TABLE OF CONSTANTS

Electron charge	e	$1\ 602 \times 10^{-19}$ coul
Electron mass	m	$9\ 1055 \times 10^{-31}$ kg
Proton mass		$1\ 6723 \times 10^{-27}$ kg
Speed of light	c	299,792 5 km/sec
Permittivity of free space	ϵ_0	$8\ 854 \times 10^{-12}$ farads/m
Permeability of free space	μ_0	$4\pi \times 10^{-7}$ henries/m

TABLE OF CONVERSION FACTORS

1 newton	=	0 225 lb = 10^5 dynes
1 joule	=	1 newton-m
1 watt	=	1 joule/sec
1 volt	=	1 joule/coul
1 ampere	=	1 coul/sec
1 volt/m	=	1 newton/coul
1 weber/m^2	=	10^4 gauss
1 amp/m	=	$4\pi \times 10^{-3}$ oersteds

TABLE OF DIMENSIONS

Charge	q	coulomb
Current	I	ampere
Volume charge density	ρ	coul/m^3
Areal current density	$\mathbf{J}$	amps/m^2
Electric field	$\mathbf{E}$	volts/m
Magnetic field	$\mathbf{B}$	webers/m^2
Electric flux density	$\mathbf{D}$	coul/m^2
Magnetic intensity	$\mathbf{H}$	amps/m
Conductivity	σ	mho/m
Permittivity	ϵ	farads/m
Permeability	μ	henries/m
Resistance	R	ohm
Capacitance	C	farad
Inductance	L	henry
Length	l	meter
Mass	m	kilogram
Time	t	second
Force	$\mathbf{F}$	newton
Energy	U	joule
Power	P	watt

ANTENNA THEORY
AND DESIGN

Revised Edition

IEEE PRESS SERIES ON ELECTROMAGNETIC WAVE THEORY

The IEEE Press Series on Electromagnetic Wave Theory consists of new titles as well as reprints and revisions of recognized classics that maintain long-term archival significance in electromagnetic waves and applications.

BOOKS IN THE IEEE PRESS SERIES ON ELECTROMAGNETIC WAVE THEORY

Christopoulos, C., *The Transmission-Line Modeling Methods. TLM*
Clemmow, P. C., *The Plane Wave Spectrum Representation of Electromagnetic Fields*
Collin. R. B., *Field Theory of Guided Waves*, Second Edition
Collin, R. E., *Foundations for Microwave Engineering*, Second Edition
Dudley, D. G., *Mathematical Foundations for Electromagnetic Theory*
Elliot, R. S., *Electromagnetics: History. Theory. and Applications*
Felsen, L. B., and Marcuvitz. N., *Radiation and Scattering of Waves*
Harrington. R. F., *Field Computation by Moment Methods*
Harrington, R. F., *Time Harmonic Electromagnetic Fields*
Hansen et al., *Plane-Wave Theory of Time-Domain Fields. Near-Field Scanning Applications*
Ishimaru, A., *Wave Propagation and Scattering in Random Media*
Jones, D. S., *Methods in Electromagnetic Wave Propagation*, Second Edition
Lindell, I. V., *Methods for Electromagnetic Field Analysis*
Peterson et al., *Computational Methods for Electromagnetics*
Tai, C. T., *Generalized Vector and Dyadic Analysis: Applied Mathematics in Field Theory*
Tai, C. T., *Dyadic Green Functions in Electromagnetic Theory*, Second Edition
Van Bladel, J., *Singular Electromagnetic Fields and Sources*
Volakis et al., *Finite Element Method for Electromagnetics: Antennas, Microwave Circuits, and Scattering Applications*
Wait, J., *Electromagnetic Waves in Stratified Media*

An IEEE Press Classic Reissue

ANTENNA THEORY AND DESIGN

Revised Edition

Robert S. Elliott
University of California
Los Angeles, California

IEEE Antennas & Propagation Society, *Sponsor*

IEEE PRESS

WILEY-INTERSCIENCE

A JOHN WILEY & SONS, INC., PUBLICATION

Published by John Wiley & Sons, Inc., Hoboken, New Jersey.
Published simultaneously in Canada.

This book was previously published by: Pearson Education Inc., formerly known as Prentice-Hall, Inc.

Library of Congress Cataloging-in-Publication Data is available.

ISBN 978-0-471-44996-6

IEEE Press
445 Hoes Lane
Piscataway, NJ 08854

To the memory of Tom Taylor

foreword to the revised edition

The purpose of the IEEE Press Series on Electromagnetic Wave Theory is to publish books of long-term archival significance in electromagnetics. Included are new titles as well as reprints and revisions of recognized classics. The book Antenna Theory and Design, by Robert S. Elliott is one such classic.

In the case of antennas and Robert S. Elliott, I should like to be personal. Much of the material that forms the basis of Antenna Theory and Design I studied as a graduate student under Bob Elliott's guidance at UCLA in the late 1950's and early 1960's. This material became the fundamental background for me during my ten-year antenna design and development career at Hughes Aircraft Company and, what was then, North American Rockwell. The notes I compiled in his courses later became the foundation for two antenna courses when I moved on to the University of Arizona in 1968.

Antenna theory can be studied, assimilated, and then written down in a textbook with little practical experience. Antenna design is another matter entirely. Bob Elliott's career has been in actuality two careers in one. He has contributed significantly to antenna and microwave component design and development at Hughes Aircraft Company and Rantec Corporation while also forming and leading a strong, internationally recognized antenna and microwave program at UCLA. The book, Antenna Theory and Design, reflects the breadth and depth of coverage that such a background would suggest. As a result, the book is useful to academics and also to practitioners in industry and government laboratories.

Professor Elliott has been an internationally well-known contributor to electromagnetics for many years. He is universally regarded among his peers and students as an electromagnetic scholar. As an example, his clear and groundbreaking exposition on electromagnetics and its relationship to the special theory of relativity appears in the widely-regarded book, Electromagnetics; History, Theory, and Applications. This scholarly work was added to the IEEE Press Series on Electromagnetic Wave Theory in 1993.

Professor Elliott is a Fellow of the IEEE (1961). Prior to his retirement from active teaching, he was the Hughes Distinguished Professor of Electromagnetics at UCLA. Among his teaching awards, he was elected Best Teacher, UCLA Campus-Wide (1983) and has been elected Best Teacher, UCLA College of Engineering, four times. Among his many professional honors, he was elected a Fellow of the National Academy of Engineering (1988). The IEEE Antennas and Propagation Society (APS) awarded him the APS

Distinguished Achievement Award (1985). In addition, he has received two APS Prize Paper awards. In 2000, he was awarded an IEEE Third Millennium Medal.

I have received many comments from Bob Elliott's colleagues and former students since beginning this reissue project. The one that most typifies this book is, "Many of the insights in his text are originally his and are still considered the fundamental way of looking at things."

It is with pleasure that I welcome this classic book into the series.

Donald G. Dudley
University of Arizona
Series Editor
IEEE Press Series on Electromagnetic Wave Theory

preface to the revised edition

This textbook first appeared in 1981 and, although it has been out of print for the past decade, a continuing demand has led to the decision that it be reissued. Like *Electromagnetics*, its predecessor in this Classic Series, it seems to have become something of a collector's item.

The primary appeal is apparently due to the fundamental treatment of both theory and design for a wide variety of antenna elements, arrays, and feeding systems. The serious reader will find in this text the basic coverage of all aspects of the antenna discipline needed as background for someone desiring to pursue a career in electronic systems, or as preparation for advanced study leading to a desired career as an antenna engineer.

The decision to reissue has provided an opportunity to eliminate errors that have been discovered in the final printing of the original text. Several colleagues and former students kindly contributed to the compiling of a list of errata. Most of the errors found were minor, a few were more serious, notably the entries in Tables 7.6 and 7.7. J.H.Anderson verified all suggested corrections and assembled the errata in a common format, thus facilitating their removal. His help is warmly acknowledged.

The opportunity was also taken to update the references in Sections 5.14 and 8.13 because of seminal advances in the design of slot arrays and in the synthesis of shaped antenna patterns.

The author wishes to thank the IEEE Press for its decision to reissue *Antenna Theory* and Design and trusts that their faith in this project will not go unrewarded.

Robert S. Elliott
Los Angeles

contents

preface

A nine-month sequence in antenna theory and design is offered on a yearly basis at the author's institution. The first and second quarters are open to seniors and first-year graduate students; the third quarter is at the graduate level. The sequence presupposes a background at the intermediate level in electromagnetic theory and a knowledge of introductory transmission line theory, including Smith charts and waveguide modal analysis. The present book has evolved from the lecture notes for the antenna sequence.

It has been the author's experience, in teaching this sequence for the past five years, that the various topics which seemed to provide a balanced treatment were not to be found at an introductory level in a single textbook currently available. Further, some recent developments, the importance of which is widely recognized, were only available in the research literature. Student frustration over nonuniformity of notation from article to article and over the economic hardship associated with buying a multiplicity of texts that would only be partially used, provided the original motivation for the lecture notes. The editing of these notes by successive groups of students is appreciated, and it is hoped that their criticisms have benefited the final product.

The topic coverage has been influenced by the author's experience and by the needs of local industry in the Los Angeles area. The reader will find emphasis on microwave antennas, particularly on arrays for use in radar and communication systems. The practical applications of such antennas have grown to occupy a major portion of the field, so it is hoped this emphasis will find wide appeal. However, other topics have not been neglected, as can be observed from the Table of Contents.

The text is divided into four parts. Part I commences with a review of electromagnetic theory and then proceeds to the establishment of integral relations between a collection of sources (the antenna) and the radiated field caused by these sources. A convenient division of antennas into two types emerges from this development. The first type, for which the actual sources are known quite well, includes dipoles, loops, and helices, and their pattern characteristics are studied in turn. The second type, for which the close-in fields are known with reasonable accuracy, can be analyzed in terms of equivalent sources. This category includes horns, slots, and patches, all of which are considered in some detail.

Part II is concerned with the analysis and synthesis of one- and two-dimensional arrays. The antenna elements studied in Part I form the constituent parts of these arrays, and focus is on the pattern characteristics. The synthesis procedures of Dolph and Taylor are introduced and extended to pattern requirements involving arbitrary side lobe topography.

In Part III the emphasis is shifted to the impedance properties of antenna elements, used either singly or in arrays. Hallén's integral equation formulation of the self-impedance of a cylindrical dipole is developed and extended to strip dipoles. Several types of solution are studied, including those obtained by the method of moments and by functional expansion. Babinet's principle is used to extend these results to slots. Mutual impedance, so important in the design of arrays, is formulated with the aid of the reciprocity theorem and then calculated for the most commonly used antenna elements. All of this information of self-impedance and mutual impedance is then employed in the design of feeding structures for single elements and for linear and planar arrays, including those which scan.

Part IV is devoted to antennas with continuous (or quasi-continuous) apertures. Long wire antennas such as the rhombic and V are studied and the properties of many surface wave structures are analyzed. These include slow wave types such as dielectric-clad and corrugated ground planes and fast wave types, notably leaky waveguides. The book concludes with an introductory treatment of reflectors and lenses, antenna types to which many of the principles of optics can be applied.

The three courses that form the antenna sequence at the author's institution span three months each, with four hours of lecture offered per week. The first course covers Chapters 1, 2, and 4 plus the first six sections of Chapter 5, the first fifteen sections of Chapter 7, the first twelve sections of Chapter 8, and the first three sections of Chapter 9. It thus concentrates on wire antennas (dipoles, monopoles, loops, and helices) after introduction of the fundamentals. The second course covers Chapter 3, the remainder of Chapters 7, 8, and 9, and all of Chapter 10. It emphasizes aperture antennas (slots, patches, reflectors, and lenses). The third course is devoted to pattern synthesis and relies on the last half of Chapter 5 and all of Chapter 6, plus some of the current literature.

For someone wishing to give a balanced offering of antenna topics in a one semester course, a combination which should prove satisfactory would contain Sections 1.1 through 1.6, Sections 1.10 through 1.18, Sections 2.1 through 2.6, Sections 3.1 through 3.6, Sections 4.1 through 4.4, Sections 5.1 through 5.3, Section 7.8, Sections 7.13 through 7.15, Sections 8.1 through 8.6, Sections 10.1 through 10.5, and Sections 10.10 through 10.11. This would provide exposure to the fundamentals, to wire antennas, to aperture antennas, to the elements of array theory, to the problem of feeding arrays in the presence of mutual coupling, and to the application of geometric optics to the design of reflector and lens antennas.

Various friends have been kind enough to read portions of the manuscript and offer their comments. The author wishes to acknowledge his indebtedness to Professors N. Alexopoulos, C. Butler, D. G. Dudley, G. Franceschetti, Y. T. Lo, C. T. Tai, and P. G. Uslenghi, and to his industrial colleagues J. Ajioka, V. Galindo-Israel, W. H. Kummer, and A. W. Love. Among the many students who have uncovered errors and assisted in modifications of the text, the efforts of D. Kim and J. Schaffner deserve explicit mention.

A special and warm expression of gratitude is reserved for my longtime colleague and friend, Alvin Clavin, Manager of the Radar Laboratory at Hughes Canoga Park. He had the confidence to offer me consulting work at Hughes when I had been away from the field for a decade, thus rekindling my interest in the subject. This tribute extends to the entire Hughes organization, which has been so generous in supporting many of the antenna research efforts which have found their way into the pages of this book. My association with the engineering staff at Hughes has been rich and valuable, and particular gratitude must be expressed for the counsel of Louis Kurtz and George Stern. The computer assistance given me at Hughes by Ralph Johnson and Annette Sato is also gratefully acknowledged.

ROBERT S. ELLIOTT
Los Angeles

I source/field relations single antenna elements

This initial part of the text, consisting of three chapters, is concerned first with establishing the general relations between a collection of sources (the antenna) and the radiated field produced by those sources (the far-field pattern). The source/field formulas are then used to deduce the pattern characteristics of the most commonly encountered antenna elements (dipole, loop, helix, horn, slot, and patch). These radiators will be seen to be ideally suited to many applications in which a single element will suffice. They have the added advantage of being useful in arrays, a subject which is discussed in Part II.

1 the far-field integrals, reciprocity, directivity

1.1 Introduction

This chapter is concerned primarily with establishing formulas for the electromagnetic field vectors **E** and **H** in terms of all the sources causing these radiating fields, but at points far removed from the sources. The collection of sources is called an *antenna* and the formulas to be derived form the basis for what is generally referred to as *antenna pattern analysis and synthesis*.

A natural division into two types of antennas will emerge as the analysis develops. There are radiators, such as dipoles and helices, on which the current distribution can be hypothesized with good accuracy; for these, one set of formulas will prove useful. But there are other radiators, such as slots and horns, for which an estimation of the actual current distribution is exceedingly difficult, but for which the close-in fields can be described quite accurately. In such cases it is possible to replace the actual sources, for purposes of field calculation, with equivalent sources that properly terminate the close-in fields. This procedure leads to an alternate set of formulas, useful for antennas of this type.

The chapter begins with a brief review of relevant electromagnetic theory, including an inductive establishment of the retarded potential functions. This is followed by a rigorous derivation of the Stratton-Chu integrals (based on a vector Green's theorem), which give the fields at any point within a volume V in terms of the sources within V and the field values on the surfaces S that bound V. This formulation possesses the virtue that it applies to either type of antenna, or to a hybrid mix of the two. Simplifications due to the remoteness of the field point from the antenna will lead to compact integral formulas, from which all the pattern characteristics of the different types of antennas can be deduced.

A general derivation of the reciprocity theorem is presented; the result is used to demonstrate that the transmitting and receiving patterns of an antenna are identical. The concept of directivity of a radiation pattern is introduced and a connection is established between the receiving cross section of an antenna and its directivity when transmitting. The chapter concludes with a discussion of the polarization of an antenna pattern.

A. REVIEW OF RELEVANT ELECTROMAGNETIC THEORY[1]

It will generally be assumed that the reader of this text is already familiar with electromagnetic theory at the intermediate level and possesses a knowledge of basic transmission line analysis (including the use of Smith charts) and of waveguide modal representations. What follows in the next several sections is a brief review of the pertinent field theory, primarily for the purposes of introducing the notation that will be adopted and highlighting some useful analogies.[2]

Throughout this text MKS rationalized units are used; the dimensions of the various source and field quantities introduced in the review are listed on the inside of the front cover.

1.2 Electrostatics and Magnetostatics in Free Space

A time-independent charge distribution

$$\rho(x, y, z) \tag{1.1a}$$

expressed in couloumbs per cubic meter, placed in what is otherwise free space, gives rise to an electrostatic field $\mathbf{E}(x, y, z)$. Similarly, a time-independent current distribution

$$\mathbf{J}(x, y, z) \tag{1.1b}$$

expressed in amperes per square meter, produces a magnetostatic field $\mathbf{B}(x, y, z)$. To heighten the analogies between electrostatics and magnetostatics, it is sometimes useful to refer to the "reduced" source distributions

$$\frac{\rho(x, y, z)}{\epsilon_0} \qquad \frac{\mathbf{J}(x, y, z)}{\mu_0^{-1}} \tag{1.2}$$

in which ϵ_0 is the permittivity of free space and μ_0^{-1} is the reciprocal of the permeability of free space.

Coulomb's law can be introduced as the experimental postulate for electrostatics and described by the equations

[1] The reader who prefers to omit this review should begin with Section 1.7.

[2] The pairing of $\mathbf{B}$ with $\mathbf{E}$ (and thus of $\mathbf{H}$ with $\mathbf{D}$), the use of μ_0^{-1}, the introduction of reduced sources, and the parallel numbering of the early equations in this review all serve to emphasize the analogies that occur between electrostatics and magnetostatics. This is done in the belief that perception of these analogies adds significantly to one's comprehension of the subject. See R. S. Elliott, "Some Useful Analogies in the Teaching of Electromagnetic Theory," *IEEE Trans. on Education*, E-22 (1979), 7–10. Reprinted with permission.

$$\mathbf{F} = q\mathbf{E} \tag{1.3a}$$

$$\mathbf{E}(x, y, z) = \int_V \frac{\rho(\xi, \eta, \zeta)\mathbf{R}\, dV}{4\pi\epsilon_0 R^3} \tag{1.4a}$$

in which $\mathbf{R}$ is the directed distance from the source point (ξ, η, ζ) to the field point (x, y, z), and $\mathbf{F}$ is the force on a charge q placed at (x, y, z), due to its interaction with the source system $\rho(\xi, \eta, \zeta)$.

Similarly, the Biot-Savart law can be introduced as the experimental postulate for magnetostatics and is represented by the equations

$$\mathbf{F} = q\mathbf{v} \times \mathbf{B} \tag{1.3b}$$

$$\mathbf{B}(x, y, z) = \int_V \frac{\mathbf{J}(\xi, \eta, \zeta) \times \mathbf{R}\, dV}{4\pi\mu_0^{-1} R^3} \tag{1.4b}$$

One can show by performing the indicated vector operations on (1.4a) that

$$\nabla \times \mathbf{E} \equiv 0 \tag{1.5a}$$

$$\nabla \cdot \mathbf{E} = \frac{\rho}{\epsilon_0} \tag{1.5b}$$

In like manner, the curl and divergence of (1.4b) yield

$$\nabla \times \mathbf{B} = \frac{\mathbf{J}}{\mu_0^{-1}} \tag{1.5c}$$

$$\nabla \cdot \mathbf{B} \equiv 0 \tag{1.5d}$$

Equations 1.5 are Maxwell's equations for static fields.

Integration of (1.5b) and use of the divergence theorem gives Gauss' law, that is,

$$\oint_S \mathbf{E} \cdot d\mathbf{S} = \int_V \left(\frac{\rho}{\epsilon_0}\right) dV = \text{total reduced charge enclosed} \tag{1.6a}$$

Similarly, integration of (1.5c) and use of Stokes' theorem yields Ampere's circuital law:

$$\oint_C \mathbf{B} \cdot dl = \int_S \left(\frac{\mathbf{J}}{\mu_0^{-1}}\right) \cdot d\mathbf{S} = \text{total reduced current enclosed} \tag{1.6b}$$

In like manner, integration of (1.5a) and (1.5d), followed by the application of Stokes' theorem or the divergence theorem results in the following relations.

$$\oint_C \mathbf{E} \cdot dl \equiv 0 \tag{1.7a}$$

$$\oint_S \mathbf{B} \cdot d\mathbf{S} \equiv 0 \tag{1.7b}$$

From (1.7a) it can be concluded that $\mathbf{E}(x, y, z)$ is a conservative field and that $\oint \mathbf{E} \cdot d\mathbf{l}$ between any two points is independent of the path. Equation 1.7b permits the conclusion that the flux lines of $\mathbf{B}$ are everywhere continuous.

Equation 1.4a can be manipulated into the form

$$\mathbf{E} = -\nabla \Phi \tag{1.8a}$$

in which

$$\Phi(x, y, z) = \int_V \frac{\rho(\xi, \eta, \zeta)\, dV}{4\pi\epsilon_0 R} \tag{1.9a}$$

is the electrostatic potential function. In like manner, Equation 1.4b can be rewritten in the form

$$\mathbf{B} = \nabla \times \mathbf{A} \tag{1.8b}$$

where

$$\mathbf{A}(x, y, z) = \int_V \frac{\mathbf{J}(\xi, \eta, \zeta)\, dV}{4\pi\mu_0^{-1} R} \tag{1.9b}$$

is the magnetostatic vector potential function. One can see that the reduced sources (1.2) play analogous roles in the integrands of the potential functions (1.8a) and (1.8b), as well as in the integrands of the field functions (1.4a) and (1.4b).

There is no compelling reason to introduce either $\mathbf{D}$ or $\mathbf{H}$ until a discussion of dielectric and magnetic materials is undertaken, but if one wishes to do it at this earlier stage, where only primary sources in what is otherwise free space are being assumed, then it is suggestive to write

$$\mathbf{D_0} = \epsilon_0 \mathbf{E} \tag{1.10a}$$

$$\mathbf{H_0} = \mu_0^{-1} \mathbf{B} \tag{1.10b}$$

with the subscripts on $\mathbf{D}$ and $\mathbf{H}$ denoting that the medium is free space. Then it follows logically from (1.5) that

$$\nabla \cdot \mathbf{D_0} = \rho \qquad \nabla \times \mathbf{H_0} = \mathbf{J} \tag{1.11}$$

and from (1.6) that

$$\oint_S \mathbf{D_0} \cdot d\mathbf{S} = \int_V \rho\, dV = \text{total charge enclosed} \tag{1.12a}$$

$$\oint_C \mathbf{H_0} \cdot d\mathbf{l} = \int_S \mathbf{J} \cdot d\mathbf{S} = \text{total current enclosed} \tag{1.12b}$$

Equations 1.12 are the forms in which one is more apt to find Gauss' law and Ampere's circuital law expressed. It is apparent from (1.12) that $\mathbf{D_0}$ and $\mathbf{H_0}$ play analogous roles in the two laws.

When flux maps are introduced, (1.12a) leads to the conclusion that the lines of $\mathbf{D}_0$ start on positive charge and end on negative charge. If one chooses to defer the introduction of **D** and **H** until materials are present, a flux map interpretation of (1.6a) includes the idea that the lines of **E** start on reduced positive charge and end on reduced negative charge.

It has already been noted in connection with equation (1.7b) that the flux lines of **B** are continuous. Since $\mathbf{H}_0$ differs from **B** only by a multiplicative constant, the flux lines of $\mathbf{H}_0$ are also continuous.

1.3 The Introduction of Dielectric, Magnetic, and Conductive Materials

The electrostatic behavior of dielectric materials can be explained quite satisfactorily by imagining the dielectric to be composed of many dipole moments of the type $\mathbf{p} = \mathbf{1}_n qd$, in which q is the positive charge of the oppositely charged pair, d is their separation, and $\mathbf{1}_n$ is a unit vector drawn from $-q$ to $+q$. If $\mathbf{P}(x, y, z)$ is the volume density of these elementary dipole moments, one can show[3] that their aggregated effect is to cause an electrostatic field given by

$$\mathbf{E}(x, y, z) = -\nabla_F\left[\oint_S \frac{\mathbf{P} \cdot d\mathbf{S}}{4\pi\epsilon_0 R} + \int_V \frac{(-\nabla_S \cdot \mathbf{P})\, dV}{4\pi\epsilon_0 R}\right] \tag{1.13a}$$

with S the dielectric surface and V its volume. In (1.13a), ∇_S operates on the source point and ∇_F operates on the field point.

Similarly, the magnetostatic behavior of magnetic materials can be explained in terms of a collection of current loops with magnetic moments of the type $m = \mathbf{1}_n \pi a^2 I$, where πa^2 is the area of the loop, I is the current, and $\mathbf{1}_n$ is a unit vector normal to the plane of the loop in the right-hand sense. If $\mathbf{M}(x, y, z)$ is the volume density of these elementary loops, one can show[4] that their aggregated effect is to cause a magnetostatic field given by

$$\mathbf{B}(x, y, z) = \nabla_F \times \left[\oint_S \frac{\mathbf{M} \times d\mathbf{S}}{4\pi\mu_0^{-1} R} + \int_V \frac{\nabla_S \times \mathbf{M}\, dV}{4\pi\mu_0^{-1} R}\right] \tag{1.13b}$$

In the more general situation that there is a primary charge distribution $\rho(x, y, z)$ somewhere in space and secondary (or *bound*) charge distributions P_n on the dielectric surface and $-\nabla \cdot \mathbf{P}$ throughout its volume, the total electrostatic field is $\mathbf{E} = \mathbf{E}_1 + \mathbf{E}_2$, with $\mathbf{E}_1$ given by (1.4a) and $\mathbf{E}_2$ given by (1.13a). No additional information would be conveyed by using $\mathbf{D}_0 = \epsilon_0 \mathbf{E}$ in this situation. However, it is extremely useful[5] to

[3]See, for example, R. S. Elliott, *Electromagnetics* (New York: McGraw-Hill Book Co., Inc., 1966), pp. 330–37.

[4]Elliott, *Electromagnetics*, pp. 404–7.

[5]Ibid., pp. 339–40.

generalize the concept of **D** through the *defining* relation

$$\mathbf{D} = \epsilon_0 \mathbf{E} + \mathbf{P} \tag{1.14a}$$

This insures the desirable feature that

$$\begin{aligned} \nabla \cdot \mathbf{D} &= \epsilon_0 \nabla \cdot \mathbf{E}_1 + \epsilon_0 \nabla \cdot \mathbf{E}_2 + \nabla \cdot \mathbf{P} \\ &= \rho - \nabla \cdot \mathbf{P} + \nabla \cdot \mathbf{P} \\ \nabla \cdot \mathbf{D} &= \rho \end{aligned} \tag{1.15a}$$

at all points in space (both within and outside the dielectric), thus permitting the assertion that the flux lines of **D** start and stop on *primary* charge alone. If there are no primary charges inside the dielectric, the **D** lines are continuous there. Outside the dielectric, (1.14a) reduces to $\mathbf{D} = \epsilon_0 \mathbf{E}$, which is consistent with (1.10a).

Since $\nabla \times \mathbf{E} = \nabla \times \mathbf{E}_1 + \nabla \times \mathbf{E}_2$, and since $\mathbf{E}_1$ and $\mathbf{E}_2$ are both expressible as the gradient of a scalar function, it follows that in this more general situation of primary and secondary charge distributions,

$$\nabla \times \mathbf{E} \equiv 0 \tag{1.15b}$$

However, one can see from the defining relation (1.14a) that $\nabla \times \mathbf{D} = \nabla \times \mathbf{P}$ and thus the generalized **D**, unlike **E**, may not be an irrotational field everywhere.

Many dielectric materials are linear (or nearly so), in the sense that $\mathbf{P} = \chi_e \epsilon_0 \mathbf{E}$ holds, where χ_e is a constant called the *dielectric susceptibility*. When this can be assumed, Equation 1.14a reduces to

$$\mathbf{D} = (1 + \chi_e)\epsilon_0 \mathbf{E} = \epsilon \mathbf{E} \tag{1.16a}$$

where ϵ is the permittivity of the dielectric medium. The quantity $\epsilon/\epsilon_0 = 1 + \chi_e$ is more useful and is known as the *relative permittivity*, or *dielectric constant*.

Similarly, in the more general situation that there is a primary current distribution $\mathbf{J}(x, y, z)$ somewhere in space and secondary (or bound) current distributions $\mathbf{M} \times \mathbf{1}_n$ on the surface of the magnetic material and $\nabla \times \mathbf{M}$ throughout its volume, the total magnetostatic field is $\mathbf{B} = \mathbf{B}_1 + \mathbf{B}_2$, with $\mathbf{B}_1$ given by (1.4b) and $\mathbf{B}_2$ given by (1.13b). No additional information would be conveyed by using $\mathbf{H}_0 = \mu_0^{-1}\mathbf{B}$ in this situation. However, it is extremely useful[6] to generalize the concept of **H** through the *defining* relation

$$\mathbf{H} = \mu_0^{-1}\mathbf{B} - \mathbf{M} \tag{1.14b}$$

This insures the desirable feature that

$$\begin{aligned} \nabla \times \mathbf{H} &= \mu_0^{-1} \nabla \times \mathbf{B}_1 + \mu_0^{-1} \nabla \times \mathbf{B}_2 - \nabla \times \mathbf{M} \\ &= \mathbf{J} + \nabla \times \mathbf{M} - \nabla \times \mathbf{M} \\ \nabla \times \mathbf{H} &= \mathbf{J} \end{aligned} \tag{1.15c}$$

[6] *op. cit.*, Elliott, *Electromagnetics*, pp. 408–10.

at all points in space (both within and outside the magnetic material) thus permitting the assertion that **H** is irrotational except at points occupied by primary sources.

Since $\nabla \cdot \mathbf{B} = \nabla \cdot \mathbf{B}_1 + \nabla \cdot \mathbf{B}_2$, and since $\mathbf{B}_1$ and $\mathbf{B}_2$ can both be expressed as the curl of a vector function, it follows that in this more general situation of primary and secondary current distributions,

$$\nabla \cdot \mathbf{B} \equiv 0 \tag{1.15d}$$

However, one can see from the defining relation (Equation 1.14b) that $\nabla \cdot \mathbf{H} = -\nabla \cdot \mathbf{M}$, and thus the generalized **H**, unlike **B**, may have discontinuous flux lines.

Most magnetic materials are nonlinear, but in the exceptional case that linearity can be assumed, **M** is linearly proportional to **B** and Equation 1.14b reduces to

$$\mathbf{H} = \frac{\mathbf{B}}{(1 + \chi_m)\mu_0} = \frac{\mathbf{B}}{\mu} \tag{1.16b}$$

in which χ_m is the magnetic susceptibility and μ is the permeability of the magnetic material.

Equations 1.15 are Maxwell's equations for static fields when dielectric and magnetic materials are present. They are supplemented by Equations 1.14, one of which links **E**, **D**, and the secondary sources **P**, with the other linking **B**, **H**, and the secondary sources **M**.

The integral forms of (1.15a) and (1.15c) lead to

$$\oint_S \mathbf{D} \cdot d\mathbf{S} = \text{primary charge enclosed} \tag{1.17a}$$

$$\oint_C \mathbf{H} \cdot d\mathbf{l} = \text{primary current enclosed} \tag{1.17b}$$

Thus the generalized **D** and **H** satisfy Gauss' law and Ampere's circuital law, respectively, in terms of the *primary* sources alone. This is their principal utility. On the other hand, **E** and **B** enter into a calculation of the force on a charge q moving through the field. In the most general static source situation (primary and secondary charge and current distributions), Equations 1.3, 1.4, and 1.13 combine to give

$$\mathbf{F} = q(\mathbf{E} + \mathbf{v} \times \mathbf{B}) \tag{1.18}$$

which is the Lorentz force law.

When conductive materials are present and Ohm's law is applicable,

$$\mathbf{J} = \sigma \mathbf{E} \tag{1.19}$$

at points occupied by the conductor, with σ the conductivity of the material.[7]

[7] *op. cit.*, Elliott, *Electromagnetics*, pp. 473–81.

1.4 Time-Varying Fields

If the sources become time-varying, represented by

$$\rho(x, y, z, t) \quad \text{coulombs per cubic meter} \tag{1.20a}$$

$$\mathbf{J}(x, y, z, t) \quad \text{amperes per square meter} \tag{1.20b}$$

and are assumed to exist in otherwise empty space, then Equations 1.5 need to be generalized. Faraday's EMF law and the continuity equation linking charge and current lead to the result that

$$\begin{aligned} \nabla \times \mathbf{E} &= -\frac{\partial \mathbf{B}}{\partial t} \\ \nabla \cdot \mathbf{E} &= \frac{\rho}{\epsilon_0} \\ \nabla \times \mathbf{B} &= \frac{\mathbf{J}}{\mu_0^{-1}} + \frac{1}{c^2}\frac{\partial \mathbf{E}}{\partial t} \\ \nabla \cdot \mathbf{B} &= 0 \end{aligned} \tag{1.21}$$

in which c is the speed of light and $\mathbf{E}(x, y, z, t)$ and $\mathbf{B}(x, y, z, t)$ are now functions of time as well as space. Equations 1.21 are Maxwell's equations in their most general form for primary sources in empty space. If one uses (1.10) and the fact that $\mu_0\epsilon_0 c^2 = 1$, these equations convert readily to the more familiar set

$$\begin{aligned} \nabla \times \mathbf{E} &= -\frac{\partial \mathbf{B}}{\partial t} \\ \nabla \cdot \mathbf{D}_0 &= \rho \\ \nabla \times \mathbf{H}_0 &= \mathbf{J} + \frac{\partial \mathbf{D}_0}{\partial t} \\ \nabla \cdot \mathbf{B} &= 0 \end{aligned} \tag{1.22}$$

If dielectric, magnetic, and conductive materials are present and are represented by time-varying dipole moments, current loops, and drifting electron clouds, respectively, if the defining relations in (1.14a) and (1.14b) are extended to apply when the fields and secondary sources are time-varying, and if Ohm's law (1.19) is still valid in the time-varying case (and all of these are good assumptions in practical situations), then Maxwell's equations become[8]

$$\begin{aligned} \nabla \times \mathbf{E} &= -\frac{\partial \mathbf{B}}{\partial t} \\ \nabla \cdot \mathbf{D} &= \rho \\ \nabla \times \mathbf{H} &= \mathbf{J} + \frac{\partial \mathbf{D}}{\partial t} \\ \nabla \cdot \mathbf{B} &= 0 \end{aligned} \tag{1.23}$$

[8] *op. cit.*, Elliott, *Electromagnetics*, pp. 393–94, 464, 509.

where now **D** and **H** have their generalized meanings, as given in the supporting Equations 1.14, and **J** is linked to **E** by (1.19) at all points occupied by conductor.

1.5 The Retarded Potential Functions

In antenna problems, one desires to find the field values at a point in terms of *all* the time-varying sources that contribute to the fields. This implies an integration of (1.22) or (1.23), a relatively difficult undertaking that will be deferred until Section 1.7. A simpler but less rigorous approach will be followed in this section, in which **E** and **B** are not found directly, but are found instead through the intermediation of potential functions whose relations to the sources are obtained intuitively.

Let the time-varying sources be given by (1.20) and be assumed to exist in a finite volume **V** in otherwise empty space. Then Maxwell's equations in the form (1.21) are point relations that connect $\mathbf{E}(x, y, z, t)$ and $\mathbf{B}(x, y, z, t)$ to the sources. Since $\nabla \cdot \mathbf{B} \equiv 0$, it is permissible to introduce a new vector function $\mathbf{A}(x, y, z, t)$ by the defining equation

$$\mathbf{B} = \nabla \times \mathbf{A} \tag{1.24}$$

Because the divergence of the curl of any vector function is identically zero, it is apparent that (1.24) automatically satisfies (1.21d).

If (1.24) is inserted in (1.21a), one obtains

$$\begin{aligned} \nabla \times \mathbf{E} &= -\frac{\partial}{\partial t}(\nabla \times \mathbf{A}) \\ \nabla \times (\mathbf{E} + \dot{\mathbf{A}}) &\equiv 0 \end{aligned} \tag{1.25}$$

where the dot over **A** implies time-differentiation. Since the curl of the gradient of any scalar function is identically zero, the most general solution to (1.25) results from the introduction of a new scalar function $\Phi(x, y, z, t)$ such that

$$\mathbf{E} = -\dot{\mathbf{A}} - \nabla\Phi \tag{1.26}$$

Equation 1.26 not only satisfies (1.21a) but, taken in conjunction with (1.24), provides a solution for **E** and **B** if the newly introduced functions **A** and Φ can be related to the sources. This can be done by forcing (1.24) and (1.26) to satisfy the two remaining Maxwell equations, that is, (1.21b) and (1.21c), notably the equations containing the sources.

If (1.24) and (1.26) are used in (1.21), the result is that

$$\begin{aligned} \nabla \times \nabla \times \mathbf{A} &= \frac{\mathbf{J}}{\mu_0^{-1}} - \frac{1}{c^2}(\ddot{\mathbf{A}} + \nabla\dot{\Phi}) \\ \nabla(\nabla \cdot \mathbf{A}) - \nabla^2\mathbf{A} &= \frac{\mathbf{J}}{\mu_0^{-1}} - \frac{1}{c^2}(\ddot{\mathbf{A}} + \nabla\dot{\Phi}) \end{aligned} \tag{1.27}$$

Equation 1.27 is a hybrid second-order differential equation (hybrid in the sense that it contains both **A** and Φ) and as a consequence would be extremely difficult to solve.

Fortunately, a simplification is possible because, up to this point, only the curl of **A** has been specified, and a vector function is not completely defined until some specification is also placed on its divergence. It is convenient in this development to choose

$$\nabla \cdot \mathbf{A} = -\frac{\dot{\Phi}}{c^2} \tag{1.28}$$

for then (1.27) reduces to

$$\nabla^2 \mathbf{A} - \frac{\ddot{\mathbf{A}}}{c^2} = -\frac{\mathbf{J}}{\mu_0^{-1}} \tag{1.29}$$

Equation 1.29 is an inhomogeneous second-order differential equation in the unknown function **A**, with the negative of the reduced current distribution (which is assumed to be known) playing the role of driving function. It is variously called the *Helmholtz equation* or the *wave equation*, the latter name arising because the solutions to (1.29) away from the sources are waves that travel at the speed of light.

The task remains to insure that (1.24) and (1.26) satisfy the remaining Maxwell equation (1.21b). Substitution gives

$$\nabla \cdot \dot{\mathbf{A}} + \nabla^2 \Phi = -\frac{\rho}{\epsilon_0}$$

This is also a hybrid differential equation, but use of (1.28) converts it to

$$\nabla^2 \Phi - \frac{\ddot{\Phi}}{c^2} = -\frac{\rho}{\epsilon_0} \tag{1.30}$$

Thus **A** and Φ satisfy the same differential equation, the only difference being the driving function; in (1.30) it is the negative of the reduced charge distribution (which is assumed to be known) which appears and governs Φ.

The development has now reached the point that if (1.29) and (1.30) can be solved for **A** and Φ, then (1.24) and (1.26) can be used to determine **E** and **B**, and the goal will have been achieved.

A solution of (1.30) can be inferred from the limiting electrostatic case. If the sources cease to vary with time so that $\rho(x, y, z, t) \longrightarrow \rho(x, y, z)$, then (1.25) and (1.30) reduce to

$$\mathbf{E} = -\nabla \Phi \tag{1.31}$$

$$\nabla^2 \Phi = -\frac{\rho}{\epsilon_0} \tag{1.32}$$

in which Φ is now a time-invariant function, that is, $\Phi(x, y, z, t) \longrightarrow \Phi(x, y, z)$. But if one returns to Section 1.2, it can be observed that (1.8a) and (1.31) are identical. Further, if the divergence of (1.8a) is taken and the result is combined with (1.5b), Equation 1.32 is reproduced, and its solution must be (1.9a), namely,

$$\Phi(x, y, z) = \int_V \frac{\rho(\xi, \eta, \zeta)\,dV}{4\pi\epsilon_0 R} \tag{1.33}$$

Thus the limiting (time-invariant) solution to (1.30) is (1.33). How can this be used to deduce the general (time-variant) solution to (1.30)?

It can be argued that a change in the charge density at a source point (ξ, η, ζ) causes a distrubance which is not immediately felt at a field point (x, y, z), since that disturbance, traveling at the speed of light, must take a time interval R/c to traverse the intervening distance R. Thus if one wishes to find the value of Φ at the point (x, y, z) at the time t, that is, $\Phi(x, y, z, t)$, one should use the charge densities at the source points (ξ, η, ζ) at the earlier times $t - (R/c)$. This suggests that a solution to (1.30) might be

$$\Phi(x, y, z, t) = \int_V \frac{\rho(\xi, \eta, \zeta, t - R/c)\, dV}{4\pi\epsilon_0 R} \tag{1.34}$$

This is admittedly a highly intuitive argument, and a rigorous solution to this problem will be presented in the development beginning in Section 1.7. However, if (1.34) is inserted in (1.30), one finds that it is indeed a solution.

By a similar argument it can be inferred that

$$\mathbf{A}(z, y, z, t) = \int_V \frac{\mathbf{J}(\xi, \eta, \zeta, t - R/c)\, dV}{4\pi\mu_0^{-1} R} \tag{1.35}$$

Equations 1.34 and 1.35 are known as *retarded potential functions* because of the use of retarded time in the integrands. In conformance with the names already given to their limiting forms in electrostatics and magnetostatics, Φ is called the *electric scalar potential function* and $\mathbf{A}$ is called the *magnetic vector potential function.*

1.6 Poynting's Theorem

One of the most useful theorems in electromagnetics concerns the power balance in a time-varying electromagnetic field. To introduce this theorem, let it be assumed that there is a system of impressed sources $\mathbf{J}^i$ that produces an electromagnetic field $\mathbf{E}^i$, $\mathbf{B}^i$, and that this impressed field causes a response system[9] of currents $\mathbf{J}^r$ to flow, creating an additional field $\mathbf{E}^r$, $\mathbf{B}^r$. If all these sources are in otherwise free space, the impressed and response fields both satisfy Maxwell's equations in the form (1.21). The total current density and field at any point are therefore

$$\mathbf{J} = \mathbf{J}^i + \mathbf{J}^r$$
$$\mathbf{E} = \mathbf{E}^i + \mathbf{E}^r$$
$$\mathbf{B} = \mathbf{B}^i + \mathbf{B}^r$$

[9]The decomposition of the total current system into impressed and response current densities is arbitrary, but often forms a natural division. For example, the currents that flow in a dipole may be considered to be a response to the impressed currents that flow in the generator and transmission line feeding the dipole.

If power is being supplied to the field, it must be at the rate[10]

$$d^3P = -\mathbf{J}^i \cdot \mathbf{E}\, dV$$

But from Maxwell's equations (1.22),

$$\mathbf{J}^i = \nabla \times \mathbf{H}_0 - \frac{\partial \mathbf{D}_0}{\partial t} - \mathbf{J}^r$$

so that

$$d^3P = \left[-\mathbf{E} \cdot \nabla \times \mathbf{H}_0 + \frac{\partial}{\partial t}\left(\frac{1}{2}\epsilon_0 E^2\right) + \mathbf{E} \cdot \mathbf{J}^r\right] dV \tag{1.36}$$

Application of the vector identity

$$\nabla \cdot (\mathbf{E} \times \mathbf{H}_0) = \mathbf{H}_0 \cdot \nabla \times \mathbf{E} - \mathbf{E} \cdot \nabla \times \mathbf{H}_0$$

coupled with the use of (1.22) gives

$$-\mathbf{E} \cdot \nabla \times \mathbf{H}_0 = \nabla \cdot (\mathbf{E} \times \mathbf{H}_0) + \mathbf{H}_0 \cdot \frac{\partial \mathbf{B}}{\partial t}$$

As a consequence, (1.36) may be rewritten as

$$d^3P = \left[\frac{\partial}{\partial t}\left(\frac{1}{2}\epsilon_0 E^2 + \frac{1}{2}\mu_0^{-1} B^2\right) + \mathbf{E} \cdot \mathbf{J}^r + \nabla \cdot (\mathbf{E} \times \mathbf{H}_0)\right] dV \tag{1.37}$$

This result gives the power balance in a volume element dV. The left side of (1.37) is the instantaneous power being supplied by the impressed sources to dV. The factor

$$\frac{\partial}{\partial t}\left(\frac{1}{2}\epsilon_0 E^2 + \frac{1}{2}\mu_0^{-1} B^2\right)$$

is the time rate of change of density of stored energy.[11] The factor $\mathbf{E} \cdot \mathbf{J}^r$ represents the power density being absorbed from the field by the response current density $\mathbf{J}^r$. If, for example, the response current is flowing in a conductor, this term accounts for ohmic loss. Alternatively, if $\mathbf{J}^r$ is due to freely moving charges, $\mathbf{E} \cdot \mathbf{J}^r$ accounts for their change in kinetic energy.

When the law of conservation of energy is invoked, it follows that the term $\nabla \cdot (\mathbf{E} \times \mathbf{H}_0)$ may be interpreted as the volume density of power leaving dV.

This conclusion can be seen from another point of view by integrating (1.37). With the aid of the divergence theorem, one is able to write

$$P = \frac{d}{dt}\int_V \left(\frac{1}{2}\epsilon_0 E^2 + \frac{1}{2}\mu_0^{-1} B^2\right) dV + \int_V \mathbf{E} \cdot \mathbf{J}^r\, dV + \oint_S \mathbf{E} \times \mathbf{H}_0 \cdot d\mathbf{S} \tag{1.38}$$

[10] *op. cit.*, Elliott, *Electromagnetics*, p. 283.

[11] *op. cit.*, Elliott, *Electromagnetics*, pp. 193–95, 283–84.

The left side of (1.38) represents the entire instantaneous power being supplied by all the sources. The first integral on the right side of this equation accounts for the time rate of change of the entire stored energy of the field. The second integral stands for the power being absorbed by the system of response currents. The last integral therefore represents the entire instantaneous power flow outward across the surface S bounding the volume V. For this reason, one may define the Poynting vector as

$$\mathcal{P} = \mathbf{E} \times \mathbf{H}_0 \tag{1.39}$$

and place upon it the interpretation that it gives in magnitude and direction the instantaneous rate of energy flow per unit area at a point. This is Poynting's theorem.

Since the units of $\mathbf{E}$ and $\mathbf{H}_0$ are volts per meter and amperes per meter, respectively, it is seen that the units of $\mathcal{P}$ are watts per square meter.

Cases in which the currents and fields are varying harmonically in time occur so frequently and have such importance as to deserve special discussion. Expressing all quantities in the form of a complex spatial vector function multiplied by $e^{j\omega t}$, such as

$$\mathbf{E}(x, y, z, t) = \mathcal{Re}\, \boldsymbol{\mathcal{E}}(x, y, z)e^{j\omega t}$$

one may write

$$\begin{aligned}
\mathcal{P} = \mathbf{E} \times \mathbf{H}_0 &= \tfrac{1}{4}(\boldsymbol{\mathcal{E}}e^{j\omega t} + \boldsymbol{\mathcal{E}}^\star e^{-j\omega t}) \times (\boldsymbol{\mathcal{H}}_0 e^{j\omega t} + \boldsymbol{\mathcal{H}}_0^\star e^{-j\omega t}) \\
&= \tfrac{1}{4}(\boldsymbol{\mathcal{E}} \times \boldsymbol{\mathcal{H}}_0^\star + \boldsymbol{\mathcal{E}}^\star \times \boldsymbol{\mathcal{H}}_0) + \tfrac{1}{4}(\boldsymbol{\mathcal{E}} \times \boldsymbol{\mathcal{H}}_0 e^{j2\omega t} + \boldsymbol{\mathcal{E}}^\star \times \boldsymbol{\mathcal{H}}_0^\star e^{-j2\omega t}) \\
&= \tfrac{1}{2}\mathcal{Re}(\mathbf{E} \times \mathbf{H}_0^\star) + \tfrac{1}{2}\mathcal{Re}(\mathbf{E} \times \mathbf{H}_0)
\end{aligned} \tag{1.40}$$

The term $\frac{1}{2}\mathcal{Re}(\mathbf{E} \times \mathbf{H}_0^\star)$ is *independent of time* and thus represents the time-average value of $\mathcal{P}$, giving

$$\bar{\mathcal{P}} = \tfrac{1}{2}\mathcal{Re}(\mathbf{E} \times \mathbf{H}_0^\star) \tag{1.41}$$

The term $\frac{1}{2}\mathcal{Re}(\mathbf{E} \times \mathbf{H}_0)$ contains the factor $e^{j2\omega t}$ and thus represents the oscillating portion of Poynting's vector. Therefore $\mathcal{P}$ may be interpreted at a point as consisting of a steady flow of energy density plus a flow which surges back and forth at frequency 2ω.

Similarly

$$\begin{aligned}
\tfrac{1}{2}\epsilon_0 E^2 &= \tfrac{1}{2}\epsilon_0 \mathbf{E} \cdot \mathbf{E} = \tfrac{1}{2}\epsilon_0[\tfrac{1}{4}(\boldsymbol{\mathcal{E}}e^{j\omega t} + \boldsymbol{\mathcal{E}}^\star e^{-j\omega t}) \cdot (\boldsymbol{\mathcal{E}}e^{j\omega t} + \boldsymbol{\mathcal{E}}^\star e^{-j\omega t})] \\
&= \tfrac{1}{4}\epsilon_0 \mathbf{E} \cdot \mathbf{E}^\star + \tfrac{1}{4}\epsilon_0 \mathcal{Re}(\mathbf{E} \cdot \mathbf{E})
\end{aligned}$$

and

$$\tfrac{1}{2}\mu_0^{-1}B^2 = \tfrac{1}{4}\mu_0^{-1}\mathbf{B} \cdot \mathbf{B}^\star + \tfrac{1}{4}\mu_0^{-1}\mathcal{Re}(\mathbf{B} \cdot \mathbf{B})$$

The terms $\frac{1}{4}\epsilon_0 \mathbf{E} \cdot \mathbf{E}^\star$ and $\frac{1}{4}\mu_0^{-1}\mathbf{B} \cdot \mathbf{B}^\star$ are independent of time and represent the time-average stored energies; their time derivatives are zero. The terms $\frac{1}{4}\epsilon_0\mathcal{Re}(\mathbf{E} \cdot \mathbf{E})$ and $\frac{1}{4}\mu_0^{-1}\mathcal{Re}(\mathbf{B} \cdot \mathbf{B})$ oscillate at a frequency 2ω and they represent the variable components of the stored energy.

Finally,

$$\mathbf{E} \cdot \mathbf{J}^r = \tfrac{1}{2}\mathcal{Re}\, \mathbf{E} \cdot \mathbf{J}^{r\star} + \tfrac{1}{2}\mathcal{Re}\, \mathbf{E} \cdot \mathbf{J}^r$$

Here again, the term $\frac{1}{2}\mathcal{R}e\, \mathbf{E} \cdot \mathbf{J}^{r*}$ represents the time-average power density being absorbed by the response currents; the term $\frac{1}{2}\mathcal{R}e\, \mathbf{E} \cdot \mathbf{J}^r$ oscillates at a frequency 2ω and represents the energy density being cyclically absorbed and released by the response currents.

With this formulation, Equation 1.38 may be rewritten in two parts. The time-average power balance is seen to be

$$\bar{P} = \tfrac{1}{2}\mathcal{R}e \int_V \mathbf{E} \cdot \mathbf{J}^{r*}\, dV + \tfrac{1}{2}\mathcal{R}e \oint_S \mathbf{E} \times \mathbf{H}_0^* \cdot d\mathbf{S} \tag{1.42}$$

while the time-variable part, oscillating at a frequency 2ω, may be written

$$\begin{aligned} P(2\omega) = \frac{d}{dt}\int_V [\tfrac{1}{4}\epsilon_0 \mathcal{R}e(\mathbf{E} \cdot \mathbf{E}) + \tfrac{1}{4}\mu_0^{-1}\mathcal{R}e(\mathbf{B} \cdot \mathbf{B})]\, dV \\ + \tfrac{1}{2}\mathcal{R}e\int_V \mathbf{E} \cdot \mathbf{J}^r\, dV + \tfrac{1}{2}\mathcal{R}e \oint_S \mathbf{E} \times \mathbf{H}_0 \cdot d\mathbf{S} \end{aligned} \tag{1.43}$$

Thus, on the time average, the sources supply power only to that component of the response currents in phase with the electric field, represented by the first integral in (1.42), and to the net energy flow out of the volume V across the surface S. In addition, the sources may have to furnish energy and take it back at the cyclic rate 2ω if the right side of (1.43) is not zero. However, in many practical circumstances, the individual integrals in (1.43) may not be in phase, but may be adjusted purposely so that they cancel each other, thus "matching" the generator.

B. INTEGRAL SOLUTIONS OF MAXWELL'S EQUATIONS IN TERMS OF THE SOURCES

The next four sections and two related appendices are devoted to a rigorous solution of Maxwell's equations in integral form, giving the fields at any point within a volume V in terms of the sources within V and the field values on the surfaces S that bound V. One advantage to this development, beyond its rigor, is that the results are in a perfect form to delineate approaches to the two types of antennas mentioned in the introduction, namely those on which the current distribution is known quite well (such as dipoles and helices), and those for which the close-in fields are known quite well (such as slots and horns). Another advantage of the development is that it delivers the retarded potential functions as an exact consequence of the central results.[12]

[12]Some authors, in contradistinction to using the Stratton-Chu formulation (which gives **E** and **B** directly as integrals involving the sources), prefer to present a rigorous proof that the retarded potential functions **A** and **Φ** are given by the integrals shown in (1.34) and (1.35). Then **E** and **B** follow from (1.24) and (1.26). That approach is comparable in complexity to the Stratton-Chu development, and suffers from the ultimate disadvantage of requiring an ad hoc introduction of fictitious magnetic sources without rigorous validation. The concept of fictitious magnetic sources arises naturally from the Stratton-Chu solution, and their results provide a sound basis for Schelkunoff's equivalence principle. See Section 1.12.

However, the reader who is not interested in delving into the complexities of this development, and who is satisfied with the intuitive introduction of the retarded potential functions given in Section 1.5, may wish to move directly to Section 1.11. This can be done without any loss of continuity.

1.7 The Stratton-Chu Solution

Since Maxwell's equations are linear in free space, no loss in generality results from assuming that time variations are harmonic and represented by $e^{j\omega t}$. The angular frequency ω may be a component of a Fourier series or a Fourier integral, thus bringing arbitrary time dependence within the purview of the following analysis. Accordingly, if $f(x, y, z, t)$ is any field component or source component, it will be assumed that $f(x, y, z, t) = f(x, y, z)e^{j\omega t}$.

Further, it will be assumed that all of the sources are in what is otherwise *free space*. This does not preclude the presence of a dielectric material if it is represented by a $\mathbf{P}$ dipole moment distribution, nor the presence of a magnetic material if it is represented by an $\mathbf{M}$ magnetic moment distribution, nor the presence of a metallic conductor if it is viewed as consisting of a positive ion lattice and an electron cloud, coexisting in free space. With dielectric or magnetic materials present, $\dot{\mathbf{P}} = \mathbf{J}_b$ and $\mathbf{J}_m = \nabla \times \mathbf{M}$ are the bound current density contributions to the total current density $\mathbf{J}$. In the case of the metallic conductor, the electrostatic fields of the lattice and cloud are assumed to cancel each other, thermal motions are assumed to be random with a null sum, and only the oscillatory motion of the electron cloud is germane, making a contribution $\sigma\mathbf{E}$ to the total current density $\mathbf{J}$, with σ the conductivity of the metal. All of these assumptions concerning the representation of electrical behavior of materials are valid in the practical realm of the actual materials used to construct most antennas. For this reason the ensuing analysis has wide applicability.

Maxwell's equations (1.21), for time-harmonic sources in otherwise free space, can be written in the form

$$\begin{aligned} \nabla \times \mathbf{E} &= -j\omega\mathbf{B} \\ \nabla \times \mathbf{B} &= \frac{\mathbf{J}}{\mu_0^{-1}} + \frac{j\omega}{c^2}\mathbf{E} \\ \nabla \cdot \mathbf{E} &= \frac{\rho}{\epsilon_0} \\ \nabla \cdot \mathbf{B} &= 0 \end{aligned} \tag{1.44}$$

Since $c^2\mu_0\epsilon_0 = 1$, the result if the divergence of the second of these equations is taken is the continuity relation

$$\nabla \cdot \mathbf{J} = -j\omega\rho \tag{1.45}$$

In all five of the above equations, the time factor $e^{j\omega t}$ is suppressed and the fields are *complex* vector functions, as is the current density. The charge density is a complex scalar function.

If the curl of either (1.44a) or (1.44b) is taken and then (1.44b) or (1.44a) is used to eliminate **E** or **B**, one obtains the vector wave equations

$$\nabla \times \nabla \times \mathbf{E} - k^2\mathbf{E} = -j\omega\left(\frac{\mathbf{J}}{\mu_0^{-1}}\right) \tag{1.46}$$

$$\nabla \times \nabla \times \mathbf{B} - k^2\mathbf{B} = \nabla \times \left(\frac{\mathbf{J}}{\mu_0^{-1}}\right) \tag{1.47}$$

in which $k = \omega/c$ is called the propagation constant, for a reason that will emerge shortly. These last two equations can be integrated through use of a technique first introduced by Stratton and Chu, and based on a vector formulation of Green's second identity.[13]

Consider a region V, bounded by the surfaces $S_1 \cdots S_N$, as shown in Figure 1.1. Let **F** and **G** be two vector functions of position in this region, each continuous and having continuous first and second derivatives everywhere within V and on the boundary surfaces S_i. Using the vector identity

$$\nabla \cdot [\mathbf{A} \times \mathbf{B}] = \mathbf{B} \cdot \nabla \times \mathbf{A} - \mathbf{A} \cdot \nabla \times \mathbf{B}$$

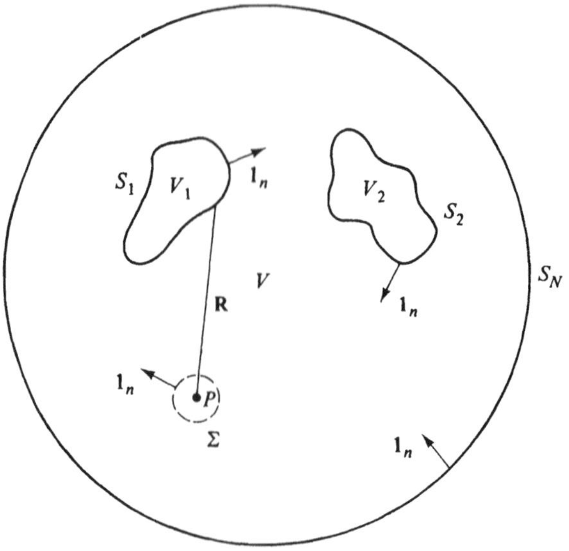

FIG. 1.1 Notation for Vector Green's Theorem.

[13] J. A. Stratton and L. J. Chu, "Diffraction Theory of Electromagnetic Waves," *Phys. Rev.*, 56 (1939), 99–107. Also, see the excellent treatment in S. Silver, *Microwave Antenna Theory and Design*, MIT Rad. Lab. Series, Vol. 12 (New York: McGraw-Hill Book Co., Inc., 1939), pp. 80–9. The present development is a reproduction, with permission, of what appears in R. S. Elliott, *Electromagnetics* (New York: McGraw-Hill Book Co., Inc., 1966), pp. 272–80 and 534–8, and differs from Silver's treatment principally in the nonuse of fictitious magnetic currents and charges.

and letting $\mathbf{A} = \mathbf{F}$ while $\mathbf{B} = \nabla \times \mathbf{G}$, one obtains

$$\nabla \cdot [\mathbf{F} \times \nabla \times \mathbf{G}] = \nabla \times \mathbf{G} \cdot \nabla \times \mathbf{F} - \mathbf{F} \cdot \nabla \times \nabla \times \mathbf{G}$$

If $\mathbf{A} = \mathbf{G}$ and $\mathbf{B} = \nabla \times \mathbf{F}$, then

$$\nabla \cdot [\mathbf{G} \times \nabla \times \mathbf{F}] = \nabla \times \mathbf{F} \cdot \nabla \times \mathbf{G} - \mathbf{G} \cdot \nabla \times \nabla \times \mathbf{F}$$

When the difference in these results is integrated over the volume V, one obtains

$$\int_V (\mathbf{F} \cdot \nabla \times \nabla \times \mathbf{G} - \mathbf{G} \cdot \nabla \times \nabla \times \mathbf{F})\, dV = \int_V \nabla \cdot [\mathbf{G} \times \nabla \times \mathbf{F} - \mathbf{F} \times \nabla \times \mathbf{G}]\, dV$$

If $\mathbf{1}_n$ is chosen to be the *inward-drawn* unit normal vector from any boundary surface S_i into the volume V, use of the divergence theorem gives

$$\int_V (\mathbf{F} \cdot \nabla \times \nabla \times \mathbf{G} - \mathbf{G} \cdot \nabla \times \nabla \times \mathbf{F})\, dV = -\int_{S_1 \cdots S_N} (\mathbf{G} \times \nabla \times \mathbf{F} - \mathbf{F} \times \nabla \times \mathbf{G}) \cdot \mathbf{1}_n\, dS \tag{1.48}$$

This result is the vector Green's theorem.

Suppose that the fields $\mathbf{E}$ and $\mathbf{B}$ of (1.46) and (1.47) both meet the conditions required of the function $\mathbf{F}$ in V; let $\mathbf{G}$ be the vector Green's function defined by

$$\mathbf{G} = \frac{e^{-jkR}}{R}\mathbf{a} = \psi \mathbf{a} \tag{1.49}$$

in which $\mathbf{a}$ is an arbitrary *constant* vector and R is the distance from an arbitrary point $P(x, y, z)$ within V to any point (ξ, η, ζ) within V or on S_i.

As defined by (1.49), $\mathbf{G}$ satisfies the conditions of the vector Green's theorem everywhere except at P. Therefore, one can surround P by a sphere Σ of radius δ and consider that portion V' of V bounded by the surfaces $S_1 \cdots S_N$, Σ. Letting $\mathbf{E} = \mathbf{F}$, one finds that

$$\int_{V'} (\mathbf{E} \cdot \nabla_S \times \nabla_S \times \psi\mathbf{a} - \psi\mathbf{a} \cdot \nabla_S \times \nabla_S \times \mathbf{E})\, dV = -\int_{S_1 \cdots S_N, \Sigma} (\psi\mathbf{a} \times \nabla_S \times \mathbf{E} - \mathbf{E} \times \nabla_S \times \psi\mathbf{a}) \cdot \mathbf{1}_n\, dS \tag{1.50}$$

in which, since ψ is a function of (x, y, z) as well as (ξ, η, ζ), it is necessary to distinguish between differentiation with respect to these two sets of variables by subscripting the operators so that

$$\nabla_S = \mathbf{1}_x \frac{\partial}{\partial \xi} + \mathbf{1}_y \frac{\partial}{\partial \eta} + \mathbf{1}_z \frac{\partial}{\partial \zeta}$$

and (1.51)

$$\nabla_F = \mathbf{1}_x \frac{\partial}{\partial x} + \mathbf{1}_y \frac{\partial}{\partial y} + \mathbf{1}_z \frac{\partial}{\partial z}$$

It is shown in Appendix A that both sides of this equation may be transformed so that **a** is brought outside the integral signs, with the following result:

$$\mathbf{a} \cdot \int_{V'} \left(j\omega\psi \frac{\mathbf{J}}{\mu_0^{-1}} - \frac{\rho}{\epsilon_0} \nabla_S \psi \right) dV - \mathbf{a} \cdot \int_{S_1 \cdots S_N, \Sigma} (\mathbf{1}_n \cdot \mathbf{E}) \nabla_S \psi \, dS$$
$$= -\mathbf{a} \cdot \int_{S_1 \cdots S_N, \Sigma} [j\omega\psi(\mathbf{1}_n \times \mathbf{B} - (\mathbf{1}_n \times \mathbf{E}) \times \nabla_S \psi] \, dS$$

Since **a** is arbitrary, it follows that the integrals on the two sides of the above equation can be equated, yielding

$$\int_{V'} \left(j\omega\psi \frac{\mathbf{J}}{\mu_0^{-1}} - \frac{\rho}{\epsilon_0} \nabla_S \psi \right) dV - \int_{S_1 \cdots S_N} [(\mathbf{1}_n \cdot \mathbf{E}) \nabla_S \psi$$
$$+ (\mathbf{1}_n \times \mathbf{E}) \times \nabla_S \psi - j\omega\psi(\mathbf{1}_n \times \mathbf{B})] \, dS \qquad (1.52)$$
$$= \int_{\Sigma} [(\mathbf{1}_n \cdot \mathbf{E}) \nabla_S \psi + (\mathbf{1}_n \times \mathbf{E}) \times \nabla_S \psi - j\omega\psi(\mathbf{1}_n \times \mathbf{B})] \, dS$$

where, for convenience, the surface integral over the sphere Σ is displayed separately.

It is further shown in Appendix A that the right side of (1.52) reaches the limit $-4\pi\mathbf{E}(x, y, z)$, with (x, y, z) the coordinates of the point P, as Σ shrinks to zero. Therefore the limiting value of (1.52) is

$$\mathbf{E}(x, y, z) = \frac{1}{4\pi} \int_V \left(\frac{\rho}{\epsilon_0} \nabla_S \psi - j\omega\psi \frac{\mathbf{J}}{\mu_0^{-1}} \right) dV$$
$$+ \frac{1}{4\pi} \int_{S_1 \cdots S_N} [(\mathbf{1}_n \cdot \mathbf{E}) \nabla_S \psi + (\mathbf{1}_n \times \mathbf{E}) \times \nabla_S \psi - j\omega\psi(\mathbf{1}_n \times \mathbf{B})] \, dS \qquad (1.53)$$

This important formula gives **E** at any point in the volume V in terms of the sources within V plus the field values on the surfaces that bound V.

By letting $\mathbf{B} = \mathbf{F}$, one may proceed in a similar fashion to deduce a companion formula for $\mathbf{B}(x, y, z)$. Alternatively, the curl of (1.53) may be taken and then (1.44a) used to obtain **B**. By either procedure, one finds that

$$\mathbf{B}(x, y, z) = \frac{1}{4\pi} \int_V \frac{\mathbf{J}}{\mu_0^{-1}} \times \nabla_S \psi \, dV$$
$$+ \frac{1}{4\pi} \int_{S_1 \cdots S_N} \left[\frac{j\omega\psi}{c^2} (\mathbf{1}_n \times \mathbf{E}) + (\mathbf{1}_n \times \mathbf{B}) \times \nabla_S \psi + (\mathbf{1}_n \cdot \mathbf{B}) \nabla_S \psi \right] dS \qquad (1.54)$$

Equations 1.53 and 1.54 comprise a solution of Maxwell's equations in terms of the time-harmonic charge and current sources within V and the field values on the boundary surfaces S_i.

1.8 Conditions at Infinity

Let it now be assumed that the surface S_N of Figure 1.1 becomes a large sphere of radius $\mathcal{R}$ centered at the point P. Initially, $\mathcal{R}$ will be taken great enough to enclose all the sources $\mathbf{J}$ and ρ of the fields; ultimately $\mathcal{R}$ will be permitted to become infinitely large. Under these circumstances, consider the contributions to (1.53) and (1.54) of the surface integrals over S_N.

If $\mathbf{1}_{\mathcal{R}}$ is a unit vector directed *outward* along the radius of the spherical surface S_N, so that $\mathbf{1}_{\mathcal{R}} = -\mathbf{1}_n$, one may write for the appropriate part of (1.54)

$$\frac{1}{4\pi}\int_{S_N}\left[\frac{j\omega\psi}{c^2}(\mathbf{1}_n \times \mathbf{E}) + (\mathbf{1}_n \times \mathbf{B}) \times \nabla_S\psi + (\mathbf{1}_n \cdot \mathbf{B})\,\nabla_S\psi\right]dS$$

$$= \frac{1}{4\pi}\int_{S_N}\left[-\frac{j\omega}{c^2}(\mathbf{1}_{\mathcal{R}} \times \mathbf{E}) + (\mathbf{1}_{\mathcal{R}} \times \mathbf{B}) \times \mathbf{1}_{\mathcal{R}}\left(jk + \frac{1}{R}\right)\right.$$

$$\left. + (\mathbf{1}_{\mathcal{R}} \cdot \mathbf{B})\mathbf{1}_{\mathcal{R}}\left(jk + \frac{1}{R}\right)\right]\frac{e^{-jkR}}{R}\,dS \tag{1.55}$$

$$= \frac{1}{4\pi}\int_{S_N}\left\{-\frac{j\omega}{c^2}(\mathbf{1}_{\mathcal{R}} \times \mathbf{E}) - \left(jk + \frac{1}{\mathcal{R}}\right)[(\mathbf{1}_{\mathcal{R}} \times \mathbf{1}_{\mathcal{R}} \times \mathbf{B}) - (\mathbf{1}_{\mathcal{R}} \cdot \mathbf{B})\mathbf{1}_{\mathcal{R}}]\right\}\frac{e^{-jk\mathcal{R}}}{\mathcal{R}}\,dS$$

$$= \frac{1}{4\pi}\int_{S_N}\left\{-\frac{j\omega}{c^2}[(\mathbf{1}_{\mathcal{R}} \times \mathbf{E}) - c\mathbf{B}] + \frac{\mathbf{B}}{\mathcal{R}}\right\}\frac{e^{-jk\mathcal{R}}}{\mathcal{R}}\,dS$$

Similarly, the appropriate part of (1.53) becomes

$$\frac{1}{4\pi}\int_{S_N}[(\mathbf{1}_n \cdot \mathbf{E})\,\nabla_S\psi + (\mathbf{1}_n \times \mathbf{E}) \times \nabla_S\psi - j\omega\psi(\mathbf{1}_n \times \mathbf{B})]\,dS$$

$$= \frac{1}{4\pi}\int_{S_N}\left\{j\omega\left[(\mathbf{1}_{\mathcal{R}} \times \mathbf{B}) + \frac{\mathbf{E}}{c}\right] + \frac{\mathbf{E}}{\mathcal{R}}\right\}\frac{e^{-jk\mathcal{R}}}{\mathcal{R}}\,dS \tag{1.56}$$

If $\mathcal{R} \longrightarrow \infty$, since the surface of the sphere increases as $\mathcal{R}^2$, the surface integral in (1.55) will vanish if

$$\lim_{\mathcal{R}\to\infty} \mathcal{R}\mathbf{B} \text{ is finite} \tag{1.57}$$

$$\mathfrak{R}[(\mathbf{1}_{\mathfrak{R}} \times \mathbf{E}) - cB] = 0 \tag{1.58}$$

Similarly, the surface integral (1.56) will vanish if

$$\lim_{\mathcal{R}\to\infty} \mathcal{R}\mathbf{E} \text{ is finite} \tag{1.59}$$

$$\lim_{\mathcal{R}\to\infty} \mathcal{R}\left[(\mathbf{1}_{\mathcal{R}} \times \mathbf{B}) + \frac{\mathbf{E}}{c}\right] = 0 \tag{1.60}$$

Relations 1.57 through 1.60 are known as the *Sommerfeld conditions at infinity*. Expressions (1.57) and (1.59) are commonly called the *finiteness conditions* (Endlichkeit Bedingungen) and Expressions 1.58 and 1.60 are customarily called *radiation conditions* (Ausstrahlung Bedingungen). The finiteness conditions require that **E** and **B** diminish as $\mathcal{R}^{-1}$, while the radiation conditions require that they bear the relation to each other found in wave propagation in regions remote from the sources. (See Section 1.11.)

It is now possible to demonstrate the extremely important result that real sources, confined to a finite volume, always give rise to fields that satisfy the Sommerfeld conditions. To see this, consider Equations 1.53 and 1.54 when the *only* boundary surface is the large sphere S_N, with radius that will be permitted to become infinitely large. It shall be assumed that the real sources **J** and ρ are finite and confined to a finite volume V_0. With the surface S_N becoming an infinite sphere, the volume V in (1.53) and (1.54) also becomes infinite, but no convergence difficulties arise with the volume integrals because the sources are all within V_0.

If one borrows from the results of Section 1.6, the fields over S_N will consist of outgoing waves with power density $\mathbf{E} \times \mathbf{H}_0$ watts per square meter. Since the surface area of S_N is increasing as $\mathcal{R}^2$, if there is even the most minute loss in V, the law of conservation of energy requires that **E** and $\mathbf{H}_0$ diminish more rapidly than $\mathcal{R}^{-1}$, and thus Conditions 1.57–1.60 are satisfied. One can then conclude that in an unbounded region, $\mathbf{B}(x, y, z)$ and $\mathbf{E}(x, y, z)$ are given solely by the volume integrals that appear in (1.53) and (1.54).

A check on this conclusion for the limiting case of no loss in V may be obtained through an ordering of the terms that comprise the volume integrals. To see this, assume that there are no bounding surfaces except the infinite sphere S_N, and that the surface integrals involving S_N in (1.53) and (1.54) are zero. Then, for this situation, Equations 1.53 and 1.54 reduce to

$$\mathbf{E}(x, y, z) = \frac{1}{4\pi}\int_V \left(\frac{\rho}{\epsilon_0} \nabla_S\psi - j\omega\psi\frac{\mathbf{J}}{\mu_0^{-1}}\right) dV = \frac{1}{4\pi}\int_V \frac{1}{j\omega\epsilon_0}[(\mathbf{J} \cdot \nabla_S) \nabla_S\psi + k^2\psi\mathbf{J}]\, dV \tag{1.61}$$

$$\mathbf{B}(x, y, z) = \frac{1}{4\pi}\int_V \frac{\mathbf{J}}{\mu_0^{-1}} \times \nabla_S\psi \, dV \tag{1.62}$$

where the second version of the integrand in (1.61) has been achieved with the aid of the continuity equation (1.45). It can now be ascertained whether or not **E** and **B**, when computed from (1.61) and (1.62), satisfy Sommerfeld's conditions at infinity.

Let an arbitrary point in V_0 be selected as the origin and let **r** be the vector drawn from the origin to the field point $P(x, y, z)$; the vector drawn from the source element to P will be labeled **R**. Then

$$\begin{aligned}(\mathbf{J} \cdot \nabla_S) \nabla_S\psi &= (\mathbf{J} \cdot \nabla_S)\left[\mathbf{1_R}\left(jk + \frac{1}{R}\right)\frac{e^{-jkR}}{R}\right] \\ &= \mathbf{1_R} J_R \frac{d}{dR}\left[\left(jk + \frac{1}{R}\right)\frac{e^{-jkR}}{R}\right] + \left(jk + \frac{1}{R}\right)\frac{e^{-jkR}}{R}\left(\frac{J_{\theta'}}{R}\frac{\partial\mathbf{1_R}}{\partial\theta'} + \frac{J_{\phi'}}{R\sin\theta'}\frac{\partial\mathbf{1_R}}{\partial\phi'}\right)\end{aligned}$$

in which spherical coordinates (r, θ', ϕ') centered at P have been used and

$$\mathbf{1_R} = -\frac{\mathbf{R}}{R}$$

Performing the indicated differentiations, one obtains

$$(\mathbf{J} \cdot \nabla_S)\nabla_S\psi = \left\{(\mathbf{J} \cdot \mathbf{1_R})\mathbf{1_R}\left[\frac{3}{R}\left(jk + \frac{1}{R}\right) - k^2\right] - \frac{\mathbf{J}}{R}\left(jk + \frac{1}{R}\right)\right\}\frac{e^{-jkR}}{R}$$

The functions ψ, $\nabla_S\psi$, and $(\mathbf{J} \cdot \nabla_S)\nabla_S\psi$ are all seen to involve polynomials in the variable R^{-1}. Retain for the moment only first-order terms; then substitution in (1.61) and (1.62) gives

$$\mathbf{E}(x, y, z) = \frac{1}{4\pi}\int_V \frac{1}{j\omega\epsilon_0}[-k^2(\mathbf{J} \cdot \mathbf{1_R})\mathbf{1_R} + k^2\mathbf{J}]\frac{e^{-jkR}}{R}\,dV \tag{1.63}$$

$$\mathbf{B}(x, y, z) = \frac{1}{4\pi}\int_V jk\,\frac{\mathbf{J}}{\mu_0^{-1}} \times \mathbf{1_R}\frac{e^{-jkR}}{R}\,dV \tag{1.64}$$

But

$$\begin{aligned} R &= [(x - \xi)^2 + (y - \eta)^2 + (z - \zeta)^2]^{1/2} \\ &= [(r \sin\theta\cos\phi - \xi)^2 + (r\sin\theta\sin\phi - \eta)^2 + (r\cos\theta - \zeta)^2]^{1/2} \end{aligned}$$

in which now conventional spherical coordinates (r, θ, ϕ) centered at the origin have been introduced. As P becomes remote, R can be expressed in the rapidly converging series

$$R = r - (\xi\sin\theta\cos\phi + \eta\sin\theta\sin\phi + \zeta\cos\theta) + 0(r^{-1}) \tag{1.65}$$

Similarly,

$$R^{-1} = r^{-1} + 0(r^{-2}) \qquad \lim_{r\to\infty} \mathbf{1_R} = \mathbf{1_r}$$

and thus as r becomes very large, Equations 1.63 and 1.64 may be written

$$\mathbf{E}(x, y, z) = \frac{j\omega e^{-jkr}}{4\pi\;\;r}\int_V \mathbf{1_r} \times \left(\mathbf{1_r} \times \frac{\mathbf{J}}{\mu_0^{-1}}\right)e^{jk\mathcal{L}}\,dV + 0(r^{-2}) \tag{1.66}$$

$$\mathbf{B}(x, y, z) = \frac{jk\,e^{-jkr}}{4\pi\;\;r}\int_V \frac{\mathbf{J}}{\mu_0^{-1}} \times \mathbf{1_r}e^{jk\mathcal{L}}\,dV + 0(r^{-2}) \tag{1.67}$$

in which $\mathcal{L} = \xi\sin\theta\cos\phi + \eta\sin\theta\sin\phi + \zeta\cos\theta$.

If one were to go back and include *all* the terms in the expressions for $\nabla_S\psi$ and $(\mathbf{J} \cdot \nabla_S)\nabla_S\psi$, they would alter the results in (1.66) and (1.67) only at the level of $0(r^{-2})$. Therefore these two expressions for **B** and **E** may be taken as exact.

In considering Expressions 1.66 and 1.67 with respect to the Sommerfeld conditions, one notices that the terms of $0(r^{-2})$ and below satisfy all four conditions and thus concern may be focused on the explicit first-order terms. But

$$\lim_{r\to\infty} r\mathbf{B} = \frac{jk}{4\pi}\lim_{r\to\infty} e^{-jkr}\int_V \frac{\mathbf{J}}{\mu_0^{-1}} \times \mathbf{1}_r e^{jk\mathfrak{L}}\, dV \tag{1.68}$$

and, since the volume integral is a function of the source coordinates and the angular direction to P, but not of r, this limit is finite. A similar argument establishes that $\lim_{r\to\infty} r\mathbf{E}$ is also finite and thus both finiteness conditions are satisfied.

Further,

$$\begin{aligned}\lim_{r\to\infty} r\left[(\mathbf{1}_r \times \mathbf{B}) + \frac{\mathbf{E}}{c}\right] \\ = \lim_{r\to\infty} \frac{e^{-jkr}}{4\pi}\int_V \left[jk\mathbf{1}_r \times \frac{\mathbf{J}}{\mu_0^{-1}} \times \mathbf{1}_r + \frac{j\omega}{c}\mathbf{1}_r \times \mathbf{1}_r \times \frac{\mathbf{J}}{\mu_0^{-1}}\right] e^{jk\mathfrak{L}}\, dV\end{aligned} \tag{1.69}$$

The integrand in (1.69) is identically zero and therefore Condition 1.60 is satisfied. In like manner, Condition 1.58 is also found to be satisfied. This supports the argument that any system of real sources confined to a finite volume V_0 gives rise to an electromagnetic field at infinity that satisfies Sommerfeld's conditions, that the surface integral over an infinite sphere S_N gives a null contribution, and that in an unbounded region the electromagnetic field at any point P, near or remote, is given precisely by (1.61) and (1.62).

Suppose now that parts of the volume V_0 are excluded from V by the finite, regular closed surfaces $S_1 \cdots S_i \cdots$. These surfaces may exclude some of the sources from V or not, but their presence does not alter the results at infinity. However, now the more general expressions in (1.53) and (1.54) apply, and one may conclude by saying that these expressions are valid even if the volume V is infinite, so long as real sources in a finite volume are assumed. If the volume V is infinite, the surface at infinity need not be considered.

This solution for **E** and **B**, given by Equations 1.53 and 1.54, is in a form that is convenient for the purpose of drawing a distinction between two types of radiators. Type I antennas will be taken to be those for which the actual current distribution is known quite well, such as dipoles and helices. Type II antennas will be those that have actual current distributions which would be difficult to deduce, but which could be enclosed by a surface over which the fields are known with reasonable accuracy. These include horns and slots.

For type I antennas, there will be no volume-excluding surfaces and (1.53) and (1.54) will contain only volume integrals. For type II antennas, the volume-excluding surfaces (usually only one) will be chosen to surround all the actual sources so that there are none to be found in the remaining part of space V. Thus for type II antennas, (1.53) and (1.54) will contain only surface integrals. In the developments that follow later in this chapter, it will be seen that it is useful to replace the field values occurring in the integrands of these surface integrals by equivalent sources. Thus for the remainder of this book, type I radiators will be referred to as *actual-source antennas* and type II radiators will be called *equivalent-source antennas*.

1.9 Field Values in the Excluded Regions

Because of its bearing on the analysis of type II (equivalent-source) antennas, it is important to consider the values of the fields **E** and **B** at points inside the excluding surfaces shown in Figure 1.1. In particular, let the field point (x, y, z) lie anywhere in the volume V_1 which has been surrounded by the closed surface S_1. A simple application of the general results in (1.53) and (1.54) gives

$$\begin{aligned}\mathbf{E}(x, y, z) &= \frac{1}{4\pi}\int_{V+V_1}\left(\frac{\rho}{\epsilon_0}\nabla_S\psi - j\omega\psi\frac{\mathbf{J}}{\mu_0^{-1}}\right)dV \\ &+ \frac{1}{4\pi}\int_{S_1\cdots S_N}[(\mathbf{1}_n\cdot\mathbf{E})\nabla_S\psi + (\mathbf{1}_n\times\mathbf{E})\times\nabla_S\psi - j\omega\psi(\mathbf{1}_n\times\mathbf{B})]\,dS\end{aligned} \tag{1.70}$$

$$\begin{aligned}\mathbf{B}(x, y, z) &= \frac{1}{4\pi}\int_{V+V_1}\frac{\mathbf{J}}{\mu_0^{-1}}\times\nabla_S\psi\,dV \\ &+ \frac{1}{4\pi}\int_{S_1\cdots S_N}\left[\frac{j\omega\psi}{c^2}(\mathbf{1}_n\times\mathbf{E}) + (\mathbf{1}_n\times\mathbf{B})\times\nabla_S\psi + (\mathbf{1}_n\cdot\mathbf{B})\nabla_S\psi\right]dS\end{aligned} \tag{1.71}$$

Another way to view this situation is to imagine that V_1 is the volume region comprising the collection of field points and that S_1 is the sole surface, performing the function of excluding all the rest of space. From this viewpoint, a second application of the general results in (1.53) and (1.54) yields

$$\begin{aligned}\mathbf{E}(x, y, z) &= \frac{1}{4\pi}\int_{V_1}\left(\frac{\rho}{\epsilon_0}\nabla_S\psi - j\omega\psi\frac{\mathbf{J}}{\mu_0^{-1}}\right)dV \\ &- \frac{1}{4\pi}\int_{S_1}[(\mathbf{1}_n\cdot\mathbf{E})\nabla_S\psi + (\mathbf{1}_n\times\mathbf{E})\times\nabla_S\psi - j\omega\psi(\mathbf{1}_n\times\mathbf{B})]\,dS\end{aligned} \tag{1.72}$$

$$\begin{aligned}\mathbf{B}(x, y, z) &= \frac{1}{4\pi}\int_{V_1}\frac{\mathbf{J}}{\mu_0^{-1}}\times\nabla_S\psi\,dV \\ &- \frac{1}{4\pi}\int_{S_1}\left[\frac{j\omega\psi}{c^2}(\mathbf{1}_n\times\mathbf{E}) + (\mathbf{1}_n\times\mathbf{B})\times\nabla_S\psi + (\mathbf{1}_n\cdot\mathbf{B})\nabla_S\psi\right]dS\end{aligned} \tag{1.73}$$

The negative signs in front of the surface integrals in (1.72) and (1.73) are occasioned by the fact that now the normal to the surface S_1 is oppositely directed.

If the difference between these two sets of formulas for the fields within V_1 is formed, one obtains

$$\begin{aligned}0 &= \frac{1}{4\pi}\int_{V}\left(\frac{\rho}{\epsilon_0}\nabla_S\psi - j\omega\psi\frac{\mathbf{J}}{\mu_0^{-1}}\right)dV \\ &+ \frac{1}{4\pi}\int_{S_1\cdots S_N}[(\mathbf{1}_n\cdot\mathbf{E})\nabla_S\psi + (\mathbf{1}_n\times\mathbf{E})\times\nabla_S\psi - j\omega\psi(\mathbf{1}_n\times\mathbf{B})]\,dS\end{aligned} \tag{1.74}$$

$$0 = \frac{1}{4\pi}\int_V \frac{\mathbf{J}}{\mu_0^{-1}} \times \nabla_S \psi \, dV$$
$$+ \frac{1}{4\pi}\int_{S_1 \cdots S_N} \left[\frac{j\omega\psi}{c^2}(\mathbf{1}_n \times \mathbf{E}) + (\mathbf{1}_n \times \mathbf{B}) \times \nabla_S \psi + (\mathbf{1}_n \cdot \mathbf{B})\nabla_S \psi\right] dS \tag{1.75}$$

The right sides of (1.74) and (1.75) are seen to be exactly the same as the right sides of (1.53) and (1.54). Therefore one can conclude that if the range of the field point (x, y, z) is unrestricted, when (x, y, z) lies within V, Equations 1.53 and 1.54 will give the true fields **E** and **B**. However, when (x, y, z) lies outside V, Equations 1.53 and 1.54 will give a null result.

1.10 The Retarded Potential Functions: Reprise

If the volume V is totally unbounded, Equations 1.53 and 1.54 give

$$\mathbf{E} = \int_V \frac{\rho \nabla_S \psi}{4\pi\epsilon_0} dV - j\omega \int_V \frac{\mathbf{J}\psi}{4\pi\mu_0^{-1}} dV \tag{1.76}$$

$$\mathbf{B} = \int_V \frac{\mathbf{J} \times \nabla_S \psi}{4\pi\mu_0^{-1}} dV \tag{1.77}$$

Since $\nabla_P \psi = -\nabla_S \psi$, and since **J** and the limits of integration are functions of (ξ, η, ζ), but not of (x, y, z), these integrals may be written in the forms

$$\mathbf{E} = -\nabla_P \int_V \frac{\rho\psi}{4\pi\epsilon_0} dV - j\omega \int_V \frac{\mathbf{J}\psi}{4\pi\mu_0^{-1}} dV \tag{1.78}$$

$$\mathbf{B} = \nabla_P \times \int_V \frac{\mathbf{J}\psi}{4\pi\mu_0^{-1}} dV \tag{1.79}$$

Therefore it is convenient to introduce two potential functions by the defining relations

$$\mathbf{A}(x, y, z, t) = \int_V \frac{\mathbf{J}(\xi, \eta, \zeta) e^{j(\omega t - kR)}}{4\pi\mu_0^{-1} R} dV \tag{1.80}$$

$$\Phi(x, y, z, t) = \int_V \frac{\rho(\xi, \eta, \zeta) e^{j(\omega t - kR)}}{4\pi\epsilon_0 R} dV \tag{1.81}$$

in which the time factor $e^{j\omega t}$ has been reinserted and e^{-jkR}/R has been substituted for ψ. The function **A** is called the *magnetic vector potential function* and Φ is called the *electric scalar potential function.*

Since $k = \omega/c$, one may write

$$\exp[j(\omega t - kR)] = \exp\left[j\omega\left(t - \frac{R}{c}\right)\right]$$

Therefore each current element in the integrand of (1.80) and each charge element in the integrand of (1.81) makes a contribution to the potential at (x, y, z) at time t which is in accord with the value it had at the earlier time $t - R/c$. But this is consistent with the idea that it takes a time R/c for a disturbance to travel from (ξ, η, ζ) to (x, y, z). For this reason, (1.80) and (1.81) are often called the *retarded potentials*.

From (1.78) and (1.79),

$$\mathbf{E} = -\nabla\Phi - \dot{\mathbf{A}} \tag{1.82}$$

$$\mathbf{B} = \nabla \times \mathbf{A} \tag{1.83}$$

in which the subscripts on the del operators have been dropped, since $\mathbf{A}$ and Φ are functions only of (x, y, z) and not also of (ξ, η, ζ).

The differential equations satisfied by $\mathbf{A}$ and Φ may be deduced by taking the divergence of (1.82) and the curl of (1.83), which leads to

$$\nabla^2\mathbf{A} - \frac{1}{c^2}\ddot{\mathbf{A}} = -\frac{\mathbf{J}}{\mu_0^{-1}} \tag{1.84}$$

$$\nabla^2\Phi - \frac{1}{c^2}\ddot{\Phi} = -\frac{\rho}{\epsilon_0} \tag{1.85}$$

These relations are valid whether $\mathbf{J}$ and ρ are harmonic functions of time or more general time functions representable by Fourier integrals. A proof may be found in Appendix B.

All of the results in this section can be seen to be consistent with those obtained in Section 1.5 by a different line of reasoning.

C. THE FAR-FIELD EXPRESSIONS FOR TYPE I (ACTUAL-SOURCE) ANTENNAS

In antenna problems, one is interested in determining the fields at points *remote from the sources*. This introduces several simplifications in the field/source relations, as can be seen in the development in the next section.

1.11 The Far-Field: Type I Antennas

The typical situation for an actual-source antenna is suggested by Figure 1.2. The sources are assumed to be oscillating harmonically with time at an angular frequency ω and to be confined to some finite volume V. There are no source-excluding surfaces S_i. For convenience, the origin of coordinates is taken somewhere in V. It is desired to find $\mathbf{E}$ and $\mathbf{B}$ at a field point (x, y, z) so remote that $R \ggg \max\,[\xi^2 + \eta^2 + \zeta^2]^{1/2}$. Said another way, the maximum dimension of the volume V that contains all the sources is very small compared to the distance from any source point to the field point.

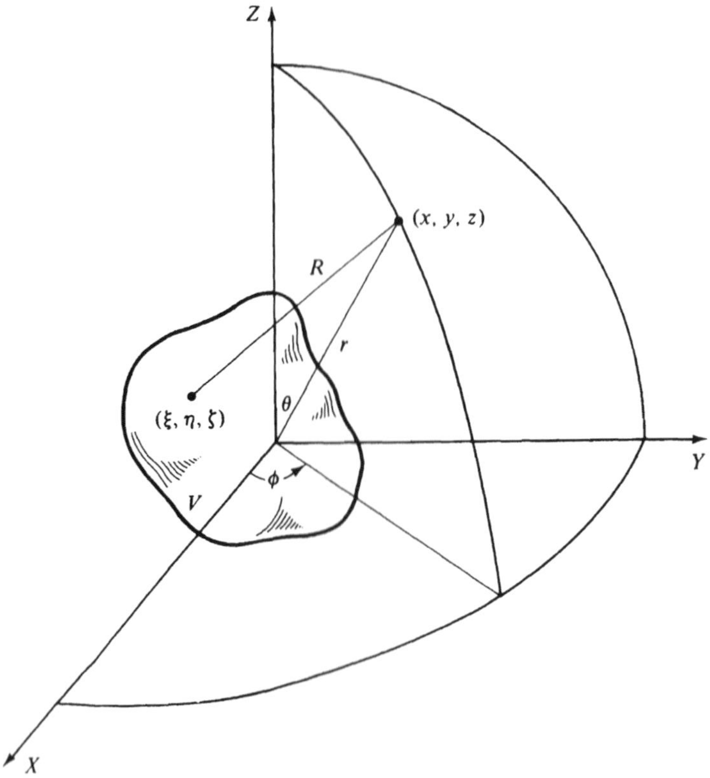

FIG. 1.2 Notation for Far-Field Analysis.

Because (x, y, z) is outside of V and thus is a source-free point, it follows that Maxwell's equations (1.44) reduce to

$$\nabla \times \mathbf{E} = -j\omega\mathbf{B} \qquad \nabla \times \mathbf{B} = \frac{j\omega}{c^2}\mathbf{E}$$
$$\nabla \cdot \mathbf{E} \equiv 0 \qquad \nabla \cdot \mathbf{B} \equiv 0 \tag{1.86}$$

As seen either in the development of Section 1.5 or Section 1.10, **B** can be related to the time-harmonic current sources by the equation

$$\mathbf{B} = \nabla \times \mathbf{A} \tag{1.87}$$

in which

$$\mathbf{A}(x, y, z, t) = \int_V \frac{\mathbf{J}(\xi, \eta, \zeta)e^{j(\omega t - kR)}}{4\pi\mu_0^{-1}R}\, dV \tag{1.88}$$

with $k = \omega/c = 2\pi/\lambda$ the wave number and R the distance from the source point (ξ, η, ζ) to the field point (x, y, z). From (1.86) and (1.87) it follows that, at source-free points (x, y, z),

$$\mathbf{E} = (c^2/j\omega)\, \nabla \times \nabla \times \mathbf{A} \tag{1.89}$$

For this reason it is not necessary to find Φ. The charge distribution on the antenna need not be known for far-field calculations; the current distribution will suffice. The procedure is reduced to finding **A** from (1.88) and **E** and **B** from (1.89) and (1.87).

The distance between source point and field point is given by

$$\begin{aligned} R &= [(x-\xi)^2 + (y-\eta)^2 + (z-\zeta)^2]^{1/2} \\ &= [(r\sin\theta\cos\phi - \xi)^2 + (r\sin\theta\sin\phi - \eta)^2 + (r\cos\theta - \zeta)^2]^{1/2} \\ &= [r^2 - 2r(\xi\sin\theta\cos\phi + \eta\sin\theta\sin\phi + \zeta\cos\theta) + \xi^2 + \eta^2 + \zeta^2]^{1/2} \\ &= r - (\xi\sin\theta\cos\phi + \eta\sin\theta\sin\phi + \zeta\cos\theta) + 0(r^{-1}) \end{aligned} \tag{1.90}$$

in which the last result is obtained via a binomial expansion. If (1.90) is inserted in (1.88) and terms of $0(r^{-2})$ are neglected, one obtains the far-field approximation

$$\mathbf{A}(x, y, z, t) = \frac{e^{j(\omega t - kr)}}{4\pi\mu_0^{-1} r} \int_V \mathbf{J}(\xi, \eta, \zeta) e^{jk\mathcal{L}}\, dV \tag{1.91}$$

in which

$$\mathcal{L} = \xi\sin\theta\cos\phi + \eta\sin\theta\sin\phi + \zeta\cos\theta \tag{1.92}$$

The distance $\mathcal{L}$ can be interpreted as the dot product of: (1) the position vector drawn from the origin to (ξ, η, ζ); and (2) a unit vector drawn from the origin toward (x, y, z). The result in (1.91) can be given the interpretation that $\mathbf{A}(x, y, z, t)$ is expressible as the product of an outgoing spherical wave

$$\frac{e^{j(\omega t - kr)}}{4\pi\mu_0^{-1} r} \tag{1.93}$$

and the directional weighting function

$$\mathcal{A}(\theta, \phi) = \int_V \mathbf{J}(\xi, \eta, \zeta) e^{jk\mathcal{L}}\, d\xi\, d\eta\, d\zeta \tag{1.94}$$

The radiated power pattern of the antenna, given by the function $\mathcal{P}(\theta, \phi)$ watts per square meter can be expressed in terms of this weighting function $\mathcal{A}(\theta, \phi)$. To see this relation, one can first perform the curl operations indicated by (1.87) and (1.89). When this is done and only the terms in r^{-1} are retained, it is found that[14]

$$\mathbf{E} = j\omega \mathbf{1}_r \times (\mathbf{1}_r \times \mathbf{A}) = -j\omega \mathbf{A}_T \tag{1.95}$$

$$\mathbf{H} = \mu_0^{-1}\mathbf{B} = -\left(\frac{j\omega}{\eta}\right)\mathbf{1}_r \times \mathbf{A} = \left(\frac{1}{\eta}\right)\mathbf{1}_r \times \mathbf{E} \tag{1.96}$$

[14]The subscript zero has been dropped on **H** as a simplification, since it is unambiguously clear that the region is free space.

in which $\mathbf{1}_r$ is a unit vector in the radial direction and $\eta = (\mu_0/\varepsilon_0)^{1/2} = 377$ ohms is the impedance of free space. The transverse part of $\mathbf{A}$ is $\mathbf{A}_T = \mathbf{1}_\theta A_\theta + \mathbf{1}_\phi A_\phi$. It can be concluded from a study of (1.95) and (1.96) that the radiated $\mathbf{E}$ and $\mathbf{H}$ fields are entirely transverse, that $\mathbf{E}$ differs from $\mathbf{A}_T$ only by a multiplicative constant, that $\mathbf{H}$ is perpendicular to $\mathbf{E}$, and that

$$\left|\frac{E}{H}\right| = \eta \tag{1.97}$$

The complex Poynting vector yields an average power density which can be written (see Section 1.6)

$$\begin{aligned}\mathcal{P}(\theta, \phi) &= \frac{1}{2}\mathcal{R}e(\mathbf{E} \times \mathbf{H}^*) \\ &= \vec{\mathbf{1}}_r\left[\frac{k^2\eta}{(4\pi r)^2}\right]\left[\frac{1}{2}\mathcal{A}_\theta\mathcal{A}_\theta^* + \frac{1}{2}\mathcal{A}_\phi\mathcal{A}_\phi^*\right]\end{aligned} \tag{1.98}$$

It is customary to call that part of the radiation pattern associated with E_θ the *θ-polarized pattern*, or the *vertically polarized pattern*, and to call that part of the radiation pattern associated with E_ϕ the *ϕ-polarized pattern*, or the *horizontally polarized pattern*. From (1.95) and (1.98), it can be seen that these two patterns are given by the functions

$$\mathcal{P}_{r,\theta}(\theta, \phi) = \frac{1}{2}\left[\frac{k^2\eta}{(4\pi r)^2}\right]|\mathcal{A}_\theta(\theta, \phi)|^2 \tag{1.99}$$

$$\mathcal{P}_{r,\phi}(\theta, \phi) = \frac{1}{2}\left[\frac{k^2\eta}{(4\pi r)^2}\right]|\mathcal{A}_\phi(\theta, \phi)|^2 \tag{1.100}$$

Often one is interested only in the *relative* power densities being radiated in different directions (θ, ϕ), in which case the factor $\frac{1}{2}[k^2\eta/(4\pi r)^2]$ can be suppressed.

Since the unit vectors in spherical and cartesian coordinates are connected by the relations

$$\begin{aligned}\mathbf{1}_\theta &= \mathbf{1}_x \cos\theta\cos\phi + \mathbf{1}_y \cos\theta\sin\phi - \mathbf{1}_z \sin\theta \\ \mathbf{1}_\phi &= -\mathbf{1}_x \sin\phi + \mathbf{1}_y \cos\phi\end{aligned}$$

it follows that the transverse components of (1.94) can be written in the forms

$$\begin{aligned}\mathcal{A}_\theta(\theta, \phi) = \int_V &[\cos\theta\cos\phi\, J_x(\xi, \eta, \zeta) + \cos\theta\sin\phi\, J_y(\xi, \eta, \zeta) \\ &- \sin\theta\, J_z(\xi, \eta, \zeta)]e^{jk\mathcal{L}}\, d\xi\, d\eta\, d\zeta\end{aligned} \tag{1.101}$$

$$\mathcal{A}_\phi(\theta, \phi) = \int_V [-\sin\phi\, J_x(\xi, \eta, \zeta) + \cos\phi\, J_y(\xi, \eta, \zeta)]e^{jk\mathcal{L}}\, d\xi\, d\eta\, d\zeta \tag{1.102}$$

These two equations are the key results of this development and form the basis of pattern analysis and synthesis for actual-source antennas. If one starts with known current distributions, $\mathcal{A}_\theta$ and $\mathcal{A}_\phi$ can be determined from (1.101) and (1.102) and then used in (1.99) and (1.100) to deduce the radiation patterns. This is the analysis problem. Conversely, if desired patterns are specified, (1.101) and (1.102) become integral equations in the sought-for current distributions. This is the synthesis problem.

The results of this section can be summarized by saying that when one is doing pattern analysis of a type I (actual-source) antenna, the steps to follow are these.

1. Place the known current distribution in (1.101) and (1.102) and determine $\mathcal{A}_\theta(\theta, \phi)$ and $\mathcal{A}_\phi(\theta, \phi)$.
2. If the far-field power patterns are desired, use (1.99) and (1.100). Then $|\mathcal{A}_\theta(\theta, \phi)|^2$ and $|\mathcal{A}_\phi(\theta, \phi)|^2$ will give the vertically and horizontally polarized relative power patterns, respectively.
3. If the **E** and **H** fields are desired, use (1.95) and (1.96).

For pattern synthesis, (1.101) and (1.102) become integral equations in the unknown current distribution with $\mathcal{A}_\theta(\theta, \phi)$ and $\mathcal{A}_\phi(\theta, \phi)$ specified.[15]

D. THE FAR-FIELD EXPRESSIONS FOR TYPE II (EQUIVALENT-SOURCE) ANTENNAS[16]

A distinction has already been made between antennas for which the actual source distribution is known to reasonable accuracy and those for which it is not. In the latter case, it is fortunately often true that the fields adjacent to the antenna are fairly well known; it is then useful to surround the antenna by surfaces that exclude all the real sources. If the Stratton-Chu formulation is used, the fields $\mathbf{E}(x, y, z)e^{j\omega t}$ and $\mathbf{B}(x, y, z)e^{j\omega t}$ can then be determined from Equations 1.53 and 1.54 with only surface integrals involved.

An alternate (and equivalent) approach that is rich in physical insight is one in which substitute sources are placed on the surfaces enclosing the antenna. These sources must be chosen so that they produce the same fields at all points exterior to the surfaces as the actual antenna does. The next two sections are concerned with developing this alternate approach.

1.12 The Schelkunoff Equivalence Principle

The concept of equivalent or substitute sources is an old and useful idea that can be traced back to C. Huyghens,[17] but the development to be presented here is patterned after S. A. Schelkunoff.[18]

[15]Often it is a vexing problem to specify the *phase* distribution of $\mathcal{A}_\theta$ and $\mathcal{A}_\phi$ since all that may really be desired is some specified $|\mathcal{A}_\theta(\theta, \phi)|$ or $|\mathcal{A}_\phi(\theta, \phi)|$. In such cases, one can search for that phase distribution of $\mathcal{A}_\theta$ and $\mathcal{A}_\phi$ which results in the simplest physically realizable current distribution. This can be a much more formidable synthesis problem.

[16]Reading the material in Part D of this chapter can be deferred without any loss in continuity until Chapter 3 is reached.

[17]C. Huyghens, *Traité de la Lumiére*, 1690 (English translation: Chicago: The University of Chicago Press, 1945).

[18]S. A. Schelkunoff, "Some Equivalence Theorems of Electromagnetics and their Application to Radiation Problems," *Bell System Tech. Jour.*, 15 (1936), 92–112.

In pursuing this idea, one finds that if the equivalent sources are to reproduce faithfully the external fields, *electric* sources alone will not suffice. It is necessary to introduce fictitious *magnetic* sources. In anticipation of this, consider the situation in which real electric sources (ρ, $\mathbf{J}$) create an electromagnetic field ($\mathbf{E}_1$, $\mathbf{B}_1$) and magnetic sources (ρ_m, $\mathbf{J}_m$) create an electromagnetic field ($\mathbf{E}_2$, $\mathbf{B}_2$). The properties of these fictitious magnetic sources are so chosen that Maxwell's equations are obeyed in the form given below. Away from the sources, no distinction can be made that would allow one to determine which type of source had given rise to either field. The two sets of sources and fields satisfy

$$\begin{aligned}
&\nabla \times \mathbf{E}_1 = -j\omega\mathbf{B}_1 && \text{(a)} \qquad && \nabla \times \mathbf{E}_2 = -\frac{\mathbf{J}_m}{\mu_0^{-1}} - j\omega\mathbf{B}_2 && \text{(e)}\\
&\nabla \times \mathbf{B}_1 = \frac{\mathbf{J}}{\mu_0^{-1}} + \frac{j\omega}{c^2}\mathbf{E}_1 && \text{(b)} && \nabla \times \mathbf{B}_2 = \frac{j\omega}{c^2}\mathbf{E}_2 && \text{(f)}\\
&\nabla \cdot \mathbf{E}_1 = \frac{\rho}{\epsilon_0} && \text{(c)} && \nabla \cdot \mathbf{E}_2 \equiv 0 && \text{(g)}\\
&\nabla \cdot \mathbf{B}_1 \equiv 0 && \text{(d)} && \nabla \cdot \mathbf{B}_2 = \frac{\rho_m}{\mu_0^{-1}} && \text{(h)}
\end{aligned} \tag{1.103}$$

The divergence of (1.103e) combined with (1.103h) reveals that $\nabla \cdot \mathbf{J}_m = -j\omega\rho_m$. In other words, the manner in which the magnetic sources have been introduced insures that the continuity equation applies for magnetic as well as electric sources.

In a development paralleling what is found in Section 1.5, it is useful once again to introduce potential functions, this time by means of the defining relations

$$\mathbf{B}_1 = \nabla \times \mathbf{A} \quad \text{(a)} \qquad \mathbf{E}_2 = -\nabla \times \mathbf{F} \quad \text{(b)} \tag{1.104}$$

As before, $\mathbf{A}$ will be called the *magnetic vector potential function*; by analogy, it is appropriate to call $\mathbf{F}$ the *electric vector potential function*. Equation 1.104a insures compliance with (1.103d); similarly, (1.104b) is in agreement with (1.103g). Equations 1.103a and 1.103f then lead to

$$\nabla \times (\mathbf{E}_1 + j\omega\mathbf{A}) \equiv 0 \qquad \nabla \times \left(\mathbf{B}_2 + \frac{j\omega}{c^2}\mathbf{F}\right) \equiv 0$$

from which

$$\mathbf{E}_1 = -j\omega\mathbf{A} - \nabla\Phi \qquad \mathbf{B}_2 = -\frac{1}{c^2}(j\omega\mathbf{F} + \nabla\Phi_m) \tag{1.105}$$

with Φ and Φ_m called the *electric and magnetic scalar potential functions*, respectively. If the total fields are $\mathbf{E} = \mathbf{E}_1 + \mathbf{E}_2$ and $\mathbf{B} = \mathbf{B}_1 + \mathbf{B}_2$, then

$$\mathbf{E} = -\nabla \times \mathbf{F} - j\omega\mathbf{A} - \nabla\Phi \tag{1.106}$$

$$\mathbf{B} = \nabla \times \mathbf{A} - \frac{1}{c^2}(j\omega\mathbf{F} + \nabla\Phi_m) \tag{1.107}$$

Equations 1.103b and 1.103e can next be converted to the forms

$$\nabla \times \nabla \times \mathbf{A} = \frac{\mathbf{J}}{\mu_0^{-1}} - \frac{j\omega}{c^2}(j\omega\mathbf{A} + \nabla\Phi) \qquad \nabla \times \nabla \times \mathbf{F} = \frac{\mathbf{J}_m}{\mu_0^{-1}} - \frac{j\omega}{c^2}(j\omega\mathbf{F} + \nabla\Phi_m)$$

If the divergences of **A** and **F** are selected to satisfy

$$\nabla \cdot \mathbf{A} = -\frac{j\omega}{c^2}\Phi \qquad \nabla \cdot \mathbf{F} = -\frac{j\omega}{c^2}\Phi_m$$

then these hybrid equations reduce to

$$\nabla^2\mathbf{A} + k^2\mathbf{A} = -\frac{\mathbf{J}}{\mu_0^{-1}} \quad \text{(a)} \qquad \nabla^2\mathbf{F} + k^2\mathbf{F} = -\frac{\mathbf{J}_m}{\mu_0^{-1}} \quad \text{(b)} \qquad (1.108)$$

Finally, (1.103c) and (1.103h) transform to

$$\nabla^2\Phi + k^2\Phi = -\frac{\rho}{\epsilon_0} \quad \text{(a)} \qquad \nabla^2\Phi_m + k^2\Phi_m = -\frac{\rho_m}{\epsilon_0} \quad \text{(b)} \qquad (1.109)$$

The solutions for **A** and Φ have already been given (see Section 1.10) and the solutions for **F** and Φ_m are obviously similar. If the electric and magnetic sources are confined to reside in surfaces, then lineal current densities **K** amperes per meter and $\mathbf{K}_m$ magnetic amperes per meter replace **J** and $\mathbf{J}_m$. In like manner, the areal charge densities ρ_s coulombs per square meter and ρ_{sm} magnetic coulombs per square meter replace ρ and ρ_m. The potential functions are then given by

$$\mathbf{A}(x, y, z, t) = \int_S \frac{\mathbf{K}(\xi, \eta, \zeta)e^{j(\omega t - kR)}}{4\pi\mu_0^{-1}R}\, dS \qquad \mathbf{F}(x, y, z, t) = \int_S \frac{\mathbf{K}_m(\xi, \eta, \zeta)e^{j(\omega t - kR)}}{4\pi\mu_0^{-1}R}\, dS$$

$$\Phi(x, y, z, t) = \int_S \frac{\rho_s(\xi, \eta, \zeta)e^{j(\omega t - kR)}}{4\pi\epsilon_0 R}\, dS \qquad \Phi_m(x, y, z, t) = \int_S \frac{\rho_{sm}(\xi, \eta, \zeta)e^{j(\omega t - kR)}}{4\pi\epsilon_0 R}\, dS$$

$$(1.110)$$

Suppose one desires to find the values that these surface sources should have in order to give a specified electromagnetic field external to S but a *null* field within S. As suggested by Figure 1.3a, let a contour C_δ be constructed such that the leg ab is just outside S and parallel to $\mathbf{B}_{\text{tang}}$; the leg cd is parallel to ab and just inside S; both legs have infinitesimal lengths dl. Since (1.103b) and (1.103f) combine to give $\nabla \times \mathbf{B} = (\mathbf{J}/\mu_0^{-1}) + (j\omega/c^2)\mathbf{E}$, integration of this result and the application of Stokes' theorem yields

$$\oint_{C_\delta} \mathbf{B} \cdot d\mathbf{l} = \int_{S_\delta} \frac{\mathbf{J}}{\mu_0^{-1}} \cdot d\mathbf{S} + \frac{j\omega}{c^2}\int_{S_\delta} \mathbf{E} \cdot d\mathbf{S} \qquad (1.111)$$

in which S_δ is the membranelike surface stretched over the infinitesimal rectangular contour C_δ.

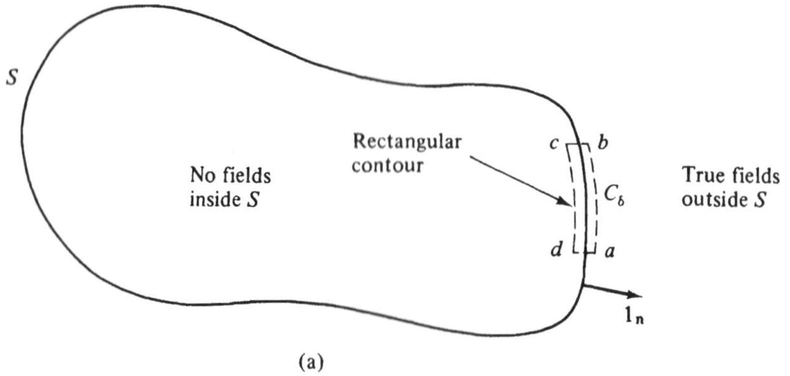

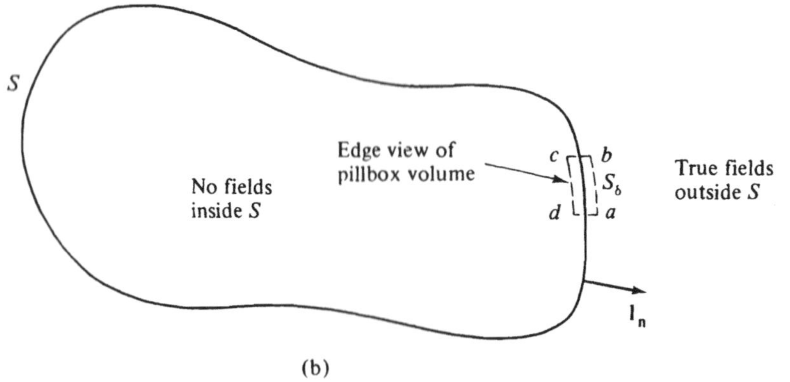

Fig. 1.3 Determination of Equivalent Surface Sources

As the legs *bc* and *da* are shrunk toward the limit zero, with *ab* always outside *S* and *cd* always inside, the electric flux enclosed goes to zero, the current enclosed is Kdl, and the line integral in (1.111) gives $B_{tang}dl$, since there is no contribution from inside. In Figure 1.3a, **K** emerges from the paper if $\mathbf{B}_{tang}$ is in the direction from *a* to *b*. One obtains the result that

$$\frac{\mathbf{K}}{\mu_0^{-1}} = \mathbf{1}_n \times \mathbf{B} \tag{1.112}$$

with $\mathbf{1}_n$ a unit outward-drawn normal vector.

Similarly, if (1.103a) and (1.103e) are added and the result integrated, with the contour taken so that its leg *ab* is parallel to $\mathbf{E}_{tang}$, one finds that

$$\frac{\mathbf{K}_m}{\mu_0^{-1}} = -\mathbf{1}_n \times \mathbf{E} \tag{1.113}$$

Next, imagine that an infinitesimal pillbox has been erected, straddling *S* as shown in Figure 1.3b. If the view in the figure were to be rotated 90°, one would see an infinitesimal disclike area dS, with the upper surface of the pillbox just outside S and the lower surface just inside S. Since (1.103c) and (1.103g) combine to give $\nabla \cdot \mathbf{E} = \rho/\epsilon_0$, integration plus use of the divergence theorem yields

$$\oint_{S_\delta} \mathbf{E} \cdot d\mathbf{S} = \int_{V_\delta} \frac{\rho}{\epsilon_0} dV$$

in which S_δ is the total surface of the pillbox, enclosing the volume V_δ.

Because there is to be no field inside S, as the height of the pillbox is reduced toward the limit zero, with one pillbox face always on each side of S, in the limit $E_n dS = (\rho_s/\epsilon_0)\, dS$, with E_n the normal component of $\mathbf{E}$. Thus

$$\frac{\rho_s}{\epsilon_0} = \mathbf{1}_n \cdot \mathbf{E} \tag{1.114}$$

In like manner, if (1.103d) and (1.103h) are combined and this process is repeated, one finds that

$$\frac{\rho_{sm}}{\mu_0^{-1}} = \mathbf{1}_n \cdot \mathbf{B} \tag{1.115}$$

Schelkunoff's equivalence principle in essence asserts that, if the equivalent sources given by (1.112) through (1.115) are inserted in the potential functions (1.110), and the results are used in (1.106) and (1.107), the calculation of **E** and **B** will give the true fields at all points external to S and null fields at all points internal to S.

It is not immediately obvious that this should be so, since all that has been done so far is to choose equivalent sources that would correspond to the situation that the true fields exist infinitesimally outside S and that no fields exist infinitesimally inside S, with no obvious indication that this will produce the proper field values at points further removed from S. However, Schelkunoff's assertion can be affirmed by following his suggested procedure. If the factor $e^{j\omega t}$ is suppressed, if ψ replaces e^{-jkR}/R, and if equations (1.112) through (1.115) are substituted in (1.110), the result is that

$$\begin{aligned} \mathbf{A} &= \frac{1}{4\pi}\int_S (\mathbf{1}_n \times \mathbf{B})\psi\, dS \qquad & \mathbf{F} &= -\frac{1}{4\pi}\int_S (\mathbf{1}_n \times \mathbf{E})\psi\, dS \\ \Phi &= \frac{1}{4\pi}\int_S (\mathbf{1}_n \cdot \mathbf{E})\psi\, dS \qquad & \Phi_m &= \frac{c^2}{4\pi}\int_S (\mathbf{1}_n \cdot \mathbf{B})\psi\, dS \end{aligned} \tag{1.116}$$

When these expressions for the potential functions are used in (1.106) and (1.107), and the vector transformations $\nabla_F \psi = -\nabla_S \psi$ and $\nabla_F \times [\psi(\mathbf{1}_n \times \mathbf{E})] = \nabla_F \psi \times (\mathbf{1}_n \times \mathbf{E}) = (\mathbf{1}_n \times \mathbf{E}) \times \nabla_S \psi$ are employed, one finds that

$$\mathbf{E}(x, y, z) = \frac{1}{4\pi}\int_S [(\mathbf{1}_n \cdot \mathbf{E})\nabla_S \psi + (\mathbf{1}_n \times \mathbf{E}) \times \nabla_S \psi - j\omega\psi(\mathbf{1}_n \times \mathbf{B})]\, dS \tag{1.117}$$

$$\mathbf{B}(x, y, z) = \frac{1}{4\pi}\int_S \left[\frac{j\omega\psi}{c^2}(\mathbf{1}_n \times \mathbf{E}) + (\mathbf{1}_n \times \mathbf{B}) \times \nabla_S \psi + (\mathbf{1}_n \cdot \mathbf{B})\nabla_S \psi\right] dS \tag{1.118}$$

where ∇_F and ∇_S are the del operators for the field point variables and source point variables respectively; they have been defined by Equations 1.51.

These integral solutions for **E** and **B** at the field point (x, y, z), in terms of the field values over the surface S, are seen to be identical to the Stratton-Chu solutions 1.53 and 1.54 for the case that all the real sources have been excluded from the exterior volume V. Since it has already been shown in Sections 1.7 and 1.9, via a direct integration of Maxwell's equations, that (1.53) and (1.54) give the true fields at all points exterior to S, whereas they give a null result at all points interior to S, it follows that Schelkunoff's equivalence principle has been established.

1.13 The Far Field: Type II Antennas

In a development paralleling what was done in Section 1.11 for actual-source antennas, the potential expressions (1.110) for equivalent-source antennas can be simplified if the field point (x, y, z) is remote from all the sources. The details need not be repeated, but the thread of the argument proceeds as follows.

Away from the sources, (1.103b) and (1.103e) give

$$\mathbf{E}_1 = \frac{c^2}{j\omega}\nabla \times \mathbf{B}_1 = \frac{c^2}{j\omega}\nabla \times \nabla \times \mathbf{A}$$

$$\mathbf{B}_2 = -\frac{1}{j\omega}\nabla \times \mathbf{E}_2 = \frac{1}{j\omega}\nabla \times \nabla \times \mathbf{F}$$

so that (1.106) and (107) simplify to

$$\mathbf{E} = -\nabla \times \mathbf{F} + \frac{c^2}{j\omega}\nabla \times \nabla \times \mathbf{A} \qquad (1.119)$$

$$\mathbf{B} = \nabla \times \mathbf{A} + \frac{1}{j\omega}\nabla \times \nabla \times \mathbf{F} \qquad (1.120)$$

As before, one can dispense with the need to know the charge distributions if the fields are only sought at source-free points; knowledge of the current distributions, which determine **A** and **F**, is sufficient.

The far-field forms of these vector potential functions can be written as the product of the outgoing spherical wave factor (1.93) with the directional weighting functions

$$\mathcal{A}(\theta, \phi) = \int_S \mathbf{K}(\xi, \eta, \zeta)e^{jk\mathcal{L}}\, dS \qquad (1.121)$$

$$\mathcal{F}(\theta, \phi) = \int_S \mathbf{K}_m(\xi, \eta, \zeta)e^{jk\mathcal{L}}\, dS \qquad (1.122)$$

When (1.119) and (1.120) are applied to the far-field forms of **A** and **F** and only the terms in r^{-1} are retained, the result is

$$\mathbf{E} = -j\omega \mathbf{A}_T + jk(\mathbf{1}_r \times \mathbf{F}_T) \tag{1.123}$$

$$\mathbf{H} = \mu_0^{-1}\mathbf{B} = -\frac{j\omega}{\eta}(\mathbf{1}_r \times \mathbf{A}_T) - j\omega\epsilon_0 \mathbf{F}_T = \frac{1}{\eta}\mathbf{1}_r \times \mathbf{E} \tag{1.124}$$

with $\mathbf{A}_T$ and $\mathbf{F}_T$ the transverse components of the vector potential functions. Once again it can be noted that the far field $\mathbf{E}$ and $\mathbf{H}$ are both transverse to the radial direction and are perpendicular to each other, and that $|E/H| = \eta$.

In this case of equivalent sources, the complex Poynting vector gives as the average power density

$$\begin{aligned}\mathcal{P}(\theta, \phi) &= \frac{1}{2}\mathcal{R}e\left\{[-j\omega\mathbf{A}_T + jk(\mathbf{1}_r \times \mathbf{F}_T)] \times \left[\left(\frac{j\omega}{\eta}\right)(\mathbf{1}_r \times \mathbf{A}_T^*) + j\omega\epsilon_0\mathbf{F}_T^*\right]\right\} \\ &= \frac{1}{2}\frac{k^2\eta}{(4\pi r)^2}\mathcal{R}e\left\{\left[\mathcal{A}_T - \frac{1}{c}(\mathbf{1}_r \times \mathcal{F}_T)\right] \times \left[(\mathbf{1}_r \times \mathcal{A}_T^*) + \frac{1}{c}\mathcal{F}_T^*\right]\right\} \\ &= \mathbf{1}_r\frac{k^2\eta}{2(4\pi r)^2}\left[\mathcal{A}_\theta\mathcal{A}_\theta^* + \mathcal{A}_\phi\mathcal{A}_\phi^* + \frac{1}{c^2}(\mathcal{F}_\theta\mathcal{F}_\theta^* + \mathcal{F}_\phi\mathcal{F}_\phi^*)\right. \\ &\quad \left. + \frac{2}{c}\mathcal{R}e(\mathcal{A}_\theta\mathcal{F}_\phi^* - \mathcal{A}_\phi\mathcal{F}_\theta^*)\right]\end{aligned} \tag{1.125}$$

A study of (1.123) reveals that the vertically polarized (E_θ) pattern is related to $\mathcal{A}_\theta$ and $\mathcal{F}_\phi$, whereas the horizontally polarized (E_ϕ) pattern is governed by $\mathcal{A}_\phi$ and $\mathcal{F}_\theta$. Thus the component patterns are given by the functions

$$\mathcal{P}_{r,\theta}(\theta, \phi) = \frac{1}{2}\left[\frac{k^2\eta}{(4\pi r)^2}\right]\left[|\mathcal{A}_\theta(\theta, \phi)|^2 + \frac{1}{c^2}|\mathcal{F}_\phi(\theta, \phi)|^2 + \frac{2}{c}\mathcal{R}e(\mathcal{A}_\theta\mathcal{F}_\phi^*)\right] \tag{1.126}$$

$$\mathcal{P}_{r,\phi}(\theta, \phi) = \frac{1}{2}\left[\frac{k^2\eta}{(4\pi r)^2}\right]\left[|\mathcal{A}_\phi(\theta, \phi)|^2 + \frac{1}{c^2}|\mathcal{F}_\theta(\theta, \phi)|^2 - \frac{2}{c}\mathcal{R}e(\mathcal{A}_\phi\mathcal{F}_\theta^*)\right] \tag{1.127}$$

Once again, the factor $\frac{1}{2}[k^2\eta/(4\pi r)^2]$ can be suppressed when only relative levels are of interest.

As before, the transverse components of $\mathcal{A}$ and $\mathcal{F}$ can be obtained by expanding (1.121) and (1.122) into components. This gives

$$\begin{aligned}\mathcal{A}_\theta(\theta, \phi) = \int_S [&\cos\theta\cos\phi\, K_x(\xi, \eta, \zeta) + \cos\theta\sin\phi\, K_y(\xi, \eta, \zeta) \\ &- \sin\theta\, K_z(\xi, \eta, \zeta)]e^{jk\mathcal{L}}\, dS\end{aligned} \tag{1.128}$$

$$\mathcal{A}_\phi(\theta, \phi) = \int_S [-\sin\phi\, K_x(\xi, \eta, \zeta) + \cos\phi\, K_y(\xi, \eta, \zeta)]e^{jk\mathcal{L}}\, dS \tag{1.129}$$

$$\begin{aligned}\mathcal{F}_\theta(\theta, \phi) = \int_S [&\cos\theta\cos\phi\, K_{xm}(\xi, \eta, \zeta) + \cos\theta\sin\phi\, K_{ym}(\xi, \eta, \zeta) \\ &- \sin\theta\, K_{zm}(\xi, \eta, \zeta)]e^{jk\mathcal{L}}\, dS\end{aligned} \tag{1.130}$$

$$\mathcal{F}_\phi(\theta, \phi) = \int_S [-\sin\phi\, K_{xm}(\xi, \eta, \zeta) + \cos\phi\, K_{ym}(\xi, \eta, \zeta)]e^{jk\mathcal{L}}\, dS \tag{1.131}$$

These four equations are the key results of this development and form the core of pattern analysis and synthesis for most equivalent-source antennas.[19] If one starts with known equivalent-current distributions, $\mathcal{A}_\theta$, $\mathcal{A}_\phi$, $\mathcal{F}_\theta$, and $\mathcal{F}_\phi$ can be determined from (1.128) through (1.131) and then used in (1.126) and (1.127) to deduce the radiation patterns. This is the analysis problem. Conversely, if desired patterns are specified, (1.128) through (1.131) become integral equations in the equivalent current distributions that are sought. This is the synthesis problem.

The results of this section can be summarized by indicating the procedure for doing pattern analysis of a type II (equivalent-source) antenna.

1. Surround the antenna with a closed surface S over which the actual fields are known, at least to a good approximation.
2. Use (1.112) and (1.113) to find the equivalent lineal current densities $\mathbf{K}(\xi, \eta, \zeta)$ and $\mathbf{K}_m(\xi, \eta, \zeta)$ on S.
3. Find $\mathcal{A}_T(\theta, \phi)$ and $\mathcal{F}_T(\theta, \phi)$ from (1.128) through (1.131).
4. If the component power patterns are needed, use (1.126) and (1.127) to determine them.
5. If the far fields $\mathbf{E}$ and $\mathbf{B}$ are required, use (1.119) and (1.120).

For pattern synthesis, (1.128) through (1.131) assume the roles of integral equations in the unknown equivalent-current distributions, with $\mathcal{A}_T(\theta, \phi)$ and $\mathcal{F}_T(\theta, \phi)$ specified.[20]

E. RECIPROCITY, DIRECTIVITY, AND RECEIVING CROSS SECTION OF AN ANTENNA

This penultimate part of Chapter 1 is concerned with the development of several concepts that have proven to be extremely useful in antenna theory. The first of these is the concept of reciprocity, based on a simple deduction from Maxwell's equations. The second (directivity) is a measure of the ability of any antenna to radiate preferentially in some directions relative to others. The last concept (receiving cross section) introduces a measure of the ability of an antenna to "capture" an incoming electromagnetic wave.

[19] Occasionally a design problem will be encountered in which the antenna is very long in one dimension and the sources are essentially independent of that dimension. It is then convenient to assume that the problem is two dimensional and use cylindrical coordinate expressions equivalent to (1.128) through (1.131). See Appendix G for the development of these expressions.

[20] The synthesis problem is actually quite a bit more complicated than this simple statement would suggest. Often it is only $\mathcal{P}_{r,\theta}(\theta, \phi)$ and $\mathcal{P}_{r,\phi}(\theta, \phi)$ that are specified. The division into $\mathcal{A}_T(\theta, \phi)$ and $\mathcal{F}_T(\theta, \phi)$ is immaterial to the desired result, but it may be critical in terms of physical realizability of a synthesized antenna. Another difficulty is that the *phase* of the far-field pattern is seldom specified. This offers the antenna designer an added degree of freedom, but complicates the synthesis problem. One should strive for a phase distribution of the far-field pattern that permits the simplest physically realizable antenna. This can be a formidable undertaking.

1.14 The Reciprocity Theorem

One of the most important and widely used relations in electromagnetic theory is the reciprocity theorem, which will be invoked many times in this text as various subjects are presented. A derivation of this theorem is based on the idea that either of two sets of sources, ($\mathbf{J}^a$, $\mathbf{J}_m^a$, ρ^a, ρ_m^a) or ($\mathbf{J}^b$, $\mathbf{J}_m^b$, ρ^b, ρ_m^b), can be established in a region, producing the fields ($\mathbf{E}^a$, $\mathbf{B}^a$) and ($\mathbf{E}^b$, $\mathbf{B}^b$), respectively. It is assumed that the two sets of sources oscillate at a common frequency. There may be dielectric, magnetic, and conductive materials present in which some or all of these sources reside, but if so the electromagnetic behavior of these materials must be *linear*. The equivalent situation of free and bound sources in free space will be used to represent the behavior of the materials, as a consequence of which Maxwell's curl equations in the free space form,

$$
\begin{aligned}
\nabla \times \mathbf{E}^a &= -\frac{\mathbf{J}_m^a}{\mu_0^{-1}} - j\omega\mathbf{B}^a \qquad & \nabla \times \mathbf{H}^a &= \mathbf{J}^a + j\omega\mathbf{D}^a \\
\nabla \times \mathbf{E}^b &= -\frac{\mathbf{J}_m^b}{\mu_0^{-1}} - j\omega\mathbf{B}^b \qquad & \nabla \times \mathbf{H}^b &= \mathbf{J}^b + j\omega\mathbf{D}^b
\end{aligned} \tag{1.132}
$$

can be used to connect the fields and current sources for each set. Equations 1.132 are a restatement of (1.103) in combined form, with $\mathbf{D} = \epsilon_0\mathbf{E}$ and $\mathbf{H} = \mu_0^{-1}\mathbf{B}$. These curl equations can be dotted as indicated to give

$$
\begin{aligned}
\mathbf{H}^b \cdot \nabla \times \mathbf{E}^a &= -\mu_0\mathbf{H}^b \cdot \mathbf{J}_m^a - j\omega\mu_0\mathbf{H}^b \cdot \mathbf{H}^a \\
\mathbf{E}^a \cdot \nabla \times \mathbf{H}^b &= \mathbf{E}^a \cdot \mathbf{J}^b + j\omega\epsilon_0\mathbf{E}^a \cdot \mathbf{E}^b \\
\mathbf{H}^a \cdot \nabla \times \mathbf{E}^b &= -\mu_0\mathbf{H}^a \cdot \mathbf{J}_m^b - j\omega\mu_0\mathbf{H}^a \cdot \mathbf{H}^b \\
\mathbf{E}^b \cdot \nabla \times \mathbf{H}^a &= \mathbf{E}^b \cdot \mathbf{J}^a + j\omega\epsilon_0\mathbf{E}^b \cdot \mathbf{E}^a
\end{aligned} \tag{1.133}
$$

Since

$$
\begin{aligned}
\nabla \cdot (\mathbf{E}^a \times \mathbf{H}^b - \mathbf{E}^b \times \mathbf{H}^a) = {} & \mathbf{H}^b \cdot \nabla \times \mathbf{E}^a - \mathbf{E}^a \cdot \nabla \times \mathbf{H}^b - \mathbf{H}^a \cdot \nabla \times \mathbf{E}^b \\
& + \mathbf{E}^b \cdot \nabla \times \mathbf{H}^a
\end{aligned}
$$

it follows from (1.133) that

$$
\begin{aligned}
\int_S (\mathbf{E}^a \times \mathbf{H}^b - \mathbf{E}^b \times \mathbf{H}^a) \cdot dS = \int_V (&\mathbf{E}^b \cdot \mathbf{J}^a - \mathbf{B}^b \cdot \mathbf{J}_m^a - \mathbf{E}^a \cdot \mathbf{J}^b \\
&+ \mathbf{B}^a \cdot \mathbf{J}_m^b)\, dV
\end{aligned} \tag{1.134}
$$

in which integration has been taken over a volume V large enough to contain all the sources of both sets, and in which the divergence theorem has been employed. Equation 1.134 is a statement of the reciprocity theorem for sources in otherwise empty space, but with the possibility that some might be bound sources representing the behavior of linear materials. Several special forms of this reciprocity relation have proven useful and can be described as follows.

1. If S is permitted to become a sphere of infinite radius, with the sources confined to a finite volume V, the fields at infinity must consist of outgoing spherical waves for which $E_\theta = \eta H_\phi$ and $E_\phi = -\eta H_\theta$. Under these conditions the surface integral in (1.134) vanishes and one obtains

$$\int_V (\mathbf{E}^b \cdot \mathbf{J}^a - \mathbf{B}^b \cdot \mathbf{J}_m^a)dV = \int_V (\mathbf{E}^a \cdot \mathbf{J}^b - \mathbf{B}^a \cdot \mathbf{J}_m^b)dV \qquad (1.135)$$

Equation 1.135 is a principal reduction of the reciprocity theorem, which is used in circuit theory to demonstrate a variety of useful relationships. It will be used in this text to establish the equality between transmitting and receiving patterns for arbitrary antennas and to develop a basic formula for the mutual impedance between antenna elements.

2. Another important reduction of the reciprocity theorem can be derived by returning to Equation 1.134 and considering the situation illustrated in Figure 1.4a.

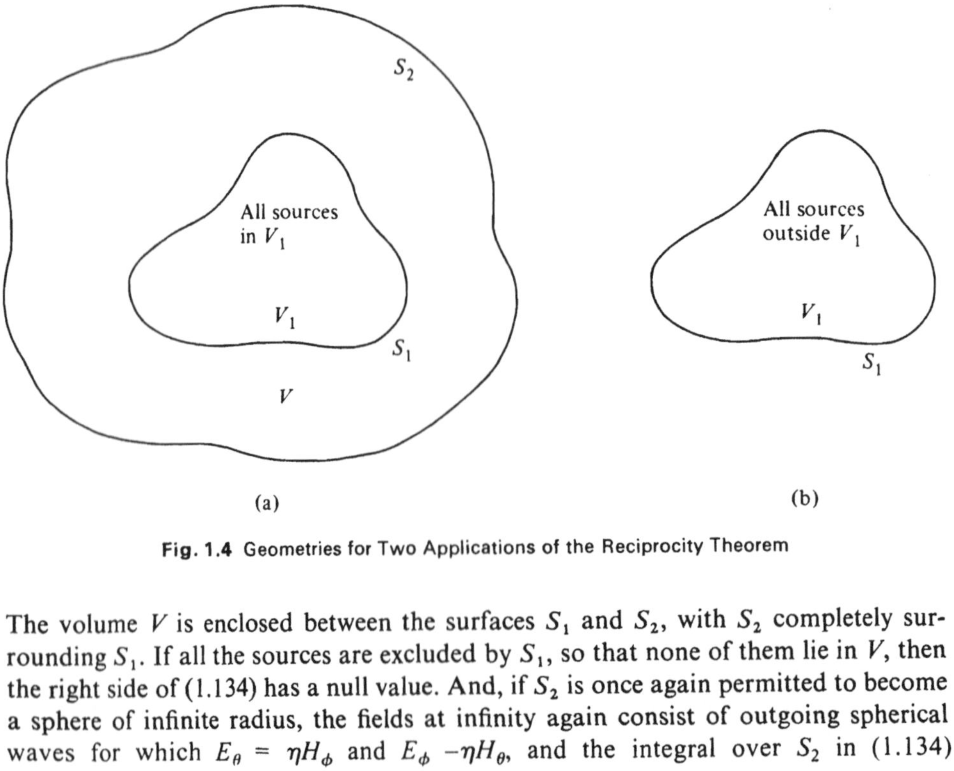

Fig. 1.4 Geometries for Two Applications of the Reciprocity Theorem

The volume V is enclosed between the surfaces S_1 and S_2, with S_2 completely surrounding S_1. If all the sources are excluded by S_1, so that none of them lie in V, then the right side of (1.134) has a null value. And, if S_2 is once again permitted to become a sphere of infinite radius, the fields at infinity again consist of outgoing spherical waves for which $E_\theta = \eta H_\phi$ and $E_\phi -\eta H_\theta$, and the integral over S_2 in (1.134) vanishes. One is left with

$$\int_{S_1} (\mathbf{E}^a \times \mathbf{H}^b - \mathbf{E}^b \times \mathbf{H}^a) \cdot d\mathbf{S} = 0 \qquad (1.136)$$

In (1.136), $d\mathbf{S}$ is drawn outward from V, but no change in (1.136) occurs if $d\mathbf{S}$ is instead drawn outward from V_1, the volume enclosed by S_1. Thus (1.136) can be interpreted by saying that if a surface S_1 is constructed to enclose *all* the sources of both sets in a finite volume V_1, then the fields caused by these sources satisfy the relation in (1.136).

Equation 1.136 will be used in Chapter 7 in the establishment of the induced EMF method for computing the self-impedance of a dipole.

3. A variant on the previous reduction is suggested by Figure 1.4b. The closed surface S_1 *excludes* all the sources, that is, the volume V_1 is source free. Application of (1.134) to this situation once again gives (1.136). This result will be used in Chapter 3 in the derivation of a formula for the scattering from a waveguide-fed slot.

1.15 Equivalence of the Transmitting and Receiving Patterns of an Antenna

The reciprocity theorem can be used to establish the very important result that the transmitting and receiving patterns of an antenna are the same. Consider the situation indicated by Figure 1.5, in which two antennas are sufficiently separated so that each

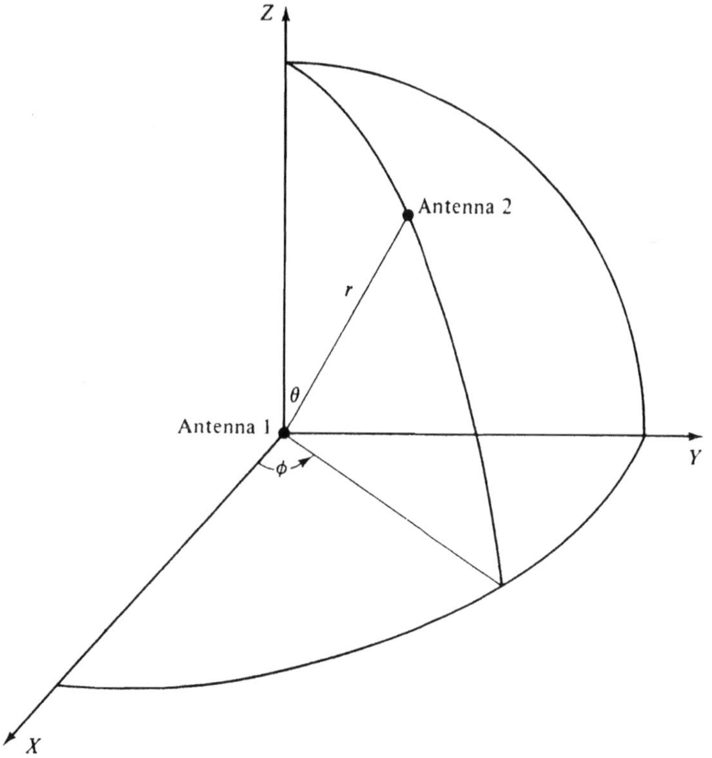

Fig. 1.5 Disposition of Two Antennas in Each Other's Far Field

is in the far-field region of the other. Spherical coordinates are arranged to place antenna 1 at the origin and antenna 2 at the point (r, θ, ϕ). Both antennas can be as simple or complicated as one wishes, so long as they are composed of linear materials. It will be assumed that a transmitter is connected to one antenna via a suitable transmission line and a receiver is connected to the other antenna, also via a suitable transmission line.[21]

In accord with the notation used in Section 1.14, let the a-set of sources occur when a transmitter is attached to antenna 1 and a receiver to antenna 2. The b-set of sources will represent the situation when the positions of transmitter and receiver are interchanged. The combination of transmitter and receiver used in the b-situation need not be the same as in the a-situation.

It will be assumed that a cross section 1 can be found in the transmission line connecting antenna 1 to the transmitter (receiver) at which a single, clean propagating mode exists, and that similarly a cross section 2 can be found in the transmission line connecting antenna 2 to the receiver (transmitter) where a single, clean propagating mode exists.[21]

For the a-situation, let electric and magnetic current sheets be placed at cross section 1 so that the fields on the antenna side are undisturbed, but so that, with the transmitter turned off, the fields on the transmitter side have been erased. From Equations 1.112 and 1.113, these port sources are given by

$$\mathbf{K}_1^a = \mathbf{1}_n \times \mathbf{H}^a \qquad \mathbf{K}_{m1}^a = -\mu_0^{-1}\, \mathbf{1}_n \times \mathbf{E}^a \tag{1.137}$$

in which $\mathbf{1}_n$ points along the transmission line toward antenna 1 and $\mathbf{E}^a$ and $\mathbf{H}^a$ are evaluated in cross section 1. In like manner, let electric and magnetic current sheets be placed at cross section 2 so that the fields on the antenna side are not altered, but so that, with the receiver turned off, the fields on the receiver side have been erased. These port sources satisfy

$$\mathbf{K}_2^a = \mathbf{1}_n \times \mathbf{H}^a \qquad \mathbf{K}_{m2}^a = -\mu_0^{-1}\, \mathbf{1}_n \times \mathbf{E}^a \tag{1.138}$$

with $\mathbf{1}_n$ pointing along the transmission line toward antenna 2, and with $\mathbf{E}^a$ and $\mathbf{H}^a$ evaluated in cross section 2.

The effective replacement of the transmitter and receiver by equivalent sources at ports 1 and 2 leaves intact all the a-sources and fields between these cross sections, including the radiation field transmitted by antenna 1 and received by antenna 2.

In precisely the same manner, equivalent electric and magnetic current sheets can be found which, when placed at cross sections 1 and 2, can serve as proxies for the transmitter and receiver in the b-situation.

These two sets of sources and the fields they produce satisfy the reciprocity theorem in the form of (1.135). The volume V over which the integration is to be performed must encompass all the original sources between the two cross sections plus the equivalent sources in the two cross sections.

[21]As a special case of this analysis, the transmission lines may be lumped circuits.

Consider any point between the cross sections that is occupied by sources. If these sources flow in conductive material,

$$\mathbf{E}^b \cdot \mathbf{J}^a = \mathbf{E}^b \cdot \sigma \mathbf{E}^a \qquad \mathbf{E}^a \cdot \mathbf{J}^b = \mathbf{E}^a \cdot \sigma \mathbf{E}^b \tag{1.139}$$

with σ the conductivity at that point. Similarly, if the material is dielectric,

$$\mathbf{E}^b \cdot \mathbf{J}^a = \mathbf{E}^b \cdot j\omega\epsilon_0\chi_e\mathbf{E}^a \qquad \mathbf{E}^a \cdot \mathbf{J}^b = \mathbf{E}^a \cdot j\omega\epsilon_0\chi_e\mathbf{E}^b \tag{1.140}$$

with χ_e the dielectric susceptibility at that point. And if the material is magnetic,

$$\mathbf{B}^b \cdot \mathbf{J}_m^a = \mathbf{B}^b \cdot j\omega\mu_0\chi_m\mathbf{B}^a \qquad \mathbf{B}^a \cdot \mathbf{J}_m^b = \mathbf{B}^a \cdot j\omega\mu_0\chi_m\mathbf{B}^b \tag{1.141}$$

in which χ_m is the magnetic susceptibility at that point.

In (1.139) through (1.141), the parameters σ, χ_e, and χ_m can be functions of position, depending on the composition and disposition of the materials that comprise the two antennas and their feeds, but, with the assumption that all materials are linear, these parameters are independent of the levels of the fields. Thus, for every source point between the cross sections, equal contributions are made to the integrals on the two sides of (1.135). What remains are the contributions made by the equivalent sources in the cross sections.

Equation 1.135 reduces to

$$\int_{S_1+S_2} (\mathbf{E}^b \cdot \mathbf{K}^a - \mathbf{B}^b \cdot \mathbf{K}_m^a)\, dS = \int_{S_1+S_2} (\mathbf{E}^a \cdot \mathbf{K}^b - \mathbf{B}^a \cdot \mathbf{K}_m^b)\, dS \tag{1.142}$$

with S_1 and S_2 the cross sectional surfaces at ports 1 and 2. By virtue of the set of relations of the type (1.137), this can be converted to the form

$$\begin{aligned}\int_{S_1+S_2} &(\mathbf{E}^b \cdot \mathbf{1}_n \times \mathbf{H}^a + \mathbf{H}^b \cdot \mathbf{1}_n \times \mathbf{E}^a)\, dS \\ &= \int_{S_1+S_2} (\mathbf{E}^a \cdot \mathbf{1}_n \times \mathbf{H}^b + \mathbf{H}^a \cdot \mathbf{1}_n \times \mathbf{E}^b)\, dS\end{aligned} \tag{1.143}$$

It is demonstrated in textbooks dealing with transmission line theory[22] that any propagating mode can be represented by a voltage wave and a current wave, defined so that

$$\mathbf{E}_{\text{tang}}(x, y, z) = V(z)\, \mathbf{g}(x, y) \tag{1.144}$$

$$\mathbf{H}_{\text{tang}}(x, y, z) = I(z)\, \mathbf{h}(x, y) \tag{1.145}$$

with Z the propagation axis and with the functions $\mathbf{g}(x, y)$ and $\mathbf{h}(x, y)$ characteristic of the given mode. The level of these characteristic functions is adjusted so that

$$\int_S \mathbf{1}_z \cdot [\mathbf{g}(x, y) \times \mathbf{h}(x, y)]\, dS = 1 \tag{1.146}$$

[22] See, for example, S. Silver, *Microwave Antenna Theory and Design*, MIT Rad. Lab. Series, Volume 12 (New York: McGraw-Hill Book Co., Inc., 1939), p. 55.

with S a cross-sectional surface. The functions $V(z)$ and $I(z)$ in (1.144) and (1.145) are called the *mode voltage* and *mode current* and are given generally by

$$V(z) = Ae^{-j\beta z} + Be^{-j\beta z} \tag{1.147}$$

$$I(z) = Y_0(Ae^{-j\beta z} - Be^{j\beta z}) \tag{1.148}$$

with A and B constants to be determined by the boundary conditions, with β the propagation constant and Y_0 the characteristic admittance of the mode.

When this representation is applied to the modes at S_1 and S_2, one finds that

$$\begin{aligned}
\mathbf{E}_1^b \cdot \mathbf{1}_n \times \mathbf{H}_1^a &= \mathbf{1}_{z_1} \cdot \mathbf{H}_1^a \times \mathbf{E}_1^b = V_1^b I_1^a \mathbf{1}_{z_1} \cdot \mathbf{h}_1 \times \mathbf{g}_1 \\
\mathbf{H}_1^b \cdot \mathbf{1}_n \times \mathbf{E}_1^a &= \mathbf{1}_{z_1} \cdot \mathbf{E}_1^a \times \mathbf{H}_1^b = V_1^a I_1^b \mathbf{1}_{z_1} \cdot \mathbf{g}_1 \times \mathbf{h}_1 \\
&\vdots \\
\mathbf{H}_2^a \cdot \mathbf{1}_n \times \mathbf{E}_2^b &= \mathbf{1}_{z_2} \cdot \mathbf{E}_2^b \times \mathbf{H}_2^a = V_2^b I_2^a \mathbf{1}_{z_2} \cdot \mathbf{g}_2 \times \mathbf{h}_2
\end{aligned}$$

Substitution in (1.143) together with use of (1.146) gives

$$\sum_{n=1}^{2} V_n^a I_n^b = \sum_{n=1}^{2} V_n^b I_n^a \tag{1.149}$$

This is a key result of the analysis and can be interpreted as saying that the mode voltages and currents at the two ports satisfy the reciprocity theorem.

Next, let Z_{11} be the impedance of antenna 1 referenced at port 1, and let Z_{22} be the impedance of antenna 2 referenced at port 2. Then

$$V_1^a = I_1^a Z_{11} \qquad V_2^b = I_2^b Z_{22} \tag{1.150}$$

Further, let Z_{R1} be the impedance of the receiver transformed to port 1 in the b-situation, and let Z_{R2} be the impedance of the receiver transformed to port 2 in the a-situation. Then

$$V_1^b = -I_1^b Z_{R1} \qquad V_2^a = -I_2^a Z_{R2} \tag{1.151}$$

When (1.150) and (1.151) are placed in (1.149), one finds that

$$\frac{V_1^a}{I_2^a}\left[1 + \left(\frac{Z_{R1}}{Z_{11}}\right)\right] = \frac{V_2^b}{I_1^b}\left[1 + \left(\frac{Z_{R2}}{Z_{22}}\right)\right] \tag{1.152}$$

The transformed receiver impedances are obviously independent of the direction (θ, ϕ) from antenna 1 to antenna 2 and, since the two antennas are in far fields of the other, so too are the driving point impedances Z_{11} and Z_{22}. Thus

$$\frac{V_1^a}{I_2^a} = K \frac{V_2^b}{I_1^b} \tag{1.153}$$

with $K = [1 + (Z_{R2}/Z_{22})]/[1 + (Z_{R1}/Z_{11})]$, a constant.

If V_1^a is held fixed and I_2^a is measured as a function of (θ, ϕ) while antenna 2 is moved along some programmed path on the spherical surface of radius r, the *transmitting* field pattern of antenna 1 is recorded. Reciprocally, if V_2^b is held fixed and I_1^b is measured as a function of (θ, ϕ) while antenna 2 is moved along the same programmed path, the *receiving* field pattern of antenna 1 is recorded. But Equation 1.153 leads to the conclusion that

$$I_1^b(\theta, \phi) = \left(\frac{KV_2^b}{V_1^a}\right) I_2^a(\theta, \phi) \tag{1.154}$$

In words, the normalized transmitting field pattern and the normalized receiving field pattern of any antenna are identical.

Some features of this proof are worth noting. No specification of the size, shape, or type of either antenna was necessary, nor were there any restrictions on the types of transmission lines feeding the two antennas, except that each should exhibit a single, clear propagating mode at the chosen ports. The materials of which the antennas and their feeds were composed were arbitrary except that they needed to be linear. It was not necessary for either antenna to be matched to its transmission line, nor was there any requirement that the transmitter or receiver be matched to either transmission line. Also, there was no restriction on the orientation of antenna 2 as it moved along its programmed path. It could be continuously reoriented to measure $E_\theta(\theta, \phi)$, or $E_\phi(\theta, \phi)$, or $E_z(\theta, \phi)$, or some arbitrarily shifting polarization. All that is needed is for antenna 2 to replicate its orientation at each point along the path after it has shifted from receive to transmit. One can conclude from this that the proof is very general.

Equation 1.154 establishes the equivalence of the transmitting and receiving *field* patterns of any antenna. A simple extension shows that this equivalence applies to the power patterns as well. If (1.154) is multiplied by its complex conjugate, the result can be used to deduce that

$$\frac{1}{2}|I_1^b(\theta, \phi)|^2 R_{R1} = \left(\frac{R_{R1}}{R_{R2}}\right)\left|\frac{KV_2^b}{V_1^a}\right|^2 \frac{1}{2}|I_2^a(\theta, \phi)|^2 R_{R2} \tag{1.155}$$

The quantities $|I_2^a|^2\ R_{R2}/2$ and $|I_1^b|^2\ R_{R1}/2$ that appear in (1.155) are the powers absorbed in the receiver when antenna 1 is transmitting and receiving, respectively. Since each is linearly proportional to the power density of the waves passing the receiving antenna, it is proper to infer that they are measures of the transmitting and receiving *power* patterns of antenna 1. With $K' = (R_{R1}/R_{R2})|KV_2^b/V_1^a|^2$, one can write

$$\mathcal{P}^{rec}(\theta, \phi) = K'\mathcal{P}^{tr}(\theta, \phi) \tag{1.156}$$

Care must be taken in interpreting (1.156). For example, if antenna 2 is linearly polarized and always oriented as it moves along its programmed path, in order to receive or transmit only θ-polarized waves, then (1.156) becomes

$$\mathcal{P}_\theta^{rec}(\theta, \phi) = K'\mathcal{P}_\theta^{tr}(\theta, \phi) \tag{1.157}$$

from which one can conclude that the normalized θ-polarized component of the power pattern of antenna 1 is the same for receive and transmit. Similarly, if antenna 2 is linearly polarized but aligned to receive or transmit only ϕ-polarized waves, (1.156) reduces to

$$\mathcal{P}_\phi^{rec}(\theta, \phi) = K'\mathcal{P}_\phi^{tr}(\theta, \phi) \tag{1.158}$$

and once again equivalence is demonstrated in the component power patterns for antenna 1. And if, for example, antenna 1 is linearly polarized with only an E_θ electric field, then (1.158) gives a null result, as it should. If antenna 1 does not radiate an E_ϕ field, it cannot detect an incoming E_ϕ field.

The sum of Equations 1.157 and 1.158 shows that the *total* power patterns are equivalent:

$$\mathcal{P}_{total}^{rec}(\theta, \phi) = K'\mathcal{P}_{total}^{tr}(\theta, \phi) \tag{1.159}$$

Acceptance of the conclusion that the normalized total power patterns of any antenna are the same for transmit and receive, and thus that one need not determine both, still leaves a measurement difficulty that should be noted. This concerns the fact that not any antenna can be chosen to play the role of antenna 2, make one traverse of the programmed path, and at each point in the path be oriented so that the received powers in (1.155) coincide with the power densities in (1.159). This will occur only if antenna 2 is *polarization-matched* to antenna 1. For example, if antenna 1 is circularly polarized, antenna 2 must be circularly polarized in the proper screw sense in order to have the received powers in the *a*- and *b*-situations that can be interpreted as the total radiated and received power patterns of antenna 1.

However, if one is content to use as antenna 2 a linearly polarized antenna, make *two* traverses of the programmed path, one with θ-orientation and the other with ϕ-orientation, and keep transmit power and receiver sensitivity stable, then the separate measurements give the component power patterns. Their sum gives the total power pattern, and Equations 1.157 through 1.159 indicate that it does not matter whether the measurements are made with antenna 1 transmitting and antenna 2 receiving, or vice versa.

1.16 Directivity and Gain

Often a principal goal in antenna design is to establish a specified radiation pattern $\mathcal{P}(\theta, \phi)$ watts per square meter through a suitable arrangement of sources. The specified pattern frequently embodies the intent to enhance the radiation in certain directions and suppress it in others. A useful measure of this is the directivity, which is simply the radiated power density in the direction (θ, ϕ) divided by the radiated power density averaged over all directions; that is,

$$D(\theta, \phi) = \frac{\mathcal{P}(\theta, \phi)}{(1/4\pi r^2)\int_0^\pi \int_0^{2\pi} \mathcal{P}(\theta', \phi') r^2 \sin\theta' \, d\theta' \, d\phi'} \tag{1.160}$$

Equation 1.160 contains the implications that the origin for spherical coordinates has been chosen somewhere in the immediate vicinity of the antenna, and that power densities are being evaluated on the surface of a sphere whose radius r is large enough to ensure being in the far field of the antenna.

If the *radiation intensity* is defined by

$$P(\theta, \phi) = r^2 \mathcal{P}(\theta, \phi) \tag{1.161}$$

then, since $\mathcal{P}(\theta, \phi)$ is measured in watts per square meter, it follows that $P(\theta, \phi)$ is measured in watts per steradian. Substitution in (1.160) gives the equivalent expression

$$D(\theta, \phi) = \frac{4\pi P(\theta, \phi)}{\int_0^{\pi} \int_0^{2\pi} P(\theta', \phi') \sin \theta' \, d\theta' \, d\phi'} \tag{1.162}$$

The value $D(\theta, \phi)$ is a pure numeric. It will have a value less than unity in directions in which radiation has been suppressed, and a value exceeding unity where the radiation has been enhanced. If (θ_0, ϕ_0) is the direction in which the radiation intensity is greatest, then D has its largest value at (θ_0, ϕ_0) and $D(\theta_0, \phi_0)$ is the *peak* directivity.

In characterizing an antenna, one must be careful to distinguish between directivity and gain. *Directivity* is used to compare the radiation intensity in a given direction to the average radiation intensity and thus pays no heed to the power losses in the materials comprising the antenna. *Gain* includes these losses, and the definition of gain is therefore

$$G(\theta, \phi) = \frac{\mathcal{P}(\theta, \phi)}{P_{acc}/4\pi r^2} \tag{1.163}$$

in which P_{acc} is the total power accepted by the antenna from the transmitter, measured in watts. The denominator of (1.163) is the value, in watts per square meter, that the radiated power density would have if all the power accepted by the antenna were radiated isotropically. Since the power accepted is greater than the actual power radiated, the denominator of (1.163) is larger than the denominator of (1.160), and, as a consequence, $G(\theta, \phi) < D(\theta, \phi)$.

Most antennas are constructed of linear materials; in this case, one may argue that

$$P_{acc} = K_L \int_0^{\pi} \int_0^{2\pi} \mathcal{P}(\theta', \phi') r^2 \sin \theta' \, d\theta' \, d\phi' \tag{1.164}$$

with K_L a pure real constant that has a value somewhat greater than unity. When this is so, Equation 1.163 becomes

$$G(\theta, \phi) = \frac{D(\theta, \phi)}{K_L} \tag{1.165}$$

The gain and directivity differ by a multiplicative factor that is independent of direction. In particular, the peak gain occurs in the same direction (θ_0, ϕ_0) as the peak directivity.

Often, gain and directivity are expressed in decibels (*dB*). From (1.165),

$$\log_{10} G(\theta, \phi) = \log_{10} D(\theta, \phi) - \log_{10} K_L \tag{1.166}$$

The gain in any direction is seen to be 10 $\log_{10} K_L$ decibels below the directivity in that direction; 10 $\log_{10} K_L$ thus represents the power losses in the materials forming the antenna.

For some applications it is useful to introduce the concept of *partial* directivity and *partial* gain. As an example of how this is done, a return to Equations 1.98 through 1.100 or 1.125 through 1.127 helps to recall that

$$\mathcal{P}(\theta, \phi) = \mathcal{P}_{r,\theta}(\theta, \phi) + \mathcal{P}_{r,\phi}(\theta, \phi) \tag{1.167}$$

If this relation is inserted in Equation 1.160, it can be seen that it is possible to write

$$D(\theta, \phi) = D'(\theta, \phi) + D''(\theta, \phi) \tag{1.168}$$

in which

$$D'(\theta, \phi) = \frac{\mathcal{P}_{r,\theta}(\theta, \phi)}{(1/4\pi r^2)\int_0^{\pi}\int_0^{2\pi} \mathcal{P}(\theta', \phi')r^2 \sin\theta' \, d\theta' \, d\phi'} \tag{1.169}$$

and

$$D''(\theta, \phi) = \frac{\mathcal{P}_{r,\phi}(\theta, \phi)}{(1/4\pi r^2)\int_0^{\pi}\int_0^{2\pi} \mathcal{P}(\theta', \phi')r^2 \sin\theta' \, d\theta' \, d\phi'} \tag{1.170}$$

are the partial directivities associated with the θ-component and ϕ-component patterns, respectively. Similar definitions follow readily for the partial gains.

An example of the utility of this concept would be when an antenna is to be designed to give peak radiation at an angle (θ_0, ϕ_0), but all the radiation should be θ-polarized; any ϕ-polarized radiation is unwanted, but for practical reasons some may be unavoidable. In such a circumstance it is the peak *partial* directivity $D'(\theta_0, \phi_0)$ that is a pertinent measure, not the peak total directivity $D(\theta_0, \phi_0)$.

The division of the total power pattern into components can be done in other ways than the θ/ϕ partition indicated above. For example, the decomposition could equally well be into right-handed and left-handed circularly polarized component power patterns. In that case one could identify right-handed and left-handed partial directivities and gains.

1.17 Receiving Cross Section

A receiving antenna will absorb energy from an incident plane wave and feed it via a transmission line to its terminating impedance. A useful measure of its ability to do this results from introducing the concept of the absorption cross section of the antenna or, as it is more commonly known, its *equivalent receiving cross-sectional area.* If S is the power density of the incoming plane wave in watts per square meter and P_r is the absorbed power in watts, then the equation

$$P_r(\theta, \phi) = SA_r(\theta, \phi) \tag{1.171}$$

serves to define the receiving cross section, in square meters, as a function of the angle of arrival of the incoming signal. In order to have $A_r(\theta, \phi)$ be a maximum measure of the capture property of the antenna, it is customary to assume that the incoming plane wave is polarization matched to the antenna, and that the antenna is terminated by a matched receiver. With these assumptions, Equations 1.155 and 1.159 are applicable and one can write

$$SA_r(\theta, \phi) = \frac{1}{2}|I_1^b(\theta, \phi)|^2 R_{R1} = P_{total}^{rec}(\theta, \phi) = K'\mathcal{P}_{total}^{tr}(\theta, \phi) \tag{1.172}$$

An integration of (1.172) gives

$$\frac{S}{4\pi}\int_0^{\pi}\int_0^{2\pi} A_r(\theta', \phi')\sin\theta'\, d\theta'\, d\phi' = \frac{K'}{4\pi r^2}\int_0^{\pi}\int_0^{2\pi} \mathcal{P}_{total}^{tr}(\theta', \phi')r^2 \sin\theta'\, d\theta'\, d\phi' \tag{1.173}$$

If the ratio of (1.172) to (1.173) is taken, one obtains

$$\frac{A_r(\theta, \phi)}{\bar{A}_r} = D(\theta, \phi) \tag{1.174}$$

in which $D(\theta, \phi)$ is the directivity of antenna 1 when it is transmitting, as given by (1.160). Then $\bar{A}_r$ is the average receiving cross section of antenna 1, defined by

$$\bar{A}_r = \frac{1}{4\pi}\int_0^{\pi}\int_0^{2\pi} A_r(\theta', \phi')\sin\theta'\, d\theta'\, d\phi' \tag{1.175}$$

It is a remarkable fact that the average receiving cross section $\bar{A}_r$ is the same for all lossless antennas that are polarization matched. This can be demonstrated as follows.

Consider again the situation of two antennas, depicted as in Figure 1.5, with antenna 1 transmitting and antenna 2 receiving in the *a*-situation and the reverse occurring in the *b*-situation. To obtain maximum power transfer, assume that in the *a*-situation the transmitter attached to antenna 1 has an internal emf V_g and an internal impedance that has been adjusted to equal Z_{11}^*, with Z_{11} the driving point impedance of antenna 1. Similarly, in the *b*-situation, let the transmitter attached to antenna 2 have an internal emf V_g and an internal impedance Z_{22}^*, with Z_{22} the driving point impedance of antenna 2.

In the *a*-situation, $I_1^a = V_g/2R_{11}$ and the power delivered to antenna 1 is $\frac{1}{2}|I_1^a|^2 R_{11} = |V_g|^2/8R_{11}$. If the losses in the antenna can be neglected, all of this power is radiated, with an average density in watts per square meter given by

$$\frac{|V_g|^2/8R_{11}}{4\pi r^2}$$

When antenna 2 is located at the point (r, θ, ϕ), the power density in the wave arriving from antenna 1 is given in watts per square meter by

$$\mathcal{P}(\theta, \phi) = \frac{|V_g|^2/8R_{11}}{4\pi r^2} D_1(\theta, \phi)$$

in which $D_1(\theta, \phi)$ is the directivity of antenna 1. If use is made of (1.171) with $S = \mathcal{P}(\theta, \phi)$, it can be argued that the power absorbed by the receiver attached to antenna 2 is given by

$$P_r = \frac{|V_g|^2/8R_{11}}{4\pi r^2} D_1(\theta, \phi) A_{r,2}(\theta, \phi) \text{ watts}$$

with $A_{r,2}(\theta, \phi)$ the receiving cross section of antenna 2. Use of (1.174) converts this to

$$P_r = \frac{|V_g|^2/8R_{11}}{4\pi r^2} D_1(\theta, \phi) D_2(\theta, \phi) \bar{A}_{r,2} \tag{1.176}$$

The power absorbed is also given by $(1/2)\mathcal{R}e\, I_2^a I_2^{a*} Z_{R2}$ but, with a matched receiver, $Z_{R2} = Z_{22}^*$, and thus

$$P_r = \tfrac{1}{2}|I_2^a|^2 R_{22} \tag{1.177}$$

When (1.176) and (1.177) are combined, the result can be written in the form

$$D_1(\theta, \phi) D_2(\theta, \phi) \bar{A}_{r,2} = 16\pi r^2 \frac{|I_2^a|^2 R_{11} R_{22}}{|V_g|^2} \tag{1.178}$$

If this analysis is repeated for the b-situation one finds that

$$D_1(\theta, \phi) D_2(\theta, \phi) \bar{A}_{r,1} = 16\pi r^2 \frac{|I_1^b|^2 R_{11} R_{22}}{|V_g|^2} \tag{1.179}$$

The currents and voltages in the two situations are related generally by Equation 1.152. In the circumstance being considered here, $V_1^a = I_1^a Z_{11} = (V_g/2R_{11})Z_{11}$, $V_2^b = I_2^b Z_{22} = (V_g/2R_{22})Z_{22}$, $Z_{R1} = Z_{11}^*$, and $Z_{R2} = Z_{22}^*$; as a consequence of this, (1.152) reduces to

$$I_2^a = I_1^b \tag{1.180}$$

Hence, upon comparing (1.178) and (1.179), one can see that

$$\bar{A}_{r,1} = \bar{A}_{r,2} \tag{1.181}$$

Since antennas 1 and 2 are completely arbitrary (except that they must be polarization-matched), Equation 1.181 is a general result.

The value of the constant $\bar{A}_r$ for linearly polarized antennas can be deduced as follows: Let antenna 1 be completely arbitrary and located at the origin, as shown in Figure 1.5, except that it is linearly polarized and has been oriented to transmit an **E** field that has only a θ-component. Antenna 2 is a single current element of length dl, located at the point (r, θ, ϕ), and oriented parallel to $\mathbf{1}_\theta$ so that the two antennas are polarization matched.

In the a-situation, let antenna 1 be transmitting with antenna 2 *absent*. In the b-situation, the current element $\mathbf{1}_\theta I_2^b\, dl$ (antenna 2) is present and radiating, and antenna 1 is receiving. Port 2 is taken to be the θ-directed line segment of length dl located at (r, θ, ϕ). In this case the reciprocity relation (1.149) becomes

$$V_1^a I_1^b + V_2^a I_2^b = V_1^b I_1^a$$

which can be rewritten in the form

$$V_1^a I_1^b - V_1^b I_1^a = E_\theta^a(r, \theta, \phi) I_2^b\, dl \tag{1.182}$$

Since $I_1^a = V_1^a/Z_{11}$ when antenna 1 is transmitting, and $I_1^b = -V_1^b/Z_{R1} = -V_1^b/Z_{11}^*$ when antenna 1 is receiving, (1.182) assumes the form

$$V_1^a V_1^b \frac{Z_{11} + Z_{11}^*}{Z_{11} Z_{11}^*} = \frac{2R_{11}}{|Z_{11}|^2} V_1^a V_1^b = E_\theta^a I_2^b\, dl \tag{1.183}$$

When (1.183) is multiplied by its complex conjugate, the result is

$$\frac{4R_{11}^2}{|Z_{11}|^4} |V_1^a V_1^b|^2 = |E_\theta^a|^2 |I_2^b\, dl|^2 \tag{1.184}$$

In the a-situation, antenna 1 accepts an amount of power given by

$$P_{acc} = \frac{1}{2} |I_1^a|^2 R_{11} = \frac{1}{2} \frac{|V_1^a|^2}{|Z_{11}|^2} R_{11}$$

from the transmitter and, if losses in antenna 1 are neglected, all of this power is radiated. The power density at (r, θ, ϕ) in the a-situation is, therefore,

$$\frac{1}{2} \frac{|E_\theta^a(r, \theta, \phi)|^2}{\eta} = \frac{|V_1^a|^2}{|Z_{11}|^2} R_{11} \frac{D_1(\theta, \phi)}{4\pi r^2} \tag{1.185}$$

When $|E_\theta^a|^2$ is eliminated from (1.184) and (1.185), one obtains the result that

$$|V_1^b|^2 = \frac{\eta |Z_{11}|^2/R_{11}}{16\pi r^2} D_1(\theta, \phi) |I_2^b\, dl|^2 \tag{1.186}$$

In the b-situation, the receiver attached to antenna 1 absorbs the power

$$P_{abs} = \frac{1}{2}|I_1^b|^2 R_{11} = \frac{1}{2}\frac{|V_1^b|^2}{|Z_{11}|^2} R_{11} \tag{1.187}$$

since $Z_{R1} = Z_{11}^*$. This absorbed power can also be expressed in terms of the power density in the waves radiated by the current element and the receiving cross section of antenna 1. From Equations 1.99 and 1.101 the maximum[23] power density radiated by a single current element is

$$\mathcal{P}_{r,\theta} = \frac{1}{2}\frac{k^2\eta}{(4\pi r)^2}|I_2^b\, dl|^2 \tag{1.188}$$

and thus the power absorbed by antenna 1 is also given by

$$P_{abs} = \frac{1}{2}\frac{k^2\eta}{(4\pi r)^2}|I_2^b\, dl|^2 A_{r,1}(\theta, \phi) \tag{1.189}$$

If (1.187) and (1.189) are combined and the result solved for $|V_1^b|^2$, further combination with (1.186) gives

$$A_{r,1}(\theta, \phi) = \frac{\lambda^2}{4\pi} D_1(\theta, \phi) \tag{1.190}$$

and thus the universal value of the average receiving cross section for linearly polarized antennas is $\lambda^2/4\pi$.

Equation 1.190 is an extremely useful result. It permits computation of the optimum power level in a receiver which is attached to an antenna of peak directivity $D(\theta_0, \phi_0)$ when the power density in the incoming signal is known. This value is diminished slightly by the losses in the antenna. It is also diminished by the multiplicative factor $(1 - |\Gamma|^2)$ when the receiver and the antenna are mismatched, with Γ the reflection coefficient.[24]

F. POLARIZATION

This concluding section of Chapter 1 is concerned with characterizing the polarization of an electromagnetic field far from the sources which produce it. Such characterization is important in many practical applications. Prominent examples include the following. (1) For purposes of optimizing propagation through a selective medium (such as the ionosphere), or optimizing back-scattering off a target, it may be desirable to specify the polarization the wave should have. This places a constraint on the design of the transmitting antenna. (2) When a sum pattern is required to have a

[23]It is the maximum value that should be used since the current element is oriented so that its maximum power density is directed at antenna 1.

[24]See, for example, Silver, *Microwave Antenna Theory, and Design*, pp. 51–53.

specified polarization and low side lobes, it is important to check that the antenna being proposed does not produce a cross-polarized pattern at a height which exceeds the desired side lobe level. This possibility exists, for example, with parabolic reflector antennas. (3) The polarization of an incoming wave may have to be accepted, which places a constraint on the design of an antenna that will receive this wave optimally. (4) The polarization of an incoming wave may be unpredictable, in which case it may be desirable to design a receiving antenna which will respond equally to all polarizations. To be equipped to deal with these and similar problems, it is important to be able to describe the polarization of an electromagnetic wave unambiguously.

1.18 Polarization of the Electric Field

It has been shown in Sections 1.11 and 1.13 that the far field of a transmitting antenna can be viewed as the product of an outgoing spherical wave and a complex directional weighting function. For the electric field (which is conventionally used as the vehicle for describing polarization), this complex directional weighting function is given by (1.95) for type I antennas and by (1.123) for type II antennas. In either case, at a far field point (r, θ, ϕ), the electric field can be represented by

$$\mathbf{E} = (\mathbf{1}_\theta E_\theta + \mathbf{1}_\phi E_\phi)e^{j\omega t} \tag{1.191}$$

when time-harmonic sources are used in the transmitting antenna.

The functions $E_\theta(r, \theta, \phi)$ and $E_\phi(r, \theta, \phi)$ that appear in (1.191) are, in general, *complex*. If this is recognized by the notation

$$E_\theta = E'_\theta + jE''_\theta \qquad E_\phi = E'_\phi + jE''_\phi \tag{1.192}$$

then it can be appreciated that what is really meant by (1.191) is that

$$\begin{aligned}\mathbf{E}(r, \theta, \phi, t) &= \Re e[\mathbf{1}_\theta(E'_\theta + jE''_\theta) + \mathbf{1}_\phi(E'_\phi + jE''_\phi)]e^{j\omega t} \\ &= \mathbf{1}_\theta(E'_\theta \cos \omega t - E''_\theta \sin \omega t) + \mathbf{1}_\phi(E'_\phi \cos \omega t - E''_\phi \sin \omega t)\end{aligned} \tag{1.193}$$

with E'_θ, E''_θ, E'_ϕ, R''_ϕ all real functions of r, θ, and ϕ.

Equation 1.193 can be rewritten in the form

$$\mathbf{E} = \mathbf{1}_\theta A \cos(\omega t + \alpha) + \mathbf{1}_\phi B \cos(\omega t + \beta) \tag{1.194}$$

in which

$$\begin{aligned} A &= \sqrt{(E'_\theta)^2 + (E''_\theta)^2} \qquad & B &= \sqrt{(E'_\phi)^2 + (E''_\phi)^2} \\ \alpha &= \arctan \frac{E''_\theta}{E'_\theta} \qquad & \beta &= \arctan \frac{E''_\phi}{E'_\phi} \end{aligned} \tag{1.195}$$

With no loss in generality, the origin of time can be selected so that $\alpha = 0$ (that is, $E''_\theta = 0$). Then (1.194) becomes

$$\mathbf{E} = \mathbf{1}_\theta A \cos \omega t + \mathbf{1}_\phi B \cos(\omega t + \beta) \tag{1.196}$$

Equation 1.196 is a particularly convenient representation of the electric field for the purpose of identifying its polarization.

(a) LINEAR POLARIZATION If $B = 0$, the electromagnetic wave is said to be *linearly polarized* in the θ-direction. Similarly, if $A = 0$, the wave is ϕ-polarized. But more generally, if $\beta = 0$ but $A, B \neq 0$, the θ and ϕ components of the electric field are *in phase*. The polarization is then tilted, but it is still linear, as can be seen from the time plots of Figure 1.6a. Therefore the most general example of linear polarization occurs when E_θ and E_ϕ are in phase.

(b) CIRCULAR POLARIZATION If $A = B$ and $\beta = -90°$, Equation 1.196 becomes

$$\mathbf{E} = A(\mathbf{1}_\theta \cos \omega t + \mathbf{1}_\phi \sin \omega t) \tag{1.197}$$

In this case the *magnitude* of $\mathbf{E}$ is constant with time. The *angle* that $\mathbf{E}$ makes with the $\mathbf{1}_\theta$ direction is ωt and this angle changes linearly with time. The locus of the tip of $\mathbf{E}$ is a circle, as indicated in Figure 1.6b. For this reason, the field is said to be *circularly polarized.*

The sequence in Figure 1.6b is drawn as though the observer were looking toward the transmitting antenna from afar, along a longitudinal line in the (θ, ϕ) direction. The progression of $\mathbf{E}$ with time is seen to be counterclockwise, which is the direction of rotation a right-hand screw would have if it were being turned to progress in the direction of propagation. For this reason, (1.197) is said to represent a *right-handed* circularly polarized wave. If one were to write

$$\mathbf{E} = A(\mathbf{1}_\theta \cos \omega t - \mathbf{1}_\phi \sin \omega t) \tag{1.198}$$

so that E_ϕ *leads* E_θ by 90°, instead of lagging by 90° as in (1.197), then a *left-handed* circularly polarized wave would be described.

(c) ELLIPTICAL POLARIZATION The most general case of (1.196) occurs when $A \neq B$, $\beta \neq 0$. The magnitude of $\mathbf{E}$ is given by

$$|\mathbf{E}(t)| = [A^2 \cos^2 \omega t + B^2 \cos^2 (\omega t + \beta)]^{1/2} \tag{1.199}$$

If the time derivative of this function is set equal to zero, the extrema of $|\mathbf{E}(t)|$ can be identified. They occur at angles $\omega t = \delta$ governed by

$$\tan 2\,\delta = -\frac{B^2 \sin 2\beta}{A^2 + B^2 \cos 2\beta} \tag{1.200}$$

If δ_1 is the angle in the first quadrant which satisfies (1.200), then $\delta_2 = \delta_1 + \pi/2$ also satisfies (1.200).

Substitution of the angles δ_1 and δ_2 in (1.196) reveals both the direction and magnitude of each of the two extrema of $\mathbf{E}(t)$. The two directions are at right angles to each other and form the principal axes of the locus. It is left as an exercise to show

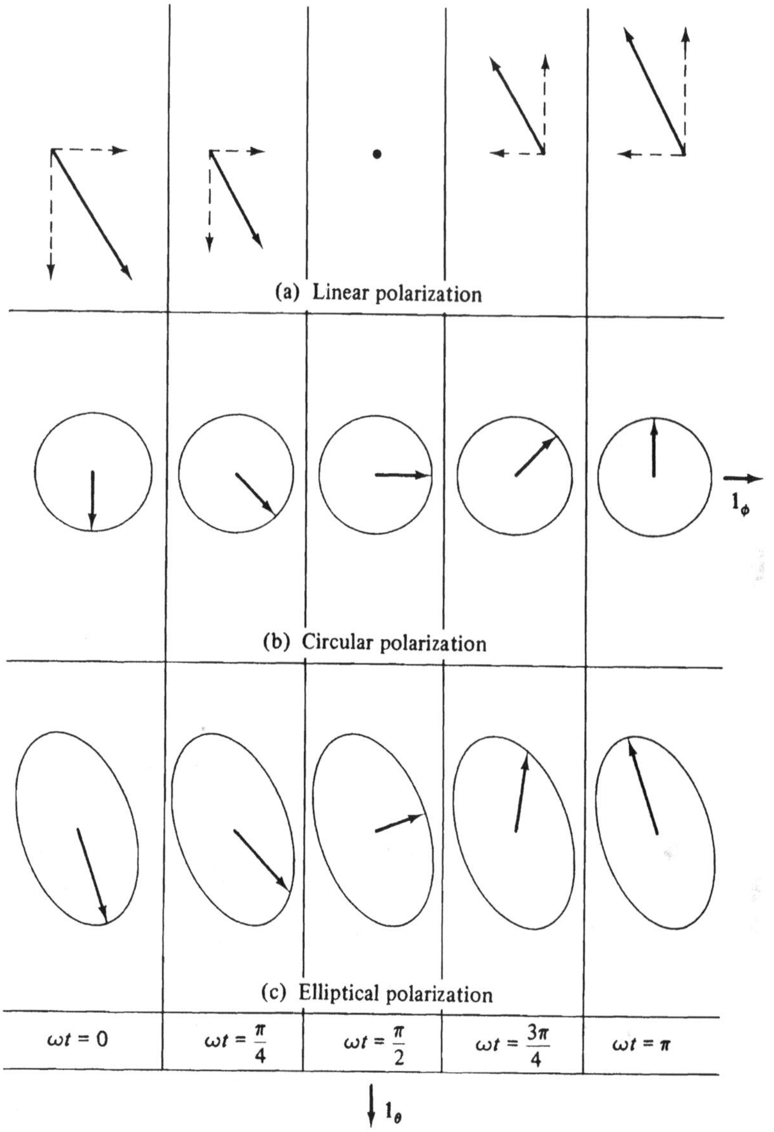

Fig. 1.6 Phasor Plots of **E** Versus Time for Electromagnetic Waves of Various Polarizations

that this locus is an ellipse.[25] Its semimajor and semiminor diameters can also be found by substituting δ_1 and δ_2 in (1.199). This gives

$$|\mathbf{E}| = \left[\frac{A^2 + B^2}{2} \mp \left(\frac{B^2}{2}\right)\frac{\sin 2\beta}{\sin 2\delta_1}\right]^{1/2} \tag{1.201}$$

[25]The components E_θ and E_ϕ that occur in (1.196) are analogous to the voltages applied to the two sets of deflecting plates in an oscilloscope in order to create a Lissajou figure on the screen.

A typical plot of (1.196) is shown in Figure 1.6c. As in the case of circular polarization, the direction of rotation of **E** can be either clockwise (left-handed elliptical polarization) or counterclockwise (right-handed elliptical polarization). This is determined by whether the phase angle β is lead or lag.

REFERENCES

COLLIN, R. E. and F. J. ZUCKER, *Antenna Theory*, Part I (New York: McGraw-Hill Book Co., Inc., 1969).

ELLIOTT, R. S., *Electromagnetics* (New York: McGraw-Hill Book Co., Inc., 1966).

ELLIOTT, R. S., "The Theory of Antenna Arrays," Chapter 1 in Volume II of *Microwave Scanning Antennas*, ed. R. C. Hansen (New York: Academic Press, 1966).

SILVER, S., *Microwave Antenna Theory and Design*, MIT Rad. Lab. Series, Volume 12 (New York: McGraw-Hill Book Co., Inc., 1939).

PROBLEMS

1.1 Complete the Stratton-Chu derivation by letting $\mathbf{F} = \mathbf{B}$ and repeating the analysis that was used in Section 1.7 to obtain **B**, thus establishing Equation 1.54 of the text.

1.2 Alternatively, take the curl of Equation 1.53 to find $-j\omega\mathbf{B}$ and in this manner verify Equation 1.54 of the text.

1.3 Use the expression for the curl of a vector in spherical coordinates and begin with Equation 1.91 in the form

$$\mathbf{A}(x, y, z, t) = \frac{e^{j(\omega t - kr)}}{4\pi\mu_0^{-1} r}\boldsymbol{\mathcal{A}}(\theta, \phi)$$

Then use (1.10b), (1.87), and (1.89) to deduce that, in the far-field

$$\mathbf{E} = -j\omega\,\mathbf{A}_T$$

$$\mathbf{H} = \mathbf{1}_r \times \left(\frac{\mathbf{E}}{\eta}\right)$$

thus confirming (1.95) and (1.96).

1.4 Demonstrate the validity of equations (1.113) and (1.115) in the text.

1.5 Use equivalent-source Equations 1.112 through 1.115 in the retarded potential functions (1.110) and show in detail that the results agree with the surface integrals in the Stratton-Chu formulation for $\mathbf{E}(x, y, z)$ and $\mathbf{B}(x, y, z)$.

1.6 Begin with the far-field expressions (1.123) and (1.124) and show that the power radiated has a density given by (1.125).

1.7 Enumerate the theorems in circuit analysis that can be proven with the aid of the reciprocity relation (1.134). Sketch the proof of each.

1.8 An antenna A, when transmitting, radiates a circularly polarized field in the direction (θ, ϕ), which is right-handed. If antenna A is receiving an elliptically polarized electromagnetic wave, incident from the direction (θ, ϕ), state the conditions of ellipticity which will maximize the received signal. State those which will minimize it.

1.9 Show that $|\mathbf{E}(t)|$, as given generally by Equation 1.199, has as its locus an ellipse with axes that occur at angles δ_1 and $\delta_2 = \delta_1 + \pi/2$, with these angles satisfying (1.200). What is the ellipticity ratio?

2 radiation patterns of dipoles, loops, and helices

2.1 Introduction

In this chapter the formulas that have been developed for the fields caused by an assumed known distribution of current will be applied to a succession of simple but practical radiators. These type I (actual-source antennas) include dipoles, loops, and helices. The center-fed dipole of length $2l$ will be taken up first, with emphasis on the two cases of greatest interest, when $2l \cong \lambda/2$, and when $2l \ll \lambda$. After a discussion of images, these results will be extended to a monopole over a ground plane and a dipole in front of a ground plane. Next, the small current loop will be examined, followed by the helix, the latter being an example of a traveling wave current distribution.

These antenna configurations have many practical applications as single radiating elements but also are widely used in arrays, a subject which will be introduced in Chapter 4.

2.2 The Center-Fed Dipole

The practical center-fed dipole usually consists of a pair of tubular conductors of diameter d aligned in tandem so that there is a small feeding gap at the center, as shown in Figure 2.1. The total length is $2l \gg d$. A voltage is applied across the gap, often by means of a two-wire transmission line. The resulting current distribution on the pair of tubular conductors gives rise to a radiating field. If a good estimation can be made of this current distribution, the formulas of Section 1.11 can be used to deduce the field.

One can gain insight to the current distribution by considering the case of a two-wire transmission line that is opened out, as shown in Figure 2.2. Without any flare, the open-circuit termination causes a standing-wave distribution of current, oppositely directed in the two conductors. Pairs of current elements, which are equal, opposite, and close together, radiate negligibly, which is the behavior of a good transmission

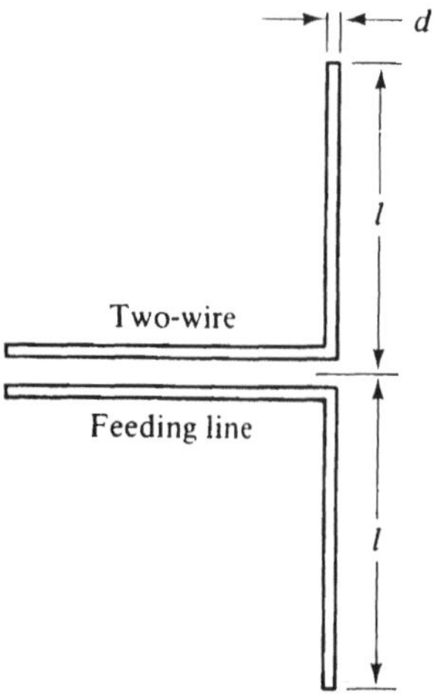

Fig. 2.1 Simple Center-Fed Dipole

lihe. If the origin of the X-axis is taken a distance l back from the end of the transmission line, the current distribution can be given by

$$I(x, t) = I_m \sin [k(l - x)]e^{j\omega t} \tag{2.1}$$

With a flare of 45°, as shown in the second panel of Figure 2.2, the inductance and capacitance per unit length change with position along the flared segment, and thus so too does the characteristic impedance; however, to first order, the wave number is still constant at the free-space value k. For this reason, one can argue that the current distribution is little altered by the flare. This is still assumed to be the case in the third panel of Figure 2.2, where the current distribution is also shown as that of a standing wave with sinusoidal spatial distribution. Note that the pair of current elements, which had canceled each other's radiation tendencies in the first panel where they were oppositely directed and close, are more widely separated and reinforcing in the third panel, which serves to illuminate why a dipole radiates.

Modern methods, pioneered by the work of E. Hallén[1] and S. A. Schelkunoff[2] and using powerful computational techniques such as the method of moments, have led to more precise knowledge of the current distribution on a cylindrical dipole, but the deviation from a sinusoidal function is found not to be great, and for pattern calculations can be ignored. (Compare with Section 7.6, and particularly Figures 7.8 and 7.9).

With the dipole diameter $d \lll \lambda$, it becomes feasible to treat the dipole as a filamentary conductor and replace $\mathbf{J}dV$ by $I d\mathbf{l}$ as the current element. The geometry is

[1]E. Hallén, "Theoretical Investigations into the Transmitting and Receiving Qualities of Antennas," *Nova Acta Upsala*, 11 (1938), 1–44.

[2]S. A. Schelkunoff, *Electromagnetic Waves* (Princeton, N.J.: D. Van Nostrand Co., Inc., 1943), pp. 441–52.

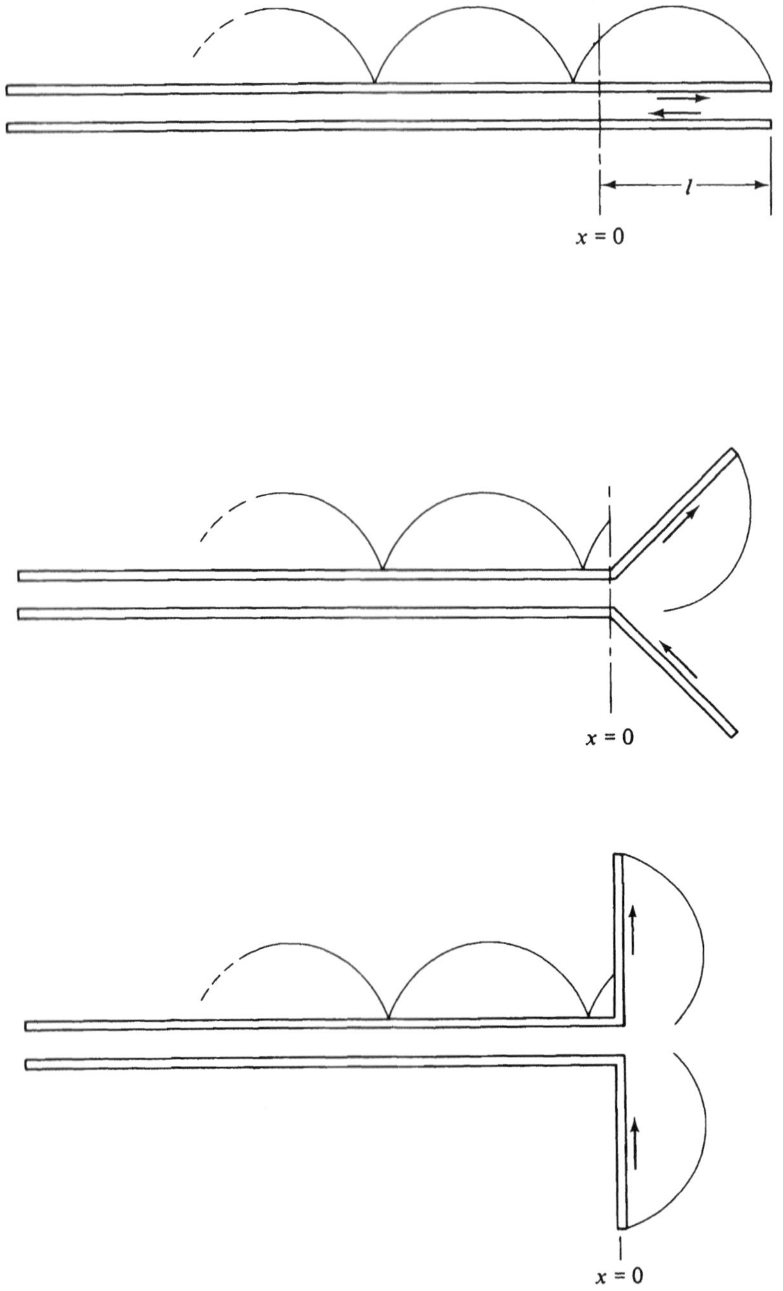

Current waveform = $I(x, t) = I_m \sin [k(l - x)] e^{j\omega t}$

Fig. 2.2 The Dipole as a Transmission Line that is Opened Out

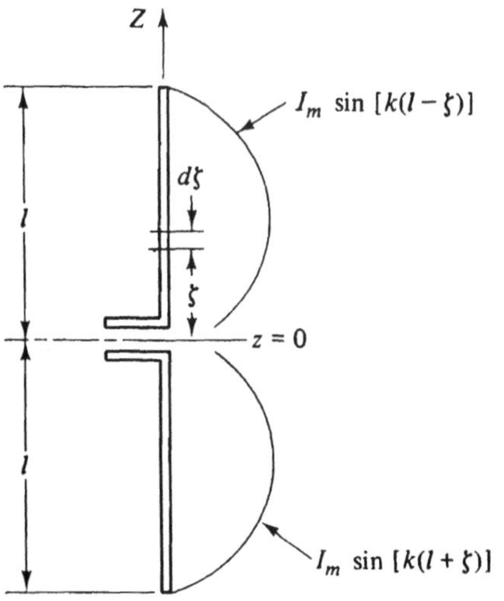

Fig. 2.3 Center-Fed Dipole with Assumed Sinusoidal Current Distribution

then as shown in Figure 2.3. The sources are $I(\zeta)d\zeta$ with

$$I(\zeta) = I_m \sin [k(l - |\zeta|)] \tag{2.2}$$

Since all the sources lie along the Z-axis,

$$\mathcal{L} = \zeta \cos \theta \tag{2.3}$$

and a return to Equations 1.101 and 1.102 reveals that, for a filamentary center-fed dipole, $\mathcal{A}_\phi \equiv 0$, whereas $\mathcal{A}_\theta$ is given by

$$\mathcal{A}_\theta(\theta) = -I_m \sin \theta \int_{-l}^{l} \sin [k(l - |\zeta|)] e^{jk\zeta \cos \theta} \, d\zeta \tag{2.4}$$

Because the current distribution is symmetrical around $\zeta = 0$, the odd part of $\exp(jk\zeta \cos \theta)$ can be discarded, yielding

$$\mathcal{A}_\theta(\theta) = -2I_m \sin \theta \int_0^l \sin [k(l - \zeta)] \cos (k\zeta \cos \theta) \, d\zeta \tag{2.5}$$

This integrates to give

$$\mathcal{A}_\theta(\theta) = -\frac{2I_m}{k \sin \theta} [\cos (kl \cos \theta) - \cos (kl)] \tag{2.6}$$

Two cases of special interest can now be considered.

1. The half-wavelength dipole, $2l = \lambda/2$.
   For this length,

$$\mathcal{A}_\theta(\theta) = -\frac{2I_m}{k} \frac{\cos [(\pi/2) \cos \theta]}{\sin \theta} \tag{2.7}$$

With the outgoing spherical wave factor of (1.93) restored, use of (1.95) and (1.96) gives

$$E_\theta = j60I_m \frac{e^{j(\omega t - kr)}}{r}\left[\frac{\cos[(\pi/2)\cos\theta]}{\sin\theta}\right] \tag{2.8}$$

$$H_\phi = j\frac{I_m}{2\pi}\frac{e^{j(\omega t - kr)}}{r}\left[\frac{\cos[(\pi/2)\cos\theta]}{\sin\theta}\right] \tag{2.9}$$

A polar plot of $E_\theta(\theta)/E_\theta(\pi/2)$ is shown in Figure 2.4. It is seen to be doughnut-shaped (the three-dimensional pattern results from rotating Figure 2.4 about the Z-axis, since E_θ is ϕ-independent), with a null along the $\theta = 0°$, $180°$ axis. This type of pattern finds wide use in omnicoverage applications when vertical polarization is required and a null can be tolerated in one direction.

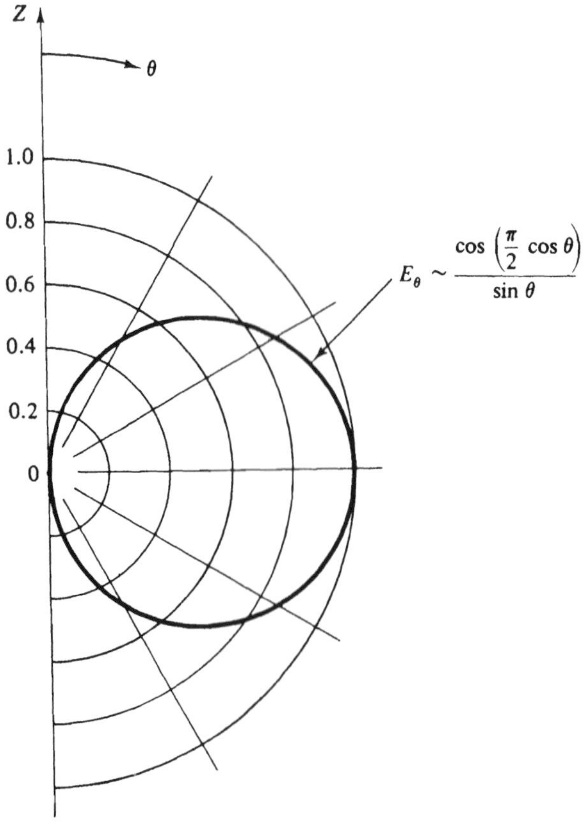

Fig. 2.4 Normalized E-Field Pattern of a Half-Wavelength Center-Fed Dipole

The power pattern can be deduced with the aid of (1.99) and is given by

$$\mathcal{P}_{r,\theta}(\theta) = \frac{2\eta I_m^2}{(4\pi r)^2}\left[\frac{\cos^2[(\pi/2)\cos\theta]}{\sin^2\theta}\right] \tag{2.10}$$

The total radiated power is

$$\begin{aligned} P_{rad} &= \int_0^{2\pi}\int_0^{\pi} \mathcal{P}_{r,\theta}(\theta, \phi)\, r^2 \sin\theta\, d\theta\, d\phi \\ &= \int_0^{\pi} \mathcal{P}_{r,\theta}(\theta)\, 2\pi r^2 \sin\theta\, d\theta \\ &= \frac{\eta I_m^2}{2\pi}\int_0^{\pi/2} \frac{\cos^2[(\pi/2)\cos\theta]}{\sin\theta}\, d\theta \end{aligned}$$

Numerical integration gives

$$P_{rad} = 0.609\frac{\eta I_m^2}{2\pi} \tag{2.11}$$

The peak directivity of a half wavelength dipole (see Section 1.16) is

$$D(peak) = \frac{\mathcal{P}_{r,\theta}(\pi/2)}{P_{rad}/4\pi r^2} = 1.64 \tag{2.12}$$

Since $l = \lambda/4$, the peak input current to the dipole (at $\zeta = 0$) is I_m and the feeding transmission line can be said to be delivering the power

$$\frac{1}{2} I_m^2 R_{rad} = (0.609)\frac{\eta I_m^2}{2\pi} \tag{2.13}$$

to a resistance R_{rad} placed across its terminus. This is called the radiation resistance of the half-wavelength dipole, and solution of (2.13) gives

$$R_{rad} = \frac{0.609\eta}{\pi} = 73 \text{ ohms} \tag{2.14}$$

Nothing has appeared in the development to indicate whether or not the current and voltage at the end of the transmission line are in phase, so there is no information at this stage about the reactance of the dipole. This subject will be explored in Part III, which is concerned with the impedance of antennas. There it will be found that the current distribution on the dipole needs to be known more accurately in order to solve for the input impedance, and that this impedance is a function of the length $2l$ and diameter d of the dipole. However, for $2l \cong \lambda/2$ and $2l \gg d$, the real part of the impedance is close to 73 ohms.

Now attention can be turned to the second special case.

2. The short dipole, $2l \ll \lambda$.

The terms $\cos(kl\cos\theta)$ and $\cos(kl)$ that appear in Equation 2.6 can be expanded in power series which converge rapidly if kl is small. One obtains

$$\mathcal{A}_\theta(\theta) = -kl^2 I_m \sin\theta\left[1 - \frac{(kl)^2}{12}(1 + \cos^2\theta) + \cdots\right] \tag{2.15}$$

The input current to the dipole is given by

$$I = I_m \sin(kl) = I_m\left[kl - \frac{(kl)^3}{3!} + \cdots\right] \tag{2.16}$$

Even for $2l$ as large as $\lambda/4$, it is seen to be a good approximation to write

$$\mathcal{A}_\theta(\theta) = -kl^2 I_m \sin\theta = -Il\sin\theta \tag{2.17}$$

Thus the short dipole also gives a vertically polarized field pattern that is doughnut-shaped, a little bit broader than Figure 2.4, but not significantly so. Where the short dipole differs radically from the half-wave dipole is in its input impedance. To see the effect on radiation resistance, (1.99) can be used to obtain

$$\mathcal{P}_{r,\theta}(\theta) = \frac{1}{2}\frac{(kl)^2\eta I^2}{(4\pi r)^2}\sin^2\theta \tag{2.18}$$

If (2.18) is integrated over a full sphere of radius r, the result is

$$P_{rad} = \frac{(kl)^2\eta I^2}{12\pi} \tag{2.19}$$

The peak directivity for a short dipole can be obtained as in Equation 2.12 and is found to be 1.5, not much less than the value for a half-wavelength dipole.

Since the radiation resistance can be defined by $P_{rad} = (1/2)I^2R_{rad}$, Equation 2.19 yields

$$R_{rad} = 20\left(\frac{\pi L}{\lambda}\right)^2 \tag{2.20}$$

in which $L = 2l$ is the length of the short dipole.

As an example, if $2l = \lambda/8$, $R_{rad} = 3$ ohms, a value considerably lower than the value of 73 ohms found for a half-wavelength dipole. The effect on the reactive component of the input impedance of a dipole is even more drastic as it is shortened. For a finite dipole diameter d, the reactance is positive at $2l = \lambda/2$, goes through zero at a dipole length slightly below $\lambda/2$, and then becomes increasingly negative as $2l$ is shortened further. For $2l = \lambda/8$, it is not unusual for X to be as much as 1000 ohms capactive. This can be tuned out by a suitable inductance placed at the feeding point, but the dipole reactance changes so rapidly with frequency at these short lengths that the combination is quite narrow band. This subject will be explored in more depth in Part III, where a study of the input impedance of a dipole is undertaken.

2.3 Images in a Ground Plane

The results of the previous section can be extended to the case of a monopole over a large highly conductive ground plane, or a dipole in front of it, by invoking the method of images. Because of these present applications and others to be encountered in later chapters, it is desirable at this point to digress and discuss the images of both electric and magnetic current elements.

Consider first an electric current element, situated at the origin, and oriented in the z-direction. The magnetic vector potential function due to this single element is

$$d\mathbf{A} = \frac{\mathbf{1}_z I dl}{4\pi\mu_0^{-1}} \frac{e^{j(\omega t - kr)}}{r} \tag{2.21}$$

The use of (1.89) in spherical coordinates gives

$$d\mathbf{E} = \frac{c^2}{j\omega} \frac{Idl}{4\pi\mu_0^{-1}} \frac{e^{j(\omega t - kr)}}{r} \left[\mathbf{1}_r \left(\frac{j2k}{r} + \frac{2}{r^2}\right)\cos\theta + \mathbf{1}_\theta\left(-k^2 + \frac{jk}{r} + \frac{1}{r^2}\right)\sin\theta\right] \tag{2.22}$$

When this expression is converted to Cartesian components, one obtains

$$d\mathbf{E} = \frac{c^2}{j\omega} \cdot \frac{Idl}{4\pi\mu_0^{-1}} \cdot \frac{e^{j(\omega t - kr)}}{r} \left\{\left[\left(-k^2 + \frac{j3k}{r} + \frac{3}{r^2}\right)\sin\theta\cos\theta\right][\mathbf{1}_x \cos\phi + \mathbf{1}_y \sin\phi] + \mathbf{1}_z\left[\left(\frac{j2k}{r} + \frac{2}{r^2}\right)\cos^2\theta + \left(k^2 - \frac{jk}{r} - \frac{1}{r^2}\right)\sin^2\theta\right]\right\} \tag{2.23}$$

Let the result of (2.23) be applied to the case of an electric current element normal to and a distance d above an infinite, perfectly conducting ground plane, as shown in Figure 2.5a. This current element will induce a current distribution in the ground plane such that $\mathbf{E}_{\text{tang}} = 0$ along the ground plane. But the same effect could be achieved if the ground plane currents were not there and an image current element were a distance d below the ground plane and in phase with the actual element. To see this, one should observe from Figure 2.5a that for any point in the ground plane $r' = r$, $\theta' = \pi - \theta$, and $\phi' = \phi$. Since $d\mathbf{E}'$ is also expressible in the form of (2.23), if $I'd\mathbf{l}' = Id\mathbf{l}$, it follows that $dE'_x = -dE_x$ and $dE'_y = -dE_y$ because $\sin(\pi - \theta) = \sin\theta$, but $\cos(\pi - \theta) = -\cos\theta$. Thus $d\mathbf{E} + d\mathbf{E}'$ has only a z-component, as required. The result is independent of d.

If this exercise is repeated, but with $Id\mathbf{l}$ parallel to the ground plane as shown in Figure 2.5b, then the proper image is oppositely directed. This can be demonstrated by noting that now, for any point on the ground plane, $r' = r$, $\theta' = \theta$, and $\phi' =$

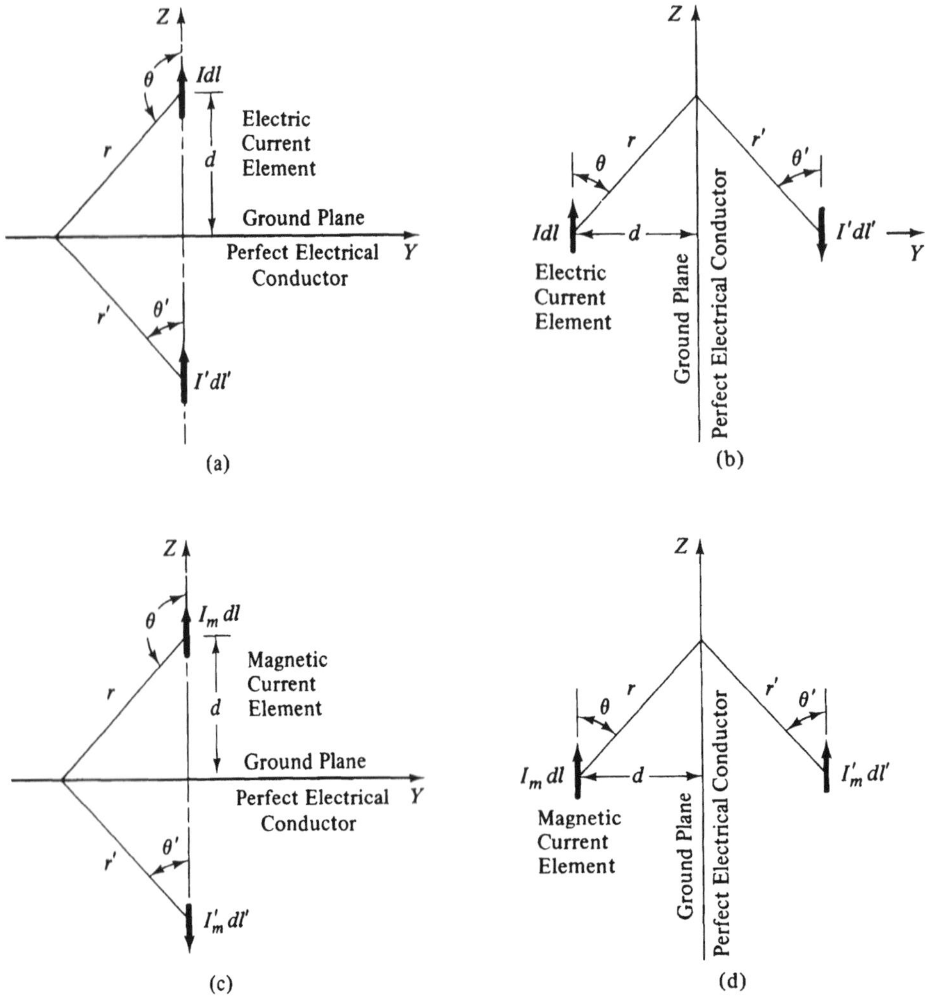

Fig. 2.5 Images in an Infinite Ground Plane

$2\pi - \phi$. Thus if $I'd\mathbf{l}' = -Id\mathbf{l}$, it follows that $dE'_x = -dE_x$ and $dE'_z = -dE_z$. But $dE'_y = dE_y$ because $\sin(2\pi - \phi) = -\sin\phi$. Therefore $d\mathbf{E} + d\mathbf{E}'$ has only a y-component, as required, and this result is independent of d.

If a magnetic current element $I_m d\mathbf{l}$ replaces the electric current element, one can write, as a special case of (1.110b),

$$d\mathbf{F} = \frac{\mathbf{1}_z I_m dl}{4\pi\mu_0^{-1}} \frac{e^{j(\omega t - kr)}}{r} \tag{2.24}$$

Use of (1.120) gives an expression identical to (2.23), except that $d\mathbf{B}$ replaces $d\mathbf{E}$ on the left and a factor c^2 is deleted on the right. Now the boundary conditions require that $\mathbf{B}_{\text{normal}} \equiv 0$ on the ground plane instead of $\mathbf{E}_{\text{tang}} \equiv 0$. For this reason, in Figure 2.5c the image is seen to be oppositely directed, whereas in Figure 2.5d the image is seen to be codirected with the current element.

These results can be summarized by saying that the image of an electric current element is codirected if the element is perpendicular to an electric ground plane, and is oppositely directed if the element is parallel to the ground plane. The image of a magnetic current element is oppositely directed if the element is perpendicular to an electric ground plane, and is codirected if the element is parallel to the ground plane.

Since any antenna can be viewed as a collection of these elementary current elements, the results just stated are also true in aggregation, that is, at the macroscopic level.

2.4 A Monopole Above a Ground Plane

When the results of the previous section on images are invoked, the development of the fields due to a center-fed dipole, considered in Section 2.2, can be extended to the case of a monopole above a ground plane. Figure 2.6 shows the arrangement; the images, taken together with the monopole, are seen to replicate the dipole. Thus the value of $\mathcal{A}_\theta(\theta)$ in $z > 0$ is given by (2.6), the fields and power pattern for a quarter-wavelength monopole are given by (2.8) through (2.10), and the corresponding results for a short monopole are given by (2.17) and (2.18).

One notable difference is that the monopole is only radiating into a half-space, so the field pattern is only the upper half of Figure 2.4. Implicit in this result is the assumption of an infinite, perfectly conducting ground plane. For a finite ground plane composed of a good conductor, diffraction at the edges causes some radiation to "spill over" into $z < 0$ with maximum radiation occurring at an angle above the horizon. For an extensive but lossy earth, radiation along the horizon is diminished due to the ohmic losses in the earth, and once again maximum radiation occurs at an angle above the horizon. The monopole fed against a ground plane has its most

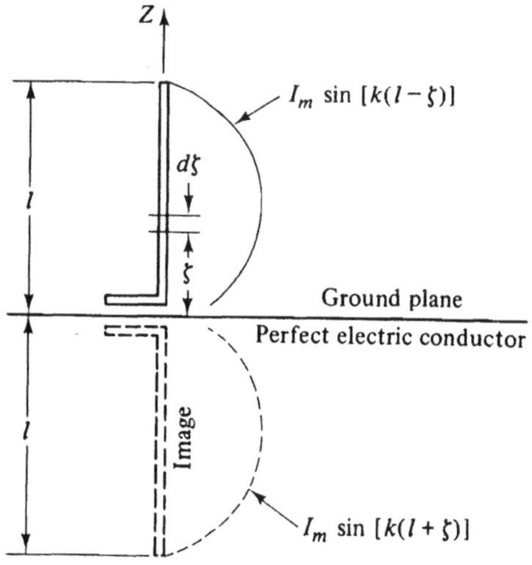

Fig. 2.6 Vertical Monopole above a Ground Plane

prominent application in broadcast antennas (such as the AM band) where omni-azimuthal coverage is desired.

Another distinction between the monopole above an infinite, perfectly conducting ground plane and a center fed dipole in a full space is that, for the monopole, P_{rad} is half that of the center-fed dipole for the same value of I_m. As a consequence, R_{rad} for a monopole is only half the value of R_{rad} for the corresponding dipole. Another way to see this is to observe from Figure 2.6 that the voltage applied between the monopole and ground is only half the voltage applied between the monopole and its image.

2.5 A Dipole in Front of a Ground Plane

The method of images can also be used to determine the pattern of a center-fed dipole which is parallel to and a distance h in front of a large ground plane, as shown in Figure 2.7. If the current in the dipole is given by (2.2), the current in the image is the negative of (2.2). A point on the dipole can be assigned the coordinates $(0, h, \zeta)$ in which case the corresponding point on the image has the coordinates $(0, -h, \zeta)$. Equation 1.102 gives $\mathcal{a}_\phi \equiv 0$ and Equation 1.101 indicates that

$$\begin{aligned}\mathcal{a}_\theta(\theta, \phi) &= -I_m \sin\theta \int_{-l}^{l} \sin[k(l - |\zeta|)] e^{jk(h \sin\theta \sin\phi + \zeta\cos\theta)}\, d\zeta \\ &\quad + I_m \sin\theta \int_{-l}^{l} \sin[k(l - |\zeta|)] e^{jk(-h \sin\theta \sin\phi + \zeta\cos\theta)}\, d\zeta \\ &= -2jI_m \sin\theta \sin(kh \sin\theta \sin\phi) \int_{-l}^{l} \sin[k(l - |\zeta|)] e^{jk\zeta\cos\theta}\, d\zeta \qquad (2.25)\end{aligned}$$

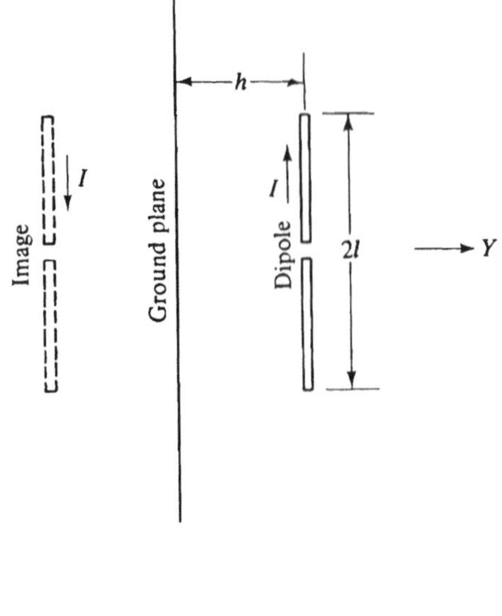

Fig. 2.7 Center-Fed Dipole Parallel to a Ground Plane Plus Image of Dipole

The integral in (2.25) is the same as the one encountered earlier in (2.4) and thus

$$\mathcal{A}_\theta(\theta, \phi) = -\frac{4jI_m}{k \sin \theta}[\cos(kl \cos \theta) - \cos kl] \sin(kh \sin \theta \sin \phi) \qquad (2.26)$$

Therefore the pattern of a dipole plus ground plane, in the half-space $y > 0$, is the pattern of an isolated dipole multiplied by the factor $2j \sin(kh \sin \theta \sin \phi)$. In the half-plane $\phi = \pi/2$, for $h = \lambda/4$, the normalized pattern is similar to Figure 2.4. In three dimensions, the pattern is ball-like.

2.6 The Small Current Loop

If a circular wire loop of radius a small compared to a wavelength is fed by a two-wire-line, as shown in Figure 2.8, it is a good approximation to assume that the current is $Ie^{j\omega t}$ everywhere on the loop, with I a constant. Since the coordinates of a point on the loop are given by

$$\xi = a \cos \psi \qquad \eta = a \sin \psi \qquad \zeta = 0$$

and since

$$I d\mathbf{l} = Ia(-\mathbf{1}_x \sin \psi + \mathbf{1}_y \cos \psi)\, d\psi$$

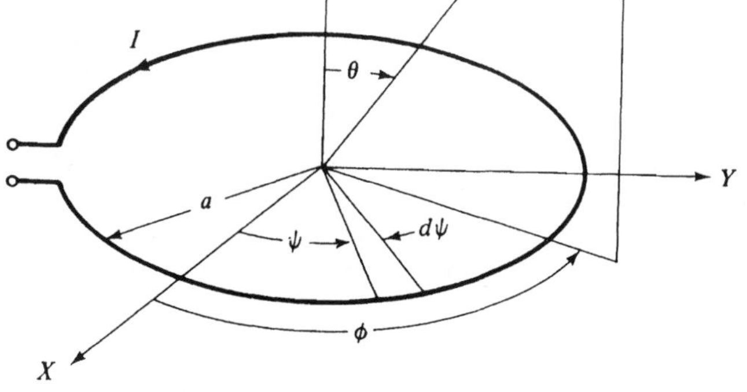

Fig. 2.8 The Small Current Loop

it follows that (1.101) becomes, for this case,

$$\mathcal{A}_\theta(\theta, \phi) = -Ia \cos\theta \int_0^{2\pi} \sin(\psi - \phi) e^{jka \sin\theta(\cos\phi\cos\psi + \sin\phi\sin\psi)}\, d\psi$$

$$= -Ia \cos\theta \int_{-\phi}^{2\pi-\phi} \sin(\psi - \phi) e^{jka \sin\theta \cos(\psi-\phi)}\, d(\psi - \phi) \qquad (2.27)$$

Since the integrand is cyclical in 2π, the limits on the integral in (2.27) can be shifted to range from 0 to 2π, which is to say that the result is independent of ϕ. When one examines the even and odd nature of the functions $\sin\psi$, $\cos(ka \sin\theta \cos\psi)$, and $\sin(ka \sin\theta\cos\psi)$ which comprise the integrand, it is a simple matter to show that

$$\mathcal{A}_\theta(\theta, \phi) \equiv 0 \qquad (2.28)$$

Proceeding similarly, one finds that, for the small loop, (1.102) takes the form

$$\mathcal{A}_\phi(\theta, \phi) = Ia \int_{-\phi}^{2\pi-\phi} \cos(\psi - \phi) e^{jka \sin\theta\cos(\psi-\phi)}\, d(\psi - \phi) \qquad (2.29)$$

Once again, the limits of integration can be shifted, indicating $\mathcal{A}_\phi$ is not a function of ϕ; elimination of the odd terms in the integrand leaves

$$\mathcal{A}_\phi(\theta) = 2jIa \int_0^{\pi} \sin(ka \sin\theta\cos\psi) \cos\psi\, d\psi \qquad (2.30)$$

If ka is assumed to be small, $\sin(ka \sin\theta\cos\psi) \cong ka \sin\theta\cos\psi$ and (2.30) becomes, to good approximation,

$$\mathcal{A}_\phi(\theta) = j(\pi a^2 I)(k \sin\theta) \qquad (2.31)$$

The far-field pattern is horizontally polarized and has a power density which, from (1.100), is

$$\mathcal{P}_{r,\phi}(\theta) = \frac{(ka)^4 I^2 \eta}{32 r^2} \sin^2\theta \qquad (2.32)$$

By comparing this result with (2.18), one can conclude that a short dipole and a small loop have similar patterns, with a difference of 90° in polarization. Thus the applications for a small loop are similar to those for a short dipole—situations in which an omni-azimuthal coverage is needed and in which a null can be tolerated along some axis. The distinction is that the short dipole gives vertical polarization, whereas the small loop gives horizontal polarization.

Integration of (2.32) over a sphere of radius r, with the result equated to $\frac{1}{2}I^2 R_{rad}$, yields

$$R_{rad} = 320\pi^6 \left(\frac{a}{\lambda}\right)^4 \qquad (2.33)$$

As an example, if $(a/\lambda) = 0.03$, $R_{rad} = 0.25$ ohms. By contrast, from (2.20), a short dipole of length $2l/\lambda = 0.06$ has a radiation resistance of 0.7 ohms. The radiation resistance of a small loop can be raised by a factor n^2 if n closely wound turns are used.

The preceding analysis is also applicable for large ka as long as one is able to assume a uniform current $Ie^{j\omega t}$ on the entire loop. For this more general case, (2.30) gives

$$\mathcal{A}_\phi(\theta) = j2\pi IaJ_1(ka \sin \theta) \tag{2.34}$$

with J_1 a Bessel function. As ka increases, more and more fine structure appears in the pattern, through the behavior of $J_1(ka \sin \theta)$. As an illustration of this, Figure 2.9 shows a polar plot of the field pattern of a loop for which $a = 2.5\lambda$, and this is contrasted to the pattern when $a = 0.05\lambda$.

The achievement of a uniform current $Ie^{j\omega t}$ in a large loop requires complicated feeding arrangements. Some examples of how this can be approximated are given by J. Blass.[3]

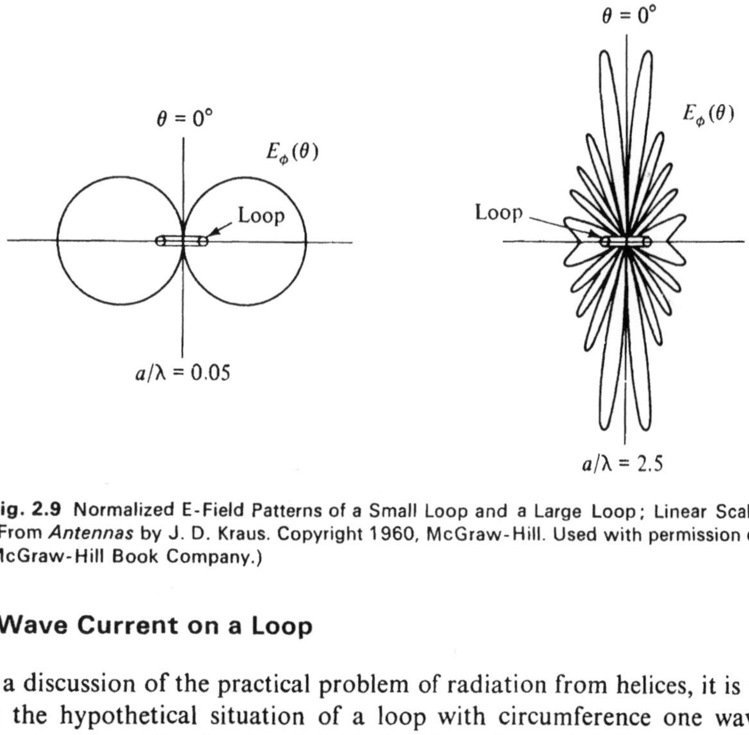

Fig. 2.9 Normalized E-Field Patterns of a Small Loop and a Large Loop; Linear Scale (From *Antennas* by J. D. Kraus. Copyright 1960, McGraw-Hill. Used with permission of McGraw-Hill Book Company.)

2.7 Traveling Wave Current on a Loop

Prior to a discussion of the practical problem of radiation from helices, it is useful to consider the hypothetical situation of a loop with circumference one wavelength, supporting a current distribution $Ie^{j(\omega t - ka\psi)}$ with ψ the angle measured from the X-axis, as shown in Figure 2.8. This can be viewed as a wave traveling along the wire at

[3] J. Blass, "Loop Antennas," *Antenna Engineering Handbook*, ed. H. Jasik (New York: McGraw-Hill Book Co., Inc., 1961), Chapter 6.

the speed of light, repeating itself every 2π radians, as required by the physical boundary conditions. For this case, because $ka = 1$, Equation 1.101 becomes

$$\mathfrak{A}_\theta(\theta, \phi) = -Ia \cos\theta \int_0^{2\pi} \sin(\psi - \phi) e^{j\sin\theta\cos(\psi-\phi)} e^{-j\psi}\, d\psi$$

$$= -Ia\cos\theta\, e^{-j\phi} \int_{-\phi}^{2\pi-\phi} \sin(\psi - \phi) e^{j\sin\theta\cos(\psi-\phi)} e^{-j(\psi-\phi)}\, d(\psi - \phi) \qquad (2.35)$$

Since the integrand is cyclical in 2π, the limits of integration can be changed to range from 0 to 2π. Thus $\mathfrak{A}_\theta(\theta, \phi)$ is a function of ϕ only in the multiplicative factor $e^{-j\phi}$. This is reasonable in view of the ϕ-symmetry of the structure and the assumed traveling wave distribution.

When the even and odd nature of the terms in the integrand of (2.35) is explored, it is found that

$$\mathfrak{A}_\theta(\theta, \phi) = (4jIa\cos\theta) e^{-j\phi} \int_0^{\pi/2} \sin^2\psi \cos(\sin\theta\cos\psi)\, d\psi \qquad (2.36)$$

In like fashion, (1.102) becomes, for this case,

$$\mathfrak{A}_\phi(\theta, \phi) = 4Iae^{-j\phi} \int_0^{\pi/2} \cos^2\psi \cos(\sin\theta\cos\psi)\, d\psi \qquad (2.37)$$

Polar plots of normalized $|\mathfrak{A}_\theta|$ and $|\mathfrak{A}_\phi|$, which are also normalized plots of $|E_\theta|$ and $|E_\phi|$, are displayed in Figure 2.10. One needs to imagine that these field patterns are sweeping azimuthally as $e^{j(\omega t-\phi)}$. It is interesting to observe that, whereas the small loop gave a doughnut pattern with nulls at $\theta = 0°, 180°$ (see Figure 2.4 as a prototype), the $ka = 1$ loop gives a "figure-eight" pattern for E_θ, with a null at $\theta° = 90°$, and an almost omnidirectional pattern for E_ϕ.

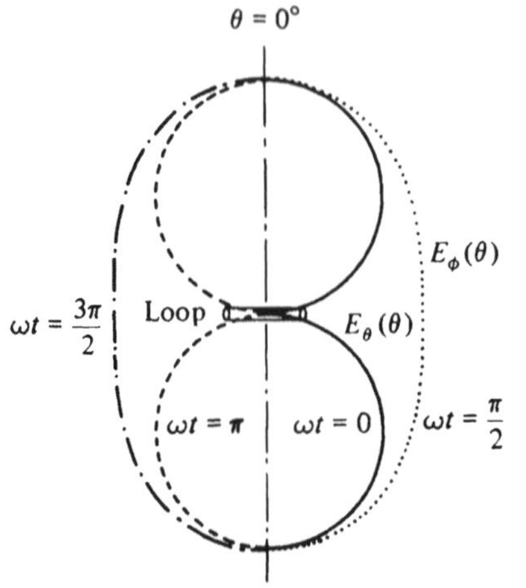

Fig. 2.10 Normalized E-Field Patterns of a Traveling Wave Loop Antenna; Linear Scale

From (1.95), (2.36), and (2.37), it is a simple matter to deduce that

$$\mathbf{E}(r, 0^\circ, \phi, t) = \mathbf{E}(r, 180^\circ, \phi, t) = \omega\mu_0 Ia e^{-j\phi} \frac{e^{j(\omega t - kr)}}{4r}(\mathbf{1}_\theta - j\mathbf{1}_\phi)$$

$$= \frac{\omega\mu_0 Ia}{4r} e^{-jkr - j\phi}[\mathbf{1}_\theta e^{j\omega t} + \mathbf{1}_\phi e^{j(\omega t - \pi/2)}] \qquad (2.38)$$

The factor $\mathbf{1}_\theta \cos \omega t + \mathbf{1}_\phi \sin \omega t$ that appears in (2.38) can be given the following interpretation: If a measurement is made of **E** at either pole of a large sphere centered on the loop, the *polarization* of **E** will rotate synchronously with a period $\tau = 2\pi/\omega$, but the *magnitude* of **E** will be independent of time.

This is an example of circular polarization. Figure 2.10 indicates that, as one departs from $\theta = 0^\circ$ or 180°, E_θ decreases more rapidly than does E_ϕ, but Equations 2.36 and 2.37 reveal that the two component polarizations are still 90° apart in time phase. The polarization becomes elliptical. In a θ-region not too far from either pole, the ellipticity is not great and the polarization remains almost circular.

2.8 The End-Fire Helix

A practical radiator of wide applicability is the helix, mounted against a ground plane and fed by a coaxial line, as shown in Figure 2.11 Experiments have shown[4] that if the circumference of the helix is approximately one wavelength and if there are several

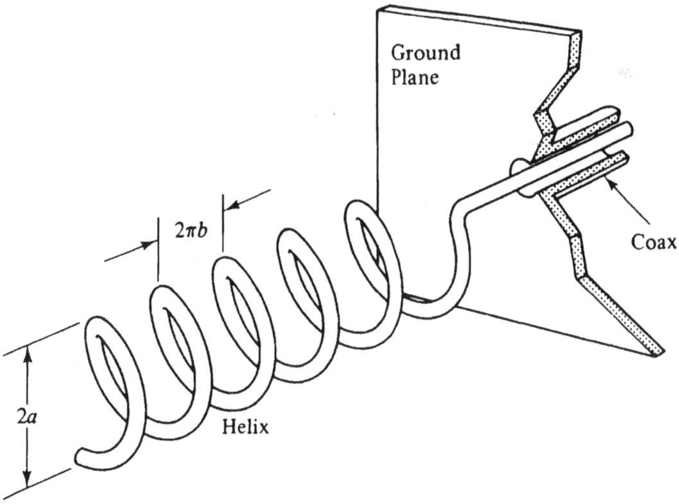

Fig. 2.11 Coaxially Fed Helical Antenna with Ground Plane

[4]See, for example, J. D. Kraus, *Antennas* (New York: McGraw-Hill Book Co., Inc., 1950), Chapter 7.

turns per wavelength, a primary component of current on the helix is a wave traveling along the wire at approximately the speed of light, and that the radiation pattern is beamlike off the end of the helix and essentially circularly polarized.

To construct a model that will explain this behavior, assume that the parametric equations of the helix are

$$\xi = a\cos\psi \qquad \eta = a\sin\psi \qquad \zeta = b\psi \tag{2.39}$$

where the notation of Figure 2.8 is once again applicable, with the helix replacing the loop. An infinitesimal length along the helix is given by

$$d\mathbf{l} = (-\mathbf{1}_x a\sin\psi + \mathbf{1}_y a\cos\psi + \mathbf{1}_z b)\,d\psi \tag{2.40}$$

Assume that there is an outgoing current wave traveling along the helix at the phase velocity $v = pc$ and decaying in amplitude to account for radiation leakage. Then

$$I(s) = I_0 e^{-(\alpha + j\beta)s} \tag{2.41}$$

where s is the distance measured along the helix from the beginning of the turn closest to the ground plane, I_0 is the input current, and $\alpha + j\beta$ is the complex propagation constant.

It is desirable to convert (2.41) to a function of the angle ψ. If one imagines that the helix is unwrapped by rolling out the cylinder on which it is wound, one turn of the helix becomes a straight line of length L, as shown in Figure 2.12. Since one turn of length L corresponds to a change of 2π in the value of ψ, it follows that $(s/L) = \psi/2\pi$. Thus

$$I(\psi) = I_0 e^{-[(\alpha + j\beta)(L/2\pi)]\psi} \tag{2.42}$$

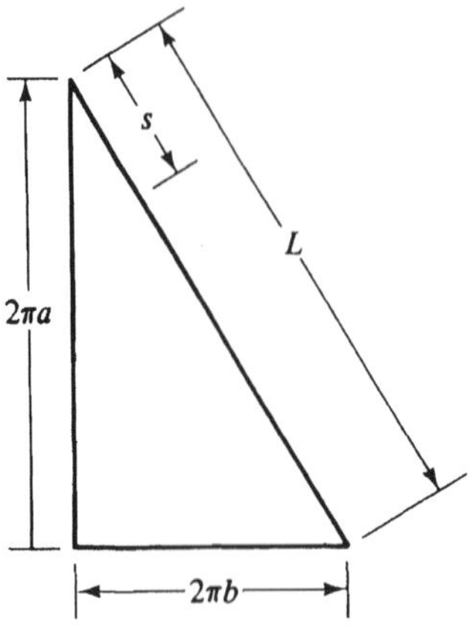

Fig. 2.12 An Unwrapped Turn of a Helix

With the use of (2.39), (2.40), and (2.42), Equations 1.101 and 1.102 can be written, for a helix of N turns, in the forms

$$\mathcal{A}_\theta(\theta, \phi) = \int_0^{2\pi N} I(\psi)[-a \cos\theta \sin(\psi - \phi) - b \sin\theta] e^{jk[a \sin\theta \cos(\psi-\phi) + \psi b \cos\theta]} \, d\psi \tag{2.43}$$

$$\mathcal{A}_\phi(\theta, \phi) = \int_0^{2\pi N} I(\psi)[a \cos(\psi - \phi)] e^{jk[a \sin\theta \cos(\psi-\phi) + \psi b \cos\theta]} \, d\psi \tag{2.44}$$

The repetitive nature of the structure can be used to convert (2.43) and (2.44) into expressions that are more easily interpreted. If (ξ_1, η_1, ζ_1) is a point on the helix in the first turn corresponding to an angle $\psi = \psi_1$, then at an angle $\psi = \psi_1 + 2\pi n$ there is a point (ξ_n, η_n, ζ_n) on the nth turn for which

$$\xi_n = \xi_1 \qquad \eta_n = \eta_1 \qquad \zeta_n = \zeta_1 + 2\pi n b \tag{2.45}$$

Further, from (2.42)

$$I(\psi_1 + 2\pi n) = I(\psi_1) e^{-(\alpha + j\beta) n L} \tag{2.46}$$

When this information is substituted in (2.43) and (2.44), the result is that

$$\mathcal{A}_\theta(\theta, \phi) = f(\theta) g_1(\theta, \phi) \tag{2.47}$$

$$\mathcal{A}_\phi(\theta, \phi) = f(\theta) g_2(\theta, \phi) \tag{2.48}$$

in which

$$f(\theta) = \sum_{n=1}^{N} e^{-j2\pi n[(\beta - j\alpha)(L/2\pi) - kb \cos\theta]} = \sum_{n=1}^{N} e^{-j2\pi n h(\theta)} \tag{2.49}$$

$$g_1(\theta, \phi) = -I_0 \int_0^{2\pi} e^{-jh\psi}[a \cos\theta \sin(\psi - \phi) + b \sin\theta] e^{jka \sin\theta \cos(\psi-\phi)} \, d\psi \tag{2.50}$$

$$g_2(\theta, \phi) = I_0 \int_0^{2\pi} e^{-jh\psi}[a \cos(\psi - \phi)] e^{jka \sin\theta \cos(\psi-\phi)} \, d\psi \tag{2.51}$$

The function $f(\theta)$, which is common to $\mathcal{A}_\theta$ and $\mathcal{A}_\phi$, is called the *array factor* of the helix and accounts for the fine structure in the field patterns. It is seen to be a sum of N phasors, one each due to the N individual turns of the helix. These phasors rotate in the complex plane as θ is varied, and in general are not coaligned. However, they will have a maximum sum at $\theta = 0°$ if

$$\frac{\beta L}{2\pi} - kb = m \tag{2.52}$$

in which m is an integer. Since $\beta = (\omega/v) = (2\pi v)/pc = k/p$, this expression can be solved for p to give

$$p = \frac{L/\lambda}{(2\pi b/\lambda) + m} \tag{2.53}$$

As seen from Figures 2.11 and 2.12, $2\pi b$ is the axial progression of one turn. Typically, helices are found to radiate end-fire (beam at $\theta = 0°$) when (L/λ) is somewhat greater than unity, $(2\pi b/\lambda) \cong 0.3$, and p is somewhat less than unity. Therefore the proper value to take for m in order to model this behavior is $m = 1$. With this choice for m,

$$h(\theta) = 1 + kb(1 - \cos\theta) - j\frac{\alpha L}{2\pi} \tag{2.54}$$

The effect of α turns out to be small in practical situations and results primarily in null-filling. For example, if the current wave is assumed to be damped to -10dB of its input value by the time it reaches the end of the helix, then $e^{-\alpha NL} = 0.316$ and $\alpha L/2\pi = 0.183/N$. For $N \geq 5$, inclusion of the term involving α in (2.52) causes only a minor change in $f(\theta)$. As an illustration of this, Figure 2.13 shows polar plots of $f(\theta)$ for $(2\pi a/\lambda) = 1$, $(2\pi b/\lambda) = 0.3$, and $N = 6$, with and without the α term in $h(\theta)$, and under the assumption that $e^{-\alpha NL} = 0.316$.

Figure 2.13 also shows that the dominant part of $f(\theta)$ is in the neighborhood of $\theta = 0°$. But in this neighborhood, with α ignored, $h(\theta) \cong 1$ and (2.50) and (2.51)

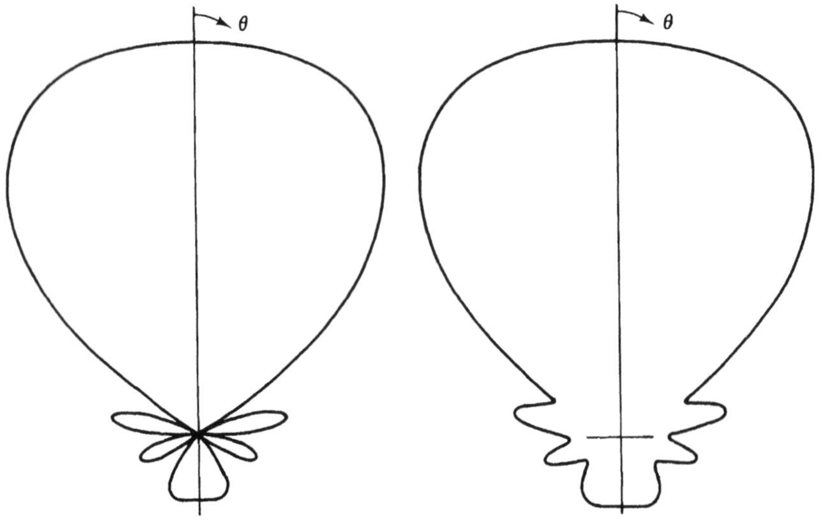

(a) Unattenuated current wave (b) 10dB attenuation in current wave

Fig. 2.13 Array Factor $f(\theta)$ for a Six-Turn Helix; Polar Plots; Linear Scale approximate the earlier equations (2.35) and (2.36), namely, the field patterns for a single loop of circumference $2\pi a = \lambda$, carrying a current wave traveling at the speed of light.

In summary, this analysis suggests that the field patterns of a helix radiating end-fire are, to good approximation, the product of $f(\theta)$ with the patterns shown in Figure 2.10. Since $g_1(\theta, \phi)$ and $g_2(\theta, \phi)$ are in time-phase quadrature, $\mathcal{E}_\theta$ and $\mathcal{E}_\phi$ for a helix combine to give a rotationally symmetric pattern, consisting of a main beam at end-fire plus sidelobes, the pattern being essentially circularly polarized in the neighborhood of $\theta = 0°$. Because $g_1(\theta, \phi)$ and $g_2(\theta, \phi)$ are broad patterns (compare with Figure 2.10), the fine structure in $\mathcal{E}_\theta$ and $\mathcal{E}_\phi$ comes from $f(\theta)$. For the example of a six-turn helix just cited, the product of Figure 2.13a with Figures 2.10a and b gives the polar plots of $\mathcal{E}_\theta$ and $\mathcal{E}_\phi$ shown in Figure 2.14. Both of these patterns are figures of rotation.

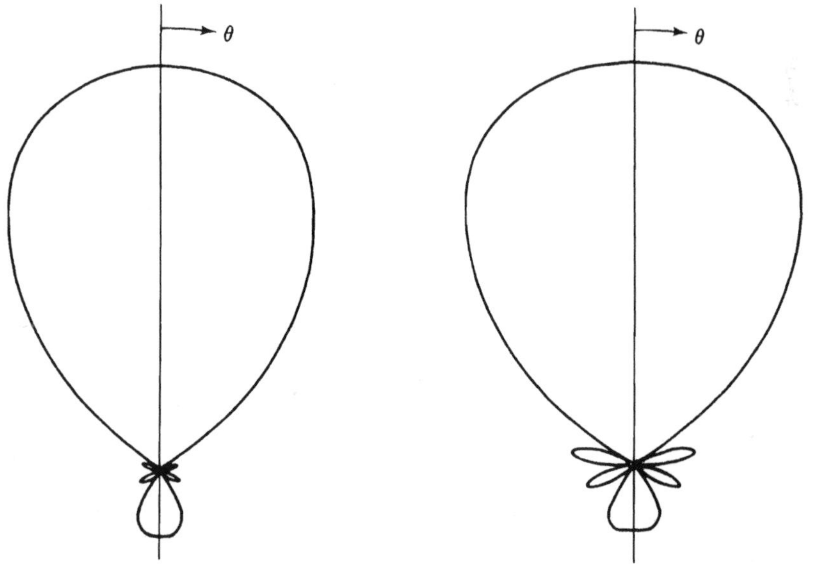

(a) $|E_\theta(\theta, \phi)/E_\theta(0°, \phi)|$ (b) $|E_\phi(\theta, \phi)/E_\phi(0°, \phi)|$

Fig. 2.14 Approximate Normalized E-Field Patterns for a Six-Turn Helix

The actual current distribution on the helix is found experimentally to be more complicated than has been assumed here. Since the total current at the end of the helix is perforce zero, there must also be a damped wave traveling back toward the ground plane. In addition, there are quasi-static mode currents corresponding to monopole-type radiation. But a current wave of the type assumed in (2.41) is the dominant factor in explaining end-fire radiation and theoretical patterns such as those shown in

Figure 2.14 are in good agreement with experiment. The interested reader should consult J. D. Kraus[5] for a detailed discussion of this topic. A different approach to the analysis of the helical antenna can be found in E. A. Wolff.[6]

REFERENCES

JASIK, H., *Antenna Engineering Handbook* (New York: McGraw-Hill Book Co., Inc., 1961).

JORDAN, E. C. and K. G. BALMAIN, *Electromagnetic Waves and Radiating Systems* (Englewood Cliffs, New Jersey: Prentice-Hall, Inc., 1968).

KRAUS, J. D., *Antennas* (New York: McGraw-Hill Book Co., Inc., 1950).

WOLFF, E. A., *Antenna Analysis* (New York: John Wiley and Sons, Inc., 1966).

PROBLEMS

2.1 Show in detail that, for a filamentary center-fed dipole with an assumed current distribution $I(\zeta) = I_m \sin[k(l - |\zeta|)]$, the magnetic vector potential function, with the outgoing spherical wave factor deleted, is given by

$$\mathcal{A}_\theta(\theta) = -\frac{2I_m}{k \sin\theta}[\cos(kl\cos\theta) - \cos(kl)]$$

2.2 With the contribution of a magnetic current element to the electric vector potential function given by (2.24), determine **dB** in Cartesian form and use the result to demonstrate that the images shown in Figures 2.4c and 2.4d are correct.

2.3 Find the expression for the power density radiated by a dipole in front of a ground plane. Numerically integrate this result over a half-space and use the answer to estimate the radiation resistance when the dipole is one-half wavelength long and $h = \lambda/4$.

2.4 If the earth is assumed to be a perfectly conducting ground plane, the radiation field of a vertical quarter wave monopole is unattenuated. The rms value of the electric field along the horizon is given by

$$E = \frac{6.17}{r}\sqrt{P_{rad}} \text{ millivolts per meter}$$

with r in miles and P_{rad} in watts. Verify this relation and derive the corresponding result for a short monopole.

2.5 A small current loop, $a = 0.02\lambda$, is to be designed to have a radiation resistance of 25 ohms. How many turns should be used?

2.6 Find the pattern of a small loop of radius $a \ll \lambda$ parallel to and a distance h above a perfectly conducting ground plane.

[5] Kraus, *Antennas*.

[6] E. A. Wolff, *Antenna Analysis* (New York: John Wiley and Sons, Inc., 1966), Chapter 9.

3 radiation patterns of horns, slots, and patch antennas

3.1 Introduction

The material in this chapter is sibling to what was presented in Chapter 2. There, various type I (actual-source) antennas were introduced and their radiation patterns deduced. Here, several simple but practical type II (equivalent-source) antennas will be analyzed. First to be treated will be the open-ended waveguide, which will then be allowed to evolve into a horn antenna. Next, a center-fed slot in a ground plane will be studied and its equivalence to a center-fed strip dipole established. Attention will then turn to waveguide-fed slots, their excitation, and their radiation patterns. Finally, a metallic patch bonded to a grounded dielectric slab will be viewed as an aperture antenna and analyzed in terms of equivalent sources along its perimeter.

As in the case of the antennas studied in Chapter 2, these configurations find many practical applications as single radiating elements, but are also used in arrays. This is particularly true of waveguide-fed slots, but arrays of horns are not unusual, and patches, stripline-fed slots, microstrip dipoles are all employed as array elements because of such desirable features as having a low profile, being lightweight, and being inexpensive to manufacture.

3.2 The Open-Ended Waveguide

A prototype for the horn antenna is a section of rectangular waveguide, open at its end and terminated in a large ground plane, as shown in Figure 3.1. It will be assumed that the dimensions a and b are chosen so that only the TE_{10} mode will propagate, and that the waveguide section is long enough so that only a TE_{10} mode is incident on the waveguide mouth. Back-scattering will be in many modes, including a TE_{10} mode, since the open-ended waveguide is not inherently matched to free space.

Efforts to deduce the complete current distribution in this structure, including currents in the ground plane, the waveguide walls, and the probe—followed by efforts to sort out which of these currents contribute to radiation and which do not—is a forbidding task. This is a classic example of a situation in which the fields can be estimated more easily than the actual sources. For this reason, a source-excluding surface S is shown surrounding the antenna in Figure 3.1. So that advantage may be taken of the image principle, this surface is chosen to lie within the ground plane, but to have a slight bulge at the mouth of the waveguide. Thus the actual sources are all excluded, except for those that have been induced in the ground plane by the electromagnetic waves emerging from the open-ended waveguide.

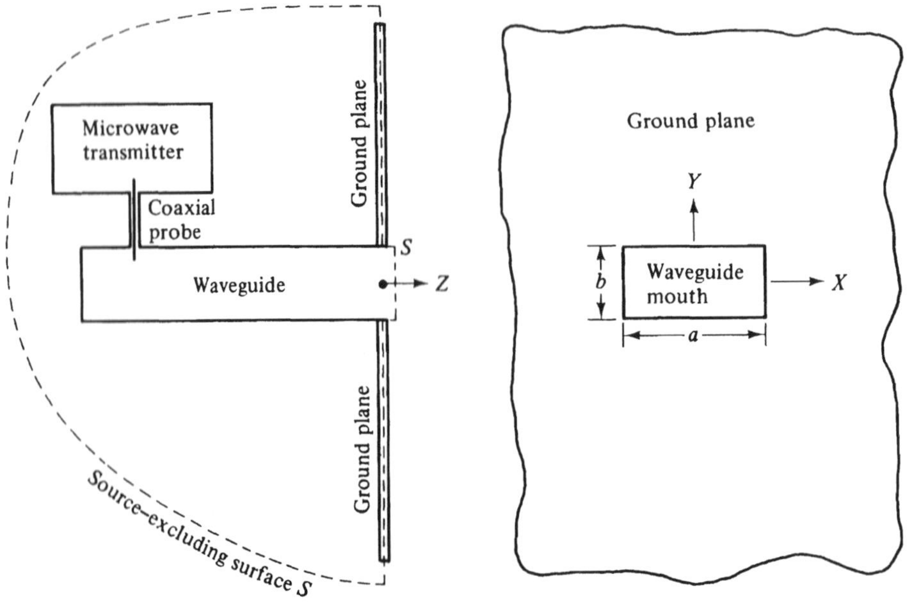

Fig. 3.1 Open-Ended Waveguide Flush-Mounted in a Large Ground Plane

The large ground plane will be modeled by assuming it is infinite in extent. Then S can be viewed as composed of a rectangular boss S_1 at the waveguide mouth plus an otherwise infinite plane S_2 lying in the ground plane and just below its outer surface, plus an infinite hemisphere S_3, which encloses the transmitter.

The radiation pattern will be determined by following the procedure outlined at the end of Section 1.13. The lineal current densities on S are given by $\mathbf{K} = \mathbf{1}_n \times \mathbf{H}$ and $\mathbf{K}_m = -\mu_0^{-1}\mathbf{1}_n \times \mathbf{E}$. But $\mathbf{E}$ and $\mathbf{H}$ are identically zero over the hemisphere S_3. Therefore equivalent sources, in the form of electric and magnetic current sheets, need to be placed only on the rectangular boss S_1. With the transmitter turned off, these current sheets—plus the actual sources in the ground plane—will maintain the actual fields in $z > 0$, while causing no fields in $z < 0$, including the region inside the waveguide.

For this reason, the mouth of the waveguide can now be closed off with conductor, causing the ground plane to become an infinite plane sheet *with no holes*. The seemingly peculiar situation emerges in which the two lineal current sheets in the boss S_1 combine to induce the actual current distribution everywhere in the hole-free ground plane except where the waveguide mouth had been. In that part of the hole-free ground plane, they combined to induce no current at all. This is a surprising effect, but one must remember that these are not physically realizable equivalent sources, but instead, convenient mathematical constructs.

With the presence of a hole-free infinite ground plane, the method of images can be invoked. The fields in $z > 0$ can be computed either from the two lineal current sheets on S_1 plus the actual currents in the ground plane, or by the lineal current sheets on S_1 plus their images. But from the results of Section 2.3, the image of $\mathbf{K} = \mathbf{1}_z \times \mathbf{H}$ is counterdirected, whereas the image of $\mathbf{K}_m = -\mu_0^{-1}\mathbf{1}_z \times \mathbf{E}$ is codirected. Therefore, as the boss is lowered so that S_1 approaches infinitesimally close to the surface of the ground plane, $\mathbf{K}$ and its image *cancel*; however, $\mathbf{K}_m$ and its image *add*. This simple formulation can be summarized by saying that the radiation pattern in $z > 0$ can be deduced solely from the magnetic lineal current distribution $-2\mu_0^{-1}\mathbf{1}_z \times \mathbf{E}$ in the mouth of the waveguide.

Next to be considered is an estimation of $\mathbf{E}_T$ in the waveguide mouth. With the origin of coordinates taken at the middle of a transverse cross section rather than at a corner, the electric field of the incident TE_{10} mode can be expressed in the form

$$E_y^i = C \cos \frac{\pi x}{a} e^{j(\omega t - \beta_{10} z)} \tag{3.1}$$

If the reflection coefficient for the TE_{10} mode is Γ, the back-scattered wave is

$$E_y^r = \Gamma C \cos \frac{\pi x}{a} e^{j(\omega t + \beta_{10} z)} \tag{3.2}$$

It will be assumed that these two fields comprise the bulk of $\mathbf{E}_T$ in the aperture. Then

$$\mathbf{E}_T = \mathbf{1}_y C' \cos \frac{\pi \xi}{a} \tag{3.3}$$

at a source point $(\xi, \eta, 0)$, with $C' = C(1 + \Gamma)$. As a consequence,

$$\mathbf{K}_m = \mathbf{1}_x 2\mu_0^{-1} C' \cos \frac{\pi \xi}{a} \tag{3.4}$$

Use of the generic integral forms (1.130) and (1.131) gives

$$\mathfrak{F}_\theta(\theta, \phi) = 2\mu_0^{-1}C' \cos\theta \cos\phi \int_{-a/2}^{a/2}\int_{-b/2}^{b/2} \cos\frac{\pi\xi}{a} e^{jk \sin\theta(\xi\cos\phi + \eta\sin\phi)}\, d\xi\, d\eta \tag{3.5}$$

$$\mathfrak{F}_\phi(\theta, \phi) = -2\mu_0^{-1}C' \sin\phi \int_{-a/2}^{a/2}\int_{-b/2}^{b/2} \cos\frac{\pi\xi}{a} e^{jk \sin\theta(\xi\cos\phi + \eta\sin\phi)}\, d\xi\, d\eta \tag{3.6}$$

The integral common to (3.5) and (3.6) reduces to

$$\int_{-a/2}^{a/2} \cos\frac{\pi\xi}{a}\cos\left(\frac{2\pi\xi}{\lambda}\sin\theta\cos\phi\right)d\xi \int_{-b/2}^{b/2} \cos\left(\frac{2\pi\eta}{\lambda}\sin\theta\sin\phi\right)d\eta \tag{3.7}$$

which readily integrates to give

$$\mathfrak{F}_\theta(\theta, \phi) = 4\pi\mu_0^{-1}abC' \cos\theta\cos\phi \frac{\cos(\pi X)}{\pi^2 - 4(\pi X)^2}\frac{\sin(\pi Y)}{(\pi Y)} \tag{3.8}$$

$$\mathfrak{F}_\phi(\theta, \phi) = -4\pi\mu_0^{-1}abC' \sin\phi \frac{\cos(\pi X)}{\pi^2 - 4(\pi X)^2}\frac{\sin(\pi Y)}{(\pi Y)} \tag{3.9}$$

in which

$$X = \frac{a}{\lambda}\sin\theta\cos\phi \tag{3.10}$$

$$Y = \frac{b}{\lambda}\sin\theta\sin\phi \tag{3.11}$$

From (1.123), since $\mathbf{A} \equiv 0$, it follows that $E_\theta = -\kappa\mathfrak{F}_\phi$ and $E_\phi = \kappa\mathfrak{F}_\theta$, with κ a common multiplier that includes the outgoing spherical wave factor. Therefore, in the XZ-plane ($\phi = 0°$, $180°$), Equations 3.8 and 3.9 indicate that there is only an E_ϕ component, given in normalized form by

$$E_\phi(\theta) = \pi^2\cos\theta \frac{\cos\left(\frac{\pi a}{\lambda}\sin\theta\right)}{\pi^2 - 4\left(\frac{\pi a}{\lambda}\sin\theta\right)^2} \tag{3.12}$$

In the YZ-plane ($\phi = 90°$, $270°$), there is only an E_θ component, given by

$$E_\theta(\theta) = \frac{\sin\left(\frac{\pi b}{\lambda}\sin\theta\right)}{\frac{\pi b}{\lambda}\sin\theta} \tag{3.13}$$

Polar plots of these two principal-plane field patterns, for the typical values $(a/\lambda) = 0.7$ and $(b/\lambda) = 0.35$, are shown in Figure 3.2. Plots for intermediate ϕ-cuts show a smooth transition, with the net polarization always parallel to the YZ-plane. The pattern is seen to be quite broad, consistent with the small size of the aperture.

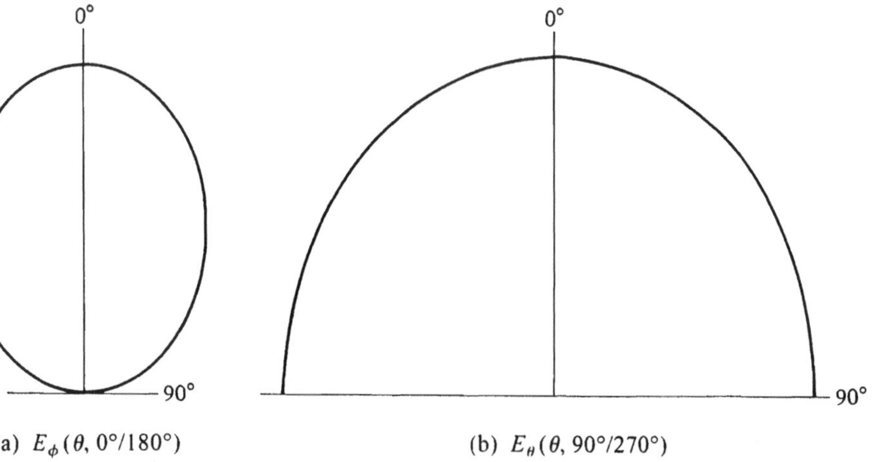

Fig. 3.2 Normalized E-Field Patterns of an Open-Ended Rectangular Waveguide with Large Ground Plane; Polar Plots; Linear Scale; $a = 0.7\lambda$, $b = 0.35\lambda$

3.3 Radiation from Horns

The functions $[\cos(\pi X)]/[\pi^2 - 4(\pi X)^2]$ and $[\sin(\pi Y)]/\pi Y$ that occur in both (3.8) and (3.9) are plotted versus X and Y in Figure 3.3. The two functions are seen to have similar features—an even symmetry and a central main lobe, with minor lobes that alternate in sign and diminish in height as X or Y is increased. From (3.10) and (3.11) one sees that the range of X and Y, as the pointing direction (θ, ϕ) varies through the half-space $z > 0$, is

$$-\frac{a}{\lambda} \leq X \leq \frac{a}{\lambda} \qquad -\frac{b}{\lambda} \leq Y \leq \frac{b}{\lambda}$$

Thus if one wants the radiation from this rectangular aperture to consist of a main beam and side lobes, clearly what is needed is to make a/λ and b/λ suitably large (how large depends on the desired narrowness of the main beam). For example, in the YZ-plane only the factor $[\sin(\pi Y)]/(\pi Y)$ is involved, and Figure 3.3 indicates that the null between the main beam and first side lobe occurs at an angle θ_1 given by

$$\sin\theta_1 = \frac{\lambda}{b} \tag{3.14}$$

Thus the larger b/λ, the smaller θ_1 and the narrower the main beam in the YZ-plane. Similarly, in the XZ-plane, only the factor $[\cos(\pi X)]/[\pi^2 - 4(\pi X)^2]$ is involved, and Figure 3.3 indicates that the null between the main beam and first side lobe occurs at an angle θ_2 given by

$$\sin\theta_2 = \frac{3}{2}\frac{\lambda}{a} \tag{3.15}$$

Here again, the larger a/λ, the narrower the main beam in the cut $\phi = 0°, 180°$.

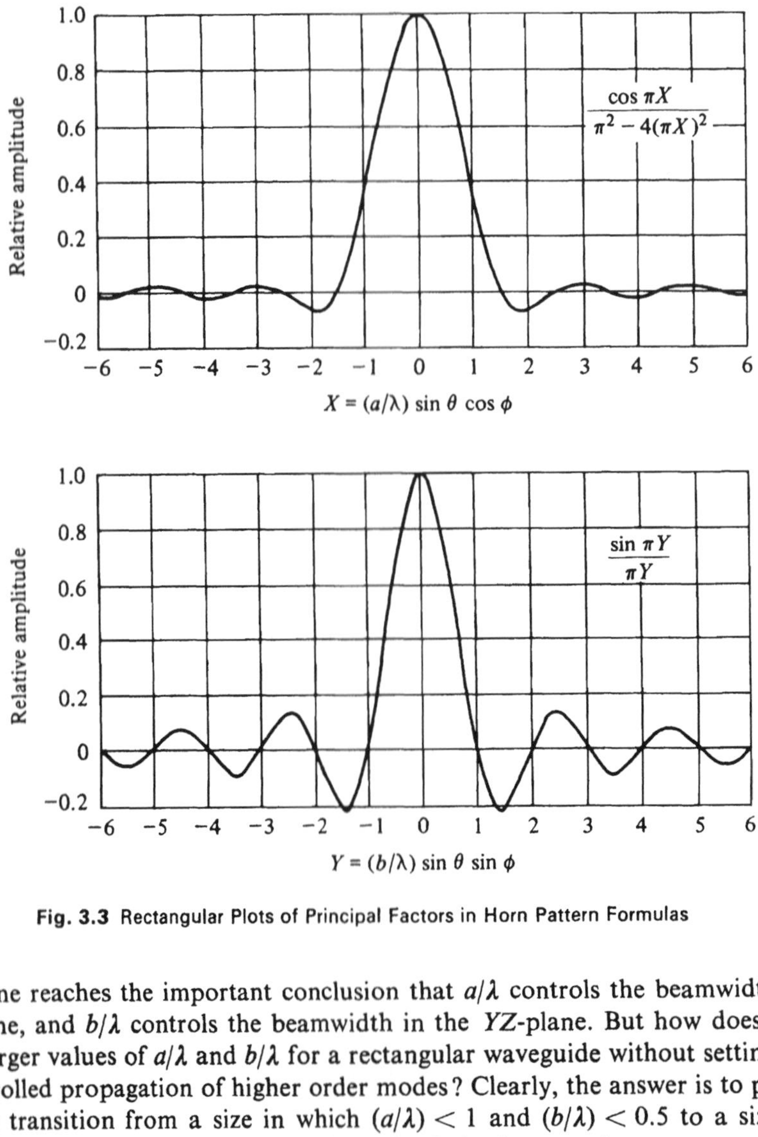

Fig. 3.3 Rectangular Plots of Principal Factors in Horn Pattern Formulas

One reaches the important conclusion that a/λ controls the beamwidth in the XZ-plane, and b/λ controls the beamwidth in the YZ-plane. But how does one get these larger values of a/λ and b/λ for a rectangular waveguide without setting up the uncontrolled propagation of higher order modes? Clearly, the answer is to provide a smooth transition from a size in which $(a/\lambda) < 1$ and $(b/\lambda) < 0.5$ to a size where a'/λ and b'/λ are large enough to produce the desired narrow beam pattern.

A common method for achieving this, because of its constructional simplicity, is to use a pyramidal horn, pictured in Figure 3.4. Ceteris paribus, the longer L, the smoother the transition. Practical limitations usually force adoption of some minimum L, below which the performance of the horn is degraded unacceptably. This typically corresponds to a flare angle of about 20°.

It should be noted that the presence of a significant flare angle has several effects on the pattern. First, forward-scattered modes of higher order are set up at the flare

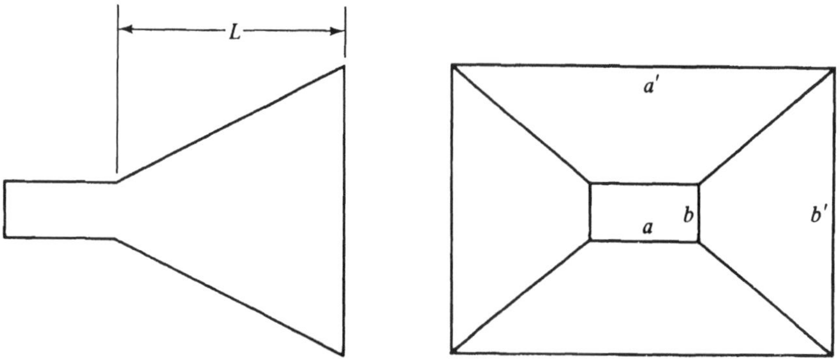

Fig. 3.4 A Pyramidal Horn Antenna

discontinuity. Some, with lower indices, do not have to travel too far before they are no longer cut off, and their presence at the ultimate a' by b' aperture affects the equivalent-source distribution and thus the pattern (though the effect is not necessarily bad). Second, the flare means that the dominant mode (expanding TE_{10}) that reaches the horn mouth has a curved phase front, which tends to broaden the main beam of the pattern somewhat and also to fill in the nulls.

Strictly speaking, the field pattern formulas in (3.8) and (3.9) apply only if TE_{10} modes with *plane* phase fronts are exclusively present in the aperture, and if the horn mouth is terminated in an infinite ground plane. In practice, horns are used without ground planes more often than with ground planes. However, if a'/λ and b'/λ are reasonably large, Figure 3.3 shows that the radiation at $\theta = 90°$ is small, and it ceases to be important whether the ground plane is there or not. Thus, despite the fact that (3.8) and (3.9) assume the presence of a ground plane and ignore the presence of higher-order modes, as well as any phase curvature to the TE_{10} mode, they provide a good first approximation to the field patterns of a pyramidal horn. As an example of this, Figure 3.5 shows the comparison of theory and experiment[1] for the principal plane cuts of a pyramidal horn for which $a'/\lambda = 1.82$ and $b'/\lambda = 1.47$.

Refinements which can account for many of the effects that have been ignored in this introductory treatment can be found in the literature. W. C. Jakes,[2] has provided a comprehensive overview of the subject and A. W. Love[3] has compiled an excellent collection of journal paper reprints.

This entire discussion could be repeated for an open-ended circular waveguide and the conical horn that evolves from it in order to produce a narrow beam of radiation.

[1]C. W. Horton, "On the Theory of the Radiation Patterns of Electromagnetic Horns of Moderate Flare Angles," *Proc. IRE*, 37 (1949), 744–49.

[2]W. C. Jakes, Jr., "Horn Antennas," *Antenna Engineering Handbook*, ed. H. Jasik (New York: McGraw-Hill Book Co., Inc., 1961), Chapter 10.

[3]*Electromagnetic Horn Antennas*, ed. A. W. Love (New York: IEEE Press, 1976).

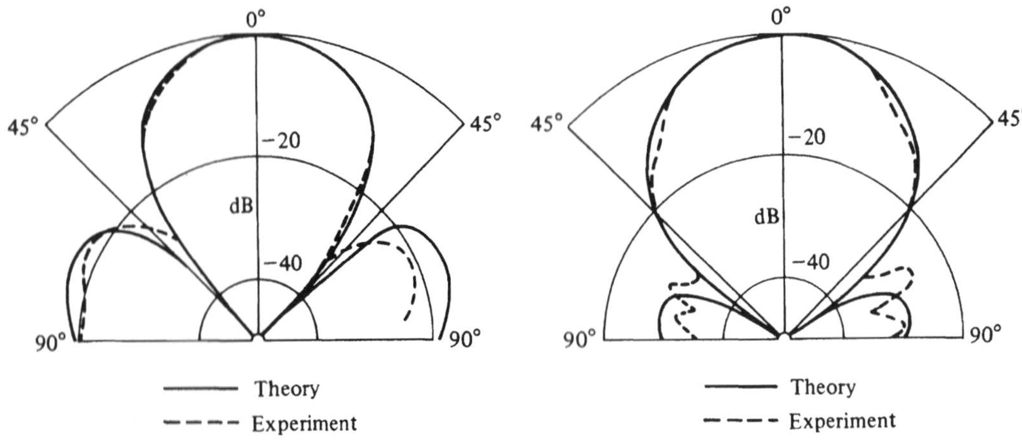

(a) E-plane pattern (b) H-plane pattern

Fig. 3.5 Normalized Power Patterns of a Pyramidal Horn Antenna; Polar Plots; Logarithmic Scale; $a' = 1.82\lambda$, $b' = 1.47\lambda$ (© 1949 IEEE. Reprinted from C. W. Horton, *Proc. IRE*, pp. 744–749, 1949.)

3.4 Center-Fed Slot in Large Ground Plane

An extremely important antenna element, not so much in its own right, but more because of its derivatives, is the narrow rectangular center-fed slot in a large ground plane, as shown in Figure 3.6. The length and width are $2l$ and w, with $2l \gg w$. A two-wire line can be imagined to be feeding the slot at the central points P_1 and P_2.

With $w \lll \lambda$, the slot itself resembles a section of two-wire line, the two "wires" being semi-infinite ground planes with adjacent edges at $x = \pm w/2$, with these "wires" shorted at $z = \pm l$. A standing wave of voltage exists on this section of line

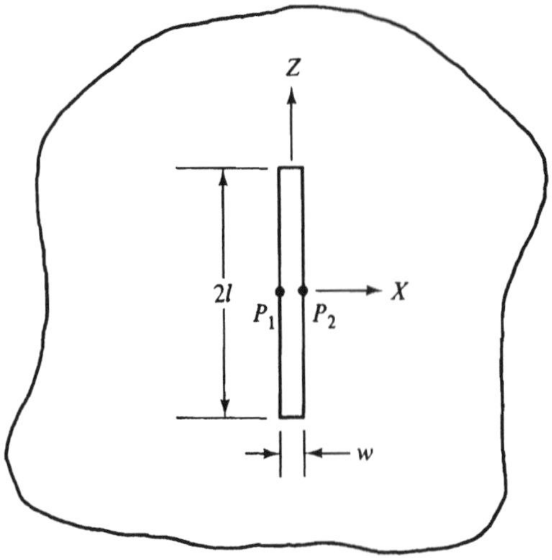

Fig. 3.6 Center-Fed Slot in a Large Ground Plane

such that the electric field in the slot is given to good approximation by

$$\mathbf{E}(\xi, \eta, \zeta) = \mathbf{1}_x E_x(\xi, 0, \zeta) = \mathbf{1}_x \frac{V_m}{w} \sin [k(l - |\zeta|)] \tag{3.16}$$

in which V_m is the peak voltage.

If the generator and two-wire line which attaches to the points P_1 and P_2 are in $y < 0$, the fields in $y > 0$ can be determined by the same technique used in connection with Figure 3.1. The actual large ground plane is modeled by an infinite ground plane. A source-excluding surface S is constructed that consists of a boss S_1 in front of the slot, plus an otherwise infinite plane S_2 inside the ground plane, plus an infinite hemisphere in $y < 0$. Use of the image principle reduces all the sources to a lineal magnetic current sheet on S_1 given by

$$\begin{aligned} \mathbf{K_m} &= -2\mu_0^{-1}\mathbf{1}_y \times \mathbf{1}_x \frac{V_m}{w} \sin [k(l - |\zeta|)] \\ &= \mathbf{1}_z \frac{2\mu_0^{-1} V_m}{w} \sin [k(l - |\zeta|)] \end{aligned} \tag{3.17}$$

Equation 3.17 is identical in form to Equation 2.2. For this reason, a narrow center-fed slot in a large ground plane is often referred to as a magnetic dipole. The analysis of Section 2.2 can be repeated with the principal result that

$$\mathfrak{F}_\theta(\theta) = \frac{-4\mu_0^{-1} V_m}{k \sin \theta} [\cos (kl \cos \theta) - \cos (kl)] \tag{3.18}$$

For a slot one-half wavelength long, use of (1.123) and (1.124) gives

$$E_\phi = -j \frac{V_m}{\pi} \frac{e^{j(\omega t - kr)}}{r} \left[\frac{\cos [(\pi/2) \cos \theta]}{\sin \theta} \right] \tag{3.19}$$

$$H_\theta = j \frac{V_m}{\pi \eta} \frac{e^{j(\omega t - kr)}}{r} \left[\frac{\cos [(\pi/2) \cos \theta]}{\sin \theta} \right] \tag{3.20}$$

which are in the same form as (2.8) and (2.9), but with the polarization rotated 90°. The field pattern shown in Figure 2.4 thus also applies for a half-wavelength slot in a large ground plane.

Since there is no **A** for this antenna, (1.125) yields

$$P_{r,\phi}(\theta) = \frac{8 V_m^2/\eta}{(4\pi r)^2} \left[\frac{\cos^2 [(\pi/2) \cos \theta]}{\sin^2 \theta} \right] \tag{3.21}$$

If the presence of the transmitter and two-wire feed in $y < 0$ can be assumed to have little influence on the far-field in $y < 0$, then (3.19) through (3.21) apply on both sides of the ground plane. Under this assumption, the total power radiated is

$$P_{rad} = (0.609) \frac{4 V_m^2/\eta}{2\pi} \tag{3.22}$$

Since, for this case of $l = \lambda/4$, the peak input voltage to the slot (at the terminals P_1/P_2) is V_m, the feeding transmission lines can be said to be delivering the power

$$\frac{1}{2}V_m^2 G_{rad} = (0.609)\frac{4V_m^2/\eta}{2\pi} \tag{3.23}$$

to a conductance G_{rad} placed across its terminus. The value $R_{rad} = 1/G_{rad}$ is called the *radiation resistance* of the center-fed half wavelength slot. From (3.23),

$$R_{rad} = \frac{\pi\eta/4}{0.609} = 486 \text{ ohms} \tag{3.24}$$

An interesting relation results if (3.24) is multiplied by (2.14), for then

$$R_{rad}^{dipole} \cdot R_{rad}^{slot} = \left(\frac{0.609\eta}{\pi}\right)\left(\frac{\pi\eta/4}{0.609}\right) = \frac{\eta^2}{4} \tag{3.25}$$

This is a special case of Booker's relation[4]. It will be shown in Chapter 7 that

$$Z^{dipole} \cdot Z^{slot} = \frac{\eta^2}{4} \tag{3.26}$$

for any length $2l$ of a narrow dipole, as long as the complementary slot in a ground plane has the same length.

The analysis undertaken in Section 2.2 for a short dipole could be repeated here for a short slot. All of the results are similar, with **E** and **H** interchanged; this is left as an exercise.

The practical applications of a two-wire fed slot, cut in a large ground plane and radiating into *both* half-spaces, are few. However, if the slot is "boxed in" on one side by a metallic-walled cavity and the dimensions of the cavity are properly chosen, the radiation in one half-space is hardly affected, whereas in the other half-space, it is virtually eliminated. The presence of the cavity affects the input impedance of the slot. This is a subject which will be treated in Chapter 8.

3.5 Waveguide-Fed Slots

Most antenna applications involving slots unify the feeding and radiating structures by placing the slots in one of the walls of a rectangular waveguide. This insures a nonradiating transmission line, permits precise machining of the slots, and provides a mechanically rigid structure. Usually the slots are arranged in arrays, which complicates the feeding because of mutual coupling. That subject will be treated in Chapter

[4]H. G. Booker, "Slot Aerials and Their Relation to Complementary Wire Aerials (Babinet's Principle)", *J.I.E.E.* (London), 93, part IIIA (1946), 620–26.

8. For now, the discussion will be limited to the behavior of a single slot, cut in one of the walls of a rectangular waveguide and excited by a TE_{10} mode.

With axes chosen as shown in Figure 3.7, the normalized field components for a TE_{10} mode, traveling in the positive z-direction, are

$$\begin{aligned} H_z &= j\cos\frac{\pi x}{a}\, e^{j(\omega t - \beta_{10} z)} \\ H_x &= \frac{-\beta_{10}}{\pi/a}\sin\frac{\pi x}{a}\, e^{j(\omega t - \beta_{10} z)} \\ E_y &= \frac{\omega\mu_0}{\pi/a}\sin\frac{\pi x}{a}\, e^{j(\omega t - \beta_{10} z)} \end{aligned} \tag{3.27}$$

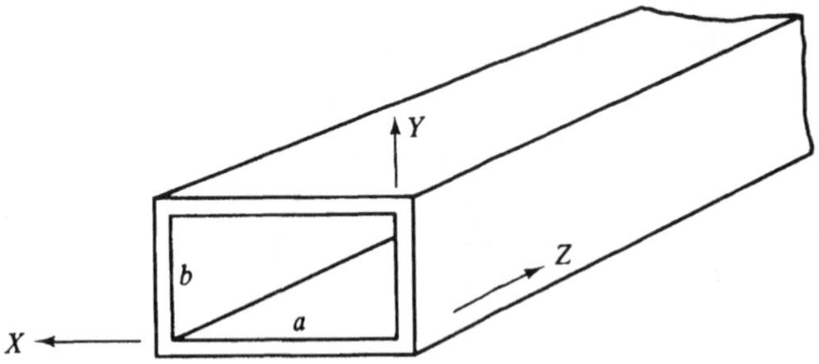

Fig. 3.7 Rectangular Waveguide

The normalization in (3.27) consists of choosing the peak magnitude of the longitudinal component to be unity, but adjusting the phase by 90° through the presence of the factor j. This causes the transverse components to have pure real amplitudes, a convenience since they enter into the Poynting vector calculation of power flow. Figure 3.8a shows the electric field distribution at a fixed time, and Figure 3.8b

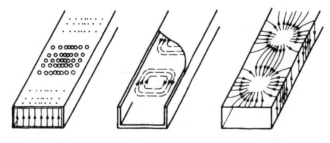

(a) Electric field and charge distribution (b) Magnetic field (c) Current flow

Fig. 3.8 Field and Source Distributions in a Rectangular Waveguide for TE_{10} Mode (From *Electromagnetics* by R. S. Elliott. Copyright 1966, McGraw-Hill. Used with permission of McGraw-Hill Book Company.)

illustrates the corresponding distribution of magnetic field. The charge and current distribution in the waveguide walls can be determined using the same procedure which led to Equations 1.112 and 1.114. Figure 3.8a shows the instantaneous distribution of charge on the upper broad wall and Figure 3.8c pictures the corresponding instantaneous current distribution. Over time, these patterns will propagate longitudinally at the phase velocity of the TE_{10} mode.

If a narrow slot is cut in one of the waveguide walls such that its long dimension runs parallel to a current line, the presence of the slot causes only a minor perturbation in the current distribution, and negligible coupling to outer space occurs. Thus the longitudinal slot on the center line in Figure 3.9a causes little disturbance, as can be seen by studying the current distribution in Figure 3.8c. Such a slot is useful for making measurements of the E-field inside the waveguide, since a vertical probe can be inserted through this slot to sample the field. If the probe is permitted to move longitudinally, VSWR data can be obtained.

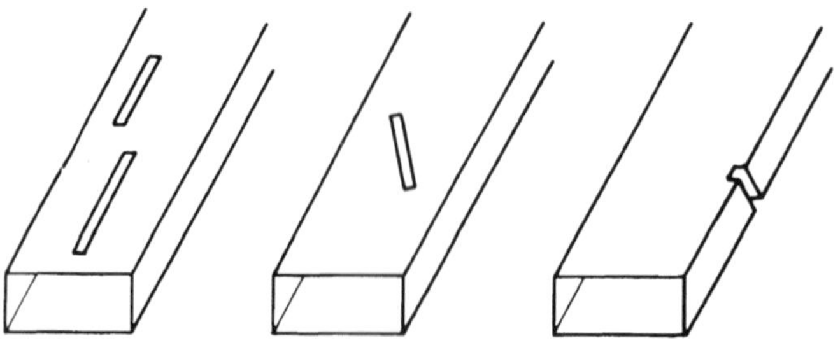

(a) Longitudinal slots in broad wall (b) Inclined slot in broad wall (c) Inclined slot in narrow wall

Fig. 3.9 Some Practical Slot Configurations in the Walls of a Rectangular Waveguide (From *Electromagnetics* by R. S. Elliott. Copyright 1966, McGraw-Hill. Used with permission of McGraw-Hill Book Company.)

However, the longitudinal slot displaced from the center line in Figure 3.9a will interrupt X-directed current; the more the displacement, the greater the interruption. The electric field developed in this slot has as one of its manifestations a displacement current which "replaces" the interrupted conduction current. This electric field can be represented by its equivalent magnetic current sheet, and can radiate into outer space.

Similarly, the inclined broad wall slot in Figure 3.9b interrupts Z-directed current, the more so the greater the inclination. This is another candidate for use as a radiating element. Finally, the inclined slot in the narrow wall, shown in Figure 3.9c, will interrupt Y-directed current, the more the inclination, the greater the interruption. This slot is a third candidate for use as a radiating element.

A desirable feature shared by all three of these radiating-type slots is that there is mechanical control over the amount of radiation, through choice of the amount of displacement or inclination. What is needed is a set of design equations which reveal this connection, a problem that is addressed in the next section.

3.6 Theory of Waveguide-Fed Slot Radiators[5]

It is assumed that the reader is familiar with waveguide mode theory, and thus it will be stated without proof that the field components in rectangular guide can be expressed in the normalized form

$$\begin{array}{ll} TE_{mn}\ mode & TM_{mn}\ mode \\ H_z = jH_{az}e^{\mp\gamma_a z} & E_z = \pm jE_{az}e^{\mp\gamma_a z} \\ \mathbf{E}_t = \mathbf{E}_{at}e^{\mp\gamma_a z} & \mathbf{E}_t = \mathbf{E}_{at}e^{\mp\gamma_a z} \\ \mathbf{H}_t = \pm\mathbf{H}_{at}e^{\mp\gamma_a z} & \mathbf{H}_t = \pm\mathbf{H}_{at}e^{\mp\gamma_a z} \end{array} \tag{3.28}$$

In (3.28) the subscript a is shorthand for the double index mn, and

$$H_{az} = \cos\frac{m\pi x}{a}\cos\frac{n\pi y}{b} \tag{3.29}$$

$$E_{az} = \sin\frac{m\pi x}{a}\sin\frac{n\pi y}{b} \tag{3.30}$$

$$\gamma_a = +\sqrt{\left(\frac{m\pi}{a}\right)^2 + \left(\frac{n\pi}{b}\right)^2 - k^2} \tag{3.31}$$

The upper signs in (3.28) need to be taken for propagation in the positive Z-direction; the lower signs for propagation in the negative Z-direction. The transverse field vectors are given by

$$TE_{mn}\ modes$$

$$\begin{aligned} \mathbf{E}_{at} &= \frac{\omega\mu_0}{\gamma_a^2 + k^2}\left(\mathbf{1}_x\frac{\partial H_{az}}{\partial y} - \mathbf{1}_y\frac{\partial H_{az}}{\partial x}\right) \\ \mathbf{H}_{at} &= -\frac{j\gamma_a}{\gamma_a^2 + k^2}\left(\mathbf{1}_x\frac{\partial H_{az}}{\partial x} + \mathbf{1}_y\frac{\partial H_{az}}{\partial y}\right) \end{aligned} \tag{3.32}$$

$$TM_{mn}\ modes$$

$$\begin{aligned} \mathbf{E}_{at} &= -\frac{j\gamma_a}{\gamma_a^2 + k^2}\left(\mathbf{1}_x\frac{\partial E_{az}}{\partial x} + \mathbf{1}_y\frac{\partial E_{az}}{\partial y}\right) \\ \mathbf{H}_{at} &= -\frac{\omega\epsilon_0}{\gamma_a^2 + k^2}\left(\mathbf{1}_x\frac{\partial E_{az}}{\partial y} - \mathbf{1}_y\frac{\partial E_{az}}{\partial x}\right) \end{aligned} \tag{3.33}$$

[5]The results to be presented in this section were first obtained by A. F. Stevenson in a classic paper "Theory of Slots in Rectangular Waveguides", *J. Appl. Phys.*, 19 (1948), 24–38. However, the development follows an approach used by J. E. Eaton, L. J. Eyges, and G. G. MacFarlane, *Microwave Antenna Theory and Design*, ed. S. Silver, vol. 12, MIT Rad. Lab Series (New York: McGraw-Hill Book Co., Inc., 1949), Chapter 9.

This information needs to be applied to scattering off a slot cut in one of the walls of the waveguide. Without at this point specifying which type of slot it might be from among those illustrated in Figure 3.8, imagine that the slot is contained in the region bounded by $z = z_1$ and $z = z_2$, with $z_2 > z_1$. If the waveguide is assumed to be infinitely long and a TE_{10} mode is launched from $z = -\infty$, traveling in the positive Z-direction, the incidence of this mode on the slot will cause a profusion of reactions. Backward and forward scattering of all TE_{mn} and TM_{mn} modes is possible. Radiation into outer space via the electric field set up in the slot is possible.

If the waveguide walls are assumed to be perfectly conducting, and if the a- and b-dimensions are chosen so that all modes except TE_{10} are cut off, then a power balance can be written that will connect the slot's excitation to its displacement or inclination. This can be done because (1) the power contained in the incident wave is calculable; (2) the power radiated is also calculable if the electric field in the slot is known; and (3) the forward and backward scattered waves in the TE_{10} mode can be determined if the electric field in the slot is known. It is this last determination that completes the linkage in the power balance equation.

To see the relation between scattering off the slot and the electric field distribution in the slot, consider two fields $(\mathbf{E}_1, \mathbf{H}_1)$ and $(\mathbf{E}_2, \mathbf{H}_2)$, both time-harmonic at the common angular frequency ω, and both satisfying Maxwell's equations in a region V bounded by a closed surface S. Let S be a rectangular parallelopiped with end faces S_1 at $z = z_1$ and S_2 at $z = z_2$; the remainder of S is a surface S_3 which is skintight against the four interior faces of the waveguide between z_1 and z_2. With S a source-free region, the reciprocity theorem in the form of (1.136) is applicable and the two fields are connected by the relation

$$\int_S (\mathbf{E}_1 \times \mathbf{H}_2 - \mathbf{E}_2 \times \mathbf{H}_1) \cdot d\mathbf{S} = 0 \tag{3.34}$$

Let $(\mathbf{E}_1, \mathbf{H}_1)$ be the scattered field due to the interaction of the slot and the incident mode. $(\mathbf{E}_2, \mathbf{H}_2)$ does not relate to the actual situation, but its use is an artifice to obtain the scattering coefficient. It will be taken to be a single normalized mode, that is, a member of either the TE or TM families given in (3.28), traveling in the positive Z-direction and designated by the subscript $b = m'n'$. The transverse components of the scattered field $(\mathbf{E}_1, \mathbf{H}_1)$ can be represented by

$$\begin{aligned}
\mathbf{E}_{1t} &= \sum_a C_a \mathbf{E}_{at} e^{-\gamma_a z} \qquad & z > z_2 \\
\mathbf{E}_{1t} &= \sum_a B_a \mathbf{E}_{at} e^{\gamma_a z} \qquad & z < z_1 \\
\mathbf{H}_{1t} &= \sum_a C_a \mathbf{H}_{at} e^{-\gamma_a z} \qquad & z > z_2 \\
\mathbf{H}_{1t} &= -\sum_a B_a \mathbf{H}_{at} e^{\gamma_a z} \qquad & z < z_1
\end{aligned} \tag{3.35}$$

in which the summation is over all TE and TM modes. The forward-scattered mode amplitudes C_a and the backward-scattered mode amplitudes B_a are yet to be determined. It should be noted that $(\mathbf{E}_1, \mathbf{H}_1)$ can*not* be represented by (3.35) in the region $z_1 < z < z_2$ because of the presence of the slot.

Since the tangential component of $\mathbf{E}_2$ is identically zero over the entire subsurface S_3, and since the tangential component of $\mathbf{E}_1$ is identically zero over all of the subsurface S_3 except that part occupied by the slot, it follows from (3.34) that

$$\int_{slot} (\mathbf{E}_1 \times \mathbf{H}_2) \cdot d\mathbf{S} = I_1 + I_2 \tag{3.36}$$

in which

$$I_1 = \int_{S_1} (\mathbf{E}_2 \times \mathbf{H}_1 - \mathbf{E}_1 \times \mathbf{H}_2) \cdot d\mathbf{S}$$

$$= \int_{S_1} \left(\mathbf{E}_{bt} e^{-\gamma_b z_1} \times \sum_a B_a \mathbf{H}_{at} e^{\gamma_a z_1} + \sum_a B_a \mathbf{E}_{at} e^{\gamma_a z_1} \times \mathbf{H}_{bt} e^{-\gamma_b z_1}\right) \cdot \mathbf{1}_z \, dS_1 \tag{3.37}$$

and

$$I_2 = \int_{S_2} (\mathbf{E}_2 \times \mathbf{H}_1 - \mathbf{E}_1 \times \mathbf{H}_2) \cdot d\mathbf{S}$$

$$= \int_{S_2} \left(\mathbf{E}_{bt} e^{-\gamma_b z_2} \times \sum_a C_a \mathbf{H}_{at} e^{-\gamma_a z_2} - \sum_a C_a \mathbf{E}_{at} e^{-\gamma_a z_2} \times \mathbf{H}_{bt} e^{-\gamma_b z_2}\right) \cdot \mathbf{1}_z \, dS_2 \tag{3.38}$$

Because of the orthogonal properties of these modes, the only contribution to I_1 comes when $a = mn$ not only equals $b = m'n'$, but also when the indices mn and $m'n'$ refer to the same *type* of mode (both *TE* or both *TM*). The proof is left as an exercise, the result being that

$$I_1 = 2B_b \int_{S_1} (\mathbf{E}_{bt} \times \mathbf{H}_{bt}) \cdot \mathbf{1}_z \, dS_1 \tag{3.39}$$

The same argument applies to the evaluation of I_2, except that in the special case $a = mn = b = m'n'$ the integrand is identically zero, and therefore $I_2 = 0$. When these results are placed in (3.36), a formula emerges from which the back-scattered mode amplitude B_b can be computed as follows.

$$B_b = \frac{\int_{slot} (\mathbf{E}_1 \times \mathbf{H}_2) \cdot d\mathbf{S}}{2 \int_{S_1} (\mathbf{E}_{bt} \times \mathbf{H}_{bt}) \cdot \mathbf{1}_z \, dS_1} \tag{3.40}$$

If this process is repeated with the only change being that $(\mathbf{E}_2, \mathbf{H}_2)$ is assumed to be propagating in the negative Z-direction, one finds that

$$C_b = \frac{\int_{slot} (\mathbf{E}_1 \times \mathbf{H}_2) \cdot d\mathbf{S}}{2 \int_{S_2} (\mathbf{E}_{bt} \times \mathbf{H}_{bt}) \cdot \mathbf{1}_z \, dS_2} \tag{3.41}$$

It is important to note that, although the denominators of (3.40) and (3.41) are equal, the numerators are not necessarily equal, because $\mathbf{H}_2$ in (3.40) is associated with a $+Z$ propagating mode, whereas the $\mathbf{H}_2$ in (3.41) is associated with a $-Z$ propagating mode.

Equations 3.40 and 3.41 have wide applicability. As an illustration of their use, consider an offset longitudinal shunt slot in the upper broad wall, depicted in Figure 3.10. It will be assumed that the waveguide walls have negligible thickness and are composed of perfect conductor. The slot is rectangular with length $2l$ and width w where $2l \gg w$. The origin of coordinates has been taken so that the XY-plane bisects the slot. The transverse dimensions of the waveguide are chosen so that only the TE_{10} mode can propagate.

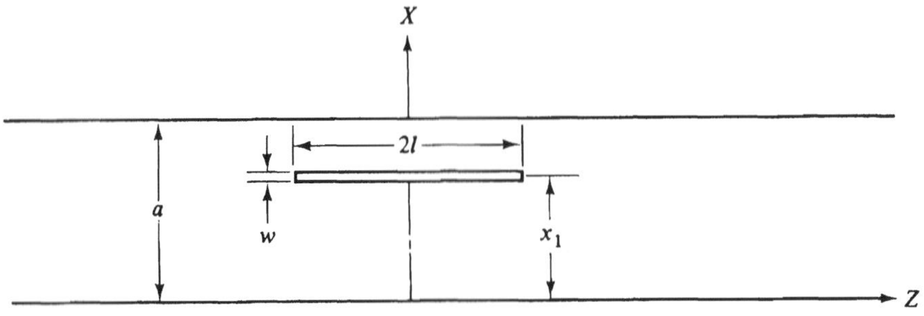

Fig. 3.10 An Offset Longitudinal Slot in the Upper Broad Wall of a Rectangular Waveguide

The forward- and backward-scattering off this slot in the TE_{10} mode will be determined with the aid of (3.40) and (3.41). First, it is a simple matter to show that

$$\int_{S_1} (\mathbf{E}_{10,t} \times \mathbf{H}_{10,t}) \cdot \mathbf{1}_z \, dS_1 = \frac{\omega\mu_0\beta_{10}}{(\pi/a)^2} \int_0^b d\eta \int_0^a \sin^2\frac{\pi\xi}{a}\, d\xi = \frac{\omega\mu_0\beta_{10}ab}{2(\pi/a)^2} \tag{3.42}$$

Second, the narrowness of the slot permits the assumption that

$$\mathbf{E}_1 = \mathbf{1}_x E_{1x}(\zeta) \tag{3.43}$$

and thus

$$B_{10} = \frac{j\int_{x_1-w/2}^{x_1+w/2} \cos\frac{\pi\xi}{a}\, d\xi \int_{-l}^{l} E_{1x}(\zeta)e^{-j\beta_{10}\zeta}\, d\zeta}{\omega\mu_0\beta_{10}ab/(\pi/a)^2}$$

$$B_{10} = \frac{-(\pi/a)^2 \cos(\pi x_1/a)}{j\omega\mu_0\beta_{10}ab} \int_{-l}^{l} V(\zeta)e^{-j\beta_{10}\zeta}\, d\zeta \tag{3.44}$$

in which $V(\zeta) = wE_{1x}(\zeta)$ is the voltage distribution in the slot.

In like manner, one can show that

$$C_{10} = \frac{-(\pi/a)^2 \cos(\pi x_1/a)}{j\omega\mu_0\beta_{10}ab}\int_{-l}^{l} V(\zeta)e^{j\beta_{10}\zeta}\,d\zeta \tag{3.45}$$

The slot voltage distribution function $V(\zeta)$ depends on the manner in which the slot is excited. Let it be assumed that a matched generator is placed at a position $z \ll -l$ and a matched load is placed at a position $z \gg l$, so that the generator launches a TE_{10} mode of amplitude A_{10} in the guide propagation in the $+Z$ direction. The TE_{10} modes scattered off the slot, of amplitudes B_{10} and C_{10}, cause no additional reflections because of the matched generator and the matched load.

Despite the fact that the generator-launched TE_{10} mode incident on the slot has a phase progression across the slot, detailed analysis shows that, if $2l \simeq \lambda_0/2$, the dominant component of $V(\zeta)$ is a symmetrical standing wave of the form

$$V(\zeta) = V_m \sin[k(l - |\zeta|)] \tag{3.46}$$

The similarity of (3.46) to (3.17) should be noted; it is as though the slot were essentially being excited at its center by a two-wire line.

With the approximation in (3.46) assumed, Equations 3.44 and 3.45 become

$$\begin{aligned} B_{10} = C_{10} &= \frac{2V_m(\pi/a)^2 \cos(\pi x_1/a)}{j\omega\mu_0\beta_{10}ab}\int_0^l \sin[k(l-\zeta)]\cos\beta_{10}\zeta\,d\zeta \\ B_{10} = C_{10} &= \frac{2V_m}{j\omega\mu_0(\beta_{10}/k)ab}(\cos\beta_{10}l - \cos kl)\cos\frac{\pi x_1}{a} \end{aligned} \tag{3.47}$$

It is important to observe that the assumed symmetry of $V(\zeta)$ resulted in the scattering off the slot being *symmetrical*, that is, $B_{10} = C_{10}$. This implies that the slot is equivalent to a shunt obstacle on a two-wire transmission line. To see this, consider the situation suggested by Figure 3.11. A transmission line of characteristic admittance G_0 is shunted at $z = 0$ by a lumped admittance Y. The voltage and current on the

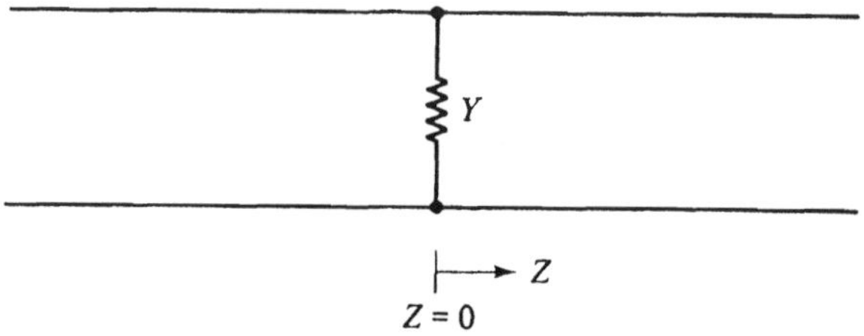

Fig. 3.11 A Shunt Obstacle on a Two-Wire Transmission Line

line are given by

$$\begin{aligned} V(z) &= Ae^{-j\beta z} + Be^{j\beta z} \\ I(z) &= AG_0e^{-j\beta z} - BG_0e^{j\beta z} \end{aligned} \quad z < 0$$
$$\begin{aligned} V(z) &= (A + C)e^{-j\beta z} \\ I(z) &= (A + C)G_0e^{j\beta z} \end{aligned} \quad z > 0 \tag{3.48}$$

The form selected for (3.48) is consistent with a matched generator being to the left of Y and a matched load to the right of Y in Figure 3.11.

The boundary conditions are

$$\begin{aligned} V(0^-) &= V(0) = V(0^+) \\ I(0^-) &= V(0)Y + I(0^+) \end{aligned} \tag{3.49}$$

which, when inserted in (3.48), give

$$A + B = A + C \tag{3.50}$$

$$(A - B)G_0 = (A + B)Y + (A + C)G_0 \tag{3.51}$$

Thus $B = C$ (the scattering is *symmetrical*), and

$$\frac{Y}{G_0} = -\frac{2B}{A + B} \tag{3.52}$$

The usefulness of (3.52) lies in the fact that, by analogy, if one can find the ratio $-2B_{10}/(A_{10} + B_{10})$ for the slot, one can then say that the slot has an equivalent normalized shunt admittance equal to that ratio.

The slot is said to be *resonant* if Y/G_0 is pure real. Since no loss in generality results from taking A_{10} to be pure real, it follows that the resonant normalized conductance of the slot is given by

$$\frac{G}{G_0} = -\frac{2B_{10}}{A_{10} + B_{10}} \tag{3.53}$$

where B_{10} is perforce pure real also. What this implies is that, for a given displacement x_1 of the slot, it is assumed in (3.53) that the length $2l$ of the slot has been adjusted so that B_{10} is either in phase with, or out of phase with, A_{10}. (It will be seen subsequently that B_{10} is out of phase with A_{10}).

Attention will now be restricted to this special case of a resonant slot.[6] The assumption of resonance permits a deduction from (3.53) via a power balance equa-

[6]The more general case of a nonresonant slot will be considered in Chapter 8.

tion. The incident power is given by

$$P_{inc} = \frac{1}{2} \mathcal{R}e \int_{S_1} (A_{10}\mathbf{E}_{10,t} \times A_{10}^* \mathbf{H}_{10,t}^*) \cdot \mathbf{1}_z \, dS_1 = \frac{\omega\mu_0\beta_{10}ab}{4(\pi/a)^2} A_{10}A_{10}^* \tag{3.54}$$

In like manner, one finds that the reflected and transmitted powers are

$$P_{refl} = \frac{\omega\mu_0\beta_{10}ab}{4(\pi/a)^2} B_{10}B_{10}^* \tag{3.55}$$

$$P_{tr} = \frac{\omega\mu_0\beta_{10}ab}{4(\pi/a)^2} (A_{10} + C_{10})(A_{10} + C_{10})^* \tag{3.56}$$

If use is made of the information that $B_{10} = C_{10}$ and that all three amplitudes are pure real, then

$$\frac{\omega\mu_0\beta_{10}ab}{4(\pi/a)^2} [A_{10}^2 - B_{10}^2 - (A_{10} + B_{10})^2] = \text{power radiated} \tag{3.57}$$

But experiment shows that resonance occurs when $2l \cong \lambda_0/2$, in which case, if the upper wall of the waveguide is imbedded in a large ground plane, the radiated power is given by one-half of (3.22). Thus

$$-\frac{\omega\mu_0\beta_{10}ab}{2(\pi/a)^2} B_{10}(A_{10} + B_{10}) = 0.609 \frac{V_m V_m^*}{\pi\eta} \tag{3.58}$$

If V_m is eliminated from (3.47) and (3.58), the result is

$$\frac{G}{G_0} = -\frac{2B_{10}}{A_{10} + B_{10}} = 2.09 \frac{(a/b)}{(\beta_{10}/k)} (\cos \beta_{10} l - \cos kl)^2 \cos^2 \frac{\pi x_1}{a} \tag{3.59}$$

When the substitution $x = x_1 - (a/2)$ is made in (3.59) and the approximation $kl \cong \pi/2$ is used, one obtains

$$\frac{G}{G_0} = \left[2.09 \frac{(a/b)}{(\beta_{10}/k)} \cos^2 \left(\frac{\beta_{10}}{k} \frac{\pi}{2}\right)\right] \sin^2 \frac{\pi x}{a} \tag{3.60}$$

in which x is the offset from the center line of the broad wall.

Equation 3.60 is a celebrated result first obtained by A. F. Stevenson[7]; it indicates that the normalized conductance of a resonant longitudinal shunt slot in the broad wall of a rectangular waveguide is approximately equal to a constant times the square of the sine of an angle proportional to its offset.

[7]Stevenson, "Theory of Slots."

Figure 3.12 gives a typical plot of experimental data showing resonant length of a longitudinal shunt slot versus offset. It can be seen that although the resonant length is offset-dependent, it stays close to the value $2l = \lambda_0/2$ assumed in (3.60).

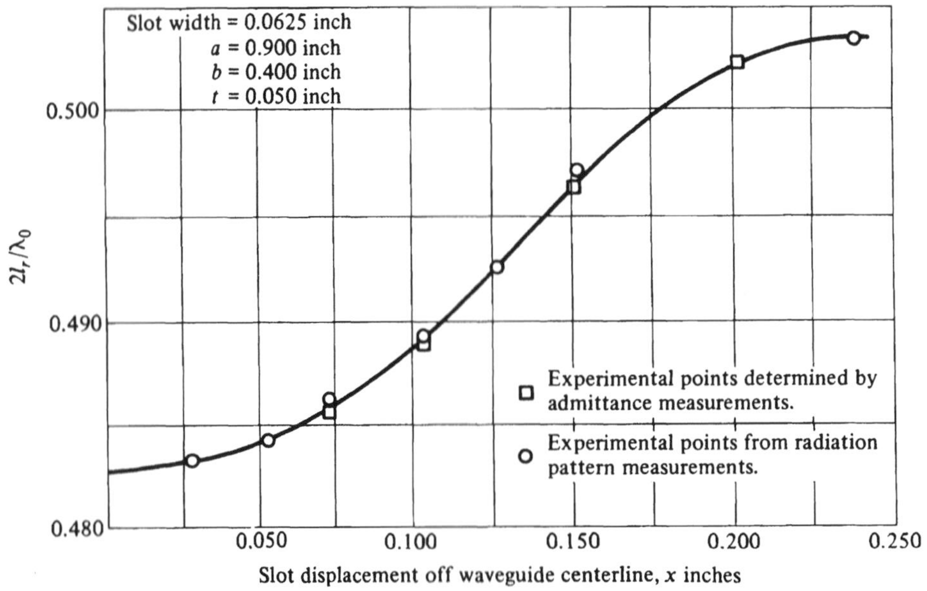

Fig. 3.12 Resonant Length versus Offset for Longitudinal Shunt Slot (After R. J. Stegen, "Longitudinal Shunt Slot Characteristics," Hughes Technical Memorandum No. 261, Nov. 1951, Hughes Aircraft Co., Culver City, California)

Figure 3.13 shows a plot of (3.60) versus experimental data. The agreement is quite good, serving to justify the approximations that were made in the theory. A more accurate analysis, which will give a better fit to the experimental data, will be presented in Chapter 8.

The assumption that the voltage distribution in the slot is given by (3.46), plus the experimental information that $2l \cong \lambda_0/2$ for a practical range of offsets, means that the radiation pattern is insensitive to offset and the same as a half-wavelength dipole (with the polarization rotated 90°). Thus when the slot is imbedded in a large ground plane, the H-plane pattern is given by Figure 2.4 and the E-plane pattern is almost semicircular. But the *power level* in these patterns is governed by the offset of the slot through the factor $\sin^2 \pi x/a$. Slot offset thus serves as a transformer, providing a means for controlling the radiation level through the amount of coupling to the incident feeding TE_{10} mode. This will prove to be a very useful feature when slot arrays are studied in Chapter 8.

The procedure followed in this section can be repeated for the cases of inclined slots in the broad and narrow walls. The analysis, though lengthy, is not difficult if the foregoing is used as a guide, and these two cases are left as exercises.

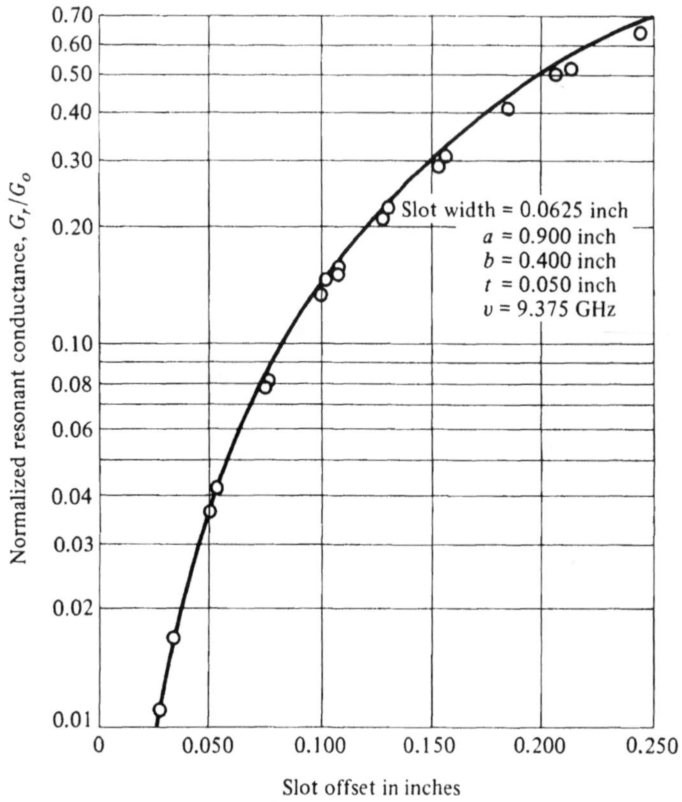

Fig. 3.13 G_r/G_0 versus Offset for Resonant Longitudinal Shunt Slot (After R. J. Stegen, "Longitudinal Shunt Slot Characteristics," Hughes Technical Memorandum No. 261, Nov. 1951, Hughes Aircraft Co., Culver City, California)

3.7 Patch Antennas

A radiating element with the attractive characteristic that it has a low profile is the *patch antenna*, illustrated in Figure 3.14. It consists of a thin metallic film bonded to a grounded dielectric substrate, and has the additional advantages of being lightweight, conformable, economical to manufacture, and easily wedded to solid state devices. The patch can be any shape, but the regular geometric shapes (such as rectangles or circular discs) are most commonly used. Feeding is achieved either via microstrip, as shown in Figure 3.15, or through use of a coaxial line with an inner conductor that terminates on the patch, as illustrated by Figure 3.16. The placement of the feed is important to the operation of the antenna.

The flow of electromagnetic power in the patch antenna can be visualized easily. Guided waves transport the energy along the microstrip or coax to the feed point. The energy then spreads out into the region under the patch; some of it crosses the boundary of the patch, to be radiated into space. If the fields in this exit region can be

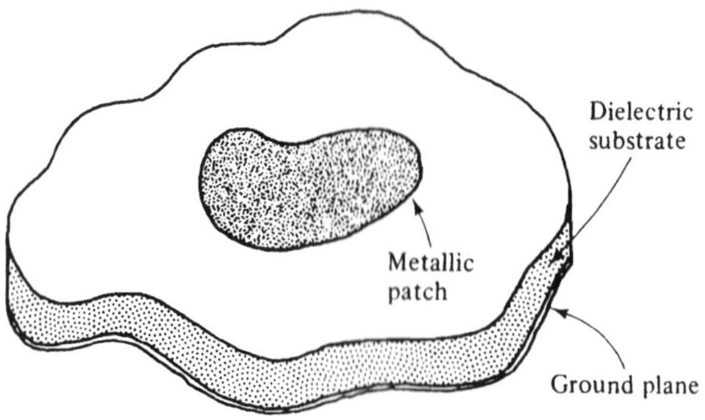

Fig. 3.14 Components of a Patch Antenna (Feed Not Shown)

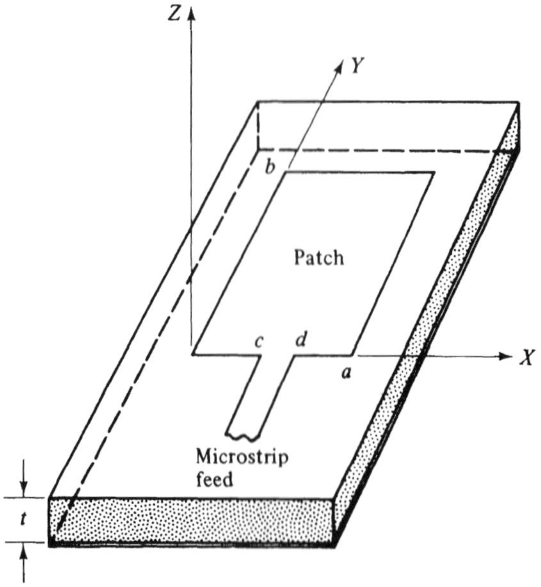

Fig. 3.15 A Microstrip-Fed Rectangular Patch

determined, equivalent sources may be placed on the boundary, from which the radiation pattern can be deduced.

In practice, the permittivity of the dielectric layer is usually not great ($\epsilon/\epsilon_0 \leq 4$) and its thickness t is small, so the region under the patch behaves very much like a portion of a parallel plate transmission line. Waves that leave the feed point see almost an open circuit when they arrive at the perimeter of the patch and considerable reflection occurs, so that the fraction of the incident energy emerging to be radiated is small. This suggests that the patch behaves more like a cavity than a radiator, and pinpoints its principal disadvantages—that it is not a highly efficient antenna and that

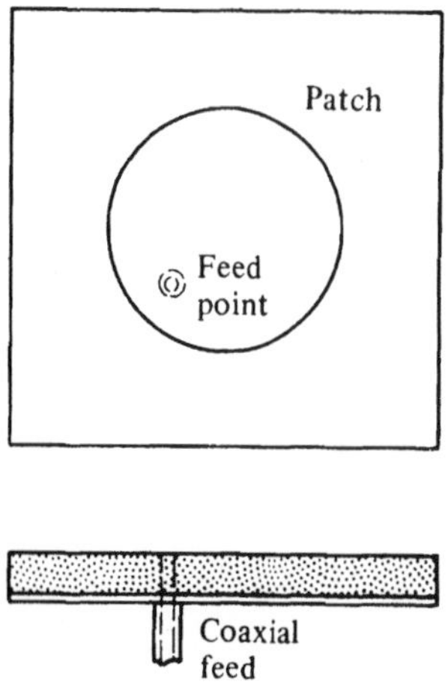

Fig. 3.16 A Coaxially Fed Circular Disc Patch

it is narrow band. The efficiency can be improved by using patches in arrays, but narrow-bandedness puts a limitation on the applications.

The assumption that the electromagnetic properties of the patch antenna can be deduced by viewing it primarily as a cavity (perhaps leaky cavity would be a better description) has been a fruitful one. Y.T. Lo and his co-workers[8] have been able to obtain good correlation between experiment and a theory based on this assumption, both for pattern and impedance when $t \ll \lambda$. Their analysis is essentially reproduced in what follows.

Let attention be confined to the region under the patch, and assume that in this region the electric field is z-directed and that the spatial variations of all field components are z-independent, with the Z-axis perpendicular to the patch. Maxwell's equations in this region take the form

$$\begin{aligned} \nabla \times \mathbf{E} &= -j\omega\mu_0\mathbf{H} \qquad & \nabla \times \mathbf{H} &= \mathbf{J} + j\omega\epsilon\mathbf{E} \\ \nabla \cdot \mathbf{E} &\equiv 0 \qquad & \nabla \cdot \mathbf{H} &\equiv 0 \end{aligned} \tag{3.61}$$

The appearance of $\mathbf{J}$ in (3.61) and its assumed nature require some explanation. Except in the subvolume occupied by the feed, $\mathbf{J} \equiv 0$ (The lower surface of the patch and the upper surface of the ground plane are excluded from the region being analyzed.) If the feed is the inner conductor of a coaxial line, then $\mathbf{J} = \mathbf{1}_z \mathbf{J}_z(u, v)e^{j\omega t}$, with u and v the transverse coordinates. If the feed is a microstrip, its presence will be modeled by a lineal current sheet $\mathbf{K} = \mathbf{1}_z \mathbf{K}_z e^{j\omega t}$ at the segment of the boundary where the patch joins to the strip line. In either case, since the current distribution is assumed to be z-independent because of the thinness of the dielectric layer, the continuity equation gives $\nabla \cdot \mathbf{J} = -j\omega\rho \equiv 0$. As a consequence of this, the volume charge

[8] Y. T. Lo, D. Solomon, and W. F. Richards, "Theory and Experiment on Microstrip Antennas" *IEEE Trans. Antennas Propagat.*, AP-27 (1979), 137–46.

density distribution is identically zero in the region, which is the reason for writing $\nabla \cdot \mathbf{E} \equiv 0$ in (3.61).

A familiar sequence of operations gives

$$\begin{aligned} \nabla \times \nabla \times \mathbf{E} = \nabla(\nabla \cdot \mathbf{E}) - \nabla^2 \mathbf{E} &= -j\omega\mu_0 \nabla \times \mathbf{H} \\ (\nabla^2 + k_d^2)E_z &= j\omega\mu_0 J_z \end{aligned} \tag{3.62}$$

with $k_d = \omega\sqrt{\mu_0 \epsilon}$ the wave number in the dielectric medium.[9] If the driving function J_z is specified, in principle (3.62) can be solved for E_z subject to the boundary conditions, after which $\mathbf{H}$ can be determined from (3.61). From this, the equivalent sources along the boundary of the patch can be deduced.

An effective approach to the solution of (3.62) is to begin by finding the characteristic solutions to the homogeneous wave equation

$$(\nabla^2 + k_{mn}^2)E_z \equiv 0 \tag{3.63}$$

To do this, assume first that the region is bounded on the top (patch) and bottom (ground plane) by perfect electric conductors and along its perimeter by a perfect magnetic conductor (to simulate an open circuit). Then if the patch is rectangular, as depicted in Figure 3.15, the solutions to (3.63) that fit the boundary conditions are given by

$$E_z = \Psi_{mn} = \cos\frac{m\pi x}{a}\cos\frac{n\pi y}{b} \tag{3.64}$$

$$k_{mn} = \sqrt{\left(\frac{m\pi}{a}\right)^2 + \left(\frac{n\pi}{b}\right)^2} \tag{3.65}$$

That (3.64) is a solution to (3.63), subject to the condition (3.65), can be verified by substitution. That it also satisfies the assumed boundary conditions can be seen by returning to (3.61) and noting that

$$\begin{aligned} \mathbf{H} = -\frac{1}{j\omega\mu_0}\left(\mathbf{1}_x\frac{\partial E_z}{\partial y} - \mathbf{1}_y\frac{\partial E_z}{\partial x}\right) &= \mathbf{1}_x\frac{n\pi/b}{j\omega\mu_0}\cos\frac{m\pi x}{a}\sin\frac{n\pi y}{b} \\ &\quad - \mathbf{1}_y\frac{m\pi/a}{j\omega\mu_0}\sin\frac{m\pi x}{a}\cos\frac{n\pi y}{b} \end{aligned} \tag{3.66}$$

This solution gives $H_x \equiv 0$ for $y = 0, b$ and $H_y \equiv 0$ for $x = 0, a$, as required.

Similarly, if the patch is a circular disc, as shown in Figure 3.16, the characteristic solutions which should be selected are

$$E_z = \psi_{mn} = J_n(k_{mn}\rho)e^{jn\beta} \tag{3.67}$$

[9]Since $\epsilon = \epsilon' + j\epsilon'' = \epsilon'(1 - j\delta)$ with δ the loss tangent (usually small), k_d will be slightly complex.

with J_n the Bessel function of the first kind and order n, and with k_{mn} chosen to satisfy

$$J'_n(k_{mn}a) = 0 \tag{3.68}$$

The solutions in (3.64) or (3.67) are seen to comprise sets of orthogonal functions, with each member satisfying the boundary conditions. With a source present that is z-directed and z-independent, these sets can be assumed to be complete and an arbitrary linear sum of the member functions can be used to represent the general solution to (3.62). This same procedure can be followed for many regular patch shapes and Y.T. Lo catalogs some of the most useful geometries.[10] Proceeding generally, one can assume that the solution to (3.62) is expressible in the form

$$E_z = \sum_m \sum_n A_{mn}\Psi_{mn} \tag{3.69}$$

The constant coefficients A_{mn} can be determined by noting that

$$\begin{aligned}\nabla^2 E_z &= \sum_m \sum_n A_{mn}\nabla^2\Psi_{mn} = -\sum_m \sum_n k_{mn}^2 A_{mn}\Psi_{mn}\\ &= j\omega\mu_0 J_z - k_d^2 E_z = j\omega\mu_0 J_z - \sum_m \sum_n k_d^2 A_{mn}\Psi_{mn}\end{aligned}$$

which can be rearranged to give

$$\sum_m \sum_n (k_d^2 - k_{mn}^2)A_{mn}\Psi_{mn} = j\omega\mu_0 J_z \tag{3.70}$$

If (3.70) is multipled by Ψ_{rs}^* and the result integrated over the domain of the patch, one obtains

$$\sum_m \sum_n (k_d^2 - k_{mn}^2)A_{mn}\int_S \Psi_{mn}\Psi_{rs}^*\,dS = j\omega\mu_0 \int_S J_z\Psi_{rs}^*\,dS \tag{3.71}$$

Since the functions Ψ_{mn} and Ψ_{rs} are orthogonal over this domain, (3.71) reduces to

$$A_{rs} = \frac{j\omega\mu_0}{k_d^2 - k_{rs}^2}\frac{\langle J_z\Psi_{rs}^*\rangle}{\langle\Psi_{rs}\Psi_{rs}^*\rangle} \tag{3.72}$$

in which

$$\langle J_z\Psi_{rs}^*\rangle = \int_S J_z\Psi_{rs}^*\,dS, \quad \langle\Psi_{rs}\Psi_{rs}^*\rangle = \int_S \Psi_{rs}\Psi_{rs}^*\,dS \tag{3.73}$$

The placement of the feed clearly influences the relative values of the excitation coefficients A_{rs}.

Insertion of (3.72) in (3.69) gives a general solution for the E-field in the region below the patch, that is,

$$E_z = j\omega\mu_0 \sum_m \sum_n \frac{1}{k_d^2 - k_{mn}^2}\frac{\langle J\Psi_{mn}^*\rangle}{\langle\Psi_{mn}\Psi_{mn}^*\rangle}\Psi_{mn} \tag{3.74}$$

[10] Lo, "Microstrip Antennas," p. 138.

Since **E** is z-directed and z-independent, the magnetic field can be expressed in the form

$$\mathbf{H} = \frac{1}{j\omega\mu_0}\mathbf{1}_z \times \nabla E_z \tag{3.75}$$

As an illustration of the use of this formulation, assume that a rectangular patch is being fed by a microstrip, as indicated in Figure 3.15. The microstrip will be assumed to be equivalent to a lineal electric current density

$$\mathbf{K} = \begin{cases} \mathbf{1}_z & c \le x \le d,\ y = 0,\ -t \le z \le 0 \\ 0 & \text{elsewhere} \end{cases} \tag{3.76}$$

When (3.76) is used in (3.74), one finds that

$$\begin{aligned} E_z(x, y) = j\omega\mu_0 \Bigg\{ & \frac{d-c}{k_d^2 ab} + \sum_{n=1}^{\infty} \frac{2(d-c)}{ab(k_d^2 - k_{on}^2)} \cos\frac{n\pi y}{b} \\ & + \sum_{m=1}^{\infty} \frac{[4\sin(m\pi(d-c)/2a)][\cos(m\pi(d+c)/2a)]}{m\pi b(k_d^2 - k_{mo}^2)} \cos\frac{m\pi x}{a} \\ & + \sum_{m=1}^{\infty}\sum_{n=1}^{\infty} \frac{[8\sin(m\pi(d-c)/2a)][\cos(m\pi(d+c)/2a)]}{m\pi b(k_d^2 - k_{mn}^2)} \cos\frac{m\pi x}{a} \cos\frac{n\pi y}{b} \Bigg\} \end{aligned} \tag{3.77}$$

with k_{mn} given by (3.65).

For a specified frequency of operation, $k_d = \omega\sqrt{\mu_0 \epsilon}$ is a constant. If the dimensions a and b of the patch are properly chosen, one of the k_{mn} wave numbers can be made almost to coincide with k_d, which for a dielectric with a small loss tangent is nearly pure real. In this case (3.77) indicates that the mnth amplitude coefficient becomes very large. The patch is then said to be resonant in the mnth mode, and this mode dominates the E_z distribution.

Let a closed surface S be chosen to bound the dielectric region under the patch, as shown in Figure 3.17. The upper face of S lies inside the metallic patch and the

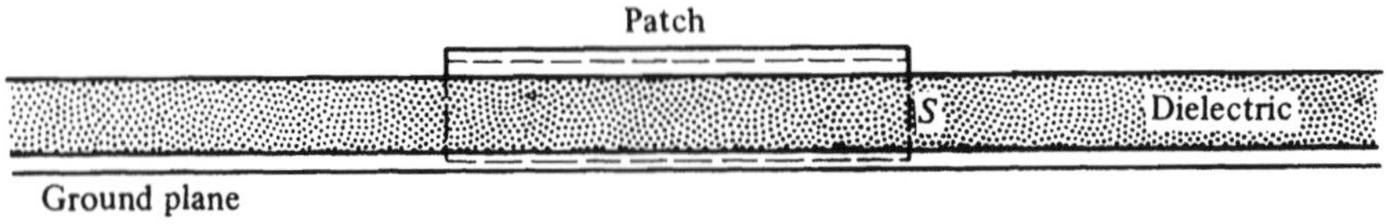

Fig. 3.17 The Choice of a Source-Excluding Surface S

lower face of S lies inside the ground plane, so the true fields on these two faces are null and no equivalent sources will appear on these faces of S. With resonance in the mnth mode postulated, the equivalent magnetic sources on the perimeter faces of S

can be deduced using (1.113) and are found (with a multiplicative constant suppressed) to be

$$\mathbf{K_m} = -\mathbf{1_y} \cos\frac{n\pi y}{b} \qquad x = 0$$

$$\mathbf{K_m} = \mathbf{1_x} \cos\frac{m\pi x}{a} \qquad y = 0$$

$$\mathbf{K_m} = \mathbf{1_y}(-1)^m \cos\frac{n\pi y}{b} \qquad x = a \tag{3.78}$$

$$\mathbf{K_m} = -\mathbf{1_x}(-1)^n \cos\frac{m\pi x}{a} \qquad y = b$$

The analysis postulates $\mathbf{H_T} \equiv 0$ along the boundary, so there are no equivalent electric sources on the perimeter faces.

To the extent that the assumptions in the analysis are valid, the equivalent sources (3.78), plus the actual sources outside S, can be used to calculate the true fields exterior to S. The actual sources outside S consist of the $\dot{\mathbf{P}}$ bound sources in the dielectric, the electric currents in the ground plane, and the electric currents on the upper face of the patch. These latter will be ignored on the argument that the fringing field is negligible. The image principle can be invoked to account for the ground plane, with the result that $-\dot{\mathbf{P}}$ image sources occur in an extra layer of thickness t and $+\mathbf{K_m}$ image sources occur in a vertical extension of the peripheral faces of S to a depth t. With t small, and ϵ/ϵ_0 not too large, the $\dot{\mathbf{P}}$ bound sources and their negatives images make a minor contribution to the radiated field, which will be ignored. What is left are the equivalent magnetic sources (3.78) extending a distance $2t$ in the Z-direction.

With the outgoing spherical wave factor in (1.93) suppressed, (1.123) indicates that the far field is given by

$$\mathbf{E} = -jk(\mathbf{1_\theta}\mathfrak{F}_\phi - \mathbf{1_\phi}\mathfrak{F}_\theta) \tag{3.79}$$

where $\mathfrak{F}_\theta$ and $\mathfrak{F}_\phi$ can be calculated from (1.130) and (1.131). When the equivalent sources of (3.78) are inserted in these integral expressions, one finds that

$$\mathfrak{F}_\theta(\theta, \phi) = (2t \cos\theta)[\cos\phi\, g_1(\theta, \phi) + \sin\phi\, g_2(\theta, \phi)] \tag{3.80}$$

$$\mathfrak{F}_\phi(\theta, \phi) = -2t[\sin\phi\, g_1(\theta, \phi) - \cos\phi\, g_2(\theta, \phi)] \tag{3.81}$$

in which

$$g_1(\theta, \phi) = [1 - (-1)^n e^{jkb \sin\theta \sin\phi}] \int_0^a \cos\frac{m\pi\xi}{a}\, e^{jk\xi \sin\theta\cos\phi}\, d\xi \tag{3.82}$$

$$g_2(\theta, \phi) = -[1 - (-1)^m e^{jka \sin\theta\cos\phi}] \int_0^b \cos\frac{n\pi\eta}{b}\, e^{jk\eta \sin\theta \sin\phi}\, d\eta \tag{3.83}$$

As a test of this theory, Lo and others constructed a rectangular patch with dimensions $a = 11.43$ centimeters and $b = 7.62$ centimeters on a copper-clad Rexolite 2200 substrate, $\frac{1}{16}$ inch thick, with $\epsilon/\epsilon_0 = 2.62$ and a loss tangent of approximately 0.001. To account for fringing, they took the effective dimensions to be

$$a_{eff} = a + \frac{t}{2} = 11.51 \text{ cm} \qquad b_{eff} = b + \frac{t}{2} = 7.70 \text{ cm}$$

From (3.65),

$$k_{10} = \frac{\pi}{0.1151} \qquad k_{01} = \frac{\pi}{0.0770}$$

so this patch should be resonant in the (1, 0) mode at a frequency such that

$$k_d = \omega\sqrt{\mu_0\epsilon} = \frac{2\pi\nu}{c}\sqrt{\frac{\epsilon}{\epsilon_0}} = \frac{\pi}{0.1151}$$

from which $\nu_{10}^{res} = 804$ MHz. Similarly, one finds that $\nu_{01}^{res} = 1202$ MHz.

Imput impedance measurements (to be discussed in Chapter 7) indicate resonant frequencies for these two modes at 804 MHz and 1197 MHz. Both values are seen to be close to the above predictions. The principal plane patterns when the patch is operating in each of these modes, and fed by a 50-ohm microstrip at the point (8.57 cm, 0), are shown in Figure 3.18. The agreement between theory and experiment is excellent. The patterns are broad, which is consistent with the fact that the resonant dimension for each mode is $\lambda_d/2$, with $\lambda_d = \lambda/(\epsilon/\epsilon_0)^{1/2}$ the wavelength of a TEM mode in the dielectric (λ is the free space wavelength). These features of small physical size and broad radiation patterns combine to make the rectangular patch, excited in one of these dominant modes, attractive for use in arrays.

Y.T. Lo and his co-workers have duplicated this analysis and its experimental validation for circular disc patches fed by microstrip.[11] The agreement between theory and experiment for the (1, 1) and (2, 1) modes was once again excellent. Their results are reproduced in Figure 3.19.

Many interesting innovations in the design of patch antennas have been discovered by various workers, including novel patch shapes and methods of feeding. As examples, if a square patch is fed by two microstrip lines, one each attached to adjacent sides of the patch, circularly polarized radiation will occur if the microstrips are connected through a 90° hybrid. The same effect can be achieved with a slightly elliptical disc patch that has an offset coaxial feed. Various scatterers can also be placed in the dielectric region under the patch in order to modify the radiation pattern. And the possibility of dual frequency operation has already been seen in the patterns of Figures 3.18 and 3.19.

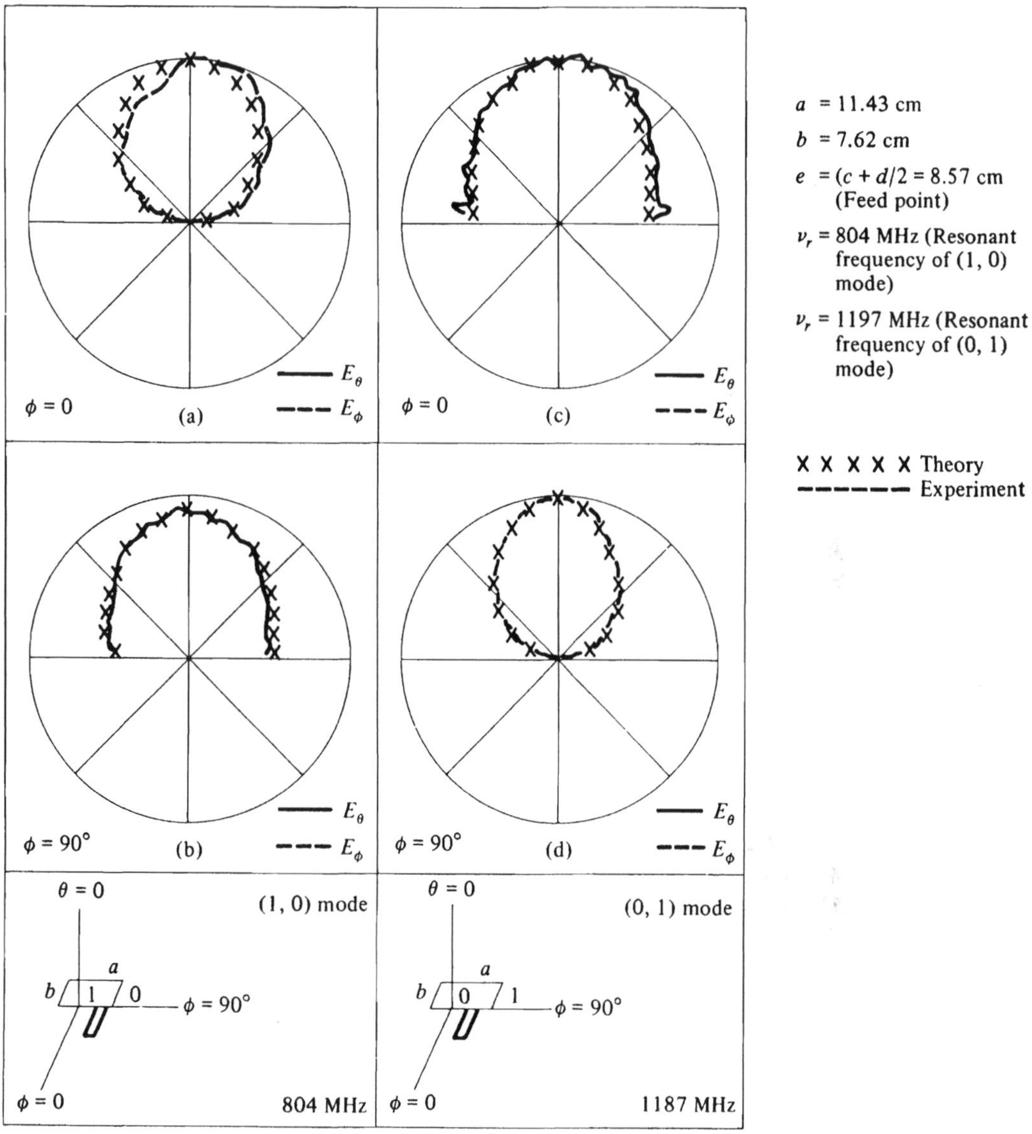

Fig. 3.18 Principal Plane Patterns for a Rectangular Patch Antenna (© 1979 IEEE. Reprinted from Lo, Solomon, and Richards, *IEEE AP Transactions*, pp. 137–146, 1979.)

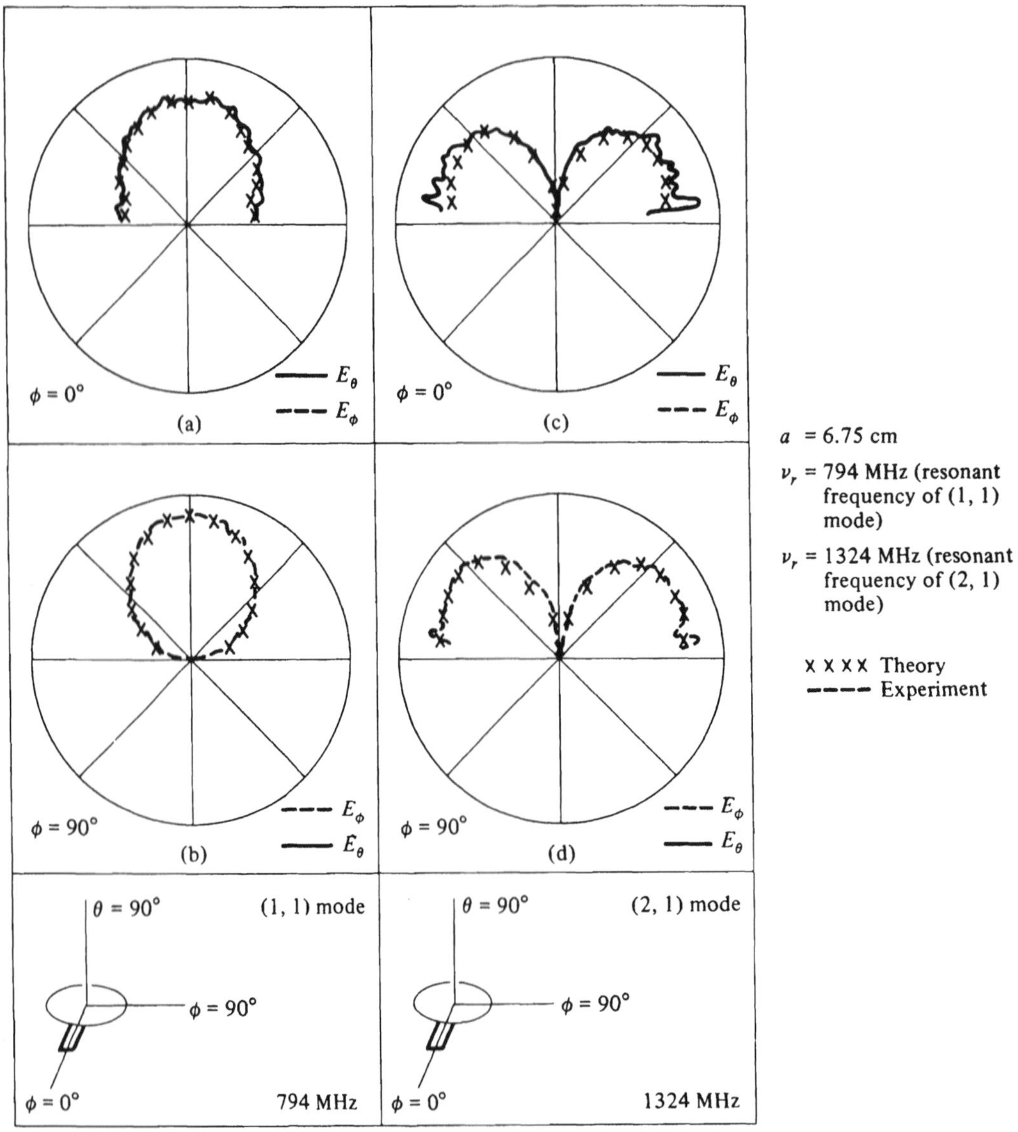

Fig. 3.19 Principal Plane Patterns for a Circular Disc Patch Antenna (© 1979 IEEE. Reprinted from Lo, Solomon, and Richards, *IEEE AP Transactions*, pp. 137–146, 1979.)

A discussion of the impedance properties of patch antennas will not be undertaken until Chapter 7, but it is significant to mention here that the input impedance level is generally higher the further the feed point is from the center of the patch. This provides the opportunity to position the feed point to match the characteristic impedance of the feed line. However, the situation is complicated when patches are used in arrays, for then mutual impedances must also be considered.

REFERENCES

JASIK, H., *Antenna Engineering Handbook* (New York: McGraw-Hill Book Co., Inc., 1961).

SCHELKUNOFF, S. A. and H. T. FRIIS, *Antennas: Theory and Practice* (New York: John Wiley and Sons, Inc., 1952).

SILVER, S., *Microwave Antenna Theory and Design*, MIT Rad. Lab. Series, Volume 12 (New York: McGraw-Hill Book Co., Inc., 1949).

PROBLEMS

3.1 Assume a TE_{mn} mode is present in the aperture of Figure 3.1 and find the expressions for the corresponding far field. Sketch the principal plane patterns for the TE_{30} mode. (This mode is often present due to forward scattering in a horn, and its presence can actually be helpful in narrowing the main beam. Do your results serve to explain how this can happen?)

3.2 Repeat the analysis of Section 3.2 for a circular waveguide, terminated in a ground plane, with a radius that is small enough to prevent all but the dominant mode from propagating.

3.3 Are the results of Section 3.2 valid when $a \longrightarrow \infty$, that is, for an infinitely long slot of width b? If not, return to first principles and solve for the fields due to the excitation $\mathbf{E}(\xi, \eta, \zeta) = \mathbf{1}_y E_0$ in a slot for which $-\infty < \xi < \infty$, $-(b/2) < \eta < (b/2)$, and $\zeta = 0$.

3.4 Assume a rectangular slot of length $2l$ and width w is cut in a perfectly conducting infinite ground plane. If $2l \gg w$ but $2l \ll \lambda$, find expressions for the far field, the power pattern, and the total power radiated. Then find a formula for the radiation resistance, R_{rad}. Does R_{rad} satisfy Booker's relation when taken in conjunction with the short dipole of Section 2.2?

3.5 Verify the scattering formulas (3.40) and (3.41).

3.6 For the inclined slot in the broad wall shown in Figure 3.9b, demonstrate that the scattering is antisymmetrical ($C_{10} = -B_{10}$) and thus that the equivalence is a series obstacle in a two-wire line. For the resonant-length case, show that

$$\frac{R}{R_0} = 0.131 \frac{\beta_{10}}{k} \frac{\lambda_0^2}{ab}\left[I(\theta) \sin\theta + \frac{\lambda_g}{2a} J(\theta) \cos\theta\right]^2$$

in which θ is the angle of tilt, and

$$\left.\begin{matrix} I(\theta) \\ J(\theta) \end{matrix}\right\} = \frac{\cos\left(\frac{\pi K_1}{2}\right)}{1 - K_1^2} \pm \frac{\cos\left(\frac{\pi K_2}{2}\right)}{1 - K_2^2}$$

$$\left.\begin{matrix} K_1 \\ K_2 \end{matrix}\right\} = \frac{\beta_{10}}{k}\cos\theta \mp \frac{\lambda_0}{2a}\sin\theta$$

This result is due to A. F. Stevenson.[12]

3.7 For the inclined slot in the narrow wall shown in Figure 3.9c, demonstrate that the scattering is symmetrical ($C_{10} = B_{10}$) and thus that the equivalence is a shunt obstacle in a two-wire line. For the resonant length case, show that

$$\frac{G}{G_0} = 0.131\frac{\lambda^4}{(\beta_{10}/k)a^3 b}\left[\frac{\sin\theta\cos\left(\frac{\beta_{10}}{k}\frac{\pi}{2}\sin\theta\right)}{1 - (\beta_{10}/k)^2\sin^2\theta}\right]^2$$

with θ the angle of tilt. This result is also due to A. F. Stevenson.[12]

3.8 Find the characteristic solutions and wave numbers for the dielectric region under a patch when the shape of the patch is a right isosceles triangle.

3.9 Verify that Equation 3.75 gives **H** in the dielectric region under a patch of general shape.

3.10 Find the modal expansion for $\mathbf{E}(\rho, \beta)$ in the dielectric region under a circular disc patch which is fed at the edge by a microstrip line. From this, deduce the equivalent sources and expressions for $\mathfrak{F}_\theta$ and $\mathfrak{F}_\phi$ when the mnth mode is dominant.

3.11 Repeat Problem 3.10 for the case that the circular disc patch is fed off center by a coaxial line intruding from below, as in Figure 3.16.

[12]Stevenson, "Theory of Slots."

II array analysis and synthesis

In Part I of this text, approximate expressions were deduced for the source distributions on various practical antenna elements (dipole, loop, helix, horn, slot, patch) and then the source/field formulas were used to determine the pattern characteristics. All of these elements have practical applications when used singly, but they also are widely used in arrays, and it is this latter class of applications which is the subject of the next three chapters. Since analysis is simpler and highly informative, it is taken up first. One- and two-dimensional arrays are studied in turn, with various relative element excitations assumed, which permits calculation of the array pattern. Conventional measures such as beamwidth, directivity, and side lobe level are introduced. Then attention is turned to array synthesis, with the desired pattern specified, the need being to find the relative element excitations which will achieve what is desired.

4 linear arrays: analysis

4.1 Introduction

Part I of this text had two principal objectives. The first was to establish formulas that would connect the radiated fields of any antenna to its sources. This was done in Chapter 1, where it was found to be desirable to divide antennas into two types, those in which the actual sources were used and those in which it was advantageous to introduce equivalent sources. Wire antennas of various shapes, notably monopoles, dipoles, loops, and helices, are practical examples of type I (actual source) antennas, and their far-field patterns were deduced in Chapter 2. Aperture antennas such as horns, slots, and patches are prominent examples of type II (equivalent-source) antennas. Their radiation patterns were determined in Chapter 3.

All of these elements can be used singly, in which case the pattern results of Chapters 2 and 3 are applicable. But they need not be used singly, and when the antenna consists of more than one element, it is called an *array*. In most practical applications, the elements will be of a common type, equispaced, and oriented to be capable of congruence through a simple translation. The discussion in this text will be limited to arrays that meet these conditions. The specialized literature should be consulted for discussions of arrays in which one or more of these restrictions is lifted.

The relative physical positioning of the elements and their relative electrical excitations are two parameters that can be used to exercise control over the shape of the radiation pattern of an array. In this chapter and the next, the positioning will be chosen so that the elements are equispaced along a straight line. Interelement spacing, the number of elements, and their relative excitations are then the principal variables available to the antenna designer. In Chapter 6 the scope of the discussion will be enlarged to include planar arrays.

There is an adage that the best way to learn synthesis is first to learn all you can about analysis. That truism certainly can be argued in the case of someone who is approaching antenna array theory for the first time and is the basis for devoting the present chapter to the analysis of linear arrays, deferring till Chapter 5 the subject of synthesis. Thus this chapter begins with a development of formulas for $\mathcal{A}_\theta$ and $\mathcal{A}_\phi$ (or $\mathcal{F}_\theta$ and $\mathcal{F}_\phi$) for the general case of arbitrarily positioned elements, but then immediately moves to the study of equispaced linear arrays with *known* excitations. A variety of excitations will be assumed and the patterns deduced. Schelkunoff's unit circle representation will be introduced and used extensively.

The analysis of the effects on pattern caused by varying the excitation will reveal many opportunities available to the antenna designer. These include the ability to form a pattern with a main beam and side lobes, to control angular placement of the main beam, to select beam sharpness by choosing the length of the array, to create a difference pattern, and to produce a shaped pattern devoid of nulls. All of these array fundamentals will form a useful basis for the synthesis procedures which follow.

4.2 Pattern Formulas for Arrays with Arbitrary Element Positions

It was established in Section 1.11 that a current density distribution $\mathbf{J}(\xi, \eta, \zeta)e^{j\omega t}$, contained in a finite volume V, causes a far-field pattern given by

$$\mathcal{A}_\theta(\theta, \phi) = \int_V [\cos\theta \cos\phi J_x(\xi, \eta, \zeta) + \cos\theta \sin\phi J_y(\xi, \eta, \zeta) - \sin\theta J_z(\xi, \eta, \zeta)]e^{jk\mathcal{L}}\, d\xi\, d\eta\, d\zeta \tag{4.1}$$

$$\mathcal{A}_\phi(\theta, \phi) = \int_V [-\sin\phi J_x(\xi, \eta, \zeta) + \cos\phi J_y(\xi, \eta, \zeta)]e^{jk\mathcal{L}}\, d\xi\, d\eta\, d\zeta \tag{4.2}$$

in which

$$\mathcal{L} = \xi \sin\theta \cos\phi + \eta \sin\theta \sin\phi + \zeta \cos\theta \tag{4.3}$$

It was also shown that $E_\theta/\mathcal{A}_\theta = E_\phi/\mathcal{A}_\phi$, so $\mathcal{A}_\theta(\theta, \phi)$ can be viewed as the vertically polarized component of the far-field and $\mathcal{A}_\phi(\theta, \phi)$ as the horizontally polarized component.

It is further evident from the development in Sections 1.12 and 1.13 that, if magnetic currents are introduced as secondary sources, $\mathcal{F}_\theta(\theta, \phi)$ is given by (4.1) with $\mathbf{J}_m$ replacing $\mathbf{J}$, and $\mathcal{F}_\phi(\theta, \phi)$ is given by (4.2) with $\mathbf{J}_m$ replacing $\mathbf{J}$. When only that part of the field due to magnetic sources is being considered, $E_\theta/\mathcal{F}_\phi = E_\phi/\mathcal{F}_\theta$ so $\mathcal{F}_\phi(\theta, \phi)$ can be viewed as the vertically polarized component of the far-field and $\mathcal{F}_\theta(\theta, \phi)$ as the horizontally polarized component. With these relations in mind, one can see that the development about to be presented applies equally well for $\mathcal{A}$ and $\mathcal{F}$, and thus equally well for type I and type II antenna arrays.

Imagine that the current distribution of an array resides in $N + 1$ identical discrete radiating elements.[1] The word *element* could mean a dipole, a helix, a horn,

[1] The analysis presented in this and the next two sections follows closely some earlier writing by the author, contained in "Beamwidth and Directivity of Large Scanning Arrays," *Microwave Journal*, 6 (1963), 53–60 and 7 (1964), 74–82, and also in *Microwave Scanning Antennas*, ed. R. C. Hansen, vol. 2, (New York: Academic Press, 1966), Chapter 1. Reprinted with joint permission.

or a slot, as examples. However, it might equally well mean a collection of helices or a hybrid collection of dipoles and slots. To say that the elements are *identical* is to impose the condition that any two of them can be made congruent by a simple translation plus rotation. To require additionally, as will be done here, that the elements are similarly oriented in space, is to impose the tighter condition that any two are capable of congruence through a translation alone. It is then possible to select a reference point in the ith element, $P_i(x_i, y_i, z_i)$, and find a point $P_j(x_j, y_j, z_j)$ that occupies the same position in the jth element. This collection of $N + 1$ reference points can serve to describe the relative positions of the different elements. It is convenient to establish local coordinate systems at each of these reference points. To do this, let

$$\xi_i = \xi - x_i \qquad \eta_i = \eta - y_i \qquad \zeta_i = \zeta - z_i \tag{4.4}$$

define any point $Q_i(\xi_i, \eta_i, \zeta_i)$ in the ith radiator, relative to its characteristic point $P_i(x_i, y_i, z_i)$. This situation is shown in Figure 4.1. Then, for example

$$J_x(\xi, \eta, \zeta) = J_x(x_i + \xi_i, y_i + \eta_i, z_i + \zeta_i)$$

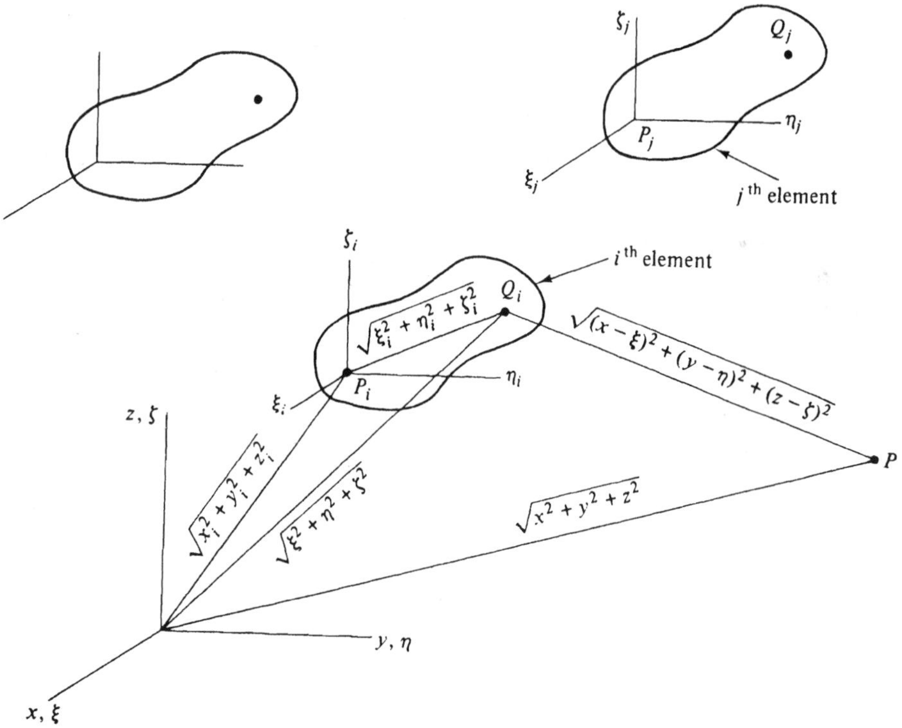

Fig. 4.1 Positional Notation for Antenna Arrays

whereas, for any point in the jth radiator,

$$J_x(\xi, \eta, \zeta) = J_x(x_j + \xi_j, y_j + \eta_j, z_j + \zeta_j)$$

Since all of the elements are assumed to be identical and similarly oriented, if I_i is the complex total current at the "terminals" of the ith radiator and I_j is the complex total current at the "terminals" of the jth radiator,[2] it follows that, if $\xi_i = \xi_j$, $\eta_i = \eta_j$, and $\zeta_i = \zeta_j$, then

$$\frac{J_x(x_i + \xi_i, y_i + \eta_i, z_i + \zeta_i)}{J_x(x_j + \xi_j, y_j + \eta_j, z_j + \zeta_j)} = \frac{I_i}{I_j} \tag{4.5}$$

because J_x is being determined at corresponding physical points in the two elements. Equation 4.5 implies that the elements are sufficiently separated to insure that the current distribution is the same (except for level) on each.

With the aid of Equation 4.5, one can rewrite (4.2) as follows:

$$\begin{aligned} \mathcal{A}_\phi(\theta, \phi) &= \sum_{i=0}^{N} \int_{V_i} [-\sin\phi J_x(x_i + \xi_i, y_i + \eta_i, z_i + \zeta_i) \\ &\quad + \cos\phi J_y(x_i + \xi_i, y_i + \eta_i, z_i + \zeta_i)] e^{jk\mathcal{L}} \, d\xi_i \, d\eta_i \, d\zeta_i \\ &= \mathcal{A}_{\phi,a}(\theta, \phi)\mathcal{A}_{\phi,e}(\theta, \phi) \end{aligned} \tag{4.6}$$

in which

$$\mathcal{A}_{\phi,a}(\theta, \phi) = \sum_{i=0}^{N} \frac{I_i}{I_0} e^{jk(x_i \sin\theta\cos\phi + y_i \sin\theta\sin\phi + z_i \cos\theta)} \tag{4.7}$$

and

$$\begin{aligned} \mathcal{A}_{\phi,e}(\theta, \phi) = \int_{V_0} &[-\sin\phi J_x(\xi_0, \eta_0, \zeta_0) + \cos\phi J_y(\xi_0, \eta_0, \zeta_0)] \\ &\cdot e^{jk(\xi_0 \sin\theta\cos\phi + \eta_0 \sin\theta\sin\phi + \zeta_0\cos\theta)} \, d\xi_0 \, d\eta_0 \, d\zeta_0 \end{aligned} \tag{4.8}$$

In the above, the origin of principal coordinates has been chosen to coincide with the characteristic point $P_0(x_0, y_0, z_0)$. This entails no loss in generality. The choice of a different origin for the principal coordinates merely introduces a phase change in $\mathcal{A}_{\phi,e}(\theta, \phi)$.

Similarly, $\mathcal{A}_\theta(\theta, \phi)$ can be recast in the form

$$\mathcal{A}_\theta(\theta, \phi) = \mathcal{A}_{\theta,a}(\theta, \phi)\mathcal{A}_{\theta,e}(\theta, \phi) \tag{4.9}$$

in which

$$\mathcal{A}_{\theta,a}(\theta, \phi) = \sum_{i=0}^{N} \frac{I_i}{I_0} e^{jk(x_i \sin\theta\cos\phi + y_i \sin\theta\sin\phi + z_i \cos\theta)} \tag{4.10}$$

[2]These terminals may be a convenient cross section in the waveguide feeding a slot, or the junction of a coaxial cable and a helix-plus-ground plane, as examples.

and

$$\mathcal{A}_{\theta,e}(\theta,\phi) = \int_{V_0} [\cos\theta\cos\phi J_x(\xi_0,\eta_0,\zeta_0) + \cos\theta\sin\phi J_y(\xi_0,\eta_0,\zeta_0) - \sin\theta J_z(\xi_0,\eta_0,\zeta_0)] \cdot e^{jk(\xi_0\sin\theta\cos\phi + \eta_0\sin\theta\sin\phi + \zeta_0\cos\theta)}\,d\xi_0\,d\eta_0\,d\zeta_0 \tag{4.11}$$

One can observe that $\mathcal{A}_{\phi,e}(\theta,\phi)$ and $\mathcal{A}_{\theta,e}(\theta,\phi)$ involve only the current distribution in one of the elements. For that reason, they are called the element patterns or element factors. It can also be noticed that $\mathcal{A}_{\phi,a}$ and $\mathcal{A}_{\theta,a}$ are identical and involve the relative current levels in the different elements as well as the relative placements of the elements. With the unnecessary ϕ and θ subscripts dropped, $\mathcal{A}_a(\theta,\phi)$ given by either (4.7) or (4.10) will be called the *array pattern* or *array factor*. It should be noted that $\mathcal{A}_a(\theta,\phi)$ is a summation of directionally weighted phasors and has no vector characteristics. The polarization of the field pattern comes from the element factors $\mathcal{A}_{\phi,e}$ and $\mathcal{A}_{\theta,e}$.

In most practical applications, the elements are small, perhaps a half-wavelength long in their maximum dimensions, in which case $\mathcal{A}_{\phi,e}(\theta,\phi)$ and $\mathcal{A}_{\theta,e}(\theta,\phi)$ are broad patterns, as has been seen for most of the elements studied in Chapters 2 and 3. When this is so, the fine structure in $\mathcal{A}_\phi(\theta,\phi)$ and $\mathcal{A}_\theta(\theta,\phi)$ comes from the array factor. This will be assumed to be the case in all subsequent developments in this chapter, but it is wise to remember the multiplication of patterns embodied in (4.6) and (4.9) and not to ignore the element factors unless it is justified.

4.3 Linear Arrays—Preliminaries

Let r_i be the distance from $P_0(x_0, y_0, z_0)$ to $P_i(x_i, y_i, z_i)$, with the line connecting P_0 to P_i having direction cosines $\cos\alpha_i$, $\cos\beta_i$, and $\cos\gamma_i$. If all the elements lie along a common line, then α, β, and γ are the same angles for every P_i. The antenna thus formed is called a *linear array*. In this case the array factor can be written

$$\mathcal{A}_a(\theta,\phi) = \sum_{n=0}^{N} \frac{I_n}{I_0} e^{jkr_n(\cos\alpha\sin\theta\cos\phi + \cos\beta\sin\theta\sin\phi + \cos\gamma\cos\theta)} \tag{4.12}$$

In the special but important and common case that the elements are *equispaced* and $2N+1$ in number,[3] the zeroth element can be taken as the central one. If one writes $r_n = nd$, in which d is the common spacing, (4.12) becomes

$$\mathcal{A}_a(\theta,\phi) = \sum_{n=-N}^{N} \frac{I_n}{I_0} e^{jknd(\cos\alpha\sin\theta\cos\phi + \cos\beta\sin\theta\sin\phi + \cos\gamma\cos\theta)} \tag{4.13}$$

[3]Equispaced arrays of an *even* number of elements $2N$ will be considered later in the development.

Equation 4.13 is the general expression for the array factor of a uniformly spaced linear array.[4] It is instructive to consider several special cases of this equation.

(a) UNIFORMLY EXCITED BROADSIDE ARRAYS Imagine first an equispaced array, laid out along the Z-axis, with radiating elements at the positions $z_i = 0, \pm d, \pm 2d, \ldots, \pm Nd$. In this case, $\cos\alpha = \cos\beta = 0$ and $\cos\gamma = 1$, so (4.13) simplifies to

$$\mathcal{A}_a(\theta) = \sum_{n=-N}^{N} \frac{I_n}{I_0} e^{jnkd\cos\theta} \tag{4.14}$$

a pattern which is rotationally symmetric (ϕ-independent). If all the currents are equal and in phase, this reduces further to

$$\mathcal{A}_a(\theta) = \sum_{n=-N}^{N} e^{jnkd\cos\theta} \tag{4.15}$$

Equation 4.15 is a sum of phasors with the common magnitude unity, possessing phase angles which depend on θ but which, for a given θ, are progressive multiples of the basic angle

$$\psi = kd\cos\theta = \frac{2\pi d}{\lambda}\cos\theta \tag{4.16}$$

A plot of these phasors for the case $(2\pi d/\lambda)\cos\theta = \pi/12$ and $2N + 1 = 15$ is shown in Figure 4.2. It is apparent that their sum is a maximum when $\cos\theta = 0$ or $\theta = \pi/2$, for then all the phasors are aligned. As θ departs from $\pi/2$ toward either 0 or π, the phasors begin to fan out, those with positive index going one way, those with negative index the other. If d/λ is large enough, when

$$(2\pi d/\lambda)\cos\theta = \pm\frac{2\pi}{2N+1}$$

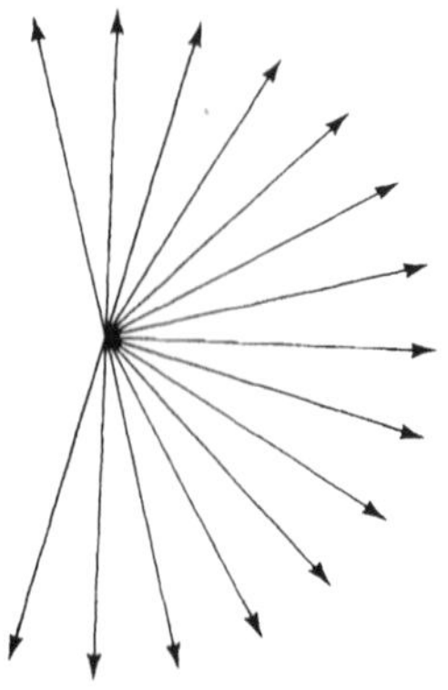

Fig. 4.2 Phasor Contributions for a Uniformly Excited Linear Array

[4]Nonuniformly spaced linear arrays have some practical applications, but are beyond the scope of this introductory treatise. The interested reader will find a starter bibliography in *Microwave Scanning Antennas*, ed. R. C. Hansen, Vol. 2 (New York: Academic Press, 1966), 53–59.

the phasors are equispaced in the complex plane and their sum is zero. This occurs at a direction θ_1 given by

$$\cos\theta_1 = \pm\frac{\lambda}{(2N+1)d}$$

If the length of the array is defined by $L = (2N+1)d$, this can be written as

$$\theta_1 = \arccos\left(\pm\frac{\lambda}{L}\right) \tag{4.17}$$

For example, if $L = \lambda$ then $\theta_1 = 0, \pi$ and a polar plot of $\mathfrak{A}_a(\theta)$ is as shown in Figure 4.3a. On the other hand, if $L \ll \lambda$, the phasors never fan out far enough to give a null value for $\mathfrak{A}_a(\theta)$ and a typical plot is almost circular, as suggested by Figure 4.3b. It should be recognized that these are polar plots in the $\phi = 0°$ half-plane. Since in this example, $\mathfrak{A}_a$ is independent of ϕ, a three-dimensional plot of $\mathfrak{A}_a$ could be obtained by rotating the patterns shown in Figure 4.3 around the Z-axis.

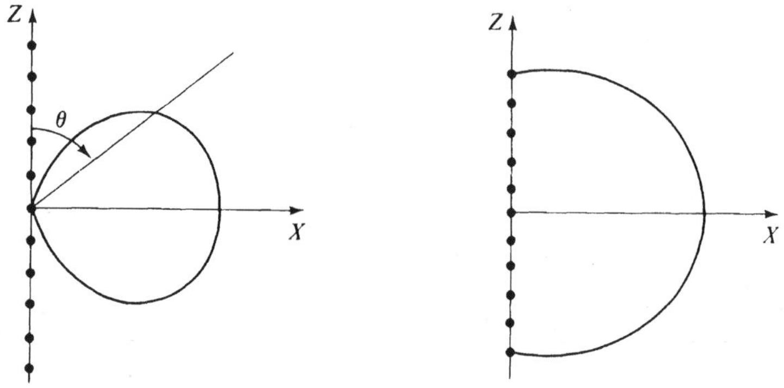

(a) $L = \lambda$ (b) $L << \lambda$

Fig. 4.3 E-Field Patterns in Polar Form for Short, Uniformly Excited Linear Arrays; Linear Scale

The most interesting case occurs when $L \gg \lambda$, in which event

$$\theta_1 = \arccos\left(\pm\frac{\lambda}{L}\right) \cong \frac{\pi}{2} \pm \frac{\lambda}{L} \tag{4.18}$$

and the phasors have fanned out to give a pair of nulls at angles only λ/L radians away from the maximum at $\theta = \pi/2$. The beamwidth between nulls is $2\lambda/L$ radians and is governed by the normalized length of the array.

In this case of uniform excitation, investigating values of θ even further removed from $\pi/2$ than θ_1 corresponds to looking at the sum of phasors that have fanned out beyond one complete rotation in the complex plane. A secondary maximum will occur when the phasors are fanned out to occupy one and one-half sheets of the complex plane, as shown in Figure 4.4a, whereas a second null will occur when they have fanned out to occupy two complete sheets of the complex plane, as shown in Figure 4.4b. The phasors lying in the second sheet are dotted to assist in identification.

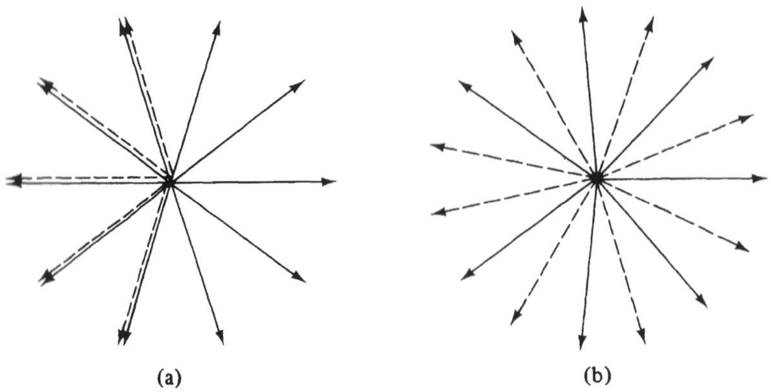

Fig. 4.4 Phasor Positions versus Polar Angle θ for a Long, Uniformly Excited Linear Array; $L = 7.5\lambda$

It can be noticed that only one-third of the phasors contribute to the net sum at the secondary maximum and that the sign of the sum is negative. This secondary maximum, or first side lobe, is thus 13.5 decibels (dB) below the primary maximum, or main lobe. The field throughout this first side lobe is out of phase with the field throughout the main lobe.

The second pair of nulls occur at angles θ_2 given by

$$\psi_2 = \frac{2\pi d}{\lambda}\cos\theta_2 = \pm\frac{4\pi}{2N+1}$$

or when

$$\theta_2 = \arccos\left(\pm\frac{2\lambda}{L}\right) \cong \frac{\pi}{2} \pm \frac{2\lambda}{L} \tag{4.19}$$

Thus the nulls near the principal maximum are approximately equispaced λ/L radians apart when $L \gg \lambda$.

If θ is varied beyond θ_2 toward either 0 or π, the phasors fan out further and encroach upon the third sheet of the complex plane. A tertiary maximum occurs when the phasors occupy two and one-half sheets, and a third null occurs when the phasors occupy a full three sheets of the complex plane. Only one-fifth of the phasors contribute to the tertiary maximum and the sign of their sum is positive. This tertiary maximum, or second side lobe, is even lower than the first side lobe, being 17.9 dB below the principal maximum. The third pair of nulls occur at angles θ_3 given by

$$\psi_3 = \frac{2\pi d}{\lambda}\cos\theta_3 = \frac{6\pi}{2N+1}$$

or when

$$\theta_3 = \arccos\frac{3\lambda}{L} \cong \frac{\pi}{2} \pm \frac{3\lambda}{L} \tag{4.20}$$

This process of letting the phasors fan out can be continued until the angle θ has reached 0° or 180°, at which extremities all of real space has been covered and the phasors have separated the maximum amount. The pattern, for this case of $L \gg \lambda$, is therefore as suggested in Figure 4.5, in which $|\mathcal{A}_a(\theta)|$ has been plotted for the half-plane $\phi = 0°$. The *phase* of the field changes by 180° in passing through each null. Put differently, the lobes alternate in sign. The three-dimensional pattern is a figure of revolution, and is seen to be an omnidirectional beam, pointing in the direction $\theta = 90°$. This is a pattern shape of practical importance in engineering applications and is commonly called a *beacon pattern.* Because the beam lies in a plane transverse to the array axis, it is also commonly referred to as a *broadside pattern.*

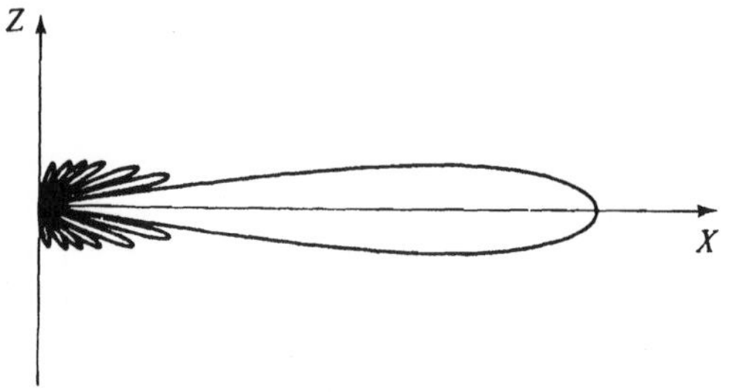

Fig. 4.5 E-Field Pattern in Polar Form for a Uniformly Excited Broadside Linear Array; $L = 7.5\lambda$; Linear Scale

Figure 4.5 was drawn for the special case $2N + 1 = 15$ elements, $d = \lambda/2$, and $L = 7.5\lambda$. It is seen to possess a single main beam and symmetrical side lobes which diminish in height as their angular distance from the main beam increases. An interesting effect would have arisen if the spacing had been greater. Imagine that the phasors are able to fan out so far before $\theta = 0°$ or 180° is reached that the angular spacing between adjacent phasors in the complex plane is 360°. In this event, all the phasors once again are aligned and sum to give a second main beam. This will occur at an angle θ' given by

$$\psi' = \frac{2\pi d}{\lambda}\cos\theta' = 2\pi$$

or when

$$\theta' = \arccos\left(\pm\frac{\lambda}{d}\right) \tag{4.21}$$

and thus will only occur if $d \geq \lambda$. By spacing the elements close enough together ($d < \lambda$), the pattern can be limited to one main beam.

If $d \gg \lambda$, the phasors can fan out to be 360°, 720°, 1080°, . . . , apart, and a sequence of main beams will occur, with the side lobe structure repeated in between. This gives what is known as an *interferometer pattern*, which has some useful applications, particularly in radio astronomy. However in most practical applications it is desirable to prevent these extra main beams by making d sufficiently small. It will be seen shortly, upon considering patterns with a single main beam which points in a direction other than $\theta = 90°$, that the requirement $d < \lambda$ is not stringent enough.

(b) BROADSIDE ARRAYS WITH TAPERED EXCITATION All of the foregoing has assumed equal, in-phase currents. Next, consider the situation when all the currents are in phase, but the amplitudes are symmetrically tapered. By this it is meant that the central element is fed by the largest current; its nearest neighbors contain currents equal but somewhat smaller than the central current, and so on, and finally the two end elements have equal currents that are the smallest in the array.

For this case, the phasor diagram of Figure 4.2 is modified as shown in Figure 4.6. Upon reflection, it can be appreciated that if $L \gg \lambda$, the phasors must fan out slightly beyond one sheet before they sum to zero and give a pattern null. Thus a tapered distribution suffers the penalty of some increase in beamwidth to the first null. However, this sacrifice is balanced by a compensating advantage. When the phasors have fanned out to occupy slightly more than one and one-half sheets a secondary maximum is reached. But this maximum is contributed to principally by that third of the phasors representing the outermost elements, that is, by the phasors with the smaller amplitudes. Thus the secondary maximum, or first side lobe, is *lower* than it was in the case of the uniform current distribution considered earlier. Similarly, the tertiary maximum is lower, because it is contributed to principally by that fifth of the phasors representing the outermost elements, and so on. One reaches the

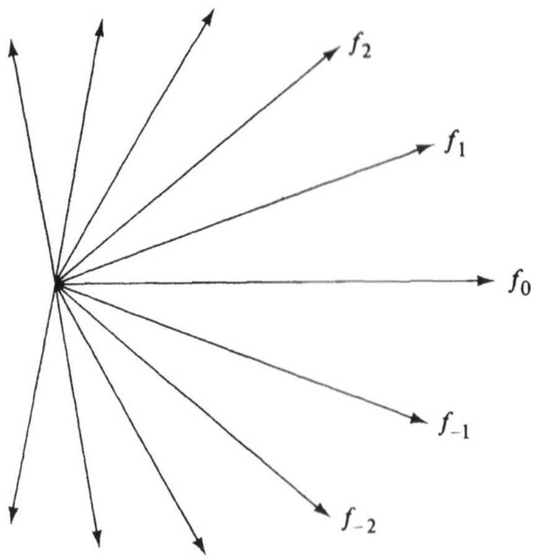

Fig. 4.6 Phasor Diagram for a Linear Array with Tapered Excitation

important conclusion that side lobe level can be controlled by tapering the array excitation, at some cost in beamwidth. Several synthesis techniques for accomplishing this will be explored in Chapter 5.

Note that if $L \ll \lambda$, tapering the excitation has negligible effect on the pattern, because the phasors never fan out far enough to give a pattern much different from Figure 4.3b. It is for this reason that the pattern of a half-wavelength dipole is essentially the same as the pattern of a very short dipole. The half-wavelength dipole can be thought of as an array of infinitesimal dipoles, touching end to end, and with $L = \lambda/2$, thus falling within the scope of the present discussion.

(c) SCANNED ARRAYS Next, consider the case of a uniformly spaced linear array laid out along the Z-axis so that Equation 4.14 continues to apply, but now assume that the currents have equal amplitudes and a uniform progressive phase, that is, let

$$I_n = I_0 e^{-jn\alpha_z} \tag{4.22}$$

in which α_z is a constant, called the *uniform progressive phase factor*. Under this assumption, (4.14) becomes

$$\mathcal{A}_a(\theta) = \sum_{n=-N}^{N} e^{jn(kd\cos\theta - \alpha_z)} \tag{4.23}$$

which differs from (4.14) only in an angular shift of origin. Whereas (4.14) gave a family of phasors that align at $\theta = \pi/2$, (4.23) gives the same family of phasors, but now they align at an angle θ_0 given by

$$kd\cos\theta_0 = \alpha_z \qquad \theta_0 = \arccos\left[\left(\frac{\alpha_z}{2\pi}\right)\left(\frac{\lambda}{d}\right)\right] \tag{4.24}$$

Thus α_z can be used as a parameter to position the main beam in space. If α_z is varied (for example, electronically), the beam will *scan*.

Once again, the position of the first null on either side of the main beam can be determined by permitting the phasors to fan out until they uniformly occupy the complex plane. This will occur when

$$(2N + 1)(kd\cos\theta_1 - \alpha_z) = \pm 2\pi \tag{4.25}$$

The two values of θ_1 that satisfy (4.25), θ_1' and θ_1'', are the two central null angles, one on each side of the beam position θ_0. Since $\alpha_z = kd\cos\theta_0$, further development of (4.25) gives

$$\cos\theta_1' - \cos\theta_0 = \frac{\lambda}{L} \qquad \cos\theta_1'' - \cos\theta_0 = -\frac{\lambda}{L}$$

If $L \gg \lambda$, one can let $\theta_1' = \theta_0 - \Delta\theta_1'$ and $\theta_1'' = \theta_0 + \Delta\theta_1''$ and write

$$\cos\theta_0 \cos\Delta\theta_1' + \sin\theta_0 \sin\Delta\theta_1' - \cos\theta_0 = \frac{\lambda}{L}$$

$$\cos\theta_0 \cos\Delta\theta_1' - \sin\theta_0 \sin\Delta\theta_1'' - \cos\theta_0 = -\frac{\lambda}{L}$$

Since $\cos\Delta\theta_1' \cong 1$, $\cos\Delta\theta_1'' \cong 1$, $\sin\Delta\theta_1' \cong \Delta\theta_1'$, and $\sin\Delta\theta_1'' \cong \Delta\theta_1''$, these last two equations sum to give

$$\theta_1'' - \theta_1' = \Delta\theta_1'' + \Delta\theta_1' \cong \left(\frac{2\lambda}{L}\right)\csc\theta_0 \tag{4.26}$$

This result is seen to embrace the earlier special case in which all the currents were equal and in phase, and which gave a beamwidth between nulls of $2\lambda/L$. (In that case, $\theta_0 = \pi/2$ and $\csc\theta_0 = 1$).

Equation 4.26 indicates that the beam broadens as it is scanned off broadside, the beamwidth between nulls being governed by the *projected* length of the array transverse to the beam direction.

Proceeding further with this case, when one permits the phasors to spread out so that they occupy one and one-half sheets of the complex plane, secondary maxima occur, giving the first side lobe on each side of the main beam, and at a relative height of -13.5 dB. When the phasors occupy two full sheets, the second pair of nulls occurs. A spread over two and one-half sheets produces tertiary maxima, and so on. Thus one finds a sequence of side lobes of steadily diminishing amplitudes, terminated when the limits of real space ($\theta = 0°, 180°$) are reached. A typical pattern computed from the magnitude of (4.23) is shown in Figure 4.7, with $2M + 1 = 15$, $d/\lambda = 1/2$, and $\alpha_z = \pi/2$. One can observe that, with the beam tilted up, the side lobe heights are still symmetrical, but that there are more side lobes below the main beam than above. Once again, this is a plot of $|\mathcal{A}_a(\theta)|$ in the half-plane $\phi = 0°$. The three-dimensional pattern can be found by rotating Figure 4.7 about the Z-axis.

Since Equation 4.23 is a series with a geometric progression, it can be summed to give

$$\mathcal{A}_a(\theta) = \frac{\sin[(\pi L/\lambda)(\cos\theta - \cos\theta_0)]}{\sin[(\pi d/\lambda)(\cos\theta - \cos\theta_0)]} \tag{4.27}$$

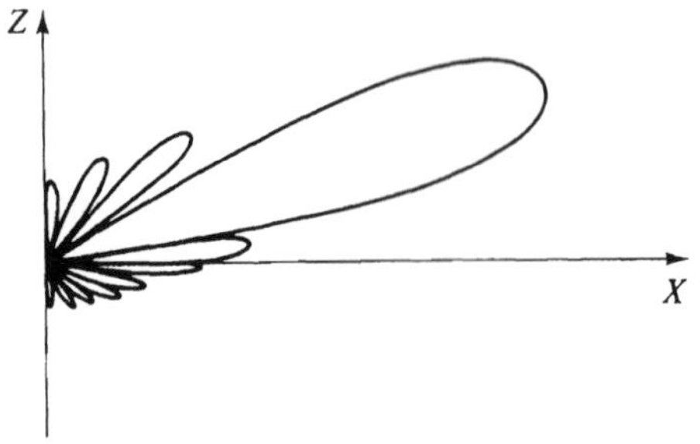

Fig. 4.7 Polar Field Plot for a Linear Array Excited with Uniform Amplitude, Uniform Progressive Phase; $L = 7.5\lambda$; Linear Scale

The proof is left as an exercise. Equation 4.27 is usually more convenient to use for calculating patterns such as the one shown in Figure 4.7.

In this case also, if the phasors can fan out so that adjacent phasors are 360° apart before the limits of real space are reached, a second main beam will occur. This will happen if

$$kd\cos\theta' - \alpha_z = \pm 2\pi, \qquad \cos\theta' = \cos\theta_0 \pm (\lambda/d) \tag{4.28}$$

Equation 4.28 is seen to be a generalization of (4.21). If $\theta_0 = \pi/2$, one obtains $\theta' = \arccos(\pm\lambda/d)$ as before. However, if θ_0 assumes another value, and one wishes to prevent the appearance of a second main beam, then it is apparent that the spacing d must be chosen so that

$$|\cos\theta_0| + (\lambda/d) > 1, \qquad |\cos\theta_0| - (\lambda/d) < -1 \tag{4.29}$$

The second of these inequalities is more demanding and requires that

$$\frac{d}{\lambda} < \frac{1}{1 + |\cos\theta_0|} \tag{4.30}$$

Equation 4.30 is the criterion for avoiding multiple beams in large scanning linear arrays. As an example of its use, one can see that if the beam is scanned close to end-fire, the elements must be spaced only one-half wavelength apart if a second main beam is to be prevented from appearing.

This discussion of a *scanned* beam can be readily enlarged to include the case of a tapered amplitude distribution together with a uniform progressive phase for the currents I_n. The main beam will still point at an angle θ_0 given by (4.24); it will be somewhat broader due to tapering, and the side lobes will be lower.

(d) EXTENSION TO AN EVEN NUMBER OF ELEMENTS—DIFFERENCE PATTERNS —If the number of elements in an equispaced linear array is even, for instance, $2N$—the array can be laid out along the Z-axis so that the elements are at the positions $\pm d/2$, $\pm 3d/2, \ldots$, and the array factor in (4.10) becomes

$$\begin{aligned}\mathcal{A}_a(\theta) = {} & \frac{I_{-N}}{I_1} e^{-j[(2N-1)/2]kd\cos\theta} + \cdots + \frac{I_{-1}}{I_1} e^{-j(1/2)kd\cos\theta} \\ & + e^{j(1/2)kd\cos\theta} + \cdots + \frac{I_N}{I_1} e^{j[(2N-1)/2]kd\cos\theta}\end{aligned} \tag{4.31}$$

From this point, the discussion of the preceding parts of this section can be repeated with little change. If all the currents are equal and in phase, $\mathcal{A}_a(\theta)$ is represented by a family of phasors lying at the angles $\pm\psi/2$, $\pm 3\psi/2, \ldots$, with $\psi = kd\cos\theta$ as before. These phasors fan out to give, in sequence, nulls and side lobe heights, and the conclusions about side lobe levels, null positions, possible multiple beams, all still prevail, now with $L = 2Nd$. Earlier arguments can be repeated for tapered excitation and when a uniform progressive phase is introduced.

What adds to the interest in linear arrays with even numbers of elements is the opportunity to excite the two halves of the array out of phase with each other, something which is neither convenient nor desirable with an odd number of elements because of the awkward presence of the middle element. As an example of this possibility, let all the currents be equal in amplitude, but let $I_{-n} = -I_n$, for all n. Then (4.31) gives a family of phasors that appear in the complex plane as shown in Figure 4.8.

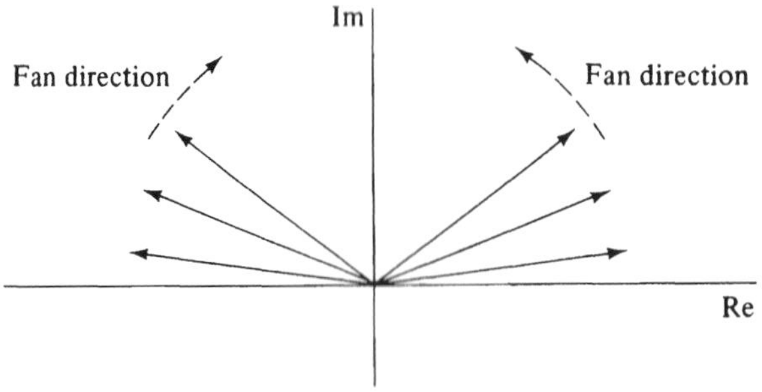

Fig. 4.8 Phasor Diagram for Difference Mode; Uniform Excitation

For $\theta = \pi/2$, half of these phasors lie along the positive real axis, the other half lie along the negative real axis, and the sum is zero; that is, there is a null in the pattern at broadside. As θ departs from $\pi/2$ toward 0 (or π), these phasors fan out into the upper (or lower) half of the complex plane. When $(2N - 1)\psi/2 \cong 3\pi/4$, their phasor sum S is a maximum, given by $S \cong \pm j\sqrt{2}\,N$. (Proof of this is left as an exercise). When $(2N - 1)\psi/2 \cong 2\pi$, their phasor sum is zero. As the phasors fan out further onto the second sheet of the complex plane, this summation process is repeated, with only one-third of the phasors effectively participating. Thus secondary maxima of approximate heights $\pm j\sqrt{2}\,N/3, \ldots$, are reached. A typical difference pattern for uniform but asymmetrical excitation is shown in Figure 4.9. Twin main lobes are found straddling $\theta = \pi/2$, with symmetrically decaying side lobes; the first pair is 9.5 dB below the level of the main lobes.

Patterns of the type found in Figure 4.9 are more readily computed from

$$\mathcal{A}_a(\theta) = \frac{\sin^2\left(\dfrac{\pi L}{2\lambda}\cos\theta\right)}{\sin\left(\dfrac{\pi d}{\lambda}\cos\theta\right)} \tag{4.32}$$

This result is obtained by summing (4.31) for the special case $I_n = -I_{-n} = I_1$, for all n.

The high side lobe level (−9.5 dB) of this pattern is due to the choice of equal amplitudes. If a tapered current distribution of the type shown in Figure 4.10 were selected, the side lobe level could be reduced. The reader might wish to confirm this by sketching the phasor diagrams as they would appear for θ values corresponding to

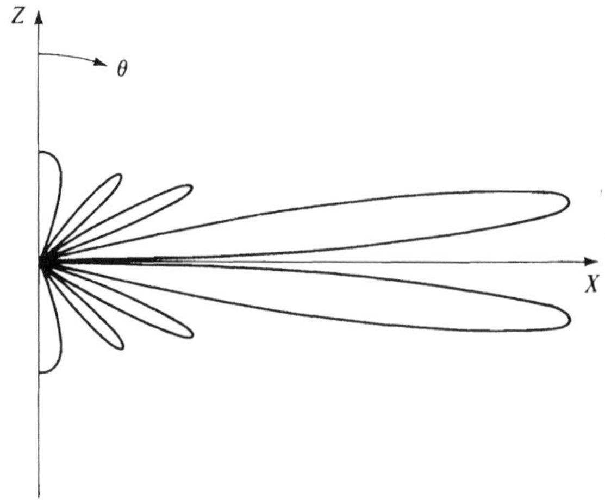

Fig. 4.9 Polar Field Plot for a 14-Element Linear Array, $d = \lambda/2$, with Uniform Asymmetric Excitation; Broadside Difference Pattern; Linear Scale

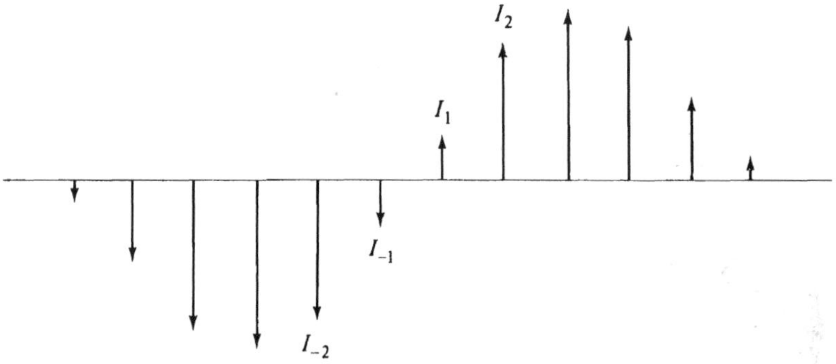

Fig. 4.10 Bar Graph of Tapered Current Distribution to Give Difference Pattern with Reduced Side Lobes

successive lobe maxima. A synthesis procedure for determining the proper taper to produce a specified side lobe level will be presented in Chapter 5.

A uniform progressive phase can be given to the current distribution, with the two halves of the array still excited out of phase with each other. The effect on Equation 4.31 is to replace $kd \cos \theta$ by $kd(\cos \theta - \cos \theta_0)$, with the current ratios pure real and θ_0 the null angle between the two principal lobes. The current distribution can, in this case, also have a taper to reduce side lobes.

Patterns such as those shown in Figures 4.5 and 4.9 form a useful pair in radar applications. With the two halves of the array fed in phase (the sum mode), one obtains a pattern with a single main beam; this is useful for acquiring a target, but not too useful for telling exactly where the target is. However, if the target is close enough and the excitation is switched to the mode in which the two halves of the aperture are excited out of phase (the difference mode), then, unless the target is exactly in the null between the two principal lobes of the difference pattern, a return signal will be detected in the radar receiver. This signal can be used either to tilt the array mechanically or to introduce a uniform progressive phase shift in the excitation, the result being to place the target in the central null of the difference pattern. The sensitivity of this process is high, so the target's angular position can be determined with considerable accuracy.

4.4 Schelkunoff's Unit Circle Representation

The development of the previous section can be reinforced and extended with the aid of an extremely useful formulation due to S. A. Schelkunoff.[5] The synthesis techniques to be considered in Chapter 5 also benefit from Schelkunoff's method of representation.

Consider an equispaced linear array of $N + 1$ elements (N can be even or odd), laid out along the Z-axis. From (4.7), the array factor can be written in the form

$$\mathcal{A}_a(\theta) = \sum_{n=0}^{N} \frac{I_n}{I_0} e^{jn(kd\cos\theta - \alpha_z)} \tag{4.33}$$

In (4.33), a uniform progressive phase factor α_z has been factored out of the current distribution and shown explicitly because many applications involve current distributions of this type. However, no loss in generality results from this, since the ratio I_n/I_0 appearing in (4.33) can still be complex.

If one lets

$$w = e^{j\psi} \tag{4.34}$$

$$\psi = kd\cos\theta - \alpha_z \tag{4.35}$$

Equation 4.33 can be converted to the form

$$\mathcal{A}_a(w) = \sum_{n=0}^{N} \frac{I_n}{I_0} w^n = \frac{I_N}{I_0} \sum_{n=0}^{N} \frac{I_n}{I_N} w^n \tag{4.36}$$

from which

$$\begin{aligned} |\mathcal{A}_a(w)| &= \left|\frac{I_N}{I_0}\right| \cdot \left| w^N + \left(\frac{I_{N-1}}{I_N}\right) w^{N-1} + \cdots + \frac{I_0}{I_N}\right| \\ &= \left|\frac{I_N}{I_0}\right| \cdot |f(w)| \end{aligned} \tag{4.37}$$

From this, by the fundamental theorem of algebra,

$$|f(w)| = |w - w_1| \cdot |w - w_2| \cdots |w - w_N| \tag{4.38}$$

[5]S. A. Schelkunoff, "A Mathematical Theory of Linear Arrays," *Bell System Tech. J.*, 22 (1943), 80–107.

Equations 4.37 and 4.38 reveal several interesting things. First, $|f(w)|$ differs from the magnitude of $\mathcal{A}_a(\theta)$ by only a multiplicative constant and thus can serve as a surrogate for the array factor. Second, the array factor can be represented as a polynomial in w of degree one less than the number of elements in the equispaced array. Third, this polynomial has N roots which are linked to the array excitation, since the current ratios comprise the coefficients of the polynomial.

The placement of these roots in the complex w-plane can be related to the field pattern in real θ space, as will be seen in what is to follow. In this manner, the analysis and synthesis of array patterns due to equispaced arrays can be tied to a study of the properties of polynomials, a distinct asset for the antenna designer.

To develop this approach further, observe that as θ varies in real space from 0 to π, the definitions in (4.34) and (4.35) require that ψ vary from $\psi_s = kd - \alpha_z$ to $\psi_f = -kd - \alpha_z$ and that w trace out a path along a *unit circle* in the complex plane, as illustrated in Figure 4.11. The total excursion of w is from $e^{j\psi_s}$ to $e^{j\psi_f}$, proceeding clockwise as θ goes from 0 to π. Thus ψ_s (read ψ-start) and ψ_f (read ψ-finish) mark the initial and terminal points of the w-excursion, the angular extent of which is $2kd$ radians.

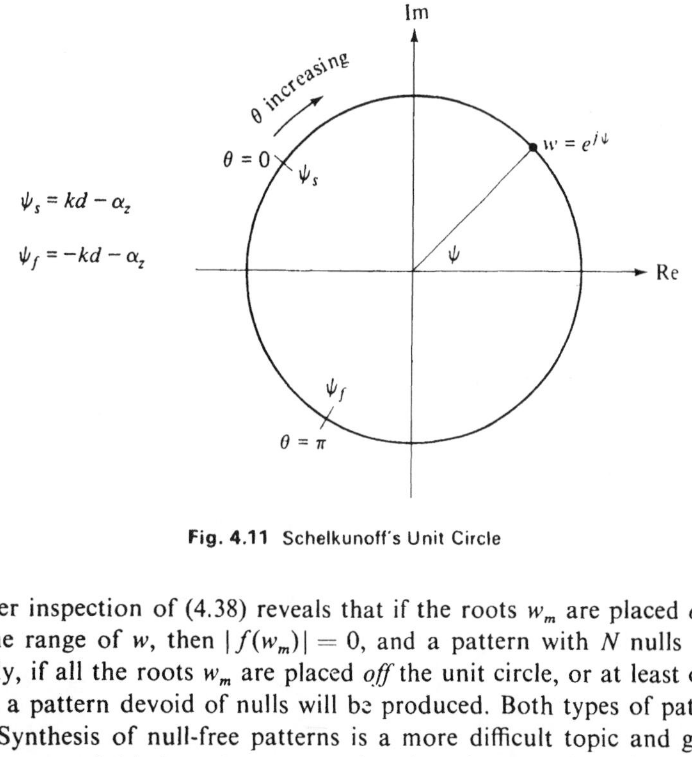

Fig. 4.11 Schelkunoff's Unit Circle

Further inspection of (4.38) reveals that if the roots w_m are placed *on* the unit circle, in the range of w, then $|f(w_m)| = 0$, and a pattern with N nulls will result. Alternatively, if all the roots w_m are placed *off* the unit circle, or at least outside the range of w, a pattern devoid of nulls will be produced. Both types of patterns have their uses. Synthesis of null-free patterns is a more difficult topic and generally is beyond the scope of this introductory treatise, though some discussion of it will be found in Chapter 5. In what follows, attention will be focused on situations in which the roots w_m have all been placed on the unit circle through proper choice of the current distribution.

(a) UNIFORMLY EXCITED BROADSIDE ARRAYS If the excitation currents have equal amplitudes and a common phase, (4.35) reduces to $\psi = kd\cos\theta$ and (4.36) simplifies to

$$\mathcal{A}_a(w) = f(w) = \sum_{n=0}^{N} w^n = \frac{1 - w^{N+1}}{1 - w} \tag{4.39}$$

DeMoivre's theorem permits the conclusion that $f(w)$ has roots at the positions

$$w_m = e^{j2\pi m/(N+1)} \qquad m = 1, \ldots, N \tag{4.40}$$

(Note that the root $w = 1$ is excluded because of the factor in the denominator). Thus a uniform amplitude/equiphase excitation of the equispaced array does put all the roots on the unit circle. As has already been seen in Section 4.3, this results in a pattern with lobes interspersed by nulls.

An illustration of (4.40) is given in Figure 4.12 for the case of a five-element array. The roots are found at the positions $\pm 2\pi/5$ and $\pm 4\pi/5$. The value of $|f(w)|$ can be found by taking the product of the four distances d_1, d_2. d_3, d_4, which is an example of the use of (4.38). As the point w moves along the unit circle (which corresponds to permitting θ to vary in real space), these four distances change, as does their product. Whenever w coincides with one of the roots w_m, the distance $d_m = 0$ and $|f(w_m)| = 0$; that is, a null has been encountered in the array factor.

This presupposes that the range of w includes the roots w_m. For example, if $d = \lambda/2$, then $\psi_s = \pi$ and $\psi_f = -\pi$; all four of the roots are within the range of w.

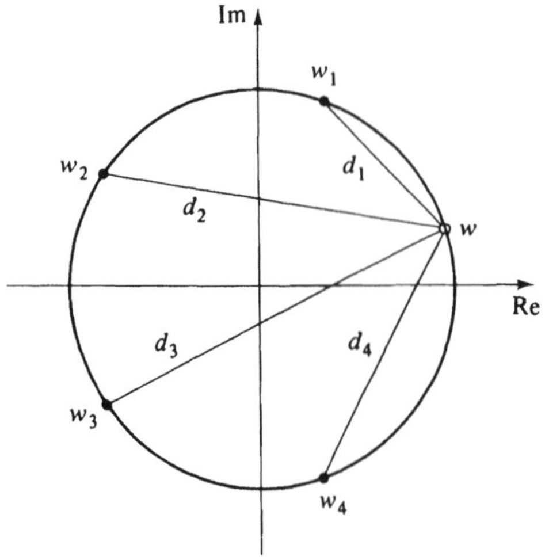

Fig. 4.12 Root Positions on a Schelkunoff Unit Circle for a Uniformly Excited Five-Element Linear Array

As w moves clockwise from $e^{j\pi}$ through e^{j0} to $e^{-j\pi}$, the product $d_1d_2d_3d_4$ traces out half a side lobe, a null at w_2, a full side lobe, a null at w_1, a main beam, a null at w_4, a side lobe, a null at w_3, and finally another half side lobe, as illustrated in Figure 4.13. The nulls in *real* space can be determined from $\psi_m = 2\pi m/(N+1) = kd\cos\theta_m$, which is in agreement with the earlier results of Section 4.3.

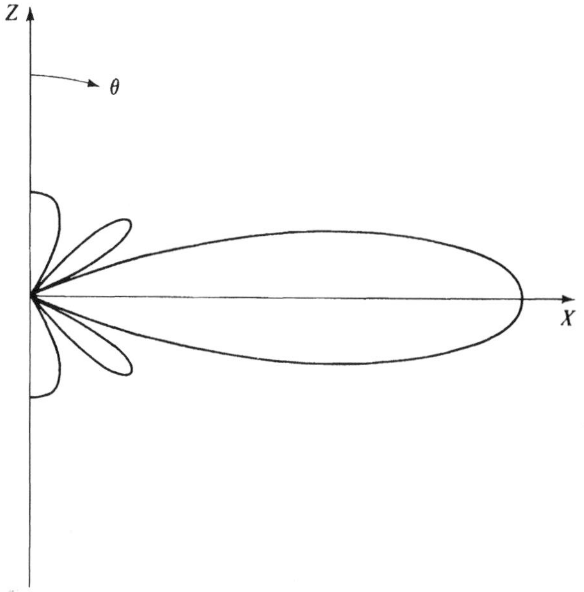

Fig. 4.13 Polar Field Plot for a Uniformly Excited Five-Element Linear Array, $d = \lambda/2$; Broadside Sum Pattern; Linear Scale

If only a quick sketch of the pattern is needed, a graphical construction using the unit circle of Figure 4.12 (suitably enlarged) can be helpful. The height of a side lobe or main lobe occurs when w is approximately halfway between roots. The product $d_1d_2 \cdots d_N$ when w is at such halfway points gives the relative lobe heights. This can be determined with reasonable accuracy if care is taken in measuring the distances. Knowledge of these lobe heights and the null positions θ_m is all that is needed to be able to produce a decent polar representation of the field pattern. The reader might wish to try this for the five-element array by constructing an enlarged version of Figure 4.12 and checking that the side lobe heights are -13.5 dB and -17.9 dB (in agreement with the conclusions of Section 4.3), and that the pattern resembles Figure 4.13.

If $d = \lambda$, then $\psi_s = 2\pi$ and $\psi_f = -2\pi$; w ranges two full revolutions around the unit circle. The result for a five-element array is the pattern shown in Figure 4.14. If one wishes to avoid these extra main beams at $\theta = 0, \pi$, a suitable restriction for a five-element array would be to have ψ_s coincide with w_4 and let w range more than

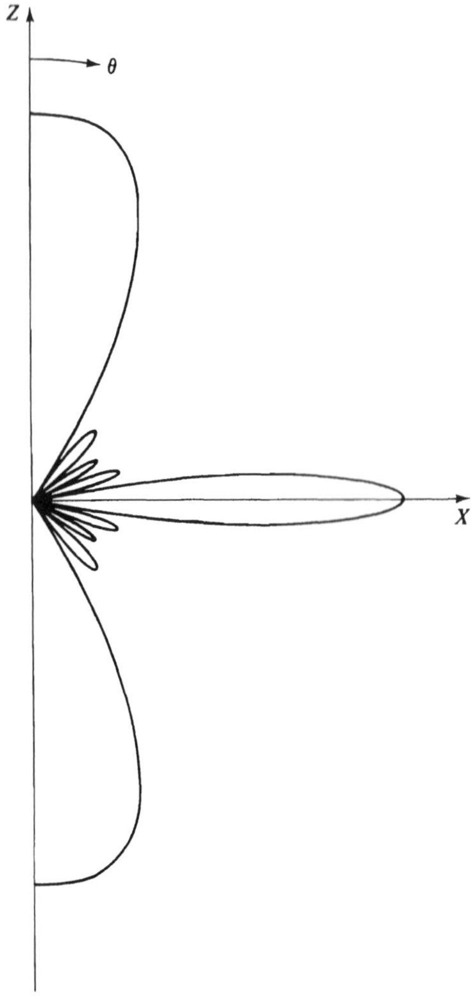

Fig. 4.14 Polar Field Plot for a Uniformly Excited Five-Element Linear Array, $d = \lambda$; Broadside Sum Pattern; Extra Main Beams at End-Fire; Linear Scale

one full revolution around the unit circle, past w_4 and on to ψ_f, which is allowed to coincide with w_1. In the more general case of $N + 1$ elements, this means choosing

$$kd = \psi_s = -\psi_f = 2\pi - \frac{2\pi}{N+1} = \frac{2\pi N}{N+1}$$

or

$$(d/\lambda)_{\max} = \frac{N}{N+1} \tag{4.41}$$

Equation 4.41 can be viewed as the maximum element separation for a uniformly excited broadside array if multiple main beams are to be avoided. For N large, it agrees with the result found in Section 4.3. For $N + 1 = 5$, the elements should be no further than 0.8λ apart. The pattern for this case is shown in Figure 4.15.

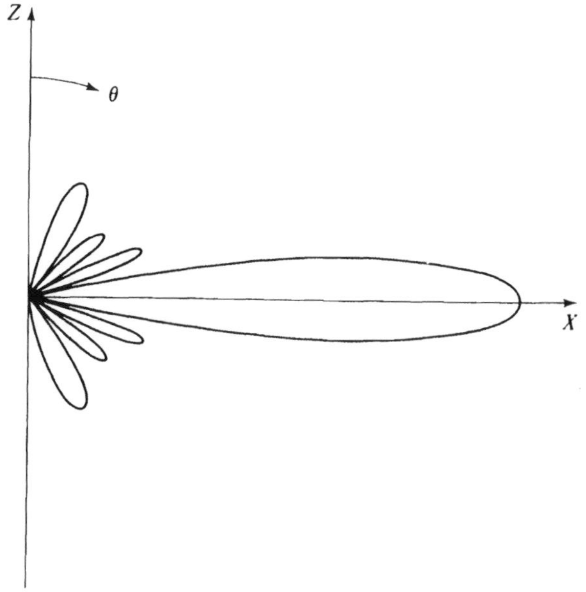

Fig. 4.15 Polar Field Plot for a Uniformly Excited Five-Element Linear Array, $d = 0.8\lambda$; Broadside Sum Pattern; Maximum Spacing for Multiple Beam Avoidance; Linear Scale

It is instructive to compare the patterns of Figures 4.13 and 4.15. Both are for uniformly excited five-element arrays, and each has a single main beam at broadside. But Figure 4.15 shows a narrower main beam and three additional side lobes. This is due to the increased length of the array (4λ versus 2.5λ).

(b) BROADSIDE ARRAYS WITH LOWERED SIDE LOBES The fact that (4.38) can be written in the form $|f(w)| = \prod_{m=1}^{N} d_m$, with d_m the distance between w and w_m, has already been noted in the discussion in this section. With w approximately halfway between the successive nulls w_m and w_{m+1}, the product of these distances gives the relative height of the mth side lobe. It follows that if these two roots are brought closer together, the height of this side lobe will be reduced. For broadside arrays, if the heights of *all* the side lobes are to be reduced, the roots must cluster closer around $-\pi$, indicating that the main beam region on the unit circle (from w_1 to w_N) must be enlarged. In other words, the main beam is broadened as the price paid to reduce the side lobes. This tradeoff has already been noted in Section 4.3.

As an example of this effect, consider again the five-element equispaced array. With uniform amplitude/equiphase excitation, this array could be represented by the Schelkunoff unit circle shown in Figure 4.12. For element spacings of $d = \lambda/2, \lambda, 0.8\lambda$, the patterns of Figures 4.13 through 4.15 were obtained. All of these patterns have the same side lobe topography, since they arise from a common unit circle diagram. The innermost side lobes are at −11.9 dB and the next set is at −13.7 dB.

Suppose it is desired to produce patterns in which all the side lobes are at −20 dB. This would require displacing the four roots of Figure 4.12 so that they are closer together and clustered closer to $-\pi$.

This design problem can be solved graphically by trial and error. At present, the roots are at ±72° and ±144°. Suppose the new positions ±87° and ±149° are tried. (This places the roots 62° apart, instead of 72°.) When a large unit circle is constructed with roots placed in these positions, the product of distance measurements gives the innermost side lobe at −18.5 dB and the other side lobes at −21.3 dB. This suggests that the roots w_2 and w_3 have been shifted too close together, but that the shift in w_1 and w_4 might be just about right. One could continue to try revised root positions, perhaps ±87° and ±147°, thus converging to the root positions that will give the desired −20 dB level for all side lobes. The interested reader might wish to pursue this and demonstrate that the proper root positions are ±89° and ±145.5°.

With the correct root positions known, one can return to (4.38) and write

$$\begin{aligned} f(w) &= (w - e^{j1.55})(w - e^{j2.54})(w - e^{-j2.54})(w - e^{-j1.55}) \\ &= (w^2 - 2w\cos 89° + 1)(w^2 - 2w\cos 145.5° + 1) \\ &= w^4 + 1.6w^3 + 1.95w^2 + 1.6w + 1 \end{aligned} \tag{4.42}$$

If this result is compared to (4.37), it can be observed that the relative current distribution is

$$1 \quad 1.6 \quad 1.95 \quad 1.6 \quad 1$$

The central element is seen to be most strongly excited. The distribution has a symmetric taper, consistent with the discussion in Section 4.3.

(c) SCANNED ARRAYS The only change, if an equispaced linear array is to have a uniform amplitude/uniform progressive phase excitation, is that one needs to return to the more general definition of ψ given in (4.35), with $\alpha_z = kd\cos\theta_0$ and θ_0 the central angle of the main beam. The Schelkunoff unit circle is unaffected; all the roots are where they were before. Only the starting and ending values of ψ are altered. They are now

$$\psi_s = kd(1 - \cos\theta_0), \qquad \psi_f = -kd(1 + \cos\theta_0) \tag{4.43}$$

The total excursion is still $2kd$, and the height of the main beam still occurs at $\psi = 0$. However, the fact that the main beam is scanned raises anew the question about multiple main beams.

As an example, one can return to the case of the five-element array, for which the unit circle of Figure 4.12 applies. If the main beam is to point at $\theta = 120°$ (that is, 30° beyond broadside) and if the element spacing is $d = \lambda/2$, then $\psi_s = 3\pi/2$, $\psi_f = -\pi/2$, and the pattern is as shown in Figure 4.16. On the other hand, if the element spacing is $d = \lambda$, then $\psi_s = \pi$, $\psi_f = -3\pi$, and the pattern assumes the shape shown in Figure 4.17. If one wishes to avoid the presence of a second main beam, it is

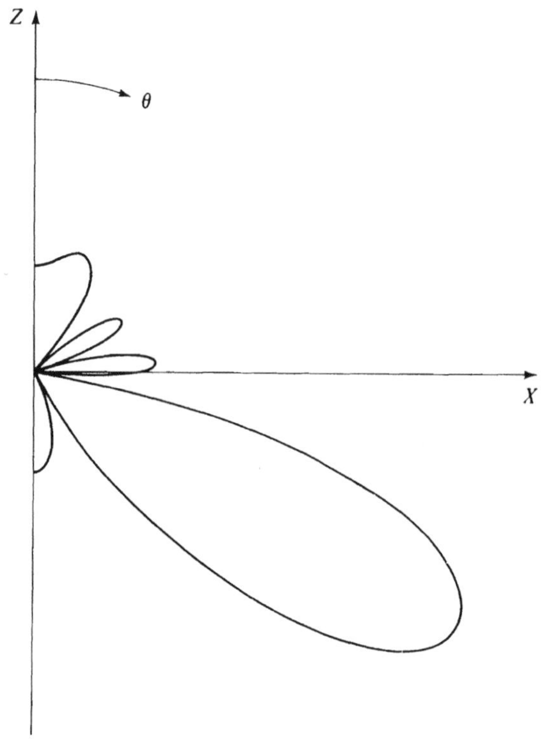

Fig. 4.16 Polar Field Plot for a Five-Element Linear Array Excited with Uniform Amplitude, Uniform Progressive Phase, $d = \lambda/2$; Sum Pattern with Main Beam Scanned to $\theta_0 = 120°$; Linear Scale

apparent that the spacing should be chosen such that

$$\psi_f \geq -\left(2\pi - \frac{2\pi}{N+1}\right) = -\frac{2\pi N}{N+1} \quad \text{if} \quad \theta_0 < \frac{\pi}{2},$$

$$\psi_s \leq 2\pi - \frac{2\pi}{N+1} = \frac{2\pi N}{N+1} \quad \text{if} \quad \theta_0 > \frac{\pi}{2}$$

as before, but now (4.43) imposes the requirement that

$$kd(1 + |\cos\theta_0|) \leq 2\pi \frac{N}{N+1}$$

or that

$$\left(\frac{d}{\lambda}\right)_{\max} = \frac{N/(N+1)}{1 + |\cos\theta_0|} \tag{4.44}$$

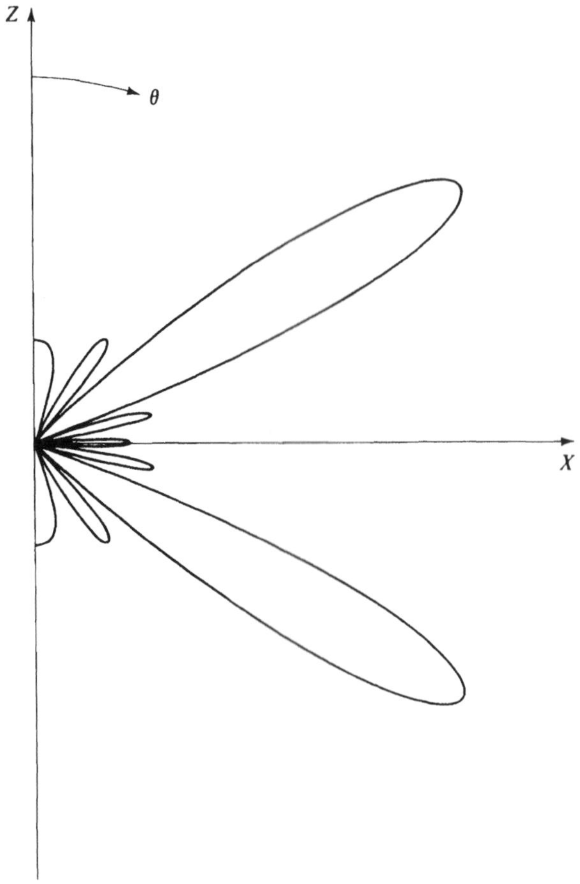

Fig. 4.17 Polar Field Plot for a Five-Element Linear Array Excited with Uniform Amplitude, Uniform Progressive Phase, $d = \lambda$; Sum Pattern with Main Beam Scanned to $\theta_0 = 120°$; Extra Main Beam at $\theta = 60°$; Linear Scale

which is the criterion for avoidance of multiple beams if a uniformly excited linear array is scanned. For N large, this agrees with (4.30). For the five-element array, with $\theta_0 = 60°$, the maximum spacing should be $8\lambda/15$.

(d) DIFFERENCE PATTERNS For equispaced linear arrays of an even number of elements, $f(w)$ has an *odd* number of roots. If one of these roots is placed at $\psi = 0$ and the others are arranged in complex conjugate pairs as suggested by Figure 4.18, a symmetrical difference pattern will result, with twin main beams straddling a null at θ_0, and with the same side lobe topography on both sides of the main beams. A special case of this occurs when $I_n = -I_{-n} = 1$, for all n, for which it has been seen that the pattern is given by (4.32). In addition to the central null at $\theta = \pi/2$, (4.32) indicates nulls at the positions

$$\frac{\pi L}{2\lambda}\cos\theta_m = m\pi$$

Since $L = 2Nd$, with $2N$ the number of elements in the array, and since $\psi = kd\cos\theta$, this result can be converted to the form

$$\psi_m = \frac{2\pi m}{N} \qquad m = \pm 1, \pm 2, \ldots, \pm N - 1 \tag{4.45}$$

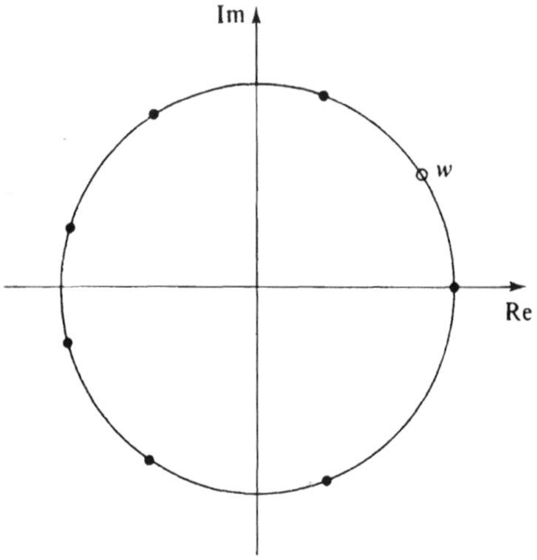

Fig. 4.18 A Schelkunoff Unit Circle with Roots Placed to Give a Symmetric Difference Pattern

Thus the roots are equispaced on the unit circle, but they are all *double* roots except for the single root at $\psi = 0$. This is an inefficient root placement. For example, the pattern shown in Figure 4.9 is for a 14-element array, $d = \lambda/2$, with uniform asymmetric excitation, and shows only two and one-half side lobes on each side of the twin main beams. It has already been remarked that this pattern does not have impressively low side lobes (the innermost pair of side lobes is only 9.5 dB below the height of the twin main beams). If the roots were repositioned appropriately on the unit circle, so that they occurred singly, there could be five and one-half side lobes on each side of the pattern, with a concomitant lowering of the side lobe level at no expense in terms of broadening the twin main beams.

Further lowering of the side lobe level could be accomplished by clustering the roots closer to $\psi = \pi$ (with one root held at $\psi = 0$ to insure a difference pattern). This would be at the expense of some broadening of the twin main beams. Such designs can be achieved graphically by trial and error in the manner already described for the sum pattern. They can also be obtained by a synthesis technique that will be presented in Chapter 5.

(e) SUPERGAIN ARRAYS Since the total excursion of w along the unit circle is $2kd$, the intriguing possibility arises that one could make the interelement spacing d

smaller and smaller, simultaneously repositioning the roots on the unit circle so that they always remained within the range of w, and in such a way that the pattern in real space was unaltered. In this manner, for example, a sum pattern with a main beam of a prescribed beamwidth and side lobes of prescribed heights could be generated by an equispaced linear array of a specified number of elements, but with the total length of the array arbitrarily small. The current distribution in this compressed array will need to be investigated. It clearly will not be uniform, because the earlier analysis of that case revealed that, for a uniformly excited array, the null-to-null beamwidth of the main beam of a sum pattern is $2\lambda/L$. (See Equation 4.18 and the related discussion.)

As a specific illustration of this possibility of a reduced-length array, consider once again the five-element equispaced linear array, excited to give a broadside sum pattern with symmetrical side lobes. If the roots are placed on the unit circle at the positions $\psi_m = \pm 72°$, $\pm 144°$, it has already been seen that the excitation will be uniform. For $d = \lambda/2$, the pattern will have a main beam plus one and one-half side lobes on each side of it. Suppose next that the interelement spacing is reduced to a fraction f of $\lambda/2$ and that the roots are simultaneously repositioned to be at $\pm\psi_1$, $\pm\psi_2$, with $\psi_1 = 72f$ degrees. This will clearly keep the roots within the range of w. As a generalization of (4.42),

$$f(w) = (w^2 - 2w\cos\psi_1 + 1)(w^2 - 2w\cos\psi_2 + 1)$$

and thus the current distribution is

$$\begin{matrix} 1 & -2(\cos\psi_1 + \cos 2\psi_1) & 2(1 + 2\cos\psi_1 \cos 2\psi_1) \\ & & -2(\cos\psi_1 + \cos 2\psi_1) \quad 1 \end{matrix} \tag{4.46}$$

When $\psi_1 = 72°$, all of these currents are unity and $[(1)^2 + (1)^2 + (1)^2 + (1)^2 + (1)^2]R$ denotes the ohmic losses, with R some appropriate ohmic representation of the resistivity and shape of an element. The field strength at the peak of the main beam is measured by the sum of the currents and therefore the total radiated power can be represented by $(5)^2K$, with K a factor that depends on pattern shape. Assume that, with $\psi_1 = 72°$, the ohmic losses are 1% of the power radiated. Then $(5)^2K = 100(5R)$ or $K = 20R$.

Now assume that $\psi_1 = 1°$, that is, there has been a 72-fold contraction in the length of the array and in the root placement on the unit circle. For this case, (4.46) gives, for the current distribution,

$$1 \qquad -3.998477 \qquad 5.996954 \qquad -3.998477 \qquad 1$$

The ohmic losses have become $70R$ and the field strength at the peak of the main beam is only 0.371×10^{-6}, so the radiated power is $0.13764 \times 10^{-12}K$. The ratio of the power radiated to the ohmic losses is

$$\frac{0.13764 \times 10^{-12}K}{70R} = 0.39326 \times 10^{-12}$$

The ohmic losses, which were assumed to be only 1% of the radiated power at $d = \lambda/2$ spacing, are in contrast a trillion times as large as the radiated power at $d = \lambda/144$ spacing.

Even with a modest reduction in spacing to $d = \lambda/4$, the ohmic losses are four times as large as the radiated power. This simple example serves to illustrate the drastic penalty one must pay in loss of efficiency if reduction of length is contemplated for linear arrays. Further study shows that mechanical and electrical tolerances become severe and frequency bandwidth is sharply curtailed as the interelement spacing is contracted. For all these reasons, supergaining (as this process is called) has proven to be impractical.

REFERENCES

Amitay, N., V. Galindo-Israel, and C. P. Wu, *Theory and Analysis of Phased Array Antennas* (New York: Wiley-Interscience, 1972), Chapter 1.

Bach, H., and J. E. Hansen, "Uniformly Spaced Arrays," in *Antenna Theory, Part I*, ed. R. E. Collin and F. J. Zucker (New York: McGraw-Hill Book Co., Inc., 1969), Chapter 5.

Elliott, R. S., "The Theory of Antenna Arrays," in *Microwave Scanning Antennas*, Vol. 2, ed. R. C. Hansen (New York: Academic Press, 1966), Chapter 1.

Jordan, E. C., and K. G. Balmain, *Electromagnetic Waves and Radiating Systems*, 2nd ed. (Englewood Cliffs, N.J.: Prentice-Hall, Inc., 1968), Chapter 12.

Wolff, E. A., *Antenna Analysis* (New York: John Wiley and Sons, Inc., 1966), Chapter 6.

PROBLEMS

4.1 Begin with Equations 4.1 and 4.5 and show that $\mathcal{A}_\theta(\theta, \phi)$ can be written as the product of the array factor (4.10) and the element factor (4.11).

4.2 Assume that a uniformly spaced linear array of $2N + 1$ elements is uniformly excited. If $2N + 1$ is very large, show that the height of the first side lobe occurs when the corresponding $2N + 1$ phasors uniformly occupy one and one-half sheets of the complex plane. Do this by letting the position of the outermost phasor, $N\psi$, be a variable. Approximate the phasor diagram by a continuum density of phasors and show that this first side lobe is 13.5 dB below the height of the main beam.

4.3 Show that Equation 4.27 is a transformation of equation (4.23).

4.4 Assume that a uniformly spaced linear array of $2N$ elements is uniformly excited, but in the difference mode. If $2N$ is very large, show that the height of the principal lobe occurs when the corresponding $2N$ phasors have fanned out such that the outermost one is at the position $(2N - 1)\psi/2 \cong 3\pi/4$. Do this by letting the position of the outermost phasor be a variable. Approximate the phasor diagram by a continuum density of phasors and show that the pair of side lobes that is closest in is approximately 9.5 dB below the level of the twin principal lobes.

4.5 Show that Equation 4.32 can be derived from (4.31) when $I_n = -I_{-n} = I_1$, for all n.

4.6 Show the equivalence of Expressions 4.27 and 4.39.

4.7 A six-element equispaced linear array is to be given uniform amplitude/equiphase excitation. Construct a suitably large Schelkunoff unit circle (say $r = 4$ inches) and mark the root positions. If $d = \lambda/2$, find the null positions in θ-space. Use the fact that $|f(w)| = d_1 \cdots d_5$ to determine the relative lobe heights, and make a rough polar plot of the field pattern.

4.8 For the six-element array discussed in the preceding problem, use trial and error to determine the proper root positions to yield an array pattern with a single main beam and all side lobes at -20 dB. Find the corresponding current distribution.

4.9 For the six-element array discussed in the two preceding problems, use trial and error to determine the proper root positions to yield a difference pattern with all side lobes at -20 dB. Find the corresponding current distribution. Note that this current distribution is substantially different from what you would get by reversing the phase of half of the array excitation found for the sum pattern in Problem 4.8.

4.10 It has been shown in the text that, for an equispaced linear array of $2N$ elements, a difference pattern will result if the nulls are placed on the unit circle at the positions $\psi_m = 2\pi m/N$ with $m = 0, \pm 1, \ldots, \pm N - 1$. Show that this gives $I_n = -I_{-n} = 1$, for all n, for the current excitation, and that the pattern is represented by Equation 4.32.

4.11 Design a six-element equispaced array to give a sum pattern with its main beam at endfire and all side lobes at -30 dB. This can be done graphically by trial and error. Then deduce the maximum spacing between elements if not even a trace of a second main beam is to be present at reverse endfire.

4.12 It has been shown in the text that if an equispaced linear array of $2N$ elements is excited uniformly but asymmetrically, a difference pattern results with the roots occurring on the unit circle at positions $\psi_m = 2\pi m/N$, $m = 0, \pm 1, \ldots, \pm N - 1$. All roots are double except ψ_0. This is an inefficient root placement. One way to correct this is to place the roots singly at $\psi_m = 2m\pi/(2N-1)$, $m = 0, \pm N - 1, \ldots, 1$. Show that if this is done, all side lobes are the same height as the twin main beams and that all currents in the array are zero except the end two, which are equal and opposite. (The result is called an *interferometer pattern.*)

4.13 With reference to the preceding problem, another possible root placement that avoids double roots is to let $\psi_m = \pi m/(N+1)$, $m = 0, \pm 2, \pm 3, \ldots, \pm N$. Show that this gives larger ψ-regions on the unit circle for the twin main beams than it does for the side lobes. Plot the polar field patterns for the case $2N = 14$, $d = \lambda/2$, and compare with Figure 4.9 of the text. Find the current distribution and compare with Figure 4.10 of text.

5 linear arrays: synthesis

5.1 Introduction

In the previous chapter the basic analysis of equispaced linear arrays was presented under the assumption that a known current distribution existed in the array and that one desired to find the resulting array pattern. A variety of practical distributions were assumed (uniform amplitude with and without uniform progressive phase, tapered amplitude with and without uniform progressive phase, the two halves of the array excited out of phase) and it was discovered that useful sum and difference patterns were caused by these distributions. In conjunction with the introduction of the Schelkunoff unit circle, the subject of synthesis was even touched on when a graphical trial-and-error technique was suggested in which root placement could be systematically altered until a desired pattern was achieved.

In the present chapter the synthesis problem will be addressed directly. In synthesis, one begins by specifying the desired array pattern. Since the discussion here is restricted to linear arrays,[1] the desired pattern must be a function of θ alone and not ϕ, that is, in the form $\mathcal{A}_a(\theta)$ or $\mathcal{F}_a(\theta)$. But the class of functions $\mathcal{A}_a(\theta)$ or $\mathcal{F}_a(\theta)$ is large. It includes sum patterns with uniform side lobes, with symmetrically tapered side lobes, and with asymmetric side lobes. It includes difference patterns with the same variety of side lobe topographies. It includes patterns with neither nulls nor side lobes. For all of these patterns, the synthesis question is basically the same: Given $\mathcal{A}_a(\theta)$ or $\mathcal{F}_a(\theta)$, what is the requisite current distribution in an equispaced array?

This question will be answered in succeeding sections of this chapter for some of the more widely used classes of prescribed patterns. Dolph's technique, which uses Chebyshev polynomials to deduce discrete current distributions that yield sum patterns with uniform side lobes, will be taken up first. Taylor's procedure, which accomplishes basically the same result but for continuous line sources, will also be

[1]This restriction will be lifted in Chapter 6 with the introduction of planar arrays.

presented. A perturbation method, which can be used to modify either a Dolph or Taylor pattern in order to produce sum patterns with arbitrary side lobe topographies, will be introduced and then extended to applications involving difference patterns. The Woodward technique, which synthesizes patterns requiring null-filling, will also be described.

The chapter is not devoted entirely to synthesis procedures. The concepts of half-power beamwidth and peak directivity of a linear array pattern are introduced and applied specifically to the case of sum patterns, since these two quantities often form a key part of the design specifications on which the pattern synthesis must be based.

5.2 Sum and Difference Patterns

Many applications of linear arrays involve the need to produce sum and difference patterns with the main beam of the sum pattern pointing at an angle θ_0, with the twin main beams of the difference pattern straddling θ_0, and with both patterns exhibiting a symmetrical side lobe structure. When there are $2N$ elements in the array, equispaced by an amount d, the array factor can be written in the form[2]

$$\mathcal{A}_a(\theta) = \sum_{n=-N}^{-1} \frac{I_n}{I_1} e^{j[(2n+1)/2]kd(\cos\theta - \cos\theta_0)} + \sum_{n=1}^{N} \frac{I_n}{I_1} e^{j[(2n-1)/2]kd(\cos\theta - \cos\theta_0)} \tag{5.1}$$

Under the above stipulations, all the current amplitudes in (5.1) can be taken as pure real. For the sum pattern, $I_n = I_{-n}$ and (5.1) becomes

$$\mathcal{S}(\theta) = 2\sum_{n=1}^{N} \frac{I_n}{I_1} \cos\left[(2n-1)\left(\frac{\pi d}{\lambda}\right)(\cos\theta - \cos\theta_0)\right] \tag{5.2}$$

For the difference pattern, $I_n = -I_{-n}$ and (5.1) takes the form

$$\mathcal{D}(\theta) = 2j\sum_{n=1}^{N} \frac{I_n}{I_1} \sin\left[(2n-1)\left(\frac{\pi d}{\lambda}\right)(\cos\theta - \cos\theta_0)\right] \tag{5.3}$$

An array with $2N + 1$ elements (an odd number) is not suitable for the creation of a difference pattern because of the awkward presence of the central element. However, as seen in Chapter 4, it can be used to produce a sum pattern. Under the assumption of symmetrical side lobes, the pattern from such an array is given, as a reduction from Equation 4.13, by

$$\mathcal{S}(\theta) = 1 + 2\sum_{n=1}^{N} \frac{I_n}{I_0} \cos\left[2n\left(\frac{\pi d}{\lambda}\right)(\cos\theta - \cos\theta_0)\right] \tag{5.4}$$

[2]For convenience $\mathcal{A}_a(\theta)$ is used, but the results apply equally well for $\mathcal{F}_a(\theta)$.

A major class of synthesis problems can be stated in terms of these equations. If a sum pattern with a specified side lobe topography is desired, how does one determine the current distribution I_n/I_1 in (5.2) or I_n/I_0 in (5.4) to achieve the desired result? Or, if a difference pattern with a certain side lobe topography is desired, how does one find the current distribution I_n/I_1 in (5.3) to bring this about? Several sections of this chapter are concerned with answers to these questions.

5.3 Dolph-Chebyshev Synthesis of Sum Patterns

The discussions of Chapter 4 revealed some useful information about the design of equispaced linear arrays, excited so as to give an array factor with one main beam plus side lobes (sum pattern). This information can be summarized as follows.

1. For $2N + 1$ elements, if the $2N$ roots are placed on the unit circle in complex conjugate pairs, a symmetrical sum pattern will result. If the positions of these root pairs are adjusted, the side lobe heights can be altered. To reduce the level of the side lobes, the root pairs need to be clustered closer to $\psi = \pi$, at the expense of broadening the main beam.

2. For $2N$ elements, if the $2N - 1$ roots are placed on the unit circle with one root at $\psi = -\pi$ and the remainder in $N - 1$ complex conjugate pairs, a symmetrical sum pattern will result. If the positions of the root pairs are adjusted, the side lobe heights can be altered. To reduce the level of the side lobes, the root pairs need to be clustered closer to $\psi = \pi$, at the expense of broadening the main beam.

3. With the roots occurring in complex conjugate pairs, $f(w)$ is a polynomial with pure real coefficients. These coefficients appear in symmetrical pairs in the polynomial, thus evidencing the fact that the current distribution in the array is symmetrical in amplitude.

With these observations as background, the problem of proper positioning of the root pairs can be addressed. If one argues that side lobes occur in spatial regions in which it is desirable to suppress radiation, and assumes that the suppression of *all* side lobes is equally important, then an optimum design is one in which all side lobes are at the same height. The reduction of a single side lobe further than this common level could only be at the expense of additional broadening of the main beam, and would deny the assumption that the region of this side lobe is no more important than the region of any other side lobe.

This problem of seeking the proper root positions (and thus the proper array excitation) to give a sum pattern with uniform side lobes at a specified height was solved by C. L. Dolph in a classic paper.[3] To do this, he took advantage of a useful property of Chebyshev polynomials, which are solutions of the differential equation

$$(1 - u^2)\frac{d^2T_m}{du^2} - u\frac{dT_m}{du} + m^2T_m = 0 \tag{5.5}$$

[3]C. L. Dolph, "A Current Distribution for Broadside Arrays Which Optimizes the Relationship between Beamwidth and Side Lobe Level", *Proc. IRE*, 34 (1946), 335–48.

It is shown in Appendix C that, if the index m is an even integer $2N$, then a solution to (5.5) is

$$T_{2N}(u) = \sum_{n=0}^{N} (-1)^{N-n} \frac{N}{N+n} \binom{N+n}{2n} (2u)^{2n} \tag{5.6}$$

If m is an odd integer $2N - 1$, then

$$T_{2N-1}(u) = \sum_{n=1}^{N} (-1)^{N-n} \frac{2N-1}{2(N+n-1)} \binom{N+n-1}{2n-1} (2u)^{2n-1} \tag{5.7}$$

with $\binom{r}{s}$ the binomial coefficient $r!/s!(r-s)!$. Both (5.6) and (5.7) can be put in the revealing form

$$\begin{aligned} T_m(u) &= \cos(m \cos^{-1} u) && -1 \leq u \leq 1 \\ &= \cosh(m \cosh^{-1} u) && u > 1 \\ &= (-1)^m \cosh(m \cosh^{-1} |u|) && u < 1 \end{aligned} \tag{5.8}$$

which is easily verified by substitution in (5.5). Thus $T_m(u)$, with m an integer, is a function that oscillates in a cosinusoidal manner in the range $|u| < 1$ and then rises hyperbolically in $|u| > 1$. It is this property of the Chebyshev polynomials that makes them so useful in antenna array design. Figure 5.1 shows the typical features of

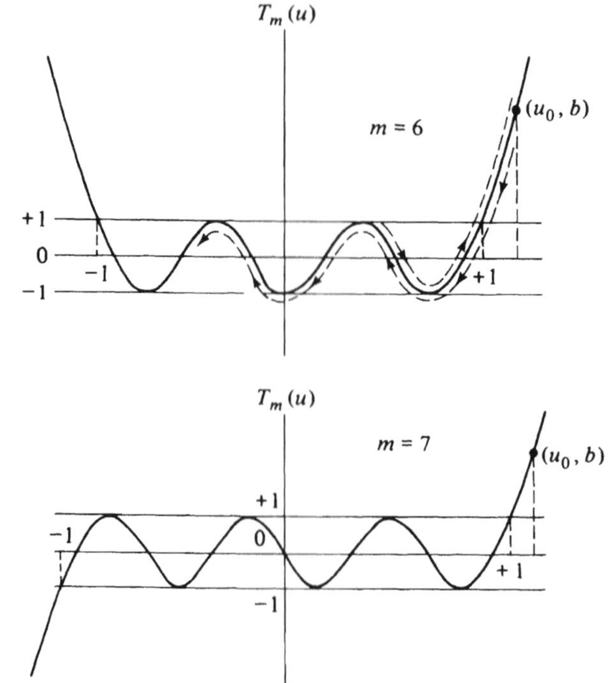

Fig. 5.1 Chebyshev Functions (Reprinted from *Microwave Scanning Antennas*, Volume 2, R. C. Hansen, Editor, Courtesy of Academic Press, Inc. © 1966 Academic Press.)

Chebyshev functions. For $m = 2N$ there is symmetry about $u = 0$, with N root pairs in $|u| < 1$. Here $T_{2N}(1) = T_{2N}(-1) = 1$. There are no roots outside $|u| = 1$, so just beyond $|u| = 1$ the function rises rapidly, with steeper slope for larger values of $2N$. For $m = 2N - 1$, there is antisymmetry about $\theta = 0$, with a root at $u = 0$ and $N - 1$ root pairs in $|u| < 1$, where $T_{2N-1}(1) = -T_{2N-1}(-1) = 1$. There are no roots beyond $|u| = 1$, so the function rises rapidly, with steeper slope for larger values of $2N - 1$.

These plots reveal why the Chebyshev polynomials were ideal for Dolph's purpose. If the variable u can be made to correspond in some manner to the real angle variable θ, so that an appropriate segment of $T_m(u)$ can be made to relate to $\mathcal{A}_a(\theta)$, then a pattern with uniform side lobes will result.

To develop this correspondence, one can return to the basic equation for the array factor, assume a uniform progressive phase and symmetrical amplitude distribution, and write as alternate forms of (5.4) and (5.2)

$$\mathcal{S}\left(\frac{\psi}{2}\right) = 1 + 2\sum_{n=1}^{N} \frac{I_n}{I_0} \cos 2n\frac{\psi}{2} \tag{5.9}$$

$$\mathcal{S}\left(\frac{\psi}{2}\right) = 2\sum_{n=1}^{N} \frac{I_n}{I_1} \cos (2n - 1)\frac{\psi}{2} \tag{5.10}$$

in which, as before,

$$\psi = kd(\cos\theta - \cos\theta_0) \tag{5.11}$$

Equation (5.9) applies for an odd number of elements and is equivalent to a polynomial of order $2N$ in the variable $\cos(\psi/2)$. Equation 5.10 applies for an even number of elements and is equivalent to a polynomial of order $2N - 1$ in the variable $\cos(\psi/2)$. Thus if one selects the transformation

$$u = u_0 \cos\frac{\psi}{2} \tag{5.12}$$

then (5.6) and (5.9) can be equated for arrays with an odd number of elements, and (5.7) and (5.10) can be equated for arrays with an even number of elements.

What this transformation accomplishes can be appreciated by returning to Figure 5.1. As θ ranges from 0 to θ_0 to π, and as ψ ranges from $\psi_s = kd(1 - \cos\theta_0)$ to zero to $\psi_f = -kd(1 + \cos\theta_0)$, u will range from $u_s = u_0 \cos(\psi_s/2)$ to u_0 to $u_f = u_0 \cos(\psi_f/2)$. The pattern trace is shown by the arrowed dotted path in Figure 5.1a. If u_0 is chosen so that $T_m(u_0) = b$, with $20 \log_{10} b$ the desired side lobe level, then a pattern will result consisting of a main beam at the relative field height b plus a family of side lobes all at the height unity.

Dolph's design procedure can now be articulated. One begins by selecting the number of elements, which determines the degree of the Chebyshev polynomial that is to be used (m is one less than the number of elements). Next, it is necessary to find u_0 from $T_m(u_0) = b$, with b fixed by the desired side lobe level. Then, from (5.8), the roots of $T_m(u)$ can be determined readily and are given by

$$u_p = \pm \cos\left[(2p - 1)\frac{\pi}{2m}\right] \tag{5.13}$$

When these values are inserted in (5.12), the corresponding root positions ψ_p on the unit circle can be computed. At this point $f(w)$ is known in factored form and can be multiplied out to give the current distribution.

As an example, consider a five-element array that is to be excited to give a sum pattern with all side lobes at -20 dB. Then it is $T_4(u)$ which should be used, and $T_4(u_0) = b = 10$. From (5.8),

$$\cosh\,(4\cosh^{-1} u_0) = 10$$

which is satisfied by $u_0 = 1.2933$. Also, from (5.8), one finds that the roots of $T_4(u)$ are ± 0.9239 and ± 0.3827. Thus, from (5.12), the roots on the unit circle are at the positions

$$\psi = \pm 88.82°, \pm 145.58°$$

If the calculation that produced Equation 4.42 is repeated (where this same problem was being solved by a graphical trial-and-error method), the present more accurate ψ values give

$$f(w) = w^4 + 1.61w^3 + 1.93w^2 + 1.61w + 1$$

with the coefficients of the various powers of w representing the relative current distribution.

With the uniform progressive phase factor α_z embedded in the definition of w (compare Equations 4.34 through 4.35), the coefficients

$$1 \quad 1.61 \quad 1.93 \quad 1.61 \quad 1$$

represent the relative *magnitudes* of the currents in the five-element array. This implies that the Dolph-Chebyshev distribution is the same in magnitude regardless of where the main beam points. All that changes when the main beam pointing direction θ_0 is altered is the uniform progressive phase, which must be attached to the amplitude distribution.

If the avoidance of extra main beams is of concern, the precautions noted in Section 4.4 must be observed. The interelement spacing should be chosen so that the ψ-excursion on the Schelkunoff unit circle only traverses the main beam region once.

The case of an end-fire Dolph-Chebyshev distribution is worth special mention, and the example cited above can serve as a typical illustration. To place the main beam at end-fire, $\psi_s = 0°$ and $\psi_f = -(360° - 88.82°) = -271.18°$. The interelement spacing should not exceed

$$d = \frac{\lambda}{2}\frac{\psi_s - \psi_f}{360°} = 0.377\lambda$$

if even a vestige of a second main beam cannot be tolerated at reverse end-fire. With this spacing, the uniform progressive phase is $\alpha_z = kd = 2.369$ radians and the end-fire Dolph-Chebyshev current distribution for this array is

$$1\underline{|271.4^\circ} \qquad 1.60\underline{|135.7^\circ} \qquad 1.93\underline{|0^\circ} \qquad 1.60\underline{|-135.7^\circ} \qquad 1\underline{|-271.4^\circ}$$

It is possible to determine the Dolph current distribution for the general case without the intermediate steps of finding the roots u_p, ψ_p, and w_p followed by multiplying out the factors of $f(w)$. To do this, one needs to insert (5.12) in (5.6) or (5.7) and make use of the relations

$$\cos^{2n}\frac{\psi}{2} = \frac{1}{2^{2n}}\sum_{q=0}^{n}\epsilon_q\binom{2n}{n-q}\cos\frac{2q\psi}{2} \tag{5.14}$$

$$\cos^{2n-1}\frac{\psi}{2} = \frac{1}{2^{2n-2}}\sum_{q=1}^{N}\binom{2n-1}{n-q}\cos(2q-1)\frac{\psi}{2} \tag{5.15}$$

A derivation of these two formulas can be found in Appendix D. In Equation 5.14, $\epsilon_q = 1$ if $q = 0$, otherwise $\epsilon_q = 2$.

With the use of (5.13) through (5.15), the expressions for the Chebyshev polynomials become

$$T_{2N}\left(\frac{\psi}{2}\right) = \sum_{n=0}^{N}\sum_{q=0}^{n}\epsilon_q(-1)^{N-n}\frac{N}{N+n}\binom{N+n}{2n}\binom{2n}{n-q}u_0^{2n}\cos 2q\frac{\psi}{2} \tag{5.16}$$

$$T_{2N-1}\left(\frac{\psi}{2}\right) = \sum_{n=1}^{N}\sum_{q=1}^{n}(-1)^{N-n}\frac{2N-1}{N+n-1}\binom{N+n-1}{2n-1}\binom{2n-1}{n-q}u_0^{2n-1}\times \cos(2q-1)\frac{\psi}{2} \tag{5.17}$$

The coefficients of $\cos 2m\psi/2$ are in the same ratio in (5.9) and (5.16) for all m, and likewise the coefficients of $\cos(2m-1)\psi/2$ are in the same ratio in (5.10) and (5.17) for all m. Therefore the relative current distribution for an array with $2N+1$ elements is

$$I_n = \sum_{p=n}^{N}(-1)^{N-p}\frac{N}{N+p}\binom{N+p}{2p}\binom{2p}{p-n}u_0^{2p} \tag{5.18}$$

and, for an array with $2N$ elements is

$$I_n = \sum_{p=n}^{N}(-1)^{N-p}\frac{2N-1}{2(N+p-1)}\binom{N+p-1}{2p-1}\binom{2p-1}{p-n}u_0^{2p-1} \tag{5.19}$$

Although they look formidable, (5.18) and (5.19) are simple to program. To use them, one still needs to start with the knowledge of the number of elements in the array and the desired side lobe level, so that u_0 can be determined. For arrays with a large number of elements, the time saved in using (5.18) or (5.19) is considerable when compared to the procedure that first determines the root positions.

5.4 Sum Pattern Beamwidth of Linear Arrays

Discussions in Chapter 4 have revealed that the angular extent of the main beam in a sum pattern is inversely related to the length of a linear array. Further, it has been seen that, for a given length array, the main beam broadens as the side lobe level is lowered. In synthesis problems, these relationships must be approached from the other end. Often the design specifications include statements about the desired beamwidth of a sum pattern, as well as the side lobe level. The designer must not only determine the requisite current distribution, but also the array length and number of elements, both chosen to avoid multiple beams (d/λ should not be too great) and supergaining (L/λ should not be too small).

Because of the importance of beamwidth as a design specification, it is desirable to sharpen these earlier discussions by introducing a more precise definition of beamwidth. The one normally used is that the beamwidth is the angular separation between θ directions at which the radiated power density is down to one-half its maximum value. For an equispaced array of $2N + 1$ elements,[4] laid out symmetrically along the Z-axis, let $\theta = \theta_2 - \theta_1$ be this beamwidth, in which θ_2 and θ_1 are the two values of θ which satisfy the relation

$$\mathcal{S}(\theta_m) = 0.707\mathcal{S}(\theta_0) = \sum_{n=-N}^{N} \frac{I_n}{I_0} e^{jn(kd\cos\theta_m - \alpha_z)} \qquad m = 1, 2 \tag{5.20}$$

In (5.20), the current amplitudes I_n are assumed to be pure real, the current phase progression is governed by α_z, d is the interelement spacing, and θ_0 is the pointing angle of the main beam.

The amplitude distribution I_n/I_0 can be described by a Fourier series, namely,

$$\frac{I_n}{I_0} = \sum_{p=-P}^{P} a_p e^{j2\pi np/(2N+1)} \tag{5.21}$$

in which P is the highest spatial harmonic needed to represent the distribution. Attention will be restricted to sum patterns in which the side lobe topography is symmetric, which means that I_n/I_0 is also symmetric, and thus that $a_p = a_{-p}$ is pure real for all p. Since $\alpha_z = kd\cos\theta_0$, insertion of (5.21) in (5.20) gives

$$\begin{aligned}\mathcal{S}(\theta_m) &= \sum_{p=-P}^{P} a_p \sum_{n=-N}^{N} e^{jnkd[\cos\theta_m - \cos\theta_0 + p\lambda/L]} \\ &= \sum_{p=-P}^{P} a_p \frac{\sin\{(\pi L/\lambda)[\cos\theta_m - \cos\theta_0 + (p\lambda/L)]\}}{\sin\{(\pi d/\lambda)[\cos\theta_m - \cos\theta_0 + (p\lambda/L)]\}}\end{aligned} \tag{5.22}$$

[4]All the results obtained in this section are equally valid for $2N$ elements.

in which $L = (2N + 1)d$ is the length of the array. It has been shown[5] that, for large arrays with conventional distributions, Equation 5.22 can be transformed to

$$\frac{\sin K\pi}{K\pi} = 0.707\left(\sum_{p=-P}^{P} \frac{a_p}{a_0}(-1)^{p+1}\frac{K^2}{p^2 - K^2}\right)^{-1} \tag{5.23}$$

where

$$K = \frac{L}{\lambda}(\cos\theta_m - \cos\theta_0) \tag{5.24}$$

is a substitution variable from which the beamwidth can be deduced. Several cases will now be considered.

CASE 1: UNIFORM DISTRIBUTION This is the simplest case of all and extremely useful as a reference. Only a_0 has a value and (5.23) yields two solutions for K such that $\sin K\pi = 0.707K\pi$ and $K\pi = \pm 1.392$, and therefore

$$\cos\theta_1 - \cos\theta_0 = 0.443(\lambda/L) \tag{5.25}$$

$$\cos\theta_2 - \cos\theta_0 = -0.443(\lambda/L) \tag{5.26}$$

from which it follows that the half-power beamwidth is given by

$$\begin{aligned}\theta = \theta_2 - \theta_1 &= \cos^{-1}\left[\cos\theta_0 - 0.443\frac{\lambda}{L}\right] \\ &\quad -\cos^{-1}\left[\cos\theta_0 + 0.443\frac{\lambda}{L}\right]\end{aligned} \tag{5.27}$$

in the range $0 < \theta_0 < \pi/2$, $\theta_1 \geq 0$.

As the main beam is scanned from broadside ($\theta_0 = \pi/2$) to end-fire ($\theta_0 = 0$), a cross section of the beam takes on a succession of positions, as indicated in Figure 5.2. As the conical beam closes toward end-fire, a position is reached at which $\theta_1 = 0$, and from this position to end-fire, there is no half-power point on one side of the beam. For this reason $\theta_1 = 0$ can be called the *scan limit*. Equation 5.25 will not give a real value for θ_1 beyond this limit.

When the end-fire position is reached, the concept of beamwidth once again takes on meaning. Equation 5.26 is still valid and one can write

$$\theta = 2\theta_2 = 2\cos^{-1}\left[1 - 0.443\frac{\lambda}{L}\right] \qquad (\theta_0 = 0, \pi) \tag{5.28}$$

The beamwidths given by (5.27) and (5.28) are plotted in Figure 5.3 as functions of array length and scan position. These curves will prove useful beyond the present case of uniform amplitude excitation, as will be seen shortly.

[5]R. S. Elliott, "Beamwidth and Directivity of Large Scanning Arrays", Appendix B, *Microwave Journal*, 6 (1963), 53–60. Also in *Microwave Scanning Antennas*, ed. R. C. Hansen, Vol. 2 (New York: Academic Press, 1966), Chapter 1.

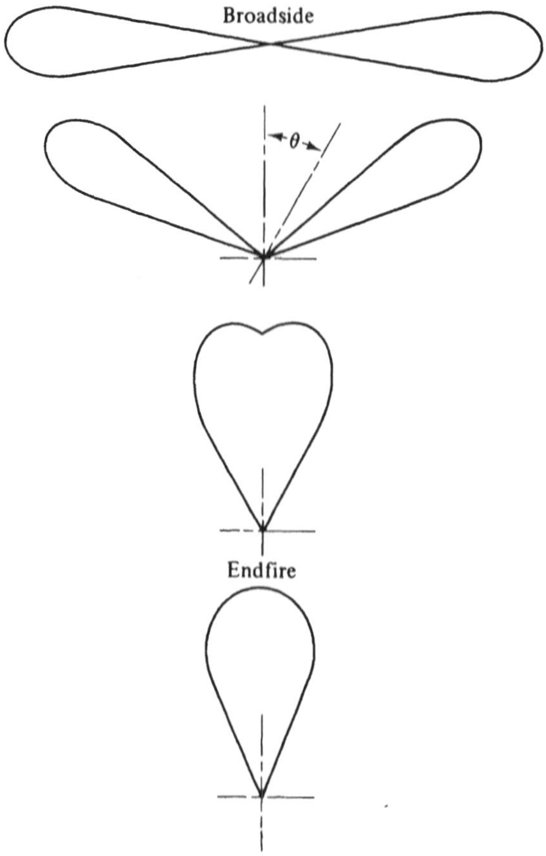

Fig. 5.2 Conical Beam Shape versus Scan

Each of these expressions for beamwidth (Equation 5.27, which is valid to within one beamwidth of end-fire, and Equation 5.28, which is valid at end-fire) has an approximate form when $L \gg \lambda$. Using small-angle expansions, one obtains

$$\theta = 0.886 \frac{\lambda}{L} \csc \theta_0 \qquad \text{(at or near broadside)} \tag{5.29}$$

$$\theta = 2\left[0.886 \frac{\lambda}{L}\right]^{1/2} \qquad \text{(at end-fire)} \tag{5.30}$$

For $L \geq 5\lambda$, Equation 5.29 is in error by less than 0.2% at broadside and is in error by less than 4% when the main beam has been scanned to within two beamwidths of end-fire. For $L \geq 5\lambda$, Equation 5.30 is in error by less than 1%.

CASE 2: DOLPH/CHEBYSHEV DISTRIBUTION It has been shown in the literature[6] that when the current distribution is chosen so that the pattern is equivalent

[6]R. S. Elliott, "An Approximation to Chebyshev Distributions," *IEEE Trans. Antennas Propagat.*, AP-11 (1963), 707–9.

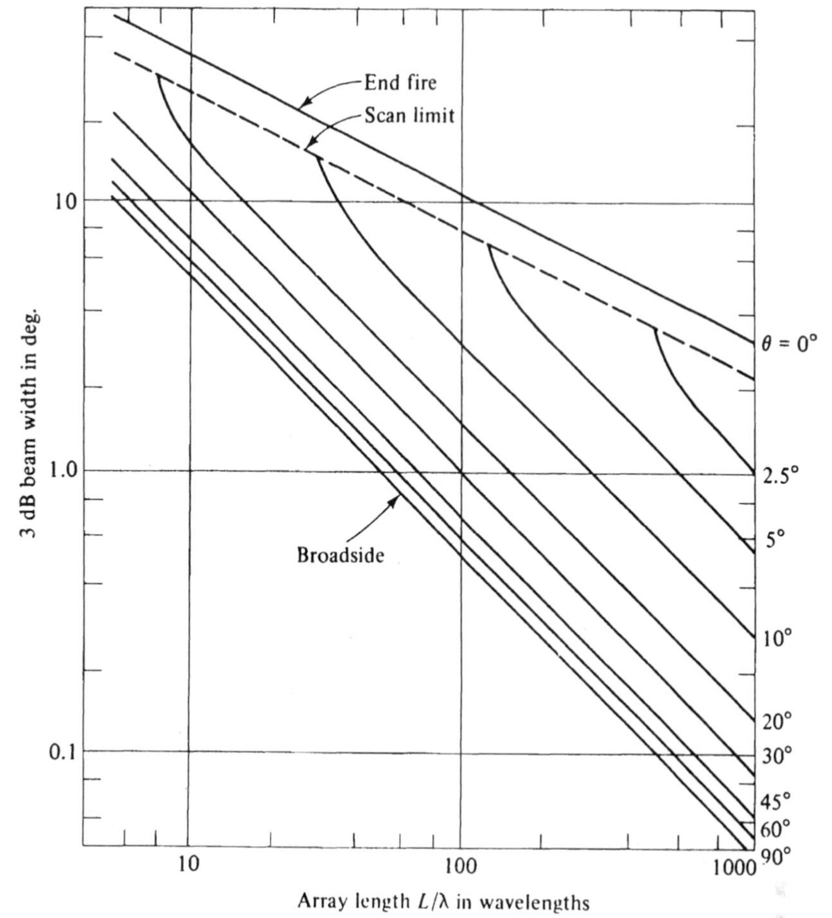

Fig. 5.3 Half-Power Beamwidth versus Linear Array Length and Scan Angle for Uniform Excitation (Reprinted from *Microwave Scanning Antennas*, Volume 2, R. C. Hansen, Editor, Courtesy of Academic Press, Inc. © 1966 Academic Press.)

to the Chebyshev polynomial $T_{2N}(u_0 \cos \psi/2)$, then the Fourier coefficients are given by

$$(2N + 1)a_p = T_{2N}\left(u_0 \cos \frac{p\pi}{2N + 1}\right) \tag{5.31}$$

It follows from (5.31) that $(2N + 1)a_0 = T_{2N}(u_0) = b$, with $20 \log_{10} b$ the side lobe level. Thus

$$a_0 = \frac{b}{2N + 1} \tag{5.32}$$

For large arrays, and for side lobe levels in the range from -20 decibels to -60 decibels, only a_0 and a_1 are significant in determining the beamwidth. If (5.8) and (5.31) are used in conjunction for $p = 1$, it is found that a_1 is given quite precisely by

$$(2N + 1)a_1 = \cosh\,[(\text{arccosh}\, b)^2 - \pi^2]^{1/2} \tag{5.33}$$

With all other Fourier coefficients insignificant in the calculation of beamwidth, (5.23) becomes for this case

$$\frac{\sin K\pi}{K\pi} = 0.707\left(1 + \frac{2a_1}{a_0}\frac{K^2}{1 - K^2}\right)^{-1} \tag{5.34}$$

Inspection of (5.34) reveals that the solution for K when $a_1 \neq 0$ differs from the solution for K when $a_1 = 0$ only through the presence of the ratio a_1/a_0. But (5.32) and (5.33) indicate that a_1/a_0 depends on side lobe level but not on scan position nor the number of elements in the array. Thus it becomes convenient to introduce a beam-broadening factor f which is simply the ratio of the half-power beamwidth when a given array is excited Dolph-Chebyshev to the half-power beamwidth when it is uniformly excited. Computations of beamwidth from (5.34), with a_0 and a_1 given by (5.32) and (5.33), can be compared to the computations that produced Figure 5.3. The result is the f-curve shown in Figure 5.4. The f-number can be interpreted as the cost in beam-broadening to convert all the side lobes to a common height and reduce them to a specified level.

The extended utility of Figure 5.3 can now be appreciated. If one wishes to determine the beamwidth of a linear array of normalized length L/λ excited to give

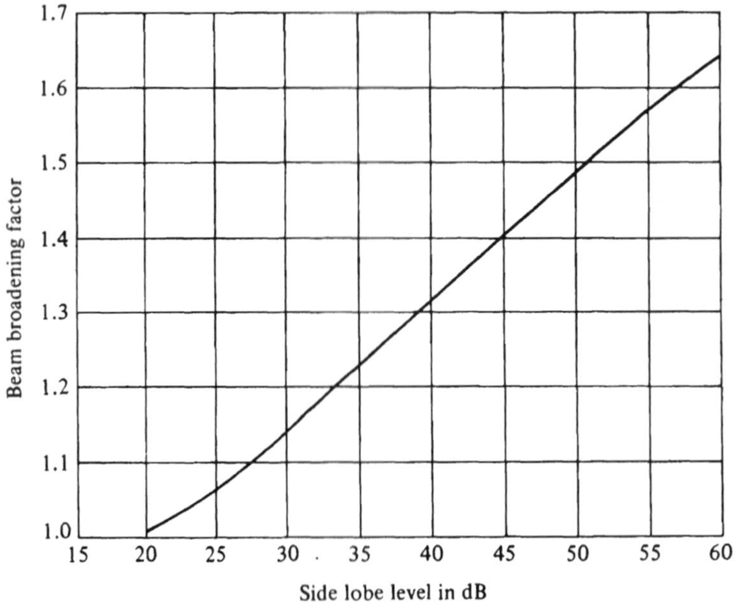

Fig. 5.4 Beam-Broadening versus Side Lobe Level for Linear Arrays with Dolph-Chebyshev Excitation

a sum pattern with the main beam at θ_0, a reading from Figure 5.3 will give the half-power beamwidth for uniform amplitude excitation. For Dolph-Chebyshev excitation with a specified side lobe level, that reading should be multiplied by the f-number read from Figure 5.4. One can also work backwards with the aid of these two figures to deduce the array length needed when the scan position, beamwidth, and side lobe level are specified for a Dolph-Chebyshev sum pattern.

5.5 Peak Directivity of the Sum Pattern of a Linear Array

The peak directivity $D(\theta_0)$ of the sum pattern produced by a linear array is a frequently encountered design specification. It can be deduced as a special case of the general definition of directivity, given as equation (1.160) and repeated here for convenience:

$$D(\theta, \phi) = \frac{\mathcal{P}(\theta, \phi)}{\dfrac{1}{4\pi r^2} \displaystyle\int_0^{\pi} \int_0^{2\pi} \mathcal{P}(\theta', \phi') r^2 \sin \theta' \, d\theta' \, d\phi'} \tag{5.35}$$

The power density in the sum pattern of a linear array is given by[7]

$$P(\theta, \phi) = \mathcal{S}_a(\theta) \mathcal{S}_a^*(\theta) [\mathcal{A}_{\theta,e}(\theta, \phi) \mathcal{A}_{\theta,e}^*(\theta, \phi) + \mathcal{A}_{\phi,e}(\theta, \phi) \mathcal{A}_{\phi,e}^*(\theta, \phi)] \tag{5.36}$$

The array and element factors that appear in (5.36) have previously been defined by Equations 4.6 through 4.11.

If the array is large, the element patterns broad, and the side lobe level of the sum pattern low, the principal contribution to the integral in the denominator of (5.35) is in the neighborhood of the main beam. When this is the case, the factor $\mathcal{A}_{\theta,e}\mathcal{A}_{\theta,e}^* + \mathcal{A}_{\phi,e}\mathcal{A}_{\phi,e}^*$ can be brought out in front of the integral and given its value at (θ_0, ϕ). If this is done,

$$D(\theta_0) = \frac{\mathcal{S}_a(\theta_0) \mathcal{S}_a^*(\theta_0)}{\dfrac{1}{4\pi} \displaystyle\int_0^{\pi} \int_0^{2\pi} \mathcal{S}_a(\theta) \mathcal{S}_a^*(\theta) \sin \theta \, d\theta \, d\phi} \tag{5.37}$$

which further simplifies to

$$D(\theta_0) = \frac{2 \mathcal{S}_a(\theta_0) \mathcal{S}_a^*(\theta_0)}{\displaystyle\int_0^{\pi} \mathcal{S}_a(\theta) \mathcal{S}_a^*(\theta) \sin \theta \, d\theta} \tag{5.38}$$

Equations 5.37 and 5.38 represent the peak directivity of the *array* factor, or what is the same thing, the peak directivity of the sum pattern when the element factor is assumed to be isotropic. They are approximate and cannot be used with good accuracy for small arrays.

[7]Equation 5.36 assumes a type I (actual-source) array. The development can be duplicated exactly when the array is represented by an equivalent magnetic current distribution.

Continuing with the assumption that the array is large ($L/\lambda \gg 1$), one can return to the definition 4.35 and write

$$\psi = kd\cos\theta - \alpha_z \qquad d\psi = -kd\sin\theta\, d\theta \tag{5.39}$$

$$\mathcal{S}_a(\theta) = \sum_{n=-N}^{N} \frac{I_n}{I_0} e^{jn\psi} \tag{5.40}$$

Use of these relations in (5.38) gives

$$D = 2\Big(\sum_{n=-N}^{N} I_n\Big)^2 \Big/ (kd)^{-1} \int_{-kd-\alpha_z}^{kd-\alpha_z} \Big(\sum_{n=-N}^{N} I_n e^{jn\psi}\Big)\Big(\sum_{m=-N}^{N} I_m e^{-jm\psi}\Big) d\psi \tag{5.41}$$

If $d = \lambda/2$ or any multiple thereof,[8] (5.41) reduces very simply to

$$D = \Big(\sum_{n=-N}^{N} I_n\Big)^2 \Big/ \sum_{n=-N}^{N} I_n^2 \tag{5.42}$$

which is a most interesting formula in several respects. The directivity given by (5.42) is a measure of the coherence of radiation from the linear array. The numerator is proportional to the total coherent field, squared, whereas the denominator is proportional to the sums of the squares of the individual fields from the various elements.

Furthermore the peak directivity, as expressed either by (5.41) or (5.42), is seen to be independent of scan angle. On the face of it this seems surprising, since it has already been observed that the main beam broadens as it is scanned away from broadside, a manifestation which usually signifies lowered directivity. However, for a *linear* array, as the conical beam is scanned toward end-fire, the "cone" occupies a smaller solid angle in space, an effect that just cancels the beam-broadening.

Although Equation 5.42 is independent of scan angle, it is not independent of current distribution. If one uses the Fourier series description of the excitation embodied in (5.21), it is evident that

$$\Big(\sum_{n=-N}^{N} \frac{I_n}{I_0}\Big)^2 = (2N+1)^2 a_0^2 \tag{5.43}$$

$$\sum_{n=-N}^{N} \Big(\frac{I_n}{I_0}\Big)^2 = (2N+1) \sum_{p=-P}^{P} a_p^2 \tag{5.44}$$

so that

$$D = \frac{2N+1}{\sum_{p=-P}^{P} (a_p/a_0)^2} \tag{5.45}$$

For half-wave spacing, $L = (2N+1)\lambda/2$, so that (5.42) can be rewritten as

$$D = \frac{2L/\lambda}{1 + 2\sum_{p=1}^{P} (a_p/a_0)^2} \tag{5.46}$$

[8]This restriction will be lifted shortly.

For element spacings in the range $\lambda/2 \leq d \leq \lambda$, if L is held fixed, computations using (5.41) show that the directivity is quite insensitive to element spacing. Since this is the range which avoids either supergaining or multiple beams, (5.46) can be adopted as a practical expression for the peak directivity of a large equispaced linear array, symmetrically excited to give a sum pattern with symmetrical side lobes, with no restriction on the pointing direction of the main beam. Several special cases can now be considered.

CASE 1: UNIFORM DISTRIBUTION Once again this is the simplest case of all and gives

$$D = \frac{2L}{\lambda} \tag{5.47}$$

which is sometimes referred to as the standard directivity of a linear array. It is the maximum directivity which can be obtained from a linear array of length L, using an aperture distribution which has uniform progressive phase.

CASE 2: DOLPH-CHEBYSHEV DISTRIBUTION If one makes use of (5.31), (5.32), and (5.46), the directivity for a Dolph-Chebyshev distribution can be written as

$$D = \frac{2L/\lambda}{\left\{1 + (2/b^2) \sum_{p=1}^{N} (T_{2N}[u_0 \cos p\pi/(2N+1)])^2\right\}} \tag{5.48}$$

Unlike the computation of beamwidth for a Dolph-Chebyshev array, in which only the first two Fourier coefficients were significant, it develops that all the Fourier coefficients that appear in the denominator of (5.48) should be included. Indeed, if the array becomes large enough, the sum of the squares of these coefficients becomes proportional to $2N + 1$ and thus the directivity tends to a limit.

It is a tedious computation to determine all the Fourier coefficients in (5.48), particularly for large arrays. Fortunately, this is not necessary, It has been shown[9] that an excellent approximation to (5.48) for large arrays is

$$D = \frac{2b^2}{1 + (b^2 - 1)(\lambda/L)f} \tag{5.49}$$

in which $20 \log_{10} b$ is the side lobe level and f is the beam-broadening factor.

Equation (5.49) has the limit

$$D_{max} = 2b^2 \tag{5.50}$$

which is reached when $L/\lambda \longrightarrow \infty$. Thus the maximum directivity for a Dolph-Chebyshev array is 3 dB more than the side lobe level. This means, for example, that if one wishes to design a linear array to have uniform side lobes and a directivity of 43 dB, it is necessary also to design it to have a side lobe level reduced at least to -40 dB.

[9]Elliott, "Beamwidth and Directivity of Large Scanning Arrays," Appendix C.

Actually, maximum directivity is approaches rather rapidly at first, as L/λ is increased, but then additional directivity is bought very dearly in terms of increased array length. This point can be appreciated by studying Figure 5.5, which is a plot of (5.49) for various values of side lobe level. An optimum directivity (and thus array length) can be selected for a given side lobe level by specifying a point on the appropriate curve of Figure 5.5 at which the curve has just barely begun to bend significantly. For example, one might not wish to design an array for a 20 dB side lobe level for which L/λ exceeded 100 and the directivity exceeded 100.

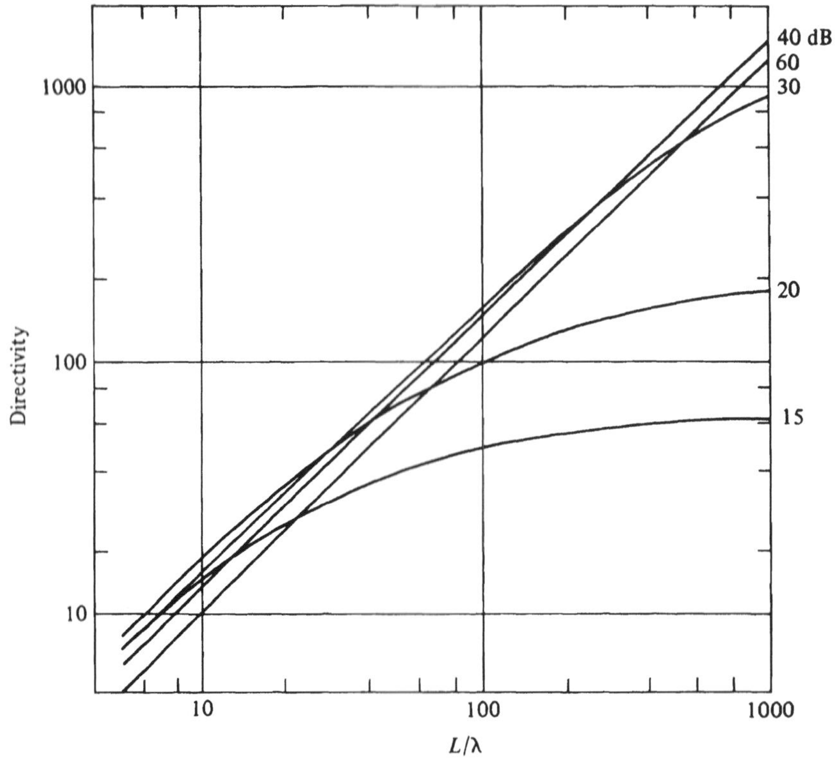

Fig. 5.5 Peak Directivity versus Linear Array Length and Side Lobe Level for Dolph-Chebyshev Excitation (Reprinted from *Microwave Scanning Antennas*, Volume 2, R. C. Hansen, Editor, Courtesy of Academic Press, Inc. © 1966 Academic Press.)

This phenomenon of a directivity limit is not observed for a uniformly excited array, with directivity given by (5.47). The difference is that the uniformly excited array has *tapered* side lobes. The Dolph-Chebyshev feature of *uniform* side lobes, while giving minimum beamwidth, is responsible for the directivity limitation. As L/λ is increased, more and more side lobes are found in real space. For a uniformly excited array, these side lobes that are farther out are at increasingly lower levels and thus make increasingly smaller contributions to the denominator of (5.38), which is not so for a Dolph-Chebyshev pattern, where all side lobes make the same contribution. However, current antenna practice is to combine large arrays with low side lobe levels, so this directivity limitation is not so serious as to preclude the use of Dolph-Chebyshev designs. For example, Figure 5.5 shows that, for arrays as long as 1000λ, very little bending has occurred in the curve for a 40 dB side lobe level.

5.6 A Relation Between Beamwidth and Peak Directivity for Linear Arrays

Equations 5.29 and 5.46 indicate that for a linear array both the peak detectivity and the reciprocal of half-power beamwidth depend linearly on the array length. .Upon eliminating L/λ from these two expressions, one obtains

$$D = \frac{1.77}{\theta_0}\left[\frac{f}{\sum_{-P}^{P}(a_p/a_0)^2}\right] \tag{5.51}$$

in which θ_0 is the *broadside* half-power beamwidth.

If the beamwidth is expressed in degrees instead of radians, and if the distribution is uniform, Equation 5.51 reduces to the simple relation

$$D = \frac{101.5}{\theta_0} \tag{5.52}$$

For a Dolph-Chebyshev distribution, until an array length is reached at which the directivity begins to limit, the factor in brackets in (5.51) is unity. Thus (5.52) is a good working relation between broadside beamwidth and peak directivity for the array excitations studied thus far, and this can be rounded off by saying that the product of broadside beamwidth in degrees and peak directivity for a linear array is approximately one hundred.

5.7 Taylor Synthesis of Sum Patterns

A horn of aperture size a by b, with $a/\lambda \gg 1$ and $b/\lambda \ll 1$ can be viewed as a *continuous* line source, as can some of the traveling wave antennas to be discussed in Chapter 9. For such structures, there is the need to develop a synthesis procedure that will permit determination of the line source distribution corresponding to a specified pattern. When the desired pattern contains a single main beam of a prescribed beamwidth and scan position, together with a family of side lobes at a common specified height, this problem parallels the one solved by Dolph for *discrete* linear arrays. The solution was achieved by T. T. Taylor[10] in an elegant paper of far-reaching importance, since Taylor distributions can be sampled and thus applied to the design of discrete arrays

[10] T. T. Taylor, "Design of Line Source Antennas for Narrow Beamwidth and Low Side Lobes," *IRE Trans. Antennas and Propagat.*, AP-7 (1955), 16–28.

as well. Further, as shall be seen in Chapter 6, Taylor was able to extend his technique to circular planar apertures. These Taylor circular distributions, as they are called, can also be sampled, and thus can give the excitation coefficients for discrete planar arrays with a circular boundary. Since there is no similar extension of Dolph's technique to circular planar apertures, it is important for the antenna designer to be cognizant of the principal features of Taylor's procedure.

If one returns to the development of the far-field vector potential functions given in Chapter 1 for a continuous line source of small cross section S stretching along the Z-axis from $-a$ to $+a$, Equations 1.101 and 1.102 take the forms

$$\mathcal{A}_\theta(\theta) = \int_{-a}^{a} S[\cos\theta\cos\phi\, J_x(\zeta) + \cos\theta\sin\phi\, J_y(\zeta) - \sin\theta\, J_z(\zeta)]e^{jk\mathcal{L}}\, d\zeta \tag{5.53}$$

$$\mathcal{A}_\phi(\theta) = \int_{-a}^{a} S[-\sin\phi\, J_x(\zeta) + \cos\phi\, J_y(\zeta)]e^{jk\mathcal{L}}\, d\zeta \tag{5.54}$$

where $\mathcal{L} = \zeta\cos\theta$.

If the *direction* of the current density is the same in every aperture element $d\zeta$, that is, if

$$J(\zeta) = (\mathbf{1}_x C_1 + \mathbf{1}_y C_2 + \mathbf{1}_z C_3)\, g(\zeta) \tag{5.55}$$

with C_1, C_2, and C_3 constants, then (5.53) and (5.54) become

$$\mathcal{A}_\theta(\theta) = (C_1\cos\theta\cos\phi + C_2\cos\theta\sin\phi - C_3\sin\theta)S\int_{-a}^{a} g(\zeta)e^{jk\zeta\cos\theta}\, d\zeta \tag{5.56}$$

$$\mathcal{A}_\phi(\theta) = (-C_1\sin\phi + C_2\cos\phi)S\int_{-a}^{a} g(\zeta)e^{jk\zeta\cos\theta}\, d\zeta \tag{5.57}$$

The factors in front of the integrals in (5.56) and (5.57) are called the *element factors* for $\mathcal{A}_\theta$ and $\mathcal{A}_\phi$, The integrals, which are seen to be common, give the array factor for the line source. This partitioning exactly parallels what has already been observed for discrete linear arrays.

Were one to deal with a continuous line source being represented by magnetic currents, the foregoing could be repeated to the point of partitioning $\mathcal{F}_\theta$ and $\mathcal{F}_\phi$ into element factors and a common array factor, the latter being an integral identical in form to the integral found in (5.56) and (5.57). For this reason, Taylor chose to start his analysis by considering the general array factor

$$\mathcal{S}(\theta) = \int_{-a}^{a} g(\zeta)e^{jk\zeta\cos\theta}\, d\zeta \tag{5.58}$$

Equation 5.58 indicates that the synthesis problem is one of finding the aperture distribution $g(\zeta)$, given the desired pattern $\mathcal{S}(\theta)$. But before proceeding to the specific class of patterns $\mathcal{S}(\theta)$ treated by Taylor, it is instructive to consider a special and idealized analysis problem, namely, given that $g(\zeta)$ has uniform amplitude, uniform progressive phase, what pattern results?

If $g(\zeta) = Ke^{-j\beta\zeta}$ with K and β constants, the integration of (5.58) is extremely simple, and gives

$$\mathcal{S}(\theta) = 2Ka\frac{\sin[ka(\cos\theta - \beta/k)]}{ka(\cos\theta - \beta/k)} \tag{5.59}$$

a result which should be compared with (4.27), a formula that was obtained for a uniformly excited *discrete* array. Since $L = 2a$, one can see that these two pattern expressions are identical (as they should be) in the limit when $d/\lambda \longrightarrow 0$.

With the substitution

$$u = \frac{2a}{\lambda}\left(\cos\theta - \frac{\beta}{k}\right) \tag{5.60}$$

a universal power pattern can be constructed from (5.59) by the definition

$$f(u) = 20\log_{10}\left|\frac{\mathcal{S}(\theta)}{\mathcal{S}(\theta_0)}\right| = 20\log_{10}\left|\frac{\sin\pi u}{\pi u}\right| \tag{5.61}$$

in which $\theta_0 = \arccos(\beta/k)$ is the pointing angle of the main beam. A decibel plot of $f(u)$ can be found in Figure 5.6. One sees a sum pattern with symmetrical side lobes

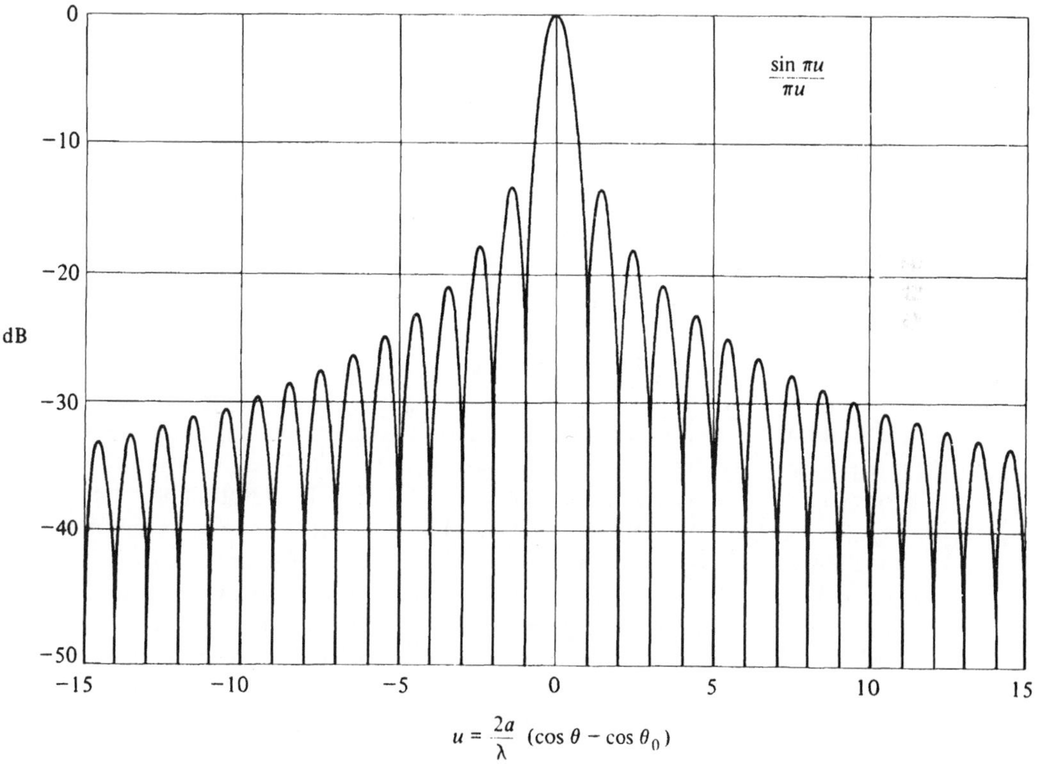

Fig. 5.6 Sum Pattern for Continuous Line Source with Uniform Excitation

whose field heights trail off as u^{-1}, the pair of side lobes closest in being down 13.5 dB. As θ moves in real space from 0 to θ_0 to π, (5.60) indicates that u moves to the left in Figure 5.6 from $(2a/\lambda)(1 - \beta/k)$ to 0 to $-(2a/\lambda)(1 + \beta/k)$. The number of side lobes in the "visible" range of u thus depends on the aperture length $2a/\lambda$. All of this is consistent with what has been learned earlier about discrete linear arrays.

What Taylor sought can be appreciated from a study of Figure 5.6. Suppose, for example, one could find a way to depress the nine innermost side lobes on each side of the main beam to a common height of -30 dB, meanwhile leaving alone all the side lobes that are further out. Clearly, this would give a satisfactory design, particularly if the further out side lobes were all in the "invisible" range of u.

An approximation to such patterns can be constructed in the following way: Select an integer $\bar{n}$ and say that for $|u| \geq \bar{n}$, the nulls of the new pattern are to occur at integral values of u, just as in Figure 5.6. But the next pair of nulls in toward the main beam will need to occur at $u = \pm u_{\bar{n}-1}$, where $|u_{\bar{n}-1}| > \bar{n} - 1$, in order to depress the intervening side lobes somewhat. Similarly, the penultimate pair of nulls needs to be shifted to $u = \pm u_{\bar{n}-2}$, where $|u_{\bar{n}-2}| > \bar{n} - 2$, and so on. The function that expresses this new pattern is

$$\mathcal{S}(u) = \frac{\sin \pi u}{\pi u} \frac{\prod_{n=1}^{\bar{n}-1} (1 - u^2/u_n^2)}{\prod_{n=1}^{\bar{n}-1} (1 - u^2/n^2)} \tag{5.62}$$

which can be seen to remove the innermost $\bar{n} - 1$ pairs of nulls from the original $\sin \pi u/\pi u$ pattern and replace them with new pairs at modified positions $\pm u_n$.

Taylor found that the new null positions should be determined from the formula

$$u_n = \bar{n}\left[\frac{A^2 + (n - \frac{1}{2})^2}{A^2 + (\bar{n} - \frac{1}{2})^2}\right]^{1/2} \tag{5.63}$$

with A a measure of the side lobe level (SLL) in that $\cosh \pi A = b$, with $20 \log_{10} b =$ SLL.

An example of a Taylor pattern is shown in Figure 5.7, with $\bar{n} = 6$ and the design side lobe level -20 dB. This plot exhibits the characteristic features of all Taylor patterns. One can observe that, for $|u| = \bar{n}$ and beyond, the nulls occur at the integers, and that the far-out side lobes decay in field value as u^{-1}. The close-in side lobes are not precisely at the design level -20 dB. The closest in pair are slightly below it, the next pair out are a bit lower, and so on, so that there is a slight droop to the envelope of the near-in side lobes. However, one finds that this droop is less if $\bar{n}$ is selected to be a larger number, and Taylor has shown that the beam broadening associated with this droop is negligible in practical circumstances.

With a Taylor pattern defined by (5.62) and (5.63), it becomes a simple matter to find the corresponding aperture distribution from (5.58). If one lets $g(\zeta) = h(\zeta)e^{-j\beta\zeta}$, with $h(\zeta)$ represented by the Fourier series

$$h(\zeta) = \sum_{m=0}^{\infty} B_m \cos \frac{m\pi\zeta}{a} \tag{5.64}$$

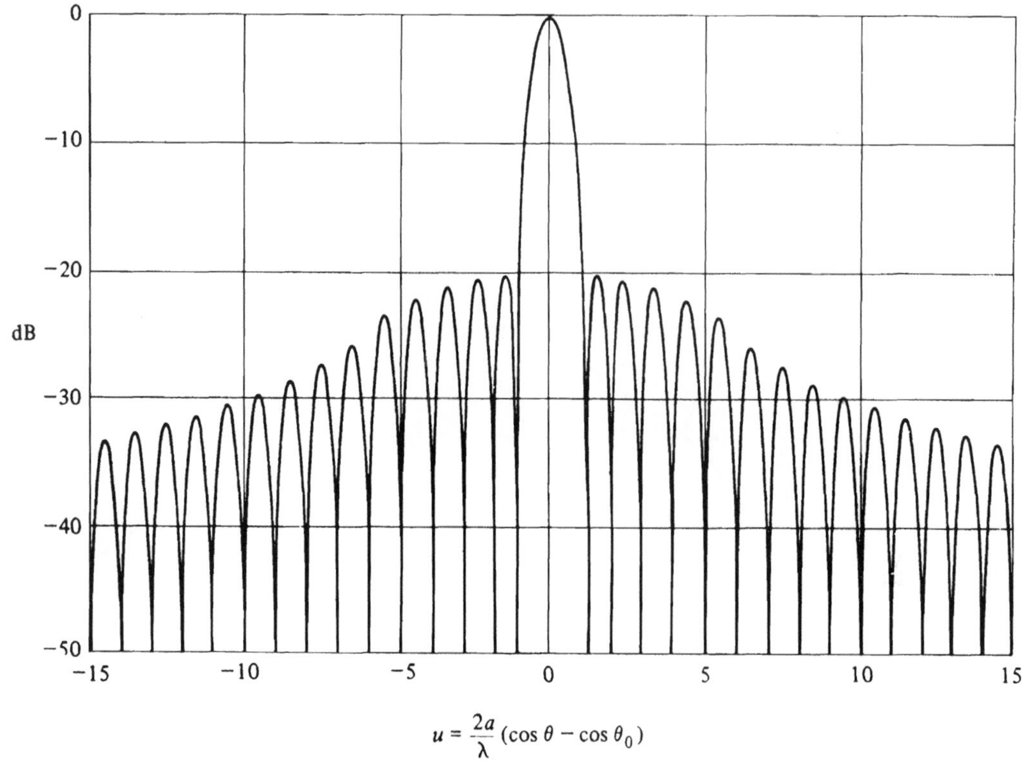

Fig. 5.7 Taylor Sum Pattern for Continuous Line Source, $\bar{n} = 6$, −20 dB SLL

then substitution in (5.58) gives

$$\mathcal{S}(u) = \sum_{m=0}^{\infty} B_m \int_{-a}^{a} \cos\frac{m\pi\zeta}{a}\, e^{ju\pi\zeta/a}\, d\zeta \tag{5.65}$$

The odd part of the integrand of (5.65) can be discarded, which leaves

$$\mathcal{S}(u) = \sum_{m=0}^{\infty} B_m \int_{-a}^{a} \cos\frac{m\pi\zeta}{a} \cos\frac{u\pi\zeta}{a}\, d\zeta \tag{5.66}$$

If u is an integer, the integral in (5.66) is zero unless $m = u$; as a consequence,

$$2aB_0 = \mathcal{S}(0), \qquad aB_m = \mathcal{S}(m), \qquad m = 1, 2, \ldots \tag{5.67}$$

However, (5.62) indicates that $\mathcal{S}(m) = 0$, $m \geq \bar{n}$, so this Fourier series truncates, and thus the continuous aperture distribution is given by

$$g(\zeta) = \frac{e^{-j\beta\zeta}}{2a}\left[\mathcal{S}(0) + 2\sum_{m=1}^{\bar{n}-1} \mathcal{S}(m) \cos\frac{m\pi\zeta}{a}\right] \tag{5.68}$$

The aperture distribution corresponding to the Taylor pattern of Figure 5.7 and computed from (5.68) is shown by the solid curve in Figure 5.8. For comparison, a bar graph is superimposed showing the Dolph excitation for a 19-element discrete array and the same design side lobe level. This reveals a common finding. The Dolph distribution typically has a wider swing near the aperture ends than does the Taylor. This can be traced to the requirement that all the side lobes in the Dolph pattern are at a common height, whereas in the Taylor pattern they droop somewhat. The Taylor distribution (discretized) is physically easier to achieve, which adds to its attractiveness.

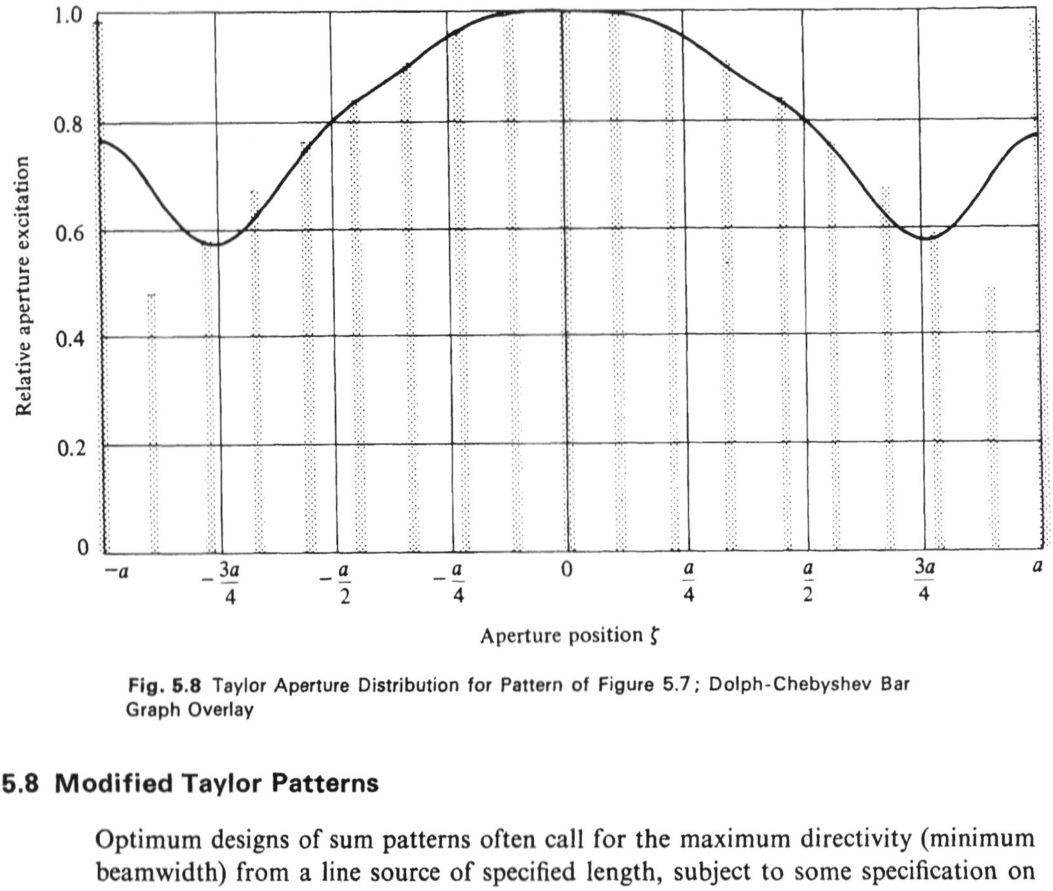

Fig. 5.8 Taylor Aperture Distribution for Pattern of Figure 5.7; Dolph-Chebyshev Bar Graph Overlay

5.8 Modified Taylor Patterns

Optimum designs of sum patterns often call for the maximum directivity (minimum beamwidth) from a line source of specified length, subject to some specification on the side lobe level. However, all directions in space may not be equally important insofar as side lobe suppression is concerned. Since every side lobe that is suppressed costs something in beam broadening, Taylor patterns (which arise when equal importance is attached to all directions) may not be optimum in some applications.[11] One

[11]These remarks are equally applicable to Dolph-Chebyshev patterns, and the development to be presented here will be extended to the discrete array case in Section 5.10.

is led to consider designs which permit high side lobes in unimportant regions, while maintaining low side lobes in critical regions, that is, patterns with *arbitrary* side lobe topography.

A perturbation procedure may be used to modify a Taylor pattern so that all the side lobes have individually arbitrary heights.[12] It is best to begin by expressing the Taylor pattern of (5.62) in the more general form

$$\mathcal{S}_0(u) = C_0 \frac{\sin \pi u}{\pi u} \frac{\prod'^{\bar{n}_R - 1}_{n=-(\bar{n}_L - 1)} (1 - u/\overset{0}{u}_n)}{\prod'^{\bar{n}_R - 1}_{n=-(\bar{n}_L - 1)} (1 - u/n)} \tag{5.69}$$

The subscripts R and L in (5.69) are used to identify the right side and the left side of the pattern. As generalizations of (5.63), the root positions are given by

$$\overset{0}{u}_n = \bar{n}_R \left[\frac{A_R^2 + (n - \frac{1}{2})^2}{A_R^2 + (\bar{n}_R - \frac{1}{2})^2} \right]^{1/2} \qquad n = 1, 2, \ldots, \bar{n}_R - 1 \tag{5.70}$$

$$\overset{0}{u}_n = -\bar{n}_L \left[\frac{A_L^2 + (n - \frac{1}{2})^2}{A_L^2 + (\bar{n}_L - \frac{1}{2})^2} \right]^{1/2} \qquad n = -1, -2, \ldots, -(n_L - 1) \tag{5.71}$$

In (5.69) through (5.71), $\bar{n}_R$ and $\bar{n}_L$ are positive integers that denote the transition roots on the two sides of the main beam (there is a root at $-\bar{n}_L$ and at each integer less than $-\bar{n}_L$; there is a root at $\bar{n}_R$ and at each integer greater than $\bar{n}_R$). The side lobe level parameters on the two sides of the pattern are A_R and A_L. The prime on each product sign in (5.69) indicates that the factor for which $n = 0$ has been excluded. C_0 is a constant.

It is readily seen that if $\bar{n}_R = \bar{n}_L = \bar{n}$ and $\bar{A}_R = \bar{A}_L = A$, Expression 5.69 reduces to the standard Taylor form of (5.62). The advantage of (5.69) is that it permits the two sides of the Taylor pattern to be treated separately, since A_R need not equal A_L, and $\bar{n}_R$ need not equal $\bar{n}_L$.

In what is to follow in Section 5.9 the root positions $\overset{0}{u}_n$ will be perturbed in order to modify individual side lobe heights. But before passing on to that development, it is interesting to pause and observe that useful patterns can be generated merely by choosing the parameters on the two sides of the main beam to be different. For example, if $\bar{n}_R = 8$, $\bar{n}_L = 3$, $A_R^2 = 1.29177$, and $A_L^2 = 0.58950$, the left side corresponds to a 15 dB Taylor and the right side to a 25 dB Taylor. This modified 15/25 Taylor, computed from (5.69), is shown in Figure 5.9. It can be seen that some averaging has taken place (the innermost side lobe on the left side, at -16.7 dB, is lower than its counterpart in the symmetrical Taylor 15/15; the innermost side lobe on the right side, at -24.3 dB, is higher than its counterpart in the symmetrical Taylor 25/25). However, this effect is systematic, and one could achieve 15/25 by designing for 13.5/26.

The asymmetry of the side lobe structure in Figure 5.9 caused a small shift in

[12] R. S. Elliott, "Design of Line-Source Antennas for Sum Patterns with Sidelobes of Individually Arbitrary Heights", *IEEE Trans. Antennas Propagat.*, 24 (1976), 76–83.

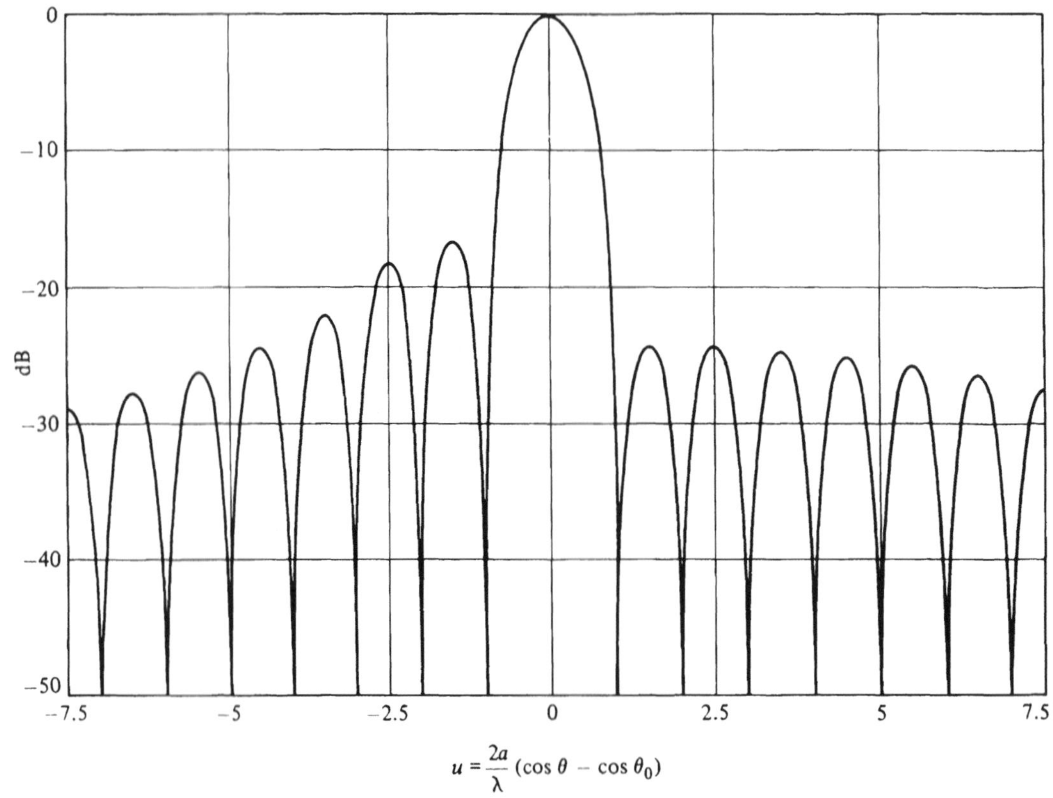

Fig. 5.9 An Asymmetric Taylor Sum Pattern (© 1975 IEEE. Reprinted from *IEEE AP Transactions*, pp. 100–107, 1975.)

the position of the main beam. This can be compensated easily by a slight change in the uniform progressive phase of the distribution.

Patterns of the type shown in Figure 5.9 are useful in applications where side lobe reduction is important on one side of the main beam but not the other. Specifically, this pattern has the advantage that the beamwidth is narrower and the directivity is higher than one finds in a symmetrical Taylor 25/25.

The aperture distributions corresponding to these modified Taylor patterns can be expressed in the Fourier form

$$g(\zeta) = h(\zeta)e^{-j\beta\zeta} = e^{-j\beta\zeta} \sum_{m=-\infty}^{\infty} B_m e^{-jm\pi\zeta/a} \tag{5.72}$$

Substitution in (5.58) gives

$$\mathcal{S}_0(\theta) = \sum_{m=-\infty}^{\infty} B_m \int_{-a}^{a} e^{-jm\pi\zeta/a} e^{jk\zeta(\cos\theta - \beta/k)}\, d\zeta$$

$$\mathcal{S}_0(u) = \frac{a}{\pi} \sum_{m=-\infty}^{\infty} B_m \int_{-\pi}^{\pi} e^{-jmp} e^{jup}\, dp \tag{5.73}$$

in which

$$p = \frac{\pi\zeta}{a} \tag{5.74}$$

Integration of (5.73) gives

$$\mathcal{S}_0(m) = 2aB_m$$

and thus

$$h(\zeta) = \frac{1}{2a} \sum_{m=-(\bar{n}_L-1)}^{\bar{n}_R-1} \mathcal{S}_0(m) e^{-jm\pi\zeta/a} \tag{5.75}$$

A plot of the distribution $h(\zeta)$ needed to produce the pattern of Figure 5.9 is shown in Figure 5.10. The amplitude distribution of the Taylor 20/20 is shown for comparison. It is well within the state of the art to achieve a discretization of either one of these distributions.

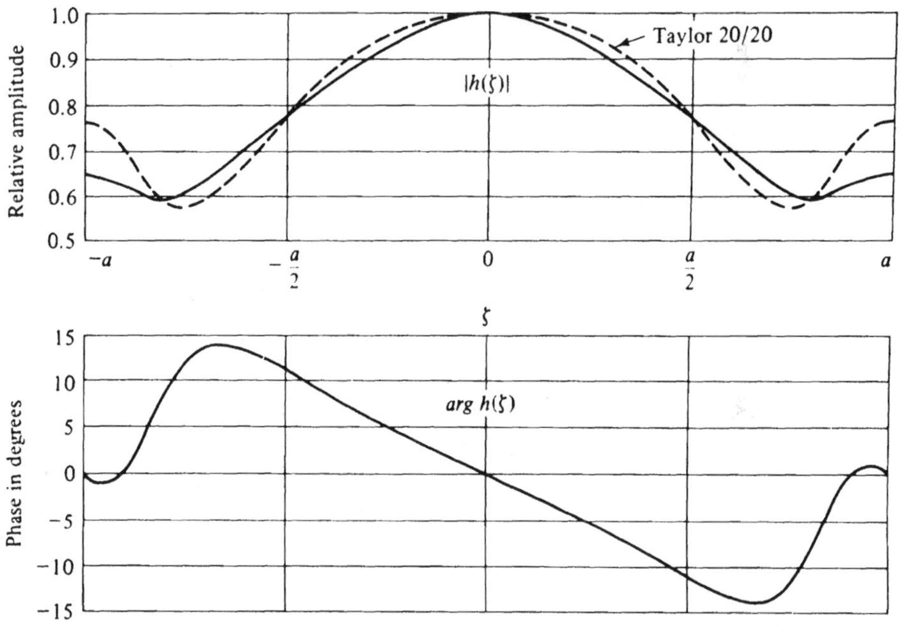

Fig. 5.10 The Aperture Distribution for the Pattern of Fig. 5.9 (© 1975 IEEE. Reprinted from *IEEE AP Transactions*, pp. 100–107, 1975.)

5.9 Sum Patterns with Arbitrary Side Lobe Topography

Imagine that a sum pattern $\mathcal{S}(u)$ has been prescribed in which the height of every side lobe is individually specified. $\mathcal{S}(u)$ can be expressed in the form of (5.69), that is,

$$\mathcal{S}(u) = Cf(u) \prod_{n=-(\bar{n}_L-1)}^{\bar{n}_R-1}{}' \left(1 - \frac{u}{u_n}\right) \tag{5.76}$$

with

$$f(u) = \frac{\sin \pi u/\pi u}{\prod'_{n=-(\bar{n}_L-1)}^{\bar{n}_R-1}\left(1 - \frac{u}{n}\right)} \tag{5.77}$$

Let a starting pattern $\mathcal{S}_0(u)$ be selected such that the side lobe levels on the two sides of its main beam separately approximate the *average* side lobe levels on the two sides of the main beam in the desired pattern. Then all the roots $\overset{0}{u}_n$ in the starting pattern are *known*, being calculable from (5.70) and (5.71).

Assume that the roots in the desired pattern are given by

$$u_n = \overset{0}{u}_n + \delta u_n \tag{5.78}$$

with the perturbations δu_n small. Then if

$$C = C_0 + \delta C \tag{5.79}$$

the desired pattern becomes

$$\mathcal{S}(u) = (C_0 + \delta C)f(u)\prod'_{n=-(\bar{n}_L-1)}^{\bar{n}_R-1}\left(1 - \frac{u}{\overset{0}{u}_n + \delta u_n}\right) \tag{5.80}$$

But

$$\begin{aligned}
1 - \left(\frac{u}{\overset{0}{u}_n + \delta u_n}\right) &= 1 - u(\overset{0}{u}_n + \delta u_n)^{-1} \\
&= 1 - u\left(\frac{1}{\overset{0}{u}_n} - \frac{\delta u_n}{\overset{0}{u}{}_n^2} + \frac{\delta u_n^2}{\overset{0}{u}{}_n^3} - \cdots\right) \\
&= 1 - \frac{u}{\overset{0}{u}_n} + \frac{u\,\delta u_n}{\overset{0}{u}{}_n^2} - \frac{u\,\delta u_n^2}{\overset{0}{u}{}_n^3} + \cdots \\
&= \left(1 - \frac{u}{\overset{0}{u}_n}\right)\left(1 + \frac{(u/\overset{0}{u}{}_n^2)}{1 - u/\overset{0}{u}_n}\delta u_n - \frac{(u/\overset{0}{u}{}_n^3)}{1 - u/\overset{0}{u}_n}(\delta u_n)^2 + \cdots\right)
\end{aligned} \tag{5.81}$$

Therefore

$$\mathcal{S}(u) = (C_0 + \delta C)f(u)\prod'_{n=-(\bar{n}_L-1)}^{\bar{n}_R-1}\left(1 - \frac{u}{\overset{0}{u}_n}\right)\left(1 + \frac{(u/\overset{0}{u}{}_n^2)}{1 - u/\overset{0}{u}_n}\delta u_n - \cdots\right) \tag{5.82}$$

and, to *first order*,

$$\mathcal{S}(u) = \mathcal{S}_0(u) + \frac{\delta C}{C_0}\mathcal{S}_0(u) + \mathcal{S}_0(u)\sum'_{n=-(\bar{n}_L-1)}^{\bar{n}_R-1}\frac{(u/\overset{0}{u}{}_n^2)}{1 - u/\overset{0}{u}_n}\delta u_n$$

which can be put in the useful form

$$\frac{\mathcal{S}(u)}{\mathcal{S}_0(u)} - 1 = \frac{\delta C}{C_0} + \sum'_{n=-(\bar{n}_L-1)}^{\bar{n}_R-1}\frac{(u/\overset{0}{u}{}_n^2)}{1 - u/\overset{0}{u}_n}\delta u_n \tag{5.83}$$

Including the main beam and all side lobes, there are $\bar{n}_R + \bar{n}_L - 1$ lobes between the anchored roots at $-\bar{n}_L$ and $+\bar{n}_R$. Let the positions of the peaks of these lobes in the starting pattern be designated by u_m^p, $m = -(\bar{n}_L - 1), \ldots, -1, 0, 1, \ldots, (\bar{n}_R - 1)$. Insertion in (5.83) gives

$$\frac{\mathcal{S}(u_m^p)}{\mathcal{S}_0(u_m^p)} - 1 = \frac{\delta C}{C_0} + \sum_{n=-(\bar{n}_L-1)}^{\bar{n}_R-1}{}' \frac{(u_m^p/\overset{0}{u}{}_n^2)}{1 - u_m^p/\overset{0}{u}_n} \delta u_n \tag{5.84}$$

If the lobe peak positions in the desired pattern are close to those in the starting pattern, $\mathcal{S}(u_m^p)$ is essentially the height of the mth lobe in the desired pattern, a *known* quantity. Then all the terms appearing in (5.84) are known except the $(\bar{n}_R + \bar{n}_L - 2)$ perturbations δu_n and the unknown $\delta C/C_0$. Since there are $(\bar{n}_R + \bar{n}_L - 1)$ values of m to insert in (5.84), this is a deterministic set of linear equations. Matrix inversion will yield the root perturbations δu_n which, with the aid of (5.78), give the new root positions. When these are used in (5.76), the new pattern may be computed and inspected to see if it is close enough to desired. If not, the process can be repeated with the new pattern used as starting pattern.

Experience shows that, for desired patterns in which the variation in heights of successive side lobes is not extreme, convergence of this process is rapid; usually several iterations are sufficient. For specified patterns of extreme variability, an interim desired pattern might need to be postulated to assure convergence.

As an example of the use of this technique, assume that a symmetrical sum pattern is desired, with the three innermost pairs of side lobes at -40 dB, the next four pairs at -20 dB, and all further outside lobes decaying as $|u|^{-1}$. It is convenient to use as starting pattern a symmetrical 30/30 Taylor with $n_R = n_L = 8$. This pattern is shown in Figure 5.11a. Three iterations yield the result displayed in Figure 5.11b. The expanded range, showing the tailing off of the outer side lobes, can be seen in Figure 5.11c. After the third iteration, all side lobes were within one quarter of a dB of specification.

The aperture distribution which will produce this desired pattern can be determined from (5.75), with $\mathcal{S}(m)$ replacing $\mathcal{S}_0(m)$, and is shown in Figure 5.12. Because the pattern is symmetrical, $h(\zeta)$ is an equiphase distribution.

A second example involves a desired pattern which is Taylor 20/20, $\bar{n} = 8$, except that the innermost three lobes on one side of the main beam are to be at -30 dB. With the unmodified Taylor used as starting pattern, three iterations produce the result shown in Figure 5.13. The corresponding aperture distribution is seen in Figure 5.14. Because of the asymmetry in the pattern, $h(\zeta)$ exhibits an asymmetric phase distribution to go with the symmetric amplitude distribution.[13]

A third example provides more of a challenge for the perturbation procedure. Suppose that the desired pattern has the innermost seven side lobes on one side of

[13] Resolution of (5.75) into its real and imaginary components indicates that, for a sum pattern, the amplitude distribution is *always* symmetrical, whereas the phase distribution is *always* asymmetrical if the pattern is asymmetrical, and is *always* zero if the pattern is symmetrical.

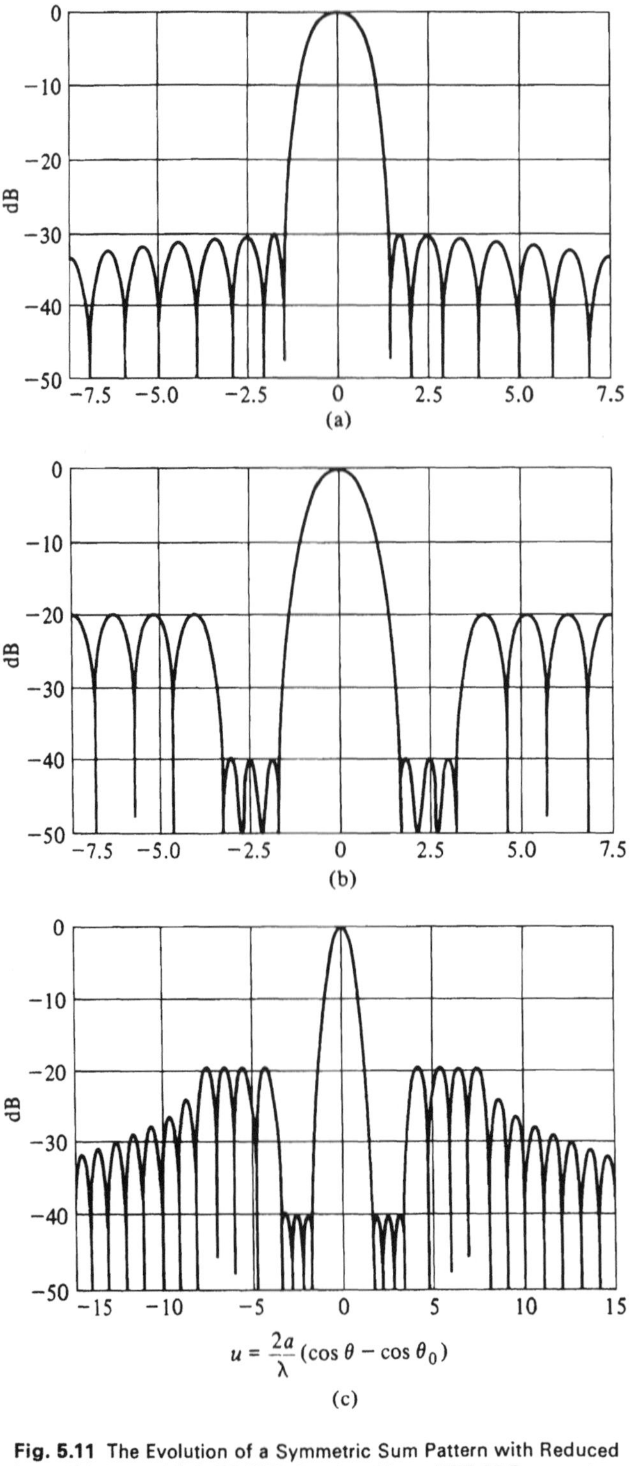

Fig. 5.11 The Evolution of a Symmetric Sum Pattern with Reduced Inner Side Lobes (© 1976 IEEE. Reprinted from *IEEE AP Transactions*, pp. 76–83, 1976.)

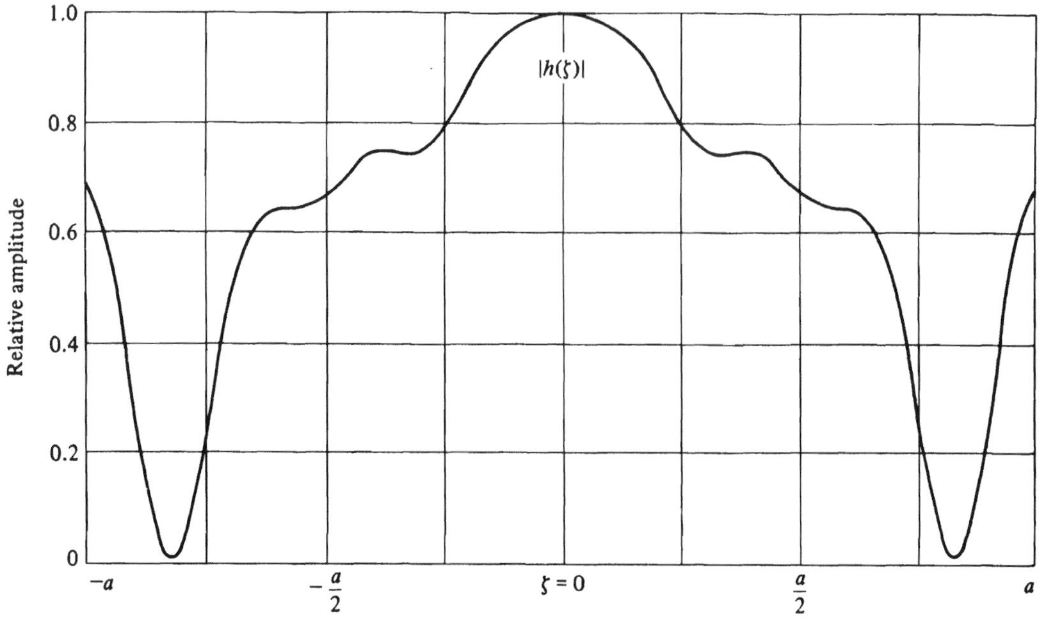

Fig. 5.12 The Aperture Distribution for the Pattern of Fig. 5.11b (© 1976 IEEE. Reprinted from *IEEE AP Transactions*, pp. 76–83, 1976.)

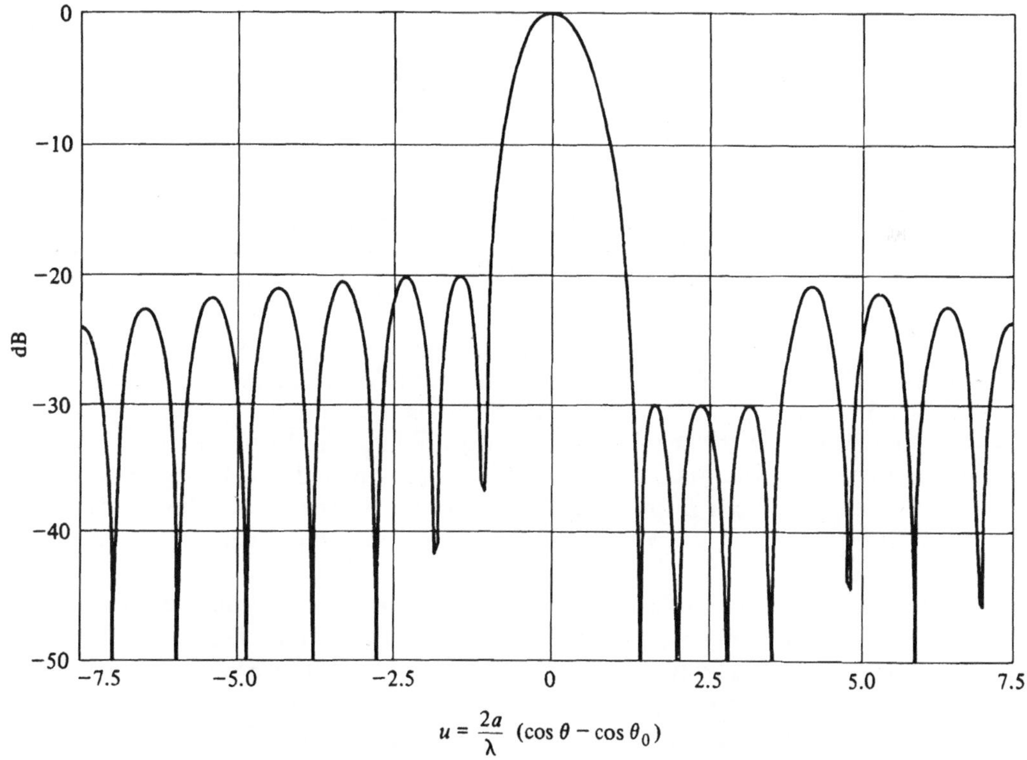

Fig. 5.13 A Taylor 20/20 Sum Pattern Modified to Contain Three Depressed Lobes (© 1977 IEEE. Reprinted from *IEEE AP Transactions*, pp. 617–621, 1977.)

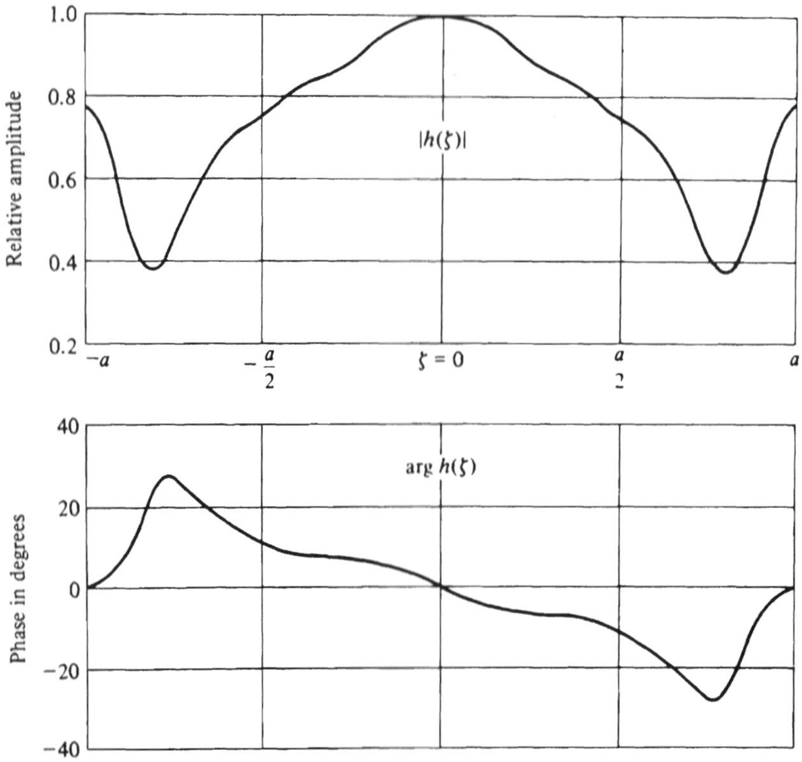

Fig. 5.14 The Aperture Distribution for the Pattern of Fig. 5.13 (© 1977 IEEE. Reprinted from *IEEE AP Transactions*, pp. 617–621, 1977.)

the main beam at a common height of −25 dB, whereas the innermost seven lobes on the other side are cascaded in 5 dB steps, the closest-in at −45 dB, the next at −40 dB, and the last at −15 dB. All further out side lobes on both sides are to decay as $|u|^{-1}$.

Though it is not quite optimum, let the pattern of Figure 5.9 serve as starting pattern. It is found that six iterations are needed to bring all side lobes within one quarter of a dB of specification. The patterns resulting from each iteration are shown in Figure 5.15 and the requisite aperture distribution in Figure 5.16. It can be observed that in this case also, because of the asymmetry in the pattern, there is an asymmetric phase distribution coupled to the symmetric amplitude distribution. Both display considerable fine structure because of the severity in the side lobe topography.

It is interesting to observe that this perturbation procedure is capable of achieving Taylor's original goal—to find a continuum equivalent to the Dolph-Chebyshev discrete excitation—a distribution that will produce a pattern with all side lobes in real space at a common specified height. However, this is a point of academic interest only. The Taylor pattern suffers an inconsequential loss in beamwidth and

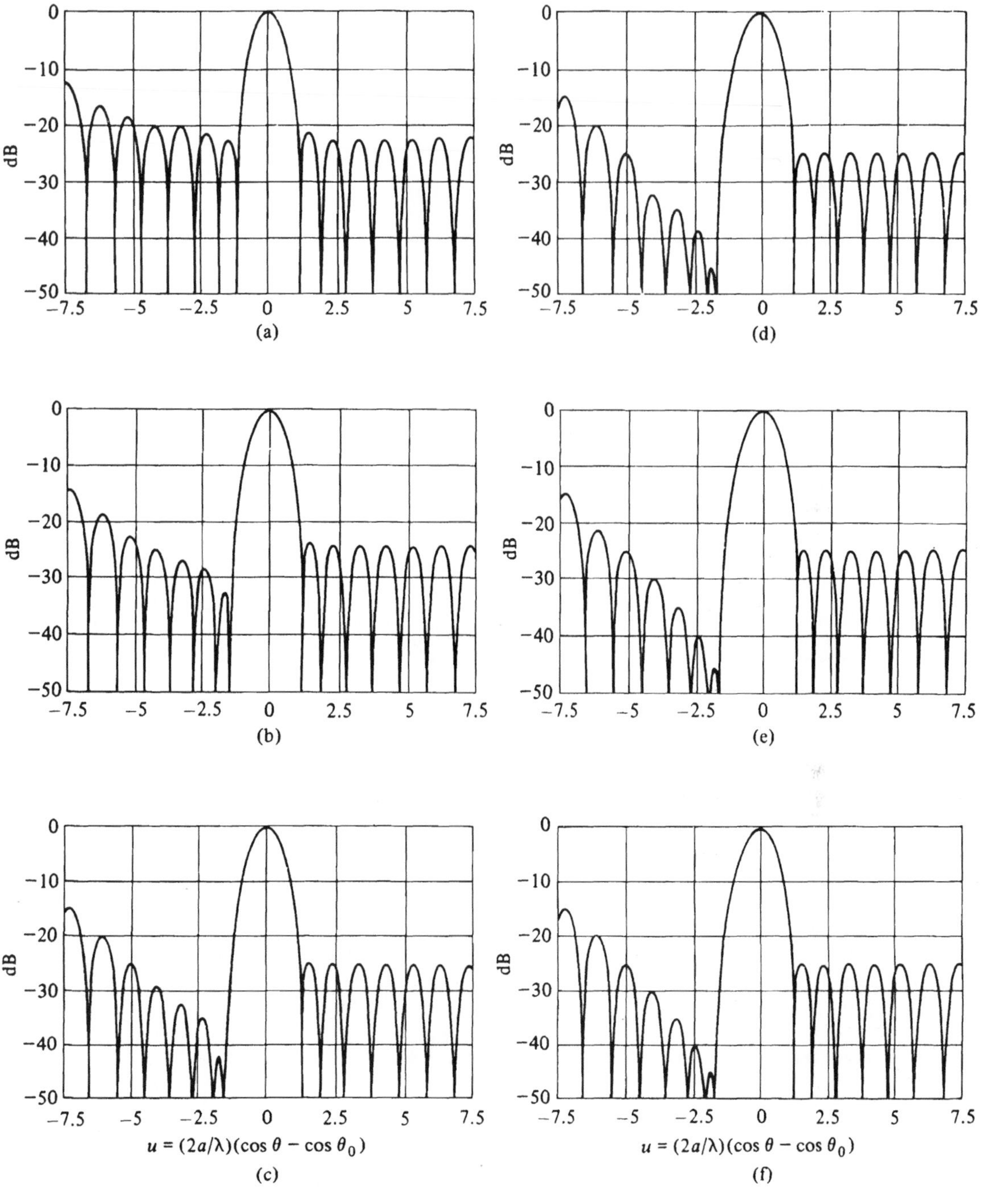

Fig. 5.15 The Evolution of a Sum Pattern with a Uniform/Cascaded Side Lobe Structure; Six Successive Iterations

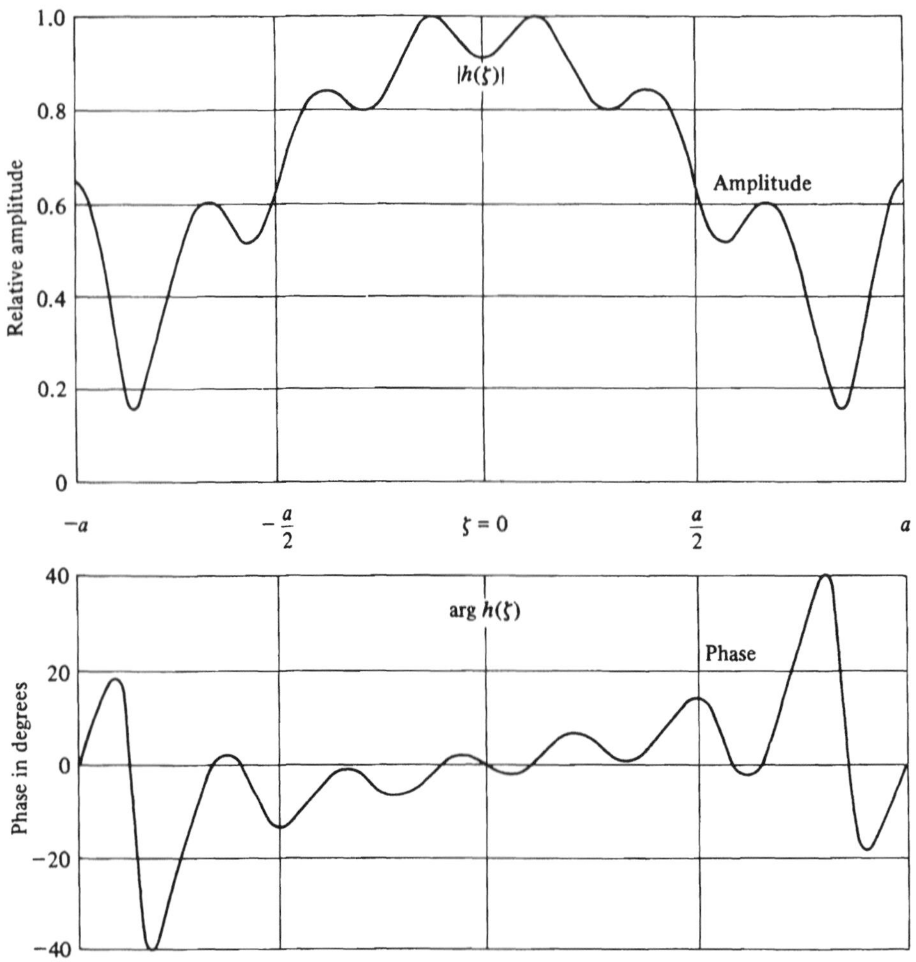

Fig. 5.16 The Aperture Distribution for the Pattern of Fig. 5.15f

directivity due to the slight droop in near-in side lobes, and requires an aperture distribution that is easier to achieve than the continuous distribution which would produce uniform side lobes.

5.10 Discretization of a Continuous Line Source Distribution

Continuous line source distributions, such as those shown in the even-numbered Figures 5.8 through 5.16 can be sampled at $N + 1$ equispaced values of ζ to determine the excitation of a linear array consisting of $N + 1$ equispaced elements. Obviously, if $N + 1$ is large enough, the sampling interval will be so small that all the fine detail in the continuous aperture distribution will be captured. Under these circumstances, the pattern from the discrete array will differ but little from the pattern due to the continuous aperture distribution. However, in many practical applications, $N + 1$ will be small enough that the sampling results in an excitation which gives a badly degraded pattern. It is possible to circumvent this difficulty by working directly with the desired pattern, rather than its continuous aperture distribution.

Suppose, as an example, one wishes to produce the pattern shown in Figure 5.13 with a linear array of $N + 1$ elements spaced $\lambda/2$ on centers. The array length is $2a = (N + 1)\lambda/2$. The full excursion of u is $(4a/\lambda) = N + 1$. Thus there are N nulls in the visible range of Figure 5.13, which exactly matches the number of roots which can be placed on a Schelkunoff unit circle for an array of $N + 1$ elements.

Since $w = e^{j\psi}$, with $\psi = kd(\cos\theta - \cos\theta_0) = \pi(\cos\theta - \cos\theta_0)$, and since $u = (2a/\lambda)(\cos\theta - \cos\theta_0)$, it follows that

$$\psi = \frac{\pi\lambda}{2a}u = \frac{2\pi u}{N+1} \tag{5.85}$$

Therefore, if u_n is a null in the pattern, then

$$w_n = e^{j2\pi u_n/(N+1)} \tag{5.86}$$

is the corresponding root on the unit circle.

The nulls u_n for the pattern of Figure 5.13 are known to good precision. (The pattern was computed from Equation 5.76 after the null positions were found using the perturbation procedure). Thus it is a simple matter to calculate the roots w_n from (5.86). Once these w_n roots are known they can be placed in (4.38). When the factors are multiplied out, the discrete current distribution is determined.

If another spacing than $\lambda/2$ is used, this does not affect the root placement, nor $f(w)$, nor the current distribution, only the extent of the w-excursion on the unit circle and the number of side lobes in visible space. Thus Equation 5.86 can be used to determine the root placement regardless of the element spacing.

As a specific example of this procedure, suppose a 19-element equispaced linear array is to be excited so as to produce the pattern of Figure 5.13. Corresponding values of u_n and w_n are listed in Table 5.1 and the normalized current distribution,

TABLE 5.1 Null positions for pattern of Figure 5.13

n	u_n	w_n	n	u_n	w_n
−9	−9.000	−0.986 − j.165	1	1.459	0.886 + j.464
−8	−8.000	−0.879 − j.476	2	2.021	0.785 + j.620
−7	−6.940	−0.663 − j.749	3	2.803	0.600 + j.800
−6	−5.888	−0.367 − j.930	4	3.561	0.383 + j.924
−5	−4.842	−0.030 − j.999	5	4.776	−0.009 + j.999
−4	−3.805	0.307 − j.952	6	5.857	−0.358 + j.934
−3	−2.789	0.604 − j.797	7	6.926	−0.659 + j.752
−2	−1.827	0.823 − j.568	8	8.000	−0.879 + j.476
−1	−1.071	0.938 − j.347	9	9.000	−0.986 + j.165

TABLE 5.2 Discretization of Figure 5.14

n	I_n Root Matching	I_n Conventional Sampling	n	I_n Root Matching	I_n Conventional Sampling
0	$1.000\,\underline{\lvert 0^\circ}$	$1.000\,\underline{\lvert 0^\circ}$	± 5	$0.730\,\underline{\lvert \pm 12.50^\circ}$	$0.714\,\underline{\lvert \pm 13.90^\circ}$
± 1	$0.986\,\underline{\lvert \pm 3.92^\circ}$	$0.981\,\underline{\lvert \pm 4.04^\circ}$	± 6	$0.673\,\underline{\lvert \pm 18.31^\circ}$	$0.596\,\underline{\lvert \pm 20.96^\circ}$
± 2	$0.914\,\underline{\lvert \pm 6.58^\circ}$	$0.906\,\underline{\lvert \pm 6.60^\circ}$	± 7	$0.448\,\underline{\lvert \pm 28.12^\circ}$	$0.387\,\underline{\lvert \pm 27.99^\circ}$
± 3	$0.852\,\underline{\lvert \pm 7.32^\circ}$	$0.847\,\underline{\lvert \pm 7.43^\circ}$	± 8	$0.373\,\underline{\lvert \pm 19.37^\circ}$	$0.518\,\underline{\lvert \pm\ 8.90^\circ}$
± 4	$0.804\,\underline{\lvert \pm 8.59^\circ}$	$0.787\,\underline{\lvert \pm 9.31^\circ}$	± 9	$0.727\,\underline{\lvert \pm\ 2.29^\circ}$	$0.782\,\underline{\lvert 0^\circ}$

found by expanding $\pi(w - w_n)$, is given in the second column of Table 5.2. For comparison, the current distribution found by conventional sampling of the continuous aperture distribution of Figure 5.14 is shown in the third column of Table 5.2. Significant differences can be noted between the two excitations, particularly in the outer elements.

The patterns produced by both current distributions listed in Table 5.2 have been computed using Equation 4.14 and are shown in Figure 5.17 for an element spacing of 0.7λ. The pattern due to conventional sampling of the continuous aperture distribution is seen to be degraded to an unacceptable level. Only one of the three innermost side lobes is depressed, and it is depressed too far, whereas the remainder of the side lobe structure does not stay below -20 dB. In contrast, the pattern resulting from root matching is an excellent approximation to the desired pattern. (The rise in the outer side lobes of both patterns in Figure 5.17 is due to the fact that they are repeats of closer-in side lobes. With $d/\lambda = 0.7$, the w-excursion is 1.4 revolutions around the Schelkunoff unit circle).

An experimental test of the current distribution obtained by root matching, and listed in Table 5.2, will be described in Chapter 8 in conjunction with the design of a 19-element waveguide-fed slot array.

This discretizing technique can be applied equally well to an unmodified Taylor pattern. For example, if a 19-element array is required to produce the Taylor 20/20, $\bar{n} = 6$ pattern of Figure 5.7, with the main beam at broadside, the 18 nulls in u-space can be computed from (5.63). It is found that

$$u_n = \pm 1.15659,\ \pm 1.91011,\ \pm 2.87579,\ \pm 3.89905,\ \pm 4.94428,\ \pm 6,\ \pm 7,\ \pm 8,\ \pm 9.$$

Thus the roots w_n occur in complex conjugate pairs at the angular positions

$$\psi_n = \pm 21.914^\circ,\ \pm 36.192^\circ,\ \pm 54.489^\circ,\ \pm 73.877^\circ,\ \pm 93.681^\circ,\ \pm 113.684^\circ,\ \pm 132.632^\circ,\ \pm 151.579^\circ,\ \pm 170.526^\circ$$

Except for a multiplicative constant, the pattern is given by

$$f(w) = \prod_{n=1}^{9} (w^2 - 2w\cos\psi_n + 1)$$

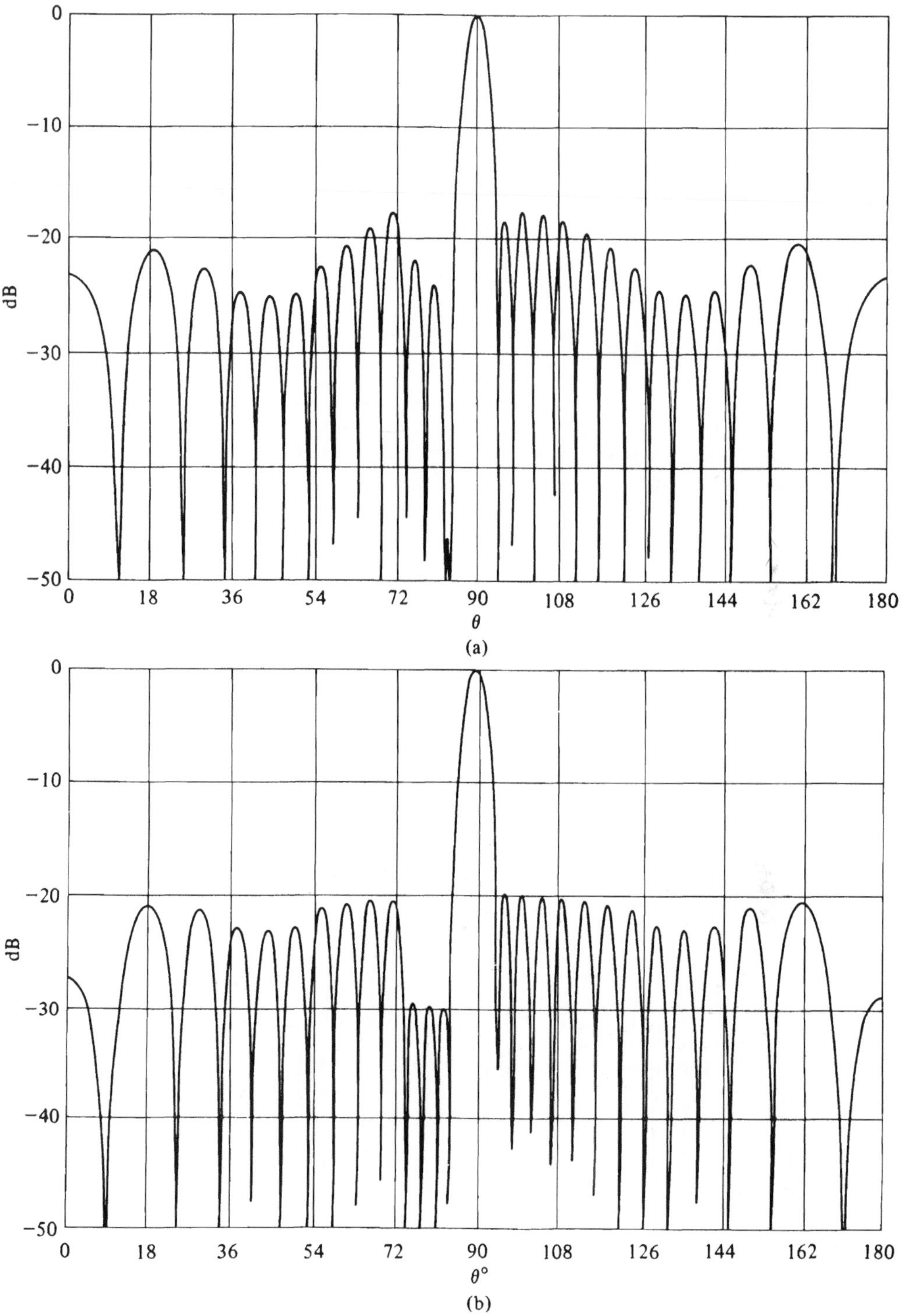

Fig. 5.17 The Sum Pattern of a 19-Element Linear Array, $d = 0.7\lambda$; Design Goal: Achieve Pattern of Fig. 5.13; Excitation: in (a), Conventional Sampling; in (b), Root Matching (© 1977 IEEE. Reprinted from *IEEE AP Transactions*, pp. 617–621, 1977.)

which, when multiplied out, gives the equiphase current distribution shown in Table 5.3.

TABLE 5.3 Current distribution for discretized Taylor 20/20, $\bar{n} = 6$

n	I_n	n	I_n
0	1.000	±5	0.769
±1	0.997	±6	0.649
±2	0.966	±7	0.563
±3	0.904	±8	0.623
±4	0.843	±9	0.749

The pattern corresponding to this discrete current distribution is shown in Figure 5.18 and is seen to be an exellent approximation to Figure 5.7. (Once again, the side lobes begin to repeat because of the element spacing.)

This determination of the discrete excitation by root matching, though superior to conventional sampling, is not sufficient if the side lobe topography becomes too

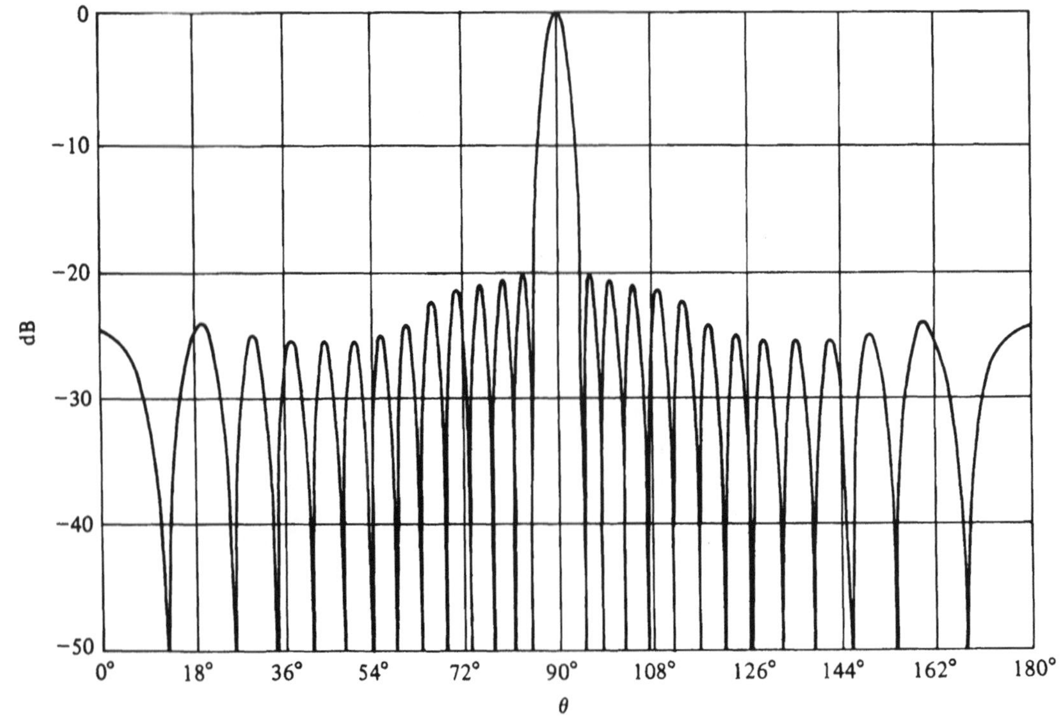

Fig. 5.18 The Sum Pattern of a 19-Element Linear Array, $d = 0.7\lambda$; Design Goal: Achieve Pattern of Fig. 5.7; Excitation Found by Root Matching

severe. To illustrate this, imagine that the pattern of Figure 5.15 is to be produced by a 15-element array with 0.475λ spacing. Conventional sampling of the aperture distribution of Figure 5.16 leads to the badly degraded and unacceptable pattern displayed in Figure 5.19a. When the current distribution is determined by root matching, considerable improvement results, as can be seen in the pattern of Figure 5.19b. However, this pattern is not an adequate approximation to the ideal of Figure 5.15. In such circumstances, a perturbation procedure can be used to improve the approximation to any degree desired.[14]

To see this, assume that—for an equispaced array of $N + 1$ elements—a current distribution $\overset{0}{I}_n$ has been found by root matching and produces the pattern (see Section 4.4)

$$f_0(w) = \sum_{n=0}^{N} \left(\frac{\overset{0}{I}_n}{\overset{0}{I}_N} \right) w^n = \prod_{n=1}^{N} (w - \overset{0}{w}_n) \tag{5.87}$$

and that this pattern is not quite a satisfactory approximation to some ideal that has been specified.

Let (5.87) be called the *starting pattern* and assume that the desired pattern can be expressed in the same form, that is,

$$f(w) = \sum_{n=0}^{N} \left(\frac{I_n}{\overset{0}{I}_N} \right) w^n \tag{5.88}$$

If the starting and desired patterns are not too disparate, one can write

$$I_n = \overset{0}{I}_n + \delta I_n \tag{5.89}$$

and expect that the perturbations δI_n will be small compared to the starting currents $\overset{0}{I}_n$. When (5.89) is used in (5.88), the result is that

$$f(w) = f_0(w) + \sum_{n=0}^{N} \left(\frac{\delta I_n}{\overset{0}{I}_N} \right) w^n \tag{5.90}$$

Let w_m^p be the positions in w-space of the lobe peaks in the starting pattern, with w_0^p the peak position of the main beam, and $w_1^p, \ldots, w_N^p$ the peak positions of the N side lobes.[15] Then

$$\frac{f(w_m^p) - f_0(w_m^p)}{f_0(w_0^p)} = \sum_{n=0}^{N} \frac{(w_m^p)^n}{\overset{0}{I}_N f_0(w_0^p)} \delta I_n \qquad m = 0, 1, \ldots, N \tag{5.91}$$

[14] R. S. Elliott, "On Discretizing Continuous Aperture Distributions", *IEEE Trans. Antennas Propagat.*, AP-25 (1977), 617–21.

[15] These peak positions can be found by a peak-finder computer routine, or with reasonable accuracy, can be taken to lie halfway between successive w roots.

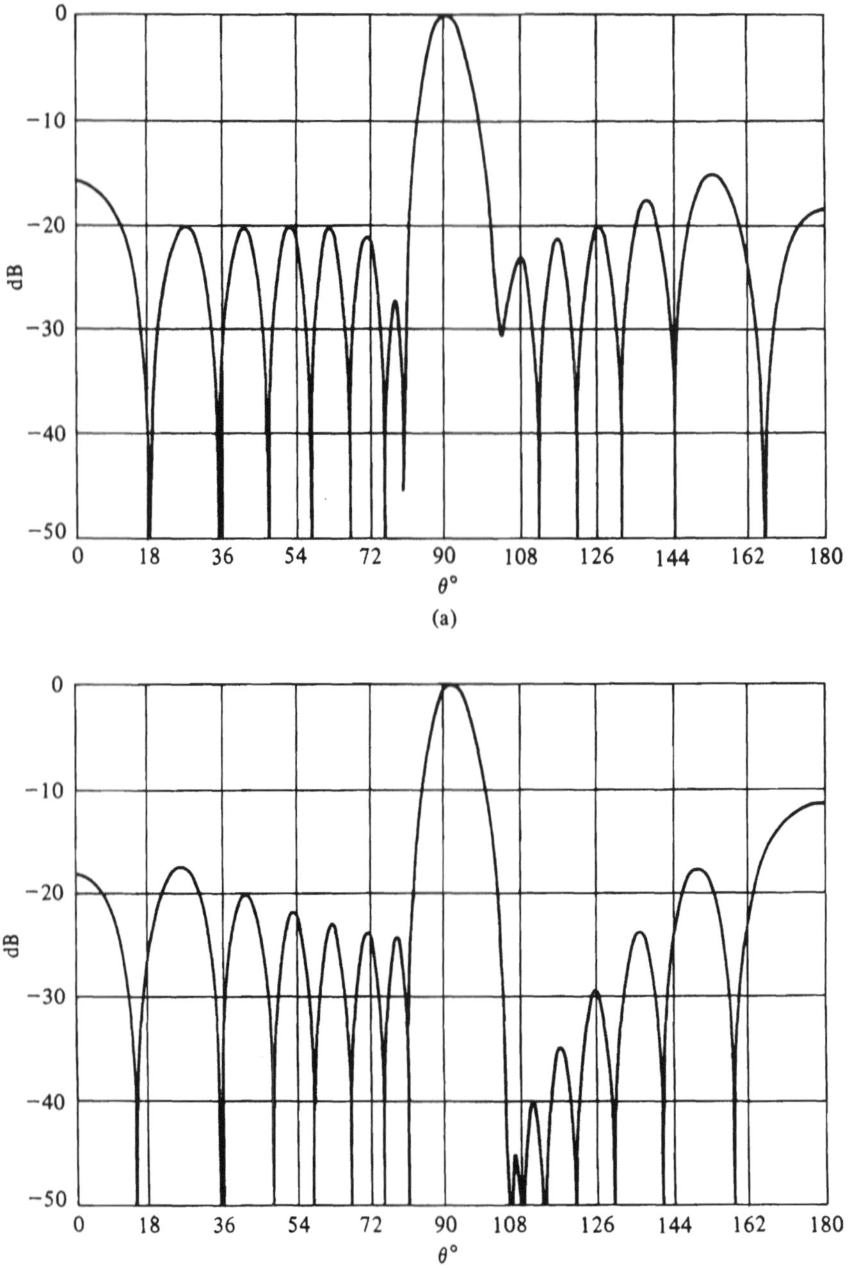

Fig. 5.19 The Sum Pattern of a 15-Element Linear Array, $d = 0.475\lambda$; Design Goal: Achieve Pattern of Fig. 5.15f; Excitation: In (a), Conventional Sampling; In (b), Root Matching (© 1977 IEEE. Reprinted from *IEEE AP Transactions*, pp. 617–621, 1977.)

If the perturbations are small, w_m^p is close to the peak position of the mth lobe in the desired pattern. And if the condition

$$f(w_0^p) - f_0(w_0^p) = 0 \tag{5.92}$$

is imposed, the two patterns will have been adjusted so that their main beam heights are essentially equal. In that case

$$\frac{f(w_m^p) - f_0(w_m^p)}{f_0(w_0^p)} = \epsilon_m \qquad m = 1, 2, \ldots, N \tag{5.93}$$

is a pure real number[16] representing the difference in height of the mth side lobe in the desired pattern and in the starting pattern. This difference is a *known* quantity.

If (5.92) and (5.93) are used in (5.91), the result is $N + 1$ simultaneous linear equations in the unknown complex quantities δI_n. Matrix inversion gives the perturbations in the relative currents, and use of (5.89) gives the new current distribution. If this is inserted in (5.88), the new pattern can be computed and compared to the desired pattern. If the agreement is satisfactory, the procedure has been completed. If not, $f(w)$ can be used as the new starting pattern, and the process repeated. Experience has shown that in practical applications only a few iterations are needed to give satisfactory convergence.

If the individual side lobes in the ultimate desired pattern have heights that vary markedly from their average value, it may be desirable to select a sequence of interim desired patterns, thus moving toward the final goal in a series of gradual steps. Since the computer program is simple, this is not a costly operation.

As an example, this perturbation procedure can be applied to the pattern of Figure 5.19b, which is not sufficiently close to the ideal of Figure 5.15. Three iterations are sufficient to bring all side lobes within $\frac{1}{4}$ dB of specification, as shown in Figure 5.20. The final normalized current distribution is listed in Table 5.4.

TABLE 5.4 Discrete current distribution for 15-element array, 0.475λ spacing, to give pattern of Figure 5.20c

n	I_n	n	I_n
0	$0.881\,\underline{|\ 0^\circ}$	± 4	$0.510\,\underline{|\pm\ 9.96^\circ}$
± 1	$1.000\,\underline{|\pm 1.72^\circ}$	± 5	$0.580\,\underline{|\pm 12.31^\circ}$
± 2	$0.778\,\underline{|\pm 4.76^\circ}$	± 6	$0.215\,\underline{|\pm\ 1.66^\circ}$
± 3	$0.828\,\underline{|\pm 5.49^\circ}$	± 7	$0.398\,\underline{|\pm 39.48^\circ}$

[16]To insure that the right side of (5.91) is also pure real, it is desirable either to index the elements from the center of the array, or to extract a factor $(w)^{N/2}$, in order to establish the phase center appropriately. Since a sum pattern with asymmetrical side lobes, interspersed by deep nulls, requires element excitations which occur as a sequence of complex conjugate pairs relative to the array center, one could parallel the development which begins with (5.87) and express the pattern in terms of sines and cosines of multiples of ψ. This leads to a real matrix equivalent of (5.91), which some users of this technique might prefer.

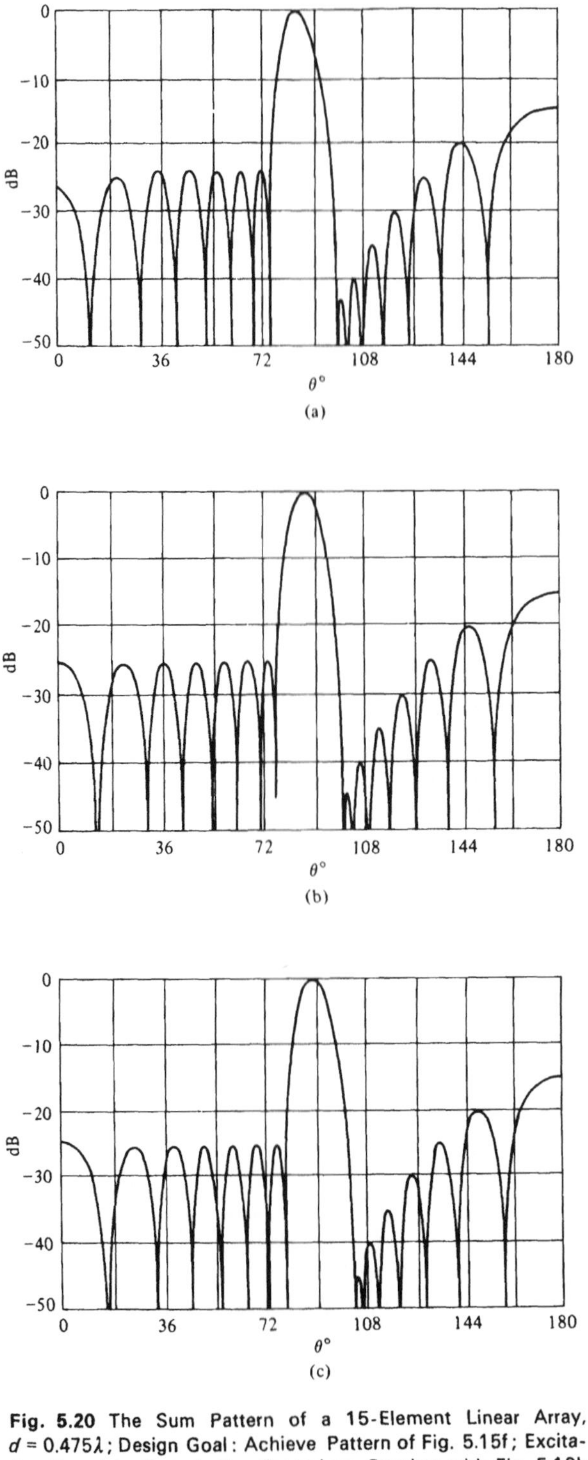

Fig. 5.20 The Sum Pattern of a 15-Element Linear Array, $d = 0.475\lambda$; Design Goal: Achieve Pattern of Fig. 5.15f; Excitation Found by Perturbation Procedure, Starting with Fig. 5.19b (© 1977 IEEE. Reprinted from *IEEE AP Transactions*, pp. 617–621, 1977.)

5.11 Bayliss Synthesis of Difference Patterns

Section 5.7 was concerned with a synthesis problem solved by Taylor, namely, the determination of a continuous line source distribution that would produce a symmetric sum pattern, with pairs of near-in side lobes at a quasi-uniform height. The requisite distribution was found to be symmetrical in amplitude and nonzero at the end points.

An analogous synthesis problem exists for difference patterns. Can one find a continuous line source distribution that will produce a symmetric difference pattern, with twin main beams surrounded by pairs of near-in side lobes at a quasi-uniform specified height?

An approach to the ultimate answer to this question can be patterned after the development in Section 5.7. In retrospect, the Taylor line source distribution $g(\zeta)$, given by Equation 5.68, is seen to be the product of two factors: (1) a uniform progressive phase term $e^{-j\beta\zeta}$ which serves the purpose of determining the pointing direction of the main beam, and (2) a pure real amplitude distribution function

$$h(\zeta) = \frac{1}{2a}\left[\mathcal{S}(0) + 2\sum_{m=1}^{\bar{n}-1} \mathcal{S}(m) \cos\frac{m\pi\zeta}{a}\right] \tag{5.94}$$

In (5.94), $h(\zeta)$ is represented by all *even* terms in a Fourier series that are nonzero at the endpoints. The special case of a *uniform* distribution $h(\zeta) =$ *constant* corresponds to taking only the first term of this Fourier series, and results in the generic pattern shown in Figure 5.6, which can be modified by root displacement to give the Taylor pattern.

By analogy, it can be anticipated that the line source distribution being sought for a Dolph-like difference pattern is representable by a Fourier series consisting of all *odd* terms that are nonzero at the end points, that is,

$$h(\zeta) = \sum_{m=0}^{\infty} B_m \sin\left[\left(m + \frac{1}{2}\right)\frac{\pi\zeta}{a}\right] \tag{5.95}$$

The generic difference pattern should result from taking only the zeroth term of this Fourier series, that is, from the aperture distribution

$$g(\zeta) = \sin\left(\frac{\pi\zeta}{2a}\right) e^{-j\beta\zeta} \tag{5.96}$$

When (5.96) is used in the array factor common to (5.56) and (5.57), it can be seen that

$$\begin{aligned}\mathcal{D}(\theta) &= \int_{-a}^{a} g(\zeta) e^{jk\zeta\cos\theta}\, d\zeta \\ &= \int_{-a}^{a} \sin\left(\frac{\pi\zeta}{2a}\right) e^{jk\zeta(\cos\theta - \beta/k)}\, d\zeta\end{aligned}$$

$$= 2j\int_0^a \sin\left(\frac{\pi\zeta}{2a}\right)\sin\left[k\zeta\left(\cos\theta - \frac{\beta}{k}\right)\right]d\zeta$$

$$= j\int_0^a \left\{\cos\left[\left(u - \frac{1}{2}\right)\frac{\pi\zeta}{a}\right] - \cos\left[\left(u + \frac{1}{2}\right)\frac{\pi\zeta}{a}\right]\right\} d\zeta \tag{5.97}$$

with u once again given by (5.60). Integration of (5.97) gives, with the suppression of an inconsequential mutiplicative constant,

$$\mathfrak{D}(u) = \frac{\pi u \cos \pi u}{(u - \frac{1}{2})(u + \frac{1}{2})} \tag{5.98}$$

This generic pattern is shown in Figure 5.21. It is seen to consist of twin main beams that straddle a null at $u = 0$, plus symmetric pairs of side lobes with heights that diminish as $|u|^{-1}$. The innermost pair is only 10 dB below the main beam. However, if the near-in null pairs could be shifted outward in some programmed manner, a useful difference pattern would result. For example, if the innermost four pairs of side lobes could be adjusted to be at a common height of —20 dB, with all further-out

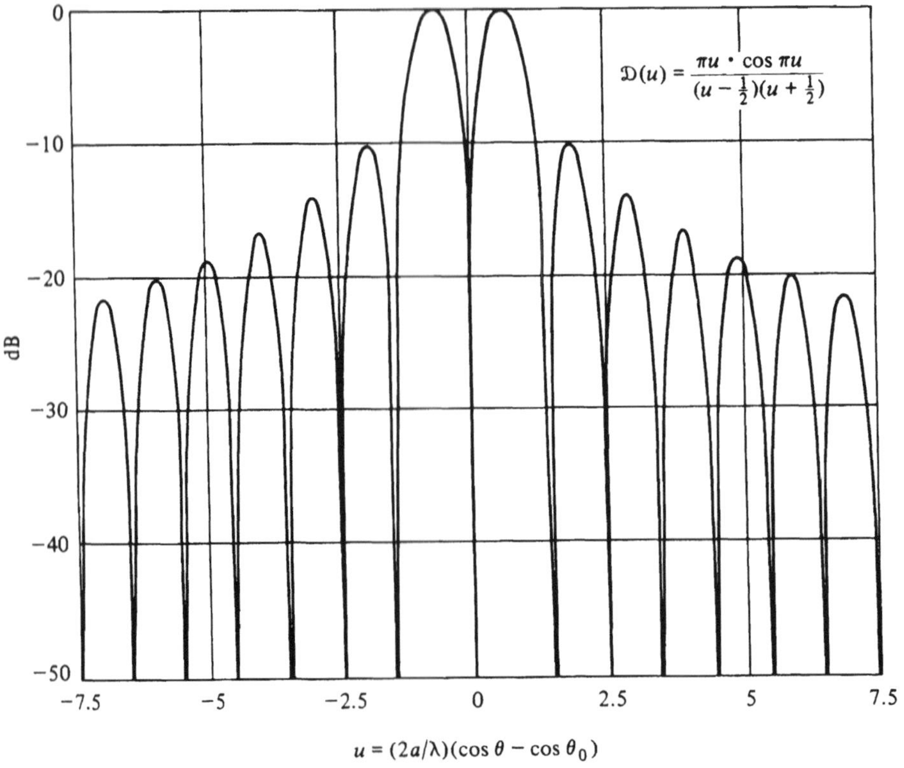

Fig. 5.21 Generic Difference Pattern for a Continuous Line Source

side lobe pairs unaffected, one would accomplish the equivalent of what Taylor did to obtain a -20 dB SLL sum pattern.

The nulls in the generic difference pattern given by (5.98) occur (1) at $u = 0$, and (2) when $\cos \pi u = 0$, that is, when

$$u = \pm(n + \tfrac{1}{2}) \qquad n = 0, 1, 2, \ldots \tag{5.99}$$

(The $n = 0$ nulls are removed by the two factors ocurring in the denominator of (5.98)). Thus if one wishes to anchor the nulls which occur at $\pm(\bar{n} + \frac{1}{2})$ and beyond, while moving the intervening null pairs so as to depress the near-in side lobes, it is clear that (5.98) should be modified to the form

$$\mathfrak{D}(u) = \pi u \cos \pi u \frac{\prod_{n=1}^{\bar{n}-1} (1 - u^2/u_n^2)}{\prod_{n=0}^{\bar{n}-1} \{1 - [u/(n + \frac{1}{2})]^2\}} \tag{5.100}$$

The synthesis problem is now focused on the need to find the new null locations $\pm u_n$ such that the near-in side lobe pairs are at a quasi-uniform specified level.

This problem was solved by E. T. Bayliss.[17] Unlike Taylor, who was able to determine his null relocation formula (5.63) through recourse to the ideal space factor $\cos \pi\sqrt{u^2 - A^2}$, Bayliss was not able to find a limiting form for the ideal difference pattern. Thus he was obliged to undertake a parametric study with the aid of a computer, the results of which have yielded the following formulas for root placement.

$$u_n = \begin{cases} 0 & n = 0 \\ \left(\bar{n} + \dfrac{1}{2}\right)\left(\dfrac{\xi_n^2}{A^2 + \bar{n}^2}\right)^{1/2} & n = 1, 2, 3, 4 \\ \left(\bar{n} + \dfrac{1}{2}\right)\left(\dfrac{A^2 + n^2}{A^2 + \bar{n}^2}\right)^{1/2} & n = 5, 6, \ldots, \bar{n} - 1 \end{cases} \tag{5.101}$$

The parameters A and ξ_n are related to the side lobe level and their appropriate values can be read from Table 5.5. A typical Bayliss pattern for $\bar{n} = 10$ and a prescribed 30 dB side lobe level, is shown in Figure 5.22. These patterns exhibit many of the

TABLE 5.5 Parameter value versus side lobe level for Bayliss difference pattern

	Side Lobe Level in dB					
	15	20	25	30	35	40
A	1.0079	1.2247	1.4355	1.6413	1.8431	2.0415
ξ_1	1.5124	1.6962	1.8826	2.0708	2.2602	2.4504
ξ_2	2.2561	2.3698	2.4943	2.6275	2.7675	2.9123
ξ_3	3.1693	3.2473	3.3351	3.4314	3.5352	3.6452
ξ_4	4.1264	4.1854	4.2527	4.3276	4.4093	4.4973

[17] E. T. Bayliss, "Design of Monopulse Antenna Difference Patterns with Low Side Lobes", *Bell System Tech. J.*, 47 (1968), 623–40.

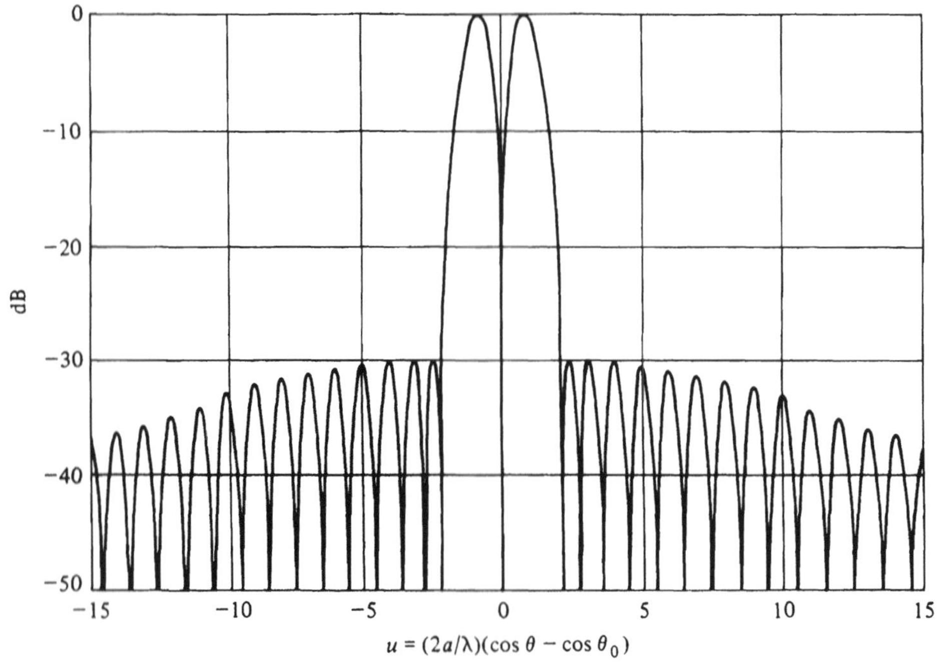

Fig. 5.22 Bayliss Difference Pattern for Continuous Line Source, $\bar{n}$ = 10, −30 dB SLL (© 1976 IEEE. Reprinted from *IEEE AP Transactions*, pp. 310–316, 1976.)

same characteristics observed in Taylor sum patterns. The near-in side lobes droop off slightly and the far-out side lobes decay as $|u|^{-1}$. The near-in nulls have been displaced outward to lower the near-in side lobes. For $|n| \geq \bar{n}$ the null positions occur at the half integers.

The aperture distribution that will produce a Bayliss pattern can be determined by multiplying (5.95) by $e^{-j\beta\xi}$ to obtain $g(\xi)$, inserting the result in the first form of (5.97), and using (5.100) for $\mathfrak{D}(u)$. One finds that

$$\mathfrak{D}(u) = 2j \sum_m B_m \int_0^a \sin\left[\left(m + \frac{1}{2}\right)\frac{\pi\zeta}{a}\right] \sin\left(u\frac{\pi\zeta}{a}\right) d\zeta \tag{5.102}$$

If u is halfway between two integers, say $n + \frac{1}{2}$, this integral is zero unless $n = m$, in which case

$$\mathfrak{D}(m + \tfrac{1}{2}) = jaB_m \tag{5.103}$$

Further, since $\mathfrak{D}(m + \frac{1}{2}) = 0$ for $m \geq \bar{n}$, as can be seen from (5.100), the Fourier series truncates. Therefore the aperture distribution for a Bayliss pattern is given by

$$g(\zeta) = e^{-j\beta\xi} \sum_{m=0}^{\bar{n}-1} \mathfrak{D}\left(m + \frac{1}{2}\right) \sin\left[\left(m + \frac{1}{2}\right)\frac{\pi\zeta}{a}\right] \tag{5.104}$$

with the factor $(ja)^{-1}$ suppressed, and with $\mathfrak{D}(m + \frac{1}{2})$ evaluated from (5.100).

For the pattern shown in Figure 5.22, use of (5.104) yields the distribution shown in Figure 5.23. One observes a symmetrical amplitude distribution which is nonzero at the end points, but which goes to zero at the midpoint, which is where a phase reversal of 180° takes place.

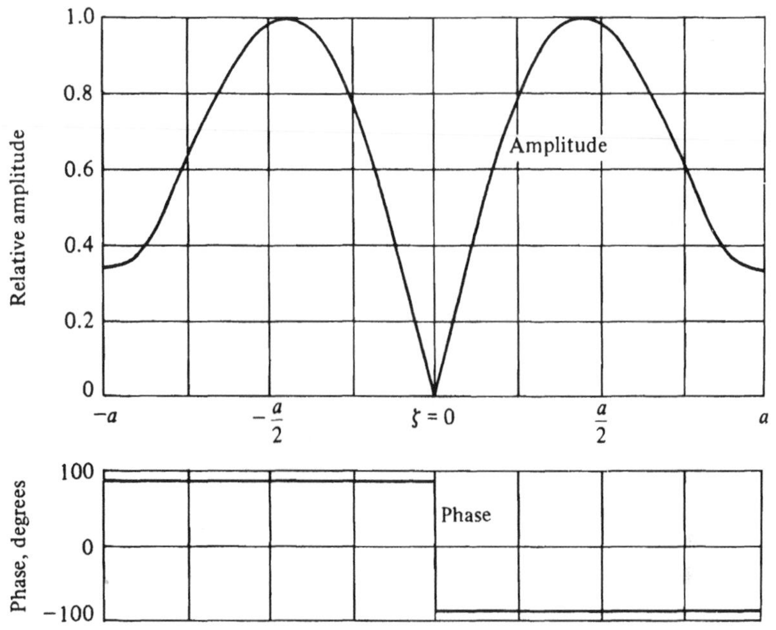

Fig. 5.23 Bayliss Aperture Distribution for Pattern of Fig. 5.22

5.12 Difference Patterns with Arbitrary Side Lobe Topography

The perturbation procedure used in Section 5.9 to shift the nulls of a Taylor pattern, so as to create a sum pattern with the heights of all side lobes individually specified, can be applied to Bayliss difference patterns for the same purpose.[18] As a generalization of (5.100), the starting pattern can be written in the form

$$\mathfrak{D}_0(u) = C_0 u f(u) \prod_{n=-(\bar{n}_L-1)}^{\bar{n}_R-1}{}' \left(1 - \frac{u}{\overset{0}{u}_n}\right) \tag{5.105}$$

with

$$f(u) = \frac{\pi \cos \pi u}{\prod_{n=-\bar{n}_L}^{\bar{n}_R-1} [1 - u/(n + \frac{1}{2})]} \tag{5.106}$$

The desired pattern can be expressed in the same form, that is,

$$\mathfrak{D}(u) = (C_0 + \delta C)(u - \delta u_0) f(u) \prod_{n=-(\bar{n}_L-1)}^{\bar{n}_R-1}{}' \left(1 - \frac{u}{\overset{0}{u}_n + \delta u_n}\right) \tag{5.107}$$

in which $u_n = \overset{0}{u}_n + \delta u_n$ is the new root position and $C = C_0 + \delta C$ is the amplitude factor of the new pattern.

[18] R. S. Elliott, "Design of Line Source Antennas for Difference Patterns with Side Lobes of Individually Arbitrary Heights", *IEEE Trans. Antennas Propagat.* AP-24 (1976), 310–16.

One new feature can be observed in the formulation of (5.107). Whereas the starting pattern $\mathfrak{D}_0(u)$ has a null at $u = 0$, the desired pattern has a null at the shifted position $u = \delta u_0$. This is necessary if one wishes to have a deterministic set of perturbations. Between the anchored roots at $u = -\bar{n}_L$ and $u = \bar{n}_R$ there are $\bar{n}_R + \bar{n}_L$ lobes with heights to be adjusted. *Including* the null between the two main lobes, there are $\bar{n}_R + \bar{n}_L - 1$ movable nulls which, combined with the adjustable amplitude factor C, provide just the proper number of degress of freedom.

When the expansion in (5.81) is used, to first order,

$$\frac{\mathfrak{D}(u)}{\mathfrak{D}_0(u)} - 1 = \frac{\delta C}{C_0} - \frac{\delta u_0}{u} + \sum_{n=-(\bar{n}_L-1)}^{\bar{n}_R-1}{}' \frac{(u/\overset{0}{u}_n^2)}{1 - u/\overset{0}{u}_n} \delta u_n \tag{5.108}$$

If the peak positions u_m^p of the lobes in the starting pattern are placed in (5.108), a set of $\bar{n}_R + \bar{n}_L$ simultaneous linear equations results and matrix inversion gives the values of the perturbations. As in the sum pattern case, $\mathfrak{D}(u_m^p)/\mathfrak{D}_0(u_m^p)$ can be identified as being essentially the ratio of the desired height of the mth lobe to its starting height.

An example of the use of this technique is the modification of a Bayliss 30/30, $\bar{n} = 10$, so that the four innermost pairs of lobes are at -40 dB. Three iterations give the pattern shown in Figure 5.24. The corresponding aperture distribution can be

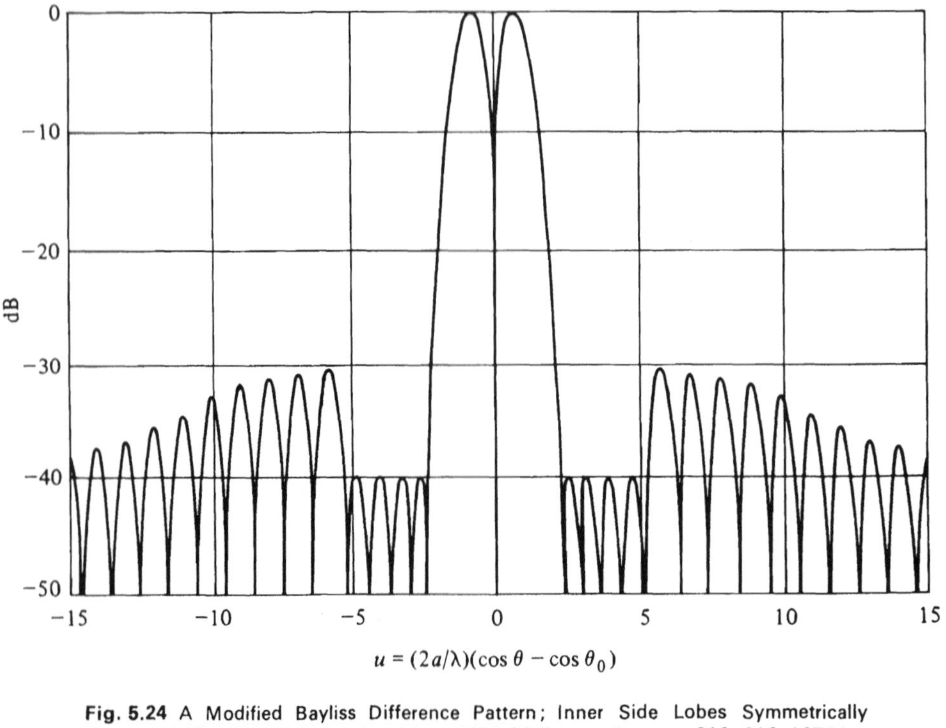

Fig. 5.24 A Modified Bayliss Difference Pattern; Inner Side Lobes Symmetrically Depressed (© 1976 IEEE. Reprinted from *IEEE AP Transactions*, pp. 310–316, 1976.)

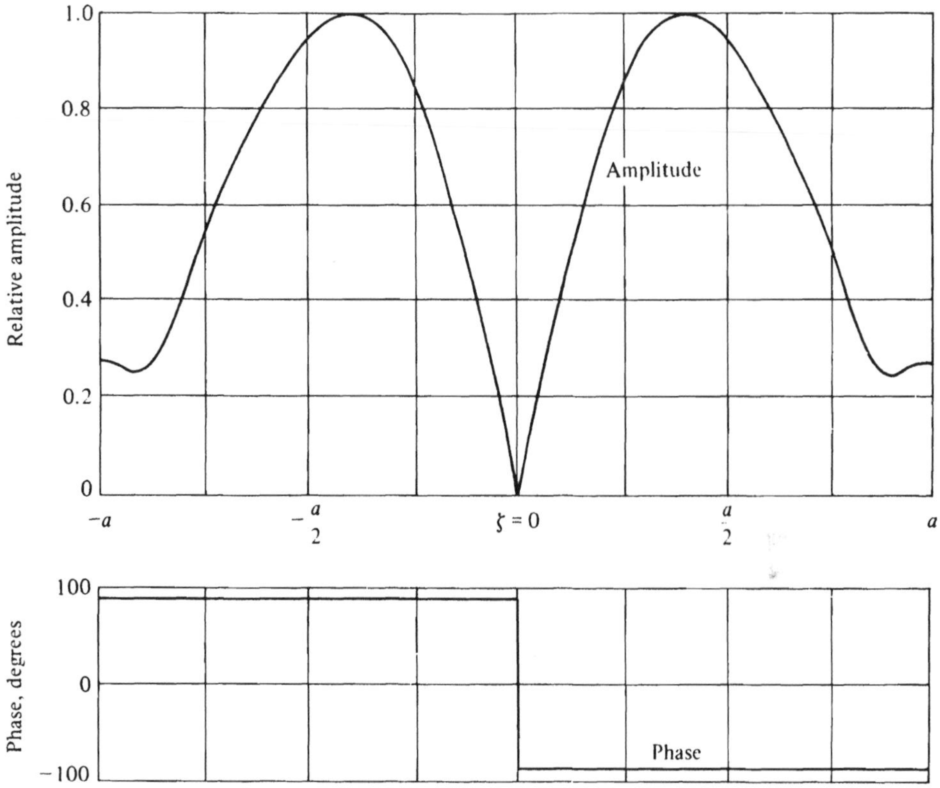

Fig. 5.25 The Aperture Distribution for the Pattern of Fig. 5.24 (© 1976 IEEE. Reprinted from *IEEE AP Transactions*, pp. 310–316, 1976.)

found in Figure 5.25. It is significantly different from the distribution for the unperturbed Bayliss 30/30 (compare with Figure 5.23), but not difficult to achieve in practice.

5.13 Discretization Applied to Difference Patterns

The discretizing technique introduced in Section 5.10 does not distinguish between sum patterns and difference patterns. If one starts with a desired pattern with known null positions, the excitation of the discrete array can be determined so that the nulls of its pattern coincide with those of the starting pattern. And, if that is not sufficient to cause the lobe heights to agree, the perturbation procedure introduced via Equations 5.87 through 5.93 can be used to effect the desired result.

As an example of the application of this technique to difference patterns, let the Bayliss 30/30, $\bar{n} = 10$, of Figure 5.22 be selected as starting pattern. The null positions can be calculated from (5.101) and are listed in Table 5.6. If a 10-element equispaced linear array is to approximate this pattern, the Schelkunoff unit circle

TABLE 5.6 Roots of the Bayliss 30/30 pattern

n	1	2	3	4	5	6	7	8	9
u_n	2.1456	2.7224	3.5553	4.4838	5.4525	6.4450	7.4494	8.4614	9.4787

should show a root at $w = 1 + j0$ plus root pairs which can be calculated from

$$w_n = e^{j2\pi u_n/10} \qquad n = \pm 1, \pm 2, \pm 3, \pm 4 \tag{5.109}$$

Thus the nine roots on the Schelkunoff unit circle occur at the angles

$$\psi_n = 0^\circ, \quad \pm 77.24^\circ, \quad \pm 98.01^\circ, \quad \pm 127.99^\circ, \quad \pm 161.42^\circ$$

The pattern is given by

$$f(w) = (w - 1) \prod_{n=1}^{4} (w^2 - 2w \cos \psi_n + 1) \tag{5.110}$$

and is shown for 0.7λ element spacing in Figure 5.26. The corresponding current distribution is found by multiplying out the factors appearing in (5.110), and is listed in the second column of Table 5.7, normalized so that the field magnitude of the peaks of the twin main beams is unity.

TABLE 5.7 Normalized currents for patterns of Figures 5.26–27

Element Number n	I_n for Figure 5.26	I_n for Figure 5.27
1	± 0.0687	± 0.0725
2	± 0.1686	± 0.1764
3	± 0.1903	± 0.1903
4	± 0.1365	± 0.1322
5	± 0.0695	± 0.0615

It can be seen from Figure 5.26 that the discretizing of the Bayliss $\bar{n} = 10$, SLL $= 30$ dB pattern is not completely satisfactory, since all the side lobes are above -30 dB, one pair being as high as -26.5 dB. One could use the perturbation procedure to place all these side lobes at -30 dB, if that were desired. But imagine instead that the desired pattern calls for the innermost pair of side lobes to be at -35 dB, all others at -30 dB. Then from the pattern of Figure 5.26, one can deduce that the lobe peaks occur at $\psi_m^p = \pm 28.1^\circ$, $\pm 87.6^\circ$, $\pm 113.0^\circ$, $\pm 144.7^\circ$, $\pm 180^\circ$. Since the desired pattern is symmetrical, the perturbations in the currents occur in equal and opposite pairs, and it is only necessary to construct a 5×5 matrix from Equation 5.91. With the starting lobe heights read from Figure 5.26, and the desired lobe heights known, the left side of (5.91) is known for each of the five values of m. Inversion of the matrix gives δI_n values which, when added to the starting current distribution, gives the new discrete currents. These currents give an inadequate approximation to the desired pattern and one finds it necessary to repeat the process by using the result

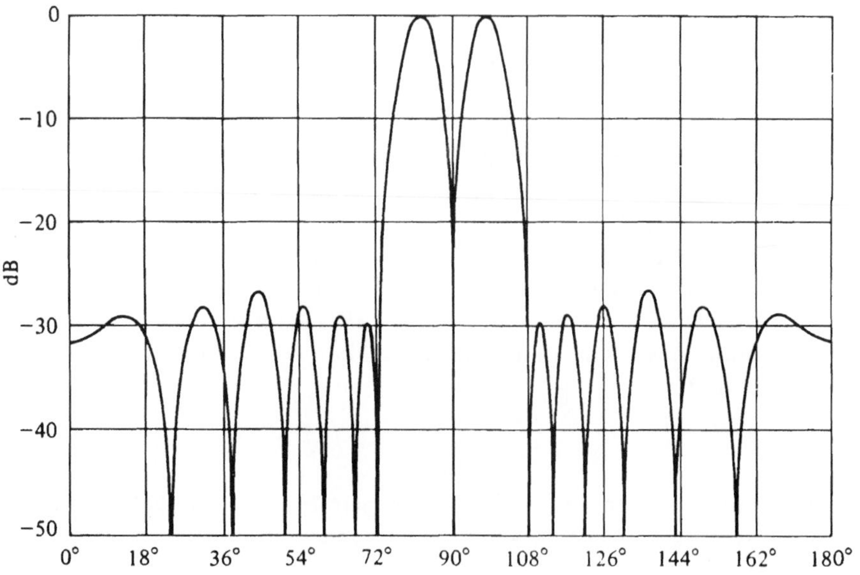

Fig. 5.26 The Difference Pattern of a 10-Element Linear Array, $d = 0.7\lambda$; Design Goal: Achieve Pattern of Fig. 5.22; Excitation Found by Root Matching

as a new starting pattern. Three successive iterations are sufficient and yield the desired current distribution listed in the third column of Table 5.7. This distribution gives the difference pattern shown in Figure 5.27.

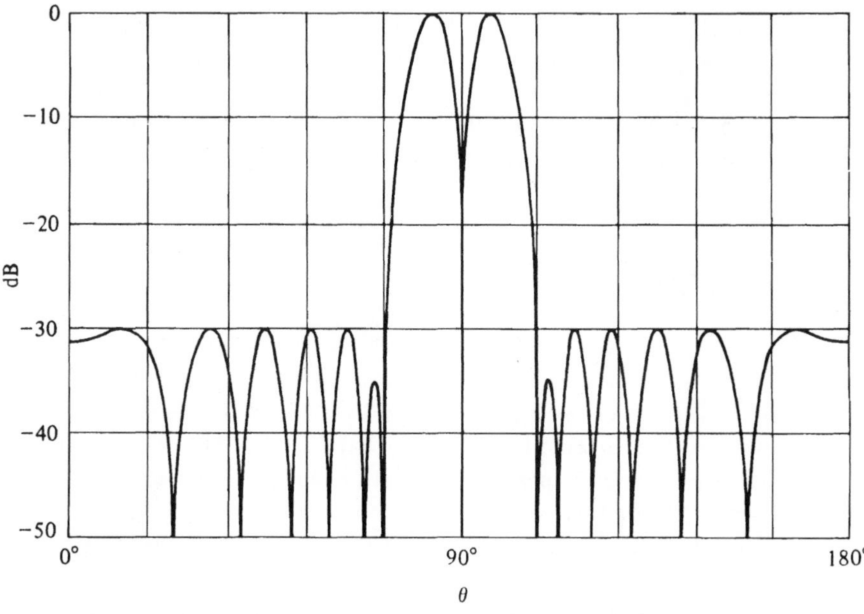

Fig. 5.27 The Difference Pattern of a 10-Element Linear Array, $d = 0.7\lambda$; Design Goal: All Side Lobes at −30 dB Except Inner Pair at −35 dB; Excitation Found by Perturbation Procedure, Starting with Fig. 5.26

5.14 Design of Linear Arrays to Produce Null-Free Patterns

Some antenna applications require patterns without nulls. An example is the airport beacon antenna which must radiate uniformly in ϕ to be able to communicate with aircraft arriving from all directions. It must also radiate without nulls in θ if it is to maintain contact with incoming aircraft which fly at constant height, and thus appear at a constantly changing angle θ with respect to the antenna.

A method due to P. M. Woodward[19] is useful in such applications. Imagine a continuous line source that is uniformly illuminated by a traveling wave distribution of the form

$$g_n(\zeta) = K_n e^{-j\beta_n\zeta} \tag{5.111}$$

It has already been seen in the development leading to (5.61) that, within a multiplicative constant, the field pattern for this distribution is given by

$$f_n(u) = K_n \frac{\sin \pi u}{\pi u} \tag{5.112}$$

in which $u = (2a/\lambda)(\cos\theta - \beta_n/k)$, with $2a$ the aperture length. This pattern is plotted in Figure 5.6 and shows a main beam at $u = 0$, or at an angle θ_n in real space given by $\theta_n = \arccos(\beta_n/k)$.

Imagine a continuous aperture distribution composed of a sum of waves of the type in (5.111) with β_n adjusted so that each partial distribution places its main beam at an angle θ_n corresponding to the closest-in null of its neighbor. If the amplitudes K_n are also properly adjusted, the peaks of the main beams can have an envelope of prescribed shape. Additionally, there will be null filling. The effect is as suggested in Figure 5.28 where only the main beams have been sketched to avoid confusion.

This type of synthesis can be accomplished if the pattern is represented by

$$F(u) = \sum_{n=0}^{N} K_n \frac{\sin \pi(u - n)}{\pi(u - n)} \tag{5.113}$$

with u defined by

$$u = \frac{2a}{\lambda}(\cos\theta - \cos\theta_0) \tag{5.114}$$

where θ_0 is the pointing angle of the main beam of the zeroth partial aperture distribution. One can observe from (5.113) that $u = 1$ is both the first null of the zeroth partial pattern and the peak of the first partial pattern. Similarly, $u = 2$ is both the first null of the first partial pattern and the peak of the second partial pattern, and so on. The values of K_n must be selected to fit the specified envelope.

[19] P. M. Woodward, "A Method of Calculating the Field Over a Plane Aperture Required to Produce a Given Polar Diagram", *J. IEEE (London)*, pt. IIIA, 93 (1947), 1554–58.

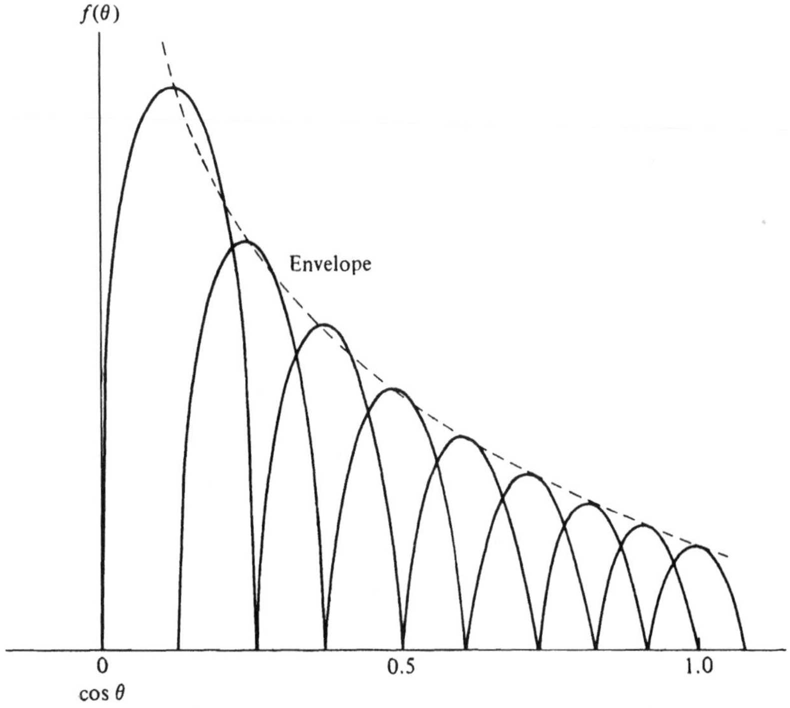

Fig. 5.28 Woodward-Type Null-Filling

As an example, consider the design of an airport beacon antenna which is to be 8λ long and have the main beam of its zeroth partial patterns pointing just far enough above the horizon (that is, above $\theta = 90°$) so that one closest-in null lies along the horizon. From (5.114) one can see that this means

$$-1 = 8\,(0 - \cos\theta_0)$$

$$\theta_0 = \arccos\left(\tfrac{1}{8}\right) = 82.82°$$

Thus in this application

$$u = 8\cos\theta - 1$$

and the zeroth partial pattern has seven nulls in the range $0° \leq \theta \leq 82.82°$.

Imagine that these nulls are to be filled by the other partial patterns such that in this range the envelope is $\csc(90° - \theta)$. This is a particularly practical selection because it would ensure that an airplane flying at a constant height would continue to receive a constant level signal from the beacon as its range changed. It follows that

$$K_n = \csc(90° - \theta_n) \quad \text{and} \quad \theta_n = \arccos[(n+1)/8] \qquad n = 1, 2, \ldots, 7$$

A tabulation of the partial beam positions and the amplitudes K_n is shown in Table 5.8. A plot of (5.113) for this case is shown in Figure 5.29. One can see a ripple around

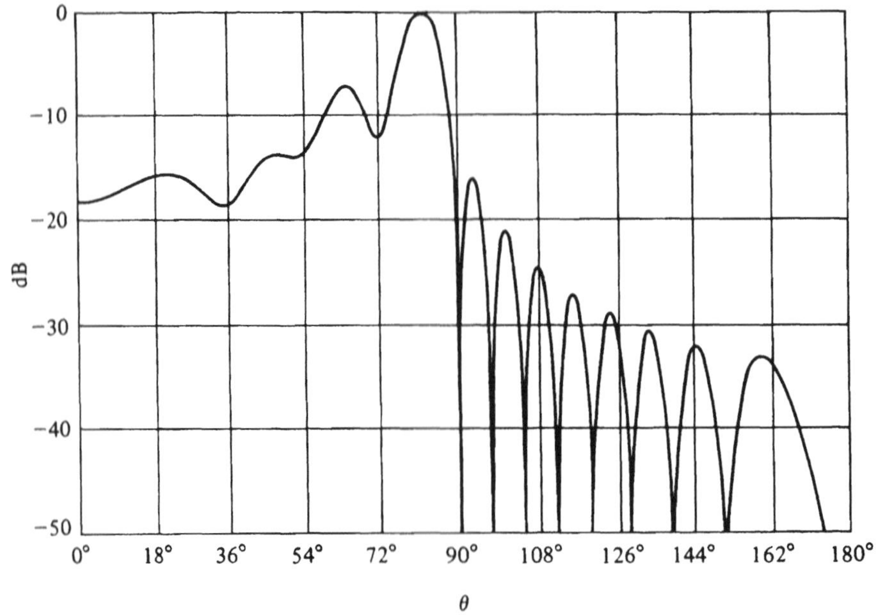

Fig. 5.29 Null-Filled Cosecant Squared Pattern

TABLE 5.8

n	0	1	2	3	4	5	6	7
θ_n	82.82°	75.52°	67.98°	60°	51.32°	41.41°	28.96°	0°
K_n	8.00	4.00	2.67	2.00	1.60	1.33	1.14	1.00

the desired envelope of about ±2.4 dB, but no nulls. The aperture distribution is

$$g(\zeta) = \sum_{n=0}^{7} K_n e^{-j\beta_n\zeta}$$

and is displayed in Figure 5.30. Note that there is considerable fine structure in both the amplitude and phase distribution. This would not be a simple aperture excitation to achieve, and some pattern degradation from the ideal would have to be anticipated.

The pattern of Figure 5.29 would be improved if the ripple could be reduced. H. J. Orchard et al.[20] have devised a synthesis procedure that can produce null-filled patterns with minimum ripple and arbitrary side lobes.

[20]H. J. Orchard, R. S. Elliott, and G. J. Stern, "Optimizing the Synthesis of Shaped Beam Antenna Patterns," *Proc. IEEE, Part H, 132* (1985), 63–68.

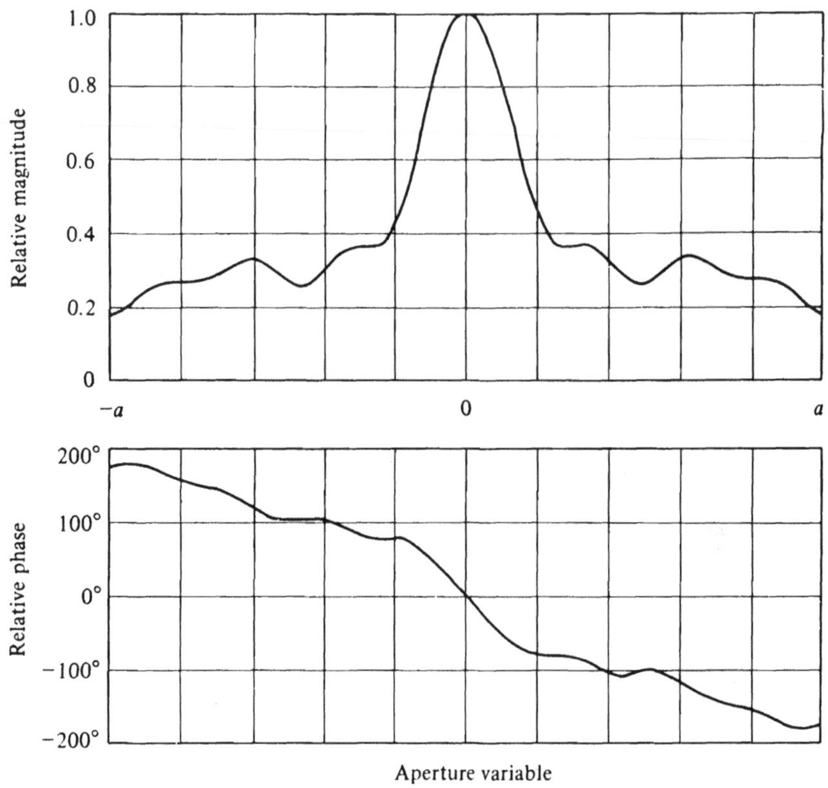

Fig. 5.30 The Aperture Distribution for the Pattern of Fig. 5.29

REFERENCES

ELLIOTT, R. S., "The Theory of Antenna Arrays", Vol. 2 of *Microwave Scanning Antennas*, ed. R. C. Hansen (New York: Academic Press, 1966), Chapter 1.

JASIK, H., *Antenna Engineering Handbook* (New York: McGraw-Hill Book Co., Inc., 1961), Chapter 2.

JORDAN, E. C. and K. G. BALMAIN, *Electromagnetic Waves and Radiating Systems* (Englewood Cliffs, New Jersey: Prentice-Hall, Inc., 1968), Chapter 12.

PROBLEMS

5.1 Show that the choice of a_0 in the development of the even-degree Chebyshev polynomial in Appendix C also insures that $T_{2N}(1) = 1$. Similarly, show that the choice of a_1 for an odd-degree Chebyshev polynomial insures that $T_{2N-1}(1) = 1$.

5.2 Design a six-element equispaced linear array to give a broadside sum pattern with all side lobes at −20 dB. Do this by using (5.12) and (5.13) to deduce the root positions on the unit circle and then the current distribution. What is the maximum interelement spacing so as to avoid completely a second main beam? If the main beam is to point at $\theta_0 = 45°$, find the maximum interelement spacing and current distribution.

5.3 Repeat the preceding problem for a seven-element array.

5.4 Use Equations 5.18 and 5.19 to obtain an independent calculation of the current distributions for Problems 5.2 and 5.3.

5.5 Design a four-element equispaced linear array to give a Dolph-Chebyshev end-fire pattern with side lobes of −15 dB. Find the maximum element spacing, current distribution, and 3 dB beamwidth.

5.6 Show that, for $2N + 1$ large, Equation 5.31 is given to good approximation by Equation 5.33 for the special case $p = 1$.

5.7 Find the 3 dB beamwidth of a 36-element linear array, with $3\lambda/4$ spacing, if it is uniformly excited and is designed to radiate a broadside sum pattern. What is your answer if the sum pattern is end-fire? If the sum pattern is broadside, but 30 dB Dolph-Chebyshev?

5.8 Find the directivity for each of the three arrays described in Problem 5.7.

5.9 With the aid of Figures 5.3 and 5.4, find the half-power beamwidth of an equispaced array consisting of 241 elements $\lambda/2$ on centers, excited to give a Dolph-Chebyshev sum pattern, with the main beam pointing at $\theta_0 = 30°$ and with side lobes of −30 dB. What is the directivity?

5.10 A continuous line source is to be designed to give a Taylor pattern at $\theta_0 = 45°$ for $\bar{n} = 6$ and a side lobe level of −20 dB. Find β/k, A, and the positions $\pm u_n$ of the five innermost pairs of pattern nulls. Write an expression for the pattern in u-space. Determine the corresponding aperture distribution $g(\zeta)$.

5.11 Use the perturbation procedure described in Section 5.9 to modify the Taylor pattern of Problem 5.10 so that the innermost side lobe on one side of the main beam is at −30 dB. Find the requisite aperture distribution.

5.12 A 10-element discrete array, 0.7λ spacing, is to be excited to produce the Taylor pattern described in Problem 5.10. Use the discretizing procedure described in Section 5.10 and determine the excitation in amplitude and phase. The pattern is given by

$$\mathcal{S}(\theta) = \sum_{n=1}^{5} I_n \cos[(2n - 1)kd(\cos\theta - \cos\theta_0)]$$

in which I_n is the amplitude distribution. If a computer plotter is available, graph $\mathcal{S}(\theta)$ and compare it to Figure 5.7. Plot I_n as a bar graph overlay of the continuous aperture distribution of Figure 5.8.

5.13 The pattern due to the discrete array of Problem 5.12 will be found to be somewhat degraded from the desired pattern shown in Figure 5.7. Use the perturbation procedure outlined in Section 5.10 to determine a modified excitation that will reproduce the desired pattern to within ±0.25 of all side lobe heights.

5.14 A continuous line source is needed to produce a Bayliss difference pattern with $\bar{n} = 7$ and a side lobe level of 20 dB. Find the null positions and plot the pattern in $|u| \leq 7.5$. If the principal null is to point at broadside determine the aperture distribution.

5.15 A 10-element equispaced linear array, 0.5λ spacing, is to be excited to produce a difference pattern all of whose lobes are at exactly -20 dB. Use the Bayliss 20/20 of Problem 5.14 as starting pattern and deduce the discrete current distribution.

6 planar arrays: analysis and synthesis

6.1 Introduction

The previous two chapters have dealt with the analysis and synthesis of equispaced linear arrays. Under certain circumstances, much of what was developed there can be carried over to apply to planar arrays. However, practical considerations will at times require the use of design techniques that are peculiar to planar arrays. Thus this chapter will be seen to consist of a mixture of extensions and new approaches.

Two basic types of planar arrays will be considered. The first consists of elements that form a rectangular grid. The second is composed of elements that lie on concentric circles. For both types of arrays it will be assumed that the elements are equispaced, though not necessarily with the same spacing in the two orthogonal directions. Often it will be assumed that the array can be divided into four symmetrical quadrants for the purpose of permitting excitations that will give sum and difference patterns. The boundary of the rectangular grid arrays wil: at different times be assumed to be square, rectangular, circular, or elliptical; the boundary of the circular grid arrays will always be taken to be circular.

For rectangular grid arrays, if the boundary is square or rectangular, and if the aperture distribution is separable, the pattern is the product of the patterns of two orthogonal linear arrays, and all of the ideas previously developed about linear arrays can be extended readily. This case will be taken up first because of its simplicity and because it reveals so many basic ideas about planar arrays. However, separable distributions suffer from some gain limitations which can be overcome by ϕ-symmetric patterns. For this reason, Taylor's extension of his line-source analyis to the case of a planar aperture with a circular boundary, containing a continuous ϕ-symmetric distribution, forms an ideal second topic. Sampling of the Taylor circular distribution can give excitation coefficients for either rectangular grid or circular grid discrete arrays.

Extension of Dolph's technique to planar arrays has not been effected for the general case, but Tseng and Cheng have shown how it can be done for rectangular grid arrays, with a rectangular boundary, if the number of elements in each direction is the same. Their design procedure will be developed in an ensuing section.

Perturbation methods have been devised that will modify a circular Taylor pattern so that the different ring side lobes have arbitrary heights, or so that the pattern is Taylor-like in every ϕ-cut, but ϕ-nonsymmetric. The resulting continuous aperture distributions can be sampled and applied to discrete arrays, and these methods will also be fully developed.

As has already been noted in the linear case, sampling continous distributions will result in some pattern degradation. For circular grid arrays, it will be shown how this degradation can be reduced considerably. The problem is much more difficult with rectangular grid arrays, but an approximate procedure will be presented which provides some improvement.

Difference patterns are amenable to many of these synthesis techniques, and some attention will be given to such applications.

A general formulation of the synthesis of a continuous planar aperture distribution needed to produce a specified far-field pattern can be given in terms of Fourier integrals. A presentation of this technique and some fundamental deductions which can be drawn from it form the concluding section of this chapter.

6.2 Rectangular Grid Arrays: Rectangular Boundary and Separable Distribution[1]

(a) PRELIMINARIES Consider a planar array in which the elements are arranged in a rectangular grid, with a rectangular boundary, as shown in Figure 6.1. Let there be $2N_x + 1$ rows of elements, each row parallel to the Y-axis, with common spacing d_x between rows. Let each row contain $2N_y + 1$ elements[2] with common spacing d_y. By the mnth element will be meant the element whose positional coordinates are $\xi_m = md_x$ and $\eta_n = nd_y$ in which $-N_x \leq m \leq N_x$ and $-N_y \leq n \leq N_y$. The current representative of the mnth element will be designated I_{mn}. With this notation, the array factor in (4.7) can be written

$$\mathcal{A}_a(\theta, \phi) = \sum_{m=-N_x}^{N_x} \sum_{n=-N_y}^{N_y} \left(\frac{I_{mn}}{I_{00}}\right) e^{jk \sin\theta(md_x \cos\phi + nd_y \sin\phi)} \tag{6.1}$$

If the representative current is magnetic, (6.1) can be replaced by an identical equation for $\mathcal{F}_a(\theta, \phi)$. Thus the following analysis applies equally well for arrays of elements which are replaced by equivalent magnetic sources.

[1]The analysis in this section follows closely some earlier writing by the author, contained in "Beamwidth and Directivity of Large Scanning Arrays: Part II," *Microwave Journal*, 7 (1964), 74–82. Also, in *Microwave Scanning Antennas*, ed. R. C. Hansen, Vol. 2, (New York: Academic Press, 1966), Chapter 1. Reprinted with joint permission.

[2]The even case of $2N_x$ by $2N_y$ elements can be treated in a completely analogous manner.

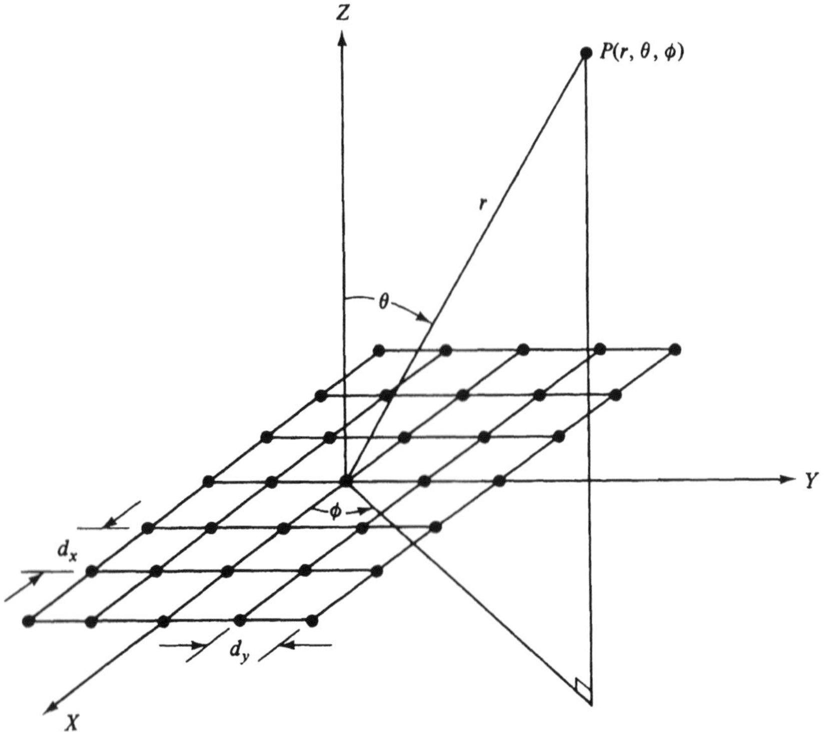

Fig. 6.1 Planar Array; Elements Arranged in a Rectangular Grid

If each row has the same current distribution—even though the current levels are different in different rows—that is, if $I_{mn}/I_{m0} = I_{0n}/I_{00}$, then the current distribution is said to be *separable* and the array factor can be expressed in the form

$$\mathcal{A}_a(\theta, \phi) = \mathcal{A}_x(\theta, \phi)\mathcal{A}_y(\theta, \phi) \tag{6.2}$$

in which

$$\mathcal{A}_x(\theta, \phi) = \sum_{-N_x}^{N_x} I_m e^{jmkd_x \sin\theta \cos\phi} \tag{6.3}$$

$$\mathcal{A}_y(\theta, \phi) = \sum_{-N_y}^{N_y} I_n e^{jnkd_y \sin\theta \sin\phi} \tag{6.4}$$

and

$$I_m = I_{m0}/I_{00}, \qquad I_n = I_{0n}/I_{00} \tag{6.5}$$

are the normalized current distributions in a row of elements parallel to the X-axis and the Y-axis, respectively.

Equation 6.2 is another example of the principle of pattern multiplication. The factors in (6.3) and (6.4) can be recognized as representing linear arrays parallel to the X- and Y-axes, since $\mathbf{1}_r \cdot \mathbf{1}_x = \cos\theta_x = \sin\theta\cos\phi$ and $\mathbf{1}_r \cdot \mathbf{1}_y = \cos\theta_y = \sin\theta\sin\phi$. Thus, under the stated restriction that the aperture distribution is separable, the array factor for a rectngular grid array with a rectangular boundary is the *product* of the array factors for two linear arrays, one laid out along the X-axis, and the other laid out along the Y-axis. Many of the results which have been developed for linear arrays in Chapters 4 and 5 can thus be interpreted to apply for this type of planar array as well.

(b) BEAM POSITION OF THE SUM PATTERN If I_{mn} differs in phase from I_{00} by the factor $\exp[-j(m\alpha_x + n\alpha_y)]$ then

$$\mathcal{A}_a(\theta, \phi) = \left[\sum_{-N_x}^{N_x} I_m e^{jm(kd_x \sin\theta\cos\phi - \alpha_x)}\right]\left[\sum_{-N_y}^{N_y} I_n e^{jn(kd_y \sin\theta\sin\phi - \alpha_y)}\right] \tag{6.6}$$

and the distribution has a uniform phase progression α_x in the X-direction and a uniform phase progression α_y in the Y-direction. The amplitude distributions I_m and I_n are now pure real. If they are also symmetric, the factor $\mathcal{A}_x$ represents a pattern that consists of a conical main beam and side lobes, rotationally symmetric about the X-axis. The main beam of $\mathcal{A}_x$ makes an angle θ'_x with the positive X-axis that satisfies the relation

$$\begin{gathered} kd_x \cos\theta'_x - \alpha_x = kd_x \sin\theta\cos\phi - \alpha_x = 0 \\ \cos\theta'_x = \frac{\alpha_x}{kd_x} = \sin\theta\cos\phi \end{gathered} \tag{6.7}$$

Similarly, the factor $\mathcal{A}_y$ gives a pattern that consists of a conical main beam and side lobes, rotationally symmetric about the Y-axis. The main beam of $\mathcal{A}_y$ makes an angle θ'_y with the positive Y-axis satisfying the relation

$$\cos\theta'_y = \frac{\alpha_y}{kd_y} = \sin\theta\sin\phi \tag{6.8}$$

The criterion developed in Chapter 4 applies to these two patterns with respect to the avoidance of multiple conical main beams in $\mathcal{A}_x$ and $\mathcal{A}_y$. Neither d_x nor d_y should exceed one-half wavelength if the two conical patterns are to be scanned to the vicinity of endfire.

The planar array factor $\mathcal{A}_a = \mathcal{A}_x\mathcal{A}_y$, since it is the product of the two linear array factors, is principally the intersection of the two conical main beams, plus those side lobes of each conical pattern which intersect with the conical main beam of the other conical pattern. Of course, it is possible to scan one conical main beam so close to the X-axis and the other conical main beam so close to the Y-axis that the two conical main beams do not intersect. This is an impractical situation, and a criterion will be developed shortly for avoiding it.

If the two conical main beams do intersect, their product gives two pencil beams, one pointing in the half-space $z > 0$, the other pointing in the half-space $z < 0$.

Almost invariably, the element pattern will be selected to give negligible radiation in the half-space $z < 0$ (through use of a ground plane, for example). There is then left a single main pencil beam, pointing in the direction (θ_0, ϕ_0), with θ_0 and ϕ_0 satisfying two equations that can be deduced from (6.7) and (6.8), namely,

$$\tan \phi_0 = \frac{\alpha_y d_x}{\alpha_x d_y} \tag{6.9}$$

$$\sin^2 \theta_0 = \left(\frac{\alpha_x}{kd_x}\right)^2 + \left(\frac{\alpha_y}{kd_y}\right)^2 \tag{6.10}$$

For given spacings d_x and d_y, and given interelement phase shifts α_x and α_y, Equations 6.9 and 6.10 give a unique pointing direction (θ_0, ϕ_0) in $z > 0$. Radiation patterns from planar arrays, exhibiting this feature of a single pencil beam, are called *sum patterns*.

Equation 6.10 can be used as the criterion for avoiding the situation that $\mathcal{A}_x$ and $\mathcal{A}_y$ contain conical main beams that do not intersect. This situation would just be reached if $\sin^2 \theta_0 = 1$. Thus, the elliptical relation

$$\left(\frac{\alpha_x}{kd_x}\right)^2 + \left(\frac{\alpha_y}{kd_y}\right)^2 = 1 \tag{6.11}$$

limits the range of α_x (or α_y) for specified values of kd_x, kd_y, and α_y (or α_x). In the remainder of this analysis, the existence of a single main pencil beam will be assumed.

(c) BEAMWIDTH OF THE SUM PATTERN Since the significant side lobes are in the two cones defined by (6.7) and (6.8), these cones are the pattern cuts which should be taken to determine the side lobe level. However, the profiles of the main pencil beam obtained in these two cuts are not, in general, due to two orthogonal slices through the pattern. (For example, if the pencil beam lies close to the XY-plane at $\phi_0 = \pi/4$, these two cuts are almost coincident.) Thus it is desirable to define beamwidth in another fashion, one which will reveal more information about the structure of the pencil beam.

In what is to follow, it will be shown that the -3 dB contour of the pencil beam is approximately elliptical. The beam cross section is suggested in Figure 6.2. At a given large distance r from the planar array, the size and shape of this elliptical contour are dependent on the pointing direction (θ_0, ϕ_0), as is the tilt of the axes of the ellipse. The two orthogonal planes which contain, respectively, one or the other of the ellipse axes, plus the origin, may be used to define the pattern cuts in which the beamwidth is measured. These two orthogonal measurements of half-power beamwidth then serve to specify the major and minor diameters of the elliptical contour, and thus give an indication of the size and shape of the beam cross section.

From (6.6), the central point in the main beam has the intensity

$$\mathcal{A}(\theta_0, \phi_0) = \sum_{-N_x}^{N_x} \sum_{-N_y}^{N_y} I_m I_n$$

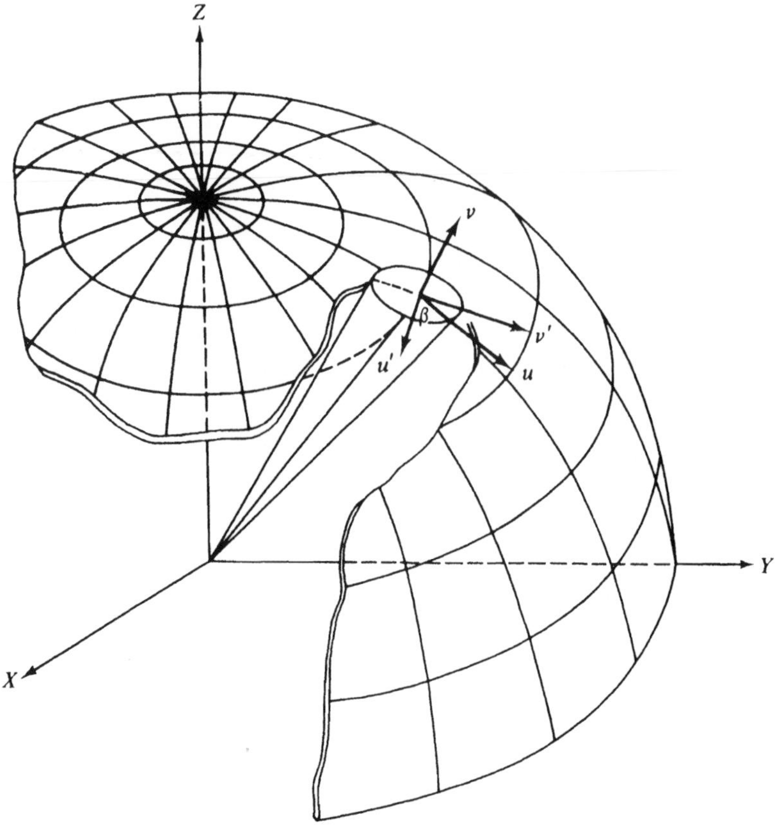

Fig. 6.2 Orthogonal Beamwidths of a Pencil Beam (Reprinted from *Microwave Scanning Antennas*, Volume 2, R. C. Hansen, Editor, Courtesy of Academic Press, Inc. © 1966 Academic Press.)

In a nearby direction $\theta_0 + \delta\theta, \phi_0 + \delta\phi$, the intensity will be down by 3 dB if

$$\begin{aligned}
\mathcal{A}(\theta_0 + \delta\theta, \phi_0 + \delta\phi) &= 0.707\mathcal{A}(\theta_0, \phi_0) = 0.707 \sum_{-N_x}^{N_x} \sum_{-N_y}^{N_y} I_m I_n \\
&= \sum_{-N_x}^{N_x} I_m \exp\{jmkd_x[\sin(\theta_0 + \delta\theta)\cos(\phi_0 + \delta\phi) - \sin\theta_0 \cos\phi_0]\} \\
&\quad \times \sum_{-N_y}^{N_y} I_n \exp\{jnkd_y[\sin(\theta_0 + \delta\theta)\sin(\phi_0 + \delta\phi) - \sin\theta_0 \sin\phi_0]\}
\end{aligned} \tag{6.12}$$

For large arrays, $\delta\theta$ and $\delta\phi$ are small, and (6.12) reduces to

$$\begin{aligned}
0.707 \sum_{-N_x}^{N_x} \sum_{-N_y}^{N_y} I_m I_n &= \sum_{-N_x}^{N_x} I_m \exp\{jmkd_x[\cos\theta_0 \cos\phi_0\, \delta\theta - \sin\theta_0 \sin\phi_0\, \delta\phi]\} \\
&\quad \times \sum_{-N_y}^{N_y} I_n \exp\{jnkd_y[\cos\theta_0 \sin\phi_0\, \delta\theta + \sin\theta_0 \cos\phi_0\, \delta\phi]\}
\end{aligned} \tag{6.13}$$

The right side of (6.13) consists of a family of phasors, symmetrically spread out in the complex plane, much as in the case of the linear array described in Section 4.3.

This sum is to be 0.707 times the sum of the phasors when they are aligned. For conventional distributions, the phasors do not need to fan out very far for this to occur. The outmost phasor normally does not reach a position of more than $\pi/2$ radians. Thus, if one lets

$$\Omega_x = kd_x[\cos\theta_0 \cos\phi_0\, \delta\theta - \sin\theta_0 \sin\phi_0\, \delta\phi]$$

and

$$\Omega_y = kd_y[\cos\theta_0 \sin\phi_0\, \delta\theta + \sin\theta_0 \cos\phi_0\, \delta\phi]$$

the phase factor $\exp[j(m\Omega_x + n\Omega_y)]$ can be expanded in a power series that will converge reasonably rapidly even for the largest values of m and n. When this is done, (6.13) becomes

$$0.707 \sum_{-N_x}^{N_x} \sum_{-N_y}^{N_y} I_m I_n = \sum_{-N_x}^{N_x} \sum_{-N_y}^{N_y} I_m I_n \times \left[1 + j(m\Omega_x + n\Omega_y) - \tfrac{1}{2}(m\Omega_x + n\Omega_y)^2 - \left(\frac{j}{3!}\right)(m\Omega_x + n\Omega_y)^3 + \cdots\right]$$

Since the distributions I_m and I_n have been assumed to be symmetrical, all summations which contain m or n to an odd power are zero. Thus, through *third order*,

$$0.586 \sum_{-N_x}^{N_x} \sum_{-N_y}^{N_y} I_m I_n = \Omega_x^2 \sum_{-N_x}^{N_x} \sum_{-N_y}^{N_y} m^2 I_m I_n + \Omega_y^2 \sum_{-N_x}^{N_x} \sum_{-N_y}^{N_y} n^2 I_m I_n \tag{6.14}$$

The two sums that appear on the right side of (6.14) can be evaluated by considering the situation in which the beam lies in either the XZ- or the YZ-plane. When the XZ-plane is chosen, $\phi_0 = 0$, and (6.14) becomes

$$\begin{aligned} 0.586 \sum_{-N_x}^{N_x} \sum_{-N_y}^{N_y} I_m I_n &= (kd_x \cos\theta_0\, \delta\theta)^2 \sum_{-N_x}^{N_x} \sum_{-N_y}^{N_y} m^2 I_m I_n \\ &\quad + (kd_y \sin\theta_0\, \delta\phi)^2 \sum_{-N_x}^{N_x} \sum_{-N_y}^{N_y} n^2 I_m I_n \end{aligned} \tag{6.15}$$

When the pencil beam lies in the XZ-plane, it is caused by the intersection of: (1) a conical beam that makes an angle $(\pi/2) - \theta_0$ with the X-axis; and (2) a conical beam that makes an angle $(\pi/2)$ with the Y-axis. The pattern cut in the XZ-plane is therefore identical to the one that would be obtained if there were only a single linear array laid out along the X-axis. But this pattern contains two points that lie in the -3 dB contour of the pencil beam, namely the points $(\theta = \theta_0 \pm \frac{1}{2}\theta_x, \phi = 0)$, where θ_x is the half-power beamwidth of the X-directed linear array when its conical main beam makes an angle $(\pi/2) - \theta_0$ with the positive X-axis. For this reason, the couplet $(\delta\theta = \frac{1}{2}\theta_x, \delta\phi = 0)$ must satisfy (6.15), which gives

$$0.586 \sum_{-N_x}^{N_x} \sum_{-N_y}^{N_y} I_m I_n = (\tfrac{1}{2}kd_x \cos\theta_0\, \theta_x)^2 \sum_{-N_x}^{N_x} \sum_{-N_y}^{N_y} m^2 I_m I_n \tag{6.16}$$

For the special case $\theta_0 = 0$, (6.16) yields

$$\sum_{-N_x}^{N_x} \sum_{-N_y}^{N_y} m^2 I_m I_n = 0.586(\tfrac{1}{2}kd_x\theta_{x0})^{-2} \sum_{-N_x}^{N_x} \sum_{-N_y}^{N_y} I_m I_n \tag{6.17}$$

in which θ_{x0} is the broadside beamwidth of the X-directed linear array.

Similarly, with the pencil beam placed in the YZ-plane, one finds that

$$\sum_{-N_x}^{N_x} \sum_{-N_y}^{N_y} n^2 I_m I_n = 0.586(\tfrac{1}{2}kd_y\theta_{y0})^{-2} \sum_{-N_x}^{N_x} \sum_{-N_y}^{N_y} I_m I_n \tag{6.18}$$

If these two results are inserted in (6.14), rearrangement gives

$$\frac{\Omega_x^2}{(\tfrac{1}{2}kd_x\theta_{x0})^2} + \frac{\Omega_y^2}{(\tfrac{1}{2}kd_y\theta_{y0})^2} = 1 \tag{6.19}$$

Since $(\delta\theta, \delta\phi)$, and thus $(\Omega_x\ \Omega_y)$, defines the set of pointing directions in which the field is 0.707 times the peak value, Equation 6.19 can be viewed as describing the -3 dB contour on the pencil beam. To see this more clearly, let u- and v-axes be erected along lines of longitude and latitude on the sphere of radius r, as shown in Figure 6.2. Then

$$u = r\,\delta\theta \qquad v = r \sin\theta_0\,\delta\phi$$

Substitution of these variables in (6.19) gives

$$\frac{(u\cos\theta_0\cos\phi_0 - v\sin\phi_0)^2}{(r\theta_{x0}/2)^2} + \frac{(u\cos\theta_0\sin\phi_0 + v\cos\phi_0)^2}{(r\theta_{y0}/2)^2} = 1 \tag{6.20}$$

This can be recognized as the equation of an ellipse in (u, v)-space. Introduction of the axes u' and v' via the rotation β (compare with Figure 6.2), such that

$$u = u'\cos\beta + v'\sin\beta, \qquad v = -u'\sin\beta + v'\cos\beta$$

permits (6.20) to be written in the form

$$\frac{(u')^2}{(d_{u'}/2)^2} + \frac{(v')^2}{(d_{v'}/2)^2} = 1 \tag{6.21}$$

In (6.21) $d_{u'}$ and $d_{v'}$ are the diameters of the ellipse measured along its two principal axes. The rotational angle β is given by

$$\tan 2\beta = \frac{2\cos\theta_0\sin 2\phi_0}{(1+\cos^2\theta_0)\cos 2\phi_0 + [(\theta_{x0}^2 + \theta_{y0}^2)/(\theta_{x0}^2 - \theta_{y0}^2)]\sin^2\theta_0} \tag{6.22}$$

At a constant zenith angle θ_0, β rotates smoothly through 90° as ϕ_0 changes through 90°. Individual expressions for $d_{u'}$ and $d_{v'}$ are unwieldy, but their product is given by the simple expression

$$d_{u'}d_{v'} = r^2 \sec\theta_0\, \theta_{x0}\theta_{y0} \tag{6.23}$$

Thus the *area* of the ellipse is independent of ϕ_0.

If the pencil beam lies in the XZ-plane in the direction $(\theta_0, 0)$, the u'- and v'-axes are aligned with the u- and v-axes and (6.20) gives

$$d_u = d_{u'} = r\sec\theta_0\,\theta_{x0}, \qquad d_v = d_{v'} = r\theta_{y0} \tag{6.24}$$

In the pattern cuts containing the u-axis or the v-axis, the half-power beamwidths are therefore

$$\theta_u = \frac{d_u}{r} = \theta_{x0}\sec\theta_0 \quad \text{and} \quad \theta_v = \frac{d_v}{r} = \theta_{y0} \qquad (\phi_0 = 0) \tag{6.25}$$

Similarly, if the pencil beam lies in the YZ-plane in the direction $(\theta, \pi/2)$, the u'-axis points in the $-v$-direction and the v'-axis is aligned with u. For this case (6.20) gives

$$d_u = d_{v'} = r\sec\theta_0\,\theta_{y0}, \qquad d_v = d_{u'} = r\theta_{x0} \tag{6.26}$$

In the pattern cuts containing the u-axis or the v-axis, the half-power beamwidths are now

$$\theta_u = \frac{d_u}{r} = \theta_{y0}\sec\theta_0 \quad \text{and} \quad \theta_v = \frac{d_v}{r} = \theta_{x0} \qquad \left(\phi_0 = \frac{\pi}{2}\right) \tag{6.27}$$

To use either Equation 6.25 or Equation 6.27, one needs first to determine θ_{x0}, θ_{y0}, and the zenith pointing angle θ_0. For uniform distributions, θ_{x0} and θ_{y0} can be determined by using L_x/λ and L_y/λ and reading the appropriate beamwidth off the broadside curve of Figure 5.3. For Dolph-Chebyshev distributions, these beamwidths need to be modified by the f-factor read from Figure 5.4. After this it is a simple matter to determine θ_u and θ_v.

It is useful to define an areal beamwidth B by the relation

$$B = \theta_{u'}\,\theta_{v'} \tag{6.28}$$

Through use of (6.23) this becomes

$$B = \left(\frac{d_{u'}}{r}\right)\left(\frac{d_{v'}}{r}\right) = \theta_{x0}\theta_{y0}\sec\theta_0 \tag{6.29}$$

The areal beamwidth, which is a measure of the area inside the -3 dB contour of the pencil beam cross section, is seen to be independent of ϕ_0. As one would expect, it has the same functional dependence on θ_0 that the projected aperture does.

The general effect of scanning a pencil beam can be constructed as suggested in exaggeration by Figure 6.3. At broadside-broadside, the cross section of the beam

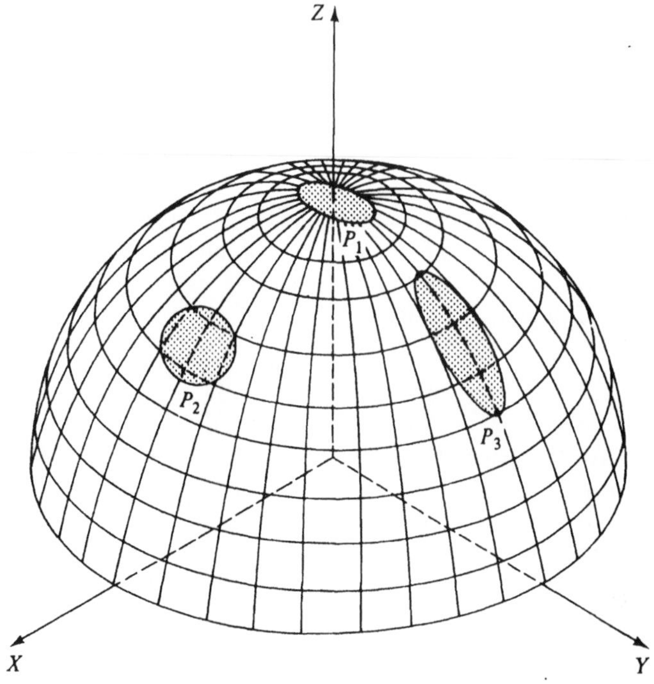

Fig. 6.3 Beam Shape versus Scan Position for a Pencil Beam (Reprinted from *Microwave Scanning Antennas*, Volume 2, R. C. Hansen, Editor, Courtesy of Academic Press, Inc. © 1966 Academic Press.)

is approximately elliptical (position P_1) with dimensions proportional to L_x^{-1} and L_y^{-1}. As the beam scans in the XZ-plane, the beam cross section elongates in that direction (position P_2). Scanning in the YZ-plane causes elongation in the other beam dimension (position P_3). For a constant angle θ_0 from the zenith, as the beam is moved from $\phi_0 = 0$ to $\phi_0 = \pi/2$, the two half-power beamwidths smoothly change and the elliptical cross section smoothly rotates, these two effects combining in such a way that the areal beamwidth remains constant.

Thus for narrow pencil beams from large rectangular grid arrays with rectangular boundaries, if the distribution is separable, the entire subject of beamwidth can be based on the results previously obtained for linear arrays. The relations derived in this section are quite good to within several beamwidths of the limiting condition of no main beam at all, defined by (6.11).

(d) PEAK DIRECTIVITY OF THE SUM PATTERN The peak directivity of this type of planar array (compare with Section 1.16) is given by

$$D = \frac{4\pi \mathcal{A}(\theta_0, \phi_0)\mathcal{A}^*(\theta_0, \phi_0)}{\int_0^{\pi/2}\int_0^{2\pi} \mathcal{A}(\theta, \phi)\mathcal{A}^*(\theta, \phi) \sin\theta \, d\theta \, d\phi} \tag{6.30}$$

in which it is assumed that the element pattern is such as to eliminate the total pattern in the half-space $\theta > \pi/2$ but is broad enough to be ignored in $\theta \leq \pi/2$. It has been shown in the literature[3] that, for large arrays which are not scanned closer than several beamwidths to end-fire, an approximate reduction of (6.30) is

$$D = \pi \cos\theta_0 \frac{2L_x/\lambda}{\sum_p (a_p/a_0)^2} \frac{2L_y/\lambda}{\sum_q (b_q/b_0)^2} \tag{6.31}$$

in which the Fourier coefficients a_p and b_p describe the aperture distribution in the X- and Y-directions, and $L_x = (2N_x + 1)d_x$ and $L_y = (2N + 1)d_y$ are the dimensions of the array. This equation has the simple interpretation that the maximum peak directivity of a planar array is

$$D = \pi D_x D_y \cos\theta_0 \tag{6.32}$$

in which D_x and D_y are the directivities of the two linear arrays. The factor $\cos\theta_0$ accounts for the decrease in projected aperture with scan. Unlike the directivity of a linear array, which was found to be independent of scan angle, the directivity of a planar array is dependent on the zenith coordinate θ_0. However, it is independent of the azimuthal coordinate ϕ_0.

Because of the form of (6.31), many of the remarks that have been made about the directivity of linear arrays can be applied as well to this type of planar array. For aperture distributions with a uniform progressive phase and a symmetric amplitude, (6.31) indicates that maximum directivity results from the choice of uniform excitation. Dolph-Chebyshev distributions suffer from a gain limit for very large arrays, and the curves of Figure 5.5 are applicable to such planar arrays.

(e) A RELATION BETWEEN BEAMWIDTH AND PEAK DIRECTIVITY As in the case of linear arrays, one finds from (6.29) and (6.31) that, for this type of planar array, peak directivity and the reciprocal of areal beamwidth depend linearly on the area of the planar aperture. Elimination of $L_x L_y/\lambda^2$ from these two expressions results in

$$D = \frac{9.87}{B}\left[\frac{f_x}{\sum_{-P}^{P} (a_p/a_0)^2} \frac{f_y}{\sum_{-Q}^{Q} (b_q/b_0)^2}\right] \tag{6.33}$$

where f_x and f_y are the beam broadening factors for the linear arrays of $2N_x + 1$ and $2N_y + 1$ elements that comprise the two dimensions of the array. The quantity in brackets is unity for a uniform distribution, and is essentially unity for a Dolph-Chebyshev distribution until gain limiting sets in. Thus, for these practical aperture distributions,

$$D = \frac{32{,}400}{B} \tag{6.34}$$

[3] R. S. Elliott, "Beamwidth and Directivity of Large Scanning Arrays", Appendix D, *Microwave Journal*, 7 (1964), 74–82. Also, *Microwave Scanning Antennas*, ed. R. C. Hansen, vol. 2 (New York: Academic Press, 1966), Chapter 1.

in which the areal beamwidth is now expressed in square degrees rather than square radians.[4] It is important to remember that, in (6.34), both quantities are measured at the same tilt angle θ_0.

(f) SUM AND DIFFERENCE PATTERNS To this point in the discussion, the array factor given by Equation 6.6 has been interpreted under the assumption that the normalized current distributions I_m and I_n were symmetrical, the result in $z > 0$ being a sum pattern consisting of a pencil beam and a family of side lobes. In such circumstances, an alternate expression for the array factor is

$$\mathcal{S}(\theta, \phi) = \left(1 + 2 \sum_{m=1}^{N_x} I_m \cos m\psi_x\right)\left(1 + 2 \sum_{n=1}^{N_y} I_n \cos n\psi_y\right) \tag{6.35}$$

in which

$$\psi_x = kd_x \sin\theta \cos\phi - \alpha_x, \qquad \psi_y = kd_y \sin\theta \sin\phi - \alpha_y \tag{6.36}$$

For the case of an even number of elements in each dimension, $2N_x$ by $2N_y$, (6.35) is replaced by

$$\mathcal{S}(\theta, \phi) = 4\left[\sum_{m=1}^{N_x} I_m \cos\left(\frac{2m-1}{2}\psi_x\right)\right]\left[\sum_{n=1}^{N_y} I_n \cos\left(\frac{2n-1}{2}\psi_y\right)\right] \tag{6.37}$$

In this latter case, two difference patterns can be generated, one by causing $I_m = -I_{-m}$, while leaving $I_n = I_{-n}$, the other by doing the reverse. The first condition gives

$$\mathfrak{D}_1(\theta, \phi) = 4j\left[\sum_{m=1}^{N_x} I_m \sin\left(\frac{2m-1}{2}\psi_x\right)\right]\left[\sum_{n=1}^{N_y} I_n \cos\left(\frac{2n-1}{2}\psi_y\right)\right] \tag{6.38}$$

while the second condition gives

$$\mathfrak{D}_2(\theta, \phi) = 4j\left[\sum_{m=1}^{N_x} I_m \cos\left(\frac{2m-1}{2}\psi_x\right)\right]\left[\sum_{n=1}^{N_y} I_n \sin\left(\frac{2n-1}{2}\psi_y\right)\right] \tag{6.39}$$

In the $\phi = 0°, 180°$ plane, $\mathfrak{D}_1(\theta, \phi)$ gives the pattern of an X-directed linear array, identical to the result in (5.3). As one examines a succession of ϕ-cuts in $0° \le \phi \le 90°$, a difference pattern is always observed, with the level of the entire pattern diminishing until, at $\phi = 90°$, there is no pattern at all. This behavior is repeated in the other three quadrants as ϕ is varied. The behavior of $\mathfrak{D}_2(\theta, \phi)$ is similar, except that its highest level occurs in $\phi = 90°, 270°$, diminishing to zero in $\phi = 0°, 180°$. In radar applications, the sum pattern of (6.37) can be used to acquire the target by proper pointing of the pencil beam, and then the difference patterns $\mathfrak{D}_1$ and $\mathfrak{D}_2$ can be used to boresight the target more accurately.

[4]Equation 6.34 is frequently encountered in the literature with the incorrect coefficient 41,253. (See, for example, J. D. Kraus, *Antennas* (New York: McGraw-Hill Book Co., Inc., 1950), p. 25.) The higher figure results from the improper assumption that the 3 dB contour is rectangular. Since the area of a rectangle is $4/\pi$ times the area of its inscribed ellipse, the ratio of these two coefficients is readily understood.

A proper selection of the current distributions in (6.37), (6.38), and (6.39) can yield sum and difference patterns with prescribed side lobe topography. With the assumption being made here of separable distributions, this selection is the same as for linear array applications, and all of the procedures discussed in Chapter 5 are applicable.

(g) AN ILLUSTRATIVE EXAMPLE Assume that the assignment has been given to design a planar array under the following specifications.

1. Rectangular grid, rectangular boundary, separable distribution.
2. Sum pattern scannable $\pm 45°$ in XZ-plane and $\pm 30°$ in YZ-plane.
3. When the sum pattern is broadside-broadside ($\theta_0 = 0°$), the 3 dB beamwidths are to be 5° in the XZ-plane and $2\frac{1}{2}°$ in the YZ-plane. Both principal cuts are to be Dolph-Chebyshev, with -30 dB side lobe levels.
4. The array should be capable of generating both difference patterns as well as the sum pattern.

The last condition implies that an even number of elements should be chosen in each dimension. One can determine the lengths L_x/λ and L_y/λ through use of Figures 5.3 and 5.4. The results are

$$\frac{L_x}{\lambda} = (1.15)(10) = 11.5$$

$$\frac{L_y}{\lambda} = (1.15)(20) = 23$$

To avoid multiple main beams when scanning in the XZ-plane, (4.30) indicates that[5]

$$\frac{d_x}{\lambda} < \frac{1}{1 + |\sin\theta_0|} = \frac{1}{1 + \sin 45°} = 0.59$$

Since $L_x = 2N_x d_x$, it follows that

$$2N_x \geq \frac{11.5}{0.59} = 19.63$$

Thus the tentative choice $2N_x = 20$ can be made, but this will need to be checked through a computation of final sum patterns at the limiting scan positions.

Similarly,

$$\frac{d_y}{\lambda} < \frac{1}{1 + \sin 30°} = 0.67$$

$$2N_y \geq \frac{23}{0.67} = 34.5$$

[5]It should be observed that, in Chapter 4, linear arrays that extended along the Z-axis were considered. Now, linear arrays along the X- and Y-axes are being considered, so $\cos\theta_0$ must be replaced by $\sin\theta_0$.

A tentative choice of $2N_y = 36$ can also be made. With $L_x/\lambda = 11.5$ and $L_y/\lambda = 23$, this also means that the tentative values

$$\frac{d_x}{\lambda} = 0.58 \quad \text{and} \quad \frac{d_y}{\lambda} = 0.64$$

are being selected.

From (6.36) one can learn that

$$\alpha_x^{\max} = 2\pi(0.58)\sin 45^\circ = (0.41)2\pi \text{ radians} = 148^\circ$$

$$\alpha_y^{\max} = 2\pi(0.64)\sin 30^\circ = (0.32)2\pi \text{ radians} = 115^\circ$$

A check of Equation 6.11 gives

$$\left(\frac{\alpha_x^{\max}}{kd_x}\right)^2 + \left(\frac{\alpha_y^{\max}}{kd_y}\right)^2 = \left(\frac{0.41}{0.58}\right)^2 + \left(\frac{0.32}{0.64}\right)^2 = 0.75 < 1$$

This indicates that, even if α_x and α_y take on their maximum values simulatneously, the two conical beams will intersect to give a pencil beam.

The normalized Dolph-Chebyshev current distributions can be computed from (5.19) for a 30 dB side lobe level with 20 and 36 elements, or can be obtained from the literature.[6] The results are shown in Table 6.1.

TABLE 6.1

$\pm m$	1	2	3	4	5	6	7	8	9	10
I_m	1.0	0.97	0.91	0.83	0.73	0.62	0.50	0.39	0.29	0.33

$\pm n$	1	2	3	4	5	6	7	8	9	10	11	12	13	14	15	16	17	18
I_n	1.0	0.99	0.97	0.95	0.91	0.87	0.82	0.77	0.71	0.65	0.59	0.53	0.47	0.40	0.35	0.29	0.24	0.49

Since the distribution is separable, $I_{mn} = I_m I_n$, and one can prepare a schedule of element excitations as suggested in Table 6.2. Only one quadrant needs to be shown since the excitation is symmetric.

One cannot take this design much further without knowing the nature of the elements making up the array. Then a feeding network would need to be devised to deliver these currents to the individual radiators. This is a complicated problem that must take into account not only the self impedance of the elements, but also their mutual impedances. The need to scan is an added complication. These problems will be addressed in Chapter 8 of this text.

[6] L. B. Brown and G. A. Scharp, "Chebyshev Antenna Distribution, Beamwidth, and Gain Tables", *Nav. Ord. Report 4629*, (California: Corona, 1958).

TABLE 6.2

m \ n	1	2	3	4	5	6	7
1	1.00	0.99	0.97	0.95	0.91	0.87	0.82
2	0.97	0.96	0.94	0.92	0.88	0.84	0.80
3	0.91	0.90	0.88	0.86	0.83	0.79	·
4	0.83	0.82	0.81	0.79	0.76	0.72	·
5	0.73	0.72	0.71	0.69	0.66	·	·
6	0.62	0.61	0.60	0.59	0.56	·	·
7	0.50	0.50	0.49	0.48	·	·	·
8	0.39	0.39	0.38	0.37	·	·	·
9	0.29	0.29	0.28	0.28	·	·	·
10	0.33	0.33	0.32	·	·	·	·

The areal beamwidth of this array can be computed from (6.29), and at broadside-broadside is given by

$$B = \theta_{x0}\,\theta_{y0} = (5)(2.5) = 12.5 \text{ square degrees}$$

From (6.34), the broadside-broadside directivity is

$$D = 2592 = 34.1 \text{ dB}$$

As the beam is scanned, the areal beamwidth broadens as $\sec\theta_0$ and the directivity decreases as $\cos\theta_0$.

(h) THE NATURE OF THE SIDE LOBE REGION It has been observed that planar arrays of this type (rectangular grid, rectangular boundary, separable distribution) when excited with a symmetrical amplitude distribution, give a sum pattern with a pencil beam and side lobes, and that this pattern can be represented as the product of two conical patterns, one each from two orthogonal linear arrays. The nulls in this sum pattern thus conicide with the nulls in either of the conical patterns. On the surface of a large sphere centered at the array midpoint, these nulls are the intersections of two families of conical surfaces with the spherical surface, the conical axes of these families being the X- and Y-axes. To a person looking down on this spherical surface from a remote point on the Z-axis, if the pencil beam is pointing broadside-broadside, the grid of null intersections looks as shown in Figure 6.4. It is clear from a study of this figure that the main beam is surrounded by *mound-type* side lobes. Those side lobes that occur in the principal planes (XZ or YZ) are depressed by an amount governed by the design of the X-oriented and Y-oriented linear arrays. Thus, for example, if the X-oriented array is to be −20 dB SLL Dolph-Chebyshev, and the Y-oriented array is to be −30 dB SLL Dolph-Chebyshev, then all the mound side lobes that occur vertically above or below the main beam in Figure 6.4 (such as the side lobe occurring in the half-tone region A) are 20 dB in power level below the main beam. Similarly, all the side lobes that occur horizontally to the right or left of the main beam in Figure 6.4 (such as the side lobe occurring in the half-tone region B)

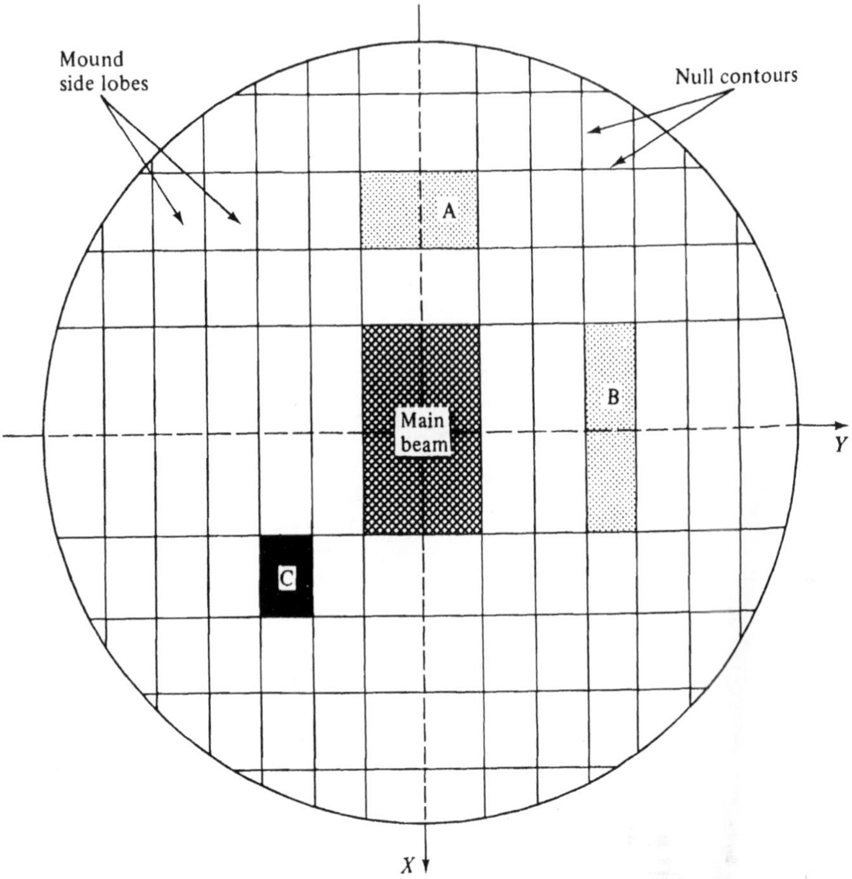

Fig. 6.4 Grid of Null Contours for the Sum Pattern of a Planar Array; Separable Aperture Distribution; Mound-Type Side Lobes

are 30 dB in power level below the main beam. However, an off-axis side lobe, such one which occupies region C in Figure 6.4, are the result of the intersection o conical side lobe of the X-directed linear array with a conical side lobe of the directed linear array. In this example, all of these off-axis mound side lobes therefore 50 dB in power level below the main beam.

This pinpoints the principal objection to separable distributions for pla arrays. Side lobe reduction is bought at the price of beam broadening (with a c comitant lowering of directivity). If the design requirement for this pattern were t side lobes be at -20 dB in the XZ-plane and at -30 dB everywhere else, then separable distribution overachieves in most of the side lobe region, at the expense resolution and directivity.

To improve on this situation, one must go to nonseparable aperture distri tions. As an example, if the entire side lobe region is of uniform importance, w ıld be most desirable is to see a Dolph-Chebyshev pattern in every ϕ-cut, that is, ıttern consisting of a pencil beam and a family of concentric ring side lobes of a ımon height, as suggested by the grid of null contours shown in Figure 6.5. But

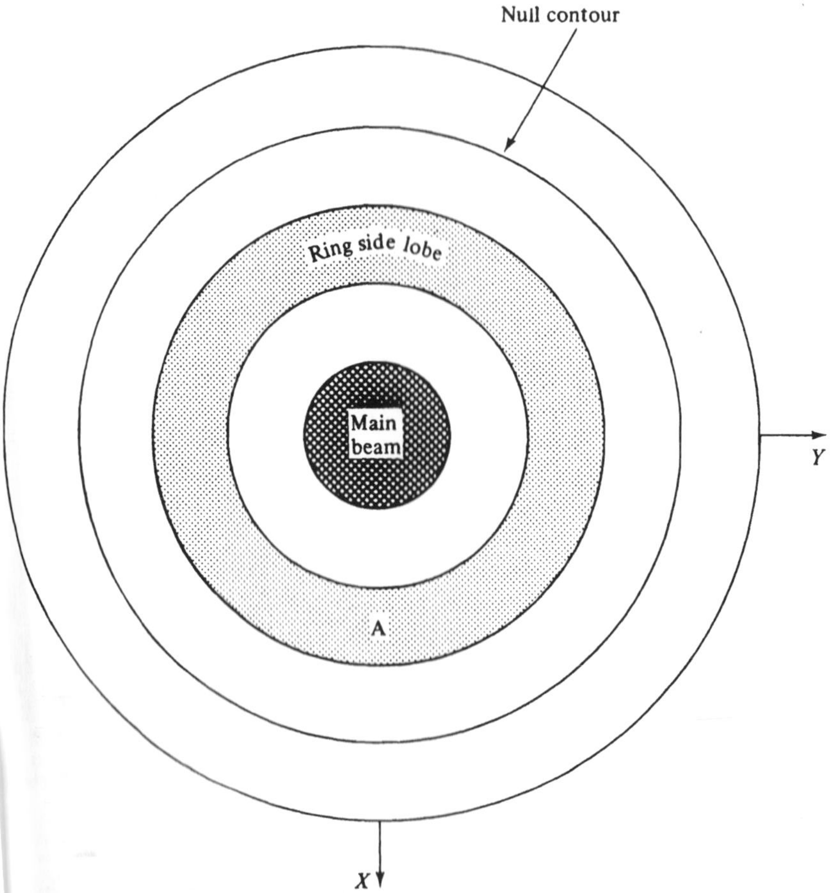

Fig. 6.5 Grid of Null Contours for a ϕ-Symmetric Sum Pattern of a Planar Array; Non-separable Aperture Distribution; Ring-Type Side Lobes

ıtain a ϕ-symmetric pattern, one must have a ϕ-symmetric aperture distribution. ·ly, a rectangular grid array, with a rectangular boundary and a separable ure distribution, does not fit this criterion. Several things are wrong. First, a ımetric aperture distribution more naturally fits a circular boundary. Second, re naturally fits a circular grid arrangement of the elements. Despite these natural backs, it will be seen in Section 6.9 that, under certain circumstances, a rectan- grid array, with a rectangular boundary and a *non*separable distribution, can a sum pattern which is Dolph-Chebyshev in all ϕ-cuts.

But first attention will be turned in Section 6.3 to the creation of a ϕ-symmetric sum pattern under the more natural condition of a circular boundary.

6.3 Circular Taylor Patterns

Consider a planar aperture with a circular boundary of radius a, as suggested by Figure 6.6. If this aperture contains a lineal current density distribution which is *unidirectional*, then for this case (1.128) and (1.129), or (1.130) and (1.131), can be written in the forms

$$\mathcal{A}_\theta(\theta, \phi) = \cos\theta \sin\phi \int_S K_y(\xi, \eta) e^{jk\mathcal{L}}\, d\xi\, d\eta \tag{6.40}$$

$$\mathcal{A}_\phi(\theta, \phi) = \cos\phi \int_S K_y(\xi, \eta) e^{jk\mathcal{L}}\, d\xi\, d\eta \tag{6.41}$$

in which $\mathcal{L}$ is reduced to

$$\mathcal{L} = \xi \sin\theta \cos\phi + \eta \sin\theta \sin\phi \tag{6.42}$$

The integral common to (6.40) and (6.41) can be viewed as the array factor for a linearly polarized planar aperture distribution. It is convenient to recast this integral in the polar coordinates illustrated in Figure 6.6 and defined by

$$\xi = \rho\cos\beta \qquad \eta = \rho\sin\beta \tag{6.43}$$

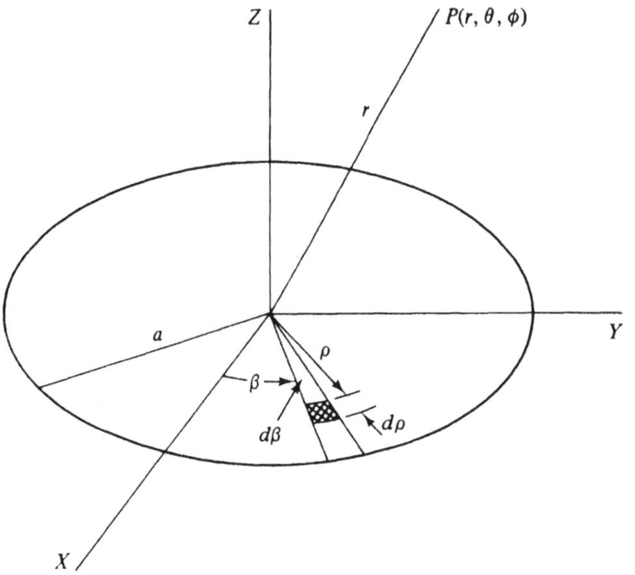

Fig. 6.6 Cartesian/Cylindrical/Spherical Coordinates

If $K_y(\xi, \eta)$ is designated as $K(\rho, \beta)$, the common integral becomes

$$F(\theta, \phi) = \int_0^a \int_0^{2\pi} K(\rho, \beta) e^{jk\rho \sin\theta \cos(\phi-\beta)} \rho \, d\rho \, d\beta \tag{6.44}$$

The aperture distribution can be represented by the Fourier series

$$K(\rho, \beta) = \sum_{n=-\infty}^{\infty} K_n(\rho) e^{jn\beta} \tag{6.45}$$

If (6.45) and the Bessel expansion

$$e^{jk\rho \sin\theta \cos(\phi-\beta)} = \sum_{m=-\infty}^{\infty} (j)^m J_m(k\rho \sin\theta) e^{jm(\phi-\beta)} \tag{6.46}$$

are inserted in (6.44), the result is that

$$F(\theta, \phi) = \sum_{m=-\infty}^{\infty} \sum_{n=-\infty}^{\infty} \int_0^{2\pi} \int_0^a K_n(\rho)(j)^m J_m(k\rho \sin\theta) e^{jm\phi} e^{j(n-m)\beta} \rho \, d\rho \, d\beta \tag{6.47}$$

The β-integration in (6.47) only has a nonzero value when $m = n$. This reduction gives

$$F(\theta, \phi) = 2\pi \sum_{n=-\infty}^{\infty} (j)^n e^{jn\phi} \int_0^a K_n(\rho) J_n(k\rho \sin\theta) \rho \, d\rho \tag{6.48}$$

It is clear from a study of (6.48) that, if a ϕ-independent pattern is desired, n should be restricted to the value zero. Returning to (6.45), one sees that this corresponds, quite logically, to choosing an aperture distribution that is β-independent.

If attention is restricted to this case, then the aperture distribution is $K_0(\rho)$ and the pattern is given by

$$F(\theta) = 2\pi \int_0^a K_0(\rho) J_0(k\rho \sin\theta) \rho \, d\rho \tag{6.49}$$

When one makes the substitutions

$$u = \frac{2a}{\lambda} \sin\theta \qquad p = \frac{\pi}{a}\rho \qquad g_0(p) = \frac{2a^2}{\pi} K_0(\rho) \tag{6.50}$$

Equation 6.49 transforms to

$$F(u) = \int_0^{\pi} p g_0(p) J_0(up) \, dp \tag{6.51}$$

It is useful to note at this point that u is a surrogate for the pointing direction in real space, and that p is a surrogate for the radial aperture coordinate.

A particular example of the use of (6.51) is of special importance. If the circular aperture is uniformly excited, a condition that can be represented by letting $g_0(p) = 1$, then integration of (6.51) gives the sum pattern

$$\mathcal{S}(u) = \frac{J_1(\pi u)}{\pi u} \tag{6.52}$$

This pattern is plotted in Figure 6.7 and shows a main beam plus a family of side lobes that decay in height as the side lobe position becomes more remote from the main beam. Since this pattern is rotationally symmetric, the main lobe is a pencil beam, surrounded by ring side lobes. How many of these side lobes are in visible space depends on the aperture size. Since $u = (2a/\lambda)\sin\theta$, the range of u corresponding to visible space is $0 \leq u \leq 2a/\lambda$.

This result should be compared to Figure 5.6 where, for the analogous case of a uniformly excited line source, the pattern was seen to be given by $\sin \pi u/\pi u$.

Continuing with the analogy, one can ask what form the function $g_0(p)$ should take in order to modify the side lobe structure of Figure 6.7 so that the near-in side

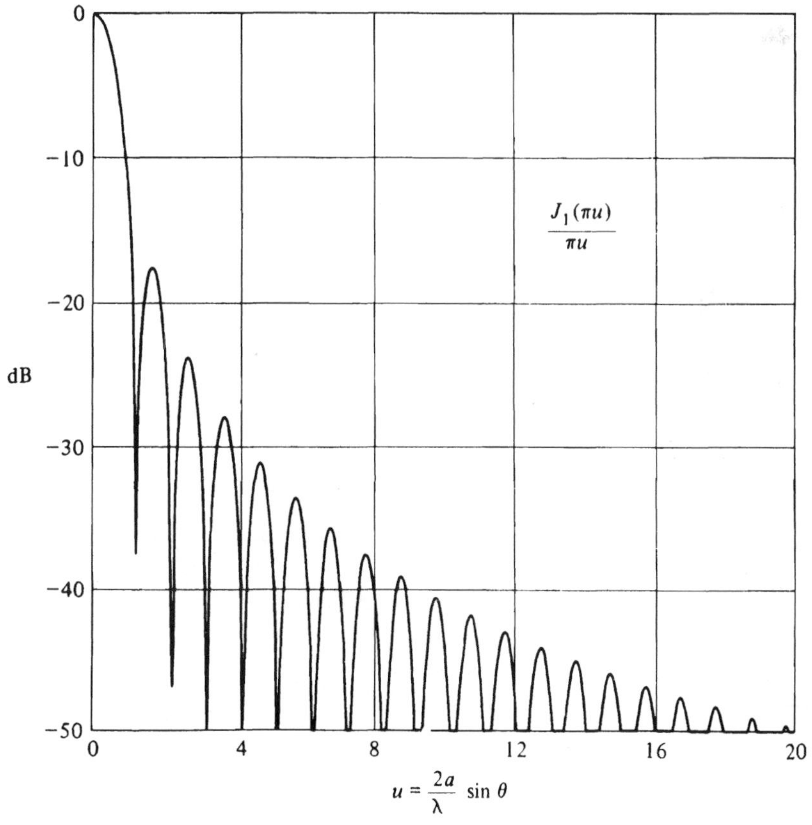

Fig. 6.7 Array Factor Pattern for a Uniformly Excited Circular Aperture

lobes are at a quasi-constant controlled height. T. T. Taylor undertook the solution to this problem in a companion paper to his earlier treatment of the line source.[7] The essence of his analysis consisted in moving the innermost $\bar{n}-1$ nulls of Figure 6.7 to achieve the desired level for the intervening side lobes. If the roots of $J_1(u)$ are defined by

$$J_1(\pi\gamma_{1n}) = 0, \qquad n = 0, 1, 2, \ldots$$

then a modification of (6.52) can be written as

$$\mathcal{S}(u) = \frac{J_1(\pi u)}{\pi u} \frac{\prod_{n=1}^{\bar{n}-1} (1 - u^2/u_n^2)}{\prod_{n=1}^{\bar{n}-1} (1 - u^2/\gamma_{1n}^2)} \tag{6.53}$$

One can see that (6.53) accomplishes the purpose of removing the first $\bar{n}-1$ root pairs of (6.52) and replacing them by $\bar{n}-1$ root pairs at the new positions $\pm u_n$. Taylor found that the new root positions should be such that

$$u_n^2 = \gamma_{1n}^2 \frac{A^2 + (n - \frac{1}{2})^2}{A^2 + (\bar{n} - \frac{1}{2})^2} \tag{6.54}$$

where once again, $-20 \log_{10} \cosh \pi A$ is the desired side lobe level.

A typical circular Taylor pattern is shown in Figure 6.8, with $\bar{n} = 6$ and a design side lobe level of -15 dB. The near-in side lobes are seen to droop somewhat. (The far-out side lobes share—with those of Figure 6.7—the property of decaying as $u^{-3/2}$.)

To find the aperture distribution $g_0(p)$ that will produce this type of pattern, it is helpful to express $g_0(p)$ as a series in the form

$$g_0(p) = \sum_{m=0}^{\infty} B_m J_0(\gamma_{1m} p) \tag{6.55}$$

When this is done, (6.51) becomes

$$\begin{aligned}\mathcal{S}(u) &= \sum_{m=0}^{\infty} B_m \int_0^{\pi} p J_0(\gamma_{1m} p) J_0(up)\, dp \\ &= \sum_{m=0}^{\infty} B_m \left[\frac{\gamma_{1m} p J_1(\gamma_{1m} p) J_0(up) - up J_0(\gamma_{1m} p) J_1(up)}{\gamma_{1m}^2 - u^2}\right]_0^{\pi}\end{aligned} \tag{6.56}$$

Since, in (6.56), $\mathcal{S}(\gamma_{1k})$ is at most contributed to by the kth term in the sum, one can write

[7] T. T. Taylor, "Design of Circular Apertures for Narrow Beamwidth and Low Side Lobes", *Trans. IRE*, AP-8 (1960), 17–22.

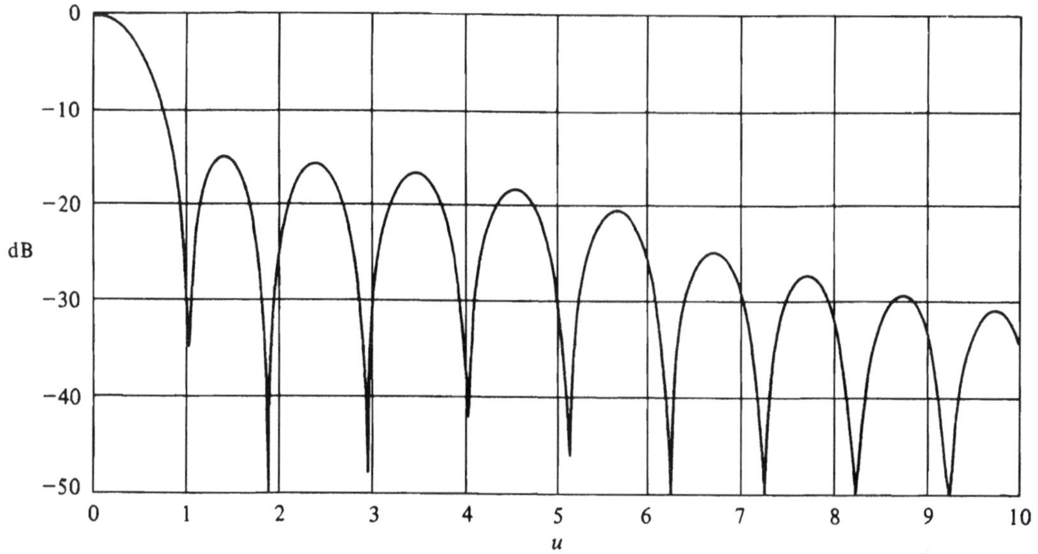

Fig. 6.8 Taylor Sum Pattern for a Circular Aperture, $\bar{n} = 6$, −15 dB SLL

$$\mathcal{S}(\gamma_{1k}) = B_k \int_0^\pi pJ_0^2(\gamma_{1k}p)\, dp$$

$$= B_k \left[\frac{p^2}{2} \{J_0^2(\gamma_{1k}p) + J_1^2(\gamma_{1k}p)\} \right]_0^\pi \tag{6.57}$$

from which

$$B_k = \frac{2}{\pi^2} \frac{\mathcal{S}(\gamma_{1k})}{J_0^2(\gamma_{1k}\pi)} \tag{6.58}$$

Because $\mathcal{S}(\gamma_{1k}) \equiv 0$ for $k \geq \bar{n}$, the series in (6.55) truncates, and the aperture distribution is given by

$$g_0(p) = \frac{2}{\pi^2} \sum_{m=0}^{\bar{n}-1} \frac{\mathcal{S}(\gamma_{1m})}{J_0^2(\gamma_{1m}\pi)} J_0(\gamma_{1m}p) \tag{6.59}$$

where $\mathcal{S}(\gamma_{1m})$ can be computed from (6.53). A plot of (6.59) for the aperture distribution corresponding to the pattern of Figure 6.8 is shown in Figure 6.9. This distribution, like the pattern, is a figure of revolution.

R. C. Hansen[8] has provided tables of the roots u_n of circular Taylor patterns,

[8] R. C. Hansen, "Tables of Taylor Distributions for Circular Aperture Antennas," Hughes Technical Memorandum No. 587, Hughes Aircraft Co. (California: Culver City, February 1959). See also *IRE Trans.*, AP-8 (1960), 22–26.

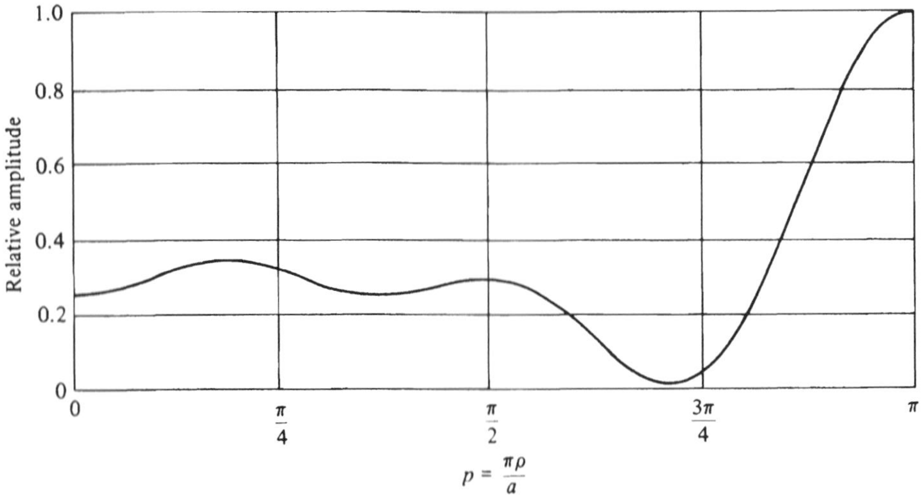

Fig. 6.9 The Continuous Aperture Distribution for the Circular Taylor Pattern of Fig. 6.8

together with the aperture distributions $g_0(p)$, for a range of side lobe levels and for a sequence of $\bar{n}$ values.

6.4 Modified Circular Taylor Patterns: Ring Side Lobes of Individually Arbitrary Heights

If one returns to Equation 6.53, which is the general expression for a circular Taylor pattern, and recalls that this equation results from replacing $\bar{n} - 1$ root pairs of $J_1(\pi u)/\pi u$ with new root pairs at the positions given by (6.54), it becomes apparent that there is nothing inviolable about these new root positions. Conceivably, one could find a set of positions that would cause the innermost $\bar{n} - 1$ ring side lobes to have individually specified heights. There are practical applications in which this is desirable.

A perturbation procedure has been devised that will determine the proper set of root positions once the height of each side lobe has been specified.[9] One begins by choosing a circular Taylor pattern whose average side lobe level in real space approximates the average side lobe level in real space of the desired pattern. Thus the starting pattern can be written in the form of (6.53), namely,

$$\mathcal{S}_0(u) = C_0 f(u) \prod_{n=1}^{\bar{n}-1} \left(1 - \frac{u^2}{{}^0u_n^2}\right) \tag{6.60}$$

[9]O. Graham, R. M. Johnson, and R. S. Elliott, "Design of Circular Apertures for Sum Patterns with Ring Side Lobes of Individually Arbitrary Heights," *Alta Frequenza*, 47 (1978), 21–25.

in which

$$f(u) = \frac{J_1(\pi u)/\pi u}{\prod_{n=1}^{\bar{n}-1} (1 - u^2/\gamma_{1n}^2)} \tag{6.61}$$

The root positions $\overset{0}{u}_n$ are known and given by (6.54).

The desired pattern $\mathcal{S}(u)$ can be expressed similarly, that is,

$$\mathcal{S}(u) = Cf(u) \prod_{n=1}^{\bar{n}-1} (1 - u^2/u_n^2) \tag{6.62}$$

The root positions u_n are unknown and will need to be determined. But if the desired and starting patterns are not too disparate, u_n and C can be given by

$$u_n = \overset{0}{u}_n + \delta u_n \tag{6.63}$$

$$C = C_0 + \delta C \tag{6.64}$$

Since, to first order,

$$1 - \frac{u^2}{u_n^2} = \left(1 - \frac{u^2}{\overset{0}{u}_n^2}\right)\left[1 + \frac{2(u^2/\overset{0}{u}_n^3)}{1 - (u^2/\overset{0}{u}_n^2)}\,\delta u_n\right] \tag{6.65}$$

if follows that (6.62) can be put in the form

$$\frac{\mathcal{S}(u)}{\mathcal{S}_0(u)} - 1 = \frac{\delta C}{C_0} + \sum_{n=1}^{\bar{n}-1} \frac{2(u^2/\overset{0}{u}_n^3)}{1 - u^2/\overset{0}{u}_n^2}\,\delta u_n \tag{6.66}$$

If u_m^p is the peak position of the mth lobe in the starting pattern, $\mathcal{S}(u_m^p)/\mathcal{S}_0(u_m^p)$ is essentially the ratio of the height of the mth lobe in the desired pattern to its height in the starting pattern. This ratio is a known quantity. Thus if $u = u_m^p$ is inserted in (6.66), all terms are known except $\delta C/C_0$ and the $\bar{n} - 1$ root perturbations δu_n. Since there are $\bar{n}$ lobes (including the main beam) in $0 \le u \le \gamma_{1\bar{n}}$, there is exactly the right number of u_m^p values to use in (6.66) in order to provide a deterministic set of simultaneous linear equations. Matrix inversion gives the perturbations δu_n from which the new root positions can be deduced. When these are inserted in (6.62) a new pattern can be computed and compared to the ideal. Iteration may be necessary, but experience has shown that convergence is usually very rapid.

As an illustration of this technique, let the starting pattern be the circular Taylor pattern already shown in Figure 6.8, and specify that the desired pattern differ from this only in that the two innermost side lobes should be at -25 dB. Three successive iterations give a pattern in which all side lobes are within $\frac{1}{4}$ dB of specification. The result is shown in Figure 6.10.

The aperture distributions for these modified Taylor patterns can be found from (6.59). For the pattern of Figure 6.10, the aperture distribution is shown in Figure 6.11. It is interesting to compare this result with Figure. 6.9. The fine structure in the two distributions is comparable, but it can be observed that $g_0(p)$ is negative in a small region for the case of two suppressed inner side lobes.

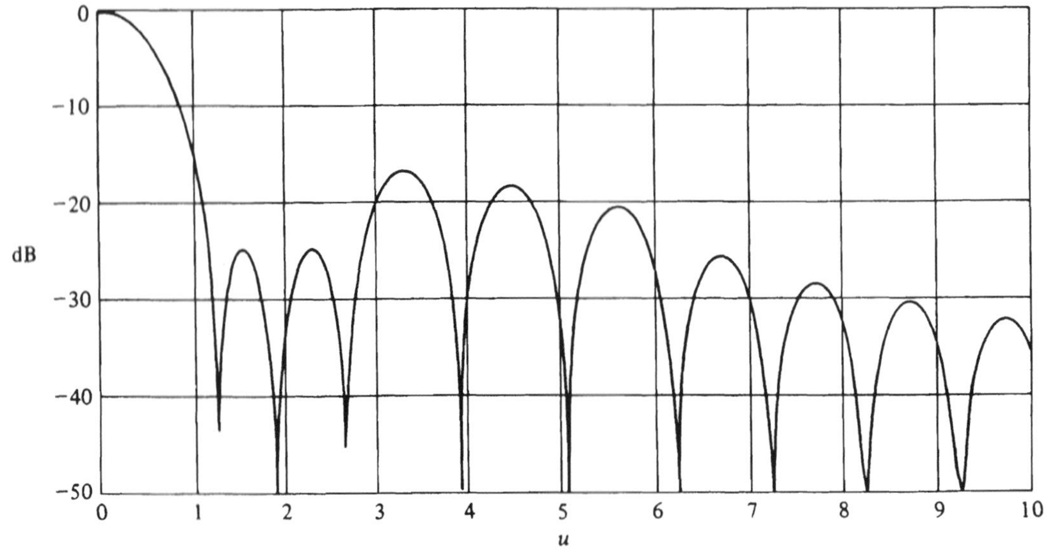

Fig. 6.10 The Circular Taylor Pattern of Fig. 6.8 Modified So That Two Innermost Ring Side Lobes Are Depressed to −25 dB (© 1978 Alta Frequenza. Reprinted from Graham, Johnson, and Elliott, *Alta Frequenza*, pp. 1–7, 1978.)

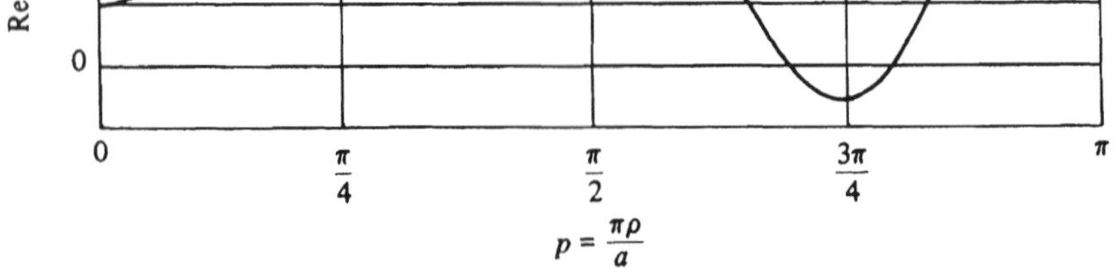

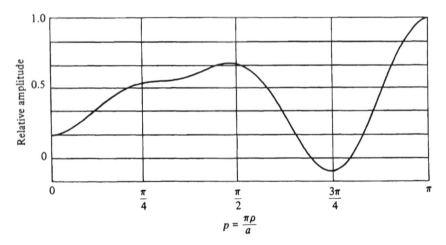

Fig. 6.11 The Continuous Aperture Distribution for the Modified Circular Taylor Pattern of Fig. 6.10 (© 1978 Alta Frequenza. Reprinted from Graham, Johnson, and Elliott, *Alta Frequenza*, pp. 1–7, 1978.)

6.5 Modified Circular Taylor Patterns: Undulating Ring Side Lobes

Just as there are practical applications in which one desires a sum pattern with ring side lobes of different heights, (θ-dependence of side lobe level), so too are there applications in which it is desirable to have a quasi-uniform side lobe level in every ϕ-cut, but with the level different in different ϕ-cuts (ϕ-dependence of side lobe level.) In this latter case, the ring side lobes undulate in height (take on a "roller-coaster" appearance) as one progresses in ϕ. A generalization of Taylor's original technique can produce such patterns.

It can be recalled from the development of Section 6.3 that a linearly polarized aperture distribution $K(\rho, \beta)$, given by (6.45), will generally create a pattern $F(\theta, \phi)$, expressed by (6.48). Taylor considered the special case $n = 0$, leading to ϕ-independent patterns. Now it will be necessary to look at the more general case.[10]

The substitutions (6.50) convert the pattern expression (6.48) to the form

$$F(u, \phi) = \sum_{n=-\infty}^{\infty} e^{jn\phi} F_n(u) \tag{6.67}$$

in which

$$F_n(u) = \int_0^{\pi} p g_n(p) J_n(up)\, dp \tag{6.68}$$

and

$$g_n(p) = \frac{2a^2}{\pi}(j)^n K_n(\rho) \tag{6.69}$$

Equation (6.67) permits the interpretation that the pattern can be represented by a Fourier series in ϕ with coefficients $F_n(u)$.

Suppose one wishes to have every ϕ-cut be a circular Taylor pattern, but with $F(u, \phi)$ displaying ϕ-asymmetry. As an example, let

$$F(u, \phi) = \begin{cases} \frac{1}{2}[F_a(u) + F_b(u)] + \frac{1}{2}[F_a(u) - F_b(u)] \cos\left(\frac{\pi\phi}{\phi_0}\right) & -\phi_0 \le \phi \le \phi_0 \\ F_b(u) & \text{otherwise} \end{cases} \tag{6.70}$$

in which $F_a(u)$ and $F_b(u)$ are circular Taylor patterns. This composition gives $F_a(u)$ at $\phi = 0°$, then a cosinusoidal transition from $F_a(u)$ to $F_b(u)$ as ϕ departs from 0° toward either $-\phi_0$ or $+\phi_0$, and then $F_b(u)$ in the regions $-\pi \le \phi \le \phi_0$ and $\phi_0 \le \phi \le \pi$. If $F_a(u)$ has low side lobes (for instance, 25 dB) and $F_b(u)$ has higher side lobes (for instance, 15 dB), then ϕ_0 controls the extent of the region in which the lower side

[10] R. S. Elliott, "Design of Circular Apertures for Narrow Beamwidth and Asymmetric Side Lobes," *IEEE Trans. Antennas and Propagat.*, AP-23, (1975), 523–27.

lobes prevail. If in a practical situation ϕ_0 need be only 30°, permitting higher side lobes in the remaining five-sixths of space can augment the directivity.

The form chosen for $F(u, \phi)$ in (6.70) is only suggestive of the kind of pattern construction which is theoretically possible. One could compose $F(u, \phi)$ out of combinations of many circular Taylor patterns, linked in many ways. Of course, the more complicated the composition, the more finely structured will be the aperture distribution, and the more difficult will be the physical realization.

It is instructive to pursue further the class of sum patterns characterized by (6.70). The partial functions $F_n(u)$ can be found by regular Fourier inversion of (6.67), namely,

$$\int_{-\pi}^{\pi} F(u, \phi)e^{-jk\phi}\, d\phi = \sum_{n=-\infty}^{\infty} F_n(u) \int_{-\pi}^{\pi} e^{j(n-k)\phi}\, d\phi = 2\pi F_k(u) \tag{6.71}$$

Since $F(u, \phi)$ has been selected in (6.70) to be an even function, its insertion in (6.71) gives, for $k = 0$,

$$F_0(u) = F_b(u) + [F_a(u) - F_b(u)]\frac{\phi_0}{2\pi} \tag{6.72}$$

and gives, for $k \neq 0$,

$$F_k(u) = F_{-k}(u) = \frac{F_a(u) - F_b(u)}{2\pi} \int_0^{\phi_0} \left[1 + \cos\left(\frac{\pi\phi}{\phi_0}\right)\right] \cos\phi\, d\phi \tag{6.73}$$

The integration indicated in (6.73) yields compact formulas when $\phi_0 = \pi/l$, with l a positive integer. Then

$$F_k(u) = \frac{F_a(u) - F_b(u)}{4l}, \qquad k^2 = l^2 \tag{6.74}$$

$$F_k(u) = \frac{F_a(u) - F_b(u)}{2l} \frac{l^2}{l^2 - k^2} \frac{\sin(\pi k/l)}{\pi k/l}, \qquad k^2 \neq l^2 \tag{6.75}$$

The partial aperture distribution corresponding to each of the partial pattern functions $F_n(u)$ can be found by taking the inverse transform of (6.68). This has already been done in Section 6.3 for $n = 0$, the result being that $g_0(p)$ is given by the truncated series (6.59).

For $n \neq 0$ one can proceed by assuming that

$$g_n(p) = \sum_{m=1}^{\infty} B_{nm} J_n(\gamma_{nm} p) \tag{6.76}$$

in which $J_n(\gamma_{nm}\pi) = 0$ defines the mth root of the nth Bessel function. Then from integration of (6.68),

$$F_n(u) = \sum_{m=1}^{\infty} B_{nm} \left[\frac{upJ_n(\gamma_{nm}p)J_{n-1}(up) - \gamma_{nm}pJ_{n-1}(\gamma_{nm}p)J_n(up)}{\gamma_{nm}^2 - u^2}\right]_0^{\pi} \tag{6.77}$$

Inspection of (6.77) reveals that $F_n(\gamma_{nk})$, with k a positive integer, is at most contributed to by the kth term of the sum. Thus, returning to (6.68), one finds that

$$\begin{aligned} F_n(\gamma_{nk}) &= B_{nk}\int_0^\pi pJ_n^2(\gamma_{nk}p)\,dp \\ &= B_{nk}\left[\frac{p^2}{2}\{J_n^2(\gamma_{nk}p) - J_{n-1}(\gamma_{nk}p)J_{n+1}(\gamma_{nk}p)\}\right]_0^\pi \\ &= -\left(\frac{\pi^2}{2}\right)B_{nk}J_{n-1}(\gamma_{nk}\pi)J_{n+1}(\gamma_{nk}\pi) \end{aligned} \tag{6.78}$$

so that

$$g_n(p) = -\left(\frac{2}{\pi^2}\right)\sum_{m=1}^{\infty}\frac{F_n(\gamma_{nm})J_n(\gamma_{nm}p)}{J_{n-1}(\gamma_{nm}\pi)J_{n+1}(\gamma_{nm}\pi)} \qquad n \neq 0 \tag{6.79}$$

The series in (6.79), unlike (6.59), do not truncate. However, for practical aperture distributions, they converge very rapidly and $g_n(p)$ is significant only for low values of n.

As a specific illustration of these results, let $l = 2$ so that $\phi_0 = \pi/2$. If $F_a(u)$ is 25 dB, $\bar{n} = 6$, and $F_b(u)$ is 15 dB, $\bar{n} = 3$, this choice gives a smooth transition from a 25 dB SLL pattern at $\phi = 0°$ to a 15 dB SLL pattern at $\phi = 90°$, then a constant 15 dB SLL to $\phi = 270°$, and another smooth transition back to a 25 dB SLL at $\phi = 360°$. A plot of (6.70) for this case is shown in Figure 6.12.

Use of (6.72) through (6.75) gives

$$F_0(u) = \tfrac{1}{4}F_a(u) + \tfrac{3}{4}F_b(u) \tag{6.80}$$

$$F_2(u) = F_{-2}(u) = \tfrac{1}{8}[F_a(u) - F_b(u)] \tag{6.81}$$

$$F_k(u) = F_{-k}(u) \equiv 0, \qquad k = 4, 6, 8, \ldots \tag{6.82}$$

$$F_k(u) = F_{-k}(u) = \frac{2(-1)^{(k-1)/2}}{\pi k(4-k^2)}[F_a(u) - F_b(u)] \qquad k = 1, 3, 5, \ldots \tag{6.83}$$

These expressions for the partial patterns can be used to determine the aperture distribution. Since $J_{-n}(up) = (-1)^n J_n(up)$, and since (6.81) through (6.83) indicate that $F_{-n}(u) = F_n(u)$, it follows from (6.68) that for this case $g_{-n}(p) = (-1)^n g_n(p)$. When this information is placed in (6.69), and then in (6.45), one obtains

$$K(p, \beta) = \frac{\pi}{2a^2}[g_0(p) - 2\cos\beta\, g_2(p) + \text{odd}\sum_{n=1}^{\infty} 2(-j)^n \cos n\beta\, g_n(p)] \tag{6.84}$$

From (6.59)

$$g_0(p) = \tfrac{1}{4}g_{0,a}(p) + \tfrac{3}{4}g_{0,b}(p) \tag{6.85}$$

in which $g_{0,a}(p)$ and $g_{0,b}(p)$ are conventional circular Taylor distribution (corresponding to $F_a(u)$ and $F_b(u)$ which can be read from the tables of R. C. Hansen.[11]

[11] Hansen, "Tables of Taylor Distributions."

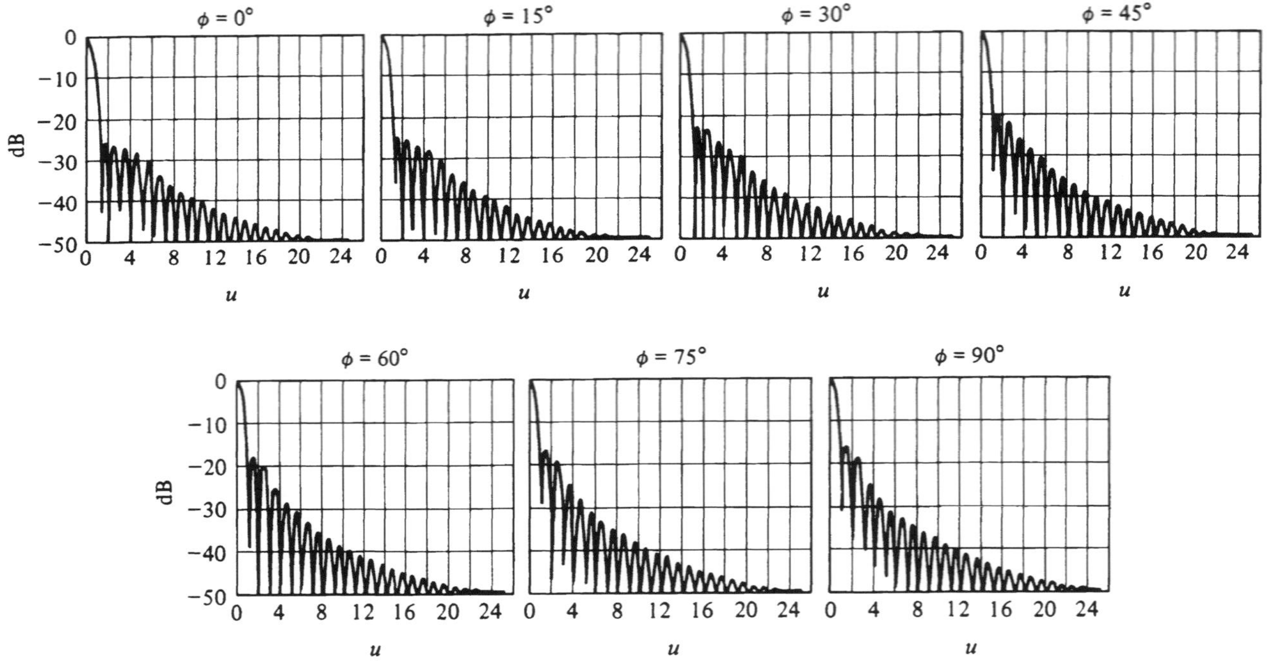

Fig. 6.12 A Sequence of ϕ-Cuts for a 15/25 dB Modified Circular Taylor Pattern (© 1975 IEEE. Reprinted from *IEEE AP Transactions*, pp. 523–527, 1975.)

The remainder of the $g_n(p)$ functions appearing in (6.84) can be determined from (6.79). Computations show for this case that the peak value of $g_1(p)$ is 11 % of $g_0(0)$ and that the peak value of $g_2(p)$ is 6 % of $g_0(0)$. The higher order modes tail off rapidly: $g_5(p)$ reaches only 0.1 % of $g_0(0)$ and no attempt was made to compute $g_7(p)$ and beyond.

The aperture distribution computed from (6.84), truncated at $n = 5$, is shown in Figure 6.13. It can be seen that, in any pair of opposing β-cuts, the amplitude distribution is symmetric and the phase distribution is asymmetric. This is typical of sum patterns with well-defined side lobes (deep nulls) of nonuniform height.

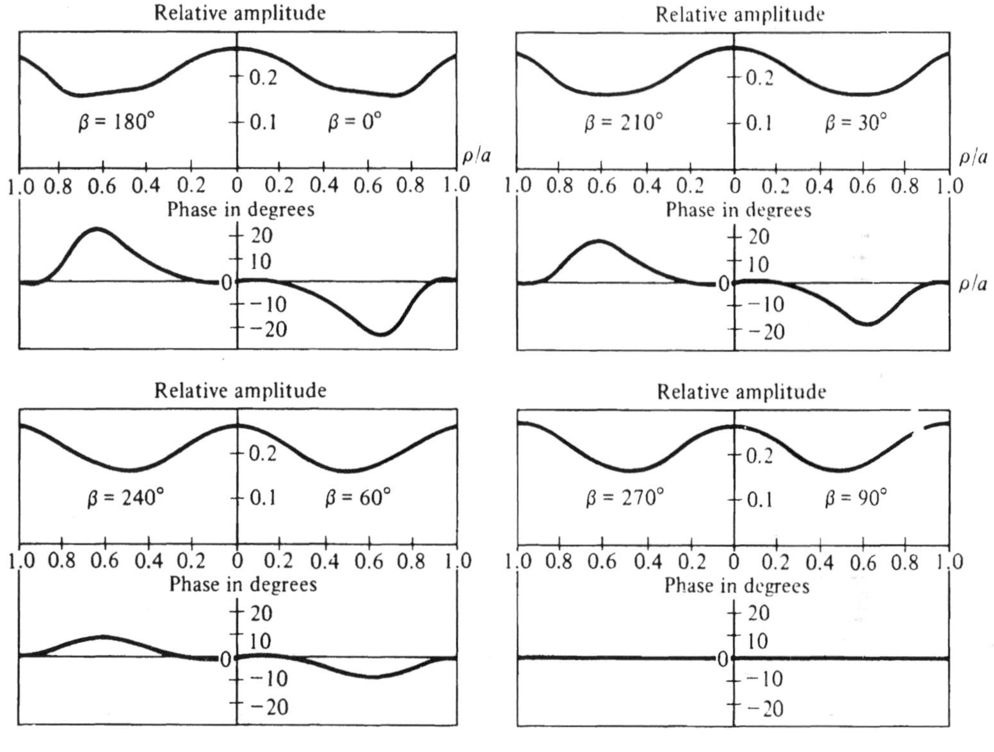

Fig. 6.13 The Continuous Aperture Distribution $K(p, \beta)$ for the Sum Pattern of Fig. 6.12 (© 1975 IEEE. Reprinted from *IEEE AP Transactions*, pp. 523–527, 1975.)

6.6 Sampling Generalized Taylor Distributions: Rectangular Grid Arrays

The practical applications of Taylor's circular patterns (or of their generalizations discussed in the previous two sections), which involve a *continuous* aperture distribution, are few. However, the excitation of *discrete* planar arrays with circular boundaries is often determined by conventional sampling of these continuous distributions. If the number of elements is large, so that the sampling interval is small, this is a satisfactory procedure. But there is always pattern degradation, and the designer must judge whether or not the result is still superior in side lobe structure and beamwidth or directivity to the pattern obtained from an easily determined discrete separable distribution. Later in this chapter, techniques will be described which improve on conventional sampling, and thus tilt the decision more strongly in favor of nonseparable distributions. However, these techniques often use sampling of the continuous distributions as a starting point, so it is useful to gain an appreciation of the conventional sampling method and what it can achieve.

A common application involves a rectangular grid array with a circular boundary, one quadrant of which is shown in Figure 6.14. For the purpose of illustration, imagine that the boundary radius is $a = 5\lambda$ and that the interelement spacing is 0.5λ in both directions. This can be viewed as a 20-by-20 array with corners that are cut off to achieve a circular boundary. If this array is to produce a sum pattern with a side lobe level of -15 dB, the circular Taylor distribution of Figure 6.9 can be sampled to obtain the discrete current distribution. For the mnth element, the distance from the origin is

$$\rho_{mn} = \left\{\left[\frac{(2m-1)d_x}{2}\right]^2 + \left[\frac{(2n-1)d_y}{2}\right]^2\right\}^{1/2} \tag{6.86}$$

which permits determination of $p_{mn} = \pi\rho_{mn}/a$ and thus $I_{mn} = g_0(p_{mn})$.

Since this is a nonseparable distribution, but one possessing quadrantal sym-

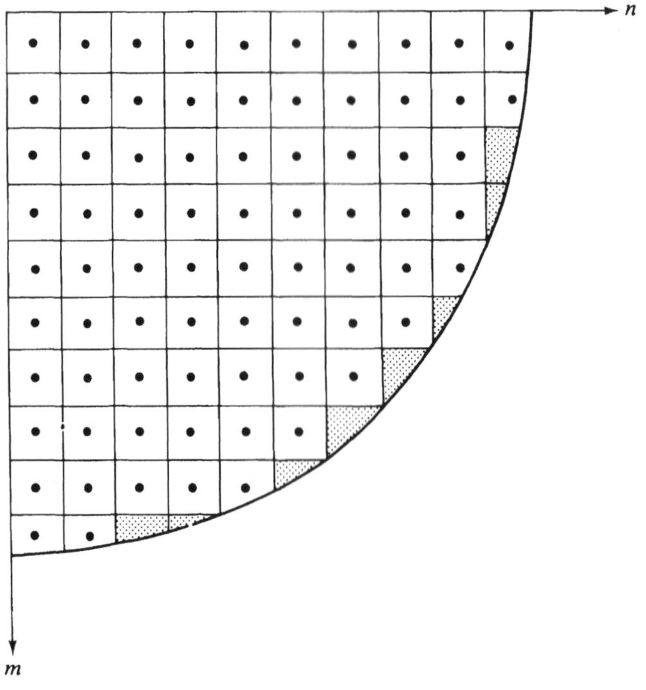

Fig. 6.14 One Quadrant of a Rectangular Grid Array with Circular Boundary; $d_x = d_y = 0.5\lambda$, $a = 5\lambda$

metry, the pattern can be calculated from

$$\mathcal{S}(\theta, \phi) = 4 \sum_{m=1}^{10} \sum_{n=1}^{10}{}' I_{mn} \cos\left[\frac{(2m-1)\psi_x}{2}\right] \cos\left[\frac{(2n-1)\psi_y}{2}\right] \tag{6.87}$$

with $\psi_x = \pi \sin\theta \cos\phi$ and $\psi_y = \pi \sin\theta \sin\phi$ because, in this example, $d_x = d_y = \lambda/2$.

Patterns computed using (6.87) are shown for a series of ϕ-cuts in Figure 6.15. Reasonable agreement with the pattern of Figure 6.8 has been achieved, although not all side lobes are under control.

Conventional sampling of *modified* Taylor patterns is also successful if the number of elements is large. As an illustration, if a discrete array of the type of Figure 6.14, but consisting of a 20-by-20 quadrant with corners that are cut off, is excited by sampling the distribution shown in Figure 6.13, the result is a pattern some of whose ϕ-cuts are shown in Figure 6.16. Agreement with the continuous aperture patterns of Figure 6.12 is seen to be reasonably good. However, some tendency to "average out" the side lobe level can be observed.

Sampling of these continuous aperture distributions, for a large number of elements, has even been found to be successful for rectangular grid arrays with elliptical boundaries. If the semimajor and semiminor axes of the boundary are a and b,

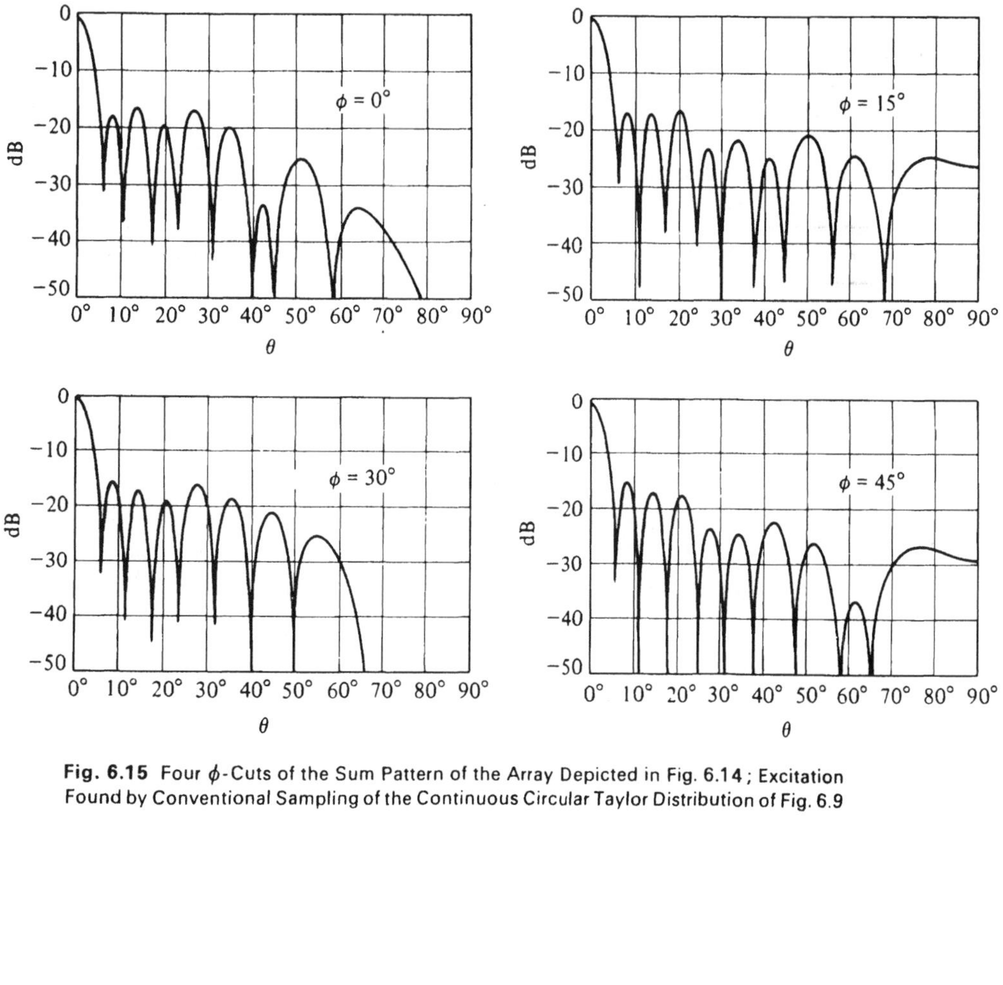

Fig. 6.15 Four ϕ-Cuts of the Sum Pattern of the Array Depicted in Fig. 6.14; Excitation Found by Conventional Sampling of the Continuous Circular Taylor Distribution of Fig. 6.9

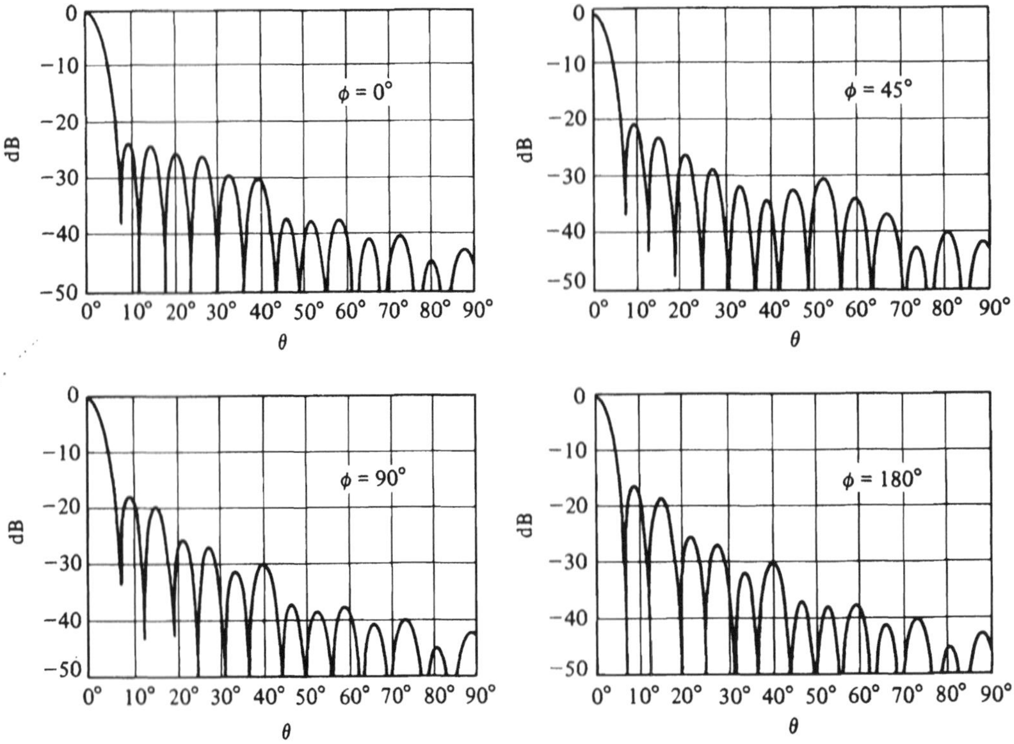

Fig. 6.16 Four ϕ-Cuts of the Sum Pattern of a Rectangular Grid Array with Circular Boundary; $d_x = d_y = 0.5\lambda$, $a = 10\lambda$; Excitation Found by Conventional Sampling of the Continuous Modified Taylor Distribution of Fig. 6.13

all that is needed is a transformation to an equivalent circular apperture via the one-way stretch

$$\xi' = \xi, \qquad \eta' = \frac{a}{b}\eta \tag{6.88}$$

This transforms the position of the mnth element from (ξ_{mn}, η_{mn}) to (ξ'_{mn}, η'_{mn}), after which the modified value

$$p'_{mn} = \frac{\pi}{a}[(\xi'_{mn})^2 + (\eta'_{mn})^2]^{1/2} \tag{6.89}$$

can be deduced, leading to the determination of the current element from $I_{mn} = g(p'_{mn})$.

Obviously, as the number of elements in the array gets smaller, conventional sampling of a continuous aperture distribution leads to more pattern degradation. As an illustration, consider the array which has only eight elements per quadrant, as suggested by Figure 6.17. If the interelement spacing is to be 0.7λ in both directions, and if a circular Taylor pattern, -22 dB, $\bar{n} = 3$ is to be approximated, sampling the continuous aperture distribution results in a pattern for which several ϕ-cuts are displayed in Figure 6.18 with the discrete current distribution shown as an inset. The

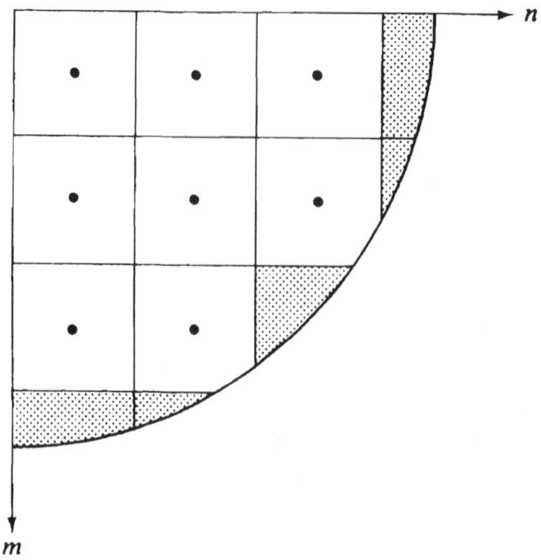

Fig. 6.17 One Quadrant of a 32-Element Planar Array; Rectangular Grid, Circular Boundary

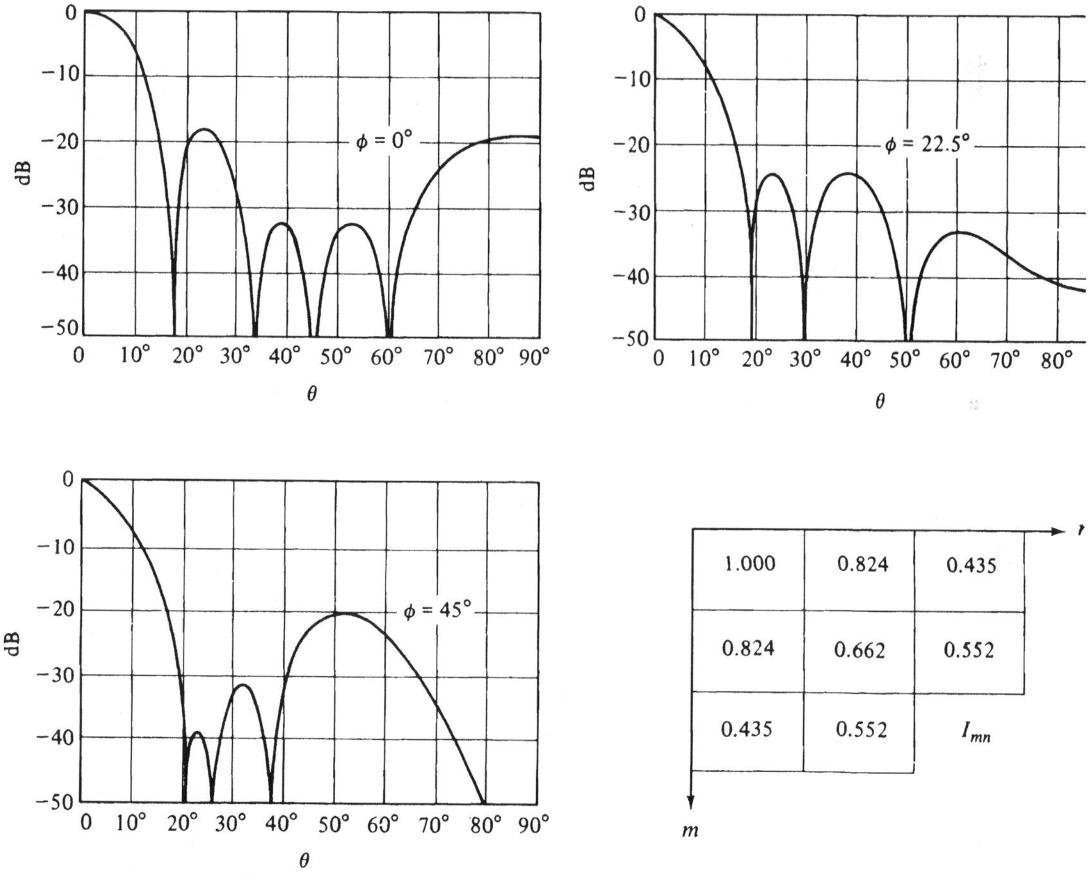

Fig. 6.18 Three ϕ-Cuts of the Sum Pattern of the Array Shown in Fig. 6.17; $d_x = d_y = 0.7\lambda$; Excitation Found by Conventional Sampling of a Continuous Circular Taylor Distribution, $\bar{n} = 6$, −22 dB SLL

degradation is clearly evident. In Section 6.10 a perturbation procedure will be presented that can improve on this situation.

6.7 Sampling Generalized Taylor Distributions: Circular Grid Arrays

It is a more natural arrangement (though not necessarily one leading to so practical an antenna array from a feeding point of view) to discretize a *circular* planar aperture by laying out a grid of concentric circles rather than using a rectangular grid. One would suspect that this might lead to less pattern degradation when a continuous aperture distribution is sampled, and this proves to be the case. As an example, let the rectangular grid of Figure 6.14 be replaced by a family of concentric circles with radii given by

$$\rho_m = \frac{(2m - 1)d}{2} \tag{6.90}$$

with d the interradial spacing. If one wishes the same spacing d between adjacent elements along any circle, then

$$2\pi\rho_m = (2m - 1)\pi d = N_m d \tag{6.91}$$

in which N_m is the number of elements on the mth circle. It is clear that the values of N_m that satisfy (6.91) are not integers, but one can round the results to the nearest integer. For example, if $a = 5\lambda$ and $b = \lambda/2$, there are 10 concentric circles, and the numbers of elements per circle are given in Table 6.3, under the restriction that N_m be

TABLE 6.3

m	1	2	3	4	5	6	7	8	9	10
N_m	4	8	16	20	28	32	40	44	52	56
I_m	0.144	0.183	0.174	0.140	0.153	0.145	0.047	0.023	0.239	0.509

divisible by 4 (so that there is quadrantal symmetry). This gives a total of 300 elements in the array, exactly the same number as were used in the rectangular grid array of Figure 6.14. The layout of one quadrant is shown in Figure 6.19. Because of the quadrantal symmetry, the sum pattern is given by

$$\mathcal{S}(\theta, \phi) = 4 \sum_{m=1}^{10} \sum_{n=1}^{N_m/4} I_{mn} \cos (k\xi_{mn} \sin \theta \cos \phi) \cos (k\eta_{mn} \sin \theta \sin \phi) \tag{6.92}$$

in which $\xi_{mn} = \rho_m \cos \beta_{mn}$ and $\eta_{mn} = \rho_m \sin \beta_{mn}$, with $\beta_{mn} = (2n - 1)\pi/N_m$. If once again the desire is to approximate the circular Taylor pattern of Figure 6.8, the currents I_{mn} that appear in (6.92) can be determined by sampling the continuous distribution shown in Figure 6.9. All the currents on a common circle are the same and

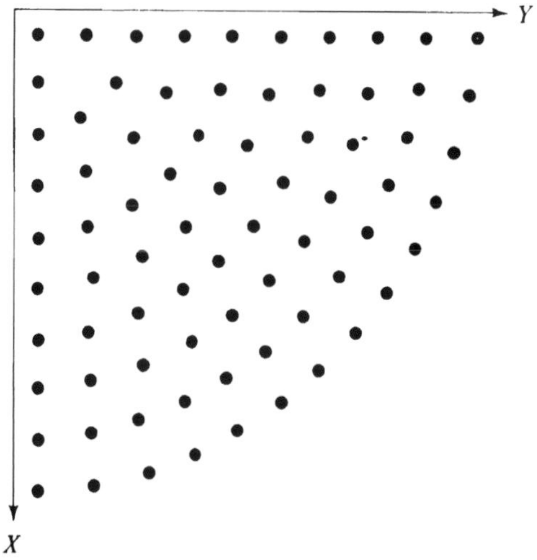

Fig. 6.19 One Quadrant of a Circular Grid Array with Circular Boundary; $d = 0.5\lambda$, $a = 5\lambda$

are listed for this application in Table 6.3. Computation gives the pattern shown in Figure 6.20, which is seen to be closer to what is desired than those due to a rectangular grid array (shown previously in Figure 6.15). (Only the $\phi = 0°$ pattern cut is shown in Figure 6.20 because it was found that, for this application, patterns for $\phi = 0°$, 15°,

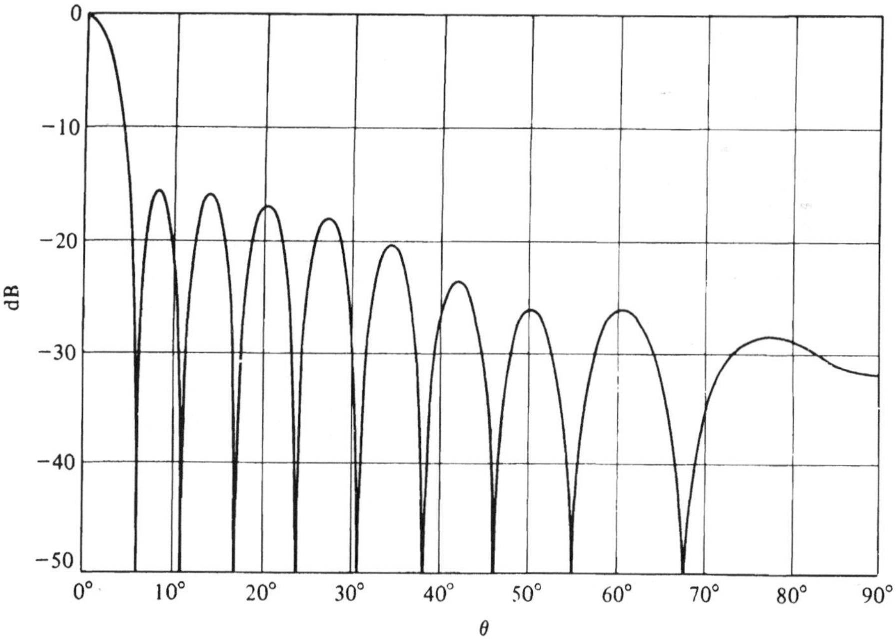

Fig. 6.20 The Sum Pattern of the Circular Grid Array of Figure 6.19; Excitation Obtained by Conventional Sampling of Fig. 6.9

30°, and 45° were virtually indistinguishable, attesting to the fact that circular grid sampling of ϕ-symmetric distributions is more natural than rectangular grid sampling).

Modified Taylor distributions can also be sampled to provide the excitation for circular grid arrays. For example, suppose five concentric rings of radii $\rho_m = ma/5$, $m = 1, 2, \ldots, 5$ contain equispaced elements $N_m = 8m$ in number, and that this circular array is intended to create the sum pattern of Figure 6.10. The reader will recall that this is a circular Taylor pattern, -15 dB SLL, $\bar{n} = 6$, except that the innermost two ring side lobes are at -25 dB. The continuous aperture distribution which would give that pattern precisely was shown in Figure 6.11. Conventional sampling gives the current distribution listed in Table 6.4, which in turn produces the pattern of Figure 6.21. Once again, only the $\phi = 0°$ cut is shown because the different ϕ-cuts are essentially the same. One can observe the general features of the desired

TABLE 6.4

m	1	2	3	4	5
N_m	8	16	24	32	40
I_m	0.225	0.292	0.189	0.002	0.472

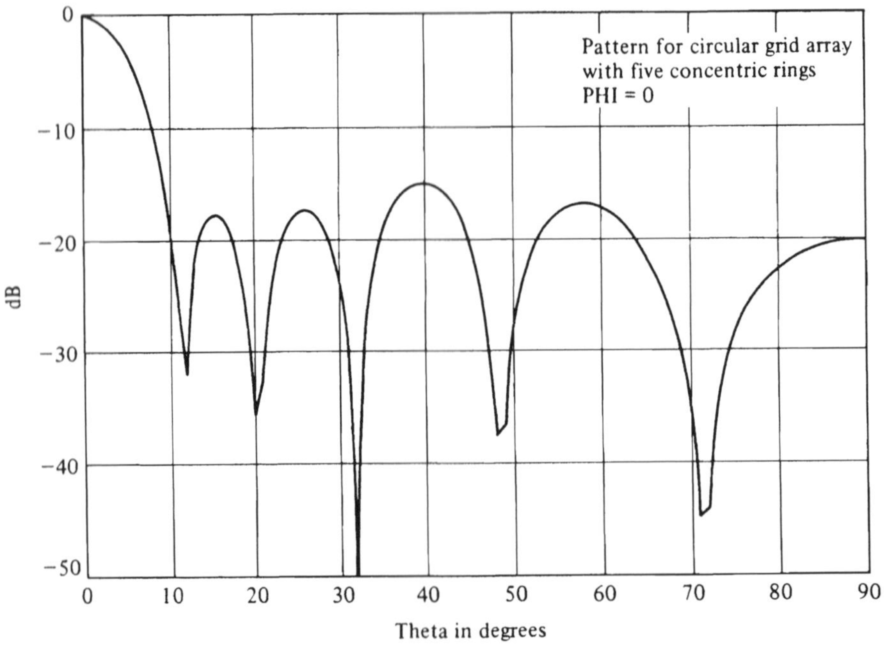

Fig. 6.21 The Sum Pattern for a Circular Grid Array of Five Equispaced Concentric Rings; $a = 2.5\lambda$; $N_m = 8m$ = Number of Equispaced Elements on mth Ring; Excitation Found by Conventional Sampling of Fig. 6.11 (© 1978 Alta Frequenza. Reprinted from Graham, Johnson, and Elliott, *Alta Frequenza*, pp. 1–7, 1978.)

pattern of Figure 6.10 though there is recognizable degradation—more so than was seen in Figure 6.20, partly because there are fewer rings and partly because the desired pattern is more intricate. In the next section, a design procedure will be presented that can improve substantially on the results of both Figures 6.20 and 6.21.

6.8 An Improved Discretizing Technique for Circular Grid Arrays

The examples of the previous section indicated that conventional sampling of generalized Taylor continuous aperture distributions causes pattern degradation which is more serious when the number of elements in the circular grid array is small. It is possible to remove most of this degradation by employing a different technique—one which focuses on the starting pattern and ignores the continuous distribution which creates it.

Assume the existence of a desired pattern $\mathcal{S}(u)$, such as Figure 6.10, which is ϕ-symmetric and which one wishes to produce with a circular grid array. As an interim step, let there be a system of concentric ring sources with normalized radii that form the sequence $0 < p_1 < p_2 \cdots < p_m \leq \pi$, with $p = \pi \rho/a$. If $\mathring{I}_m$ is the current level in the mth ring, the finite sum equivalent of (6.51) gives, for the starting pattern,

$$\mathcal{S}_0(u) = 2a \sum_{m=1}^{M} \mathring{I}_m p_m J_0(up_m) \tag{6.93}$$

The currents $\mathring{I}_m$ in (6.93) should be selected so that the nulls of $\mathcal{S}_0(u)$ coincide with the nulls of $\mathcal{S}(u)$. With $\mathring{I}_1$ arbitrarily set equal to unity, if $u_1, u_2, \ldots, u_{M-1}$ are the innermost $M - 1$ nulls of $\mathcal{S}(u)$, then

$$\mathcal{S}_0(u_n) = 2a \sum_{m=1}^{M} \mathring{I}_m p_m J_0(u_n p_m) = 0 \tag{6.94}$$

Equations 6.94 comprise $M - 1$ simultaneous linear equations in the $M - 1$ unknown currents $\mathring{I}_2, \mathring{I}_3, \ldots, \mathring{I}_M$. Matrix inversion gives what will be called the *starting ring current distribution*.

As an illustration, consider again the second example of the previous section. If $p_m = m\pi/5$, and if $\mathcal{S}_0(u)$ is to have the same null positions as the pattern shown in Figure 6.10, solution of (6.94) gives the current distribution listed in the second column of Table 6.5. Use of these sources in (6.93) produces the pattern displayed in

TABLE 6.5 Ring current distributions

Ring Number m	$\mathring{I}_m$ (for Figure 6.22a)	I_m (for Figure 6.22b)
1	1.000	1.000
2	1.305	1.479
3	0.633	1.079
4	0.483	0.037
5	0.738	1.303

Figure 6.22a. Reasonable agreement with the desired pattern of Figure 6.10 has been achieved, but the outermost side lobes are too low. This situation can be improved.

Let the desired pattern $\mathcal{S}(u)$ be expressed in the same form as (6.93), that is,

$$\mathcal{S}(u) = 2a \sum_{m=1}^{M} I_m p_m J_0(up_m) \tag{6.95}$$

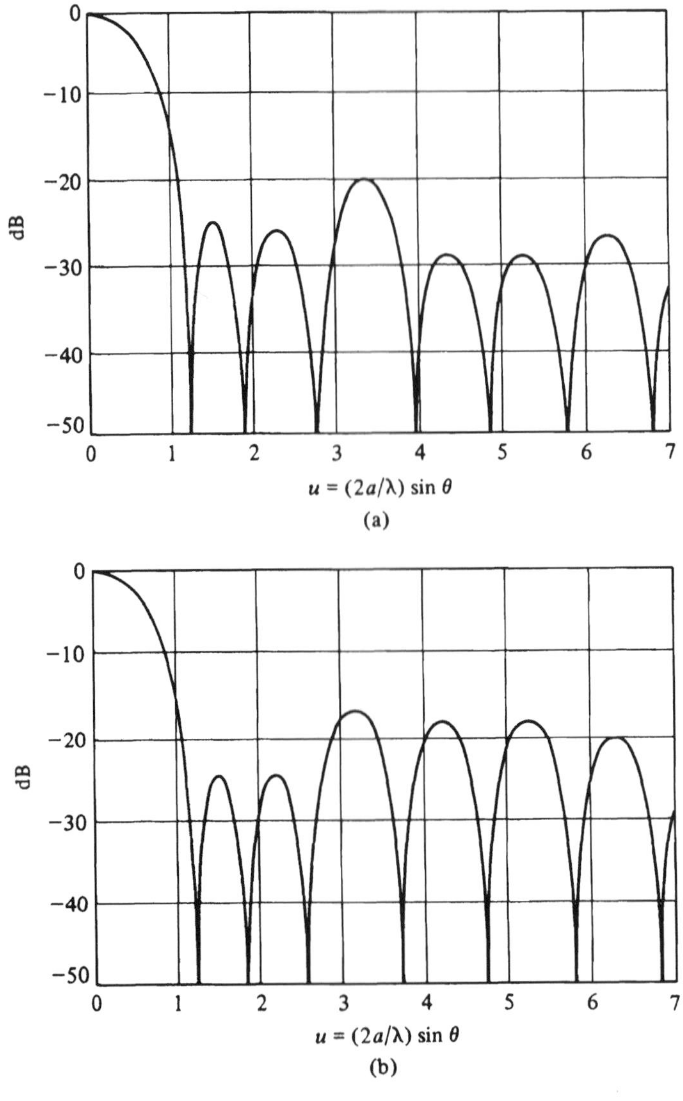

Fig. 6.22 The Sum Pattern for a Planar Array of Five Equispaced Concentric Continuous Ring Currents; $a = 2.5\lambda$; Currents are Selected (a) to Match Pattern Nulls, and (b) After Perturbation to Match Side Lobe Heights of Fig. 6.10 (© 1978 Alta Frequenza. Reprinted from Graham, Johnson, and Elliott, *Alta Frequenza*, pp. 1–7, 1978.)

with

$$I_m = \overset{0}{I}_m + \delta I_m \tag{6.96}$$

Then

$$\frac{1}{2a}[\mathcal{S}(u) - \mathcal{S}_0(u)] = \sum_{m=1}^{M} \delta I_m p_m J_0(up_m) \tag{6.97}$$

Let u_n^p be the position of the nth lobe peak in the pattern $\mathcal{S}_0(u)$, with u_0^p referring to the main beam, u_1^p to the first side lobe, and so on. Insertion of u_n^p in (6.97) creates a set of M simultaneous linear equations in the M unknown current perturbations. The left side of (6.97) involves $\mathcal{S}(u_n^p) - \mathcal{S}_0(u_n^p)$, which is approximately the difference between desired lobe level and starting lobe level. This approximate difference is a *known* quantity. Matrix inversion gives δI_m which, when added to $\overset{0}{I}_m$, gives the new current distribution I_m. When I_m is placed in (6.95), $\mathcal{S}(u)$ can be computed. If $\mathcal{S}(u)$ is sufficiently close to ideal, this part of the design procedure is completed. If not, $\mathcal{S}(u)$ can be used as a new starting pattern, with the process repeated.

Returning to the example, one finds that a sequence of iterations leads to the ring current distribution listed in the third column of Table 6.5. The corresponding pattern is shown in Figure 6.22b, and has all side lobes within $\frac{1}{4}$ dB of specification.

These ring currents are not physically realizable, but they prove to be a valuable aid in the determination of the excitation of a circular grid array. To see this connection, refer first to Figure 6.23, which shows the arrangement of discrete radiators on the mth ring. In order to insure quadrantal symmetry, the mnth radiator should be at the angular position $\beta_{mn} = (2n - 1)\pi/N_m$, with $1 \leq n \leq N_m$, and with N_m the number of radiators equispaced along the mth ring.

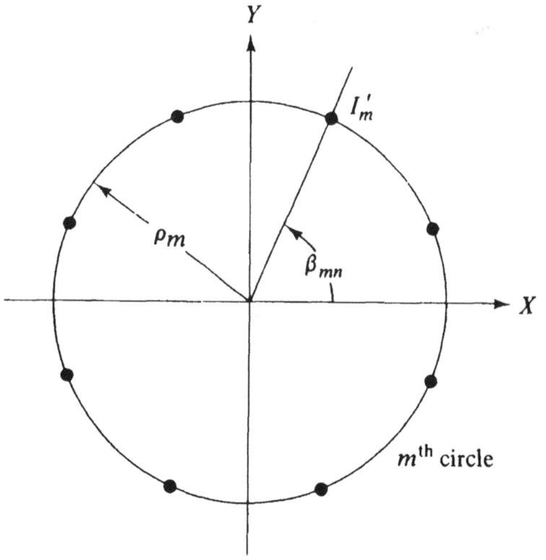

Fig. 6.23 Discrete Radiating Elements in a Circular Grid Arrangement

Since the desired pattern is ϕ-symmetric, all radiators on a common ring should have the same excitation, which will be designated by I'_m. The pattern for this circular grid array is therefore

$$\begin{aligned}\mathcal{S}(\theta,\phi) &= \sum_{m=1}^{M}\sum_{n=1}^{N_m} I'_m e^{jk\rho_m(\cos\beta_{mn}\sin\theta\cos\phi+\sin\beta_{mn}\sin\theta\sin\phi)} \\ &= \sum_{m=1}^{M}\sum_{n=1}^{N_m} I'_m e^{ju\rho_m\cos(\phi-\beta_{mn})} \\ &= \sum_{m=1}^{M}\sum_{n=1}^{N_m}\sum_{q=-\infty}^{\infty} I'_m(j)^q J_q(u\rho_m)e^{jq(\phi-\beta_{mn})}\end{aligned} \tag{6.98}$$

Since

$$\sum_{n=1}^{N_m} e^{-jq\beta_{mn}} = \sum_{n=1}^{N_m} e^{-jq(2n-1)\pi/N_m} = \begin{cases} N_m e^{js\pi} & \text{if } q = sN_m \\ 0 & \text{if } q \neq sN_m \end{cases}$$

with $s = 0, \pm 1, \pm 2, \ldots$, Equation 6.98 reduces to

$$\mathcal{S}(\theta,\phi) = \sum_{m=1}^{M} N_m I'_m J_0(u\rho_m) + 2\sum_{m=1}^{M}\sum_{s=1}^{\infty}(-1)^s N_m I'_m J_{sN_m}(u\rho_m)\cos(sN_m\phi) \tag{6.99}$$

An interesting interpretation can be placed on Equation 6.99. Each ring of the circular grid array contributes to a ϕ-independent part of the pattern, as desired. Additionally, each ring of the array causes a ϕ-harmonic series of supplemental patterns with N_m the fundamental component; this part is undesirable. However, for a given argument $u\rho_m$, the Bessel function J_{sN_m} diminishes as the order sN_m increases. Quite logically, if N_m is made large enough, one can expect that $\mathcal{S}(\theta, \phi)$ will be essentially ϕ-independent. When this is so, a comparison of (6.95) and (6.99) indicates that the discrete element currents should be related to the previously obtained ring currents by the equation

$$I'_m = 2a\frac{\rho_m}{N_m}I_m \tag{6.100}$$

Consider again the illustrative example, and assume that the five ring currents are discretized so that $N_m = 4lm$, with $l = 1, 2, 3, \ldots$. Then, since ρ_m has been chosen to equal $m\pi/5$, for this illustration $I'_m = (\pi a/10l)I_m$.

The patterns computed from the complete expression in (6.99) are shown in Figure 6.24 for the cases $l = 1, 2$. Though only the patterns for $\phi = 0°, 30°$ are shown, other ϕ-cuts display the same features. One can see that with $l = 1$, the element density is too sparse to reduce the ϕ-variable component of the field to a negligible value. However, with $l = 2$ the field is seen to be essentially the same as for the earlier cases of the ring currents and the continuous planar distribution. This is a significant improvement over the patterns of Figure 6.21, which were due to a sampling of the continuous distribution.

A specific practical example of this result would be a planar array in which collinear dipoles or slots were arranged so that their centers lay on concentric circles,

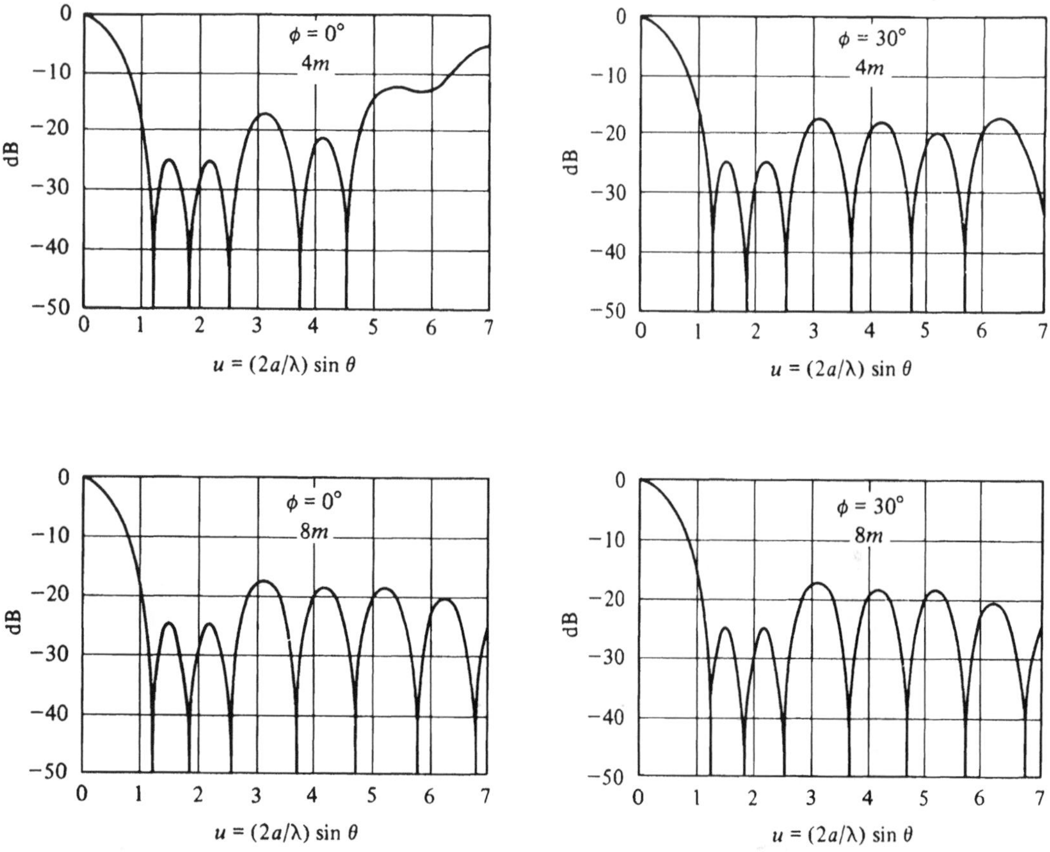

Fig. 6.24 Two ϕ-Cuts of the Sum Pattern of a Circular Grid Array of Five Equispaced Concentric Rings; $a = 2.5\lambda$; Excitation as in Table 6.5; Upper: $N_m = 4m$ = Number of Equispaced Elements on mth Ring; Lower: $N_m = 8m$ (© 1978 Alta Frequenza. Reprinted from Graham, Johnson, and Elliott, *Alta Frequenza*, pp. 1–7, 1978.)

the radii of these circles being 0.7λ, 1.4λ, . . . , 3.5λ. Eight equispaced radiators would lie on the innermost circle, 16 on the next, and so on. The relative currents would be as shown in the last column of Table 6.5. The theoretical sum array pattern would then be typified by Figures 6.24c and d with $u = 7 \sin \theta$.

6.9 Rectangular Grid Arrays with Rectangular Boundaries: Nonseparable Tseng-Cheng Distributions

The analyses presented heretofore in this chapter have shown the following.

1. If a rectangular grid array with a rectangular boundary is given a *separable* distribution, sum and difference patterns can be generated. These patterns are the product of two conical linear array patterns, associated with the row and column directions of the planar array. If the two linear array distributions are Dolph-Chebyshev,

a sum pattern will be produced with uniform height side lobes in each principal plane. However, these are mound side lobes, and those which occur off the principal planes are suppressed far below the design level, at a cost in beamwidth and/or directivity. (Though the subject was not pursued further, the two linear array distributions that comprise the separable discrete planar distribution could be such that arbitrary side lobe topography is created in each principal plane. The nature of the pattern is still to have mound side lobes, reduced in level off axis. Also, *continuous* separable rectangular aperture distributions could have been considered, but if this were only for the purpose of being sampled, it was better to begin with discrete separable distributions.)

2. If a circular aperture contains a linearly polarized *continous* distribution which is ϕ-symmetric, a sum pattern is produced which, for the proper radial dependence of the distribution (Taylor), has quasi-uniform height ring side lobes. There is a beamwidth and/or directivity advantage over a sum pattern with mound side lobes. Modified continuous distributions can yield sum patterns with ring side lobes of individually arbitrary heights, or can cause sum patterns with ring side lobes that undulate in height through a sequence of ϕ-cuts.

3. Conventional sampling of the continuous aperture distributions described in Paragraph 2 is a useful procedure when the number of elements in the discrete array is large. It results in patterns with little degradation, and such arrays enjoy the beamwidth/directivity advantage due to ring side lobes. This method of discretizing can be applied successfully to circular grid arrays with circular boundaries, and to rectangular grid arrays with either circular or elliptical boundaries. However, as smaller and smaller arrays are considered, the pattern degradation worsens and at some point becomes unacceptable.

4. An alternate discretizing technique is available for circular grid arrays that overcomes the pattern degradation due to conventional sampling, even when the array is small.

There remains the problem of rectangular grid arrays—with either rectangular or circular boundaries—for which one seeks a nonseparable distribution that will result in a sum pattern with nondegraded ring side lobes. This problem will be addressed in this section for rectangular boundaries, and in the next section for circular boundaries.

A technique due to F. I. Tseng and D. K. Cheng[12] is applicable to rectangular grid arrays with rectangular boundaries, with the one restriction that the number of elements in a row equals the number of elements in a column. The interelement spacings d_x and d_y need not be equal, and thus the array need not be square. The distribution is discrete and nonseparable and, in the original Tseng-Cheng formulation, gives a Dolph-Chebyshev pattern in every ϕ-cut (and thus ring side lobes).

The technique will be developed for an array of $2N$ by $2N$ elements. (An equivalent analysis applies for an odd number of elements per row, but the even case has more applications, since it also permits a difference pattern.) With quadrantal symmetry of the aperture distribution assumed, the expression for the sum pattern is similar to

[12] F. I. Tseng and D. K. Cheng, "Optimum Scannable Planar Arrays with an Invariant Side Lobe Level," *Proc. IEEE*, 56 (1968), 1771–78.

(6.87), that is,

$$\mathcal{S}(\theta, \phi) = 4 \sum_{m=1}^{N} \sum_{n=1}^{N} I_{mn} \cos\left(\frac{2m-1}{2}\psi_x\right) \cos\left(\frac{2n-1}{2}\psi_y\right) \tag{6.101}$$

with ψ_x and ψ_y given generally by (6.36).

Since the main beam pointing direction is defined by $\psi_x = \psi_y = 0$, if the substitutions

$$u = \frac{\pi d_x}{\lambda}(\sin\theta\cos\phi - \sin\theta_0\cos\phi_0) \tag{6.102}$$

$$v = \frac{\pi d_y}{\lambda}(\sin\theta\sin\phi - \sin\theta_0\sin\phi_0) \tag{6.103}$$

are made, (6.101) becomes

$$\mathcal{S}(u, v) = 4 \sum_{m=1}^{N} \sum_{n=1}^{N} I_{mn} \cos[(2m-1)u] \cos[(2n-1)v] \tag{6.104}$$

In the manner of Tseng and Cheng, one can introduce the transformation

$$w = w_0 \cos u \cos v \tag{6.105}$$

Then, since it has been shown in Appendix D that

$$\cos^{2s-1} u = \frac{1}{2^{2s-2}} \sum_{p=1}^{s} \binom{2s-1}{s-p} \cos(2p-1)u$$

with the same formula applying for $\cos^{2s-1} v$, it follows that a general odd polynomial of order $2N - 1$ can be written in the form

$$\begin{aligned} P_{2N-1}(w) &= \sum_{s=1}^{N} a_{2s-1} w^{2s-1} \\ &= \sum_{s=1}^{N} a_{2s-1} w_0^{2s-1} \cos^{2s-1} u \cos^{2s-1} v \\ &= \sum_{s=1}^{N} \sum_{p=1}^{s} \sum_{q=1}^{s} \frac{a_{2s-1}}{2^{2s-3}} \binom{2s-1}{s-p} \binom{2s-1}{s-q} \left(\frac{w_0}{2}\right)^{2s-1} \cos(2p-1)u \cos(2q-1)v \end{aligned} \tag{6.106}$$

If one wishes the pattern $\mathcal{S}(u, v)$ to have the characteristics of the polynomial $P_{2N-1}(w)$, then comparison of (6.104) and (6.106) indicates that

$$I_{mn} = \sum_{s=(m,n)}^{N} \frac{a_{2s-1}}{2^{2s-1}} \binom{2s-1}{s-m} \binom{2s-1}{s-n} \left(\frac{w_0}{2}\right)^{2s-1} \tag{6.107}$$

in which $(m, n) = m$ if $m \geq n$ and $(m, n) = n$ if $m < n$.

The specific case treated by Tseng and Cheng was to choose $P_{2N-1}(w)$ to be the Chebyshev polynomial

$$T_{2N-1}(w) = \sum_{s=1}^{N} \mathring{a}_{2s-1} w^{2s-1}$$
$$= \sum_{s=1}^{N} (-1)^{N-s} \frac{2^{2s-2}(2N-1)}{N+s-1} \binom{N+s-1}{2s-1} w^{2s-1} \qquad (6.108)$$

(see Appendix C). Identification of $\mathring{a}_{2s-1}$ from (6.108) and its insertion in (6.107) gives

$$I_{mn} = \sum_{s=(m,n)}^{N} (-1)^{N-s} \frac{2N-1}{2(N+s-1)} \binom{N+s-1}{2s-1} \binom{2s-1}{s-m} \binom{2s-1}{s-n} \left(\frac{w_0}{2}\right)^{2s-1} \qquad (6.109)$$

Equation 6.109 is the Tseng-Cheng formula for the excitation of an even-numbered planar array whose pattern has the features of a Chebyshev polynomial.

Despite its formidable appearance, (6.109) is a simple formula to program. As an example of its use, consider a 10 by 10 array in which $d_x = \lambda/2$ and $d_y = 3\lambda/4$. Assume that it is desired to obtain a pencil beam pointing broadside-broadside ($\theta_0 = 0°$), with 20 dB ring side lobes. Then w_0 is determined in the usual way, such that $T_{2N-1}(w_0) = T_9(w_0) = 10$ in this case. If follows that $w_0 = 1.0558$. Equation 6.109 then gives the current distribution listed in the second column of Table 6.6.

TABLE 6.6

I_{mn}	20 dB Tseng-Cheng	Innermost Side Lobe—30 dB
I_{11}	0.773	1.000
$I_{21} = I_{12}$	0.569	0.708
$I_{31} = I_{13}$	0.796	0.778
$I_{41} = I_{14}$	0.029	−0.050
$I_{51} = I_{15}$	1.000	0.846
I_{22}	0.946	0.980
$I_{32} = I_{23}$	0.119	0.100
$I_{42} = I_{24}$	0.618	0.477
$I_{52} = I_{25}$	0.667	0.564
I_{33}	0.486	0.369
$I_{43} = I_{34}$	0.777	0.642
$I_{53} = I_{35}$	0.286	0.242
I_{44}	0.387	0.325
$I_{54} = I_{45}$	0.071	0.060
I_{55}	0.008	0.007

When these currents are used in (6.104), the patterns shown in Figure 6.25 result. One can see the typical feature of Tseng-Cheng cuts, namely that they are all Dolph patterns. One can also infer the ringlike nature of individual side lobes.

The analysis presented in this section does not need to be restricted to Chebyshev polynomials. If the design requirement is a sum pattern with ring side lobes of indi-

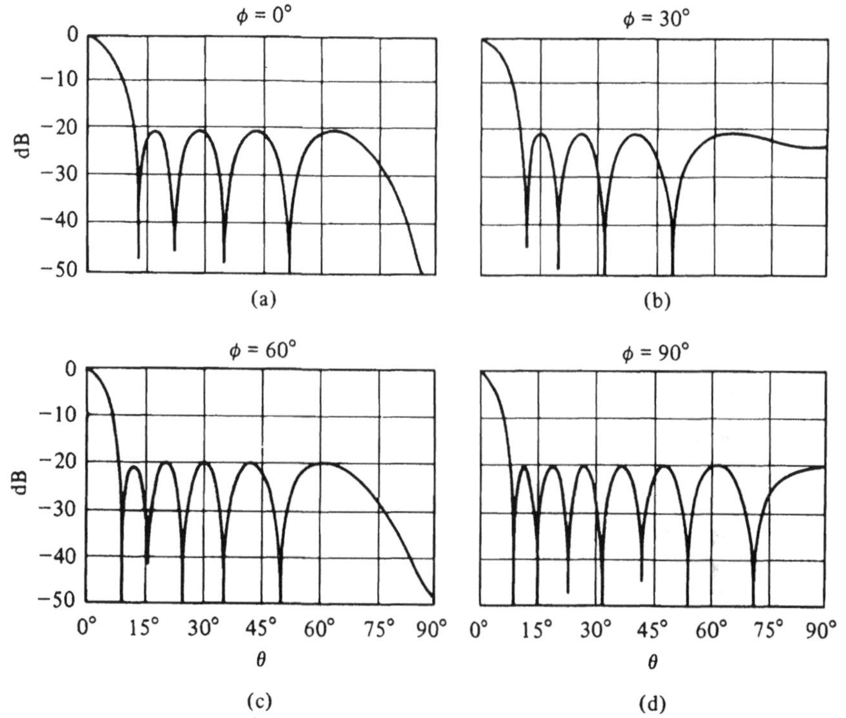

Fig. 6.25 Four ϕ-Cuts of the Sum Pattern of a 10 by 10 Rectangular Grid Array with Rectangular Boundary; $d_x = 0.5\lambda$, $d_y = 0.75\lambda$; Tseng-Cheng Distribution, −20 dB SLL (Reprinted from *Radio Science*, vol. 12, pp. 653–657, 1977, Copyrighted by American Geophysical Union.)

vidually arbitrary heights, a more general polynomial $P_{2N-1}(w)$ must be selected. It is often possible to determine this generalized polynomial through a perturbation of a suitable Chebyshev polynomial.[13]

Assume that the coefficients a_{2s-1} of $P_{2N-1}(w)$, as defined in (6.106), differ only slightly from the coefficients $\mathring{a}_{2s-1}$ of the corresponding $T_{2N-1}(w)$, defined in (6.108). That is, let

$$a_{2s-1} = \mathring{a}_{2s-1} + \delta_{2s-1} \tag{6.110}$$

Then

$$P_{2N-1}(w) - T_{2N-1}(w) = \sum_{s=1}^{N} \delta_{2s-1} w^{2s-1} \tag{6.111}$$

Let w_n^p be the position of the nth peak in the Chebyshev polynomial, as illustrated in Figure 6.26a for the case $T_9(w)$. One notes that there are $N - 1$ such peaks in $w > 0$. Further, let w_0 be that value of w which gives $T_{2N-1}(w_0) = b$, with $-20 \log_{10} b$ the side lobe level in the Dolph pattern. If one inserts successively $w_1^p, w_2^p, \ldots, w_{N-1}^p$,

[13] R. S. Elliott, "Synthesis of Rectangular Planar Arrays for Sum Patterns with Ring Side Lobes of Arbitrary Topography," *Radio Science*, 12 (1977), 653–57.

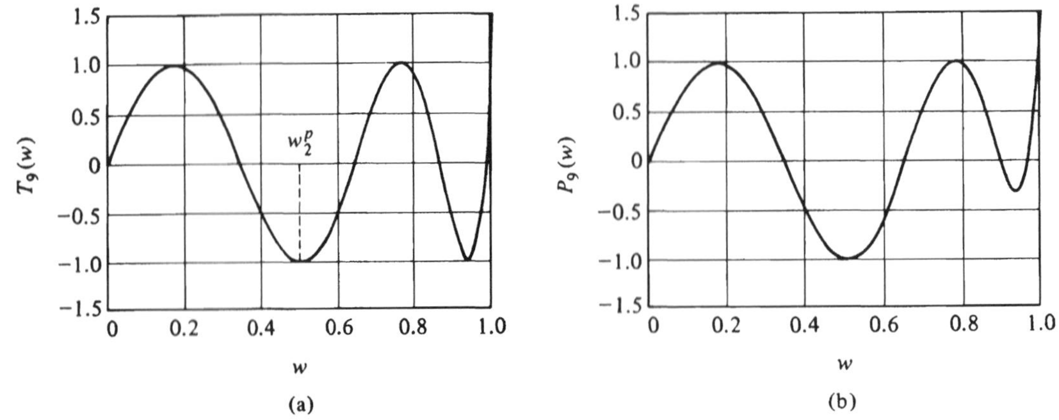

Fig. 6.26 Plots of the Chebyshev Polynomial of Ninth Order and the Corresponding Modified Polynomial (Reprinted from *Radio Science*, vol. 12, pp. 653–657, 1977, Copyrighted by American Geophysical Union.)

and w_0 in (6.111), the result is N simultaneous linear equations in the N unknowns δ_{2s-1}. This is so because the condition $P_{2N-1}(w_0) = T_{2N-1}(w_0)$ can be imposed, in which case $P_{2N-1}(w_n^p) - T_{2N-1}(w_n^p)$ is approximately the difference in levels of the nth side lobe of the desired pattern and the starting (Chebyshev) pattern, a known quantity.

After one solves for the perturbations δ_{2s-1} by matrix inversion, the results can be placed in (6.110) and then in the first form of (6.106) to see if the resulting polynomial $P_{2N-1}(w)$ is close enough to specification. If it is not, the process can be iterated until the designer is satisfied. The final set of values of a_{2s-1} can be used in (6.107) to determine the current distribution. Experience has shown that this process converges rapidly; usually two or three iterations are sufficient.

As an illustration, suppose that the requirement is to design a 10-by-10 array, with $d_x = \lambda/2$ and $d_y = 3\lambda/4$, and that the excitation is to produce a sum pattern with concentric ring side lobes, all of which are at −20 dB except the innermost which, due to noise considerations, needs to be at −30 dB. When the procedure just outlined is followed, one iteration moves from the Chebyshev plot of Figure 6.26a to the modified polynomial plotted in Figure 6.26b. The coefficients of this polynomial appear in

$$P_9(w) = 236.4w^9 - 537.1w^7 + 409.6w^5 - 116.1w^3 + 8.9w \tag{6.112}$$

which can be contrasted to the Chebyshev polynomial

$$T_9(w) = 256w^9 - 576w^7 + 432w^5 - 120w^3 + 9w \tag{6.113}$$

When the coefficients contained in (6.112) are used in (6.107), the result is the current distribution listed in the third column of Table 6.6. That current distribution causes the patterns shown in Figure 6.27. One can observe that these patterns exhibit

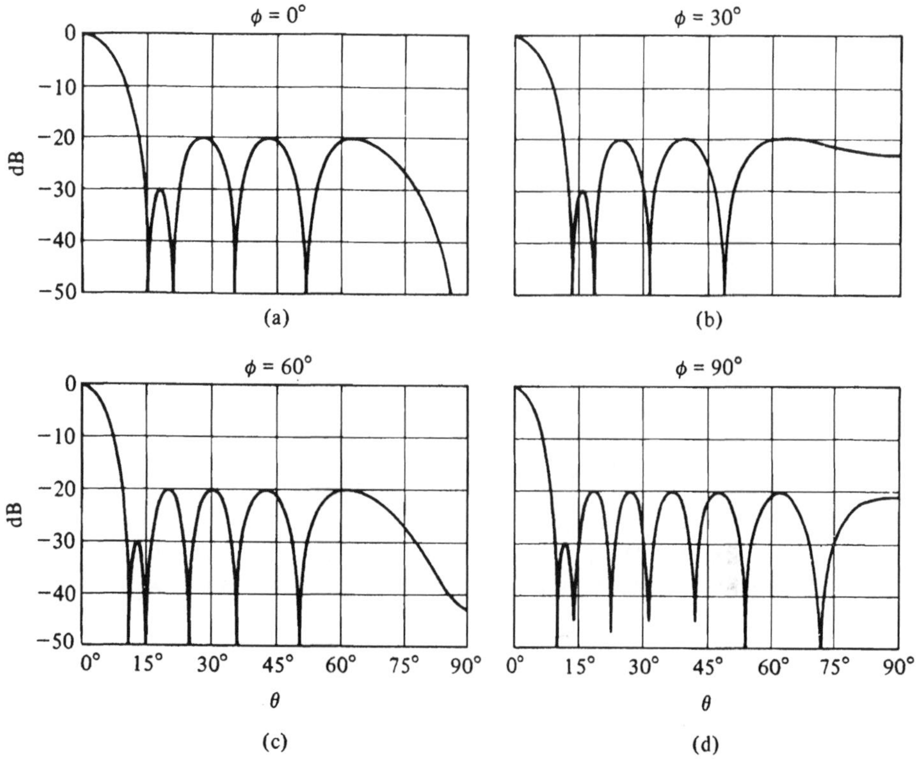

Fig. 6.27 Four ϕ-Cuts of the Sum Pattern of a 10 by 10 Rectangular Grid Array with Rectangular Boundary; $d_x = 0.5\lambda$, $d_y = 0.75\lambda$; Modified Tseng-Cheng Distribution (Reprinted from *Radio Science*, vol. 12, pp. 653–657, 1977, Copyrighted by American Geophysical Union.)

all the desired features. It is also worth noting that the current distribution is not demanding in terms of the present state of the art.

6.10 A Discretizing Technique for Rectangular Grid Arrays

In Section 6.6 conventional sampling of circular Taylor distributions was introduced and applied to rectangular grid arrays with circular boundaries. It was seen that the pattern degradation was small for arrays with many elements, but became unacceptable as the array size was reduced. In Section 6.9 a nonseparable Tseng-Cheng distribution was discussed that yields Dolph (or modified Dolph) patterns in every ϕ-cut, and is applicable to small as well as large arrays. However, the Tseng-Cheng technique requires rectangular boundaries and equal numbers of elements in the two directions. If the corners of the array are cut off to fit a circular or elliptical boundary and the remaining elements are excited Tseng-Cheng, serious pattern degradation occurs.

There remains the need to improve on conventional sampling for situations not covered by the Tseng-Cheng distribution. This problem parallels the one already encountered with linear arrays and overcome by the perturbation technique described in Sections 5.8 through 5.10. However, in the linear array case, with N elements there were $N-1$ ϕ-symmetric conical side lobes needing to be controlled and $N-1$ current ratios that could be adjusted to accomplish the task. This neat deterministic relationship does not carry over to the planar array case. For example, if one is attempting to produce a sum pattern consisting of a pencil main beam and ring side lobes, the problem is essentially one of trying to control the side lobe structure in an infinitude of ϕ-cuts with the excitation of a finite number of two-dimensionally positioned elements. The best one can hope for is to minimise the deviation of the actual side lobe structure from what is desired.

A technique that has proven useful in problems of this type starts with a current distribution $\overset{0}{I}_{mn}$, which has been determined by conventional sampling and which yields the sum pattern

$$\mathcal{S}_0(\theta, \phi) = \sum_m \sum_n \overset{0}{I}_{mn} \cos(k\xi_{mn} \sin\theta \cos\phi) \cos(k\eta_{mn} \sin\theta \sin\phi) \tag{6.114}$$

Implicit in (6.114) is the assumption that there is quadrantal symmetry in element placement and excitation. However, there is no restriction on the shape of the boundary, nor do the elements need to be arranged in a regular grid. If the grid *is* rectangular, $\xi_{mn} = (2m-1)d_x/2$ and $\eta_{mn} = (2n-1)d_y/2$, but what follows has a more general applicability.

Imagine that the starting pattern given by (6.114) is not acceptable and needs to be improved. Let an achievable pattern be given by

$$\mathcal{S}(\theta, \phi) = \sum_m \sum_n I_{mn} f_{mn} \tag{6.115}$$

in which $f_{mn} = \cos(k\xi_{mn} \sin\theta \cos\phi) \cos(k\eta_{mn} \sin\theta \sin\phi)$ and I_{mn} is a current distribution which will improve the pattern. If one can assume that

$$I_{mn} = \overset{0}{I}_{mn} + \delta I_{mn} \tag{6.116}$$

then the difference between (6.115) and (6.114) is simply

$$\mathcal{S}(\theta, \phi) - \mathcal{S}_0(\theta, \phi) = \sum_m \sum_n \delta I_{mn} f_{mn} \tag{6.117}$$

Let a ϕ-cut of the starting pattern be designated by ϕ_q and let the peak of the pth lobe in this ϕ-cut occur at the angle θ_{pq}. Then

$$\mathcal{S}(\theta_{pq}, \phi_q) - \mathcal{S}_0(\theta_{pq}, \phi_q) = \sum_m \sum_n \delta I_{mn} f_{mnpq} \tag{6.118}$$

in which $f_{mnpq} = \cos(k\xi_{mn} \sin\theta_{pq} \cos\phi_q) \cos(k\eta_{mn} \sin\theta_{pq} \sin\phi_q)$.

Suppose that a number of pointing directions (θ_{pq}, ϕ_q) is chosen to equal exactly the number of elements in one quadrant of the array. If one of these directions is at the peak of the main beam, and if $\mathcal{S}$ is equated to $\mathcal{S}_0$ in this direction, then for the other directions the left side of (6.118) is approximately the difference between desired and starting side lobe level, a known quantity. Equation 6.118 then becomes a deterministic set that can be solved for the current perturbations δI_{mn} by matrix inversion. The remainder of the procedure follows in the usual way. One places δI_{mn} in (6.116) to obtain the new currents, which in turn are used in (6.115) to give the new pattern. If the result is not satisfactory, iteration can be undertaken.

The success of this method hinges on judicious selection of the ϕ-cuts and the particular lobes in those ϕ-cuts for use as the pointing directions (θ_{pq}, ϕ_q). Experience has shown that the most easily controlled side lobes are those closest to the main beam and that the ϕ-cuts should be chosen to divide angle space into roughly equal regions.

As an illustration, one can return to the rectangular grid array with circular boundary and eight elements per quadrant, depicted in Figure 6.17. Conventional sampling of the circular Taylor distriubtion for a -22 dB, $\bar{n} = 3$ pattern led to the current distribution shown as an inset to Figure 6.18 and the accompanying unacceptable patterns. If it is desired to move as close as possible to the ideal Taylor pattern (ϕ-independent, with ring side lobes at a quasi-constant height of -22 dB), then clearly the discrete aperture distribution should be as ϕ-symmetric as possible, implying that $I_{mn} = I_{nm}$. Thus, for this case, (6.118) reduces to

$$\begin{aligned} \mathcal{S}(\theta_{pq}, \phi_q) - \mathcal{S}_0(\theta_{pq}, \phi_q) = {} & \delta I_{11} f_{11pq} + \delta I_{22} f_{22pq} \\ & + \delta I_{21}(f_{21pq} + f_{12pq}) \\ & + \delta I_{31}(f_{31pq} + f_{13pq}) \\ & + \delta I_{32}(f_{32pq} + f_{23pq}) \end{aligned} \tag{6.119}$$

and there are only five unknown current increments to determine.

If the pattern cuts $\phi_q = 11.25°, 33.75°$ are chosen, the regions "belonging" to these ϕ-cuts are equal. And if, in addition to the main beam peak position $(0°, \phi)$, the positions of the two innermost side lobes in the 11.25° and 33.75° cuts are chosen, five independent linear equations arise from (6.119) which can be solved simultaneously for the values of the δI's. When this is done, a sequence of iterations leads to the current distribution shown in the inset of Figure 6.28. The accompanying patterns are seen to be a significant improvement over those of Figure 6.18. Though not shown, intervening ϕ-cuts differ but little from those displayed in Figure 6.28. The current distribution is also seen to be reasonable, though distinct from the starting excitation.

In some applications a better starting pattern can be found than was obtained by conventional sampling if use is made of the idea of collapsed distributions. The concept of a collapsed distribution can be understood if the coordinate rotation

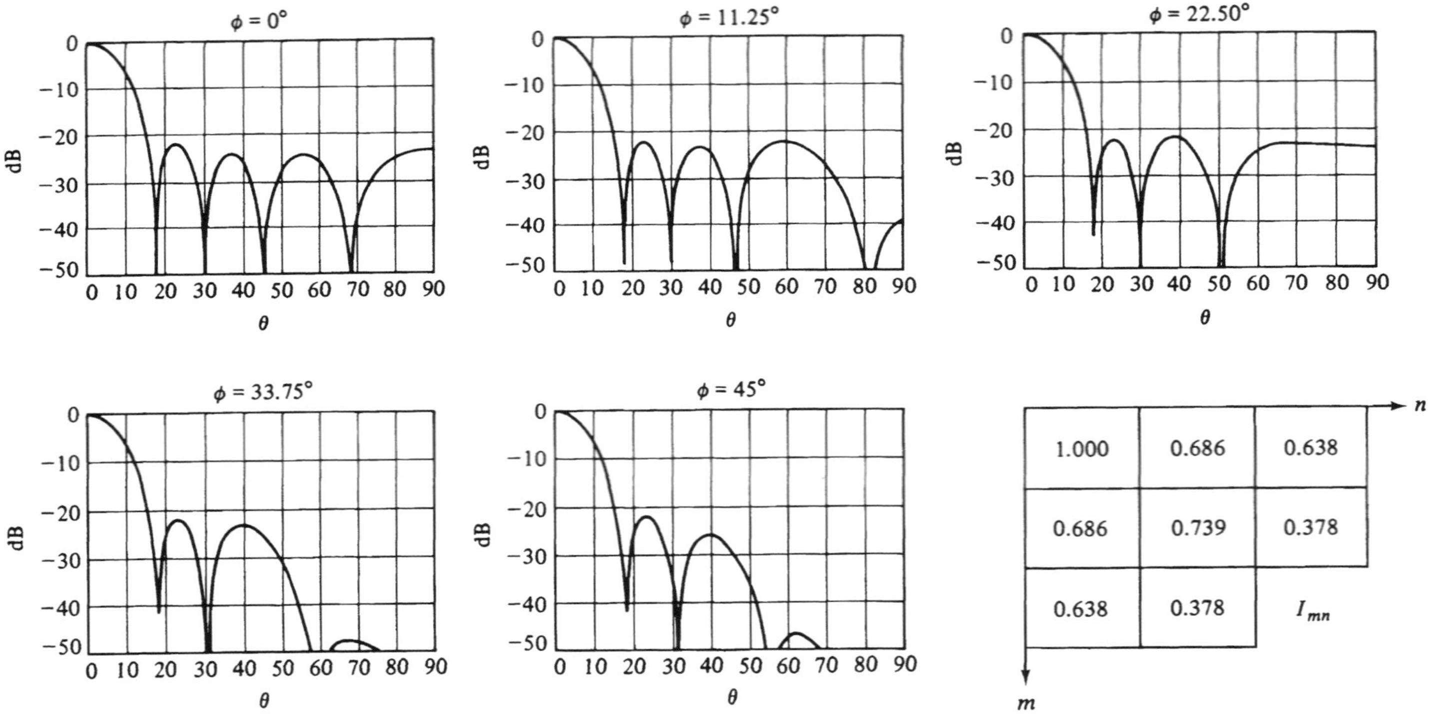

Fig. 6.28 Five ϕ-Cuts of the Sum Pattern of a 32-Element Planar Array; Rectangular Grid, Circular Boundary; $d_x = d_y = 0.7\lambda$; Currents Obtained by the Perturbation Procedure of Equation 6.119

$$\begin{aligned} \xi_{mn} &= \xi'_{mn} \cos \beta - \eta'_{mn} \sin \beta \\ \eta_{mn} &= \xi'_{mn} \sin \beta + \eta'_{mn} \cos \beta \end{aligned} \tag{6.120}$$

is used to transform the general formula for the planar array factor

$$\mathcal{A}_a(\theta, \phi) = \sum_m \sum_n I_{mn} e^{jk\sin\theta(\xi_{mn}\cos\phi + \eta_{mn}\sin\phi)} \tag{6.121}$$

to the form

$$\mathcal{A}_a(\theta, \phi) = \sum_m \sum_n I_{mn} e^{jk\sin\theta[\xi'_{mn}\cos(\phi-\beta) + \eta'_{mn}\sin(\phi-\beta)]} \tag{6.122}$$

If the special condition $\beta = \phi$ is imposed, (6.122) reduces to

$$\mathcal{A}_a(\theta, \phi) = \sum_m \sum_n I_{mn} e^{jk\xi'_{mn} \cos \theta_{x'}} \tag{6.123}$$

in which $\theta_{x'}$ is the polar angle measured from the X'-axis (that is, in the $X'Z$-plane, $\theta_{x'} = (\pi/2) - \theta$ and $\sin \theta = \cos \theta_{x'}$).

In words, (6.123) says that if all elements in the planar array are projected onto the X'-axis and given their original excitations I_{mn}, the pattern of the resulting linear array, *in the $X'Z$-plane* is the same as the pattern of the actual planar array in that same plane (that is, the ϕ-plane). This result is true whether one is dealing with sum patterns or difference patterns, and whether or not the elements are regularly spaced.

Let this concept be applied to the illustrative example of this section, namely the array shown in Figure 6.17. With $\overset{0}{I}_{mn} = \overset{0}{I}_{nm}$, the normalized starting current distribution can be represented simply, as shown in Figure 6.29. When this distribution

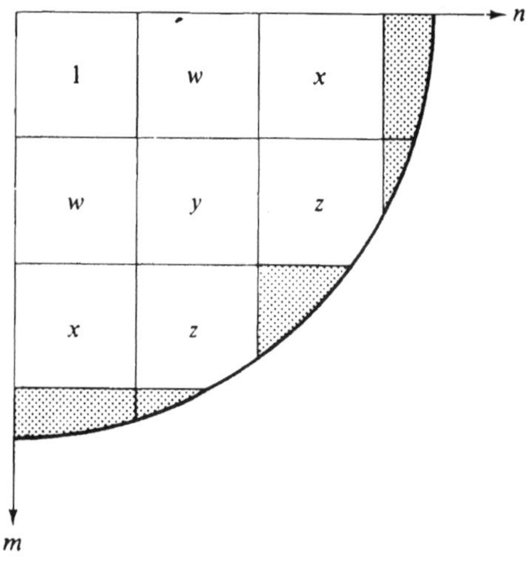

Fig. 6.29 Notation for the Starting Current Distribution of a 32-Element Planar Array; Sum Mode

is collapsed onto the X-axis, one obtains the six-element linear array of relative excitation

$$(x + z) \quad (w + y + z) \quad (1 + w + x) \quad (1 + w + x) \quad (w + y + z) \quad (x + z)$$

However, if it is collapsed onto the 45°-line, a nine-element linear array of relative excitation

$$(2z) \quad (2x + y) \quad (2w + 2x) \quad (1 + 2w + 2z) \quad (2 + 2y) \quad (1 + 2w + 2z) \quad (2w + 2x) \quad (2x + y) \quad (2z)$$

is obtained.

If one finds the linear array distributions

$$\begin{array}{ccccccccc} & & a_2 & a_1 & 1 \quad 1 & a_1 & a_2 & & \\ b_4 & b_3 & b_2 & b_1 & 1 & b_1 & b_2 & b_3 & b_4 \end{array}$$

which give conical sum patterns with a -22 dB side lobe level, then the planar array pattern would be forced to be correct in the $\phi = 0°$ and $\phi = 45°$ planes if

$$\begin{aligned} w + y + z &= a_1(1 + w + x) \\ x + z &= a_2(1 + w + x) \\ 1 + 2w + 2z &= b_1(2 + 2y) \\ 2w + 2x &= b_2(2 + 2y) \\ 2x + y &= b_3(2 + 2y) \\ 2z &= b_4(2 + 2y) \end{aligned} \tag{6.124}$$

One could then hope that the pattern in between these two ϕ-cuts would not wander too far from what was desired.

Unfortunately, it would take six controllable currents to satisfy all of these conditions, and there are only four relative currents available to be adjusted. So let us be content to satisfy the first four equations of (6.124).

The Brown and Scharp[14] tables give, for a -22 dB side lobe level,

$$a_1 = 0.7559 \qquad a_2 = 0.4683 \qquad b_1 = 0.9447 \qquad b_2 = 0.7924$$

When these values are inserted in the first four equations of (6.124), matrix inversion gives the starting current distribution shown in the table inset of Figure 6.30. The accompanying patterns are a considerable improvement over those of Figure 6.18 and are therefore more desirable for use as starting patterns, since they cut down on the number of iterations needed in the process associated with Equations 6.119.

[14] Brown, "Tables."

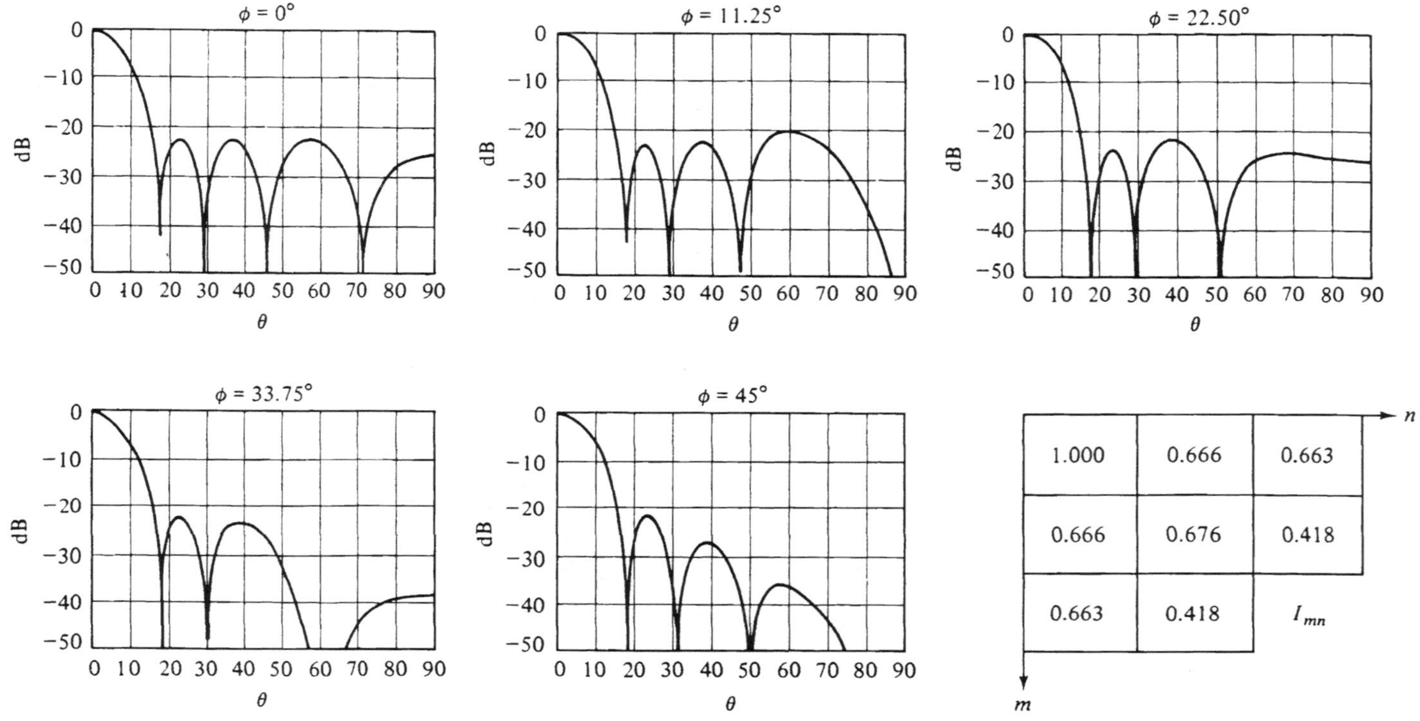

Fig. 6.30 Five ϕ-Cuts of the Sum Pattern of a 32-Element Planar Array. Rectangular Grid, Circular Boundary; $d_x = d_y = 0.7\lambda$; Currents Obtained by Fitting to Collapsed Distribution

6.11 Circular Bayliss Patterns

In the general analysis that introduced Section 6.3, it was shown that, for a planar aperture with a circular boundary of radius a, if the aperture contained a continuous, unidirectional lineal current density distribution $K(\rho, \beta)$, then the array factor could be represented by the function

$$F(\theta, \phi) = 2\pi \sum_{n=-\infty}^{\infty} (j)^n e^{jn\phi} \int_0^a K_n(\rho) J_n(k\rho \sin\theta)\rho \, d\rho \tag{6.125}$$

in which $K_n(\rho)$ is the nth partial aperture distribution, contained in

$$K(\rho, \beta) = \sum_{n=-\infty}^{\infty} K_n(\rho) e^{jn\beta} \tag{6.126}$$

In the remainder of the analysis of Section 6.3, attention was focused on the special case $n = 0$, which led to ϕ-independent sum patterns of the circular Taylor type. Now, if attention is turned to the special case $n = \pm 1$, Equation 6.125 becomes

$$\mathfrak{D}(\theta, \phi) = 2\pi j \int_0^a [e^{j\phi} K_1(\rho) J_1(k\rho \sin\theta) - e^{-j\phi} K_{-1}(\rho) J_{-1}(k\rho \sin\theta)]\rho \, d\rho \tag{6.127}$$

Since $J_{-1}(k\rho \sin\theta) = -J_1(k\rho \sin\theta)$, if $K_{-1}(\rho) = K_1(\rho)$, Equations 6.126 and 6.127 reduce to

$$K(\rho, \beta) = 2K_1(\rho) \cos\beta \tag{6.128}$$

$$\mathfrak{D}(\theta, \phi) = 4\pi j \cos\phi \int_0^a K_1(\rho) J_1(k\rho \sin\theta)\rho \, d\rho \tag{6.129}$$

In words, this pair of equations indicates that if one excites the aperture with a continuous distribution of the type $K_1(\rho) \cos\beta$, the resulting pattern will be in the form $f(\theta)\cos\phi$, in which

$$f(\theta) = \int_0^n K_1(\rho) J_1(k\rho \sin\theta)\rho \, d\rho \tag{6.130}$$

Clearly, the shape of $K_1(\rho)$ will affect the shape of $f(\theta)$.

In any plane that contains the Z-axis, one finds two opposed ϕ-cuts (half-planes). If one half-plane is defined by the angle ϕ, the other will correspond to the angle $\phi + \pi$. From (6.129) and (6.130)

$$\mathfrak{D}(\theta, \phi) = 4\pi j \cos\phi \, f(\theta) \tag{6.131}$$

$$\mathfrak{D}(\theta, \phi + \pi) = -4\pi j \cos\phi \, f(\theta) \tag{6.132}$$

Thus the patterns in these two half-planes are mirror images, except for a change of sign, and taken together they consitute a *difference* pattern. As one examines the pattern in all the planes that contain the Z-axis, it can be seen from (6.131) and (6.132) that the *shape* of this pattern is governed by $f(\theta)$ in all the planes, with the relative height in different planes controlled by $\cos\phi$. In the plane $\phi = 0°/180°$, the pattern is at its greatest level, and in the plane $\phi = 90°/270°$ the pattern is at a null level. Therefore this pattern is useful to provide resolution in the XZ-plane.

Similarly, if $K_{-1}(\rho) = -K_1(\rho)$, Equations 6.126 and 6.127 reduce to

$$K(\rho, \beta) = 2jK_1(\rho)\sin\beta \tag{6.133}$$

$$\mathfrak{D}(\theta, \phi) = -4\pi \sin\phi\, f(\theta) \tag{6.134}$$

The entire argument of the previous paragraph can be repeated, except for a shift of 90° in ϕ. The distribution in (6.133) will give rise to a difference pattern, which can be used to provide resolution in the YZ-plane. Many modern radar antenna systems make use of a sum pattern and both of these difference patterns.

Since the shape of both difference patterns is due to $f(\theta)$, which in turn is governed by the radial aperture distribution $K_1(\rho)$, attention can be directed to Equation 6.130. It is convenient once again to introduce the substitutions

$$u = (2a/\lambda)\sin\theta \qquad p = \pi\rho/a$$

which convert (6.130) to the form

$$f(u) = (a/\pi)^2 \int_0^\pi K_1(p)J_1(up)p\,dp \tag{6.135}$$

It will prove desirable to express $K_1(p)$ as an orthogonal expansion in the form

$$K_1(p) = \sum_{m=0}^{\infty} A_m J_1(\mu_m p) \tag{6.136}$$

in which the μ_m coefficients are eigenvalues, to be defined shortly. Substitution of (6.136) in (6.135) gives

$$\begin{aligned} f(u) &= \left(\frac{a}{\pi}\right)^2 \sum_{m=0}^{\infty} A_m \int_0^\pi J_1(\mu_m p)J_1(up)\,p\,dp \\ &= \left(\frac{a}{\pi}\right)^2 \sum_{m=0}^{\infty} A_m \left[\frac{\mu_m p J_1(up)J_0(\mu_m p) - upJ_0(up)J_1(\mu_m p)}{u^2 - \mu_m^2}\right]_0^\pi \end{aligned} \tag{6.137}$$

Since $vJ_0(v) = J_1(v) + vJ_1'(v)$, the preceding result can be converted to

$$\begin{aligned} f(u) &= \left(\frac{a}{\pi}\right)^2 \sum_{m=0}^{\infty} A_m \left[\frac{\mu_m p J_1'(\mu_m p)J_1(up) - upJ_1'(up)J_1(\mu_m p)}{u^2 - \mu_m^2}\right]_0^\pi \\ &= \left(\frac{a}{\pi}\right)^2 \sum_{m=0}^{\infty} A_m \frac{\pi\mu_m J_1'(\pi\mu_m)J_1(\pi u) - \pi u J_1'(\pi u)J_1(\pi\mu_m)}{u^2 - \mu_m^2} \end{aligned} \tag{6.138}$$

A study of (6.138) reveals that the condition $f(\mu_n) = 0$, $n \neq m$, can be obtained if either μ_m is defined by $J_1(\pi\mu_m) = 0$ or by $J_1'(\pi\mu_m) = 0$. A return to (6.136) shows that, if the first option is chosen, $K_1(\pi)$ must equal zero. Since the desirable aperture distributions will be found to be nonzero at the boundary $p = a$, it is appropriate to select the second option and define the eigenvalues μ_m by the equation

$$J_1'(\pi\mu_m) = 0 \tag{6.139}$$

The first twenty of these eigenvalues are listed in Table 6.7.

TABLE 6.7 Bessel function zeros, $J_1'(\pi\mu_m) = 0$

m	μ_m	m	μ_m	m	μ_m	m	μ_m
0	0.5860670	5	5.7345205	10	10.7417435	15	15.7443679
1	1.6970509	6	6.7368281	11	11.7424475	16	16.7447044
2	2.7171939	7	7.7385356	12	12.7430408	17	17.7450030
3	3.7261370	8	8.7398505	13	13.7435477	18	18.7452697
4	4.7312271	9	9.7408945	14	14.7439856	19	19.7455093

With the selection of (6.139), Equation 6.138 reduces to

$$f(u) = \left(\frac{a}{\pi}\right)^2 \sum_{m=0}^{\infty} A_m J_1(\pi\mu_m) \frac{\pi u J_1'(\pi u)}{\mu_m^2 - u^2} \tag{6.140}$$

It is instructive to consider the special case that the aperture distribution consists solely of the $m = 0$ (or fundamental) term. Then

$$f(u) = \left(\frac{a}{\pi}\right)^2 \frac{A_0 J_1(0.586\pi)}{(0.586)^2}\left[\frac{\pi u J_1'(\pi u)}{1 - (u/0.586)^2}\right] \tag{6.141}$$

The aperture distribution $J_1(0.586p)$ is shown in Figure 6.31 and the pattern is plotted in Figure 6.32. The typical features of a difference pattern are evident—a null at $u = 0$, then one of the twin main peaks, followed by a sequence of side lobes which steadily decay in height. (Because $|J_1'(\pi u)| \sim u^{-1/2}$ as $u \longrightarrow \infty$, $|f(u)| \sim u^{-3/2}$ for u large). However, for many practical applications, the innermost side lobes are too high, and the beamwidth of the twin main beams is enlarged because the further-out side lobes are lower than required. Thus a more complicated aperture distribution is needed in order to get an improved side lobe structure.

This problem parallels the one already encountered for the sum pattern in Section 6.3. There it was noted that Taylor was able to modify the generic sum pattern $J_1(\pi u)/\pi u$ by an appropriate shift of its innermost $\bar{n} - 1$ null pairs and achieve a quasi-uniform side lobe level of specified height. E. T. Bayliss[15] has shown how to accomplish the same result for circular difference patterns.

[15] E. T. Bayliss, "Design of Monopulse Antenna Difference Patterns with Low Sidelobes," *Bell System Tech. J.*, 47 (1968), 623–50.

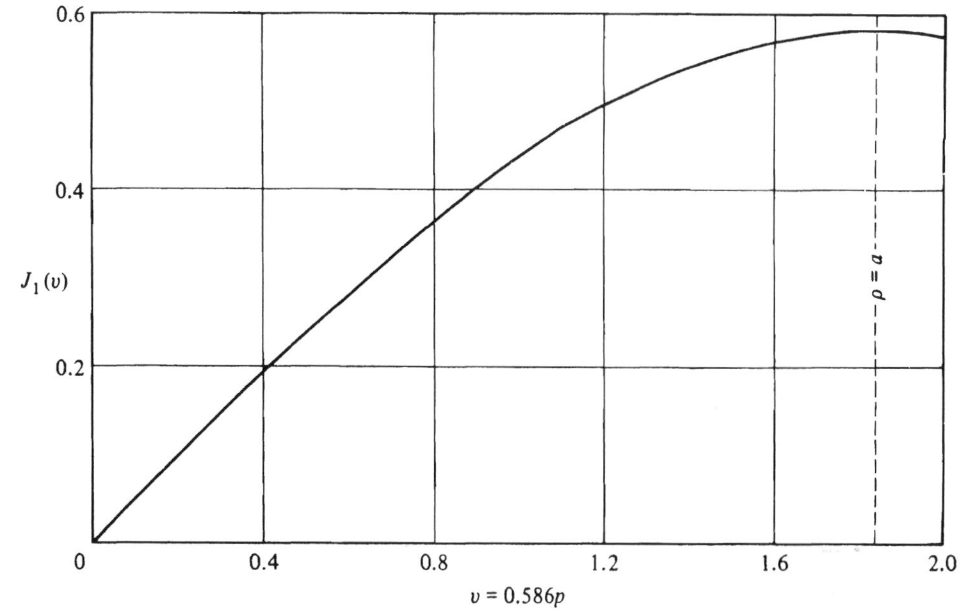

Fig. 6.31 The Aperture Distribution for the Generic Circular Difference Pattern

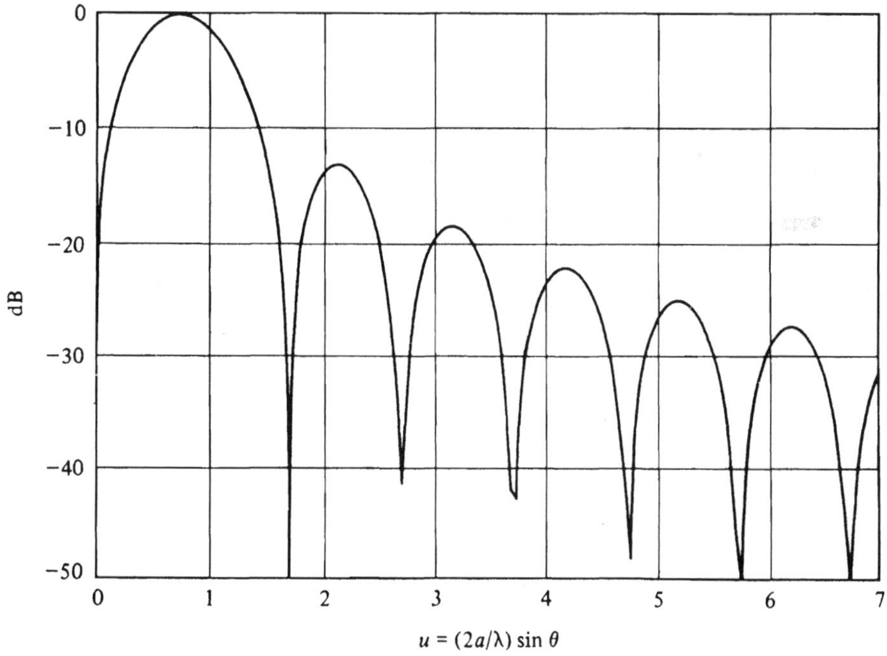

Fig. 6.32 The Difference Pattern for a Circular Planar Aperture Excited by the Generic Aperture Distribution of Fig. 6.31

Bayliss' pattern function is given by the expression

$$\mathfrak{D}(u, \phi) = \cos\phi \; \pi u J_1'(\pi u) \frac{\prod_{n=1}^{\bar{n}-1} [1 - (u/u_n)^2]}{\prod_{n=0}^{\bar{n}-1} [1 - (u/\mu_n)^2]} \tag{6.142}$$

Except for an inconsequential multiplicative constant, the θ-factor of (6.142) is seen to be a modification of (6.141), with the innermost $\bar{n}$ null pairs of $J_1'(\pi u)$ removed and $\bar{n} - 1$ new null pairs placed at the positions $\pm u_n$. Unlike Taylor, Bayliss was not able to find a simple formula from which the roots u_n could be computed. He used fitted polynomials to determine that

$$u_n = \begin{cases} \mu_{\bar{n}} \left(\dfrac{\xi_n^2}{A^2 + \bar{n}^2} \right)^{1/2} & \text{for } n = 1, 2, 3, 4 \\ \mu_{\bar{n}} \left(\dfrac{A^2 + n^2}{A^2 + \bar{n}^2} \right)^{1/2} & \text{for } n \geq 5 \end{cases} \tag{6.143}$$

The parameters $A, \xi_1, \ldots, \xi_4$ have already been given in Table 5.5 as functions of side lobe level.

As an illustration of a circular Bayliss pattern, Figure 6.33 shows a plot of Equation 6.142 for the case of a -30 dB SLL with $\bar{n} = 4$. A characteristic Taylor-like droop is seen in the envelope of the close-in side lobes, and then the envelope reverts to a $u^{-3/2}$ decay.

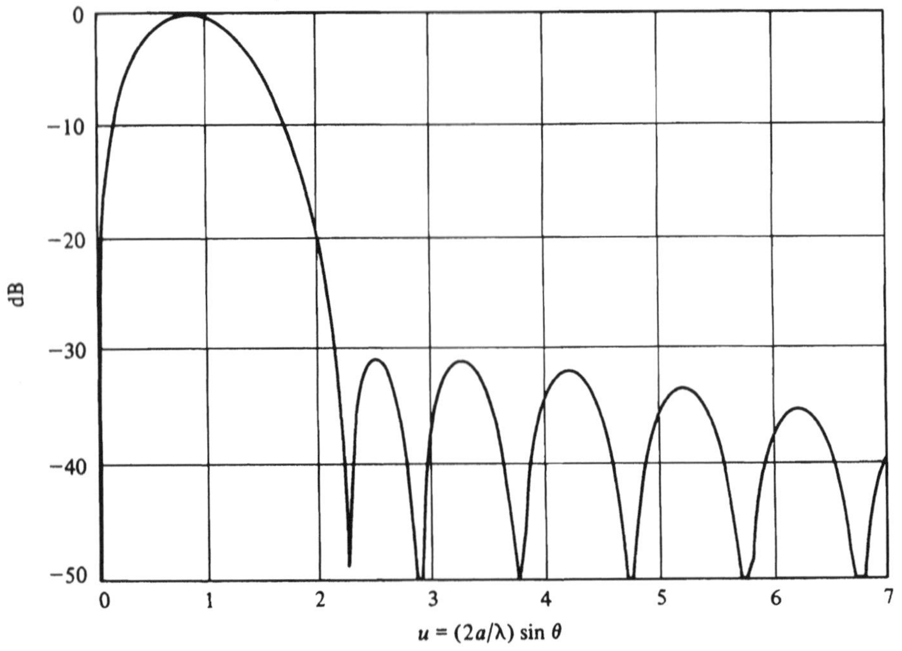

Fig. 6.33 Typical ϕ-Cut of a Bayliss Circular Difference Pattern; $\bar{n} = 4$, -30 dB SLL

The aperture distribution that causes a circular Bayliss difference pattern can be deduced by returning to Equation 6.138. Since $J_1'(\pi\mu_m) = 0$, the only possible nonzero contribution to this sum occurs when $u = \mu_n$, and then only for the term $m = n$. From (6.137),

$$f(\mu_n) = \left(\frac{a}{\pi}\right)^2 A_n \int_0^\pi J_1^2(\mu_n p)p\,dp$$
$$= \left(\frac{a}{\pi}\right)^2 A_n \frac{\pi^2}{2} J_1^2(\pi\mu_n) \tag{6.144}$$

Since $f(\mu_n) = 0$ for $n \geq \bar{n}$, the series in (6.136) truncates, and

$$K_1(p) = \frac{2}{a^2} \sum_{m=1}^{\bar{n}-1} \frac{f(\mu_m)}{J_1^2(\pi\mu_m)} J_1(\mu_m p) \tag{6.145}$$

in which $f(\mu_m)$ can be taken as the difference pattern evaluated at $(\mu_m, 0)$. It is calculable, with the aid of L'Hospital's rule, from (6.142).

For the Bayliss circular difference pattern shown in Figure 6.33, the aperture distribution computed from (6.145) is shown in Figure 6.34, and is considerably changed from the generic case of Figure 6.31.

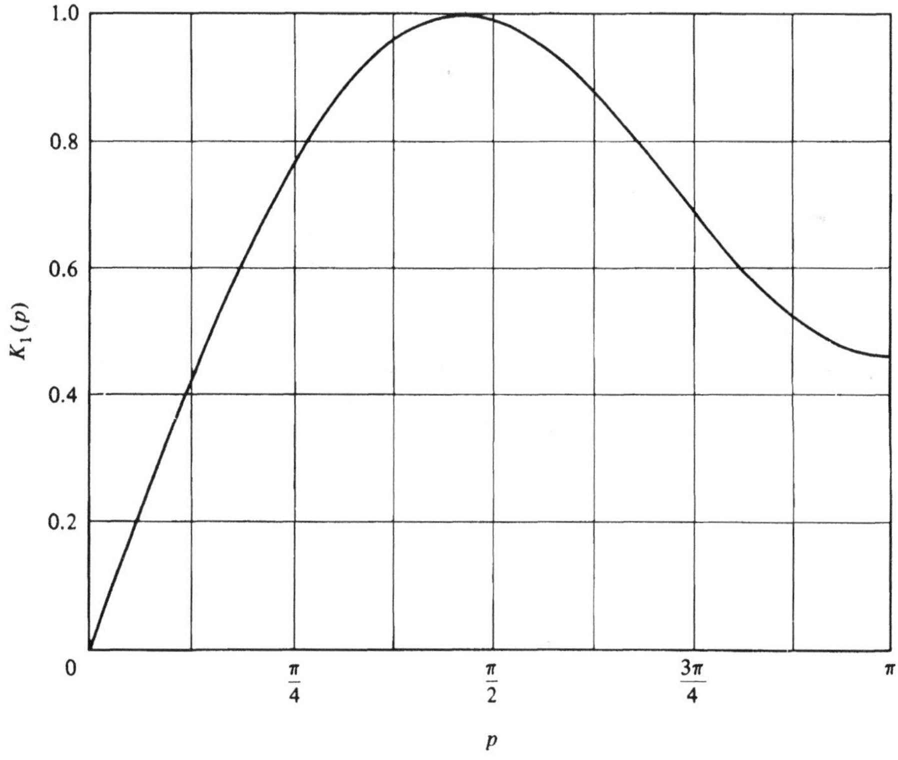

Fig. 6.34 The Aperture Distribution for the Bayliss Circular Difference Pattern of Fig. 6.33

6.12 Modified Circular Bayliss Patterns

The side lobe structure of the difference pattern produced by a circular aperture need not be quasi-uniform in height. The perturbation procedure that has already been used for sum patterns from line sources, for ϕ-symmetric sum patterns from circular apertures, and for difference patterns from line sources, can be applied as well to the problem of modifying a circular Bayless pattern to yield an arbitrary side lobe topography.

Let the starting pattern be given in the form (6.142), namely,

$$\mathfrak{D}_0(u, \phi) = C_0 f(u, \phi) \prod_{n=1}^{\bar{n}-1} \left(1 - \frac{u^2}{\overset{0}{u}{}_n^2}\right) \tag{6.146}$$

in which the roots $\overset{0}{u}_n$ are known and

$$f(u, \phi) = \frac{(\cos \phi)\pi u J_1'(\pi u)}{\prod_{n=1}^{\bar{n}-1} (1 - u^2/\mu_n^2)} \tag{6.147}$$

The desired pattern can be expressed similarly:

$$\mathfrak{D}(u, \phi) = C f(u, \phi) \prod_{n=1}^{\bar{n}-1} \left(1 - \frac{u^2}{u_n^2}\right) \tag{6.148}$$

The assumptions that $u_n = \overset{0}{u}_n + \delta u_n$ and $C = C_0 + \delta C$, together with the expansion in (6.65), lead to

$$\frac{\mathfrak{D}(u_m^p, \phi)}{\mathfrak{D}_0(u_m^p, \phi)} - 1 = \frac{\delta C}{C_0} + \sum_{n=1}^{\bar{n}-1} \frac{2(u_m^p)^2/\overset{0}{u}{}_n^3}{1 - u^2/\overset{0}{u}{}_n^2} \delta u_n \tag{6.149}$$

in which u_m^p is the peak position of the mth lobe in the starting pattern. Since there are $\bar{n}$ lobes to be adjusted and $\bar{n}$ unknowns, (6.149) is a deterministic set of simultaneous linear equations for which $\mathfrak{D}(u_m^p, \phi)/\mathfrak{D}_0(u_m^p, \phi)$ plays the role of driving function, being essentially the ratio of desired to starting height for the mth lobe. Matrix inversion gives the perturbations δu_n, which permit computation of a new pattern. Comparison with the ideal determines whether or not further iterations are needed.

An application of this technique to circular difference patterns is posed in Problem 6.21 at the end of this chapter.

6.13 The Discretizing Technique Applied to Planar Arrays Excited to Give a Difference Pattern

With some minor modifications, the technique described in Section 6.10, which was used there in application to sum patterns, can also be used for difference patterns. Imagine that a starting current distribution $\overset{0}{I}_{mn}$ has been determined, either by conventional sampling of a continuous Bayliss distribution (such as the one shown in Figure 6.34) or by a procedure to be described later in this section. Then the starting pattern can be expressed as

$$\mathfrak{D}_0(\theta, \phi) = \sum_m \sum_n \overset{0}{I}_{mn} g_{mn} \tag{6.150}$$

in which

$$g_{mn}(\theta, \phi) = \sin (k\xi_m \sin \theta \cos \phi) \cos (k\eta_{mn} \sin \theta \sin \phi)$$

if the difference pattern is XZ-oriented, or

$$g_{mn}(\theta, \phi) = \cos (k\xi_{mn} \sin \theta \cos \phi) \sin (k\eta_{mn} \sin \theta \sin \phi)$$

if the difference pattern is YZ-oriented.

One assumes that the achievable pattern can also be given in the same form as (6.150); that is,

$$\mathfrak{D}(\theta, \phi) = \sum_m \sum_n I_{mn} g_{mn} \tag{6.151}$$

where, once again, $I_{mn} = \overset{0}{I}_{mn} + \delta I_{mn}$. Then

$$\mathfrak{D}(\theta_{pq}, \phi_q) - \mathfrak{D}_0(\theta_{pq}, \phi_q) = \sum_m \sum_n \delta I_{mn} g_{mnpq} \tag{6.152}$$

and the problem is to choose a judicious set of pointing directions equal to the number of elements in a quadrant in order to solve for δI_{mn} by matrix inversion.

It has already been observed, in connection with the illustrative example of Section 6.10, that—for small arrays—conventional sampling of continuous aperture distributions is inferior to working with collapsed distributions as a method for determining a starting set of currents $\overset{0}{I}_{mn}$. Therefore, continuing with that example, let it be assumed that the starting current distribution is in the form shown in Figure 6.35. (This distribution lacks the $\overset{0}{I}_{mn} = \overset{0}{I}_{nm}$ symmetry of Figure 6.29 because, unlike the desired sum pattern, the desired difference pattern does not have symmetry about the 45°-axis.

From the collapsed distribution in the $\phi = 0°$ plane, one obtains

$$t + w + z = a_1(1 + v + y) \tag{6.153}$$

$$u + x = a_2(1 + v + y) \tag{6.154}$$

whereas the collapsed distribution in the $\phi = 45°$-plane yields

$$t + u + v - y = b_1(1 + t + x - v - z) \tag{6.155}$$

$$u + w + y = b_2(1 + t + x - v - z) \tag{6.156}$$

$$x + z = b_3(1 + t + x - v - z) \tag{6.157}$$

in which the a_i and b_j are relative currents in six- and nine-element linear arrays excited to give a difference pattern.

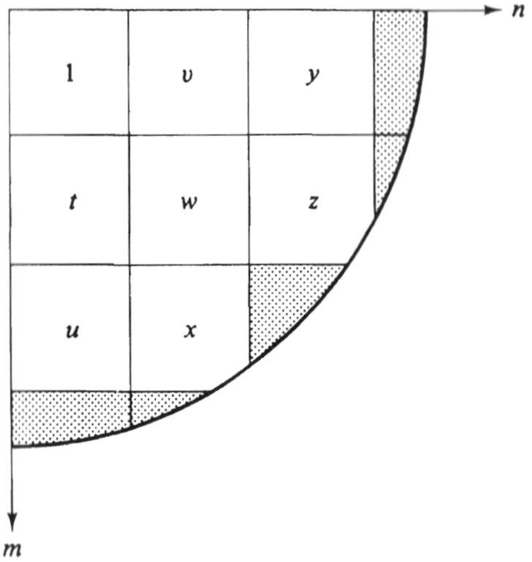

Fig. 6.35 Notation for the Starting Current Distribution of a 32-Element Planar Array; Difference Mode

Equations 6.153 through 6.157 give five conditions on the seven unknown currents $t, u, \ldots, z$. Two more conditions can be found as follows: Let (ρ_{mn}, β_{mn}) be the position of the *mn*th element, expressed in polar coordinates. Then for the rectangular grid array of Figure 6.35,

$$\frac{\cos\beta_{nm}}{\cos\beta_{mn}} = \frac{\cos(\pi/2 - \beta_{mn})}{\cos\beta_{mn}} = \tan\beta_{mn}$$

Thus, if one is trying to approximate a Bayliss distribution (which varies as $\cos\beta$) for the example of Figure 6.35,

$$t = 3v \tag{6.158}$$

$$u = 5y \tag{6.159}$$

Equations 6.158 and 6.159 can be used as the two additional conditions on the current distribution.

If root positions are adjusted graphically on a Schelkunoff unit circle for six- and nine-element linear arrays, in order to give difference patterns with uniform 20 dB side lobe levels, one finds that

$$a_1 = 2.04 \qquad a_2 = 1.35 \qquad b_1 = 1.51 \qquad b_2 = 1.44 \qquad b_3 = 0.91$$

When these values of a_i and b_j are placed in (6.153) through (6.159), simultaneous solution gives the current distribution shown in the table inset of Figure 6.36. The corresponding patterns, computed from (6.150), are also shown in Figure 6.36 and comprise an acceptable starting point for the iterative procedure.

For this illustrative example, efforts to improve on the patterns of Figure 6.36 through use of Equation (6.152) proved fruitless when the desired pattern was prescribed to behave in ϕ as $\cos\phi$. However, when the design goal was changed to permit the main beam peak to subside as $\cos\phi$, but to allow the side lobes to stay at -20 dB relative to the *highest* main beam peak (that is, the one seen in the $\phi = 0°$ cut), the iterative procedure yielded the improvement shown in Figure 6.37. The need to alter

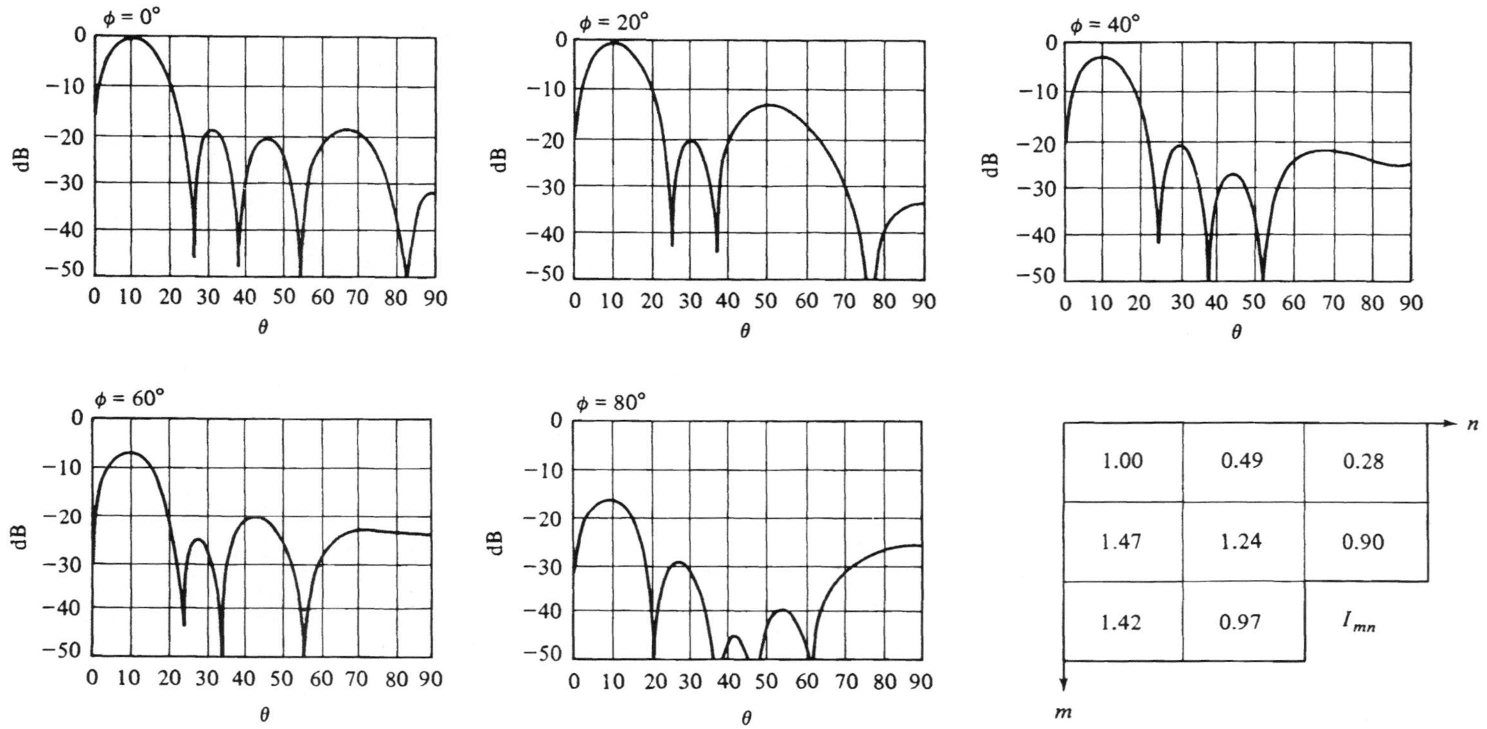

Fig. 6.36 Five ϕ-Cuts of the Difference Pattern of a 32-Element Planar Array; Rectangular Grid, Circular Boundary; $d_x = d_y = 0.7\lambda$; Design Goal: −20 dB SLL; Currents Obtained through Use of Equations 6.153 through 6.159

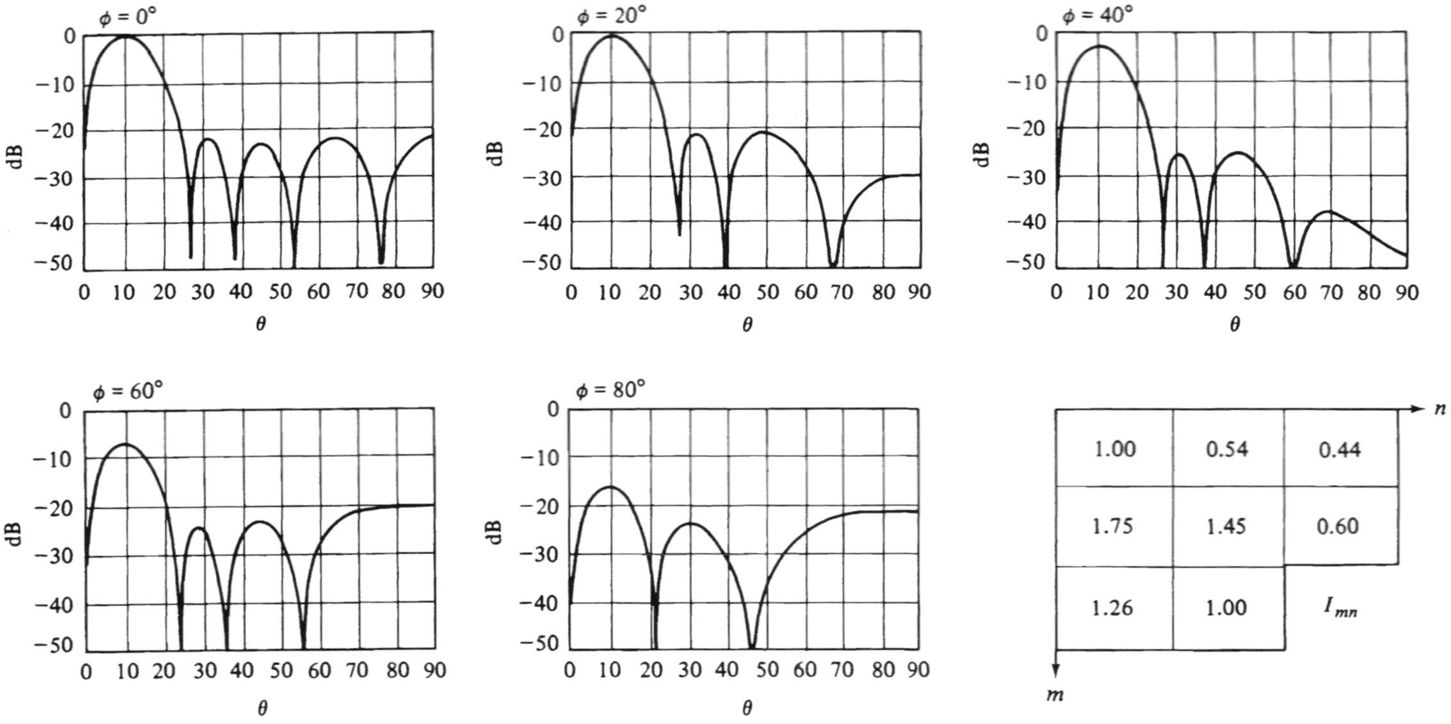

Fig. 6.37 Improvement on Patterns of Fig. 6.36; Currents Modified by Perturbation Procedure

the design goal can probably be attributed to the small size of the array, since the problem has not been encountered with larger arrays. However, this altered goal is actually more in keeping with typical design specifications, in which all regions of space are given equal weight in so far as side lobe suppression is concerned.

6.14 Comparative Performance of Separable and Nonseparable Excitations for Planar Apertures

Many of the results that were obtained in earlier sections of this chapter can be brought together to provide a comparison of the performance of planar apertures, excited continuously or discretely to produce a sum pattern with a specified side lobe level. The principal criteria in this comparative study will be peak directivity and beamwidth, and it is advantageous to normalize these measures as follows.

Let a planar aperture bounded by an arbitrary contour C be excited by a continuous distribution that is uniform in amplitude and uniform progressive in phase. This causes a sum pattern whose main beam points in a direction (θ_0, ϕ_0). If the element pattern is assumed to be hemispherically isotropic, it is a simple matter[16] to show that the peak directivity is

$$D_0 = \frac{4\pi A}{\lambda^2} \tag{6.160}$$

in which A is the projected area of the aperture in a plane transverse to the direction (θ_0, ϕ_0) and λ is the free-space wavelength. The areal beamwidth of this sum pattern will be designated by the symbol B_0.

Added significance can be attached to the result (6.160) because no other amplitude distribution, combined with the given uniform progressive phase distribution, can produce this high a peak directivity from the given aperture. Thus D_0 can be used as a figure of merit and the peak directivity D of a sum pattern caused by another amplitude distribution can be compared to D_0. Concurrently, the areal beamwidth B of the sum pattern caused by another amplitude distribution can be contrasted to B_0.

With an assumed hemispherically isotropic element pattern, (1.160) and (6.160) combine to give for the normalized peak directivity

$$\frac{D}{D_0} = \frac{\mathcal{S}_a(\theta_0, \phi_0)\mathcal{S}_a^*(\theta_0, \phi_0)}{(A/\lambda^2)\int_0^{\pi/2}\int_0^{2\pi} \mathcal{S}_a(\theta, \phi)\mathcal{S}_a^*(\theta, \phi)\sin\theta\, d\theta\, d\phi} \tag{6.161}$$

in which $\mathcal{S}_a(\theta, \phi)$ is the array pattern, and could be produced by either a continuous or discrete planar aperture distribution.

As an illustration of the use of (6.161), first consider an array of 18 by 18 elements, with $d_x = d_y = 0.492\lambda$, and with the elements uniformly excited in amplitude and equiphase. The aperture is 8.86λ square and $\mathcal{S}_a(\theta, \phi)$ can be determined from

[16]See, for example, S. Silver, *Microwave Antenna Theory and Design*, MIT Rad. Lab. Series, Vol. 12 (New York: McGraw-Hill Book Co., Inc., 1939), pp. 177–78.

(6.37). Insertion in (6.161) gives $D/D_0 = 0.965$. Interestingly, there is a 3.5% loss in peak directivity just due to discretization.

Next, imagine that these 324 elements are rearranged into a rectangular grid with a *circular* boundary. If they are placed so that $d_x = d_y = 0.500\lambda$, one finds that the circular aperture has the same area as the previous square aperture, namely $(8.86\lambda)^2$. With uniform excitation, the normalized directivity is now $D/D_0 = 0.924$. There has been an additional 4.1% loss in peak directivity due to the inefficient use of aperture space along the periphery.

Now, picture a 20-by-20 square array, with $d_x = d_y = 0.500\lambda$, excited in a *separable* Dolph-Chebyshev distribution to give 15 dB side lobes in both principal planes. For this case, it is found that $D/D_0 = 0.621$. If some of these elements are cut off[17] so that the remaining 324 fit in a circular boundary of 5λ radius, then $D/D_0 = 0.748$. This figure is 17.6% lower than the previous value of 0.924, found when the same array was uniformly excited, even though the side lobe levels are not substantially different (−15 dB in the Dolph-Chebyshev case, −13.5 dB in the uniform case).

To see the improvement that can be obtained by going to a nonseparable distribution, consider a circular aperture of radius 5λ, excited by a continuous circular Taylor distribution such that the pattern is −15 dB SLL, $\bar{n} = 3$. In this situation, $D/D_0 = 0.967$. If this continuous distribution is sampled in order to find the discrete excitation for a rectangular grid of 324 elements, the resulting pattern has a normalized peak directivity of 0.940. The pattern has ring side lobes (slightly undulating because of the discretization), at a quasi-uniform height of −15 dB. The value $D/D_0 = 0.940$ is clearly superior to the value 0.748, found when the same 324 elements were separably excited Dolph-Chebyshev, −15 dB SLL. In that case, mound side lobes at −30 dB existed outside the principal planes, broadening the main beam, and lowering the directivity. This penalty for using a separable distribution is less severe as the array is made smaller (because the ratio of the number of off-axis mound side lobes to on-axis mound side lobes goes down), but there is always a penalty. It can, and should, be avoided whenever directivity is an important consideration.

Because of its clear superiority when judged by the directivity criterion, the Taylor circular distribution deserves further attention. The sum pattern produced by this distribution is given by Equation 6.53 and insertion in (6.161) will give the normalized peak directivity for a planar aperture with a circular boundary of radus a. The normalized half-power beamwidth can be found by the following procedure.

If the aperture is uniformly excited, the pattern is given by (6.52), that is,

$$\mathcal{S}_0(u) = \frac{J_1(\pi u)}{\pi u} \tag{6.162}$$

and L'Hospital's rule can be invoked to determine that $\mathcal{S}_0(0) = 0.500$. The 3-decibel beamwidth can therefore be determined from

$$\mathcal{S}_0(u_1) = 0.707\,\mathcal{S}_0(0) = 0.3535$$

[17]This will result in some pattern degradation.

When this result is used in (6.158), it is found that

$$\pi u_1 = 1.617 = ka \sin \theta_1$$

from which θ_1 can be deduced. The reference half-power beamwidth is $2\theta_1$ and the reference areal beamwidth is

$$B_0 = 4\theta_1^2$$

The half-power beamwidth for the Taylor circular pattern can be computed from (6.53) by seeking the value u_1 at which $\mathcal{S}(u_1) = 0.707\mathcal{S}(0)$. From this, the normalized areal beamwidth B/B_0 can be deduced.

The calculations just described lead to Figures 6.38 and 6.39, which show the normalized peak directivity and normalized beamwidth of a Taylor circular pattern

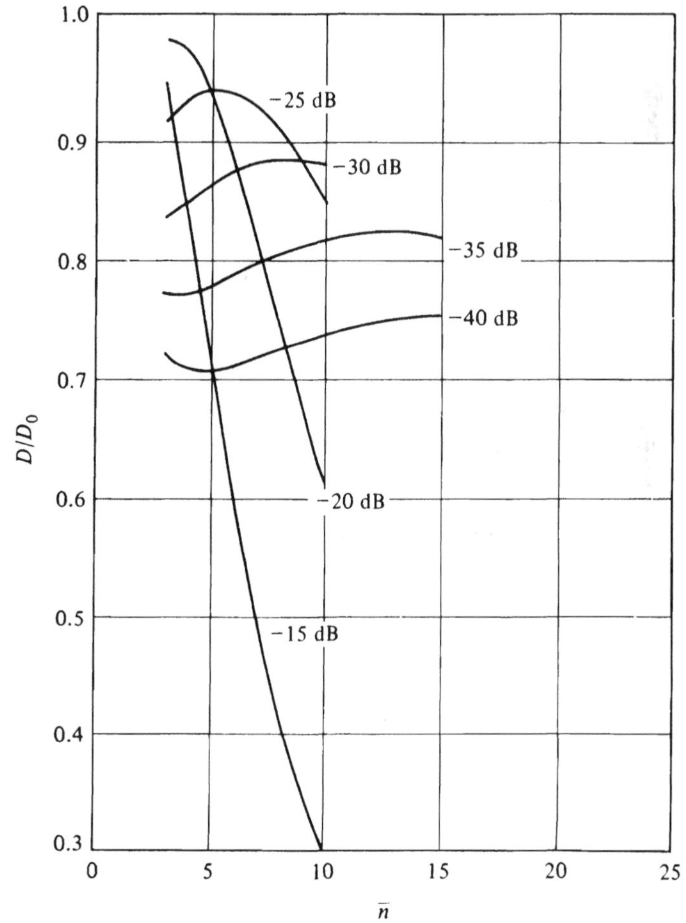

Fig. 6.38 Normalized Peak Directivity of Circular Taylor Sum Patterns

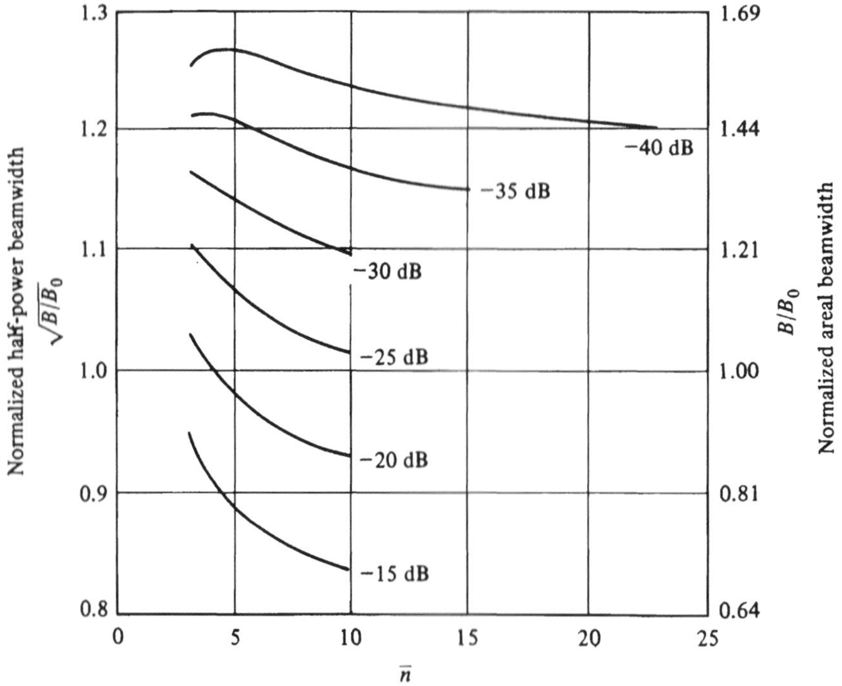

Fig. 6.39 Normalized Areal Beamwidth of Circular Taylor Sum Patterns

as functions of side lobe level and the transition integer $\bar{n}$. Several significant conclusions can be drawn from a study of these figures. Foremost, at a given side lobe level, the normalized beamwidth narrows as $\bar{n}$ is increased, but not dramatically; however, the normalized directivity peaks and then plunges precipitously. For this reason, it is generally wise to pick that value of $\bar{n}$ that maximizes the normalized directivity for the specified side lobe level. When this is done, the values of D/D_0 and B/B_0 can be tabulated versus side lobe level, as has been done in Table 6.8. This table points up the other significant conclusion, that the price paid for a lowered side lobe level is a substantial reduction in directivity and a concomitant increase in beamwidth.

The last column of Table 6.8 indicates that the product of peak directivity and areal beamwidth is not constant as the side lobe level is varied. However, it is constant for a given side lobe level as the aperture size is permitted to vary.

For the antenna designer who intends to sample a continuous circular Taylor distribution to obtain the excitation of a discrete planar array, it is extremely important to be aware of the penalty for not choosing the optimum value of $\bar{n}$. As an illustration, for an array of 324 elements in a rectangular grid with a circular boundary, it was found that when a Taylor 15 dB SLL, $\bar{n} = 3$ was sampled, a normalized peak directivity of 0.940 was achieved. Had one chosen to sample the Taylor 15 dB SLL, $\bar{n} = 6$, the result would have been $D/D_0 = 0.500$. This would represent a loss of nearly 3 dB merely because of an improper choice of $\bar{n}$.

TABLE 6.8 Normalized peak directivity and normalized areal beamwidth versus side lobe level at optimum $\bar{n}$ for circular Taylor patterns

Side Lobe Level, dB	Optimum $\bar{n}$	$\frac{D}{D_0}$	$\frac{B}{B_0}$	$\left(\frac{D}{D_0}\right)\left(\frac{B}{B_0}\right)$
−20	3	0.982	1.065	1.046
−25	5	0.941	1.134	1.067
−30	9	0.884	1.212	1.071
−35	14	0.821	1.325	1.088
−40	22	0.758	1.445	1.095

6.15 Fourier Integral Representation of the Far Field

Various earlier sections of this chapter have been concerned with the development of specific techniques for synthesizing planar aperture distributions to produce desired far field patterns. For continuous apertures with *circular* boundaries, these developments included the techniques of Taylor and Bayliss and their extensions. The *rectangular* boundary case was treated principally for discrete arrays. The planar aperture with an *arbitrary* boundary has not heretofore been considered. A powerful technique which is applicable to the general boundary case involves the use of Fourier transforms. This technique permits many general deductions to be made about the properties of planar aperture antennas without specifying their shapes and is widely used in the theoretical research literature.

To appreciate this technique, consider a thin perfectly conducting infinite plane which contains a collection of arbitrarily shaped holes $S_1, S_2, \ldots, S_N$, as suggested by Figure 6.40. Regardless of the method of excitation, the currents in this planar antenna are constrained to lie in the XY-plane. The fields due to these currents must therefore have the following properties in the surface $z = 0$:

$$\mathbf{H}_x \equiv \mathbf{H}_y \equiv \mathbf{E}_z \equiv 0 \qquad \text{in } S_1, S_2, \ldots, S_N$$

$$\mathbf{E}_x \equiv \mathbf{E}_y \equiv \mathbf{H}_z \equiv 0 \qquad \text{except in } S_1, S_2, \ldots, S_N$$

If the collective area of the holes $S_1, S_2, \ldots, S_N$ is small compared to the area of the metallic part of the conducting plane, the fields in $z > 0$ can be conveniently formulated by the procedure introduced in Section 3.2 as an application of the Schelkunoff equivalence principle. Secondary sources can be placed above the holes (in the plane $z = 0+$) after which the holes can be covered over with perfect conductor. Use of the image principle then results in removal of the infinite ground plane and a new source system consisting of a doubled magnetic current sheet in $S_1, S_2, \ldots, S_N$ and no electric currents whatsoever. Since the magnetic current sheet is related to the value of the true tangential electric field in $S_1, S_2, \ldots, S_N$ by Equation 1.113, it follows that the far field in $z > 0$ is uniquely determined by $\mathbf{E}_{\text{tang}}$ in $z = 0$.

It can be shown (see Appendix F) that the far field in $z > 0$ is also uniquely determined by $\mathbf{H}_{\text{tang}}$ in $z = 0$. This alternate formulation is useful when the collective

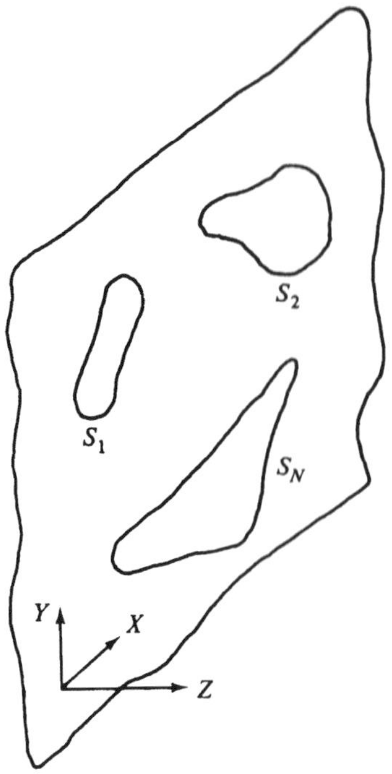

Fig. 6.40 Infinite Perfectly Conducting Screen with Arbitrarily Shaped Holes

area of the holes $S_1, S_2, \ldots, S_N$ is large compared to the area of the metallic part of the conducting plane. The Fourier integral formulation to be introduced in this section will be in terms of $\mathbf{E}_{tang}$ in $z = 0$, but it can be developed with equal facility in terms of $\mathbf{H}_{tang}$ in $z = 0$.

Let the electric field in $z > 0$ be represented by $\mathbf{E}(x, y, z)e^{j\omega t}$. With the time factor suppressed, a two-dimensional Fourier transform of this field function can be defined by the integral

$$\mathbf{E}(k_x, k_y, z) = \frac{1}{2\pi}\int_{-\infty}^{\infty}\int_{-\infty}^{\infty} \mathbf{E}(x, y, z)e^{jk_x x + jk_y y}\, dx\, dy \tag{6.163}$$

with the inverse transform given by

$$\mathbf{E}(x, y, z) = \frac{1}{2\pi}\int_{-\infty}^{\infty}\int_{-\infty}^{\infty} \mathbf{E}(k_x, k_y, z)e^{-jk_x x - jk_y y}\, dk_x\, dk_y \tag{6.164}$$

$\mathbf{E}(x, y, z)$ satisfies the homogeneous wave equation $(\nabla^2 + k^2)\mathbf{E}(x, y, z) \equiv 0$ in the source-free region $z > 0$. When this operation is performed on (6.164), the result is that

$$\left[\frac{\partial^2}{\partial z^2} + k^2 - (k_x^2 + k_y^2)\right]\mathbf{E}(k_x, k_y, z) \equiv 0 \tag{6.165}$$

A general solution of (6.165) can be represented by

$$\mathbf{E}(k_x, k_y, z) = \mathbf{g}(k_x, k_y)e^{-jk_z z} \tag{6.166}$$

in which

$$k_z = [k^2 - (k_x^2 + k_y^2)]^{1/2} \tag{6.167}$$

The positive real root must be chosen in (6.167) when $k^2 > k_x^2 + k_y^2$ and the negative imaginary root must be chosen when $k^2 < k_x^2 + k_y^2$, in order to satisfy the radiation condition at infinity in $z > 0$.

Returning to (6.164), one can see that the electric field in $z > 0$ can be expressed in the form

$$\mathbf{E}(x, y, z) = \frac{1}{2\pi}\int_{-\infty}^{\infty}\int_{-\infty}^{\infty} \mathbf{g}(k_x, k_y)e^{-j\mathbf{k}\cdot\mathbf{r}}\, dk_x\, dk_y \tag{6.168}$$

in which $\mathbf{r}$ is the position vector drawn from the origin to the point (x, y, z) and $\mathbf{k} = \mathbf{1}_x k_x + \mathbf{1}_y k_y + \mathbf{1}_z k_z$. Equation 6.168 permits the interpretation that $\mathbf{E}(x, y, z)$ in $z > 0$ can be viewed as the superposition of plane waves with amplitudes $\mathbf{g}(k_x, k_y)dk_x\, dk_y$ and directions of propagation given by $\mathbf{k}$.

In analogy to the time/frequency use of Fourier transforms in electric circuit theory, the components k_x and k_y of the propagation vector $\mathbf{k}$ are often called spatial frequencies because of their conjugate relationship to the spatial variables x and y.

Because $\nabla\cdot\mathbf{E}(x, y, z) = 0$ in $z > 0$, if the divergence of (6.168) is taken, one finds that

$$\nabla \cdot (\mathbf{g}e^{-j\mathbf{k}\cdot\mathbf{r}}) = \mathbf{g} \cdot \nabla(e^{-j\mathbf{k}\cdot\mathbf{r}}) = -j\mathbf{g}\cdot\mathbf{k}e^{-j\mathbf{k}\cdot\mathbf{r}} \equiv 0$$

which requires that $\mathbf{g} \cdot \mathbf{k} \equiv 0$. Thus only two of the components of $\mathbf{g}$ are independent. In particular,

$$g_z(k_x, k_y) = -\frac{k_x g_x(k_x, k_y) + k_y g_y(k_x, k_y)}{k_z} \tag{6.169}$$

This conclusion is consistent with the earlier argument that $\mathbf{E}(x, y, z)$ is uniquely determined in $z > 0$ if only $\mathbf{E}_{\text{tang}}$ is specified in the aperture. An important consequence of this result is that the Fourier transform in (6.168) can be simplified by suppressing the z-component. When additionally the point (x, y, z) is restricted to lie in the aperture, that is, when $(x, y, z) \to (\xi, \eta, 0+)$, one can further reduce (6.168) to

$$\mathbf{E}_t(\xi, \eta, 0+) = \mathbf{E}_t(\xi, \eta) = \frac{1}{2\pi}\int_{-\infty}^{\infty}\int_{-\infty}^{\infty} \mathbf{g}_t(k_x, k_y)e^{-jk_x\xi - jk_y\eta}\, dk_x\, dk_y \tag{6.170}$$

with the t subscripts indicating that only the transverse (x- and y-) components are being used.

The inverse transform of (6.170) is

$$\mathbf{g}_t(k_x, k_y) = \frac{1}{2\pi}\int_{-\infty}^{\infty}\int_{-\infty}^{\infty} \mathbf{E}_t(\xi, \eta)e^{jk_x\xi + jk_y\eta}\, d\xi\, d\eta \qquad (6.171)$$

The utility of the transform pair (9.170) and (6.171) in problems involving pattern synthesis can be appreciated if the substitutions

$$k_x = k \sin\theta \cos\phi \qquad k_y = k \sin\theta \sin\phi \qquad (6.172)$$

are introduced[18] and specified to apply for $k_x^2 + k_y^2 \leq k^2$. When this is done, (6.171) becomes

$$\mathbf{g}_t(k_x, k_y) = \frac{1}{2\pi}\iint_{S_1, S_2, \ldots, S_N} \mathbf{E}_t(\xi, \eta)e^{jk(\xi\sin\theta\cos\phi + \eta\sin\theta\sin\phi)}\, d\xi\, d\eta \qquad (\text{for } k_x^2 + k_y^2 \leq k^2) \qquad (6.173)$$

Equation 6.173 is in the exact same form as the electric vector potential function $\mathfrak{F}(\theta, \phi)$ expressed as Equation 1.122. Therefore $\mathbf{g}_t(k_x, k_y)$, with k_x and k_y given by (6.172) and with $k_x^2 + k_y^2 < k^2$, can be identified as the far-field $\mathfrak{F}$-function of the planar antenna.

If $\mathbf{g}_t(k_x, k_y)$ is *fully* specified, the inverse transform of (6.170) can be used to determine the required aperture distribution. In principle, this solves the synthesis problem. However, close inspection uncovers some serious difficulties. When $\mathfrak{F}(\theta, \phi)$ is specified in the far field in both amplitude and phase, $\mathbf{g}_t(k_x, k_y)$ is known exactly for $k_x^2 + k_y^2 < k^2$. However, one must choose $\mathbf{g}_t(k_x, k_y)$ for $k_x^2 + k_y^2 > k^2$ so that, when (6.170) is used, $\mathbf{E}_t(\xi, \eta) = 0$ except over $S_1, S_2, \ldots, S_N$. To complicate the situation further, $\mathfrak{F}(\theta, \phi)$ is often specified in magnitude but not in phase. When this situation prevails, there is no unique solution to the synthesis problem. Ideally, one should choose a phase distribution for $\mathfrak{F}(\theta, \phi)$ which results in a simple, physically realizable aperture distribution over $S_1, S_2, \ldots, S_N$ and which minimizes the aggregate effect of $\mathbf{g}_t(k_x, k_y)$ in the range $k_x^2 + k_y^2 > k^2$. This latter condition arises because the plane waves in this range are evanescent (k_z is imaginary) and do not contribute to radiation but rather to reactive stored energy. (This point will be elaborated shortly). The general synthesis problem is thus seen to be formidable. It can be raised to its ultimate level of difficulty if one also seeks the optimum planar antenna shape to produce a specified pattern with the simplest physically realizable aperture distribution.

Despite these difficulties, Equations 6.170 through 6.173 constitute a useful formulation of the synthesis problem for planar apertures. Indeed, one of the virtues of this formulation is that it permits a penetrating perception of the difficulties of

[18] This linking of (k_x, k_y) with the real space angle variables (θ, ϕ) is consistent with results obtained by applying the method of stationary phase to the integral transform (6.168). See, e.g., R. E. Collin and F. J. Zucker, *Antenna Theory: Part I*, (New York: McGraw-Hill Book Co., Inc., 1969), pp. 62–9.

synthesis. But beyond this, it permits other important deductions to be made which have wide validity. One of these involves the calculation of radiated power and average reactive stored energy.

Since $\nabla \times \mathbf{E} = -j\omega\mu_0\mathbf{H}$ in $z > 0$, it follows from (6.168) that

$$\mathbf{H}(x, y, z) = -\frac{1}{2\pi j\omega\mu_0}\int_{-\infty}^{\infty}\int_{-\infty}^{\infty} \nabla \times (\mathbf{g}e^{-j\mathbf{k}\cdot\mathbf{r}})\, dk_x\, dk_y$$

$$= \frac{1}{2\pi j\omega\mu_0}\int_{-\infty}^{\infty}\int_{-\infty}^{\infty} \mathbf{g} \times \nabla(e^{-j\mathbf{k}\cdot\mathbf{r}})\, dk_x\, dk_y$$

$$= \frac{1}{2\pi j\omega\mu_0}\int_{-\infty}^{\infty}\int_{-\infty}^{\infty} \mathbf{k} \times \mathbf{g}e^{-j\mathbf{k}\cdot\mathbf{r}}\, dk_x\, dk_y \tag{6.174}$$

The complex power flow (see Section 1.6) across the aperture plane into the half space $z > 0$ is given by

$$P = \bar{P} + j2\omega(W_m - W_e) = \tfrac{1}{2}\int_{-\infty}^{\infty}\int_{-\infty}^{\infty} \mathbf{E} \times \mathbf{H}^* \cdot \mathbf{1}_z\, d\xi\, d\eta \tag{6.175}$$

in which $\bar{P}$ is the real power flow and W_m and W_e are the time-average magnetic and electric stored energies. Use of (6.168) and (6.174) in (6.175) yields

$$P = \frac{1}{8\pi^2\omega\mu_0}\int_{-\infty}^{\infty} \cdots \int_{-\infty}^{\infty} [\mathbf{g}(k_x', k_y')e^{-j\mathbf{k}'\cdot\mathbf{r}} \times \mathbf{k}^* \times \mathbf{g}^*(k_x, k_y)e^{j\mathbf{k}^*\cdot\mathbf{r}}]$$
$$\cdot\, \mathbf{1}_z\, d\xi\, d\eta\, dk_x\, dk_y\, dk_k'\, dk_y' \tag{6.176}$$

This sixfold integral can be reduced through use of the orthogonality relation

$$\int_{-\infty}^{\infty} e^{-jk_x'\xi + jk_x\xi}\, d\xi = 2\pi\delta(k_x - k_x') \tag{6.177}$$

with δ the Dirac delta function. Since the same reduction applies for the η integration, one finds that

$$P = \frac{1}{2\omega\mu_0}\int_{-\infty}^{\infty}\int_{-\infty}^{\infty} [\mathbf{g}(k_x, k_y) \times \mathbf{k}^* \times \mathbf{g}^*(k_x, k_y)] \cdot \mathbf{1}_z\, dk_x\, dk_y \tag{6.178}$$

With the aid of (6.169), this result can be put in the revealing form

$$P = \frac{1}{2\omega\mu_0}\int_{-\infty}^{\infty}\int_{-\infty}^{\infty} [(k^2 - k_t^2)\mathbf{g}_t \cdot \mathbf{g}_t^* + |\mathbf{k}_t \cdot \mathbf{g}_t|^2]\frac{dk_x\, dk_y}{k_z^*} \tag{6.179}$$

A study of the integrand of (6.179) reveals that the expression inside the square brackets is always pure real. However, k_z^* is pure real or pure imaginary according to whether or not $k_t^2 = k_x^2 + k_y^2$ is less than or greater than k^2. Therefore the real radiated power comes from that portion of the plane wave spectrum for which $k_x^2 + k_y^2 \le k^2$, whereas the average stored reactive energy comes from the remainder of the plane wave spectrum (the evanescent waves) for which $k_x^2 + k_y^2 > k^2$. This result is consistent with the comments made earlier about the synthesis of a physically realizable aperture distribution.

This formulation of complex power flow in terms of the plane wave spectrum is useful in the determination of the input impedance or admittance of various planar antennas. As an example, consider the case of an infinitely long narrow slit in a perfectly conducting ground plane, as suggested by Figure 6.41. The electric field in the slit is assumed to be uniform, X-directed, and given by

$$\mathbf{E}_t(\xi, \eta, 0+) = \frac{\mathbf{1}_x V}{a} \tag{6.180}$$

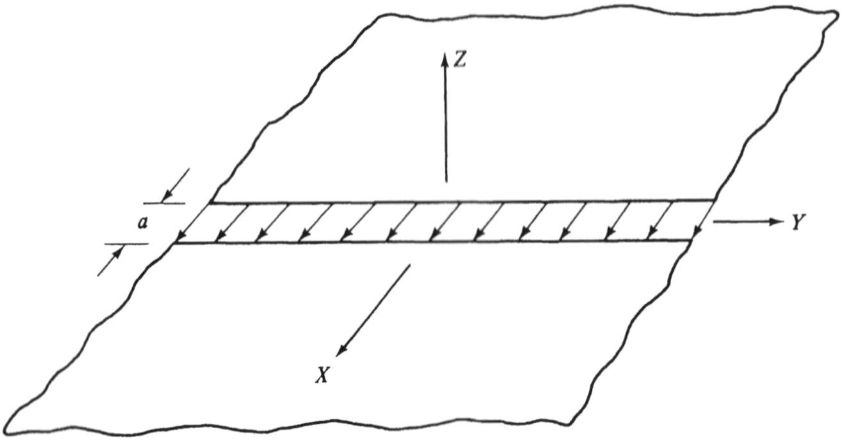

Fig. 6.41 An Infinitely Long Narrow Slit in an Infinite Perfectly Conducting Ground Plane

in which a is the slot width and V is the voltage across the gap. From (6.171), one finds that in this case

$$\begin{aligned}\mathbf{g}_t(k_x, k_y) &= \frac{1}{2\pi}\int_{-a/2}^{a/2} d\xi \int_{-\infty}^{\infty} \mathbf{1}_x \frac{V}{a} e^{jk_x\xi + jk_y\eta}\, d\eta \\ &= \mathbf{1}_x \frac{V}{a}\,\delta(k_y - 0)\int_{-a/2}^{a/2} e^{jk_x\xi}\, d\xi \\ &= \mathbf{1}_x \frac{2V}{a}\,\delta(k_y - 0)\frac{\sin(k_x a/2)}{k_x}\end{aligned} \tag{6.181}$$

It is seen that $\mathbf{g}$ does not have a y-component, which is consistent with the fact that $E_y \equiv 0$ in $z > 0$. However, there is a z component which, with the aid of (6.169), is found to be

$$g_z(k_x, k_y) = -\frac{2V}{a}\,\delta(k_y - 0)\frac{\sin(k_x a/2)}{k_z} \tag{6.182}$$

Because this is a two-dimensional problem, resulting in a delta function dependency of g on k_y, it is convenient to compute the complex power flow into the half-space $z > 0$ by returning to (6.176). If in this case P represents the flow per unit length of the slit; then

$$P = \frac{1}{8\pi^2\omega\mu_0}\int_{-\infty}^{\infty}\cdots\int_{-\infty}^{\infty}[k_z^* g_x(k_x', k_y')g_x^*(k_x, k_y) - k_x g_x(k_x', k_y')g_z^*(k_x, k_y)]$$

$$\cdot\, e^{j(k_x-k_x')\xi}e^{j(k_y-k_y')\eta}\, d\xi\, dk_x\, dk_y\, dk_x'\, dk_y'$$

$$= \frac{1}{4\pi\omega\mu_0}\int_{-\infty}^{\infty}\int_{-\infty}^{\infty}\delta(k_x - k_x')[k_z^* g_x(k_x')g_x^*(k_x) - k_x g_x(k_x')g_z^*(k_x)]\, dk_x\, dk_x'$$

$$= \frac{1}{4\pi\omega\mu_0}\int_{-\infty}^{\infty}\frac{k_x k_x^* + k_z^* k_z^*}{k_z^*}\, g_x(k_x)g_x^*(k_x)\, dk_x$$

$$= \frac{2VV^*}{\lambda\eta a^2}\int_{-\infty}^{\infty}\frac{\sin^2(k_x a/2)}{k_x^2 k_z^*}\, dk_x \tag{6.183}$$

The input admittance per unit length can be defined by the relation $(1/2)VV^*Y_a^* = P$, as a result of which

$$G_a = \frac{4}{\lambda\eta a^2}\int_{-k}^{k}\frac{\sin^2(k_x a/2)}{k_x^2\sqrt{k^2 - k_x^2}}\, dk_x \tag{6.184}$$

$$B_a = \frac{4}{\lambda\eta a^2}\left(\int_{-\infty}^{-k} + \int_{k}^{\infty}\right)\frac{\sin^2(k_x a/2)}{k_x^2\sqrt{k_x^2 - k^2}}\, dk_x \tag{6.185}$$

Plots of the real and imaginary components of the input admittance are shown in Figure 6.42. The susceptance is positive and this is a capacitance-type aperture. It

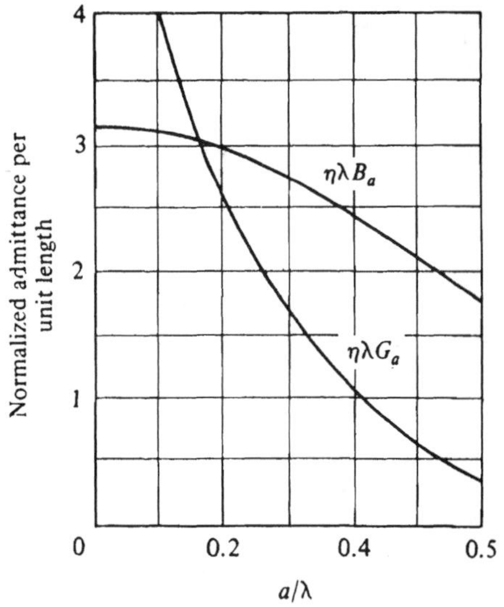

Fig. 6.42 Admittance per Unit Length of the Slit Radiator Shown in Fig. 6.41

can be seen that resonance does not occur in the range

$$0 < a/\lambda \leq 0.5$$

There are other simple planar antennas for which formulas for the input admittance can be obtained readily using this plane wave spectrum approach. (See, for example, Problem 6.24 at the end of this chapter.) For a full exposition of the utility of a Fourier integral representation of the fields produced by planar antennas, the interested reader is referred to the pertinent literature.[19]

REFERENCES

AMITAY, N., V. GALINDO, and C. P. WU, *Theory and Analysis of Phased Array Antennas* (New York: Wiley-Interscience, 1972), Chapter 1.

COLLIN, R. E., and F. J. ZUCKER, *Antenna Theory: Part I* (New York: McGraw-Hill Book Co., Inc., 1969), Chapter 5.

ELLIOTT, R. S., "The Theory of Antenna Arrays," Vol. 2 of *Microwave Scanning Antennas*, ed. R. C. Hansen (New York: Academic Press, 1966), Chapter 1.

MA, M. T., *Theory and Application of Antenna Arrays* (New York: Wiley-Interscience, 1974), Chapter 3.

RHODES, D. K., *Synthesis of Planar Antenna Sources* (London: Clarendon Press, Oxford, 1974).

PROBLEMS

6.1 Find the separable current distribution for a rectangular grid array with a rectangular boundary if $d_x = 5\lambda/8$, $d_y = 3\lambda/4$, $2N_x = 8$, and $2N_y = 12$, and if 25 dB and 35 dB Dolph-Chebyshev patterns are desired in the XZ- and YZ-planes, respectively. Assume the main beam points at $\theta_0 = 0°$ and write an equation for the -3 dB contour of the main beam. What is the height of the off-axis side lobes?

6.2 In Problem 6.1, if the element pattern is hemispherically isotropic in $z > 0$ and is zero in $z < 0$, find the peak directivity. What is the areal beamwidth? Find the changes in directivity and areal beamwidth if the beam is scanned to the position $\theta = 30°$, $\phi = 45°$.

6.3 Suppose the array of Problem 6.1 is to be used for conical scanning, that is, the position of the main beam is to be given by $\theta =$ constant, $\phi = Kt$, with K a rate constant and t

[19]For the original exposition, see H. G. Booker and P. C. Clemmow, "The Concept of an Angular Spectrum of Plane Waves and Its Relation to That of Polar Diagram and Aperture Distribution," Proc. IEEE, 97 (1950), 11–17. The books by D. R. Rhodes, *Synthesis of Planar Antenna Sources*, (London: Clarendon Press, Oxford, 1974), and by J. R. Goodman, *Introduction to Fourier Optics*, (New York: McGraw-Hill Book Co., Inc., 1968) are also recommended.

the time. Find the uniform progressive phase factors α_x and α_y as functions of time in order to achieve this effect.

6.4 If, instead of the conical scan of Problem 6.3, a raster scan is desired for the array, find α_x and α_y as functions of time. [A *raster scan* can be defined by $\theta = K_1$ while ϕ changes linearly in time from $-\phi_0$ to $+\phi_0$; then $\theta = K_2$ while ϕ changes linearly in time from $+\phi_0$ to $-\phi_0$; and so on, with $K_1, K_2, \ldots$ a monotonic sequence of constants.]

6.5 A more practical application of a raster scan is one in which the planar array sits in a vertical plane over a horizontal earth, with θ measured from the earth's zenith and ϕ measured in the earth plane. Repeat the analysis of Problem 6.4 for this case and find the sequence $K_1, K_2, \ldots$ if the main beam positions on successive legs of the raster are to overlap at the -3 dB points.

6.6 For the array of Problem 6.1, sketch the -3 dB contour of the main beam when it is at the pointing directions (0°, 0°), (30°, 0°), (30°, 30°), (30°, 60°), and (30°, 90°).

6.7 Assume that the sum pattern current distribution found for the array of Problem 6.1 is retained, except that the sign of the excitation is reversed for the two quadrants in which $x < 0$. Write an expression for the resulting difference pattern. If a computer is available, plot this difference pattern in $\phi = 0°/180°$ and observe the side lobe level. If not, collapse the distribution onto the X-axis (that is, let $\phi = 0°$) and determine the side lobe level by trial and error.

6.8 The side lobe level found for the difference pattern in Problem 6.7 will be seen to be poor. Use the perturbation technique described in Section 6.12 to find a current distribution that will give no side lobe in any ϕ-cut higher than -25 dB relative to the twin main beam peak of the $\phi = 0°/180°$ cut.

6.9 Design an equispaced planar array under the following specifications.
(a) Rectangular grid, rectangular boundary, separable distribution.
(b) Sum and difference pattern capability.
(c) Sum pattern scannable out to $\theta = 30°$ in any ϕ-cut.
(d) $\theta_{x0} = 14°$, $\theta_{y0} = 20°$.
(e) Both principal cuts are Dolph-Chebyshev, -20 dB in XZ and -15 dB in YZ.

6.10 A circular Taylor pattern, -20 dB SLL, $\bar{n} = 3$, is desired from a continuous circular aperture for which $a = 3\lambda$. Find A^2, σ, and the modified root positions u_1 and u_2. Write the explicit expression for this Taylor pattern. Determine the orthogonal components in the expression for the aperture distribution. If a computer is available, plot the pattern in $0° \leq \theta \leq 90°$ and the aperture distribution in $0 \leq \rho \leq 3\lambda$. Note the characteristic droop of the side lobe structure in the pattern plot.

6.11 Use the perturbation technique described in Section 6.4 to determine the modified Taylor circular aperture distribution which will give a pattern identical to the one found in Problem 6.10 except that the innermost side lobe is depressed to -30 dB. If available, use a computer to plot the aperture distribution and pattern.

6.12 If undulating ring side lobes are desired for the pattern caused by the circular aperture described in Problem 6.10, find the modified aperture distribution if the side lobe level is to vary smoothly from -25 dB at $\phi = 0°$ to -15 dB at $\phi = 180°$ and back again. Assume all ϕ-cuts are to exhibit Taylor-like patterns with $\bar{n} = 3$.

6.13 A rectangular grid array with $d_x = d_y = 0.7\lambda$ has a circular boundary for which $a = 3\lambda$. Because of the cut-off corners, there are only 13 elements per quadrant. Find the excitation of this array if one uses Dolph-Chebyshev separable -20 dB SLL and merely sets the excitation of the three cut-off elements equal to zero. Use a computer to plot the pattern in the cuts $\phi = 0°, 15°, 30°$, and $45°$.

6.14 Repeat Problem 6.13, except use a Tseng-Cheng nonseparable -20 dB SLL excitation.

6.15 Repeat Problem 6.13, except use an excitation found by conventional sampling of the continuous circular Taylor distribution determined in Problem 6.10.

6.16 Use the perturbation technique described in Section 6.10 to bring the degraded patterns of Problem 6.15 as close as you can to the desired pattern found in Problem 6.10.

6.17 Determine the beamwidths and directivities of the patterns found in Problems 6.13 through 6.16.

6.18 A circular grid array with four concentric rings at radii $\rho/\lambda = 0.7, 1.4, 2.1$, and 2.8 is to be excited with ring currents that will give a pattern approximating a circular Taylor -20 dB SLL, $\bar{n} = 3$. Determine the ring currents (a) by conventional sampling of the Taylor distribution; (b) by matching to the nulls of the pattern found in Problem 6.10; (c) by perturbation of either of the foregoing patterns, using the technique described in Section 6.8.

6.19 Use the ring currents determined in Part (c) of the previous problem to determine the discrete element excitation if the inner ring has (a) four elements; (b) eight elements. Assume other rings have the same element separation. Compute the patterns for these two cases and plot the ϕ-cuts for $\phi = 0°, 15°, 30°$, and $45°$.

6.20 For a 6λ diameter continuous circular aperture, find the excitation that will produce a -20 dB SLL, $\bar{n} = 3$, Bayliss difference pattern. Write the explicit expressions for pattern function and aperture distribution. If a computer is available, plot both the pattern and the distribution.

6.21 Use the perturbation technique outlined in Section 6.12 to modify the Bayliss circular pattern shown in Figure 6.33 so that the innermost side lobe is at -40 dB, with all others unchanged. Find the requisite aperture distribution.

6.22 For the array of Problem 6.13 use conventional sampling of the distribution found in Problem 6.20 to produce an approximation to the Bayliss difference pattern. Plot the resulting pattern for $\phi = 0°, 20°, \ldots, 80°$.

6.23 Use the perturbation procedure detailed in Section 6.13 to improve on the patterns of the previous problem.

6.24 For the infinite slit shown in Figure 6.41, assume that the E-field in the aperture is y-directed and given by $E_y(\xi) = K \cos(\pi\xi/a)$. Use the plane wave spectrum approach to find the complex radiated power and from this deduce an expression for the input admittance per unit length.

III self-impedance and mutual impedance, feeding structures

Part I of this text dealt with pattern analysis of individual antenna elements (such as a single dipole, helix, or slot), and Part II dealt with pattern analysis and synthesis of linear and planar arrays of these elements. The treatment of arrays culminated, in Chapters 5 and 6, with a variety of procedures for the determination of array element excitations which will produce specified patterns.

A different class of practical engineering questions now needs to be addressed. What is the best way to feed a single element? How does one provide a match between the element and its feeding line at the design frequency? How does one minimize the mismatch over a frequency band? And, for arrays, how does one actually achieve the desired excitation of the elements in order to produce the specified pattern? Further, how does one achieve it, and at the same time provide a match to the transmitter (or receiver) at the design frequency? More difficult still, how does one minimize pattern degradation and mismatch over a frequency band? Most difficult of all, how is this done if the pattern is to scan?

Before answers to these questions can be attempted, knowledge about the input impedance of a single antenna element and about the mutual impedances among elements when they are used in arrays must be acquired. This information is vital when a feed line is to be designed to connect to a single element, or when a network of feed lines is to be designed to connect together the elements of an array. Thus this part of the text is devoted first to the determination of the self-impedance of various types of isolated antenna elements and to the determination of self-impedances and mutual impedances among elements in an array. This is followed by an introduction to the design of various feeding structures which have the purpose of providing a match and, in the case of arrays, do this in concert with establishing the array excitation which will yield a specified pattern.

7 self-impedance and mutual impedance of antenna elements

7.1 Introduction

In this chapter the reader will find a sequence of analyses leading to the determination of the self-impedance and mutual impedance (or admittance) of cylindrical dipoles, strip dipoles, monopoles, and slots, these being among the most common antenna elements used singly or in arrays. A theoretical formulation of the self-impedance of a patch antenna is also presented.[1] The chapter begins with the derivation of an integral equation relating the current density distribution on an arbitrarily shaped antenna element to the sources which excite it. Specific application of this integral equation is then made to the center-fed cylindrical dipole of circular cross section, using an approach pioneered by Hallén.[2] The method of moments is introduced and used to solve for the current distribution on the cylindrical dipole for a known applied voltage. The input current is then used to compute the self-impedance.

When the inquiry is focused on self-impedance, one needs to find only the input current and not the entire current distribution. A variety of techniques is available for doing this, and the ones developed in this chapter, all applied to the cylin-

[1]Theoretical analyses of the mutual impedance between patches, and of the self-impedances and mutual impedances of helices are extremely difficult and of dubious value. However, these two elements are often used in arrays, with the necessary data on self-impedance and mutual impedance obtained experimentally.

[2]An independent and highly original approach to the determination of the current distribution on a center-fed dipole and its self-impedance has been provided by Schelkunoff. He started with the biconical antenna as a prototype and provided an approximate solution to the relevant differential equations. Compare with S. A. Schelkunoff, *Electromagnetic Waves* (Princeton, New Jersey: D. Van Nostrand Co., Inc., 1943), pp. 446–69. For an excellent summary of Schelkunoff's approach, see E. C. Jordan and K. G. Balmain, *Electromagnetic Waves and Radiating Systems* 2nd Ed. (Englewood Cliffs, New Jersey: Prentice-Hall, Inc., 1968), pp. 572–88. The integral equation approach of Hallén will be followed in this text because of certain computational advantages and because it provides a convenient framework for several related developments.

drical dipole, are the induced EMF method, Storer's variational approach, and Hallén's method. The results are summarized in sets of curves showing the real and imaginary parts of the self-impedance versus cylindrical dipole length and radius.

A proof is then presented that a slender strip dipole (rectangular cross section) is equivalent to a cylindrical dipole of the same length and appropriately chosen radius, so that all of the preceding developments and results for cylindrical dipoles including current distribution and input impedance, can be carried over intact to the strip dipole.

The reciprocity theorem is used next to develop a formula for the mutual impedance between parallel cylindrical (or strip) dipoles that are arbitrarily positioned relative to each other. Computations based on this formula lead to a family of curves relating mutual impedance to the two dipole lengths and their relative positions. A discussion is undertaken about the meaning of self-impedance and mutual impedance in dipole arrays consisting of many parallel elements, and it is indicated under what conditions isolated self-impedance and one-on-one mutual impedance can be used as approximations in this general case.

The simple extension of all the foregoing results for vertical monopoles fed above a horizontal ground plane is indicated.

Babinet's principle and Booker's extension of it to complementary arrays of slots and strip dipoles are introduced and used to establish the equivalence between the field distribution in a slot and the current distribution on a strip dipole. This permits deduction of the equivalence between the input impedances of a single slot and dipole, and the equivalences between (1) self-impedance and mutual impedance of strip dipoles in an array, and (2) self-impedance and mutual admittances of slots in the complementary array. Thus the utility of all the results on cylindrical dipoles is extended still further, beyond monopoles and strip dipoles, to two-wire-fed slots. In Chapter 8, a further extension to waveguide-fed slots will be presented. Because of these equivalences, the subject of the self-impedance and mutual impedance of cylindrical dipoles takes on added importance.

The chapter concludes with a formulation of an expression for the self-impedance of a patch antenna (metallic film of rectangular or circular shape bonded to a grounded dielectric slab).

7.2 The Current Distribution on an Antenna: General Formulation

Consider an electromagnetic system consisting of a transmitter (or receiver), a feeding network, and an antenna element (such as a dipole, a slot, or a helix). Some simple examples are shown in Figure 7.1. More complicated examples could be created by using an ensemble of these elements in arrays.

In the analysis that follows, the feeding network will be idealized by assuming that it can be replaced by a generator contained in a small feeding volume. As an illustration, the two-wire-fed dipole of Figure 7.1a will be modeled as shown in

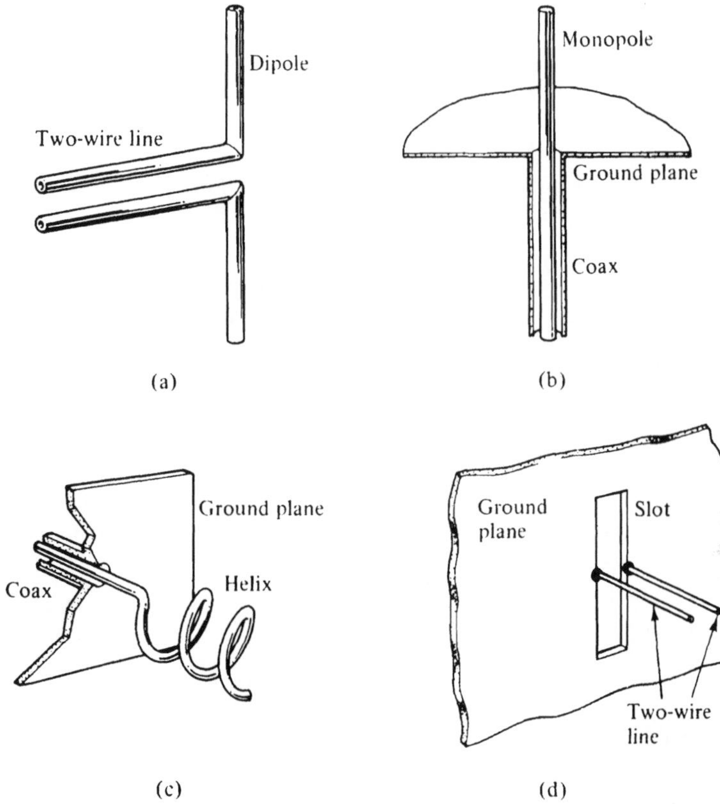

Fig. 7.1 Some Simple Antenna Elements and Their Feeding Structures

Figure 7.2, with the generator confined to a cylindrical region in the gap between the two arms of the dipole.

If $\mathbf{J}(\xi, \eta, \zeta)e^{j\omega t}$ is the current density distribution of the antenna plus idealized generator, then the fields at any point (x, y, z) are related to the sources at (x, y, z) by the differential relations (compare with Chapter 1)

$$\mathbf{B} = \nabla \times \mathbf{A} \tag{7.1}$$

$$\nabla \times \mathbf{B} = \nabla(\nabla \cdot \mathbf{A}) - \nabla^2 \mathbf{A} = \frac{\mathbf{J}}{\mu_0^{-1}} + \frac{j\omega}{c^2}\mathbf{E} \tag{7.2}$$

$$(\nabla^2 + k^2)\mathbf{A} = \frac{-\mathbf{J}}{\mu_0^{-1}} \tag{7.3}$$

The elimination of $\mathbf{J}$ from (7.2) and (7.3) provides a connection between $\mathbf{E}$ and $\mathbf{A}$, namely,

$$\nabla(\nabla \cdot \mathbf{A}) + k^2 \mathbf{A} = \frac{j\omega}{c^2}\mathbf{E} \tag{7.4}$$

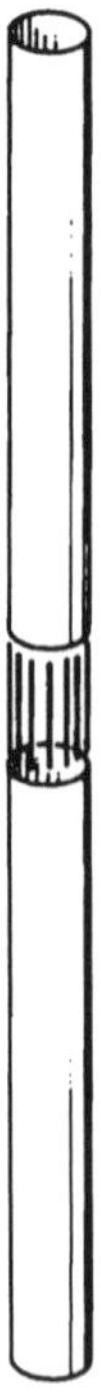

Fig. 7.2 A Cylindrical Dipole Center-Fed by an Idealized Gap Generator

with **A** given by

$$\mathbf{A}(x, y, z) = \int_V \frac{\mathbf{J}(\xi, \eta, \zeta)e^{-jkR}}{4\pi\mu_0^{-1}R}\,dV \tag{7.5}$$

where $R = [(x-\xi)^2 + (y-\eta)^2 + (z-\zeta)^2]^{1/2}$ is the distance from the source point (ξ, η, ζ) to the field point (x, y, z). Equation 7.5 is a restatement of (1.80) with the time factor suppressed.

In the analysis of most electromagnetic problems that involve use of the magnetic vector potential function (7.5), the collection of points (ξ, η, ζ) is defined to be all points at which there are sources. The collections of points (x, y, z) is usually a larger set. It may include points occupied by sources, but often refers principally to all other points in space. The analyses to be presented shortly differ in that the collection of points (x, y, z) will be the same set as the collections of points (ξ, η, ζ). To emphasize the equivalence of these two sets, a field point will still be characterized by the triplet (x, y, z), but a source point will hereafter be identified as (x', y', z') whenever a distinction needs to be made between source points and field points.

With this explanation as background, Equations 7.4 and 7.5 can be combined to give

$$(\nabla_F \nabla_F \cdot + k^2) \int_V \frac{\mathbf{J}(x', y', z')e^{-jkR}}{4\pi R}\,dV' = j\omega\epsilon_0 \mathbf{E}(x, y, z) \tag{7.6}$$

in which

$$R = [(x-x')^2 + (y-y')^2 + (z-z')^2]^{1/2} \tag{7.7}$$

and ∇_F retains its normal meaning, that is, differentiation with respect to the variables x, y, and z. (When needed, ∇_S will imply differentiation with respect to the variables x', y', and z').

Equation 7.6 is a cornerstone for what is to follow. If, for a given antenna, $\mathbf{E}(x, y, z)$ is known at all points occupied by the antenna, (7.6) is an integral equation in the unknown and sought current distribution $\mathbf{J}(x', y', z')$.[3] Since most antennas are composed of good conductors, it is usually an excellent assumption to take $\mathbf{E}(x, y, z) \equiv 0$, except in the region where the antenna is being fed. If a good estimation of $\mathbf{E}(x, y, z)$ can be made in the feeding region, modern numerical techniques applied to (7.6) will yield satisfactory solutions for the current distribution. The resulting knowledge of the value of the current density at the interface between the antenna and the idealized generator, together with the specified initial value of the terminal voltage across the generator, also permits a calculation of the input impedance. This will first be demonstrated for cylindrical dipoles.

7.3 The Cylindrical Dipole: Arbitrary Cross Section

A principal application of Equation 7.6 is to the center-fed dipole. As shown in Figure 7.3, this is an antenna element consisting of two identical arms, each a cylinder of length $l - \delta$, with the two arms axially aligned and separated by a gap of length

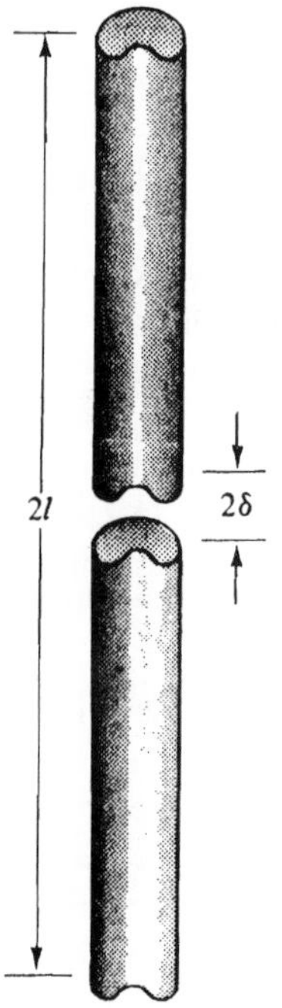

Fig. 7.3 The Center-Fed Cylindrical Dipole of Arbitrary Cross Section

[3]Alternatively, if $\mathbf{J}(x', y', z')$ is known at all points occupied by the antenna, (7.6) can be used to determine the electric field distribution $\mathbf{E}(x, y, z)$ throughout the antenna.

2δ. All transverse cross sections of the two cylinders are identical. The cross section may have an arbitrary shape, as suggested by Figure 7.4a. A perimeter coordinate s can be used to identify points on the contour C via the parametric equations

$$x = g_1(s), \qquad y = g_2(s) \tag{7.8}$$

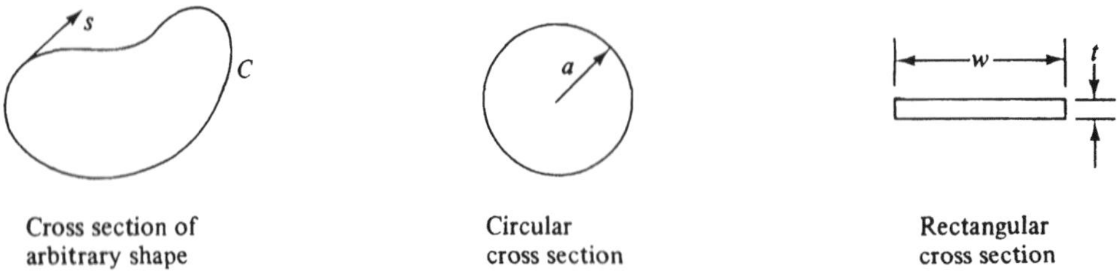

Fig. 7.4 Cylindrical Dipoles of Various Cross-Sectional Shapes

When the need exists to distinguish between the points (x, y, z) and (x', y', z') the transverse coordinates of the latter will be identified by

$$x' = g_1(s'), \qquad y' = g_2(s') \tag{7.9}$$

In practice, the two most commonly encountered cross sections are circular and rectangular. For the former, shown in Figure 7.4b, $g_1 = a \cos \phi$ and $g_2 = a \sin \phi$, with a the radius. For the latter, shown in Figure 7.4c, $x = \pm t/2$ and $y = \pm w/2$, with w the width and t the thickness.

Loose usage of the word *cylindrical*, plus historical precedence and the desire for word economy, has led to the tradition that the phrase *cylindrical dipole* refers exclusively to the circular cross section of Figure 7.4b.

The advent of printed circuits has occasioned an increased interest in thin dipoles of rectangular cross section which can be laid or printed on grounded dielectric substrates and fed by stripline. But even prior to this practical application, theoretical interest in such dipoles was stimulated by recognition of their complementarity to rectangular slots cut in large ground planes. To provide a distinction between dipoles of circular and rectangular cross section, since both will be treated in subsequent developments, the name *strip dipole* will always be used to refer to the rectangular shape of Figure 7.4c.

The analysis leading to the determination of the current distribution on a dipole and its input impedance can be carried a good distance before one is forced to particularize the cross section. Therefore, attention will continue to focus for the remainder of this section on the general shape of Figure 7.4a. If $2u$ is the maximum lineal dimension of this arbitrary cross section, it will be assumed in all that follows that $ku \ll 1$ and $u \ll l$. Thus what ensues is a theory of slender dipoles. This is not a serious restriction, since most practical applications fit this condition.

The gap 2δ between the two arms of the dipole will be taken to be an infinitesimal.[4] The equivalent idealized generator will be assumed to be confined within the cylindrical volume of height 2δ and contour C.

Since almost without exception the practical applications of dipoles involve the use of arms made of good conductors, it is customary to assume in the model that the arms are perfectly conducting. Consideration should then be given to the question whether the arms are tubular or solid. If they are solid, currents will flow on the disclike surfaces at $z = \pm\delta$ and $z = \pm l$ and this complicates the analysis. If they are hollow tubes (which is often the case in practice), currents will flow primarily on the outer surfaces of the arms, but can penetrate up inside the tubes at $z = \pm\delta$, and can lapover and down into the tubes at $z = \pm l$. However, if $ku \ll 1$, as is being assumed, these currents inside the tubes do not penetrate very far and can usually be ignored.

The present analysis will be confined to the case in which the dipole arms are tubular, with negligibly thin walls, and composed of perfect conductors. For this reason, the equivalent idealized generator will be taken to lie in the cylindrical shell of contour C and height 2δ. Because of the slenderness assumption ($u \ll l$), the spatial disposition of the idealized generator, and the assumption of perfect conductivity, the current density is lineal, Z-directed, and flows on a surface S, with S a cylinder of contour C extending from $z = -l$ to $z = +l$. For this case, Equation 7.6 becomes

$$\left(\frac{\partial^2}{\partial z^2} + k^2\right) \int_S \frac{K_z(s', z')e^{-jkR}}{4\pi R} dS = j\omega\epsilon_0 E_z(s, z) \tag{7.10}$$

With the dipole composed of perfectly conducting tubes, $E_z(s, z) \equiv 0$ except for $-\delta \leq z \leq \delta$. But the voltage measured at the dipole terminals is given by

$$V = -\int_{-\delta}^{\delta} E_z(s, z)\, dz \tag{7.11}$$

a result that is independent of s. This implies that E_z is not a function of s. With δ an infinitesimal, it is appropriate to take E_z as a Dirac delta function, that is,

$$E_z(z) = 0, \quad z \neq 0 \tag{7.12}$$

$$\int_{-l}^{l} E_z(z)\, dz = -1 \tag{7.13}$$

This is equivalent to saying that positive unit voltage is impressed on the dipole.

[4]This is not always an assumption that models physical reality with sufficient accuracy. Considerable discussion of the "gap problem" can be found in the literature, and the interested reader may particularly wish to consult King and Thiele among the references listed at the end of this chapter. The monopole protruding vertically from a large ground plane, as in Figure 7.1b, with the monopole an extension of the inner conductor of a coax, is perhaps the easiest real situation to model in terms of an equivalent idealized generator. It is also the configuration for which theory and experiment are most often compared. Thiele argues for the use of a magnetic frill as the equivalent feeding source for the dipole. The present treatment, which is introductory, will be confined to use of the infinitesimal gap model.

With these elaborations, (7.10) becomes

$$\left(\frac{\partial^2}{\partial z^2} + k^2\right) \int_{-l}^{l} \oint_C \frac{e^{-jkR}}{4\pi R} K_z(s', z')\, ds'\, dz' = j\omega\epsilon_0 E_z(z) \tag{7.14}$$

It is interesting to observe that, although the integrand in (7.14) is a function of s (since R is a function of x and y), the integral itself is not a function of s, since E_z depends only on z.

Equation 7.14 is a generalized version of an integral equation first applied to dipoles by H.C. Pocklington.[5] It can be written in the form

$$\iint G(s, z, s', z') K_z(s', z')\, ds'\, dz' = j\omega\epsilon_0 E_z(z) \tag{7.15}$$

and thus identified as a Fredholm integral equation of the first kind. (Were there an additive factor on the right side, consisting of a constant times $K_z(s, z)$, it would be of the second kind.) The function G is called the kernel and in the present case is given by

$$G(s, z, s', z') = \left(\frac{\partial^2}{\partial z^2} + k^2\right) \frac{e^{-jkR}}{4\pi R} \tag{7.16}$$

The general problem of finding solutions to (7.15) has been widely studied. Several techniques useful for the particular kernel indicated by (7.16) will be presented in subsequent sections of this chapter. The appropriate solution of (7.14) is quite obviously influenced by the shape of the contour C, and attention will first be focused on the important and common case when C is circular.

7.4 The Cylindrical Dipole: Circular Cross Section, Hallén's Formulation

If the contour C is circular, by symmetry the lineal current density distribution K_z on the dipole is only a function of z, and the integral appearing in (7.14) is also only a function of z. Thus for a dipole whose cross section is a circle of radius a, (7.14) can be written in the form

$$\left(\frac{d^2}{dz^2} + k^2\right) \mathcal{A}(z) = j\omega\epsilon_0 E_z(z) \tag{7.17}$$

in which

$$\mathcal{A}(z) = \int_{-l}^{l} \int_0^{2\pi} \frac{e^{-jkR}}{4\pi R} K_z(z') a\, d\phi\, dz' \tag{7.18}$$

In (7.18), $R = [(x - x')^2 + (y' - y')^2 + (z - z')^2]^{1/2}$, with both points lying in the

[5] H. C. Pocklington, "Electrical Oscillations in Wire", *Cambridge Phil. Soc. Proc.*, Vol. 9 (1897), pp. 324–32.

cylindrical surface. The function $\mathcal{A}(z)$ is in the form of a magnetic vector potential function (lacking only the factor μ_0^{-1}) and satisfies the inhomogeneous wave equation.

It is useful to pause at this stage in the analysis and appreciate that if $\mathcal{A}(z)$ can be found by solving the differential equation in (7.17), the solution can then be used as the driving function in (7.18), which is a simpler integral equation to solve for the unknown current density distribution $K_z(z')$ than is the earlier Equation 7.14, which contains a more complicated kernel.

Solution of (7.17) is not difficult. Since E_z is a Dirac delta function, (7.17) is homogeneous except in the neighborhood of $z = 0$. Thus for $z > 0$,

$$\mathcal{A}(z) = A \cos kz + B \sin kz \tag{7.19}$$

whereas for $z < 0$,

$$\mathcal{A}(z) = C \cos kz + D \sin kz \tag{7.20}$$

with A, B, C, and D constants. One can match these solutions across $z = 0$ by noting from (7.17) that, if $\mathcal{A}(z)$ is to be finite everywhere, the singularity in E_z at $z = 0$ must be matched by a singularity in $d^2\mathcal{A}/dz^2$ at $z = 0$, since it cannot be accommodated by $k^2\mathcal{A}(0)$. Therefore

$$j\omega\epsilon_0 \int_{0-\delta}^{0+\delta} E_z(z)\, dz = -j\omega\epsilon_0 = \int_{0-\delta}^{0+\delta} \frac{d^2\mathcal{A}}{dz^2}\, dz = \left.\frac{d\mathcal{A}}{dz}\right|_{0-\delta}^{0+\delta} \tag{7.21}$$

Thus there is a jump of $-j\omega\epsilon_0$ in the first derivative of $\mathcal{A}(z)$ as the origin is traversed. Differentiation of (7.19) and (7.20), followed by letting $z \to 0$, yields the relation

$$kB - kD = -j\omega\epsilon_0 \tag{7.22}$$

By symmetry, $K_z(z')$ is even in z', and a study of Equation 7.18 indicates that this forces $\mathcal{A}(z)$ to be even also. Thus $A = C$, $B = -D$, and

$$\mathcal{A}(z) = C \cos kz - \frac{j\omega\epsilon_0}{2k} \sin k|z| \tag{7.23}$$

a solution which is valid for all z.

Insertion of (7.23) in (7.18) gives

$$\int_{-l}^{l} \int_{0}^{2\pi} \frac{e^{-jkR}}{4\pi R} K_z(z')a\, d\phi\, dz' = C \cos kz - \frac{j\omega\epsilon_0}{2k} \sin k|z| \tag{7.24}$$

Since the assumption has been made that $a \ll \lambda$, the magnetic vector potential function $\mathcal{A}(z)$, that is, the left side of (7.24), can be computed on the dipole surface S by assuming that the total current is concentrated along the Z-axis (see Appendix E). For this reason, (7.24) can be replaced by

$$\int_{-l}^{l} \frac{e^{-jkr}}{4\pi r} I(z')\, dz' = C \cos kz - \frac{j\omega\epsilon_0}{2k} \sin k|z| \tag{7.25}$$

where $I(z') = 2\pi a\, K_z(z')$ is the total current, and $r = [a^2 + (z - z')^2]^{1/2}$ is the distance from an *equivalent* source point $(0, 0, z')$ on the Z-axis to a field point (x, y, z) which is on the cylindrical surface. Equation 7.25 is Hallén's integral equation.[6] The constant C in (7.25) can be evaluated by requiring that $I(l) = 0$.

Hallén's equation can be solved either for the complete current distribution $I(z')$, or merely for the input current $I(0)$, with the latter equal to the self-admittance, since unit voltage is being applied to the cylindrical dipole. A powerful technique for solving (7.25) for the complete current distribution is the method of moments, which is described in the next section.

7.5 The Method of Moments

With the advent of high-speed, large-capacity computers, integral equations such as (7.25) can be solved rapidly, inexpensively, and with good accuracy by a variety of numerical techniques. The approach to be described here is a special case of the method of moments[7] and is often called a *point-matching technique*.

Consider the general one-dimensional Fredholm integral equation of the second kind:

$$\int_a^b G(z, z')f(z')\,dz' = g(z) + \gamma f(z) \tag{7.26}$$

with $\gamma = 0$, one obtains a Fredholm integral equation of the first kind, of which Hallén's equation in (7.25) is an example.

In (7.26) it is assumed that the kernel $G(z, z')$ is a known function, that $g(z)$ is a known driving function, that γ is a known constant, and that $z \in [a, b]$. The problem is to determine the unknown function $f(z)$.

The point-matching technique begins with the assumption that the unknown function $f(z)$ can be approximated by a linear combination of known functions $f_n(z)$, called basis functions, as follows.

$$f(z) \cong c_1 f_1(z) + c_2 f_2(z) + \cdots + c_N f_N(z) \tag{7.27}$$

In (7.27), the functions $f_n(z)$ are linearly independent and the constants c_n are unknown at this stage of the analysis.

If (7.27) is substituted in (7.26) one obtains

$$\sum_{n=1}^{N} c_n \left\{\int_a^b G(z, z')f_n(z')\,dz' - \gamma f_n(z)\right\} \cong g(z) \tag{7.28}$$

in which, because of the linearity assumption, it has been permissible to interchange the order of the summation and the integration.

[6]E. Hallén, "Theoretical Investigations into Transmitting and Receiving Qualities of Antennas," *Nova Acta Regiae Soc. Sci. Upsaliensis*, (January 1938), 1–44.

[7]R. F. Harrington, *Field Computation by Moment Methods*, (New York: The Macmillan Co., 1968).

The true function $f(z)$ insures that the two sides of (7.26) are equal for every value of $z \in [a, b]$. The approximate solution in (7.27) cannot similarly guarantee equality of the two sides of (7.28) for all $z \in [a, b]$, which is why the approximately equals sign was used. However, one can force the two sides of (7.28) to be equal at specified match points z_m by appropriate choice of the constants c_n.[8] If there are M such match points, then

$$\sum_{n=1}^{N} c_n \left\{ \int_a^b G(z_m, z') f_n(z')\, dz' - \gamma f_n(z_m) \right\} = g(z_m) \qquad m = 1, 2, \ldots, M, \quad z_m \in [a, b] \tag{7.29}$$

Since $G(z_m, z')$, $f_n(z')$, γ, and $f_n(z_m)$ are all known, the quantity within the braces in (7.29), which shall be called a_{mn}, can be computed for every value of m and every value of n. Similarly, $g(z_m) = b_m$ is known, and thus (7.29) can be represented by

$$\sum_{n=1}^{N} a_{mn} c_n = b_m \tag{7.30}$$

which is recognizable as a system of M linear equations in the N unknowns $c_1, c_2, \ldots, c_N$. If $M \geq N$, matrix inversion will yield values for the coefficients c_n which, when placed in (7.27), will give an approximation to the function $f(z)$ that is sought.

When one reviews this procedure, it is clear that the calculation of a_{mn} is influenced by the choice of partial functions $f_n(z)$ and by the choice of matching points z_m. Judicious selection of both can reduce the computational difficulties and enhance the prospect of getting a good approximation to $f(z)$. But judicious selection requires skill based on experience. A burgeoning body of knowledge is now available concerning solutions to problems of this type and the interested reader is urged to consult the current literature.[9] The application of this point matching technique to Hallén's integral equation will be described in the next two sections.

7.6 Solution of Hallén's Integral Equation: Pulse Functions[10]

Since, in the center-fed cylindrical dipole problem formulated in Section 7.4, the current distribution $I(z)$ is even, Hallén's integral equation (7.25) can be rewritten in the form

$$\int_0^l G(z, z') I(z')\, dz' = C \cos kz - \frac{j}{2\eta} \sin kz \tag{7.31}$$

[8]More generally, the method of moments involves the selection of a set of known weighting functions $w_m(z)$; the inner products of these functions with (7.28) are computed. In the point matching technique being described here, the $w_m(z)$ are Dirac delta functions centered at the points z_m. The interested reader should consult Reference 7 for the general development.

[9]See, for example, *Computer Techniques for Electromagnetics*, ed. R. Mittra (Oxford: Pergamon Press Ltd., 1973).

[10]The procedures outlined in this and the next section follow closely the development presented by C. M. Butler in Chapter 2 of *Supplementary Notes for a Short Course in the Application of Moment Methods to Field Problems* (University of Mississippi, May 1973).

with $z \in [0, l]$ and $\eta = k/\omega\epsilon_0 = \sqrt{\mu_0/\epsilon_0} = 377$ ohms. In (7.31),

$$G(z, z') = \frac{e^{-jkr}}{4\pi r} + \frac{e^{-jkr'}}{4\pi r'} \tag{7.32}$$

in which

$$r = [a^2 + (z - z')^2]^{1/2} \tag{7.33}$$

$$r' = [a^2 + (z + z')^2]^{1/2} \tag{7.34}$$

An approximation to $I(z)$, the unknown current distribution in (7.31), can be expressed in the form of (7.27). One selection for the basis functions $f_n(z)$ can be made as follows.

Construct a sequence of equispaced points $z_1, z_2, \ldots, z_{N+1}$ in $[0, l]$ such that $z_1 = 0$ and $z_{N+1} = l$, thus dividing the interval into N subintervals of equal length. The nth point is located by

$$z_n = \frac{(n-1)l}{N} \tag{7.35}$$

and the nth subinterval can be described by the relation

$$\Delta z_n = z_{n+1} - z_n \tag{7.36}$$

Let the pulse function $p_n(z)$ be defined as

$$p_n(z) = 1 \quad \text{if} \quad z \in \Delta z_n, \qquad p_n = 0 \quad \text{if} \quad z \notin \Delta z_n \tag{7.37}$$

If $I(z)$ is represented by

$$I(z) \cong \sum_{n=1}^{N} c_n p_n(z) \tag{7.38}$$

(which is seen to be in the form of (7.27)), then in the interval Δz_n, $I_n(z)$ is approximated by c_n. This situation is suggested by Figure 7.5.

If the matching points z_m are selected to be N in number and to occur at the midpoints of the subintervals, that is, if

$$z_m = \frac{(2m-1)l}{2N} \tag{7.39}$$

then for this case Equation 7.29 gives N simultaneous linear equations, in the form

$$\sum_{n=1}^{N} c_n \int_0^l G(z_m, z')p_n(z')\,dz' = C \cos kz_m - \frac{j}{2\eta} \sin kz_m \tag{7.40}$$

Because of the characteristics of the pulse functions, the matrix element a_{mn} can be identified from (7.40) as

$$a_{mn} = \int_{z_n}^{z_{n+1}} G(z_m, z')\,dz' \tag{7.41}$$

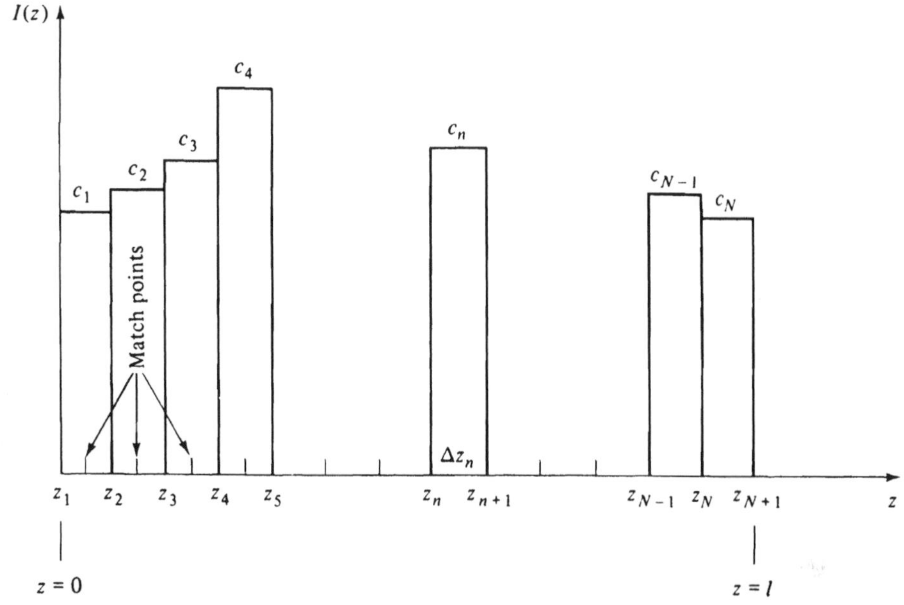

Fig. 7.5 The Use of Pulse Functions to Approximate the Current Distribution $I(z)$ on a Center-Fed Dipole

In the original formulation of Hallén's integral equation (see Section 7.4), it was stated that the constant C was to be determined from the boundary condition $I(l) = 0$. In the present formulation, that boundary condition is satisfied if one lets $c_N = 0$. Then C must be treated as an unknown in the system of linear equations in (7.40), in effect taking the place of c_N. The matrix to be solved assumes the form

$$\begin{bmatrix} a_{11} & a_{12} & \cdots & a_{1,N-1} & d_1 \\ a_{21} & a_{22} & \cdots & a_{2,N-1} & d_2 \\ \cdot & \cdot & & \cdot & \cdot \\ \cdot & \cdot & & \cdot & \cdot \\ \cdot & \cdot & & \cdot & \cdot \\ a_{N1} & a_{N2} & \cdots & a_{N,N-1} & d_N \end{bmatrix} \begin{bmatrix} c_1 \\ c_2 \\ \cdot \\ \cdot \\ \cdot \\ c_{N-1} \\ C \end{bmatrix} = \begin{bmatrix} b_1 \\ b_2 \\ \cdot \\ \cdot \\ \cdot \\ b_{N-1} \\ b_N \end{bmatrix} \tag{7.42}$$

with $d_m = -\cos kz_m$ and $b_m = -(j/2\eta)\sin kz_m$.

A computer solution to (7.42) has been obtained for the eight combinations of $2l/\lambda = 0.25, 0.50, 0.75, 1.00$ and $a/\lambda = 0.01, 0.0001$. The magnitude of $I(z)$ is plotted for these eight cases in Figure 7.6; the phase is plotted in Figure 7.7. These graphs permit an assessment of the assumption made in Chapter 2 for the purpose of computing the dipole radiation pattern, namely, that the current distribution is sinusoidal and given by $I(z) = I_m \sin[k(l - |z|)]$. If the dipole is very slender, $|I(z_m)|$ is seen to fit this assumption quite well. To emphasize this point, the case $2l/\lambda = 0.5$ and $a/\lambda =$

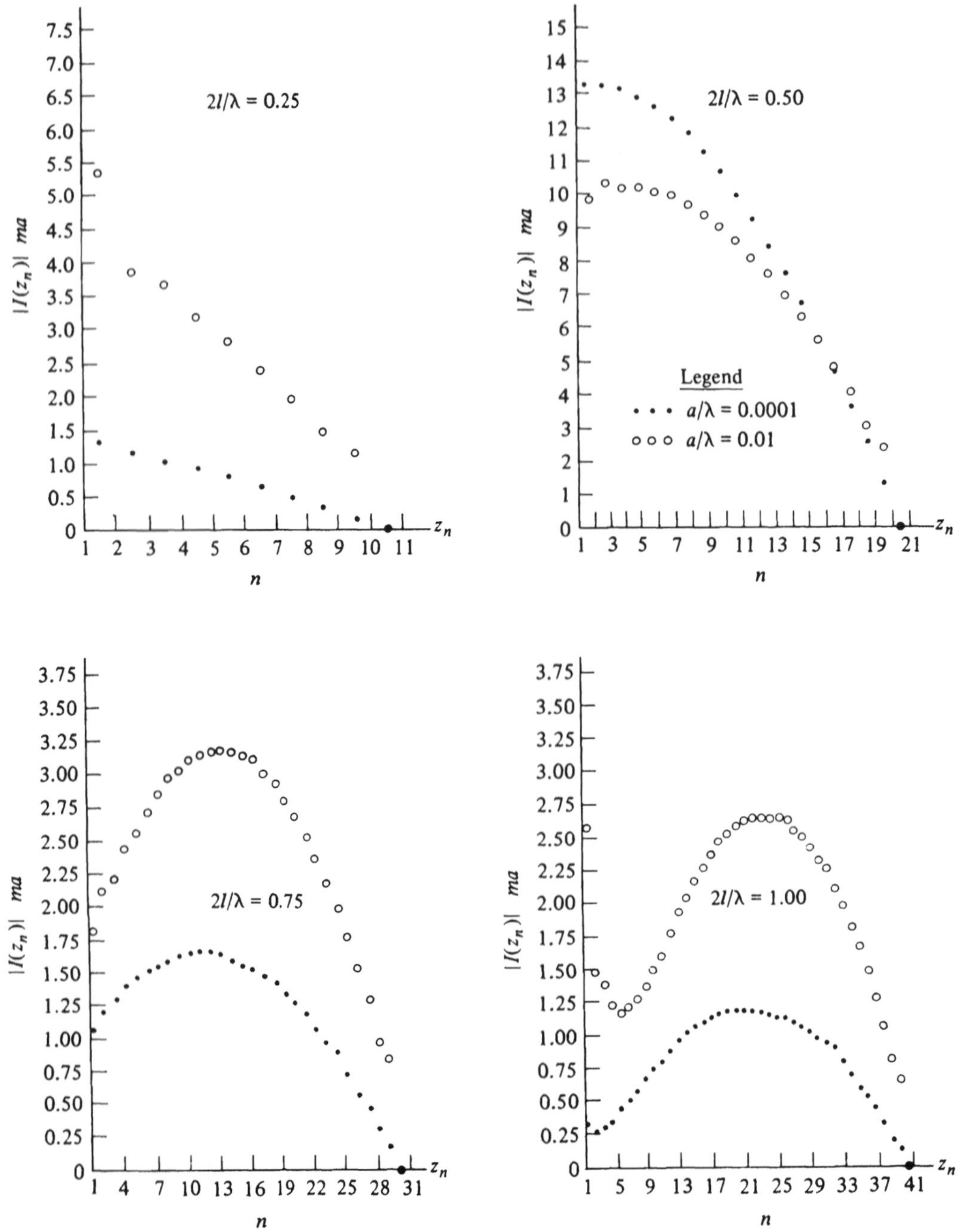

Fig. 7.6 The Magnitude of $I(z)$ for Center-Fed Cylindrical Dipoles of Various Lengths and Diameters; Pulse Function Solution

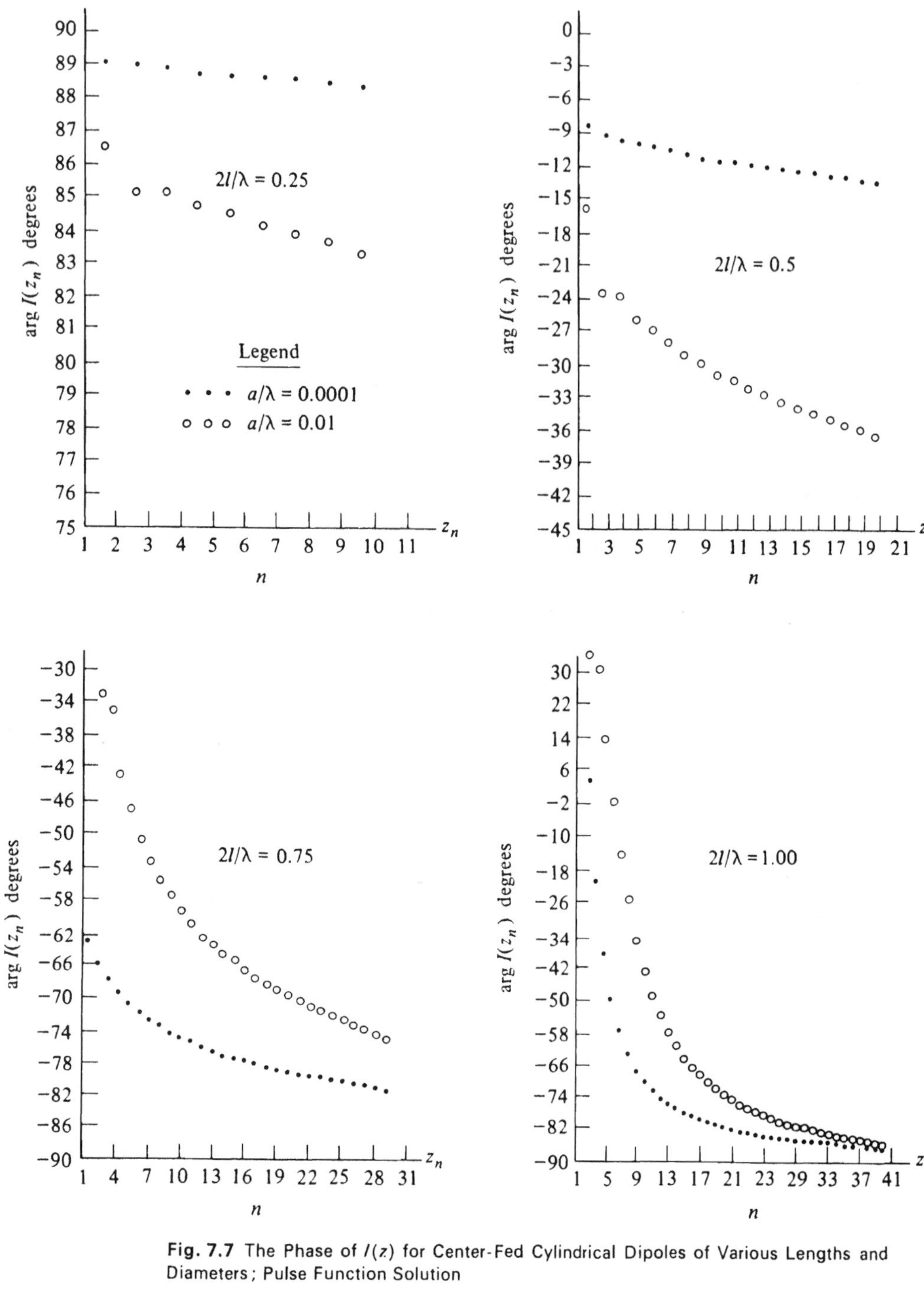

Fig. 7.7 The Phase of $I(z)$ for Center-Fed Cylindrical Dipoles of Various Lengths and Diameters; Pulse Function Solution

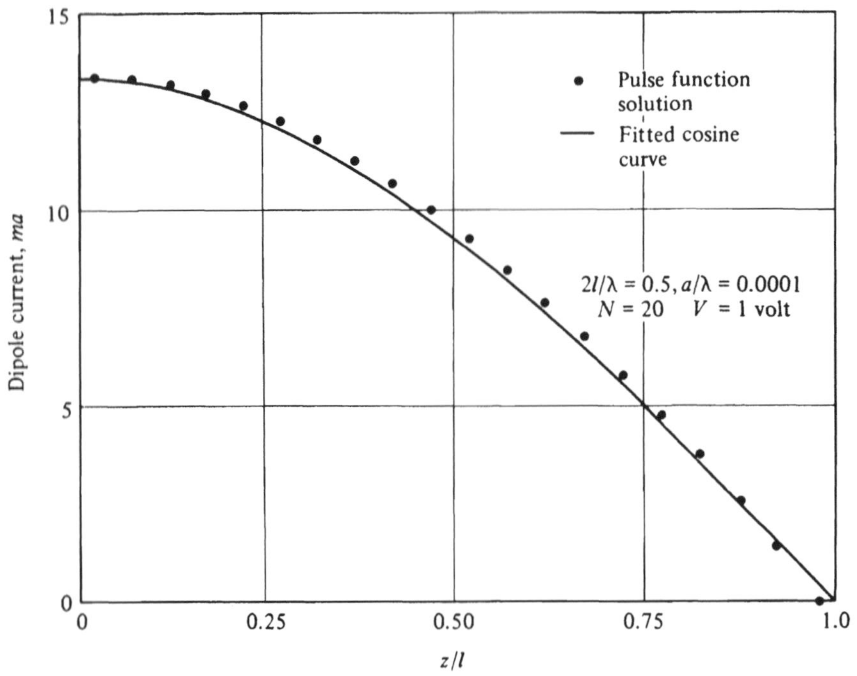

Fig. 7.8 The Current Distribution on a Half-Wavelength Long Center-Fed Cylindrical Dipole

0.0001 has been replotted in Figure 7.8 and compared to a fitted sine curve. For the fatter dipole ($a/\lambda = 0.01$), the assumptions made in going from (7.24) to (7.25) are less valid (Compare with Appendix E), and some erratic behavior can be observed in the plots of Figure 7.6 and 7.7, particularly at the ends of the interval.

Also, if the dipole is very slender and not too long, the phase plots of Figure 7.7 indicate that the assumption of constant phase for the current distribution is not a bad one. But even for slender dipoles, at the longer lengths this is no longer a valid assumption, as can be seen from a study of the case $2l/\lambda = 1.00$, and $a/\lambda = 0.0001$. However, for dipoles of the most commonly used lengths ($2l/\lambda \cong 0.5$ or less), these departures from the assumption made in Chapter 2 do not influence the element pattern significantly. As an example, the field amplitudes for the case $2l/\lambda = 0.5$ and $a/\lambda = 0.01$ are shown in Table 7.1 for the current distribution found by the pulse function method and for the idealized current distribution used in Chapter 2.

Current distributions computed using pulse functions and the point-matching procedure agree quite well with experimental results. An illustration of this is offered in Figure 7.9 where the experimental data of T. Morita[11] for a half-wavelength dipole

[11]T. Morita, "Current Distributions on Transmitting and Receiving Antennas," *Proc. IRE*, 38 (1950), 898–904.

TABLE 7.1 $\mathcal{A}_\theta(\theta)$ versus θ for idealized current distribution and for current distribution found using pulse functions

	$\mathcal{A}_\theta(\theta)$	
θ°	$I(z) = I_m \sin k(l - z)$	$I(z)$ Found by Using Pulse Functions
0	0.000	0.000 |1.163°
6	0.082	0.081 |1.141°
12	0.165	0.162 |1.098°
18	0.249	0.244 |1.029°
24	0.333	0.327 |0.940°
30	0.418	0.411 |0.833°
36	0.503	0.496 |0.717°
42	0.587	0.580 |0.596°
48	0.668	0.662 |0.475°
54	0.746	0.740 |0.361°
60	0.816	0.812 |0.258°
66	0.879	0.876 |0.169°
72	0.930	0.928 |0.096°
78	0.968	0.967 |0.043
84	0.992	0.992 |0.011°
90	1.000	1.000 |0°

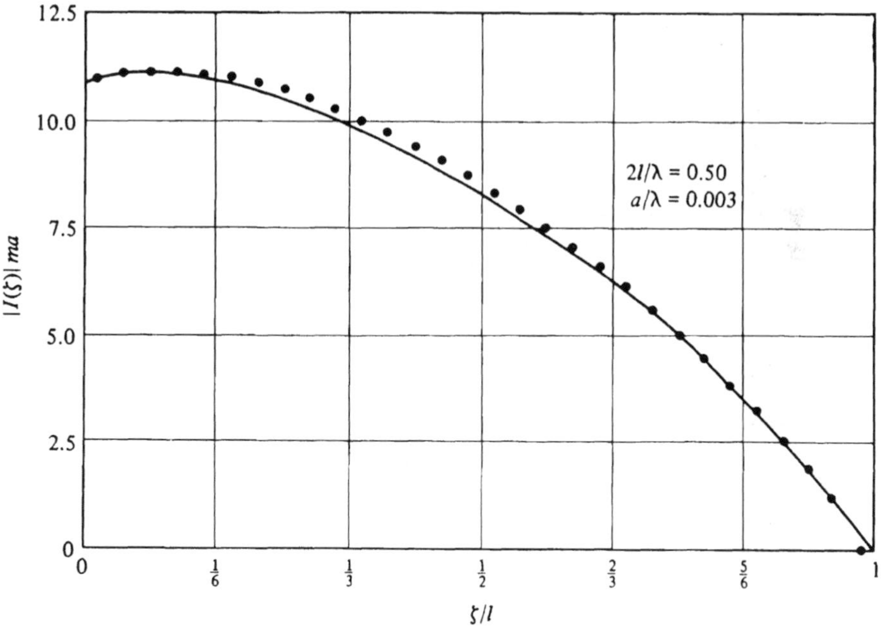

Fig. 7.9 Comparison of Theory and Experiment: Current Distribution on a Half-Wavelength Center-Fed Cylindrical Dipole (Solid Curve Experimental Results of T. Morita, *Proc. IRE*, vol. 38, pp. 898–904. © 1950 IEEE. Dots Computer Results Using Pulse Functions.)

with $a/\lambda = 0.003$ is represented by the solid curve. The computer printout is shown as a sequence of points.

Since the applied voltage has been taken as unitary in computing the current distributions shown in Figures 7.6 and 7.7, the input current is numerically equal to the self-admittance. In the pulse function formulation, the input current is approximated by c_1, and therefore $1/c_1$ is a measure of the self-impedance. The fourth column of Table 7.2 lists the values of $1/c_1$ for the eight cases under study.

TABLE 7.2 Approximations to input impedance for various cylindrical dipoles using pulse functions or sinusoidal functions as basis functions

Normalized Length $2l/\lambda$	Normalized Radius a/λ	Number of Pulse Functions N	Self-Impedance in Ohms $(c_1)^{-1}$ Pulse Functions	Self-Impedance in Ohms Sinusoidal Basis Functions
0.25	0.01	10	$11.3 - j186$	$10.2 - j185$
0.25	0.0001	10	$12.9 - j737$	$12.5 - j739$
0.50	0.01	20	$97.3 + j27.8$	$90.2 + j22.2$
0.50	0.0001	20	$74.0 + j11.3$	$74.2 + j26.4$
0.75	0.01	30	$534 + j79.9$	$477 + j180$
0.75	0.0001	30	$424 + j827$	$403 + j882$
1.00	0.01	40	$178 - j344$	$40 - j255$
1.00	0.0001	40	$2724 - j1067$	$439 - j1445$

7.7 Solution of Hallén's Integral Equation: Sinusoidal Basis Functions[12]

The observation gleaned in the previous section—that the current distribution on a cylindrical dipole is approximately sinusoidal—suggests that a judicious choice for the primitive functions $f_n(z)$ might be the spatially harmonic sequence

$$f_n(z) = \sin\left[\frac{n\pi}{2l}(l - z)\right] \tag{7.43}$$

which permits the current distribution to be approximated by

$$I(z) = \sum_{n=1}^{N} c_n \sin\left[\frac{n\pi}{2l}(l - z)\right] \tag{7.44}$$

with the anticipation that the coefficients c_n will be complex. In this case

[12]This solution technique was first introduced by H. P. Neff, C. A. Siller, and J. D. Tillman, "Simple Approximation to the Current on the Surface of an Isolated Thin Cylindrical Center-Fed Dipole Antenna of Arbitrary Length, *IEEE Trans. Antennas Propagat.*, AP-18 (1970), 399–400.

$$a_{mn} = \int_0^l G(z_m, z') \sin\left[\frac{n\pi}{2l}(l - z')\right] dz' \tag{7.45}$$

and the integration is seen to extend over the entire interval, in contrast to what was found in the case of the pulse function, where (7.41) called for integration only over the subinterval Δz_n. A tradeoff is evident in contrasting the two approaches. Because the true current distribution $I(z)$ is quasi-sinusoidal, it should take fewer terms in the series of (7.44) to obtain an approximation of a given level of quality than in the series (7.38), particularly if the dipole length is close to a multiple of a half-wavelength. On the other hand, the computation involved in (7.45) is more extensive than in (7.41) and thus the overall computer cost is often comparable in the two approaches.

Because each of the primitive functions in (7.43) separately satisfies the condition that $I(l) = 0$, once again the constant C in (7.27) needs to be treated as an unknown and one must select $N + 1$ matching points, rather than N. If the matching points are equispaced and one of them is placed at $z = l$,

$$z_m = \frac{(2m - 1)l}{2N + 1}, \qquad m = 1, 2, \ldots, N + 1 \tag{7.46}$$

The matrix set (7.30) becomes, for this case,

$$\begin{bmatrix} a_{11} & a_{12} & \cdots & a_{1N} & d_1 \\ a_{21} & a_{22} & \cdots & a_{2N} & d_2 \\ \cdot & \cdot & & \cdot & \cdot \\ \cdot & \cdot & & \cdot & \cdot \\ \cdot & \cdot & & \cdot & \cdot \\ & & & & \\ a_{N+1,1} & a_{N+1,2} & \cdots & a_{N+1,N} & d_{N+1} \end{bmatrix} \begin{bmatrix} c_1 \\ c_2 \\ \cdot \\ \cdot \\ \cdot \\ c_N \\ C \end{bmatrix} = \begin{bmatrix} b_1 \\ b_2 \\ \cdot \\ \cdot \\ \cdot \\ b_N \\ b_{N+1} \end{bmatrix} \tag{7.47}$$

where once again $d_m = -\cos kz_m$ and $b_m = -(j/2\eta)\sin kz_m$.

As an illustration, (7.47) has been inverted for the eight cases $2l/\lambda = 0.25, 0.50, 0.75, 1.00$ and $a/\lambda = 0.01, 0.0001$. The magnitude of the resulting $I(z)$ is plotted in Figure 7.10. The results are seen to be very close to those found using pulse functions.

From (7.44) the input current is

$$I(0) = \sum_{n=1}^{N} c_n \sin(n\pi/2) = \sum_{n=1}^{N} (-1)^{n-1} c_n \tag{7.48}$$

For the cases shown in Figure 7.10, the summation in (7.48), whose reciprocal is also an approximation to the self-impedance, gives the set of values shown in the last column of Table 7.2. The agreement with the earlier results found using pulse functions is seen to be quite good at the shorter dipole lengths. An improvement on both of these computations of self-impedance can be obtained using the methods outlined in the next two sections.

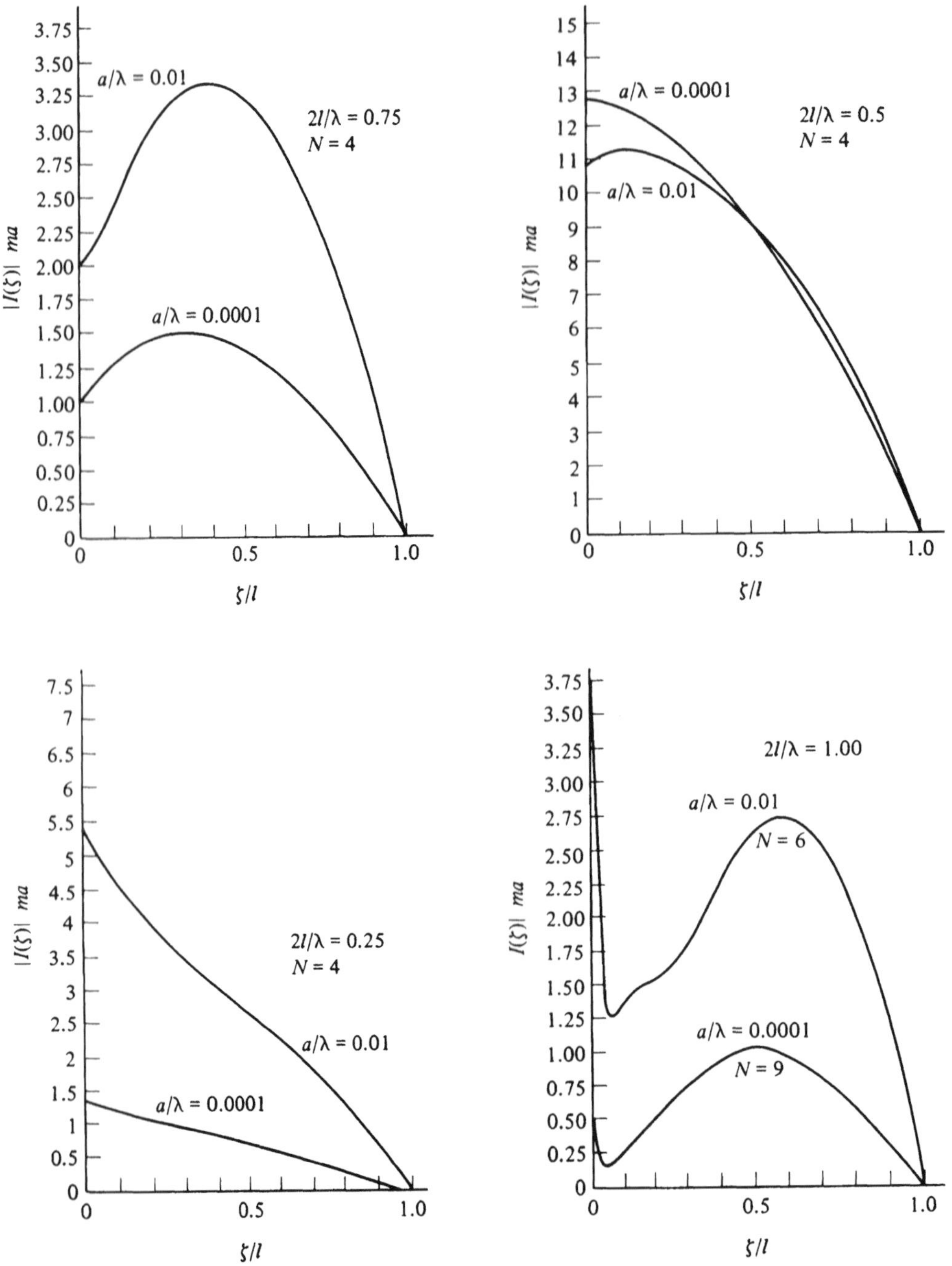

Fig. 7.10 The Magnitude of $I(z)$ for Center-Fed Cylindrical Dipoles of Various Lengths and Diameters; Solution Using Sinusoidal Basis Functions

7.8 Self-Impedance of Center-Fed Cylindrical Dipoles: Induced EMF Method

The method of moments, as illustrated in the previous two sections, is a useful computational technique when applied to the problem of determining the current distribution on a cylindrical dipole. This can in principle be done with increased accuracy by choosing more basis functions and more matching points, but at greater computer cost, with the amount of computer storage available ultimately setting a level on the accuracy. A complexity which arises is that, for a given a/λ, as the interval is more finely divided (thus increasing the number of basis functions and matching points needed) (7.25) becomes a poorer approximation to (7.24). The ultimate solution is to apply the method of moments directly to (7.24). But this is a much costlier computer operation. Additionally, the method of moments as applied here delivers the entire current distribution. If one is only interested in the *input* current (to obtain the self-impedance), it is advantageous to carry the analysis further before embarking on a computational program. For this reason attention will now be turned to techniques which focus exclusively on the problem of finding the input current.

The first of these techniques is the so-called induced EMF method, introduced by L. Brillouin[13] in 1922 and elaborated by A. A. Pistolkors[14] and P. S. Carter[15]. It involves a self-impedance formula which can be derived with the aid of the reciprocity theorem. Consider again the cylindrical dipole of length $2l$ and radius a, center-fed across a gap of infinitesimal height 2δ by an idealized generator, as suggested in Figure 7.2. Let a surface S be constructed that would just enclose this dipole without touching it. Then S consists of a section of a circular cylinder of radius $a + \epsilon_1$, capped by circular discs at $z = \pm(l + \epsilon_2)$, with ϵ_1 and ϵ_2 infinitesimals.

As before, the arms of the dipole will be assumed to consist of perfectly conducting tubular material of negligible wall thickness. Let $K_z(z)$ be the lineal current density distribution along the dipole when the generator is adjusted so that one volt exists across the gap. (The distribution $K_z(z)$ includes the current which flows through the generator.) If the perfectly conducting tubes are removed and a source distribution $K_z^a(z) = K_z(z)$ is established in free space in the exact location of the previous source system, the fields caused by the original system will be duplicated at every point by the new system.

Next, let $I^b(z)$, with $z \in [-l, l]$, be any filamentary current distribution along the axis of the cylinder S, also established in free space. Since the two source systems $K_z^a(z)$ and $I^b(z)$ are both entirely contained within S, the reciprocity theorem in the form of Equation 1.36 applies, and one can write

[13] L. Brillouin, "Origin of Radiation Resistance", *Radioélectricité*, 3 (1922), 147–52.

[14] A. A. Pistolkors, "Radiation Resistance of Beam Antennae", *Proc. IRE*, 17 (1929), 562–79.

[15] P. S. Carter, "Circuit Relations in Radiating Systems and Applications to Antenna Problems", *Proc. IRE*, 20 (June 1932) 1004–41.

$$\int_S (\mathbf{E}^a \times \mathbf{H}^b - \mathbf{E}^b \times \mathbf{H}^a) \cdot d\mathbf{S} = 0 \tag{7.49}$$

in which $(\mathbf{E}^a, \mathbf{H}^a)$ is the electromagnetic field caused by $K_z^a(z)$ and $(\mathbf{E}^b, \mathbf{H}^b)$ is the electromagnetic field caused by $\mathrm{I}^b(z)$.

Since $\mathbf{E}_{\text{tang}}^a$ and $\mathbf{E}_{\text{tang}}^b$ are both Z-directed along the cylindrical portion of S, if one neglects the contributions to (7.49) from the disclike end surfaces (justifiable when $a \ll \lambda$), the result is that

$$\int_{-l}^{l} \int_0^{2\pi} [E_z^a(a, z')H_\phi^b(a, z') - E_z^b(a, z')H_\phi^a(a, z')]a \, d\phi \, dz' = 0 \tag{7.50}$$

Because $E_z^a(a, z')$ is a Dirac delta function (compare with Equations 7.12 and 7.13) and because all quantities in (7.50) are ϕ-independent, integration gives

$$2\pi a H_\phi^b(a, 0) = -\int_{-l}^{l} E_z^b(a, z')H_\phi^a(a, z')2\pi a \, dz' \tag{7.51}$$

However, $H_\phi^a(a, z') = K_z^a(z')$, and $2\pi a K_z^a(z') = I^a(z')$, with $I^a(z')$ the total current distribution for the dipole. These substitutions convert (7.51) to

$$2\pi a H_\phi^b(a, 0) = -\int_{-l}^{l} E_z^b(a, z')I^a(z') \, dz' \tag{7.52}$$

For the disclike surface of radius a which lies in the XY-plane and is centered at the origin, the integral form of the appropriate Maxwell equation gives

$$\oint_C \mathbf{H}^b \cdot d\mathbf{l} = \int_S \mathbf{J}^b \cdot d\mathbf{S} + j\omega\epsilon_0 \int_S \mathbf{E}^b \cdot d\mathbf{S}$$

$$2\pi a H_\phi^b(a, 0) = I^b(0) + 2\pi j\omega\epsilon_0 \int_0^a E_z^b(\rho, 0)\rho \, d\rho \tag{7.53}$$

For $a \ll \lambda$ and $a \ll l$, the second term on the right side of (7.53) is negligible compared to $I^b(0)$ and one can write $2\pi a H_\phi^b(a, 0) \cong I^b(0)$. When this approximation is placed in (7.52), the result is that

$$I^b(0) = -\int_{-l}^{l} E_z^b(a, z')I^a(z') \, dz' \tag{7.54}$$

Since $I^b(z)$ is completely arbitrary, one is at liberty to let $I^b(z) = I^a(z)$. When this is done, there is no longer any need to retain the superscripts, and (7.54) becomes

$$I(0) = -\int_{-l}^{l} E_z(a, z')I(z') \, dz' \tag{7.55}$$

Finally, one can write for the self-impedance

$$Z = \frac{V}{I(0)} = \frac{VI(0)}{I^2(0)} = -\frac{1}{I^2(0)}\int_{-l}^{l} E_z(a, z')I(z')\,dz' \tag{7.56}$$

since $V = 1$ volt.

Equation 7.56 is a key result of the induced EMF method for determining the self-impedance of a center-fed cylindrical dipole. It is a peculiar result in some respects, since it contains $I^2(0)$ in the denominator, and one could argue that if $I(0)$ is known the problem is solved, since $I(0)$ is numerically equal to the input admittance. However, that line of reasoning is based on the logic sequence that one volt is applied across the gap and that the problem is to find the resulting current distribution, or at least the input current. The reasoning that is used in the induced EMF method is almost the reverse. One *assumes* a current distribution for the dipole and thus "knows" $I(0)$. One then computes $E_z(a, z')$ in response to this current distribution and uses this $E_z(a, z')$ in the integrand of (7.56), together with the assumed $I(z')$, in order to compute the self-impedance Z. Quite obviously, the accuracy of the value computed for Z depends on the quality of the assumption for $I(z')$.

An alternate derivation for the central equation of the induced EMF method is based on power relations. With the aid of Poynting's theorem, one can argue that

$$\tfrac{1}{2}I(0)I^*(0)Z = \tfrac{1}{2}\oint_S \mathbf{E} \times \mathbf{H}^* \cdot d\mathbf{S} \tag{7.57}$$

With power flow across the end caps of the dipole ignored, this becomes

$$|I(0)|^2 Z = -\int_{-l}^{l}\int_0^{2\pi} E_z(a, z')H_\phi^*(a, z')a\,d\phi\,dz'$$

$$= -\int_{-l}^{l} E_z(a, z')I^*(z')\,dz'$$

from which the self-impedance is given by

$$Z = -\frac{1}{|I(0)|^2}\int_{-l}^{l} E_z(a, z')I^*(z')\,dz' \tag{7.58}$$

If a real trial function is chosen for $I(z)$, Formulas 7.56 and 7.58 give the same result for Z. C. T. Tai[16] discusses the implications of the differences in these two formulas. The developments in this text will be based on (7.56).

It was shown in Section 7.4 that, for $a \ll \lambda$ and $a \ll l$, $E_z(a, z)$ can be determined from

$$j\omega\epsilon_0 E_z(a, z) = \left(\frac{d^2}{dz^2} + k^2\right)\int_{-l}^{l} \frac{e^{-jkr}}{4\pi r} I(z')\,dz' \tag{7.59}$$

[16]C. T. Tai, "A Variational Solution to the Problems of Cylindrical Antennas" Technical Report No. 12, (Palo Alto, CA: Stanford Research Institute, August, 1950). See also his article "A Study of the EMF Method" *Jour. App. Phys.*, 20 (1949), 717–23.

in which $r = [a^2 + (z - z')^2]^{1/2}$. If one defines a kernel function by

$$G(z, z') = -\frac{1}{4\pi j\omega\epsilon_0}\left(\frac{\partial^2}{\partial z^2} + k^2\right)\frac{e^{-jkr}}{r} \tag{7.60}$$

then (7.59) can be written in the compact form

$$E_z(a, z) = -\int_{-l}^{l} G(z, z')I(z')\,dz' \tag{7.61}$$

When this result is placed in (7.56), one obtains

$$Z = \frac{1}{I^2(0)}\int_{-l}^{l}\int_{-l}^{l} G(\zeta, \zeta')I(\zeta)I(\zeta')\,d\zeta\,d\zeta' \tag{7.62}$$

which is a formula for the self-impedance of a center-fed dipole that is particularly suitable as the input to a computer program.

When the induced EMF method was first introduced, it was natural to assume that the current distribution was sinusoidal, since techniques did not yet exist for determining the distribution more accurately. The results of Section 7.6, 7.7 indicate that this is not at all a bad assumption. But early workers also assumed initially that the radius a of the dipole had negligible effect on the input impedance and used $r = |z - z'|$ in (7.60), thus in effect setting $a = 0$. This proved to be a valid assumption insofar as computing the real part of Z in the range $0 \leq 2l/\lambda \leq 0.6$, but gave an infinite value for the imaginary part of Z except for the particular lengths $2l/\lambda = (2n + 1)/2$, with n an integer. Thus one should avoid this simplification and use $r = [a^2 + (z - z')^2]^{1/2}$ in (7.60). This presents no difficulties for a modern electronic computer.

When it is assumed that

$$I(\zeta) = I_m \sin k(l - |\zeta|) \tag{7.63}$$

Equation 7.62 becomes

$$Z = \frac{1}{\sin^2 kl}\int_{-l}^{l}\int_{-l}^{l} G(\zeta, \zeta')\sin[k(l - |\zeta|)]\sin[k(l - |\zeta'|)]\,d\zeta\,d\zeta' \tag{7.64}$$

By performing the differentiation indicated in (7.60) and then putting the expanded form for G in (7.64), one is able to show that[17]

$$Z = \frac{j60}{\sin^2 kl}\{4\cos^2 kl \cdot S(kl) - \cos 2kl \cdot S(2kl) - \sin 2kl[2C(kl) - C(2kl)]\} \tag{7.65}$$

in which

$$C(ky) = ln\frac{2y}{a} - \frac{1}{2}\mathrm{Cin}(2ky) - \frac{j}{2}\mathrm{Si}(2ky) \tag{7.66}$$

[17]See, for example, E. C. Jordan and K. G. Balmain, *Electromagnetic Waves and Radiating Systems*, 2nd Ed. (Englewood Cliffs, New Jersey: Prentice-Hall, 1968), pp. 540–47.

$$S(ky) = \frac{1}{2}\,\mathrm{Si}(2ky) - \frac{j}{2}\,\mathrm{Cin}(2ky) - ka \tag{7.67}$$

with Si(x) and Cin(x) tabulated functions.[18] The function Si(x) is called the *sine integral*, and is defined by

$$\mathrm{Si}(x) = \int_0^x \frac{\sin u}{u}\,du \tag{7.68}$$

whereas the function Cin(x) is sometimes called the *modified cosine integral* and is given by

$$\mathrm{Cin}(x) = \int_0^x \frac{1 - \cos u}{u}\,du \tag{7.69}$$

The real and imaginary parts of the self-impedance of a center-fed cylindrical dipole, as computed from (7.65), are plotted versus $2l/\lambda$ in Figure 7.11 for a sequence of a/λ values. The single resistance plot reflects the fact that the real part of (7.65) is independent of a/λ. (This would not be true if a current distribution more complicated than a pure sinusoid were used in (7.62).) The reactance *is* seen to be sensitive to a/λ, and one can note that resonance ($X = 0$) occurs at shorter lengths as a/λ is increased. Also, the X-curves are more gently sloped for larger values of a/λ. Fat dipoles are less frequency sensitive than skinny dipoles.

The curves of Figure 7.11 are in a useful form if one wishes to find, for a given a/λ, the length needed to produce a dipole impedance with a specified reactance. In Chapter 8, the design of dipole arrays will be seen to involve such deductions. However, another useful form in which to present Equation 7.65 graphically results from fixing $2l/a$ and then finding $Z(kl)$. The conventional method for doing this is to define a parameter Ω by the equation

$$\Omega = 2 ln(2l/a)$$

and then to plot $Z(kl)$ for a fixed Ω. Figure 7.12 gives a family of curves covering the practical range of Ω values. These curves are useful if one wishes to determine the behavior of self-impedance with frequency for a specific dipole ($2l$ and a fixed). (Since the real part of Z in (7.65) is independent of a/λ, $R(kl)$ as it appears in Figure 7.12 is merely a replotting of part of Figure 7.11 to a logarithmic scale. This is done for later comaprison with the results of Hallén and King.)

C. T. Tai[19] has shown that the values of Z computed from (7.65) are fitted extremely well in the range $0 \leq 2l/\lambda \leq \pi/2$ by the expression

$$Z = R(kl) - j\left[120\left(ln\frac{2l}{a} - 1\right)\cot kl - X(kl)\right] \tag{7.70}$$

[18] See, for example, R. W. P. King, *The Theory of Linear Antennas*, (Cambridge, Massachusetts: Harvard University Press, 1956), pp. 857–64.

[19] C. T. Tai, "Characteristics of Linear Antenna Elements," *Antenna Engineering Handbook*, ed. H. Jasik (New York: McGraw-Hill Book Co., Inc., 1961), Chapter 3.

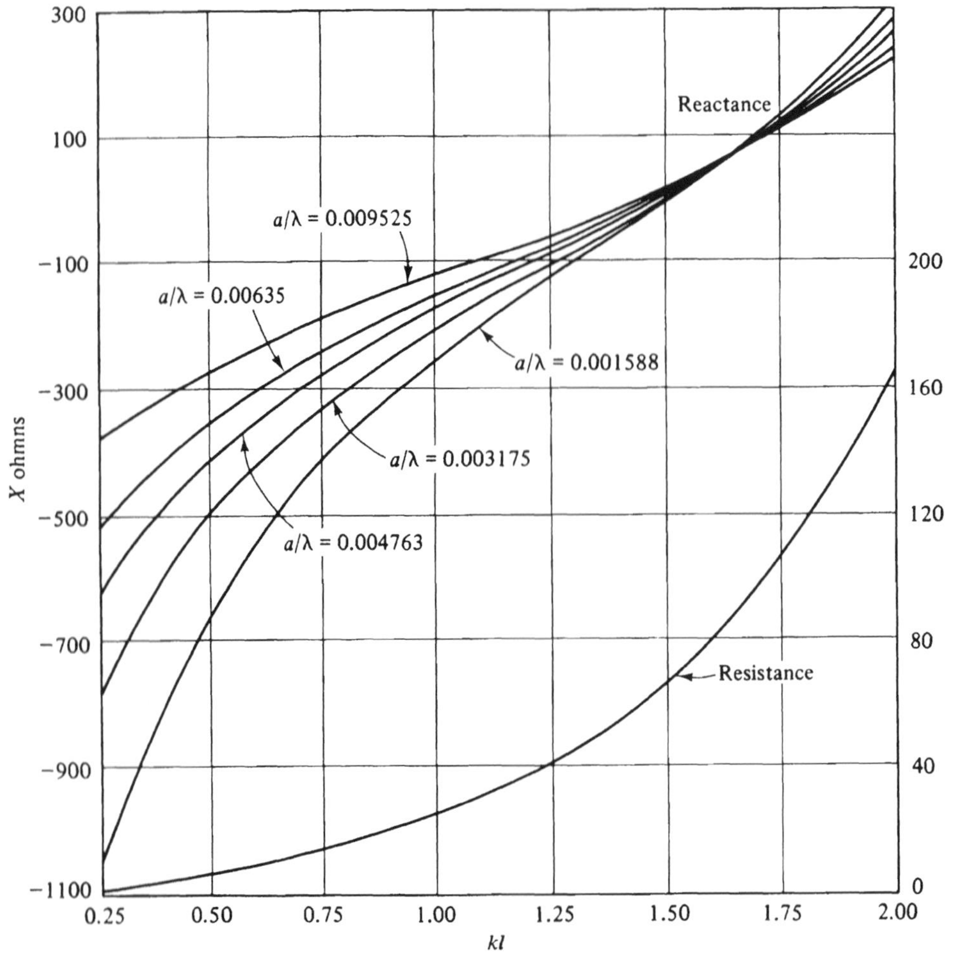

Fig. 7.11 The Resistance and Reactance of a Center-Fed Dipole versus kl and a/λ; Values Computed by the Induced EMF Method Using Equation 7.65

with $R(kl)$ and $X(kl)$ smooth, simple functions which he tabulates and graphs. If one represents Tai's functions $R(kl)$ and $X(kl)$ by second-degree polynomials with coefficients chosen to fit data computed from (7.65) in the range $1.3 \leq kl \leq 1.7$ and $0.001588 \leq a/\lambda \leq 0.009525$, Equation 7.70 takes on the specific form

$$Z = [122.65 - 204.1kl + 110(kl)^2] - j\left[120\left(ln\frac{2l}{a} - 1\right)\cot kl - 162.5 + 140kl - 40(kl)^2\right] \tag{7.71}$$

For the specified range of dipole lengths and diameters, the real part of (7.71) does

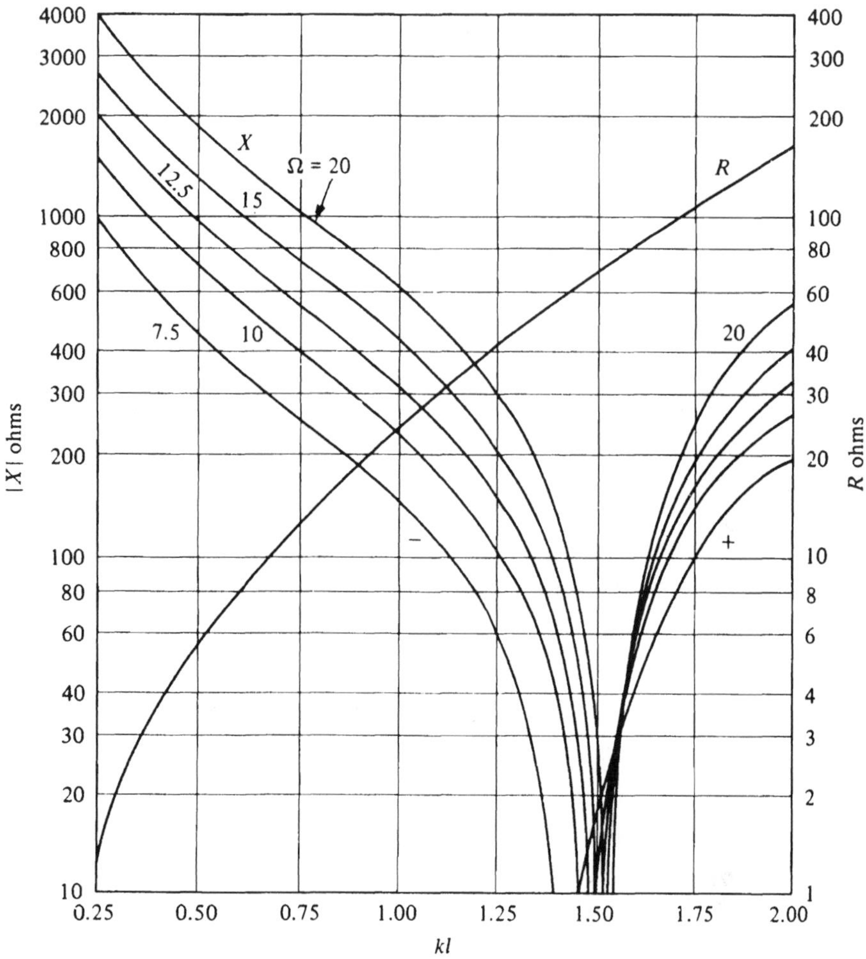

Fig. 7.12 The Resistance and Reactance of a Center-Fed Dipole versus kl and Ω; Values Computed by the Induced EMF Method Using Equation 7.65

not deviate from the induced EMF impedance expression in (7.65) by more than 0.42 ohms, with an rms error of 0.14 ohms. The imaginary part of (7.71) stays within 2.33 ohms of (7.65), with an rms error of 0.20 ohms. Equation 7.71, which can be used with a pocket calculator, is a much simpler formula to use than is (7.65).

The resonant length l_r of the center-fed cylindrical dipole can be deduced from (7.65) by setting the reactance equal to zero. When this is done, one finds a relation between $2l_r/\lambda$ and a/λ which, when plotted, appears as shown in Figure 7.13.

The resonant resistance of a cylindrical dipole is also of some interest and can be found easily by inserting kl_r in the real part of (7.65). This results in the curve shown in Figure 7.14. One can observe that the resonant resistance is in the neighborhood of 73 ohms for very thin dipoles, but falls off from this value steadily as a/λ is

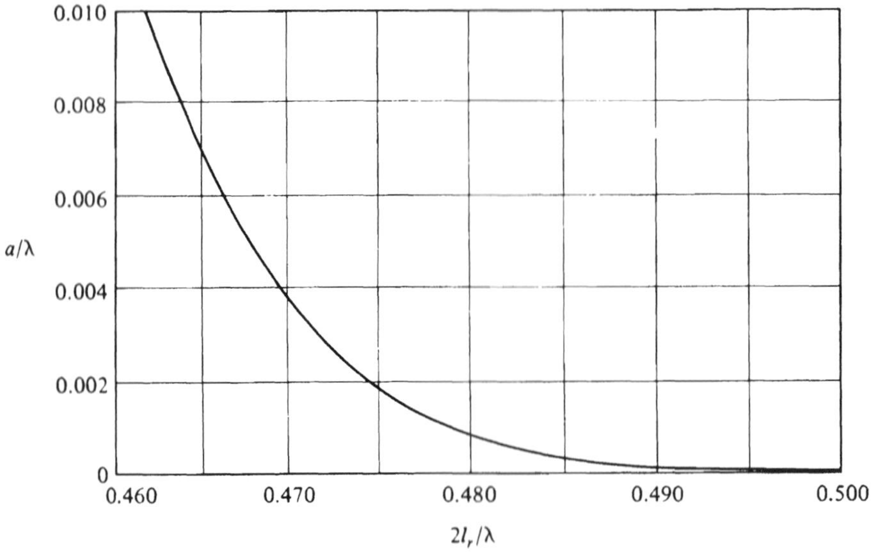

Fig. 7.13 Resonant Length versus Radius for Center-Fed Cylindrical Dipoles

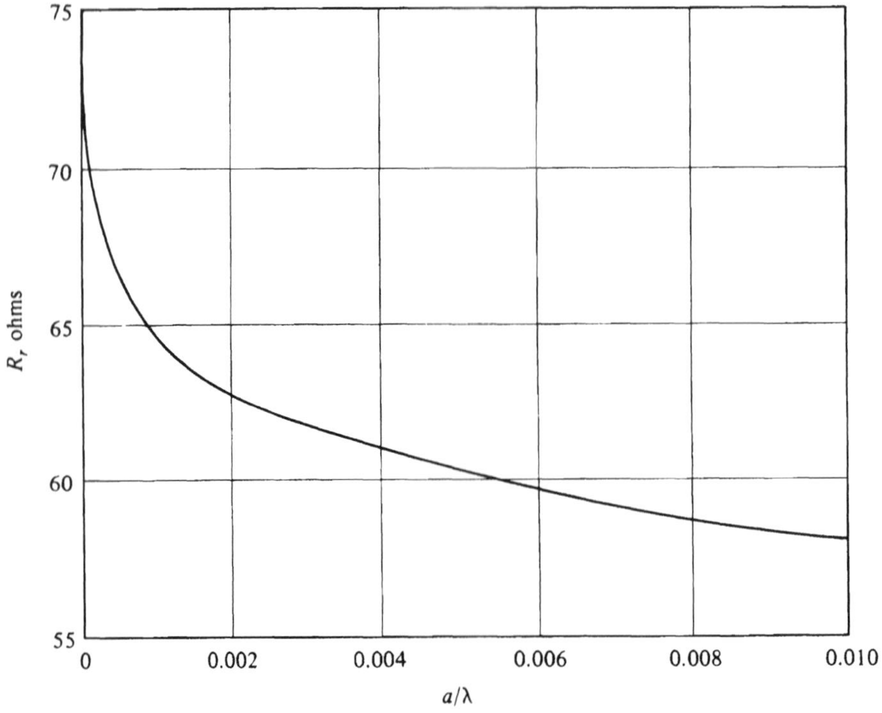

Fig. 7.14 Resonant Resistance versus Radius for Center-Fed Cylindrical Dipoles

increased. The limiting figure of 73 ohms agrees with the value obtained by power pattern integration in Section 2.2.

7.9 Self-Impedance of Center-Fed Cylindrical Dipoles: Storer's Variational Solution

A refinement of the induced EMF method has been provided by J. E. Storer[20] and rests on the stationary property of Equation 7.62. The reader will recall that this equation permits computation of a value for the self-impedance Z of a center-fed cylindrical dipole once an assumption has been made about the current distribution $I(\zeta)$. It is a pleasant fact that Z, as given by (7.62), is stationary with respect to variations in $I(\zeta)$. This means that if the true current distribution is $\overset{0}{I}(\zeta)$, if the true input impedance is $\overset{0}{Z}$, and if one uses a trial current distribution $I(\zeta) = \overset{0}{I}(\zeta) + \delta I(\zeta)$ in (7.62), then one will compute an incorrect impedance $Z = \overset{0}{Z} + \delta Z$, but that to first order $\delta Z = 0$. In other words, a certain degree of inaccuracy in "guessing" the current distribution will result in a much smaller degree of inaccuracy in the computed value of the input impedance.

This assertion of the stationary nature of (7.62) is of sufficient importance to call for a proof. For the trial current distribution $I(\zeta)$ and the true current distribution $\overset{0}{I}(\zeta)$, Equation 7.62 takes the forms

$$Z = \overset{0}{Z} + \delta Z = \frac{1}{[\overset{0}{I}(0) + \delta I(0)]^2} \int_{-l}^{l} \int_{-l}^{l} G(\zeta, \zeta')[\overset{0}{I}(\zeta) + \delta I(\zeta)][\overset{0}{I}(\zeta') + \delta I(\zeta')]\, d\zeta\, d\zeta' \tag{7.72}$$

$$\overset{0}{Z} = \frac{1}{\overset{0}{I}{}^2(0)} \int_{-l}^{l} \int_{-l}^{l} G(\zeta, \zeta')\overset{0}{I}(\zeta)\overset{0}{I}(\zeta')\, d\zeta\, d\zeta' \tag{7.73}$$

The difference between these two equations is

$$\begin{aligned} \delta Z = {} & \frac{1}{\overset{0}{I}{}^2(0)} \left\{1 - 2\frac{\delta I(0)}{\overset{0}{I}(0)} + 3\left[\frac{\delta I(0)}{\overset{0}{I}(0)}\right]^2 - \cdots\right\} \int_{-l}^{l} \int_{-l}^{l} G(\zeta, \zeta')[\overset{0}{I}(\zeta)\delta I(\zeta') \\ & \qquad + \overset{0}{I}(\zeta')\delta I(\zeta) + \delta I(\zeta) \cdot \delta I(\zeta')] d\zeta\, d\zeta' \\ & + \frac{1}{\overset{0}{I}{}^2(0)} \left\{-2\frac{\delta I(0)}{\overset{0}{I}(0)} + 3\left[\frac{\delta I(0)}{\overset{0}{I}(0)}\right]^2 - \cdots\right\} \int_{-l}^{l} \int_{-l}^{l} G(\zeta, \zeta')\overset{0}{I}(\zeta)\overset{0}{I}(\zeta') d\zeta\, d\zeta' \end{aligned} \tag{7.74}$$

If one returns to Equation 7.61, which applies for any current distribution, including the true one, multiplication of both sides by $\delta I(\zeta)\, d\zeta$, followed by integration, gives

[20] J. E. Storer, "Variational Solution to the Problem of the Symmetrical Cylindrical Antenna," Cruft Laboratory Report No. 101 (Harvard University, 1950).

$$\int_{-l}^{l} E_z(a, \zeta)\delta I(\zeta)\, d\zeta = -V\delta I(0) = -\overset{0}{Z}\overset{0}{I}(0)\delta I(0)$$

$$= -\int_{-l}^{l}\int_{-l}^{l} G(\zeta, \zeta')\overset{0}{I}(\zeta')\delta I(\zeta)\, d\zeta\, d\zeta' \qquad (7.75)$$

Since $G(\zeta, \zeta')$ is symmetrical, it follows that (7.75) is also valid if ζ and ζ' are interchanged. Substitution of both forms of (7.75) in (7.74) results in

$$\delta Z = \frac{1}{\overset{0}{I^2}(0)}\left\{1 - \frac{\delta I(0)}{\overset{0}{I}(0)} + 3\left[\frac{\delta I(0)}{\overset{0}{I}(0)}\right]^2 - \cdots\right\}\int_{-l}^{l}\int_{-l}^{l} G(\zeta, \zeta')\delta I(\zeta)\delta I(\zeta')\, d\zeta\, d\zeta'$$
$$+ \frac{1}{\overset{0}{I^2}(0)}\left\{-\left[\frac{\delta I(0)}{\overset{0}{I}(0)}\right]^2 + 2\left[\frac{\delta I(0)}{\overset{0}{I}(0)}\right]^3 - \cdots\right\}\int_{-l}^{l}\int_{-l}^{l} G(\zeta, \zeta)\overset{0}{I}(\zeta)\overset{0}{I}(\zeta')\, d\zeta\, d\zeta' \qquad (7.76)$$

If all terms in δI which are second order or smaller are ignored in (7.76), the right side reduces to zero, which is to say that $\delta Z = 0$ to first order in variations in the current distribution, as has been alleged. For this reason, Formula 7.62 is attractive for the purpose of computing input impedance; it is somewhat forgiving of imprecise knowledge about the current distribution. Indeed, the stationary property of (7.62) serves to explain why the induced EMF method gives such satisfactory results.

Storer has elected to attempt to improve on the results of the induced EMF method by assuming that the current distribution can be expressed in the form

$$I(\zeta) = Af_1(\zeta) + Bf_2(\zeta) \qquad (7.77)$$

where

$$f_1(\zeta) = \sin[k(l - |\zeta|)] \qquad (7.78)$$

$$f_2(\zeta) = 1 - \cos[k(l - |\zeta|)] \qquad (7.79)$$

The particular selection of $f_1(\zeta)$ is justified by the knowledge that, in the limit as $a/\lambda \longrightarrow 0$, the current distribution becomes truly sinusoidal. The form of $f_2(\zeta)$ is cusplike and permits an even perturbation on $f_1(\zeta)$, with either bulging or indenting near the middle of the interval $[-l, l]$.

C. T. Tai[21] has pointed out that Storer's choice for $f_2(\zeta)$ is only useful in the range $l < \lambda$ and proposes instead the function

$$f_2(\zeta) = k(l - |\zeta|) \cos k(l - |\zeta|) \qquad (7.80)$$

which is applicable for all values of l/λ. The trial current distribution (7.77) could also be enlarged to consist of the linear sum of three or more functions. However, since Storer's choice of (7.77) through (7.79) is valid in the length interval of principal interest and is fully illustrative of the method, what follows will be based on his formulation.

[21] C. T. Tai, "A New Interpretation of the Integral Equation Formulation of Cylindrical Antennas," *IRE Trans. Antennas and Propagat.*, AP-3, (1955), 125–27.

When (7.77) is placed in the impedance expression of (7.62), one finds that

$$Z = \frac{A^2\gamma_{11} + 2AB\gamma_{12} + B^2\gamma_{22}}{[Af_1(0) + Bf_2(0)]^2} \tag{7.81}$$

in which

$$\gamma_{mn} = \int_{-l}^{l}\int_{-l}^{l} G(\zeta, \zeta')f_m(\zeta)f_n(\zeta')\,d\zeta\,d\zeta' \tag{7.82}$$

Because of the stationary nature of the impedance expression, if B is held fixed and A is changed by an amount δA, to first order, δZ will be zero. Or, what is the same thing,

$$\frac{\partial Z}{\partial(A/B)} = 0 \tag{7.83}$$

If this differentiation is applied to (7.81), the result is that

$$\frac{A}{B} = \frac{\gamma_{22}f_1(0) - \gamma_{12}f_2(0)}{\gamma_{11}f_2(0) - \gamma_{12}f_1(0)} \tag{7.84}$$

Substitution of this result in (7.81) gives

$$Z = \frac{\gamma_{11}\gamma_{22} - \gamma_{12}^2}{\gamma_{11}[f_2(0)]^2 - 2\gamma_{12}f_1(0)f_2(0) + \gamma_{22}[f_1(0)]^2} \tag{7.85}$$

Storer found that for his choice of trial function, the double integral (7.82) could be expressed in terms of sine and cosine integrals. The interested reader is referred to the original report. Calculations of Z from (7.85) for the eight cases $a/\lambda = 0.01$, 0.0001 and $2l/\lambda = 0.25$, 0.50, 0.75, 1.00 are listed in the fourth column of Table 7.3.

If one chooses the simpler trial function $I(\zeta) = Af_1(\zeta)$, Equation 7.85 gives

$$Z = \frac{\gamma_{11}}{\sin^2 kl} \tag{7.86}$$

This result is identical with (7.64), as it should be, and the values of Z for the eight cases under study, as computed from (7.86), are listed in the third column of Table 7.3. The values shown in the third and fourth columns are quite close for short dipoles, but they begin to deviate from each other as the dipole length is increased. In particular, Storer's formula gives a resistive component of Z which is dependent on a/λ, unlike the result obtained using the induced EMF method. Storer's formula also gives a finite impedance for $2l/\lambda = 1$.

The reader may wish to compare the entries in Table 7.3 to the earlier results arising from use of the method of moments, and listed in Table 7.2.

7.10 Self-Impedance of Center-Fed Cylindrical Dipoles: Zeroth and First Order Solutions to Hallén's Integral Equation

The previous two sections have dealt with the computation of the input impedance of a cylindrical dipole using a formula derived with the aid of the reciprocity theorem. That formula, (7.62), was seen to be stationary with respect to variations in the trial function chosen to represent the current distribution. The induced EMF method consisted of using a one-term sinusoidal function in (7.62) to approximate the current distribution. Storer's variational method permitted a linear sum of arbitrary known functions to be used for the current distribution, with the relative levels of these known functions determinable because of the stationary nature of (7.62).

TABLE 7.3 Approximations to the input impedance of a center-fed cylindrical dipole using Storer's variational formulation

Normalized Length $2l/\lambda$	Normalized Radius a/λ	Input Impedance: Induced EMF Method	Input Impedance: Two-Term Trial Function
0.25	0.01	$13.44 - j185.75$	$11.63 - j184.86$
0.25	0.0001	$13.44 - j723.45$	$12.93 - j722.62$
0.50	0.01	$73.13 + j38.78$	$101.13 + j32.82$
0.50	0.0001	$73.13 + j42.51$	$80.15 + j42.61$
0.75	0.01	$371.62 + j502.35$	$565.84 + j3.10$
0.75	0.0001	$371.62 + j1069.90$	$521.15 + j1019.24$
1.00	0.01	∞	$290.13 - j363.46$
1.00	0.0001	∞	$2370.31 - j2128.60$

A fundamentally different approach to this problem has been pioneered by E. Hallén[22] and exploited extensively by R. W. P. King and his co-workers.[23] Hallén's development, up to the establishment of his basic integral equation (7.25), has already been traced in Section 7.4. That equation, which links the unknown current distribution on the cylindrical dipole to a Dirac delta function distribution of longitudinal electric field along the dipole, was solved earlier in this chapter using the method of moments to determine the current distribution (compare with Sections 7.5 through 7.7). If one is interested in obtaining the input impedance without finding the entire current distribution in the process, Hallén's integral equation (7.25) can be manipulated to accomplish this.

The development begins with the addition and subtraction of a supplementary integral to (7.25), which will serve to convert it to a Fredholm integral equation of the second kind.

[22]Hallén, "Investigations into Transmitting and Receiving Qualities of Antennas."
[23]King, *Theory of Linear Antennas*, Chapter 2.

$$C \cos kz - \frac{j}{2\eta} \sin k|z| = \int_{-l}^{l} \frac{I(z')e^{-jkr} - I(z)}{4\pi r} dz' + \int_{-l}^{l} \frac{I(z)}{4\pi r} dz' \tag{7.87}$$

in which $r = [a^2 + (z - z')^2]^{1/2}$. Integration of the last term in (7.87) gives

$$\int_{-l}^{l} \frac{I(z)}{4\pi r} dz' = \frac{I(z)}{4\pi} \ln \frac{z + l + [a^2 + (z + l)^2]^{1/2}}{z - l + [a^2 + (z - l)^2]^{1/2}} \tag{7.88}$$

If the factor for which the logarithm is being computed in (7.88) is multiplied by the unitary ratio

$$\frac{z - l - [a^2 + (z - l)^2]^{1/2}}{z - l - [a^2 + (z - l)^2]^{1/2}} \cdot \frac{4l^2}{4l^2} \cdot \frac{l^2 - z^2}{(l - z)(l + z)} \cdot \frac{a^2}{a^2} \tag{7.89}$$

some rearrangenent leads to

$$\int_{-l}^{l} \frac{I(z)}{4\pi r} dz' = \frac{I(z)}{4\pi} \left\{\Omega + \ln\left[1 - \left(\frac{z}{l}\right)^2\right] + \Delta\right\} \tag{7.90}$$

in which

$$\Omega = 2 \ln \frac{2l}{a} \tag{7.91}$$

and

$$\Delta = \ln\left[\frac{1}{4}\left\{1 + \left[1 + \frac{a^2}{(l - z)^2}\right]^{1/2}\right\}\left\{1 + \left[1 + \frac{a^2}{(l + z)^2}\right]^{1/2}\right\}\right] \tag{7.92}$$

Ω, which can be called the *slenderness index*, has already been encountered in Section 7.8 in connection with the construction of Figure 7.12. It will be seen to be a measure of the rate of convergence of the iterative procedure that will be introduced shortly.

The placement of (7.90) in (7.87) gives

$$\begin{aligned} I(z) = \frac{4\pi}{\Omega}\bigg[&C \cos kz - \left(\frac{j}{2\eta}\right) \sin k|z|\bigg] \\ &- \frac{1}{\Omega}\left[I(z)\left\{\ln\left[1 - \left(\frac{z}{l}\right)^2\right] + \Delta(z)\right\} + \int_{-l}^{l} \frac{I(z')e^{-jkr} - I(z)}{r} dz'\right] \end{aligned} \tag{7.93}$$

Since

$$\lim_{z \to l} \left\{\ln\left[1 - \left(\frac{z}{l}\right)^2\right] + \Delta\right\} = \ln\left\{\frac{a}{2l}\left(1 + \left[1 + \left(\frac{a}{2l}\right)^2\right]^{1/2}\right)\right\} \tag{7.94}$$

is finite, it follows from (7.93) that

$$\begin{aligned} I(l) = 0 = \frac{4\pi}{\Omega}&\left[C \cos kl - \left(\frac{j}{2\eta}\right) \sin kl\right] \\ &- \frac{1}{\Omega} \int_{-l}^{l} \frac{I(z')e^{-jkr'}}{r'} dz' \end{aligned} \tag{7.95}$$

in which $r' = [a^2 + (l - z')^2]^{1/2}$.

An expression for $I(z)$ which insures a null current at $z = l$ results when (7.95) is subtracted from (7.93). One obtains

$$I(z) = \frac{4\pi}{\Omega} H_0(z) - \frac{I(z)}{\Omega}\left\{\ln\left[1 - \left(\frac{z}{l}\right)^2\right] + \Delta(z)\right\} - \frac{1}{\Omega}\left[\int_{-l}^{l} \frac{I(z')e^{-jkr} - I(z)}{r}\,dz' - \int_{-l}^{l} \frac{I(z')e^{-jkr'}}{r'}\,dz'\right] \tag{7.96}$$

in which

$$H_0(z) = CF_0(z) - \frac{j}{2\eta} G_0(z) \tag{7.97}$$

$$F_0(z) = \cos kz - \cos kl \tag{7.98}$$

$$G_0(z) = \sin k\,|z| - \sin kl \tag{7.99}$$

A careful study of (7.96) reveals that, except in the neighborhood of $z = l$, $ln[1 - (z/l)^2]$ and $\Delta(z)$ are both small, while near $z = l$, their sum is approximately $ln(a/l)$. Since $I(z)$ itself is small near $z = l$, the second term on the right side of (7.96) is dominated by $(4\pi/\Omega)H_0(z)$. So, too, is the difference of the two integrals. Thus an initial estimate of the current distribution on the cylindrical dipole, called the *zeroth order approximation*, is

$$I_0(z) = \frac{4\pi}{\Omega} H_0(z) \tag{7.100}$$

which, by virtue of (7.97) through (7.99), is seen to be a spatially sinusoidal distribution. This is consistent with the findings of Sections 7.6 and 7.7, where it was discovered (particularly for a/λ small) that moment method solutions were quasi-harmonic.

Successive approximations to (7.96) can be obtained by an iterative procedure. Let the first order solution $I_1(z)$ be generated by a modification of (7.96) with the modification consisting of the replacement of $I(z)$ by $I_0(z)$ in the right side of (7.96). That is, let

$$I_1(z) = \frac{4\pi}{\Omega} H_0(z) - \frac{4\pi}{\Omega^2} H_0(z)\left\{\ln\left[1 - \left(\frac{z}{l}\right)^2\right] + \Delta(z)\right\} - \frac{4\pi}{\Omega^2}\left[\int_{-l}^{l} \frac{H_0(z')e^{-jkr} - H_0(z)}{r}\,dz' - \int_{-l}^{l} \frac{H_0(z')e^{-jkr'}}{r'}\,dz'\right] \tag{7.101}$$

If the function $H_1(z)$ is defined by

$$H_1(z) = -H_0(z)\left\{\ln\left[1 - \left(\frac{z}{l}\right)^2\right] + \Delta(z)\right\} - \int_{-l}^{l} \frac{H_0(z')e^{-jkr} - H_0(z)}{r}\,dz' \tag{7.102}$$

then the expression for $I_1(z)$ can be written in the form

$$I_1(z) = \frac{4\pi}{\Omega} H_0(z) + \frac{4\pi}{\Omega^2}[H_1(z) - H_1(l)] \tag{7.103}$$

The current distribution $I_1(z)$ is seen to be comprised of $I_0(z)$ plus correction terms, which rank at a level Ω^{-1} compared to $I_0(z)$.

Similarly, a second-order approximation $I_2(z)$ can be generated by using $I_1(z)$ for $I(z)$ in the right side of (7.96). This second-order approximation contains additional correction terms which rank at a level Ω^{-2} compared to $I_0(z)$.

Proceeding in this manner, one can generate an nth-order approximation to the current distribution, $I_n(z)$, which contains correction terms at the levels $\Omega^{-1}, \Omega^{-2}, \ldots, \Omega^{-n}$ relative to $I_0(z)$. Since the value of the slenderness index Ω is typically 10 or greater, a sequence of approximations obtained in this manner ostensibly should converge with reasonable rapidity.

A procedure by which one can obtain an expression for the input admittance of a center-fed cylindrical dipole without the need to find the complete current distribution is outlined by the following: (a) Decide on the order of the approximation n and then develop the formula for $I_n(z)$. Equations 7.100 and 7.101 are examples of this for $n = 0$ and $n = 1$. (b) Place this expression for $I_n(z)$ in (7.95) and compute the value of the constant C. Note that C will appear repeatedly in the formula for $I_n(z)$, so this computation will become increasingly more complicated for larger n. (c) Use the computed value of C in the formula for $I_n(z)$ and then solve for the nth-order approximation to the input impedance from the relation $Z_n = I_n^{-1}(0)$.

As an illustration of this procedure, suppose that a zeroth-order approximation to the input impedance is desired. If $I_0(z)$, given by (7.100), is placed in (7.95), the result is that

$$C \cos kl - \frac{j}{2\eta} \sin kl = \frac{1}{\Omega} \int_{-l}^{l} \frac{H_0(z')e^{-jkr'}}{r'} dz' \tag{7.104}$$

from which

$$C = \frac{j}{(2\eta)} \frac{\sin kl + G_1(l)/\Omega}{\cos kl + F_1(l)/\Omega} \tag{7.105}$$

where $F_1(z)$ and $G_1(z)$ are defined by

$$H_1(z) = CF_1(z) - \frac{j}{2\eta} G_1(z) \tag{7.106}$$

so that, from (7.102),

$$F_1(l) = -\int_{-l}^{l} \frac{F_0(z')e^{-jkr'}}{r'} dz' \tag{7.107}$$

$$G_1(l) = -\int_{-l}^{l} \frac{G_0(z')e^{-jkr'}}{r'} dz' \tag{7.108}$$

When the value of C given by (7.105) is inserted in (7.100), one finds that

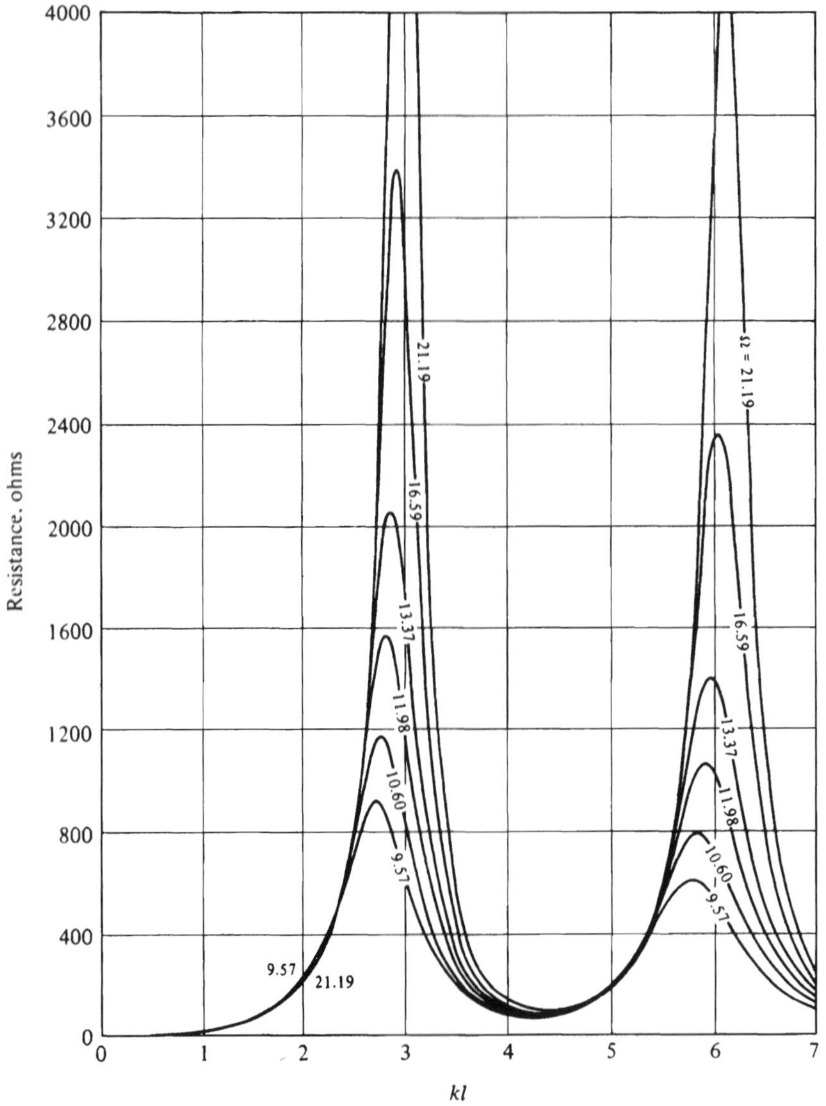

Fig. 7.15a Hallén's Curves of Resistance of a Center-Fed Cylindrical Dipole versus kl and Ω (Reprinted from E. Hallén, Cruft Laboratory Report No. 46, 1946, Courtesy of Harvard University.)

$$Z_0 = \frac{1}{I_0(0)} = -j60\Omega \frac{\cos kl + \Omega^{-1}F_1(l)}{\sin kl + \Omega^{-1}[(1 - \cos kl)G_1(l) + \sin klF_1(l)]} \tag{7.109}$$

Computations from (7.109) for the eight cases of $(2l/\lambda, a/\lambda)$ studied in the previous four sections give the zeroth-order values for self-impedance, listed in the third column of Table 7.4. These values are at considerable variance with the corresponding entries in Tables 7.2 and 7.3, indicating that the zeroth-order approximation is not sufficiently accurate.

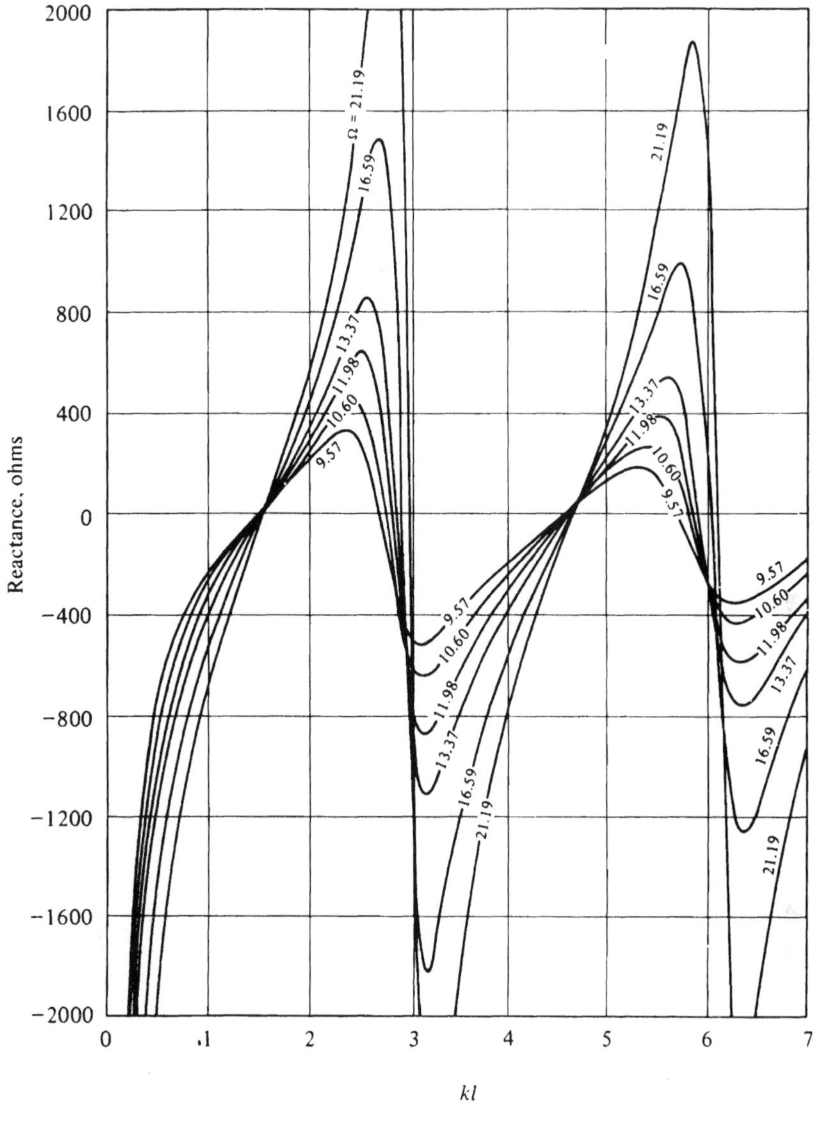

Fig. 7.15b Hallén's Curves of Reactance of a Center-Fed Cylindrical Dipole versus kl and Ω (Reprinted from E. Hallén, Cruft Laboratory Report No. 46, 1946, Courtesy of Harvard University.)

If (7.101) is used, together with the appropriate value of C, the first-order results shown in the fourth column of Table 7.4 are obtained. These values are in better agreement with the corresponding data listed in Tables 7.2 and 7.3, particularly at the shorter lengths.

Hallén, working before the advent of electronic computers with a mechanical desk calculator, laboriously calculated first-order values for $Z = R + jX$ in the range $0 \leq kl \leq 7$, for $\Omega = 2 ln(2l/a) = 9.57$, 10.60, 11.98, 13.37, 16.59, and 21.19. His curves are reproduced in Figure 7.15.

TABLE 7.4 Approximations to the input impedance of a center-fed cylindrical dipole using Hallén's integral equation

Normalized Length $2l/\lambda$	Normalized Radius a/λ	Input Impedance in Ohms: Zeroth-Order Approximation	First-Order Approximation
0.25	0.01	$19.35 - j359$	$15.99 - j240$
0.25	0.0001	$17.46 - j934$	$14.67 - j756$
0.50	0.01	$80.36 + j35.73$	$87.34 + j35.68$
0.50	0.0001	$76.81 + j41.57$	$79.08 + j43.52$
0.75	0.01	$268 + j526$	$437 + j318$
0.75	0.0001	$250 + j1120$	$433 + j1018$
1.00	0.01	$1685 - j1357$	$559 - j594$
1.00	0.0001	$6923 - j5385$	$3052 - j2626$

7.11 Self-Impedance of Center-Fed Cylindrical Dipoles: King-Middleton Second-Order Solution

For the benefit of the reader who has been following the development throughout this chapter, it is desirable to stop and make an assessment. Hallén's integral equation (7.25), which links the unknown current distribution on a center-fed cylindrical dipole to the unit voltage delta generator that excites it, was first solved (in Sections 7.6 and 7.7) using a method of moments approach. The input current provided a measure of the dipole's self-impedance, and representative values were listed in Table 7.2 for two different types of basis functions used in the computations.

A stationary expression for the self-impedance, Equation 7.62, was derived in Section 7.8 and used, together with one-and two-term trial functions, to obtain the representative values shown in Table 7.3.

Finally, a return to Hallen's integral equation (7.25) led, in Section 7.10, to the development of an nth order approximation to the self-impedance. Table 7.4 displayed representative values for the zeroth- and first-order approximations to Z.

It is disconcerting to see that the six sets of self-impedance values listed in these three tables, though showing general and qualitative agreement, cannot really be said to corroborate each other in a quantitative sense. One can excuse the entries in Table 7.2 on the valid argument that either (1) a sufficient number of basis functions had not been chosen to provide high accuracy, or (2) improved accuracy would require return to the more accurate integral equation of (7.24). (The purpose of that exercise was to illustrate use of the method of moments and to show the nature of the entire current distribution, but not to determine the input current with precision.) Further, one can argue that the two-term Storer solutions listed in the last column of Table 7.3 should be more accurate than the one-term induced EMF solutions shown in the third column of that table. By a similar argument, one can state a preference for the first-order solutions over the zeroth-order solutions in Table 7.4. By elimination, the comparison is reduced to the two sets of impedance values given in the third and fourth columns of Table 7.5.

TABLE 7.5 Comparison of the input impedance of a center-fed cylindrical dipole using different computational methods

		Input Impedance in Ohms		
Normalized Length $2l/\lambda$	Normalized Radius a/λ	Storer's Two-Term Approximation	First-Order Approximation to Hallén's Equation	King-Middleton Second-Order Approximation
0.25	0.01	$11.63 - j185$	$15.99 - j240$	$13.98 - j166$
0.25	0.0001	$12.93 - j723$	$14.67 - j756$	$12.90 - j811$
0.50	0.01	$101 + j32.82$	$87.34 + j35.68$	$92.51 + j38.30$
0.50	0.0001	$80.15 + j42.61$	$79.08 + j43.52$	$79.89 + j43.47$
0.75	0.01	$566 + j3.10$	$437 + j318$	$543 + j32.2$
0.75	0.0001	$521 + j1019$	$433 + j1018$	$540 + j1016$
1.00	0.01	$290 - j363$	$559 - j594$	$177 - j339$
1.00	0.0001	$2370 - j2129$	$3052 - j2626$	$2233 - j2150$

When one considers the difficulty involved in making these calculations, some satisfaction can be taken in the general agreement between the Storer-type values and the first order Hallén-type values. But which set is closer to the truth? And how far from the truth?

It can be argued that if either approach is carried to a more refined level of approximation, the accuracy of the calculations should improve. In the case of the Storer method, part of the difficulty is in knowing how to compose the functions which will serve as three-term, four-term, and n-term trial expressions. The complexity of calculation increases drastically as more terms are added to the trial function. The situation is less complicated in Hallén-type solutions. No choice of trial functions needs to be made, and the computational procedure for higher order solutions can be organized into a repetitive format.

King and Middleton have given full development to a second-order approximate solution of a version of Hallén's integral equation. Their curves of self impedance for a center-fed cylindrical dipole are shown as Figures 30.5a and b in R. W. P. King's text[24] and can be compared to the values obtained by Hallén (Figure 7.15). One finds general qualitative agreement. The tabulated data which accompanies the King-Middleton curves can be linearly interpolated to provide impedance values for the eight cases under study here. This gives the entries shown in the fifth column of Table 7.5. One can observe reasonable agreement between the King-Middleton results and the two-term Storer values, particularly at the longer dipole lengths. This impression is reinforced by a study of Figure 7.16, which gives a graphical comparison of Storer's results and the King-Middleton calculations for $\Omega = 15$ and a 2π range in kl.[25] All of this would suggest that higher-order approximations to Hallén's integral equation and higher-order Storer variational solutions might be converging to the true values.

[24] R. W. P. King, *Theory of Linear Antennas*, pp. 158–59.

[25] Storer, "Symmetrical Cylindrical Antennas," Figure 3 in particular.

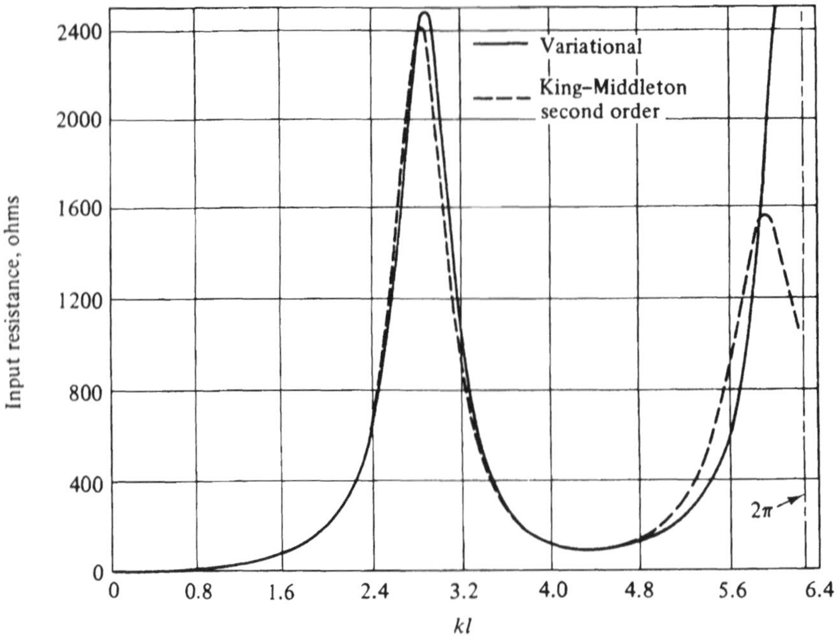

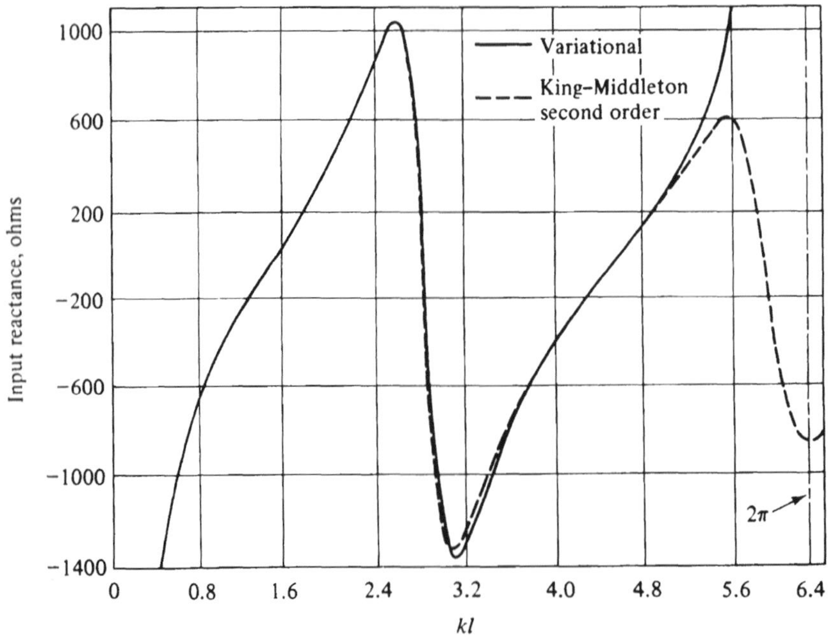

Fig. 7.16 A Comparison of Storer's Variational Solution and King-Middleton Second Order Values for the Input Impedance of a Center-Fed Cylindrical Dipole; $\Omega = 2 ln(2l/a) = 15$ (Reprinted from J. E. Storer, Cruft Laboratory Report No. 101, 1950, Courtesy of Harvard University.)

King and his coworkers have investigated the convergence question by computing the input admittance $Y = G + jB$ of center-fed cylindrical dipoles for selected lengths and diameters out to the 30th iteration.[26] They found that, for $k_a \leq 0.02$, G converged to a stable value at the second (or at most third) order approximation, but that B diverged. Since one would expect that the real and imaginary components of the admittance should exhibit the same convergence properties, this was a surprising result. The cause was traced to the assumption of a delta function generator feeding two hollow tubes of negligible thickness across an infinitesimal gap, thus creating a discontinuity in scalar potential in parallel with a knife-edge capacitance at $z = 0$.[27] The nature of the iterative process is to provide an additional contribution to B, at each iterative step, proportional to the susceptance of this infinitesimal gap knife-edge capacitance. Thus the intrinsic susceptance values, attributable to the dipole itself, actually converge, but the overall values of B grow linearly. This growth can be represented by $(2\pi a)nk$, in which K is a proportionality constant, n is the level of the iteration, and $2\pi a$ is the gap circumference. (The gap capacitance is proportional to $2\pi a$). Unfortunately, no theoretical method has been found to determine the value of K and thus remove the effect of the gap capacitance. However, the removal can be accomplished if one accurate value of B is obtained experimentally for each thickness of the antenna at a convenient value of kl (such as the value yielding the first antiresonance).

This has been done using the very precise experimental results of R. B. Mack[28] and an illustration of the correction is shown in Figure 7.17. The uncorrected King-Middleton second-order values are indicated by the crosses. With a constant susceptance of -0.7 millimhos removed, the corrected King-Middleton values are shown by the solid lines. Mack's experimental data give the dotted curves. The agreement is seen to be quite good.

R. W. P. King and others[29], proceeding in this fashion for a sequence of dipole radii, have deduced an improved King-Middleton second-order solution and have provided a table of impedance values versus kl and a/λ. In the important practical range $0.0016 \leq a/\lambda \leq 0.01$ and $1.3 \leq kl \leq 1.7$, an empirical double polyfit to their data yields the equations

$$R\left(kl, \frac{a}{\lambda}\right) = \sum_{m=0}^{4} \sum_{n=0}^{4} a_{mn}(kl)^m \left(\frac{a}{\lambda}\right)^n \tag{7.110}$$

$$X\left(kl, \frac{a}{\lambda}\right) = \sum_{m=0}^{4} \sum_{n=0}^{4} b_{mn}(kl)^m \left(\frac{a}{\lambda}\right)^n \tag{7.111}$$

[26]For a review of this work see R. W. P. King, "The Linear Antenna—Eighty Years of Progress," IEEE Proceedings 55 (1967), pp. 2–16.

[27]The divergence in susceptance values disappears if the gap is finite. See G. E. Albert and J. L. Synge, "General Problem of Antenna Radiation and Fundamental Integral Equation with Application to Antennas of Revolution," *Quart. App. Math.*, 6 (1948), 117–56.

[28]R. B. Mack, "A Study of Circular Arrays," Cruft Laboratory Technical Reports Nos. 381–386, (Harvard University, May 1963).

[29]R. W. P. King, E. A. Aronson, and C. W. Harrison, Jr., "Determination of the Admittance and Effective Length of Cylindrical Antennas," *Radio Science*, 1 (1966), 835–50.

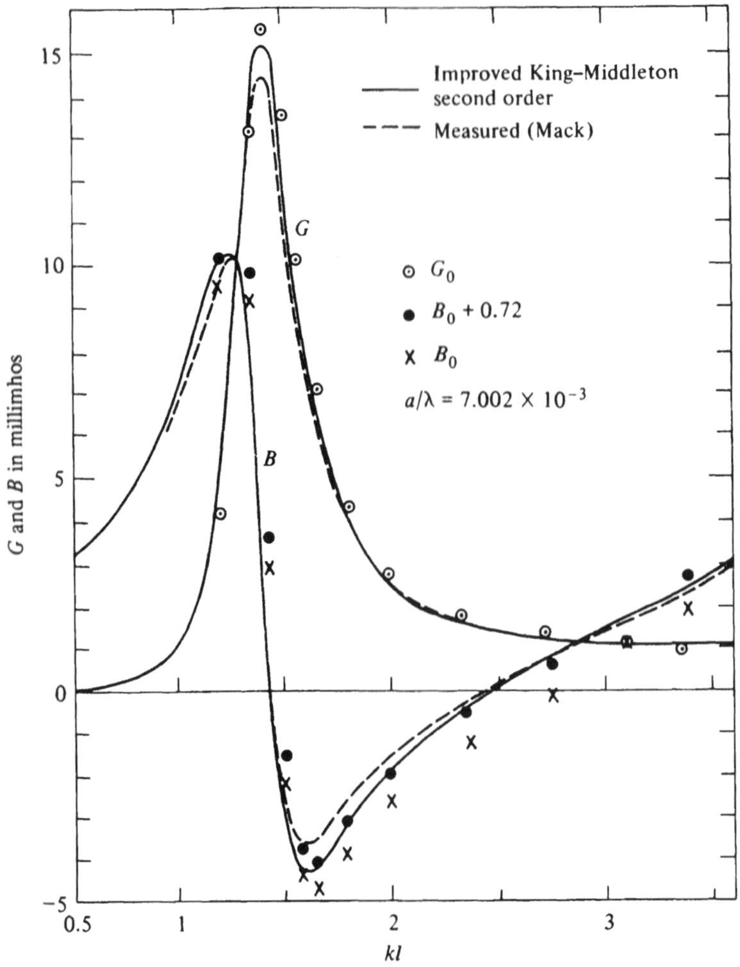

Fig. 7.17 A Comparison of the Improved King-Middleton Second-Order Admittance and the Measured Admittance of a Center-Fed Cylindrical Dipole (Measurements by Mack[28]) (© 1967 IEEE. Reprinted from R. W. P. King, *IEEE Proceedings*, pp. 2–16, 1967.)

The coefficients a_{mn} and b_{mn} are listed in Tables 7.6 and 7.7. Plots of $R(kl, a/\lambda)$ and $X(kl, a/\lambda)$ for the five values of a/λ tabulated in Reference 21 are shown in Figure 7.18. For the interested reader, the tabulation of the improved King-Middleton

TABLE 7.6 a_{mn} coefficients for use in Equation 7.110

m \ n	0	1	2	3	4
0	1.8115E + 02	3.9100E + 05	−4.2139E + 07	1.5370E + 09	6.1253E + 10
1	−6.3433E + 02	−1.0232E + 06	9.5644E + 07	−1.6606E + 09	−2.8222E + 11
2	8.3517E + 02	1.0004E + 06	−7.7690E + 07	−8.8929E + 08	4.0597E + 11
3	−4.6128E + 02	−4.3749E + 05	2.7195E + 07	1.3870E + 09	−2.2850E + 11
4	1.0222E + 02	7.3332E + 04	−3.6772E + 06	−3.3416E + 08	4.3131E + 10

TABLE 7.7 b_{mn} coefficients for use in Equation 7.111

m \ n	0	1	2	3	4
0	$-8.7489E+02$	$-1.4335E+05$	$1.1955E+08$	$-2.0911E+10$	$9.5064E+11$
1	$2.7551E+01$	$9.4225E+05$	$-4.1973E+08$	$6.6190E+10$	$-2.9289E+12$
2	$9.6056E+02$	$-1.2014E+06$	$4.6413E+08$	$-7.0841E+10$	$3.0937E+12$
3	$-5.9137E+02$	$5.8176E+05$	$-2.1273E+08$	$3.2124E+10$	$-1.3886E+12$
4	$1.3101E+02$	$-1.0342E+05$	$3.6025E+07$	$-5.3901E+09$	$2.3005E+11$

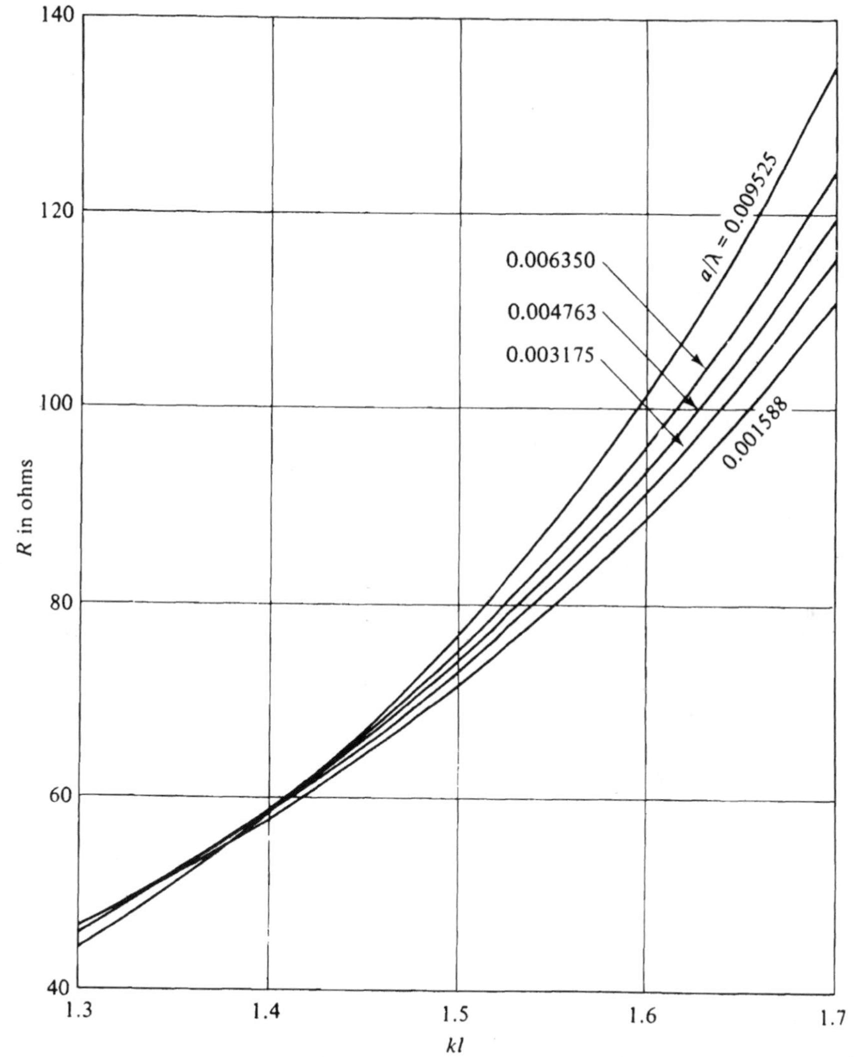

Fig.7.18a The Resistance of a Center-Fed Cylindrical Dipole versus kl and a/λ; Improved King-Middleton Second-Order Approximation

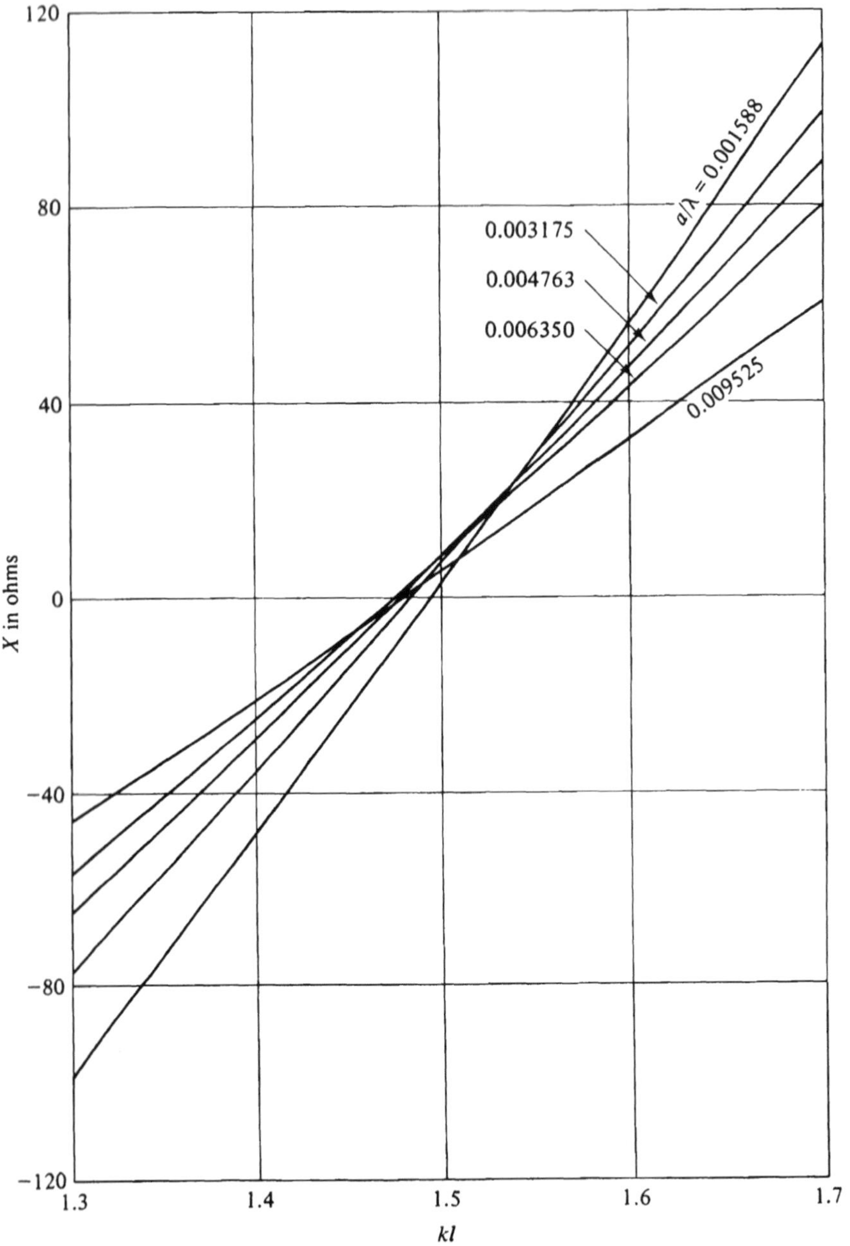

Fig. 7.18b The Reactance of a Center-Fed Cylindrical Dipole versus *kl* and *a*/λ; Improved King-Middleton Second-Order Approximation

second-order approximation to self-impedance appearing in the reference of footnote 29 covers the range $0 \leq kl \leq 4$. (The above formulas have only been fitted in the subrange $1.3 \leq kl \leq 1.7$.) This data is the most accurate available for the self-impedance of an idealized center-fed dipole. The idealization consists of picturing the dipole as composed of perfectly conducting thin-walled tubes with an infinitesimal feeding gap. The normalized radius a/λ is assumed to be small enough that end effects at $z = \pm l$ can be ignored. When this idealization is inappropriate, one can resort to experimentation to determine the input impedance versus parameters of interest (such as frequency, length, radius). But the trend in experimental data should always conform to theoretical curves such as those shown in Figure 7.18, which are therefore useful as a guide even in situations in which they do not strictly apply. Most practical applications for which the idealization is valid fall in the range for which Equations 7.110 and 7.111 or Figure 7.18 may be used.

Comparison of Figures 7.11 and 7.18 indicates that the results using the induced EMF method are in better and better agreement with King-Middleton as the dipole becomes thinner. For $a/\lambda \leq 0.001$, the agreement is sufficient to make the use of Tai's simple formula in (7.71) adequate for most purposes.

7.12 Self-Impedance of Center-Fed Strip Dipoles

A dipole shape which is finding widespread practical use is one with transverse cross section that is a rectangle of width w and thickness t, with $w \gg t$. Strip dipoles, as such radiators are called, can be fabricated on dielectric substrates and used in linear and planar arrays at microwave frequencies. Knowledge of the impedance properties of strip dipoles is needed in the design of the feeding structures for such arrays. Additionally, strip dipoles in free space are complementary radiators to slots cut in thin ground planes, and Babinet's principle (compare with section 7.16) links the electrical characteristics of the two types of radiators. For such applications, a determination of the impedance properties of strip dipoles provides knowledge which can be transferred simply to the complementary slot problem. For these reasons, it is desirable to study the behavior of dipoles with a rectangular cross sectional shape.

If the strip is slender ($kw \lll 1$ and $w \lll l$) it is possible to find an equivalent cylindrical dipole of radius a and the same length $2l$, which has a current distribution and input impedance that are essentially the same as those of the strip dipole. Thus all the knowledge developed in the preceding sections about cylindrical dipoles can be carried over to apply to slender strip dipoles. But first one must establish a relation connecting the radius of the equivalent cylindrical dipole to the dimensions w and t of the strip dipole.

Actually, this equivalence can be established for a dipole of more arbitrary transverse cross section than a rectangle.[30] The general situation has been depicted in Figure 7.2a and the development carried out in Section 7.3 led to the conclusion

[30]The development in the remainder of this section is patterned after a treatment which can be found in King, *The Theory of Linear Antennas*, pp. 16–20.

that, whatever the transverse shape, the integral equation for the current distribution on the dipole could be connected to the Z-directed magnetic vector potential $\mathcal{A}(z)$, the latter being given by

$$\mathcal{A}(z) = \int_{-l}^{l} \oint_C \frac{e^{-jkR}}{4\pi R} K_z(s', z') ds' \, dz' \tag{7.112}$$

in which $K_z(s', z')$ is the lineal current density on the dipole, and $R = [(x - x')^2 + (y - y')^2 + (z - z')^2]^{1/2}$. Because of the assumption of perfect conductivity, both the field point (x, y, z) and the source point (x', y', z') lie on the surface of the dipole.

If $2u$ is the maximum lineal extent of the transverse cross section, it will be assumed that $ku \lll 1$ and that $u \lll l$. It can then be argued that the lineal current density $K_z(s', z')$ can be represented as the product of two functions:

$$K_z(s', z') = f(s')I(z') \tag{7.113}$$

In (7.113), $I(z')$ is the total current, and thus

$$\int_C K_z(s', z') \, ds' = I(z') = I(z') \oint_C f(s') \, ds' \tag{7.114}$$

so that $f(s')$ is the normalized lineal current density. When (7.113) is substituted in (7.112), the result can be expanded into the form

$$\mathcal{A}(z) = \left[\int_{-l}^{z-b} + \int_{z-b}^{z+b} + \int_{z+b}^{l}\right] I(z') \oint_C f(s') \frac{e^{-jkR}}{4\pi R} ds' \, dz' \tag{7.115}$$

and, if $b \cong 10u$, this can be approximated by

$$\begin{aligned} \mathcal{A}(z) \cong & \left[\int_{l}^{z-b} + \int_{z+b}^{l}\right] I(z') \frac{e^{-jk|z-z'|}}{4\pi |z - z'|} dz' \\ & + I(z) \int_{z-b}^{z+b} \oint_C \frac{f(s')}{4\pi R} ds' \, dz' \end{aligned} \tag{7.116}$$

To obtain (7.116), use has been made of the knowledge that $kb \ll 1$ and $b^2 \gg u^2$. The mean value theorem has been used to place $I(z)$ in front of the last integral, which in turn is given by

$$\int_{z-b}^{z+b} \oint_C \frac{f(s')}{4\pi R} ds' \, dz' = 2 \int_0^b \oint_C \frac{f(s')}{4\pi R} ds' \, d\zeta' \tag{7.117}$$

wherein $\zeta' = z' - z$ so that $R = [(x - x')^2 + (y - y')^2 + (\zeta')^2]^{1/2}$.

Since $\mathcal{A}(z)$ is not a function of the transverse coordinates (x, y), nor are the first two integrals in (7.116), it follows that neither is the third integral in (7.116). But this implies, together with (7.117), that

$$\int_0^b \oint_C \frac{f(s')}{R}\, ds'\, d\zeta' = \text{constant} \tag{7.118}$$

The ζ' integration of (7.118) gives

$$\begin{aligned}\int_0^b \frac{d\zeta'}{R} &= \ln \frac{b + [b^2 + (x - x')^2 + (y - y')^2]^{1/2}}{[(x - x')^2 + (y - y')^2]^{1/2}} \\ &\cong \ln \frac{2b}{[(x - x')^2 + (y - y)^2]^{1/2}}\end{aligned} \tag{7.119}$$

which means that, approximately,

$$\oint_C f(s') \ln \frac{2b}{[(x - x')^2 + (y - y')^2]^{1/2}}\, ds' = \text{constant} \tag{7.120}$$

Because the constant on the right side of (7.120) is independent of the shape of the cross section, its value may be determined by considering the special case of a circular cylinder of radius a.

It has already been noted (see Appendix E) that $\mathfrak{a}(z)$ can be computed accurately, under the present assumptions, by taking the source point to be on the Z-axis. Thus for the circular cylinder case, $[(x - x')^2 + (y - y')^2]^{1/2}$ can be replaced in (7.120) by $[x^2 + y^2]^{1/2} = a$. This allows one to conclude, because of (7.114), that the constant in (7.120) has the value $\ln(2b/a)$. As a result, for *any* shape of the cross section, (7.120) reduces to

$$\oint_C f(s') \ln[(x - x')^2 + (y - y')^2]^{1/2} ds' = \ln a \tag{7.121}$$

in which (x, y) and (x', y') are constrained to lie on the contour of the cross section.

Equation 7.121 may be used to determine the equivalent radius a of a dipole of circular cross section that gives the same $\mathfrak{a}(z)$, and thus the same input impedance, as the dipole of arbitrary cross section, with contour as specified by the parametric equations of (7.8). As an illustration, consider a transverse cross section of elliptical shape with major and minor diameters $2a_e$ and $2b_e$, as shown in Figure 7.19. A point (x, y) on the contour is given by

$$x = a_e \cos\theta \qquad y = b_e \sin\theta \tag{7.122}$$

Similarly, $x' = a_e \cos\theta'$ and $y' = b_e \sin\theta'$, and in this instance

$$ds' = [d\xi^2 + d\eta^2]^{1/2} = [a_e^2 \sin^2\theta' + b_e^2 \cos^2\theta']^{1/2}\, d\theta'$$

so that (7.121) becomes

$$\begin{aligned}\int_0^{2\pi} f(\theta') \ln\, &[a_e^2(\cos\theta - \cos\theta')^2 + b_e^2(\sin\theta - \sin\theta')^2] \\ &\cdot [a_e^2 \sin^2\theta' + b_e^2 \cos^2\theta']^{1/2} d\theta' = 2 \ln a\end{aligned} \tag{7.123}$$

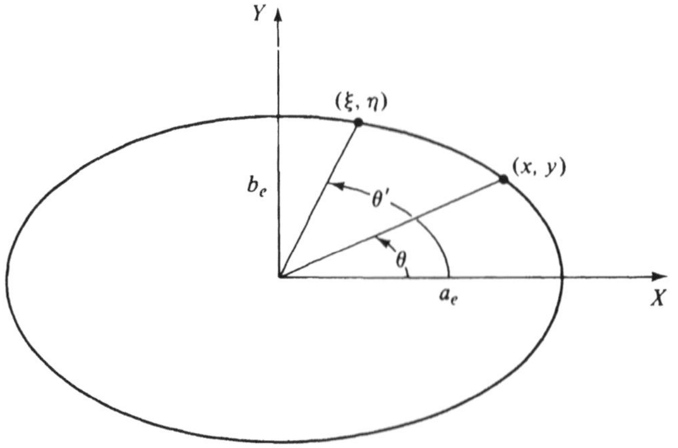

Fig. 7.19 A Dipole with Elliptical Transverse Cross Section

subject to the condition in (7.114) which, for the elliptical case, takes the form

$$\int_0^{2\pi} f(\theta')[a_e^2 \sin^2\theta' + b_e^2 \cos^2\theta']^{1/2} d\theta' = 1 \tag{7.124}$$

Equation 7.124 is clearly satisfied by

$$f(\theta') = \frac{1}{2\pi[a_e^2 \sin^2\theta' + b_e^2 \cos^2\theta']^{1/2}} \tag{7.125}$$

If the substitution of (7.125) in (7.123) leads to a result that is independent of θ, then (7.125) is the true normalized distribution of the surface current on the elliptic cylinder dipole. To test this, let $\phi = (\theta + \theta')/2$ and $\psi = (\theta - \theta')/2$, so that

$$\cos\theta - \cos\theta' = -2 \sin\tfrac{1}{2}(\theta + \theta') \sin\tfrac{1}{2}(\theta - \theta') = -2 \sin\phi \sin\psi$$
$$\sin\theta - \sin\theta' = 2 \sin\tfrac{1}{2}(\theta - \theta') \cos\tfrac{1}{2}(\theta + \theta') = 2 \sin\psi \cos\phi$$

As a consequence of this, (7.123) becomes

$$\begin{aligned} 4\pi \ln a &= \int_0^{2\pi} \ln(4 \sin^2\psi)(a_e^2 \sin^2\phi + b_e^2 \cos^2\phi)\, d\theta' \\ &= -2 \int_0^{\pi} \ln(4 \sin^2\psi)\, d\psi + 2 \int_0^{\pi} \ln(a_e^2 \sin^2\phi + b_e^2 \cos^2\phi)\, d\phi \end{aligned} \tag{7.126}$$

Through the use of standard trigonometric identities, (7.126) can be converted to

$$\begin{aligned} 4\pi \ln a = &-2 \int_0^{\pi} \ln(2 - 2\cos 2\phi)\, d(2\phi) \\ &+ 2 \int_0^{\pi} \ln\left\{\frac{a_e^2}{2}\left[\left(1 + \frac{b_e^2}{a_e^2}\right) - \left(1 - \frac{b_e^2}{a_e^2}\right)\cos 2\phi\right]\right\} d(2\phi) \end{aligned} \tag{7.127}$$

Both integrals in (7.127) can be evaluated from

$$\int_0^\pi \ln (a + b \cos x)\, dx = \pi \ln \tfrac{1}{2}[a + (a^2 + b^2)^{1/2}]$$

The first integral has a null value. The second serves to reduce (7.127) to

$$2\pi \ln \frac{a}{a_e} = \int_0^\pi \ln \left\{ \frac{1}{2}\left[\left(1 + \frac{b_e^2}{a_e^2}\right) - \left(1 - \frac{b_e^2}{a_e^2}\right) \cos x \right] \right\} dx$$

$$= 2\pi \ln \left[\frac{1}{2}\left(1 + \frac{b_e}{a_e}\right)\right] \tag{7.128}$$

as a consequence of which

$$a = \tfrac{1}{2}(a_e + b_e) \tag{7.129}$$

Equation 7.129 is a key result. It says that the equivalent circular cylinder has a radius a which is the arithmetic mean of the major and minor radii of the elliptical cylinder. Since a highly eccentric ellipse is a good approximation to a rectangle, if one lets $2a_e = w$ and $2b_e = t$, then

$$a = \tfrac{1}{4}(w + t) \tag{7.130}$$

with w and t the dimensions of the rectangular contour. Equation 7.130 can be used to find the equivalent cylindrical dipole for a specified strip dipole, after which Equations 7.110 and 7.111 can be used to determine the input impedance of the strip dipole.

7.13 The Derivation of a Formula for the Mutual Impedance Between Slender Dipoles

The previous eleven sections of this chapter have been concerned with the self-impedance of isolated dipoles, that is, a single dipole in otherwise empty space. If a dipole is to be used in conjunction with a ground plane or in an array of dipoles, it is necessary also to be able to determine the mutual impedance between dipoles. This section is concerned with formulating an expression from which the mutual impedance can be calculated.

Consider two center-fed dipoles, as shown in Figure 7.20. Without any loss in generality, the first dipole can be centered at the origin and placed to coincide with the Z-axis. Complete generality in the placement of the second dipole would have its center at an arbitrary point (x, y, z) and would have its orientation arbitrary as well. However, for almost all practical applications, the two dipoles will be parallel, and that assumption will be made here. It is then sufficient to locate the second dipole in the YZ-plane, that is, with its center at the point $(0, y, z)$.

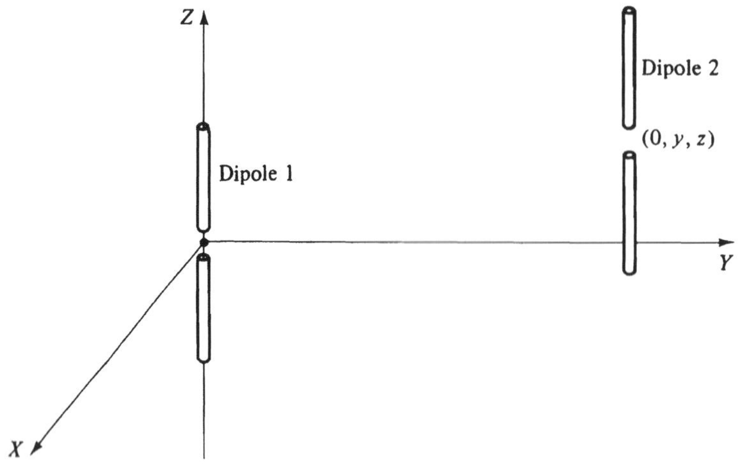

Fig. 7.20 Two Parallel Dipoles

The dipoles will be assumed to have lengths and radii $(2l_1, a_1)$ and $(2l_2, a_2)$, which are generally not the same. The values a_1 and a_2 may be the actual radii of cylindrical dipoles or the equivalent radii of dipoles of other cross-sectional shapes (rectangular in the case of strip dipoles).

If voltages V_1 and V_2 are applied across the central gaps of these two dipoles, input currents I_1 and I_2 will flow into the dipoles. This is a linear bilateral network, so one may write

$$\begin{aligned} V_1 &= I_1 Z_{11} + I_2 Z_{12} \\ V_2 &= I_1 Z_{21} + I_2 Z_{22} \end{aligned} \tag{7.131}$$

If dipole 2 is *present* but open-circuited, $I_2 = 0$ and the first of Equations 7.131 indicates that, under those conditions, the ratio of V_1 to I_1 is Z_{11}. In many practical applications, this ratio of V_1 to I_1 is negligibly different from what would occur if dipole 2 were *absent*. But in the latter case, the ratio of V_1 to I_1 is the isolated self-impedance of dipole 1, a subject which was extensively investigated in the first half of this chapter. We can therefore conclude that when the presence of a second (open-circuited) dipole has negligible effect on the input impedance of the driven dipole, Z_{11} (and Z_{22}) can be determined from the curves of Figure 7.18 or Equations 7.110 through 7.111.

The reciprocity theorem can be used to demonstrate that $Z_{12} = Z_{21}$; this is a standard proof in circuit theory that will not be repeated here. But once again, if dipole 2 is present but open-circuited, the second of Equations 7.131 indicates that Z_{21} is the ratio of the open-circuit voltage V_2 to the input current I_1 in the driven dipole. The reciprocity theorem in the form (1.135) can be used to develop a formula from which Z_{21} can be computed.

To see this, consider first the situation in which dipole 1 is present and energized but dipole 2 is absent. If perfect conductivity is assumed and end effects as well as gap effects are idealized, one can picture a source-and-response arrangement as suggested in Figure 7.21. A cylindrical sheath generator occupies the surface $r = a$, and extends from $z = -\delta$ to $z = +\delta$, with δ an infinitesimal. Over this sheath, E_z is uniform, $2\delta E_z(0)$ is the value of the applied voltage. Because of the assumption of perfect conductivity, $E_z \equiv 0$ over the cylindrical surface $r = a_1$, which extends from $z = -l$ to $z = -\delta$, and also over the cylindrical surface $r = a_1$, which extends from $z = \delta$ to $z = l_1$. A Z-directed surface current of lineal density $K_z(z)$ flows over the entire cylindrical surface $r = a_1$ from $z = -l_1$ to $z = +l_1$ and produces an electromagnetic field distribution (**E**, **H**) throughout space.

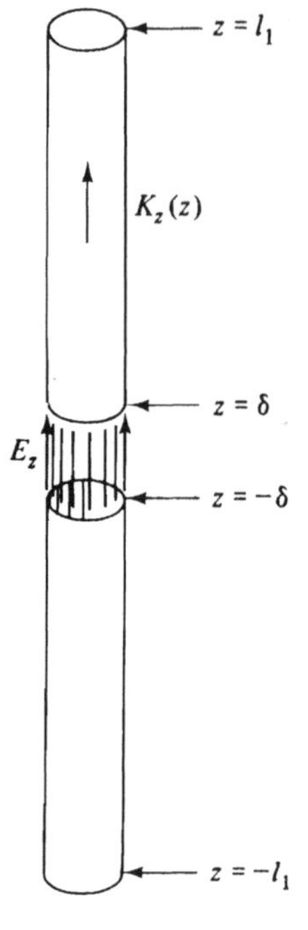

Fig. 7.21 A Cylindrical Sheath Generator Energizing a Cylindrical Dipole

If one removes the perfect conductor of which dipole 1 is assumed to be composed, but establishes in free space the same lineal current density $K_z(z)$ over the cylindrical surface $r = a_1$, extending from $z = -l_1$ to $z = +l_1$, then (**E**, **H**) will have the same distribution throughout space as before, including the values of $E_z(z)$ along the surface $r = a_1$. This new situation fits the assumption of sources in otherwise free space that was invoked in the derivation of the reciprocity theorem, and will be designated as containing the a-source system.

Next, imagine that both dipoles are present, with dipole 2 energized and dipole 1 open circuited. Now $K_z(z)$ will have a value on the cylindrical surface $r = a_2$ over the full length $2l_2$ and on the cylindrical surface $r = a_1$ over the full length $2l_1$, except for the central gap of length 2δ. Also, E_z will be identically zero along either cylindrical arm of either dipole, but will have a value in the gap of each dipole.

Once again, the perfect conductor can be removed. If the source distribution $K_z(z)$ is re-established in free space on the surfaces $r = a_1$ and $r = a_2$, the field distribution will be replicated. This new situation will be designated as containing the b-source system.

Since no magnetic sources are involved and since all the electric sources flow on surfaces, for this application (1.135) becomes

$$\int_S \mathbf{E}^b \cdot \mathbf{K}^a \, dS = \int_S \mathbf{E}^a \cdot \mathbf{K}^b \, dS \tag{7.132}$$

When the particular information just developed is placed in (7.132), one obtains

$$\int_{-\delta}^{\delta} \int_0^{2\pi} E_z^b K_z^a a_1 \, d\phi_1 \, d\zeta_1 = \int_{-l_2}^{l_2} \int_0^{2\pi} E_z^a K_z^b a_2 \, d\phi_2 \, d\zeta_2 \tag{7.133}$$

which reduces to

$$V_1^b I_1^a(0) = -\int_{-l_2}^{l_2} E_z^a(\zeta_2) I_2^b(\zeta_2) \, d\zeta_2 \tag{7.134}$$

in which V_1^b is the open circuit voltage at dipole 1 in the b-situation, $I_1^a(0)$ is the input current to dipole 1 in the a-situation, $I_2^b(\zeta_2)$ is the current distribution on dipole 2 in the b-situation, and $E_z^a(\zeta_2)$ is the *free-space* longitudinal field distribution on the surface $r = a_2$, which dipole 2 will occupy in the b-situation but does not occupy in the a-situation. It has been assumed in making the reduction from (7.134) that a_1/λ and a_2/λ are so small that E_z^b over $r = a_1$ is the same as though $a_1 = 0$ and that E_z^a over $r = a_2$ is the same as though $a_2 = 0$.

The minus sign in (7.134) requires an explanation. In terms of the notation of (7.131), if dipole 2 is energized and a load Z_L is placed across the terminals of dipole 1, the equivalent circuit is as shown in Figure 7.22. Because of the assumed positive direction of I_1, it follows that $V_1 = -I_1 Z_L$. This is true even when $Z_L \to \infty$ and V_1

Fig. 7.22 The Equivalent Circuit with Dipole 2 Transmitting and Dipole 1 Receiving

is the open circuit voltage. Therefore, if the positive direction of I_1 is taken as upward in Figure 7.22, the positive direction of the open-circuited V_1 will be downward. For this reason, in the evaluation of the left side integral of (7.133), one needs to write

$$\int_{-\delta}^{\delta} E_z^b d\zeta_1 = -V_1^b$$

Since Z_{12} is the open circuit voltage at dipole 1 in the b-situation, divided by $I_2^b(0)$, it follows from (7.134) that

$$Z_{12} = -\frac{1}{I_1^a(0)I_2^b(0)} \int_{-l_2}^{l_2} E_z^a(\zeta_2)I_2^b(\zeta_2)\, d\zeta_2 \tag{7.135}$$

If one knows the current distribution on a driven dipole and the field it produces, (7.135) can be used to determine the mutual impedance.

7.14 The Exact Field of a Dipole: Sinusoidal Current Distribution

In most practical applications for which one desires to know the mutual impedance between dipoles, they are not sufficiently separated to be in each other's far field. Indeed, they may be only a small fraction of a wavelength apart. Therefore $E_z^a(\zeta_2)$, as it appears in the integrand of (7.135), needs to be calculated in the near-field region of dipole 1. Fortunately, if one assumes that a sinusoidal current distribution exists on a driven dipole, it is possible to get exact expressions for the fields produced that are valid in both the near and far fields.

The assumption of a sinusoidal current distribution has already been seen to be justified for an isolated dipole if a/λ is sufficiently small. In the present application the additional assumption must be made (for the b-situation) that the presence of a nearby open-circuited dipole does not distort the current distribution of the driven dipole.

With these assumptions one can write

$$I_1^a(\zeta_1) = I_{m1} \sin k(l_1 - |\zeta_1|) \tag{7.136}$$

$$I_2^b(\zeta_2) = I_{m2} \sin k(l_2 - |\zeta_2|) \tag{7.137}$$

and (7.135) becomes

$$Z_{12} = -\frac{1}{\sin kl_1 \cdot \sin kl_2} \int_{-l_2}^{l_2} \frac{E_z^a(\zeta_2)}{I_{m1}} \sin k(l_2 - |\zeta_2|)\, d\zeta_2 \tag{7.138}$$

There remains the problem of finding $E_z^a(\zeta_2)$ before the integration in (7.138) can be performed.

In terms of the coordinate system arrangement of Figure 7.23, the magnetic vector potential function due to the current distribution (7.136) on dipole 1 is given by

$$A_z(x, y, z, t) = \int_{-l_1}^{l_1} \frac{I_m \sin k(l_1 - |\zeta_1|)e^{j(\omega t - kR)}}{4\pi\mu_0^{-1}R} d\zeta_1 \tag{7.139}$$

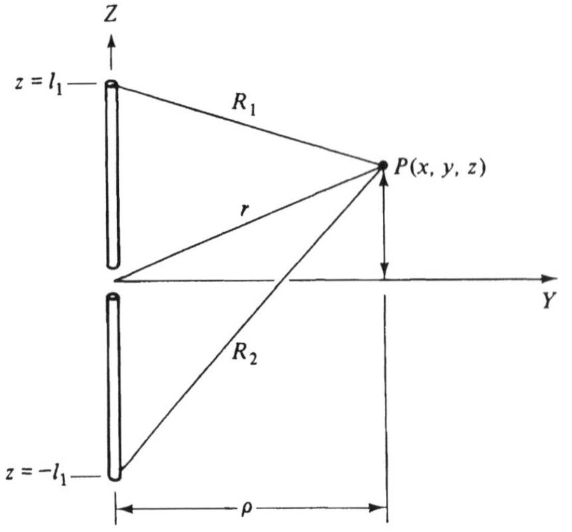

Fig. 7.23 Geometric Notation for Dipole 1

in which

$$R = [x^2 + y^2 + (z - \zeta_1)^2]^{1/2} \tag{7.140}$$

with (x, y, z) the field point and $(0, 0, \zeta_1)$ the source point. If $\sin k(l_1 - |\zeta_1|)$ is replaced by its exponential equivalent and $e^{j\omega t}$ is suppressed, (7.140) becomes

$$A_z(x, y, z) = \frac{I_m}{j8\pi\mu_0^{-1}}\left[e^{jkl_1}\int_{-l_1}^{0}\frac{e^{-jk(R-\zeta_1)}}{R}d\zeta_1 - e^{-jkl_1}\int_{-l_1}^{0}\frac{e^{-jk(R+\zeta_1)}}{R}d\zeta_1 \right.$$
$$\left. + e^{jkl_1}\int_{0}^{l_1}\frac{e^{-jk(R+\zeta_1)}}{R}d\zeta_1 - e^{-jkl_1}\int_{0}^{l_1}\frac{e^{-jk(R-\zeta_1)}}{R}d\zeta_1\right] \tag{7.141}$$

In cylindrical coordinates, the only component of the magnetic field will be B_ϕ, given by

$$B_\phi = (\nabla \times \mathbf{A})_\phi = -\frac{\partial A_z}{\partial \rho} \tag{7.142}$$

with $\rho = [x^2 + y^2]^{1/2}$. If the indicated differentiation is performed on the first integral of (7.141), the result is that

$$\mathcal{I}_1 = -\frac{\partial}{\partial\rho}\left[\frac{I_m}{j8\pi\mu_0^{-1}} e^{jkl_1}\int_{-l_1}^{0}\frac{e^{-jk(R-\zeta_1)}}{R}\,d\zeta_1\right]$$

$$= -\frac{I_m}{j8\pi\mu_0^{-1}} e^{jkl_1}\int_{-l_1}^{0} e^{jk\zeta_1}\frac{\partial}{\partial\rho}\left(\frac{e^{-jkR}}{R}\right)d\zeta_1$$

$$= \frac{I_m}{j8\pi\mu_0^{-1}} e^{jkl_1}\int_{-l_1}^{0} e^{-jk(R-\zeta_1)}\left[\frac{jk\rho}{R^2}+\frac{\rho}{R^3}\right]d\zeta_1 \qquad (7.143)$$

This integrand is a perfect differential, and it is not difficult to show that

$$\mathcal{I}_1 = \frac{I_m}{j8\pi\mu_0^{-1}} e^{jkl_1}\int_{-l_1}^{0}\rho\frac{\partial}{\partial\zeta_1}\left\{\frac{e^{-jk(R-\zeta_1)}}{R[R+z-\zeta_1]}\right\}d\zeta_1 \qquad (7.144)$$

Integration and substitution of the limits gives

$$\mathcal{I}_1 = \frac{\rho I_m}{j8\pi\mu_0^{-1}} e^{jkl_1}\left[\frac{e^{-jkr}}{r(r+z)} - \frac{e^{-jk(R_2+l_1)}}{R_2(R_2+z+l_1)}\right]$$

$$= \frac{\rho I_m}{j8\pi\mu_0^{-1}} e^{jkl_1}\left\{\frac{(r-z)e^{-jkr}}{r(r^2-z^2)} - \frac{[R_2-(z+l_1)]e^{-jk(R_2+l_1)}}{R_2[R_2^2-(z+l_1)^2]}\right\} \qquad (7.145)$$

But $R_2^2-(z+l_1)^2 = r^2 - z^2 = \rho^2$ and thus one may write

$$\mathcal{I}_1 = \frac{I_m e^{jkl_1}}{j8\pi\mu_0^{-1}\rho}\left[\frac{r-z}{r}e^{-jkr} - \frac{R_2-(z+l_1)}{R_2}e^{-jk(R_2+l_1)}\right] \qquad (7.146)$$

The other three integrals in (7.141) may be evaluated by the same procedure. When the four results are combined, one finds that

$$B_\phi = -\frac{I_m}{j4\pi\mu_0^{-1}\rho}[e^{-jkR_1} + e^{-jkR_2} - (2\cos kl_1)e^{-jkr}] \qquad (7.147)$$

The electric field can be found from Maxwell's curl equation, $\nabla \times \mathbf{H} = j\omega\epsilon_0\mathbf{E}$, which in cylindrical coordinates means that

$$E_\rho = -\frac{1}{j\omega\epsilon_0}\frac{\partial}{\partial z}(\mu_0^{-1}B_\phi)$$

$$= j30I_m\left(\frac{z-l_1}{\rho}\frac{e^{-jkR_1}}{R_1} + \frac{z+l_1}{\rho}\frac{e^{-jkR_2}}{R_2} - (2\cos kl_1)\frac{z}{\rho}\frac{e^{-jkr}}{r}\right) \qquad (7.148)$$

and

$$E_z = \frac{1}{j\omega\epsilon_0\rho}\frac{\partial}{\partial\rho}(\rho\mu_0^{-1}B_\phi)$$

$$= -j30I_m\left(\frac{e^{-jkR_1}}{R_1} + \frac{e^{-jkR_2}}{R_2} - 2\cos kl_1\frac{e^{-jkr}}{r}\right) \qquad (7.149)$$

It is this last result, the expression for the vertical component of electric field, which is needed in the present analysis. As many others have remarked, it is truly an extraordinary result. The E_z field of the dipole can be viewed as being composed of three contributions, each of which is an isotropic spherical wave, one each emanating from the middle and two ends of the dipole. The result is critically dependent on the assumption of a spatially sinusoidal current distribution on the dipole, but for slender dipoles this is not at all a bad assumption. Equation 7.149 will be used in the next section to establish an integral formula from which the mutual impedance between parallel slender dipoles can be computed.

7.15 Computation of the Mutual Impedance Between Slender Dipoles

If one returns to Figure 7.20, it is clear that a point on the axis of dipole 2 has the coordinates $(0, y, z + \zeta_2)$ with the central point of dipole 2 at the arbitrary position $(0, y, z)$ in the YZ-plane. From (7.149), the vertical component of electric field at $(0, y, z + \zeta_2)$ due to dipole 1 in the a-situation can be written as

$$E_z = -j30I_m\left(\frac{e^{-jkr_1}}{r_1} + \frac{e^{-jkr_2}}{r_2} - 2\cos kl_1 \frac{e^{-jkr}}{r}\right) \tag{7.150}$$

with

$$r = [y^2 + (z + \zeta_2)^2]^{1/2} \tag{7.151}$$

$$r_1 = [y^2 + (z + \zeta_2 - l_1)^2]^{1/2} \tag{7.152}$$

$$r_2 = [y^2 + (z + \zeta_2 + l_1)^2]^{1/2} \tag{7.153}$$

When (7.150) is substituted in (7.138) the result is that

$$Z_{12} = \frac{j30}{\sin kl_1 \cdot \sin kl_2} \int_{-l_2}^{l_2} \left(\frac{e^{-jkr_1}}{r_1} + \frac{e^{-jkr_2}}{r_2} - 2\cos kl_1 \frac{e^{-jkr}}{r}\right) \cdot \sin k(l_2 - |\zeta_2|)\, d\zeta_2 \tag{7.154}$$

It is customary to normalize (7.154) to the wavelength. When this is done, the real and imaginary components are given by

$$R_{12} = \frac{30}{\sin kl_1 \cdot \sin kl_2} \int_{-l_2/\lambda}^{l_2/\lambda} \left(\frac{\sin kr_1}{r_1/\lambda} + \frac{\sin kr_2}{r_2/\lambda} - 2\cos kl_1 \frac{\sin kr}{r/\lambda}\right) \cdot \sin k(l_2 - |\zeta_2|)\, d\left(\frac{\zeta_2}{\lambda}\right) \tag{7.155}$$

$$X_{12} = \frac{30}{\sin kl_1 \cdot \sin kl_2} \int_{-l_2/\lambda}^{l_2/\lambda} \left(\frac{\cos kr_1}{r_1/\lambda} + \frac{\cos kr_2}{r_2/\lambda} - 2 \cos kl_1 \frac{\cos kr}{r/\lambda} \right) \cdot \sin k(l_2 - |\zeta_2|)\, d\left(\frac{\zeta_2}{\lambda}\right) \tag{7.156}$$

Figure 7.24 shows plots of R_{12} and X_{12} when the two dipoles are the same length,

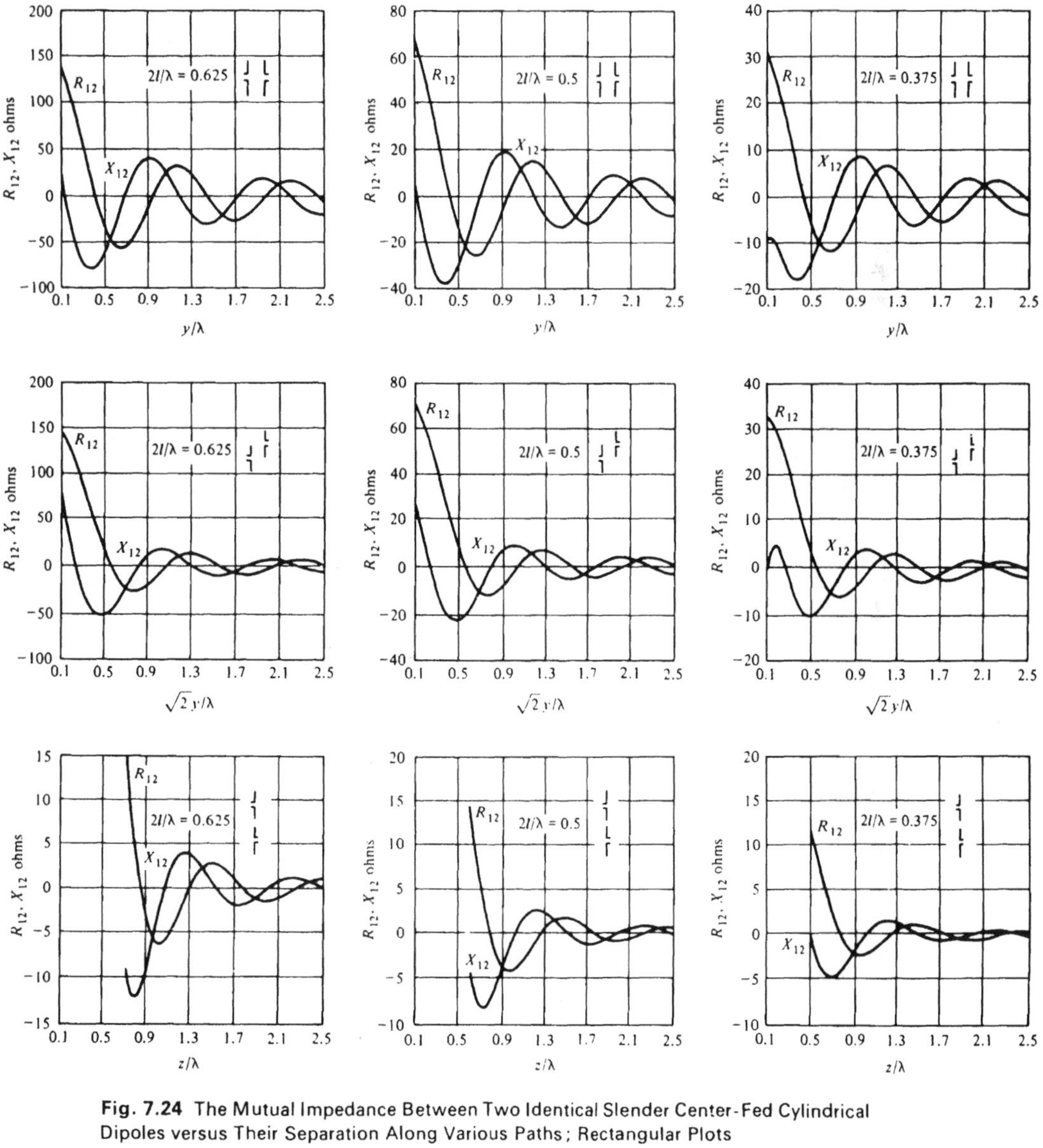

Fig. 7.24 The Mutual Impedance Between Two Identical Slender Center-Fed Cylindrical Dipoles versus Their Separation Along Various Paths; Rectangular Plots

for lengths $2l/\lambda = \frac{3}{8}, \frac{1}{2}$, and $\frac{5}{8}$, and for three directions of separation: broadside ($z = 0$, y variable), echelon ($z = y$, both variable), and end-fire ($y = 0$, z variable). One can observe a cyclical, decaying behavior to both R_{12} and X_{12}, with the peaks and nulls of one essentially coinciding with the nulls and peaks of the other. End-fire separation exhibits the most rapid decay, but at close spacings Z_{12} for the end-fire arrangement is just as strong as for the broadside arrangement. These observations will prove significant when the design of feeding networks for dipole arrays is undertaken.

Because of the cyclical variations of R_{12} and Z_{12} with separation distance between dipoles, it is illuminating to plot Z_{12} in polar form. This is done in Figure 7.25 for the same data that was used to construct Figure 7.24. One can see clearly that $|Z_{12}|$ decays most rapidly for the end-fire case, least rapidly for the broadside case, and that the phase angle of $|Z_{12}|$ retards almost linearly with separation at a rate corresponding to the speed of light, this effect being essentially independent of the direction of separation. The value of $|Z_{12}|$ is clearly influenced by the lengths of the dipoles.

Under the assumptions that the dipoles are slender and not too close to each other, the field of one in the vicinity of the other is negligibly different from what one would compute by collapsing the current distribution onto the dipole axis. Also, the variation of this field over the surface of the other dipole is negligibly different from the variation of this field along the axis of the other dipole. For these reasons, under the stated assumptions it does not matter what the cross-sectional shapes of the dipoles are. Thus the formulas (7.155) and (7.156) can be used to compute Z_{12} between strip dipoles as well as between cylindrical dipoles, as long as the slenderness criteria are met.

The case of vertical monopoles fed against a horizontal ground plane corresponds, via the image principle, to the broadside separation case for dipoles. The one difference that affects the computation is that it only takes half the voltage between the monopole and ground to establish a given current level that it does between the two halves of a dipole. Thus, for monopoles, one needs to take half the R_{12} and X_{12} values calculated from (7.155) and (7.156).

Several of the assumptions that have been made in the development of the past three sections can be tested for the special case of two parallel dipoles, each $\lambda/2$ long, separated by a distance b in the broadside position. C. T. Tai[31] has investigated this case rigorously, using coupled integral equations. By exciting the dipoles equally, either in phase or out of phase, he was able to compute both Z_{11} and Z_{12} versus separation distance. Tai's results are reproduced in Figure 7.26. One can observe that, for a separation of $\lambda/2$ or more, Z_{11} has settled down to the value of the isolated self-impedance. It is also clear that Z_{12} is but little affected by the slenderness index of the dipoles for $\Omega \geq 10$.

[31]C. T. Tai, "Coupled Antennas," *Proc. I.R.E.*, 36 (1948), 487–500. See particularly Figure 16.

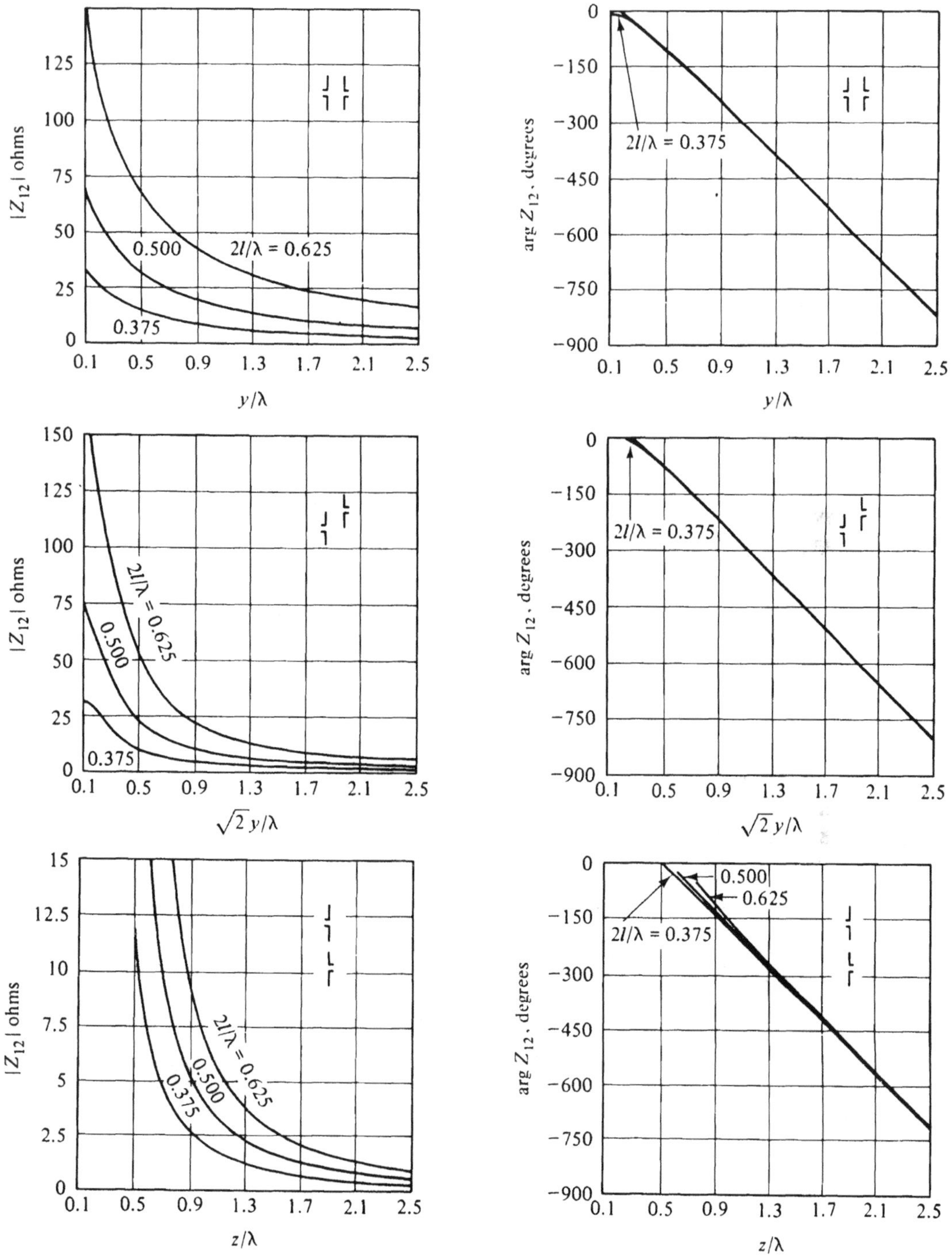

Fig. 7.25 The Mutual Impedance Between Two Identical Slender Center-Fed Cylindrical Dipoles versus Their Separation Along Various Paths; Polar Plots

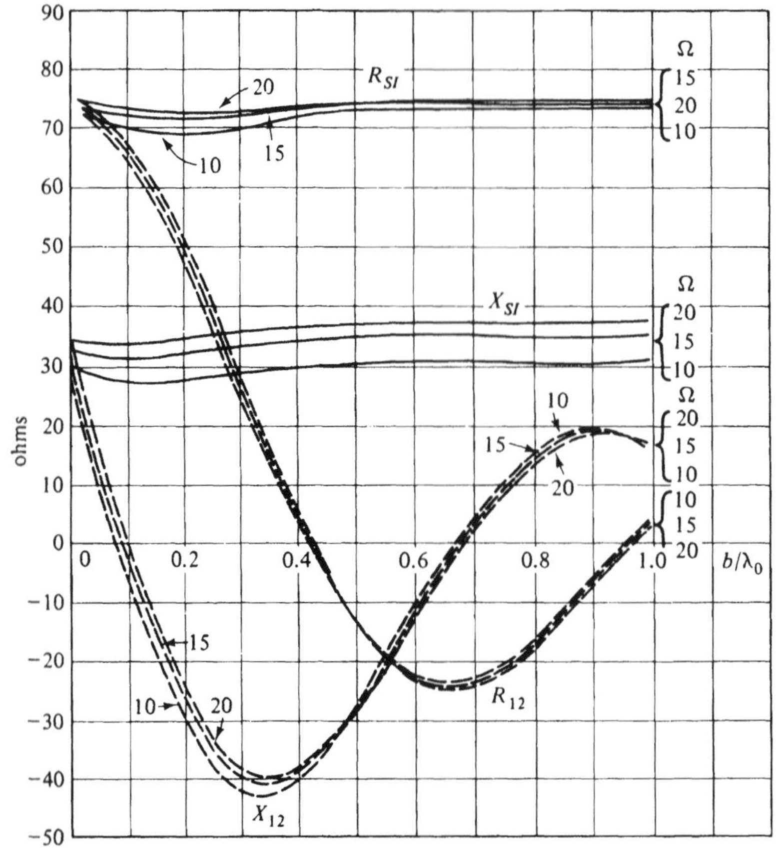

Fig. 7.26 The Self-Impedance and Mutual Impedance of Two Parallel Center-Fed Half-Wavelength Long Cylindrical Dipoles Versus Their Broadside Separation Distance (© 1948 IEEE. Reprinted from C.T. Tai, *Proc. IRE*, vol. 36, pp. 487–500, 1948.)

7.16 The Self-Admittance of Center-Fed Slots in a Large Ground Plane: Booker's Relation

Assume that two thin-walled tubes of perfect *magnetic* conductor form a dipole and are fed by an idealized magnetic generator across a central infinitesimal gap. The shape of this antenna has already been pictured in Figure 7.2, except that now what is being described is a magnetic dipole, not an electric one.

How could one determine the current distribution on such an antenna? One way to proceed would be to develop an analysis that is the exact dual of what appears in Section 7.2. An integral equation like (7.6) would emerge, with the unknown magnetic current density contained in the integrand and the axial magnetic field playing the role of driving function. For slender cylindrical magnetic dipoles, Hallen's integral equation would apply, and one would conclude that, for the same dimensions

$(2l/\lambda, a/\lambda)$, the magnetic current distribution on the magnetic dipole is the same as the electric current distribution on the electric dipole.

Further, the analysis contained in Section 7.12 could be emulated, thus extending the duality to noncircular cross sections. In particular, one can argue that center-fed strip dipoles have the same current distribution, whether they be electric or magnetic.

With this duality in mind, consider the antenna shown in Figure 7.27, consisting of a rectangular slot of width w and length $2l$, cut in a large ground plane, and energized at its center by a two-wire line. If $w/\lambda \ll 1$, the electric field in the slot is constrained to be essentially transverse, that is, y-directed, and perforce must vanish at the ends of the slot. Since $\partial E_y/\partial x \equiv 0$ in the slot, it follows that $\mathbf{B}_{\text{tang}} \equiv 0$ in the slot.

If the two-wire line is modeled by a generator placed in the plane of the slot and attached to the feed points, and if the ground plane is modeled by a "zero-thickness" perfect conductor of infinite extent, the technique of equivalent sources described in Section 3.2 can be used to compute the fields in either half of space.

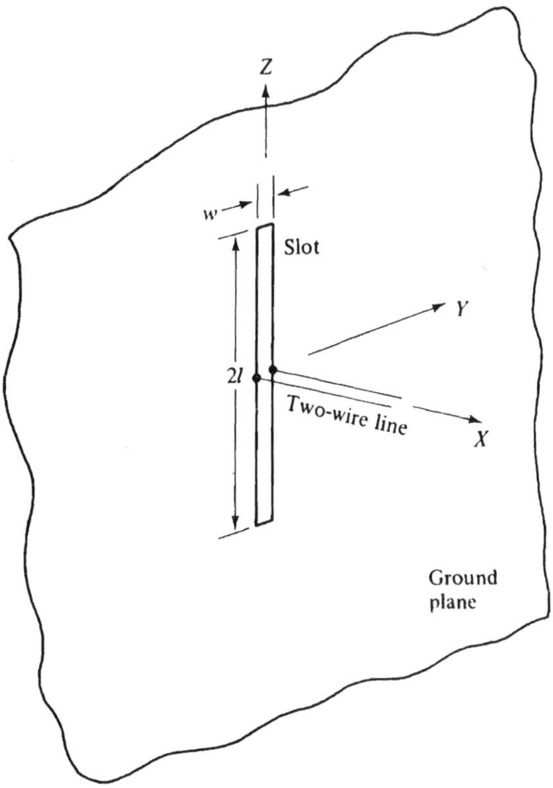

Fig. 7.27 A Rectangular Slot in a Large, Thin Ground Plane, Center-Fed by a Two-Wire Transmission Line

For the half-space $x > 0$, the equivalent sources consist of a lineal magnetic current distribution

$$K_{mz}(z) = -2\mu_0^{-1}E_y(z) \tag{7.157}$$

occupying the strip region vacated by the slot. Since $\mathbf{B}_{\text{tang}} \equiv 0$ in the original problem, these magnetic sources can be assumed to be flowing on a zero-thickness strip of material composed of perfect magnetic conductor.

The actual antenna can be modeled further by assuming that an infinitesimal transverse slit is cut centrally in the magnetic dipole, across which a delta-function magnetic generator is placed as the energizing source for the magnetic dipole. If the magnetic voltage of this generator is adjusted so that the input magnetic current is

$$wK_{mz}(0) = -2\mu_0^{-1}wE_y(0) \tag{7.158}$$

then the fields everywhere in $x > 0$ will be essentially the same as for the actual antenna.

One can determine the fields caused by the magnetic strip dipole with the aid of the relations developed in Section 7.5. In particular,

$$\mathbf{H}_2 = -j\omega\epsilon_0\mathbf{F} - \epsilon_0\nabla\Phi_m \tag{7.159}$$

in which

$$\mathbf{F} = \int_{-l}^{l}\int_{-w/2}^{w/2} \frac{\mathbf{K}_m e^{j(\omega t - kR)}}{4\pi\mu_0^{-1}R}\, dS \tag{7.160}$$

$$\Phi_m = \int_{-l}^{l}\int_{-w/2}^{w/2} \frac{\sigma_m e^{j(wt - kR)}}{4\pi\epsilon_0 R}\, dS \tag{7.161}$$

Away from the sources, the electric field can be found from

$$j\omega\epsilon_0\mathbf{E}_2 = \nabla \times \mathbf{H}_2 \tag{7.162}$$

It is interesting to compare these results with what one would obtain for the fields from a center-fed *electric* dipole of the same dimensions. Suppose that the sources on the electric dipole are related to those on the magnetic dipole by

$$\mathbf{K} = -\kappa\epsilon_0\mathbf{K}_m \qquad \rho_s = -\kappa\epsilon_0\rho_{sm} \tag{7.163}$$

Equations 7.163 are consistent with the deduction that the current distributions on the two dipoles must be the same; the multiplicative constant $-\kappa\epsilon_0$(with $\kappa = 1$ ohm), causes the two sides of (7.163) to be dimensionally consistent. The electric field caused by the electric dipole is given by

$$\mathbf{E}_1 = -j\omega\mathbf{A} - \nabla\Phi \tag{7.164}$$

where, by virtue of (1.43) and (7.163),

$$\mathbf{A} = -\kappa\epsilon_0\mathbf{F} \qquad \Phi = -\kappa\epsilon_0\Phi_m \tag{7.165}$$

Thus $\mathbf{E}_1 = \kappa(j\omega\epsilon_0\mathbf{F} + \epsilon_0\nabla\Phi_m)$ which, when compared to (7.159), indicates that

$$\mathbf{E}_1 + \kappa\mathbf{H}_2 = 0 \tag{7.166}$$

Also, away from the sources,

$$-j\omega\mu_0\mathbf{H}_1 = \nabla \times \mathbf{E}_1 = -\kappa\nabla \times \mathbf{H}_2 = -j\omega\epsilon_0\kappa\mathbf{E}_2 \tag{7.167}$$

so that

$$\mathbf{H}_1 - \frac{\kappa}{\eta^2}\mathbf{E}_2 = 0 \tag{7.168}$$

Equations 7.166 and 7.168, which link the fields $(\mathbf{E}_1, \mathbf{H}_1)$ of the electric dipole to the fields $(\mathbf{E}_2, \mathbf{H}_2)$ of the magnetic dipole, are a particular illustration of Babinet's principle, which is discussed in detail in Appendix F. The essence of Babinet's principle is that, if two complementary screens (the holes of one are the metallic part of the other) are excited by conjugate sources, the resulting total fields are related by

$$\mathbf{E}_1 + \kappa\mathbf{H}_2 = \mathbf{E}_1^i \qquad \mathbf{H}_1 - \frac{\kappa}{\eta^2}\mathbf{E}_2 = \mathbf{H}_1^i \qquad x > 0 \tag{7.169}$$

$$\mathbf{E}_1 - \kappa\mathbf{H}_2 = \mathbf{E}_1^r \qquad \mathbf{H}_1 + \frac{\kappa}{\eta^2}\mathbf{E}_2 = \mathbf{H}_1^r \qquad x < 0 \tag{7.170}$$

Equations 7.169 and 7.170 are derived in Appendix F. The sources are assumed to be in $z < 0$ and $(\mathbf{E}_1^i, \mathbf{H}_1^i)$ is the incident field on screen 1. The field that would be reflected from screen 1 if it contained no holes is $(\mathbf{E}_1^r, \mathbf{H}_1^r)$.

In the present application, the two-wire-fed slot of Figure 7.27 and the center-fed electric strip dipole can be viewed as complementary screens. The magnetic dipole was used as a surrogate for the slot in the ground plane; its fields are therefore $(\mathbf{E}_2, \mathbf{H}_2)$. Under the assumption of infinitesimal gap generators, $(\mathbf{E}_1^i, \mathbf{H}_1^i)$ is negligible except at the feed point, and (7.169) is seen to be consistent with (7.166) and (7.168).

The reader may wonder about the change in sign evident in (7.170). With infinitesimal gap generators, $(\mathbf{E}_1^r, \mathbf{H}_1^r)$ is also negligible except at the feed point. Why then does (7.170) not agree with (7.166) and (7.168)? The reason for this is that the equivalent magnetic current distribution (7.157) was deduced in order to compute the fields in $x > 0$. To determine the fields in $x < 0$, one would use a current sheet given by $K_{mz}(z) = 2\mu_0^{-1}E_y(z)$. Stated another way, the magnetic dipole causes an E-field with flux lines that are circular, while the slot in a ground plane causes an E-field with flux lines that are opposing "semicircles" Thus the magnetic dipole models the slot's fields on one side of the ground plane, but introduces a phase shift of 180° in modeling the slot's fields on the other side of the ground plane.

Equations 7.166 and 7.168 establish the fact that the two-wire-fed slot in an infinite ground plane and the complementary center-fed electric strip dipole have the same radiation pattern (with a 90°-rotation in polarization). But these two equations apply in the near-field region as well, and this enables one to deduce a connection between the input admittance of the slot and the input impedance of the strip dipole.

The connection between slot admittance and dipole impedance was discovered by H. G. Booker[32], and his analysis is essentially reproduced in what follows. With reference to Figure 7.28, imagine two complementary radiators, a center-fed rectangular slot in a large ground plane and a center-fed strip electric dipole, each of dimensions $2l$ by w. The coordinate axes are placed centrally as shown and two small circular contours are constructed, each with its center at the origin, one in the XY-plane and the other in the XZ-plane, as suggested in the projections.

If (V_1, I_1) and (V_2, I_2) are the applied voltages and input currents to the dipole and slot, respectively, then

$$\begin{aligned} &\oint_{ABCDA} \mathbf{H}_1 \cdot d\mathbf{l} = 2\oint_{ABC} \mathbf{H}_1 \cdot d\mathbf{l} = I_1 \qquad \oint_{abc} \mathbf{E}_1 \cdot d\mathbf{l} = V_1 \\ &\oint_{abcda} \mathbf{H}_2 \cdot d\mathbf{l} = 2\oint_{abc} \mathbf{H}_2 \cdot d\mathbf{l} = -I_2 \qquad \oint_{ABC} \mathbf{E}_2 \cdot d\mathbf{l} = V_2 \end{aligned} \tag{7.171}$$

When the excitation levels are adjusted to conform to (7.163) so that (7.166) and (7.168) apply, one finds that

$$V_1 = -\kappa \int_{abc} \mathbf{H}_2 \cdot d\mathbf{l} = \frac{\kappa}{2} I_2$$

$$V_2 = \frac{\eta^2}{\kappa} \int_{ABC} \mathbf{H}_1 \cdot d\mathbf{l} = \frac{\eta^2}{2\kappa} I_1$$

and thus that $V_1/I_1 = (\eta^2/4)(I_2/V_2)$, or

$$\frac{Z_1}{Y_2} = \frac{\eta^2}{4} \tag{7.172}$$

Equation 7.172 is Booker's relation and is often written in the form $Z_1 Z_2 = \eta^2/4$, which in words says that $\eta/2 = 188.5$ ohms is the geometric mean between the input impedance of a slender strip dipole and the input impedance of the complementary slender slot. However, (7.172) is the preferred form since it can be generalized to the case of complementary *arrays* of slots and dipoles, as shall be seen in the next section of this chapter.

Booker's relation is one of the more useful results in antenna theory. It extends the entire body of knowledge that has been gathered on the self-impedance of center-fed slender dipoles to apply to a center-fed slender slot in a large ground plane. Admittedly, the practical applications for a slot which radiates into *both*

[32]H. G. Booker, "Slot Aerials and Their Relation to Complementary Wire Aerials (Babinet's Principle)", *JIEE*, 93, pt. IIIA (1946), 620–26.

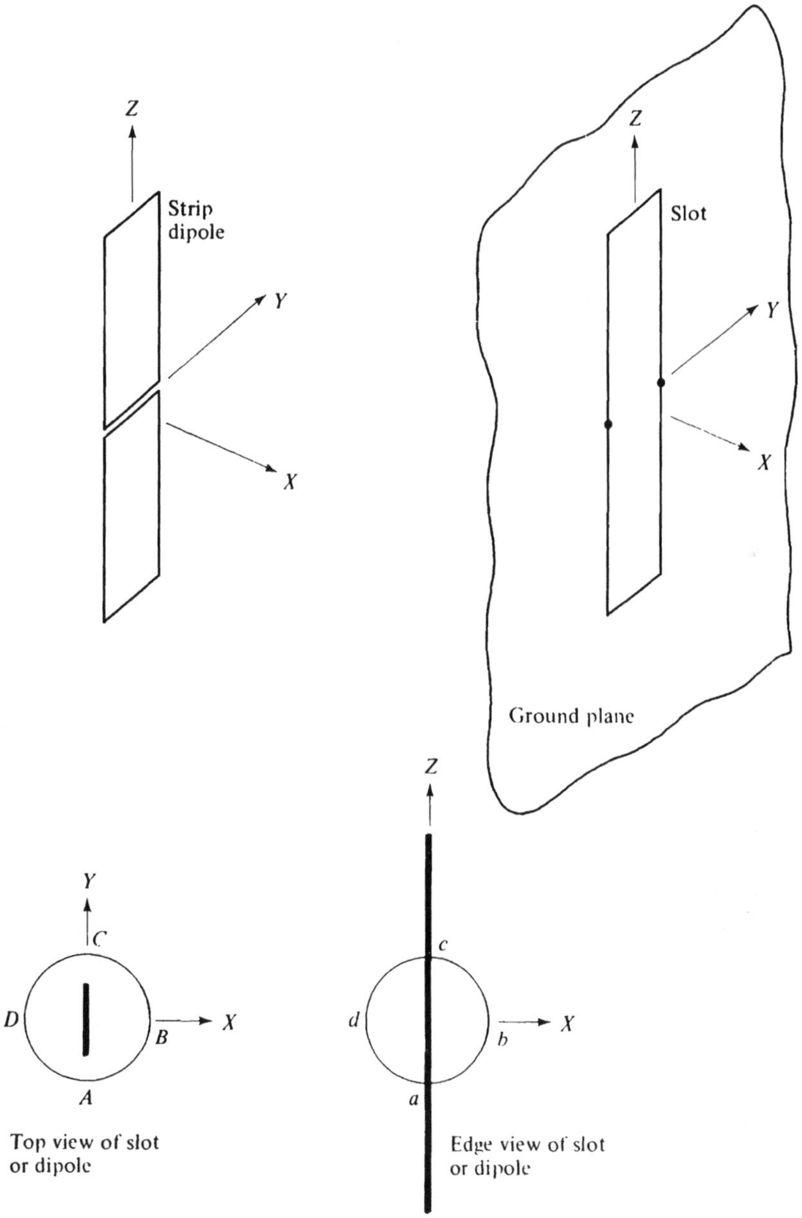

Fig. 7.28 Complementary Slot and Strip Dipole Radiators

half-spaces are few, but Booker's relation provides insight to the design of cavity-backed slots, and will be seen to play a significant role in the theory of waveguide-fed slot arrays. Both of these topics are treated in Chapter 8.

7.17 Arrays of Center-Fed Slots in a Large Ground Plane: Self-Admittance and Mutual Admittance

Rectangular arrays of parallel slender slots (usually waveguide-fed) find wide application in radar and communication systems and their proper design requires an understanding of mutual coupling. This problem can be introduced by considering a set of N arbitrarily placed (but parallel) center-fed slots in a common large ground plane, as suggested by Figure 7.29. This array can be viewed as an N-port linear bilateral system. If (V_m^s, I_m^s) are the applied voltage and input current at the mth slot, then one can write

$$I_m^s = \sum_{n=1}^{N} V_n^s Y_{mn}^s \tag{7.173}$$

with Y_{mm}^s the self-admittance of the mth slot and Y_{mn}^s the mutual admittance between the mth and nth slots.

The complementary array of center-fed strip dipoles is also an N-port linear bilateral system. If (V_m^d, I_m^d) are the applied voltage and input current at the mth

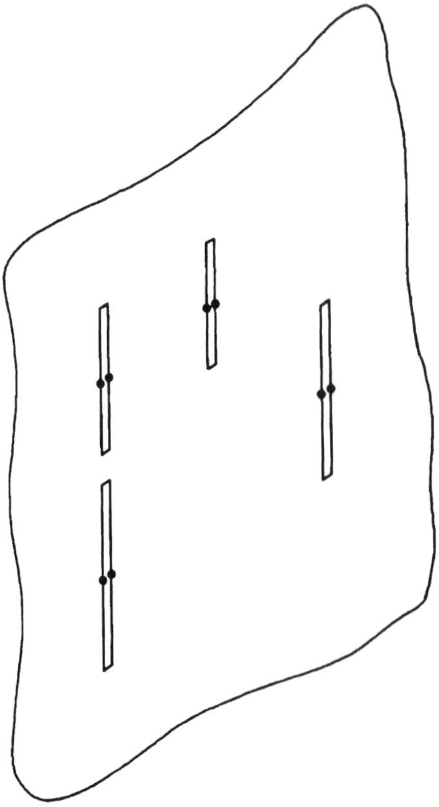

Fig. 7.29 An Arbitrary Array of Parallel Center-Fed Slots in a Large, Thin Ground Plane

dipole, then

$$V_m^d = \sum_{n=1}^{N} I_n^d Z_{mn}^d \tag{7.174}$$

in which Z_{mm}^d is the self-impedance of the mth dipole and Z_{mn}^d is the mutual impedance between the mth and nth dipoles.

Suppose that all slots but the mth are short-circuited at their feed points, and that all dipoles but the mth are open-circuited. Equations 7.173 and 7.174 give

$$I_m^s = V_m^s Y_{mm}^s \qquad V_m^d = I_m^d Z_{mm}^d \tag{7.175}$$

In this condition the two screens are exactly complementary in the Babinet sense (including shorting wires across all but the mth slot and open slits across all but the mth dipole). With the excitation levels adjusted in conformance with (7.163), the relations in (7.166) and (7.168) apply. Repetition of the analysis embodied in Equations 7.171 leads to the conclusion that

$$\frac{Z_{mm}^d}{Y_{mm}^s} = \frac{\eta^2}{4} \tag{7.176}$$

Thus Booker's relation applies to the dipole self-impedance/slot self-admittance ratio in the generalized case of two or more elements.

If the input impedance to the mth dipole with all other dipoles present but open-circuited is negligibly different from its input impedance with all other dipoles absent, then Z_{mm}^d can be taken to be the isolated self-impedance of the mth dipole, and all of the results of Sections 7.10 through 7.13 are pertinent. This is a good assumption if the interelement spacing is $\lambda/2$ or greater. By the same token, the self-admittance of the mth slot can be approximated by its isolated self-admittance.

Continuing with the assumption that all slots but the mth are short-circuited and that all dipoles but the mth are open-circuited, one can also see from (7.173) and (7.174) that

$$I_n^s = V_m^s Y_{mn}^s \qquad V_n^d = I_m^d Z_{mn}^d \quad n \neq m \tag{7.177}$$

with I_n^s the short-circuit current at the nth slot and V_n^d the open-circuit voltage at the nth dipole.

Let the *ABCD* contours shown in Figure 7.28 be erected at the mth slot (and mth dipole), but let the *abcd* contours, also shown in that figure, be erected at the nth slot (and nth dipole). Then

$$\begin{aligned} &\oint_{ABCDA} \mathbf{H}_1 \cdot d\mathbf{l} = 2\int_{ABC} \mathbf{H}_1 \cdot d\mathbf{l} = I_m^d \qquad \int_{abc} \mathbf{E}_1 \cdot d\mathbf{l} = V_n^d \\ &\oint_{abcda} \mathbf{H}_2 \cdot d\mathbf{l} = 2\int_{abc} \mathbf{H}_2 \cdot d\mathbf{l} = -I_n^s \qquad \int_{ABC} \mathbf{E}_2 \cdot d\mathbf{l} = V_m^s \end{aligned} \tag{7.178}$$

Use of (7.166) and (7.168), in conjunction with (7.177), leads to the conclusion that

$$\frac{Z^d_{mn}}{Y^s_{mn}} = \frac{\eta^2}{4} \tag{7.179}$$

Therefore Booker's relation also applies to the ratio of the dipole mutual impedance to the slot mutual admittance in the generalized case of two or more elements.

These results can be summarized by the equation

$$[Y^s] = \frac{4}{\eta^2}[Z^d] \tag{7.180}$$

in which $[Y^s]$ is the admittance matrix of the slot array and $[Z^d]$ is the impedance matrix of the complementary dipole array. It is important to note that the product $Z^d_{mn}Z^s_{mn}$ does *not* equal $\eta^2/4$.

The diagonal terms of the $[Y^s]$ matrix (self-admittance terms) can be computed by using Equations 7.110 and 7.111, modified by the multiplicative factor $(4/\eta^2)$. The off-diagonal terms of the $[Y^s]$ matrix (mutual admittance terms) can be determined through use of Equations 7.155 and 7.156, also modified by the multiplicative factor $(4/\eta^2)$.

7.18 The Self-Impedance of a Patch Antenna

The patch antenna was described in 3.7, with its generic form pictured in Figure 3.14. Simply stated, it consists of a metallic film bonded to a grounded dielectric substrate. The boundary of the film may be any shape, but rectangular and circular patches are most common. The maximum dimension of the patch seldom exceeds one-half of a free-space wavelength. Feeding is usually by means of a microstrip or coaxial line, as suggested by Figures 3.15 and 3.16.

It was seen in Section 3.7 that when the patch antenna was viewed as a slightly leaky cavity, approximate expressions for the field distribution could be readily deduced. From this, secondary sources could be calculated for placement along the perimeter of the patch-cavity, permitting calculation of the far field pattern. Comparison with experimental patterns was seen to be excellent, as evidenced for rectangular and circular patches by Figures 3.18 and 3.19.

The viewing of a patch antenna as a leaky cavity is also fruitful when one wishes to develop an expression for its self impedance. W. F. Richards, Y. T. Lo, and D. D. Harrison[33] have adopted this model and idealized the feed region in order to provide a development whose essentials are reproduced in what follows.

If the feed is coaxial, as in Figure 3.16, it can be represented by a cylindrical

[33] W. F. Richards, Y. T. Lo, and D. D. Harrison, "Improved Theory for Microstrip Antennas," *Electronic Letters*, 15 (1979), 42–44.

band of electric current flowing from the ground plane to the patch, plus an annular ring of magnetic current at the coaxial opening in the ground plane. The latter can be neglected with little error, and the former can be idealized by assuming that it is equivalent to a uniform current ribbon of some effective width d, centered on the feed axis and oriented (for instance) in the X-direction. The choice of a proper value for d will be considered later.

If a microstrip feed is used, as in Figure 3.15, the idealization consists of replacing the microstrip by a uniform current ribbon of some effective width d, placed at the boundary between patch and microstrip. The value of d may be somewhat larger than the physical width of the microstrip due to fringing.

The foregoing idealizations permit both types of feed to be modeled by a uniform current ribbon. If the fields in the leaky cavity are assumed to be insignificantly different from the fields that would exist if the patch were surrounded by a perfect magnetic wall (compare with Section 3.7), then the electric field beneath the patch can be represented in the form of Equation 3.69, with the constant coefficients A_{mn} calculable from (3.72).

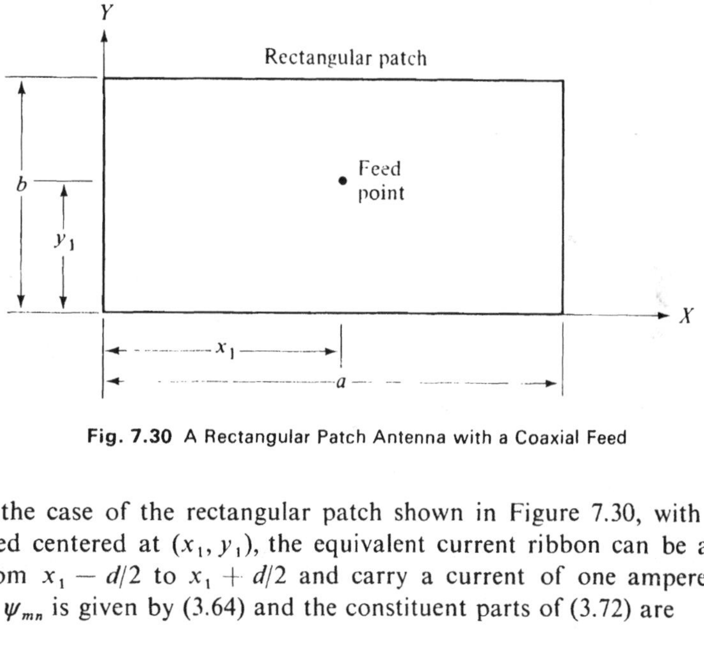

Fig. 7.30 A Rectangular Patch Antenna with a Coaxial Feed

For the case of the rectangular patch shown in Figure 7.30, with the actual coaxial feed centered at (x_1, y_1), the equivalent current ribbon can be assumed to stretch from $x_1 - d/2$ to $x_1 + d/2$ and carry a current of one ampere. For this geometry, ψ_{mn} is given by (3.64) and the constituent parts of (3.72) are

$$\langle \Psi_{rs} \Psi_{rs}^* \rangle = \int_0^a \int_0^b \cos^2 \frac{r\pi x}{a} \cos^2 \frac{s\pi y}{b} \, dx \, dy = \frac{ab}{\epsilon_r \epsilon_s} \tag{7.181}$$

$$\langle J_z \psi_{rs}^* \rangle = \int_{x_1 - d/2}^{x_1 + d/2} \frac{1}{d} \cos \frac{r\pi x}{a} \cos \frac{s\pi y}{b} \, dx \tag{7.182}$$

In (7.181), $\epsilon_t = 1$ if $t = 0$, and $\epsilon_t = 2$ if $t > 0$. The substitution $\xi = x - x_1$ converts (7.182) to

$$\langle J_z \psi_{rs}^* \rangle = \frac{1}{d} \cos \frac{s\pi y_1}{b} \int_{-d/2}^{d/2} \left(\cos \frac{r\pi\xi}{a} \cos \frac{r\pi x_1}{a} - \sin \frac{r\pi\xi}{a} \sin \frac{r\pi x_1}{a} \right) d\xi \tag{7.183}$$

If the second term in the integrand of (7.183) is dropped on the basis of the argument that $\sin(r\pi\xi/a)$ is very small in $-(d/2) \le \xi \le (d/2)$,[34] then

$$\langle J_z \psi_{rs}^* \rangle = \frac{\sin(r\pi d/2a)}{(r\pi d/2a)} \cos \frac{r\pi x_1}{a} \cos \frac{s\pi y_1}{b} \tag{7.184}$$

When this information is placed in (3.72) and the resultant used in (3.69), an approximate expression for the electric field in the cavity is given by

$$E_z = j\omega\mu_0 \sum_m \sum_n \frac{\phi_{mn}(x, y)\phi_{mn}(x_1, y_1)}{k_d^2 - k_{mn}^2} j_0\left(\frac{m\pi d}{2a}\right) \tag{7.185}$$

in which

$$\phi_{mn}(x, y) = \left(\frac{\epsilon_m \epsilon_n}{ab}\right)^{1/2} \cos \frac{m\pi x}{a} \cos \frac{n\pi y}{b} \tag{7.186}$$

and

$$j_0(u) = \frac{\sin u}{u} \tag{7.187}$$

The mode wave numbers which appear in (7.185) are defined by Equation 3.65 and the square of the wave number in the dielectric medium can be expressed as

$$k_d^2 = \epsilon_r(1 - j\delta)k_o^2 \tag{7.188}$$

with ϵ_r the relative permittivity, with δ the loss tangent of the dielectric, and with $k_0 = 2\pi/\lambda_0$, where λ_0 is the free-space wavelength.

Near resonance the factor $k_d^2 - k_{mn}^2$ becomes very small for the dominant mode, even with k_d slightly complex, and the field E_z is contributed to principally by the dominant mode term. This being the case, with all other modes neglected, E_z is given by a single term from (7.185). If this simplified field expression is used to deduce the equivalent magnetic sources at the periphery of the patch, computations can be made of the radiative loss P_{rad} and of the surface wave loss P_{sw}. Also, within the cavity region, the power loss P_{met} in the metallic walls can be estimated, as can the power loss P_d in the dielectric. If the sum of these four losses is represented by P, the loss

[34]This assumption loses its validity for large values of the index r. However, the dominant mode in practical applications occurs for r, s small.

tangent of an equivalent nonleaky cavity can be determined from

$$\delta_{eff} = \frac{P}{2\omega W_e} \tag{7.189}$$

in which $\omega = 2\pi\nu$ is the resonant radian frequency, W_e is the stored electric energy at resonance, and δ_{eff} is the loss tangent of the equivalent dielectric. Under the reasonable assumption that δ_{eff}, while greater than δ, is still quite small compared to unity, W_e can also be computed easily by using the dominant term of (7.185).

If next k_{eff} is defined by

$$k_{eff}^2 = \epsilon_r(1 - j\delta_{eff})k_0^2 \tag{7.190}$$

and used in place of k_d^2 in (7.185), then an improved calculation of E_z is possible. Since $tE_z(x_1, y_1)$ is the voltage at the feed, with t the dielectric thickness, and since a feeding current of one ampere has been assumed, it follows that the self-impedance is given by

$$Z = j\omega\mu_0 t \sum_m \sum_n \frac{\phi_{mn}^2(x_1, y_1)}{k_{eff}^2 - k_{mm}^2} j_0\left(\frac{m\pi d}{2a}\right) \tag{7.191}$$

Near resonance this series is contributed to mainly by the dominant mode term.

Computations using (7.191) can only be carried out after assuming some value for the equivalent current ribbon width d. Richards et al.[33] adjusted the value of d so that agreement was obtained at one frequency between the theoretical calculation of Z from (7.191) and the experimental value of Z. They then proceeded to compare theory and experiment for the rectangular patch shown in Figure 7.31a, using three different feed points. The results are shown in the Smith chart of Figure 7.31b. The correlation can be seen to be extraordinarily good. They were also pleased to find that the same effective ribbon width d was applicable to all three loci.

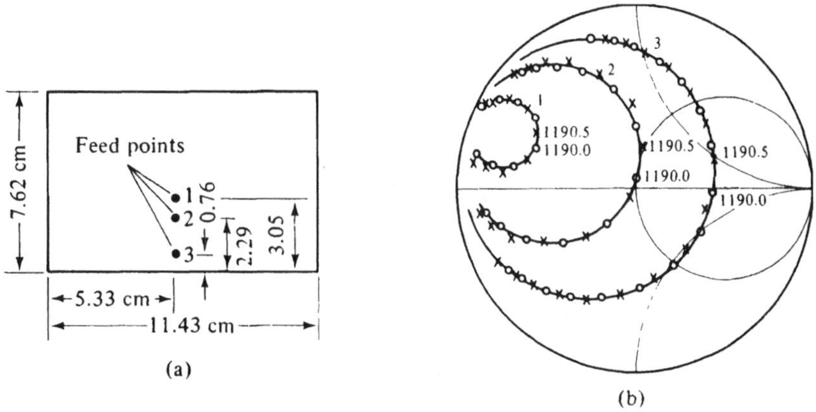

Fig. 7.31 The Input Impedance versus Frequency and Coaxial Feed Position for a Rectangular Patch Antenna (© 1979 IEE, London. Reprinted from Richards, Lo, and Harrison, *Electronic Letters*, vol. 15, pp. 42–44, 1979.)

This same approach was used successfully by Richards et al.[33] for a circular patch, as can be seen from a study of Figure 7.32.

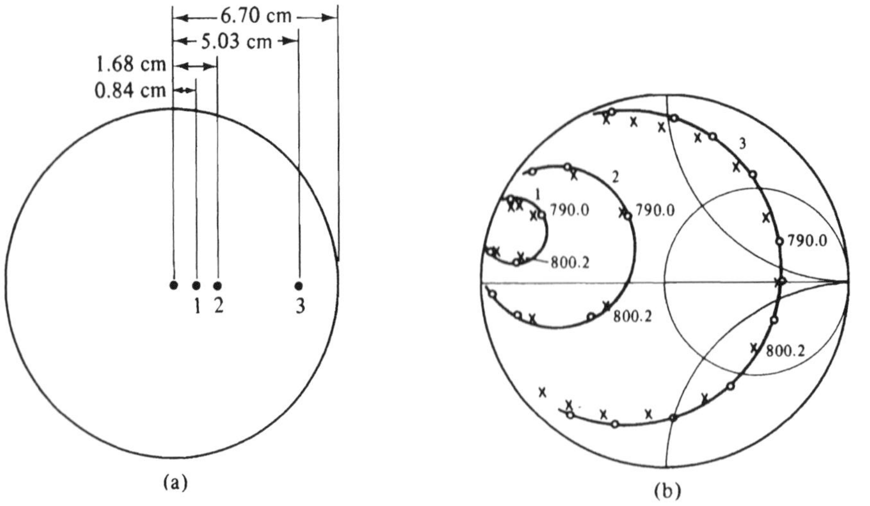

Fig. 7.32 The Input Impedance versus Frequency and Coaxial Feed Position for a Circular Patch Antenna (© 1979 IEE, London. Reprinted from Richards, Lo, and Harrison, *Electronic Letters*, vol. 15, pp. 42–44, 1979.)

REFERENCES

AHARONI, J., *Antennae* (Oxford: Clarendon Press, 1946).

BUTLER, C. M., "Integral Equation Solution Methods," *Supplementary Notes for a Short Course in the Application of Moment Methods to Field Problems*, (University of Mississippi, May 1973), Chapter 2.

GALEJS, J., *Antennas in Inhomogeneous Media* (Oxford: Pergamon Press, 1969).

HARRINGTON, R. F., *Field Computation by Moment Methods* (New York: The Macmillan Company, 1968).

JORDAN, E. C. and K. G. BALMAIN, *Electromagnetic Waves and Radiating Systems*, 2nd Ed. (Englewood Cliffs, New Jersey: Prentice-Hall, Inc., 1968).

KING, R. W. P., *Theory of Linear Antennas* (Cambridge, Mass.: Harvard University Press, 1956).

KING, R. W. P., *Tables of Antenna Characteristics* (New York: IFI/Plenum Publishing Co., 1971).

TAI, C. T., "Characteristics of Linear Antenna Elements," *Antenna Engineering Handbook*, ed. H. Jasik (New York: McGraw-Hill Book Co., Inc., 1961), Chapter 3.

THIELE, G. A., "Wire Antennas," *Computer Techniques for Electromagnetics*, ed. R. Mittra (Oxford: Pergamon Press, 1973), Chapter 2.

PROBLEMS

7.1 Repeat the analysis of Section 7.2 through 7.4 for the case of a magnetic dipole of arbitrary cross section and show that, under the same assumptions of slenderness, Hallén's integral equation (7.25) is obtained for the distribution of magnetic current on the dipole, with the roles of **E** and **B** interchanged.

7.2 Retrace the development of Section 7.12 to show that, for slender magnetic dipoles of arbitrary cross section, there is an equivalent magnetic dipole of circular cross section that has the same total current distribution and the same input impedance. Thus show the complementarity of electric and magnetic strip dipoles and establish Booker's relation that $\eta/2$ is the geometric mean between the input impedance of the two dipoles.

7.3 Use the method of moments and pulse functions spread over 20 equal intervals to find the current distribution on a slender center-fed cylindrical dipole for $a/\lambda = 0.01$ and $2l/\lambda = 0.45\ (.01)\ 0.55$. Tabulate the input impedance versus dipole length and compare your results to those found by the induced EMF method and by the improved King-Middleton second-order approximation.

7.4 Repeat Problem 7.3, but use four sinusoidal basis functions.

7.5 Use the curves of Figure 7.12 (or the accompanying fitted equations) to determine the impedance bandwidth of a dipole for which $\Omega = 10$. Define bandwidth as $(f_2 - f_1)/f_0$, in which f_0 is the lowest frequency at which the dipole is resonant and f_2 and f_1 are the upper and lower frequencies straddling f_0 at which the input VSWR has risen to a value of 2: 1, assuming a match at f_0.

7.6 Storer's two-term trial function for the current distribution on a cyclindrical dipole can be expressed in a form that contains only one arbitrary constant by combining (7.77) and (7.84). Do this for his choice of partial functions in (7.78) and (7.79) and then determine the current distribution explicitly for $a/\lambda = 0.01$ and $2l/\lambda = 0.5$. Compare your result with the method of moments solutions displayed in Figures 7.6 and 7.10.

7.7 Develop an expression for the constant C in Hallén's integral equation suitable for the first order approximation. Your result should be analogous to (7.105), which was obtained for the zeroth-order approximation. Use this value of C in Equation (7.99) and find $I_1(0)$ for $a/\lambda = 0.01$ and $2l/\lambda = 0.45\ (.01)\ 0.55$. How do the impedance values found by this method compare with those found in Problem 7.3?

7.8 Use Equations 7.110 and 7.111 to find the resonant length of a center-fed cylindrical dipole as a function of radius in the interval $0.0016 \leq a/\lambda \leq 0.01$, according to the improved King-Middleton second-order approximation. Also find the input resistance at resonance as a function of a/λ. Compare these results to those found by the induced EMF method and shown in Figures 7.13 and 7.14.

7.9 Find the input impedance of a center-fed strip dipole in free space if the operating frequency is 300 MHz. The dipole is 1.25 in. wide, negligibly thick, and 17.70 in. long. Use the improved King-Middleton second-order approximation to determine your answer.

7.10 Estimate the input impedance of a center-fed rectangular slot in a large ground plane if the slot is the complement of the strip dipole described in Problem 7.9, and if the center frequency is the same.

7.11 Find the mutual impedance between two center-fed dipoles, parallel and in the broadside position, when they are 0.7λ apart and of common length $2l/\lambda = 0.40\ (.01)\ 0.52$.

7.12 Repeat Problem 7.11 if the dipoles are in the echelon position; if they are end-fire.

7.13 What is the mutual admittance between two rectangular slots, parallel and in the broadside position, if they are in a large thin ground plane and each is 0.01λ wide, 0.47λ long, and they are 0.7λ apart? What is their mutual impedance? Their self-impedance? What would be the self-impedance of one of the slots if the other were absent?

7.14 Develop an expression for the input impedance of a circular patch antenna with an offset coaxial feed.

8 the design of feeding structures for antenna elements and arrays

8.1 Introduction

The results of the previous chapter, which was concerned with the self-impedances and mutual impedances of various antenna elements, will now be utilized in the design of transmission line systems (feeding structures) to connect these antennas to a transmitter or receiver. In the case of a single radiating element, the most common criteria are that it be matched to its feed at some specified frequency, and that the input impedance and pattern stay within some prescribed limits over a certain frequency band. The impedance specification may require that the feed contain some frequency compensating features. In the case of an array of radiating elements, the feed may be a network of transmission lines, with signal division at each junction designed so that the excitation of all elements, in amplitude and phase, is exactly what is required to produce the specified pattern in the presence of mutual coupling among elements. The feed port connected to the transmitter (receiver) is usually required to be matched.

Individual sections of this chapter will be concerned with the design of a coaxially fed monopole above a large ground plane, a single dipole parallel to a large ground plane and fed by a balun, a cavity-backed slot, and a coaxially fed helix backed by a ground plane. The study of feeding structures for arrays will begin with the design of a two-wire harness for an end-fire array of driven dipoles in free space. Yagi-Uda and frequency independent arrays will be considered, as well as one-and two-dimensional arrays of balun-fed dipoles that are parallel to and in front of a large ground plane. For the linear dipole array, an introductory treatment of scanning a sum pattern in the presence of mutual coupling will be undertaken. The equivalent problem of waveguide-fed slot arrays will then be treated, and the chapter will conclude with an analysis of feeding structures for two-dimensional slot arrays designed to produce both sum and difference patterns.

8.2 Design of a Coaxially Fed Monopole with Large Ground Plane

A relatively simple example of the problem of feed design for an elementary radiator occurs when a vertical monopole is to be fed against a large horizontal ground plane. Assume that the feed is a coaxial line with an inner conductor that is extended to form the monopole, and an outer conductor that is terminated in the ground plane, as shown in Figure 8.1. It is desired to determine (1) if it is feasible to select a radius for the monopole that will cause it to be matched to the coaxial line, and (2) how its input impedance varies with frequency.

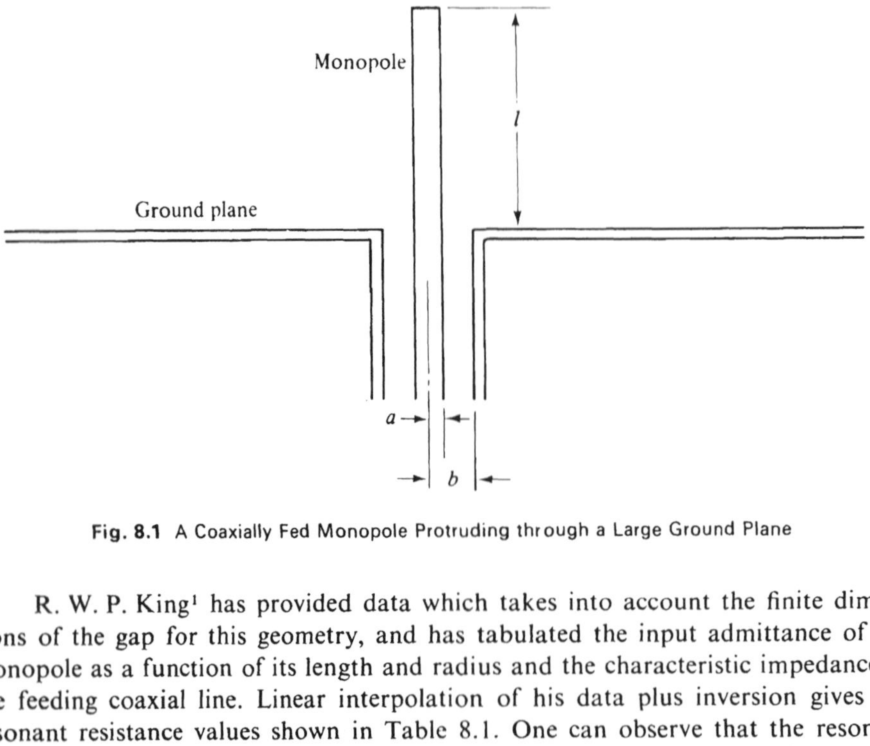

Fig. 8.1 A Coaxially Fed Monopole Protruding through a Large Ground Plane

R. W. P. King[1] has provided data which takes into account the finite dimensions of the gap for this geometry, and has tabulated the input admittance of the monopole as a function of its length and radius and the characteristic impedance of the feeding coaxial line. Linear interpolation of his data plus inversion gives the resonant resistance values shown in Table 8.1. One can observe that the resonant resistance is quite insensitive to monopole radius and to characteristic impedance. Clearly, to obtain a match, one should choose as feed a coaxial line with a characteristic impedance of about 37 ohms.

Since King gives admittance data for $Z_0 = 25$ ohms and 50 ohms, and since the two sets of data are very close, it is convenient to compute the average and invert this value, thereby obtaining the input impedance of the monopole versus its length and radius for a coaxial feed whose characteristic impedance is 37.5 ohms. The results are shown in Table 8.2. Further linear interpolation gives the dependence of resonant length and resonant resistance on monopole radius. This data is collected in Table 8.3.

[1] R. W. P. King, *Tables of Antenna Characteristics* (New York: IFI/Plenum, 1971), pp. 29–32.

TABLE 8.1 Resonant resistance of coaxially-fed thin tubular monopoles

	Resonant Resistance R_r			
	b/a (Z_0 in ohms)			
a/λ	1.517 (25)	2.301 (50)	3.49 (75)	5.30 (100)
0.001588	36.82	36.80	36.78	36.76
0.003175	37.09	37.01	36.94	36.84
0.004763	37.36	37.20	37.05	36.88
0.006350	37.64	37.38	37.12	36.89

TABLE 8.2 Impedance of thin tubular monopoles in ohms†

	$a/\lambda = 0.001588$	$a/\lambda = 0.003175$	$a/\lambda = 0.004763$	$a/\lambda = 0.006350$
l/λ	$R + jX$	$R + jX$	$R + jX$	$R + jX$
0.06250	$1.49 - j385.35$	$1.77 - j297.61$	$1.26 - j250.62$	$1.45 - j220.01$
0.09375	$3.58 - j267.69$	$3.10 - j210.26$	$3.20 - j178.83$	$3.50 - j158.03$
0.12500	$6.23 - j191.37$	$6.23 - j151.72$	$6.43 - j129.89$	$6.51 - j115.04$
0.15625	$10.85 - j131.82$	$10.90 - j104.91$	$11.05 - j\ 89.80$	$11.27 - j\ 79.47$
0.18750	$17.60 - j\ 79.58$	$17.95 - j\ 62.63$	$18.26 - j\ 53.06$	$18.63 - j\ 46.52$
0.21875	$27.87 - j\ 29.29$	$28.75 - j\ 21.11$	$29.52 - j\ 16.57$	$30.20 - j\ 13.55$
0.25000	$43.79 + j\ 22.82$	$45.88 + j\ 22.47$	$47.52 + j21.86$	$48.87 + j\ 21.09$
0.28125	$69.53 + j\ 80.10$	$74.08 + j\ 70.22$	$77.42 + j\ 63.31$	$79.94 + j\ 57.65$
0.31250	$113.40 + j145.58$	$122.85 + j122.55$	$129.05 + j105.92$	$132.87 + j\ 92.20$
0.34375	$193.49 + j218.63$	$210.52 + j171.14$	$218.36 + j135.24$	$220.08 + j105.86$
0.37500	$347.46 + j278.96$	$361.31 + j177.42$	$351.73 + j105.11$	$331.48 + j\ 53.56$

†Monopole is extension of inner conductor of coaxial line for which $b/a = 1.868$. Characteristic impedance of TEM mode in coaxial line is $Z_0 = 60 \ln(b/a) = 37.5$ ohms. Table entries have been calculated by linear interpolation and inversion of data found in King, *Tables of Antenna Characteristics*, pp. 29–32.

TABLE 8.3 Resonant length and resonant resistance of coaxially-fed tubular monopoles. $Z_0 = 37.5$ ohms

	$a/\lambda = 0.001588$	$a/\lambda = 0.003175$	$a/\lambda = 0.004763$	$a/\lambda = 0.006350$
l_r/λ	0.236	0.234	0.232	0.231
R_r (ohms)	36.82	37.05	37.28	37.50

A perusal of Table 8.3 indicates that, for an exact match, one should choose $a/\lambda = 0.00635$ and $l/\lambda = 0.231$. However, in order to use the available data maximally in terms of finding the frequency response of the impedance of the monopole, let the selection $a/\lambda_0 = 0.00397$, $l/\lambda_0 = 0.233$, and $R_r = 37.17$ ohms be made, with λ_0 the central wavelength.

Next, define four wavelengths by

$$
\begin{aligned}
(a/\lambda_1) &= 0.001588 = (a/\lambda_0)(\lambda_0/\lambda_1) = 0.00397(\lambda_0/\lambda_1), &\quad \lambda_1 &= 2.500\lambda_0 \\
(a/\lambda_2) &= 0.003175 = (a/\lambda_0)(\lambda_0/\lambda_2) = 0.00397(\lambda_0/\lambda_2), &\quad \lambda_2 &= 1.250\lambda_0 \\
(a/\lambda_3) &= 0.004763 = (a/\lambda_0)(\lambda_0/\lambda_3) = 0.00397(\lambda_0/\lambda_3), &\quad \lambda_3 &= 0.833\lambda_0 \\
(a/\lambda_4) &= 0.006350 = (a/\lambda_0)(\lambda_0/\lambda_4) = 0.00397(\lambda_0/\lambda_4), &\quad \lambda_4 &= 0.625\lambda_0
\end{aligned}
\tag{8.1}
$$

The corresponding normalized monopole lengths are

$$l/\lambda_1 = 0.093 \qquad l/\lambda_2 = 0.186 \qquad l/\lambda_3 = 0.280 \qquad l/\lambda_4 = 0.373$$

When these normalized lengths are used in conjunction with the appropriate columns of data in Table 8.2, linear interpolation yields the input impedance of the monopole at the four wavelengths. These data points, plus the resonant resistance value at λ_0, can be connected by the smooth curves shown in Figure 8.2.

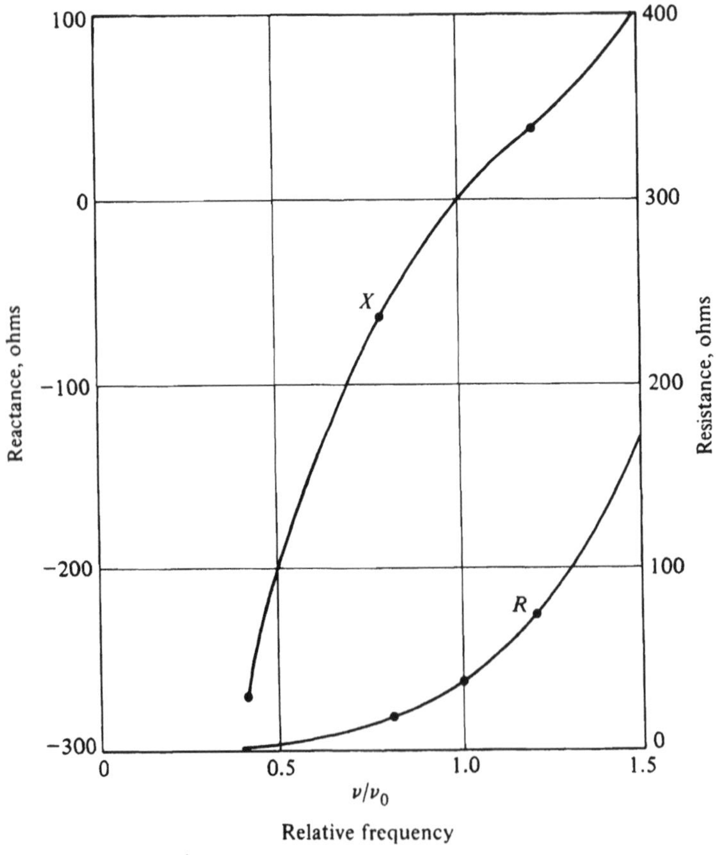

Fig. 8.2 Input Impedance versus Frequency for a Coaxially Fed Monopole Protruding through a Large Ground Plane

This antenna element is seen to be quite narrow band. One way to characterize the bandwidth is to state the extreme values of VSWR which occur in a specified frequency range. For this monopole-plus-ground-plane, the VSWR rises from 1.01 at ν_0 to 2.39 at $0.9\nu_0$, and to 1.82 at $1.1\nu_0$. A 20% bandwidth can thus only be contained within a 2.4 VSWR circle.

8.3 Design of a Balun-Fed Dipole Above a Large Ground Plane

A commonly used antenna element consists of a center-fed dipole parallel to and a distance h above a large ground plane, as shown in Figure 8.3a. One could feed the dipole with a two-wire line from the upper half of space, but this is awkward and seldom acceptable. Usually, it is desirable to have the transmitter (receiver) behind the ground plane, in which case it becomes difficult to pass a two-wire line up to the dipole. It must go through a hole in the ground plane that is large enough not to affect the TEM mode. The presence of the hole is undesirable, and a good *mechanical* connection of dipole, feed, and ground plane is not achieved.

Another possibility is to have a rigid coaxial line emerge vertically from the ground plane, as shown in Figure 8.3b, with its inner and outer conductors connected to the two arms of the dipole. This is a better design from a mechanical point of view, but is undesirable electrically. The reason for this is that the coax is an *unbalanced* feed for the dipole. In effect, the outer surface of the coax becomes part of one of the dipole arms and will be excited, thus contributing to the radiation pattern, introducing an unwanted cross-polarized field component.

This difficulty can be overcome by the design shown in Figure 8.3c. The dipole is supported by a pair of metal tubes of length h which are electrically connected to the ground plane at one end and to the arms of the dipole at the other. A center conductor is brought up inside one of these tubes and looped over to connect electrically to the junction of the other tube with the other dipole arm. The resulting coax is seen to feed two elements in parallel: (a) the dipole, and (b) a two-wire line of length h, shorted at its other end by the ground plane. The system is now electrically balanced. It is for this reason that feeds of this type are called *baluns*, the word being a contracted form of *balanced/unbalanced*.

If $h = \lambda/4$, the input impedance of the two-wire line is very high, and negligible current flows on it. For all practical purposes, only the dipole is being fed. And with $h = \lambda/4$, another beneficial effect is achieved. Because a large ground plane has been assumed, the method of images may be invoked, with the ground plane replaced by an image dipole for the purpose of computing the pattern in the upper half of space. This image dipole carries a current equal and opposite to the driven dipole (compare with Section 2.3). Being $\lambda/2$ away, it reinforces the field of the driven dipole maximally in the zenith direction, which is usually desired.

It will be assumed in what follows that $h = \lambda/4$. Still to be determined are the dipole length $2l/\lambda$ and the transverse dimensions of the coaxial line in order to achieve a match at the coax input. These quantities can be deduced by first considering the equivalent situation of two dipoles and no ground plane, for which the equations

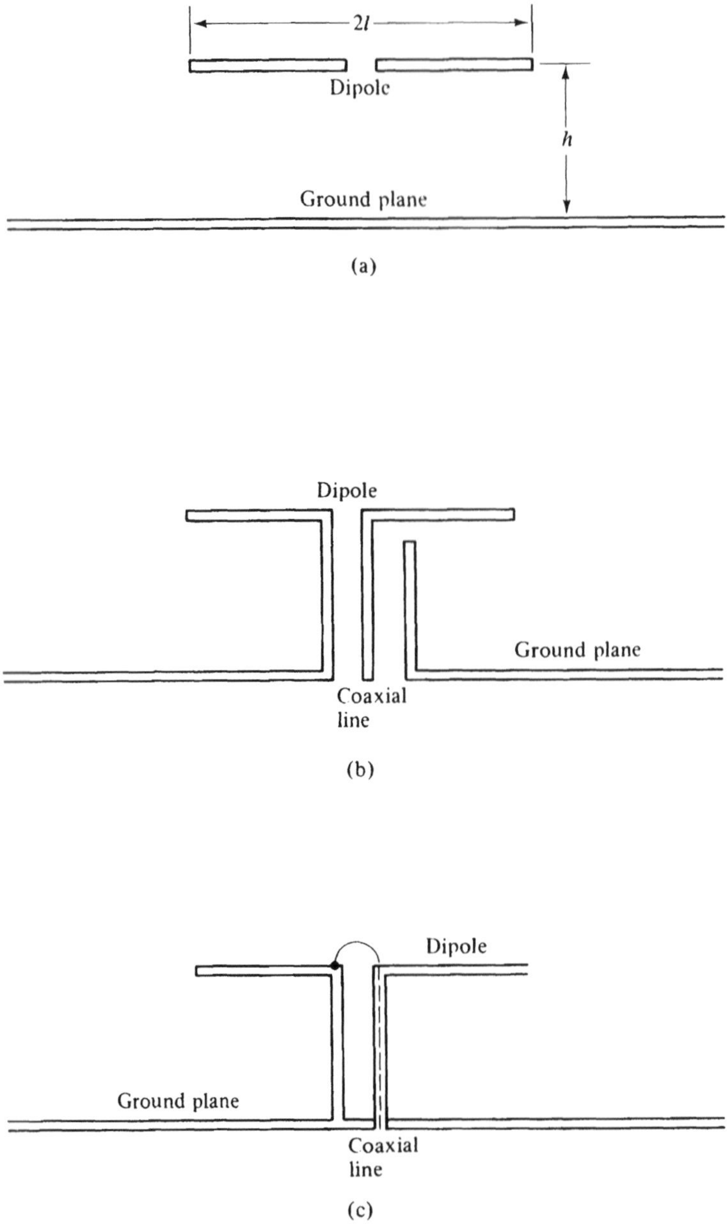

Fig. 8.3 A Center-Fed Dipole Above and Parallel to a Large Ground Plane; Various Feeding Arrangements

$$\begin{aligned} V_1 &= I_1 Z_{11} + I_2 Z_{12} \\ V_2 &= I_1 Z_{21} + I_2 Z_{22} \end{aligned} \tag{8.2}$$

can be written. In this case, in order to represent the image dipole properly, $Z_{11} = Z_{22}$ and $V_1 = -V_2$, as a consequence of which $I_1 = -I_2$. The first of Equations

8.2 can be rewritten as

$$Z_1^a = \frac{V_1}{I_1} = Z_{11} - Z_{12} \tag{8.3}$$

Z_1^a, called the *active impedance of the first dipole*, is its input impedance when the two dipoles are contraexcited. It is also approximately the input impedance of the single driven dipole above its large ground plane. (The degree of approximation is governed by how well the actual ground plane is modeled by an infinite ground plane of perfect conductivity.)

The mutual impedance term Z_{12} that appears in (8.3) represents the coupling between two parallel dipoles of the same length a distance $\lambda/2$ apart. Equations 7.155 and 7.156 can be used to compute Z_{12} versus the common normalized length $2l/\lambda$. The results of such calculations are displayed in Figure 8.4.

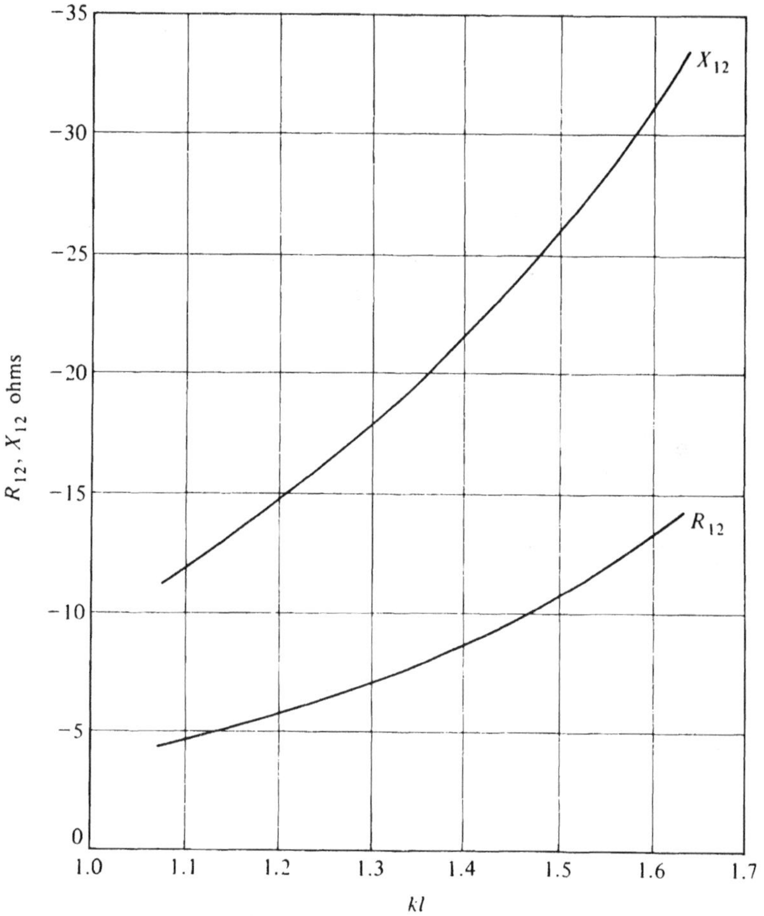

Fig. 8.4 The Mutual Impedance versus Their Common Length for Two Parallel Dipoles with a Broadside Separation of a Half-Wavelength

Assume that an infinitesimal gap generator is an adequate model for the center-feeding of the dipole, and that Z_{11} is essentially the same as the isolated self-impedance of the dipole. If the formula in (7.71) is used to compute the self impedance Z_{11}, then for every a/λ, a curve of X_{11} versus kl can be constructed. Where it crosses the X_{12} curve of Figure 8.4 defines a resonant length for the dipole of that radius plus ground plane, since Z_1^a, as defined by (8.3), will be pure real. This procedure permits the determination of a curve of resonant length versus dipole radius, with the result shown in Figure 8.5. With this relationship determined, one is able to deduce $R_1^a = R_{11} - R_{12}$ versus either resonant length or dipole radius. This result is also shown in Figure 8.5. One can observe that both the resonant length and input resistance

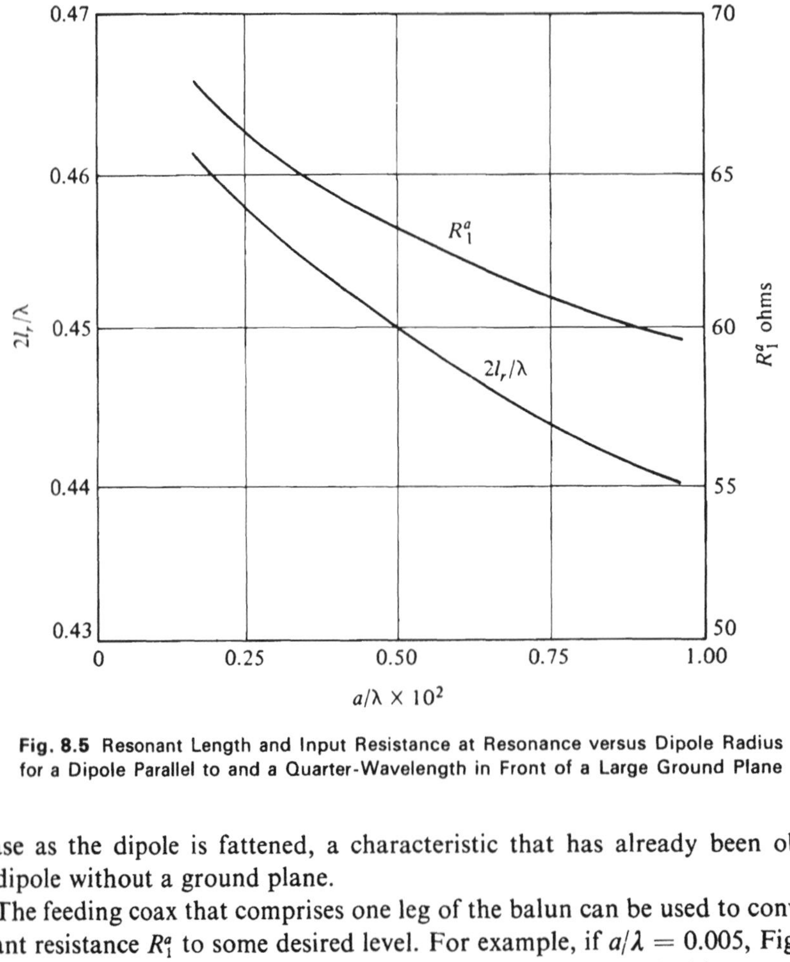

Fig. 8.5 Resonant Length and Input Resistance at Resonance versus Dipole Radius for a Dipole Parallel to and a Quarter-Wavelength in Front of a Large Ground Plane

decrease as the dipole is fattened, a characteristic that has already been observed for a dipole without a ground plane.

The feeding coax that comprises one leg of the balun can be used to convert the resonant resistance R_1^a to some desired level. For example, if $a/\lambda = 0.005$, Figure 8.5 indicates that $R_1^a = 63$ ohms. Imagine that it is desired to match this to a 50-ohm coax which runs along behind the ground plane. If the balun coax is air-filled, it becomes a quarter wave transformer, and should have a characteristic impedance given by $Z_0 = [(63)(50)]^{1/2} = 56$ ohms. The inner and outer conductors of the balun coax should thus be in a ratio that satisfies the equation $Z_0 = 60 \ln(b/a)$. This means that $b/a = 2.55$, which is a reasonable ratio, easy to achieve in a practical design.

8.4 Two-Wire-Fed Slots: Open and Cavity-Backed

If a slender rectangular slot is cut in a large ground plane and center-fed by a two-wire line, as suggested by Figure 7.1d, a radiator that is essentially the complement of the center-fed dipole will result. Booker's extension of Babinet's principle (compare with Section 7.16) provides the information that

$$Y^s = \frac{4}{\eta^2} Z^d \tag{8.4}$$

in which Y^s is the self-admittance of the slot, Z^d is the self-impedance of the equivalent dipole, and $\eta = 377$ ohms is the impedance of free space. If the slot has a length $2l$ and a width $w \ll l$, and if the ground plane has a negligible thickness, then the equivalent dipole has a length $2l$ and a radius $a = w/4$ (compare with Section 7.12). Equation 8.4 indicates that the slot admittance is pure real when the impedance of the equivalent dipole is pure real. It follows that the resonant length of the slot can be deduced from Figure 7.18 if a/λ is replaced by $w/4\lambda$.

If the slot is tuned to resonance, its input conductance can be computed by using (8.4) in conjunction with the information contained in Figure 7.18. For example, if $w/\lambda = 0.0064$ then $kl_{res} = 1.493$ and $G^s = (2/377)^2(70.75) = 1.99$ millimhos. Since the characteristic impedance of a two-wire line is given by

$$Z_0 = 120 \ln \left\{ \frac{D}{d} + \left[\left(\frac{D}{d} \right)^2 - 1 \right]^{1/2} \right\} \tag{8.5}$$

a match of this resonant slot with its feed can be achieved by choosing the wire diameter d and center-to-center spacing D of the two-wire line so that $Z_0 = R_s = 500$ ohms.

The slot in a ground plane, center-fed by a two-wire line, and radiating into both halves of space, gives approximately the same pattern as an electric dipole of the same length, but with **E** and **H** interchanged, that is, with the polarization rotated 90°. The practical applications of such radiators are limited, but a useful antenna emerges if the slot is forced to radiate only into a half-space by the introduction of a cavity that "boxes in" the slot on one side, as shown in Figure 8.6. The combination of slot and rectangular cavity can be fed by a coax, as shown, and if the cavity dimensions are large enough, the electric field distribution in the slot is approximately the same as before introduction of the cavity.

Assume that this is the case, and that the a-dimension of the cavity is in the range to permit propagation only of the TE_{10} mode. If additionally $b \cong w$, the field distribution in the cavity primarily consists of a standing wave of the TE_{10} type, with

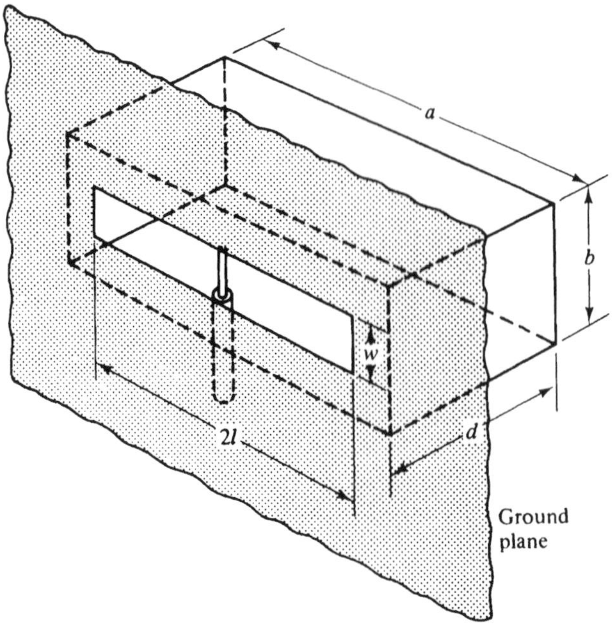

Fig. 8.6 A Cavity-Backed Slot in a Large Ground Plane

field components given by

$$
\begin{aligned}
E_y &= C \sin \frac{\pi x}{a} \sin \beta (d - z) \\
H_x &= -\left(\frac{\beta}{j\omega\mu_0}\right) C \sin \frac{\pi x}{a} \cos \beta (d - z) \\
H_z &= -\left(\frac{\pi}{j\omega\mu_0 a}\right) C \cos \frac{\pi x}{a} \sin B (d - z)
\end{aligned}
\tag{8.6}
$$

The complex power flow into the cavity is then

$$
\begin{aligned}
VI &= \int_0^a \int_0^b \mathbf{E} \times \mathbf{H} \cdot \mathbf{1}_z \, dx \, dy \\
&= \left(\frac{\beta}{j\omega\mu_0}\right) C^2 \sin \beta d \cos \beta d \int_0^a \int_0^b \sin^2 \frac{\pi x}{a} \, dx \, dy \\
&= \left(\frac{\beta ab}{2j\omega\mu_0}\right) C^2 \sin \beta d \cos \beta d
\end{aligned}
\tag{8.7}
$$

Since $V = E_y b = bC \sin \beta d$, the input admittance of the cavity is

$$
Y^c = \frac{I}{V} = \frac{VI}{V^2} = -j\left(\frac{a}{b}\right)\left(\frac{\beta}{2\omega\mu_0}\right) \cot \beta d \tag{8.8}
$$

With the current distribution in the radiating face of the ground plane the same as though the cavity were not present, the admittance of the slot is half what it was before, or

$$Y^s = \frac{2Z^d}{\eta^2} \tag{8.9}$$

The total admittance seen by the coaxial feed is therefore

$$Y = Y^s + Y^c = \frac{2Z^d}{\eta^2} - j\left(\frac{a}{b}\right)\left(\frac{\beta}{2\omega\mu_0}\right) \cot \beta d \tag{8.10}$$

If the cavity dimensions and the slot length are properly adjusted, the susceptive part of Y^s can cancel Y^c so that the coax sees a resistive load of amount $\eta^2/2R^d$. This is a high resistance, typically in the range of 1000 ohms. Coaxial feeds with their characteristic impedances in this range are impractical, and stepdown impedance transformations are narrow band. However, a T-bar transition, in which the center conductor of the coax terminates in a transverse bar rather than the upper side of the slot, has proven very effective in overcoming this impedance level problem. The cavity dimensions must be adjusted experimentally, but an exact match at the design frequency can be achieved, with the input VSWR held under 1.5 over a 30% bandwidth.[2]

8.5 Coaxially Fed Helix Plus Ground Plane

A helical antenna with circumference C_λ that is approximately one free-space wavelength will radiate in the axial mode, producing an end-fire beam that is circularly polarized. The helix is usually mounted over a ground plane and excited by a coax, as shown in Figure 2.11. The ground plane should be at least one-half wavelength in diameter, but all dimensions of this antenna are surprisingly noncritical, and good operation can be obtained over an extremely broad band of frequencies.

A typical example is provided by J. D. Kraus[3] who describes the performance of a six-turn helix with a 14° pitch angle. The helix diameter was $0.31\lambda_0$ at the center frequency of 400 MHz. Kraus used tubing of $0.02\lambda_0$ for the helix but comments that tubing with diameters ranging from $0.006\lambda_0$ to $0.05\lambda_0$ have little effect on the antenna characteristics. His measured patterns in the frequency range 300–500 MHz are shown in Figure 8.7. (Outside this range the patterns deteriorated).

The measured input impedance at 400 MHz was 130 ohms and this was transformed to 53 ohms via a quarter-wave section. Kraus measured the input VSWR versus frequency referred to a 53-ohm line, and the results are given in Figure 8.8. Also shown is the axial ratio ($|E_\theta|/|E_\phi|$ at end-fire) and the half-power beamwidth for the E_θ and E_ϕ pattern components. On all counts, this is seen to be a highly

[2] Radio Research Laboratory Staff, *Very High Frequency Techniques*, (New York: McGraw-Hill Book Co., Inc., 1947), Chapter 7.

[3] J. D. Kraus, *Antennas*, (New York: McGraw-Hill Book Co., Inc., 1950), pp. 208–12.

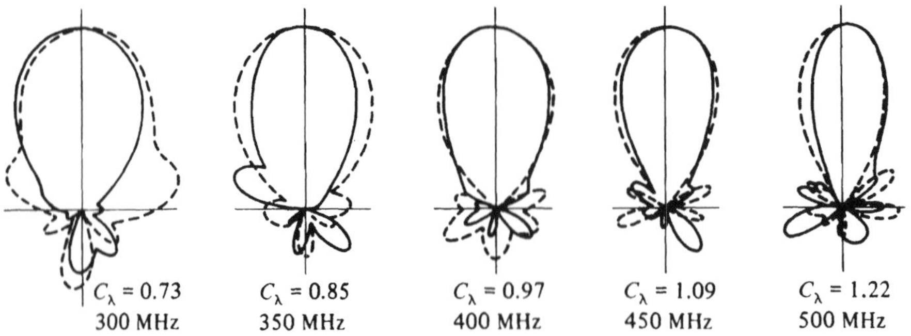

Fig. 8.7 Measured E-Field Patterns of a Six-Turn Helix with 14° Pitch; Polar Plots; Linear Scale (From *Antennas* by J. D. Kraus. Copyright 1950 McGraw-Hill. Used with permission of McGraw-Hill Book Company.)

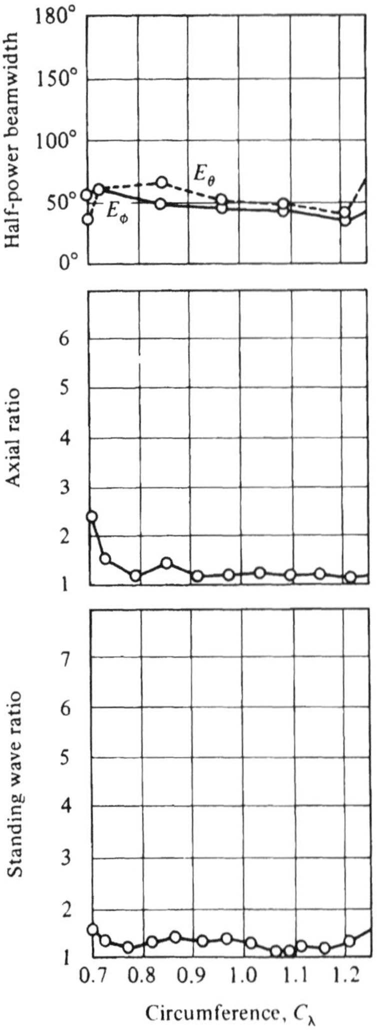

Fig. 8.8 Performance Data for a Six-Turn Helix with 14° Pitch (From *Antennas* by J. D. Kraus. Copyright 1950 McGraw-Hill. Used with permission of McGraw-Hill Book Company.)

satisfactory antenna for circularly polarized applications requiring a beam-type pattern.

8.6 The Design of an Endfire Dipole Array

A linear array of transverse, parallel dipoles, excited to produce an endfire beam, finds many applications, both as a transmitting antenna and as a receiving antenna. Chapters 4 and 5 dealt with the questions of current distribution, array length, and interelement spacing if a linear array is to produce an endfire pattern with a specified beamwidth and side lobe level. There remains the problem of determining how to deliver the desired currents to the individual radiators. Often this is done by means of a properly designed transmission line network. An example of an endfire dipole array fed in this manner will be presented in this section.

The reader will find that the design of endfire dipole arrays is a rich and diverse subject. In some designs the dipoles are not equispaced (log-periodic) and in some designs they may not all be fed by the transmission line (Yagi-Uda). These approaches will be treated in ensuing sections.

As an example of an equispaced endfire array with all elements directly fed, consider the antenna system shown in Figure 8.9. The dipoles are spaced $\lambda/3$ on centers and series-coupled to a two-wire transmission line which can be assumed air-filled. The currents are to be $I_1 = 1\underline{|-120^\circ}$, $I_2 = 1.5\underline{|0^\circ}$, and $I_3 = 1\underline{|120^\circ}$, resulting in an array factor with -17 dB side lobes, as shown in Figure 8.10.

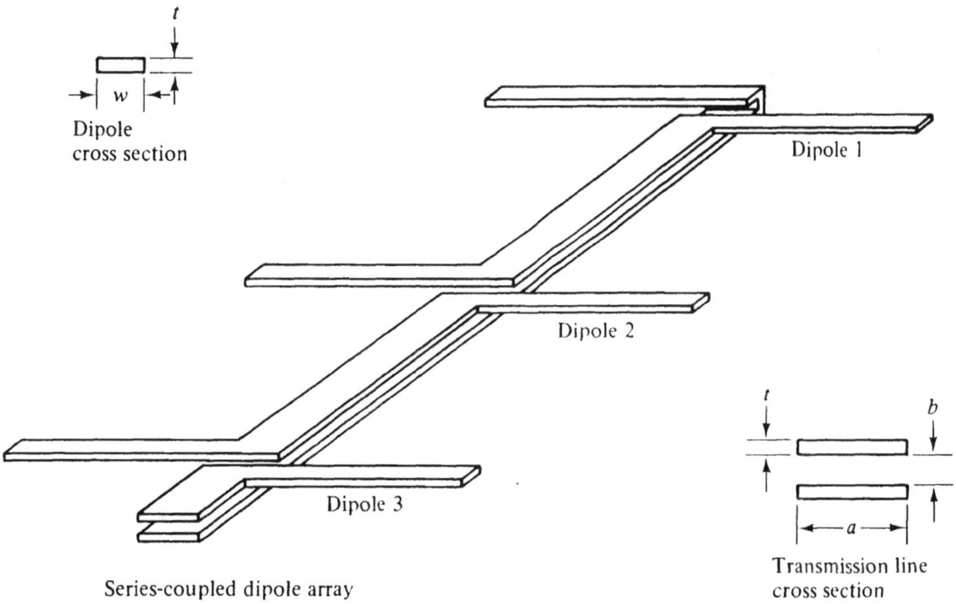

Fig. 8.9 An Endfire Array of Three Dipoles, Series-Fed by a Two-Wire Transmission Line

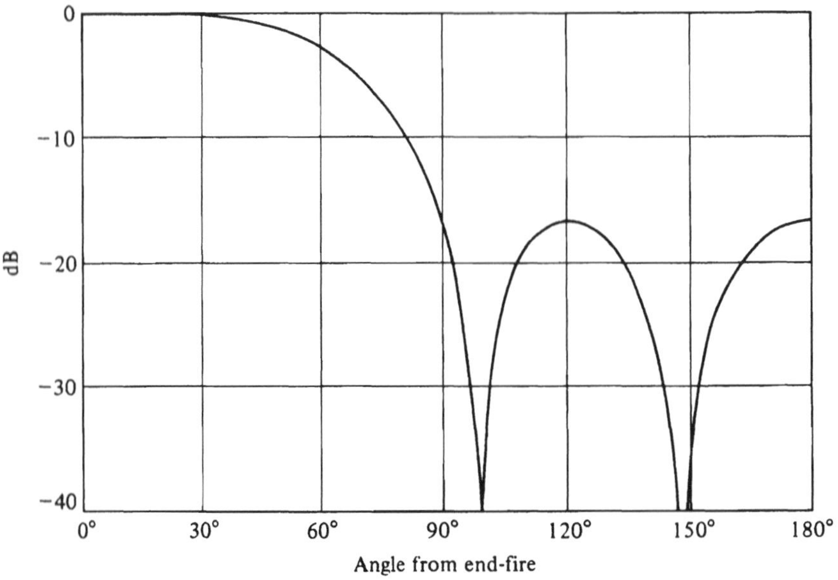

Fig. 8.10 Principal Plane Pattern; End-Fire Array of Three Dipoles, One-Third Wavelength on Centers

If (V_m, I_m) is the voltage/current pair at the terminals of the mth dipole, then

$$V_m = \sum_{n=1}^{3} I_n Z_{mn} \tag{8.11}$$

and thus the active (or input) impedances at the three terminals are

$$\begin{aligned} Z_1^a &= Z_{11} + \frac{I_2}{I_1} Z_{12} + \frac{I_3}{I_1} Z_{13} \\ &= Z_{11} + 1.5e^{j2\pi/3} Z_{12} + e^{j4\pi/3} Z_{13} \\ Z_2^a &= Z_{22} + 0.67e^{-j2\pi/3} Z_{21} + 0.67e^{j2\pi/3} Z_{23} \\ Z_3^a &= Z_{33} + 1.5e^{-j2\pi/3} Z_{32} + e^{-j4\pi/3} Z_{31} \end{aligned} \tag{8.12}$$

For the purpose of making a first calculation of the mutual impedances, assume that all three dipoles are $\lambda/2$ long. Equations 7.155 and 7.156 can be used to compute Z_{ij} for $\lambda/2$ dipoles spaced $\lambda/3$ on centers. The results are

$$\begin{aligned} Z_{12} = Z_{21} = Z_{23} = Z_{32} &= 21.40 - j36.76 = 42.54\underline{/-59.8^\circ} \\ Z_{13} = Z_{31} &= -25.34 - j5.32 = 25.90\underline{/-168.1^\circ} \end{aligned} \tag{8.13}$$

When this information is placed in (8.12), one finds that

$$
\begin{aligned}
Z_1^a &= Z_{11} + 38.76 + j79.98 \\
Z_2^a &= Z_{22} - 14.28 + j24.52 \\
Z_3^a &= Z_{33} - 46.54 - j19.52
\end{aligned} \tag{8.14}
$$

Consider the feeding problem from the vantage point of port 2, looking along the transmission line toward port 1. The equivalent circuit of this part of the antenna system is suggested by Figure 8.11. Since the input and output currents of a section of transmission line of length l are related by the equation

$$I_{IN} = I_{OUT}\left[\cos \beta l + j\frac{Z_L}{Z_0}\sin \beta l\right] \tag{8.15}$$

with Z_L the load impedance through which the output current flows, and Z_0 the characteristic impedance of the transmission line, it follows that in this case

$$1.5\underline{|0^\circ} = 1.0\underline{|-120^\circ}\left(\cos 120^\circ + j\frac{Z_1^a}{Z_0}\sin 120^\circ\right]$$

or that

$$\frac{Z_1^a}{Z_0} = 1.5 + j0.29 = 1.53\underline{|10.9^\circ} \tag{8.16}$$

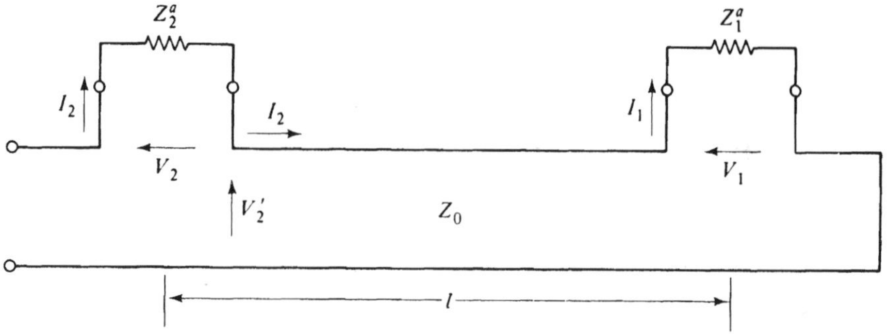

Fig. 8.11 The Equivalent Circuit of the Rearward Part of the Antenna System

If one makes the tentative assumption that the mutual impedance terms are insensitive to small length adjustments, Equations 8.14a and 8.16 taken in concert indicate that $2l_1$, the length of dipole 1, should be adjusted so that $Z_{11} + 38.76 + j79.98$ has a phase angle of 10.9°. Assume that Tai's empirical formula (7.71) gives an adequate measure of Z_{11}. If the dipole cross section (see the inset of Figure 8.9) is such that $w/\lambda = 0.0056$ and $t/\lambda = 0.0008$, then the equivalent cylindrical dipole (see Section 7.12) has the radius $a = (w + t)/4$ and thus $a/\lambda = 0.0016$. When this normalized radius is used in (7.71), it is found that if $2l_1/\lambda = 0.439$, then $Z_{11} = 50.28 - j62.54$ and $Z_1^a = 90.04 + j17.44 = 91.71\underline{|10.96^\circ}$. When this information is placed in Equation 8.16, one finds that the stretch of two-wire line connecting dipoles 1 and 2 should have a characteristic impedance of 60 ohms. With the transverse dimensions of the feed line as indicated in the second inset of Figure 8.9, the characteristic impedance is given by $Z_0 = \eta(b/a)$. For this section of line, $(b/a) = 60/377 = 0.16$, which is quite reasonable.

This process can be repeated by drawing the equivalent circuit of the antenna system from the vantage point of port 3. This is illustrated by Figure 8.12, which shows that the load impedance consists of Z_2^a in series with $Z_1^{a'}$, where $Z_1^{a'}$ is the active impedance Z_1^a transformed through an electrical length of 120°, that is,

$$Z_1^{a'} = Z_0 \frac{(Z_1^a/Z_0)\cos 120^\circ + j\sin 120^\circ}{\cos 120^\circ + j(Z_1^a/Z_0)\sin 120^\circ}$$

$$= 39.96 + j11.54 = 41.60\underline{|16.1^\circ} \tag{8.17}$$

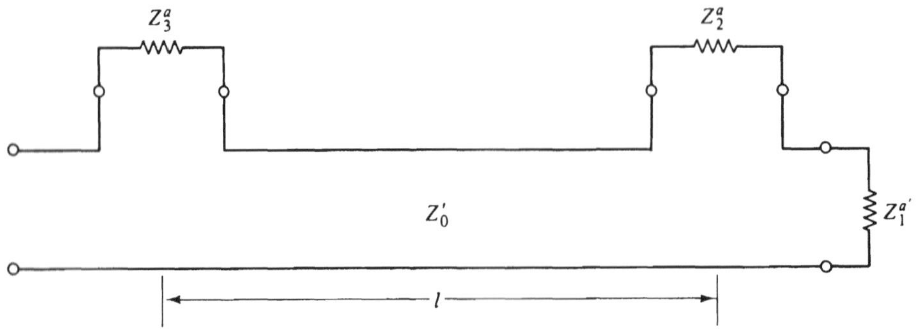

Fig. 8.12 The Equivalent Circuit of the Forward Part of the Antenna System

A second use of (8.15) gives

$$1\underline{|120^\circ} = 1.5\underline{|0^\circ}\left[\cos 120^\circ + j\frac{Z_2^a + Z_1^{a'}}{Z_0'}\sin 120^\circ\right] \tag{8.18}$$

in which Z_0' is the characteristic impedance of the length of two-wire line connecting dipoles 2 and 3.

Solution of (8.18) yields

$$\frac{Z_2^a + Z_1^{a'}}{Z_0'} = 0.667 - j0.192 = 0.694\underline{|-16.1^\circ} \tag{8.19}$$

From (8.14) and (8.17) it is known that

$$Z_2^a + Z_1^{a'} = Z_{22} + 25.68 + j36.06$$

and from (8.19) this should be an impedance with a phase angle of $-16.1°$. Another use of Tai's formula (7.71) indicates that, if $2l_2/\lambda = 0.441$, then $Z_{22} = 51.08 -$

$j58.32$ and $Z_2^a + Z_1^{a\prime} = 76.76 - j22.26 = 79.92\underline{|-16.2^\circ}$. A return to (8.19) reveals that $Z_0' = 115$ ohms should be the characteristic impedance of this stretch of two-wire line.

The input impedance to the entire antenna system is given by

$$Z_{IN} = Z_3^a + Z_0' \frac{[(Z_2^a + Z_1^{a\prime})/Z_0'] \cos 120^\circ + j \sin 120^\circ}{\cos 120^\circ + j[(Z_2^a + Z_1^{a\prime})/Z_0'] \sin 120^\circ}$$
$$= Z_{33} + 126.54 - j52.84 \tag{8.20}$$

If this total input impedance is to be pure real, then a third use of Tai's formula (7.71) indicates that the length of dipole 3 should be $2l_3/\lambda = 0.507$; then $Z_{33} = 76.58 + j52.82$ and

$$Z_{IN} = 203 \text{ ohms} \tag{8.21}$$

This input resistance can be matched to the transmitter (receiver) by a quarter-wave transformer or a tapered line.

The design procedure just described was predicated on the assumption that Z_{ij} is insensitive to these small dipole length changes. One can check this by repeating the process, using the new dipole lengths to compute the mutual impedances. When this is done, one finds that

$$Z_{12} = Z_{21} = 15.16 - j26.12$$
$$Z_{13} = Z_{31} = -21.50 - j4.66$$
$$Z_{23} = Z_{32} = 18.42 - j31.66$$

With these values used in the design procedure, it can be determined that

$$\frac{2l_1}{\lambda} = 0.449 \qquad \frac{2l_2}{\lambda} = 0.442 \qquad \frac{2l_3}{\lambda} = 0.507$$
$$Z_0 = 55 \qquad Z_0' = 125 \qquad Z_{IN} = 225$$

One can observe that $2l_1/\lambda$ is 2% higher, $2l_2/\lambda$ is changed but little, and $2l_3/\lambda$ is unchanged. Also, Z_0 is 8% lower, Z_0' is 8% higher, and Z_{IN} has been raised by 10%. Another iteration would show almost negligible further change and will not be undertaken.

If the gap problem is such that neither the Tai empirical formula (7.71) nor the more accurate King-Middleton equations (7.110) and (7.111) give a valid representation of the self-impedance, experimental data can be taken (at a modeled frequency if that is more convenient) and then formula-fitted. The design procedure is unchanged except for the substitution of the fitted formula for (7.71). Mutual impedance is not so seriously affected by the gap problem and thus (7.155) and (7.156) should still be applicable.

8.7 Yagi-Uda Type Dipole Arrays: Two Elements

If the end-fire array described in the previous section (three parallel dipoles in free space, all driven) were enlarged to become an array of four or more driven elements, an increase in directivity could be achieved. The same design procedure could be used to determine dipole lengths and transmission line characteristics. However, the design quickly becomes complicated by the addition of more elements, and the complexity of feed construction becomes onerous. A nice way out of this difficulty would result if it were to prove possible to eliminate the feed network, to short all dipoles save one, and to adjust the lengths and spacings so that the currents induced in the shorted dipoles (by the field of the driven dipole) would contribute to the creation of an end-fire pattern.

This possibility was first investigated by S. Uda[4] in the 1920's and reported in an English-language journal by his colleague H. Yagi[5], as a result of which such antennas have come to be known as Yagi-Uda arrays. They have many practical applications, including wide use by amateur radio enthusiasts.

One can gain considerable insight about such arrays by considering first the case of two parallel dipoles a distance d apart, one driven and the other parasitic (shorted). This situation is suggested by Figure 8.13. The mesh equations for this

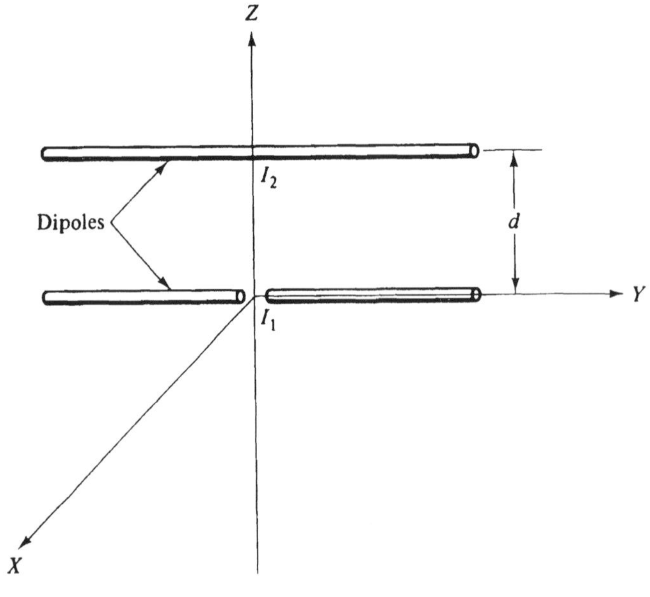

Fig. 8.13 An Array of Two Parallel Dipoles, One Driven, One Parasitic

[4] S. Uda, "Wireless Beam of Short Electric Waves," *J. IEEE (Japan)*, (1926), pp. 273–82 and (1927), pp. 1209–19.

[5] H. Yagi, "Beam Transmission of Ultra Short Waves," *IRE Proceedings.*, 16 (1928), 715–41.

array are

$$\begin{aligned} V_1 &= I_1 Z_{11} + I_2 Z_{12} \\ O &= I_1 Z_{21} + I_2 Z_{22} \end{aligned} \tag{8.22}$$

and thus

$$\frac{I_2}{I_1} = -\frac{Z_{12}}{Z_{22}} \tag{8.23}$$

Since the array factor is given by

$$\mathcal{A}(\theta) = 1 + \frac{I_2}{I_1} e^{jkd\cos\theta} \tag{8.24}$$

it is clear that the shape of the pattern is controlled by the spacing d/λ and by $-Z_{12}/Z_{22}$.

If the lengths of the two dipoles are near first resonance, the phase of the mutual impedance as a function of d/λ is quite insensitive to the values of $2l_1/\lambda$ and $2l_2/\lambda$. (Even the magnitude of Z_{12} does not show much sensitivity.) As an illustration of this, let $2l_1/\lambda = 0.475$, and let $2l_2/\lambda = 0.450$, 0.475, and 0.500, successively. If Equations 7.155 and 7.156 are used to compute the mutual impedance, one obtains the values entered in Table 8.4.

Because of this insensitivity, the *phase* of I_2/I_1, at a given spacing, is governed primarily by the phase of Z_{22}, as can be seen from (8.23). Continuing with the present illustration, one can determine an approximation to Z_{22} by using either Equations 7.110 and 7.111 or Figure 7.18. With $a_2/\lambda = 0.0032$, this gives the values shown in Table 8.5. When these values of Z_{22} are used in conjunction with the entries of Table 8.4 and Equation 8.23, the current ratios shown in Table 8.6 are obtained.

Since the objective is to produce an end-fire pattern with this two-dipole array, one can scan the entries of Table 8.5 to see if there is some combination of d/λ and $2l_2/\lambda$ which will enhance end-fire radiation. For the beam to be at $\theta = 0°$, enhance-

TABLE 8.4 Mutual impedance versus spacing between two parallel dipoles: $2l_1/\lambda = 0.475$

	Z_{12} ohms		
d/λ	$2l_2/\lambda = 0.450$	$2l_2/\lambda = 0.475$	$2l_2/\lambda = 0.500$
0.10	$53.94\,\underline{\vert 1.52°}$	$58.19\,\underline{\vert 3.22°}$	$62.78\,\underline{\vert 4.98°}$
0.15	$49.08\,\underline{\vert -9.38°}$	$52.73\,\underline{\vert -8.45°}$	$56.62\,\underline{\vert -7.50°}$
0.20	$44.42\,\underline{\vert -21.93°}$	$47.67\,\underline{\vert -21.42°}$	$51.12\,\underline{\vert -20.90°}$
0.25	$40.23\,\underline{\vert -35.53°}$	$43.18\,\underline{\vert -35.28°}$	$46.30\,\underline{\vert -35.02°}$
0.30	$36.55\,\underline{\vert -49.91°}$	$39.26\,\underline{\vert -49.82°}$	$42.11\,\underline{\vert -49.73°}$
0.35	$33.35\,\underline{\vert -64.87°}$	$35.84\,\underline{\vert -64.89°}$	$38.47\,\underline{\vert -64.91°}$
0.40	$30.57\,\underline{\vert -80.32°}$	$32.86\,\underline{\vert -80.40°}$	$35.31\,\underline{\vert -80.48°}$
0.45	$28.14\,\underline{\vert -96.14°}$	$30.28\,\underline{\vert -96.26°}$	$32.56\,\underline{\vert -96.38°}$
0.50	$26.04\,\underline{\vert -112.25°}$	$29.02\,\underline{\vert -112.40°}$	$30.16\,\underline{\vert -112.56°}$

TABLE 8.5 Self-impedance of a cylindrical dipole versus length (King-Middleton corrected second-order approximation; ($a_2/\lambda = 0.0032$)

$2l_2/\lambda$	Z_{22} ohms
0.450	$60.56 - j29.58 = 67.40 \lfloor -26.03^\circ$
0.475	$72.06 + j\,4.04 = 72.17 \lfloor 3.21^\circ$
0.500	$83.60 + j41.34 = 93.26 \lfloor 26.31^\circ$

TABLE 8.6 Relative current versus spacing for two parallel dipoles, one driven, one parasitic: $2l_1/\lambda = 0.475$

	$I_2/I_1 = -Z_{12}/Z_{22}$		
d/λ	$2l_2/\lambda = 0.450$	$2l_2/\lambda = 0.475$	$2l_2/\lambda = 0.500$
0.10	$0.800 \lfloor -152.45^\circ$	$0.806 \lfloor 180.01^\circ$	$0.673 \lfloor 158.67^\circ$
0.15	$0.728 \lfloor -163.35^\circ$	$0.731 \lfloor 168.34^\circ$	$0.607 \lfloor 146.19^\circ$
0.20	$0.659 \lfloor -175.90^\circ$	$0.661 \lfloor 155.37^\circ$	$0.548 \lfloor 132.79^\circ$
0.25	$0.597 \lfloor 170.50^\circ$	$0.598 \lfloor 141.51^\circ$	$0.496 \lfloor 118.67^\circ$
0.30	$0.542 \lfloor 156.12^\circ$	$0.544 \lfloor 126.97^\circ$	$0.452 \lfloor 103.96^\circ$
0.35	$0.495 \lfloor 141.16^\circ$	$0.497 \lfloor 111.90^\circ$	$0.413 \lfloor 88.78^\circ$
0.40	$0.454 \lfloor 125.71^\circ$	$0.455 \lfloor 96.39^\circ$	$0.379 \lfloor 73.21^\circ$
0.45	$0.418 \lfloor 109.89^\circ$	$0.420 \lfloor 80.53^\circ$	$0.349 \lfloor 57.31^\circ$
0.50	$0.386 \lfloor 93.78^\circ$	$0.388 \lfloor 64.39^\circ$	$0.323 \lfloor 41.13^\circ$

ment will occur if I_2 lags I_1 by kd radians. Clearly, none of the values in Table 8.6 fit this condition. However, cancellation in the direction $\theta = 180^\circ$ also corresponds to a beam at $\theta = 0^\circ$. Such cancellation results if I_2 lags I_1 by $\pi - kd$ radians. It can be observed that if $d/\lambda = 0.10$ and $2l_2/\lambda = 0.450$, this condition is almost satisfied.

Similarly, if one wishes to produce an end-fire beam at $\theta = 180^\circ$, this goal will be helped if I_2 leads I_1 by kd radians. Inspection of the entries in Table 8.6 indicates that this will occur if $d/\lambda = 0.30$ and $2l_2/\lambda = 0.500$.

The criterion of attempting to match the phase of I_2/I_1 with either kd or its supplement is actually too crude, since a forward optimum and a rearward optimum cannot be achieved at the same spacing. A more useful way to go about determining the optimum spacing is to compute the directivity corresponding to each entry in Table 8.6. When this is done, one finds that for $2l_2/\lambda = 0.450$, an end-fire beam occurs at $\theta = 0^\circ$, and that the directivity versus spacing is given by the solid curve in Figure 8.14. Similarly, for $2l_2/\lambda = 0.500$, an end-fire beam occurs at $\theta = 180^\circ$, with the directivity versus spacing indicated by the dashed curve in Figure 8.14.

An important conclusion can be drawn from this exercise. If a shorted dipole is spaced an appropriate distance from the driven dipole, an end-fire beam can be produced. When the parasitic dipole is shorter than the driven dipole, it is called a *director*, and the end-fire beam is in the direction from the driven element past the parasite. When the parasitic dipole is longer than the driven dipole, it is called a

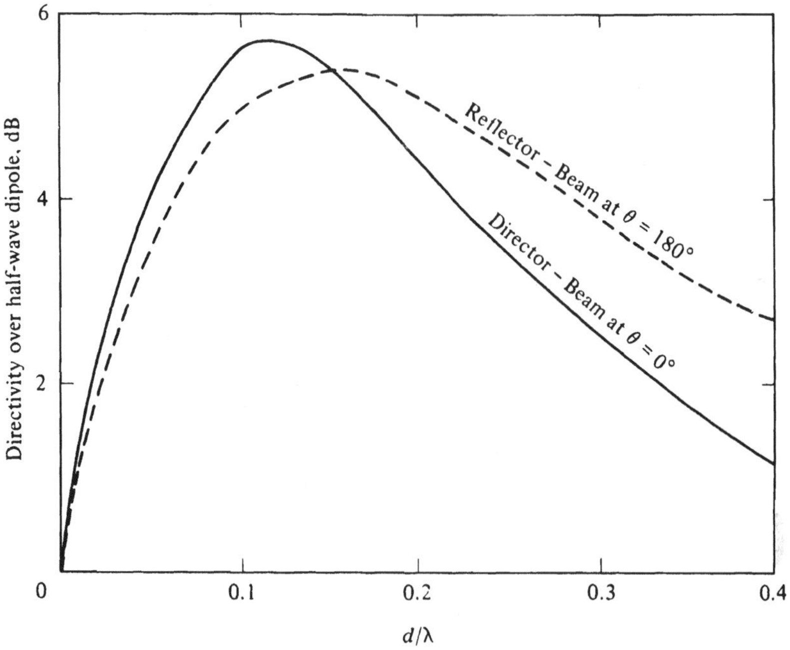

Fig. 8.14 Peak Directivity versus Element Spacing for an Array of Two Parallel Dipoles; One Element Driven, the Other Parasitic

reflector, and the end-fire beam is in the direction from the parasite past the driven element. From Figure 8.14, the optimum spacing (in the sense of maximizing directivity) is seen to be about 0.12λ for the director case and 0.16λ for the reflector case. Pattern cuts in the XZ-plane are shown in Figure 8.15 for these two optimum designs.

Under the assumption that Z_{12} is insensitive to small changes in $2l_1/\lambda$. all of the foregoing is still valid if $2l_1/\lambda$ is no longer exactly 0.475. Adjustment of $2l_1/\lambda$ can make the input impedance pure real, which is often desired. From (8.22),

$$Z_{IN} = \frac{V_1}{I_1} = Z_{11} + \frac{I_2}{I_1} Z_{12} = Z_{11} - \frac{Z_{12}^2}{Z_{22}} \tag{8.25}$$

For the optimum director case, one can deduce from Tables 8.4 and 8.6 that

$$-\frac{Z_{12}^2}{Z_{22}} = -37.59 - j13.95 \tag{8.26}$$

Therefore, $2l_1/\lambda$ should have a value such that the reactive component of Z_{11} is $+13.95$ ohms. Use of (7.110) and (7.111) or Figure 7.19 leads to the conclusion that, with $a_1/\lambda = 0.0032$, $2l_1/\lambda$ should be 0.482, and then $Z_{11} = 75.82 + j13.95$ and $Z_{IN} = 38$ ohms.

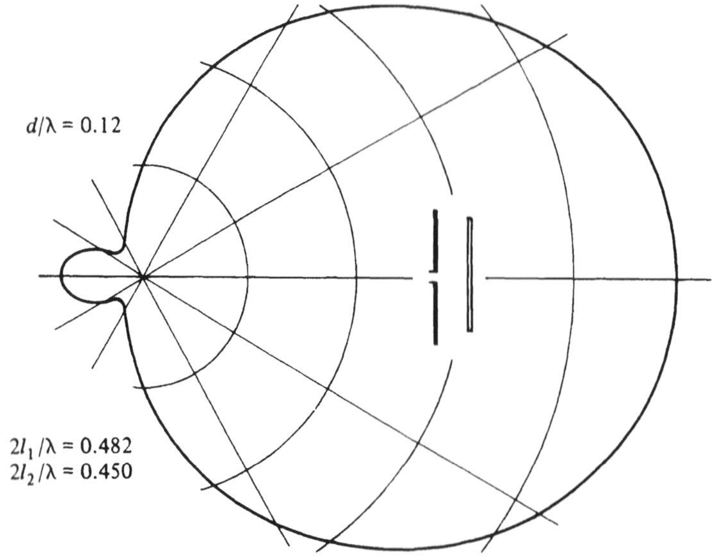

(a)

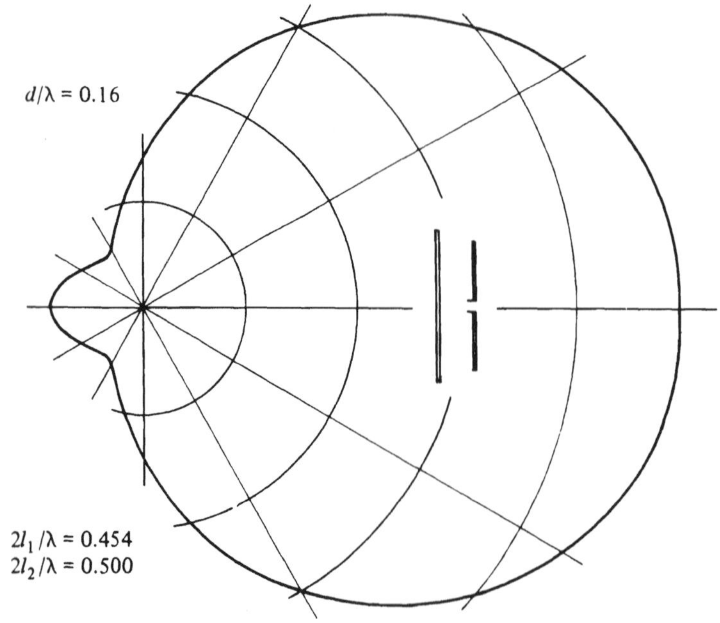

(b)

Fig. 8.15 *H*-Plane Power Patterns for Two-Element Yagi-Uda Arrays

Similarly, for the optimum reflector case, Tables 8.4 and 8.6 yield the information that

$$-\frac{Z_{12}^2}{Z_{22}} = -22.67 + j24.03 \tag{8.27}$$

For this situation, $2l_1/\lambda$ should be chosen so that Z_{11} has a reactive component equal to -24.03 ohms. One finds that $2l_1/\lambda$ should be 0.454, and this gives $Z_{11} = 62.46 - j24.03$, so that $Z_{IN} = 40$ ohms.

Improved accuracy would result if this design procedure were iterated, but the values obtained are indicative of the method, and probably of sufficient accuracy for most purposes.

8.8 Yagi-Uda Type Dipole Arrays: Three or More Elements

A natural extension of the development in the previous section is the three-element dipole array, with one driven element flanked on each side by a director and a reflector, as shown in Figure 8.16. The mesh equations for this array are

$$V_m = \sum_{n=1}^{3} I_n Z_{mn} \tag{8.28}$$

with $V_1 = V_3 = 0$. Simultaneous solution of the first and third equations of (8.28) gives

$$I_1/I_2 = \frac{Z_{13}Z_{23} - Z_{12}Z_{33}}{Z_{11}Z_{33} - Z_{13}^2} \tag{8.29}$$

$$I_3/I_2 = \frac{Z_{13}Z_{12} - Z_{23}Z_{11}}{Z_{11}Z_{33} - Z_{13}^2} \tag{8.30}$$

whereas the second equation of (8.28) gives the input impedance

$$\begin{aligned} Z_{IN} = \frac{V_2}{I_2} &= Z_{22} + \left(\frac{I_1}{I_2}\right)Z_{21} + \left(\frac{I_3}{I_2}\right)Z_{23} \\ &= \frac{Z_{11}Z_{22}Z_{33} - Z_{23}^2Z_{11} - Z_{13}^2Z_{22} - Z_{12}^2Z_{23} + 2Z_{12}Z_{13}Z_{23}}{Z_{11}Z_{33} - Z_{13}^2} \end{aligned} \tag{8.31}$$

Under the assumption that the presence of the director does not materially affect the proper length of the reflector and vice versa, one can initiate a computer search in the neighborhood of the two separate designs of the previous section to find optimum values of $2l_i/\lambda$ and d_i/λ. When this is done, a fairly broad range of dimensions gives good results. Both the reflector and the director can be spaced 0.15λ to 0.20λ from the driven element without much effect on the pattern. As an example, if

$$d_1/\lambda = d_3/\lambda = 0.20, \qquad 2l_1/\lambda = 0.450, \qquad 2l_2/\lambda = 0.475, \qquad 2l_3/\lambda = 0.500$$

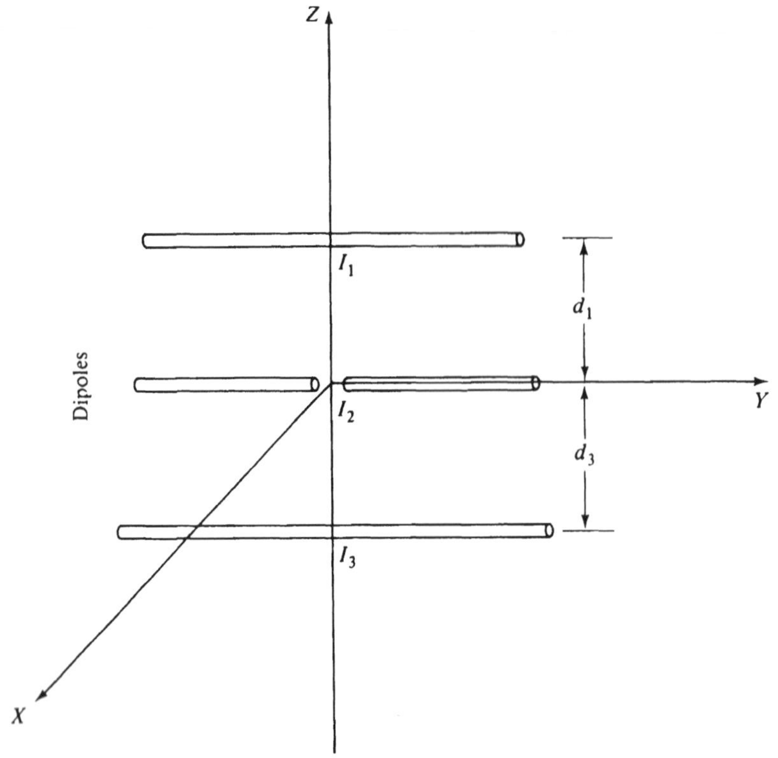

Fig. 8.16 A Linear Array of Three Parallel Equispaced Dipoles

then it is found that

$$\frac{I_1}{I_2} = 0.635\underline{|-143^\circ}, \qquad \frac{I_3}{I_2} = 0.389\underline{|143^\circ}, \qquad Z_{IN} = 25.6\underline{|10^\circ}$$

A small adjustment in the length of the driven element will tune the input impedance to resonance.

An XZ-plane plot of the power pattern for this current distribution is shown in Figure 8.17. The directivity of this three-element Yagi-Uda array is 7.5 dB above that of a single dipole.

Experience shows that attempts to place more reflectors behind the driven element are ineffectual because the total field reaching them is small, but a string of directors can be placed in front of the driven element, with each additional director resulting in an increase in directivity. For example, with 20 directors, one driven element, and one reflector, and with the overall length of the Yagi-Uda array 6.5λ, the directivity over a single half-wave dipole is 19 dB. In such designs, successive directors are about 0.5% shorter, and the interelement spacing increases to about

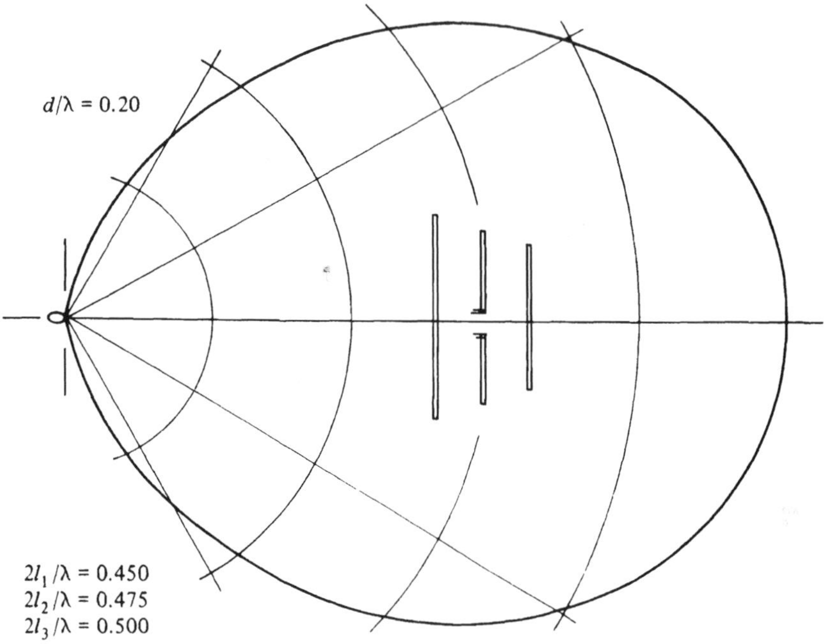

Fig. 8.17 *H*-Plane Power Pattern for a Three-Element Yagi-Uda Array

0.35λ at the fifth director and then remains constant. The reader interested in practical details should consult the Radio Amateur's Handbook.[6]

The multi-element Yagi-Uda array has been analyzed as a slow-wave structure, and a summary of this approach, with bibliography, can be found in E. A. Wolff.[7] It has also been treated by G. A. Thiele,[8] using a point-matching technique within the framework of the method of moments. His theoretical results for the pattern of a 15-element array are in extraordinary agreement with experiment.

8.9 Frequency-Independent Antennas: Log-Periodic Arrays

Yagi-Uda dipole arrays, described in the previous two sections, have the advantage that there is only one fed element. At a single frequency, with many dipoles in the array, it is possible to get a good end-fire pattern and a real input impedance at a convenient level. However, such arrays are not particularly broadband, either in terms of pattern performance or input VSWR.

[6]See the Radio Amateur's VHF Manual, 3rd Edition (Newington, Conn: ARRL, Inc., 1972), pp. 153–55.

[7]E. A. Wolff, *Antenna Analysis* (New York: John Wiley and Sons, Inc., 1966), pp. 405–9.

[8]G. A. Thiele, "Wire Antennas," *Computer Techniques for Electromagnetics*, ed. R. Mittra (Oxford: Pergamon Press, 1973), Chapter 2, pp. 43–48. See particularly Figure 2.21.

There exists a type of dipole array, with a simple feeding structure, which will produce patterns similar to those of a Yagi-Uda array, but over a remarkably broad band of frequencies (6:1 or more), with the added virtue of a low input VSWR throughout the same frequency range. Such arrays are called *log-periodic*, and a satisfactory way to understand their performance is to start with a general analysis of frequency independent antennas.

It was recognized by V. H. Rumsey[9] that an antenna with shape specified entirely in terms of angles will have pattern and impedance characteristics that are independent of frequency. His analysis of such antennas is reproduced in what follows, except that the treatment of the three dimensional case is simplified.[10,11]

Consider an antenna, with both terminals indefinitely close to the origin of a spherical coordinate system, being symmetrically disposed along the $\theta = 0°$, $180°$ axis. Assume that the antenna consists of perfect conductors and is surrounded by an infinite homogeneous and isotropic medium. Let the surface of this antenna be described by

$$r = F(\theta, \phi) \tag{8.32}$$

Equation 8.32 does not necessarily imply that the material composing the antenna is indefinitely thin. There may be several branches to the function $F(\theta, \phi)$, corresponding to inner and outer surfaces.

Suppose that one wishes to scale this antenna to a new frequency that is K times lower than the original frequency. The antenna must be made K times bigger, resulting in a surface

$$r' = KF(\theta, \phi) \tag{8.33}$$

in which K depends neither on θ nor on ϕ.

Imagine that when this is done the new surface is found to be identical to the old, that is, the surfaces are not only similar, but they can actually be made *congruent*. (This implies, of course, that both surfaces are infinite.) A little thought will convince the reader that congruence, if it occurs, can only be established through a rotation in ϕ. (Translation is barred because both antennas have their terminals at the same origin. Rotation in θ is barred because both pairs of terminals are symmetrically disposed along the $\theta = 0°$, $180°$ axis.) Thus, for congruence,

$$KF(\theta, \phi) = F(\theta, \phi + C) \tag{8.34}$$

[9]V. H. Rumsey, "Frequency Independent Antennas," *IRE National Convention Record, Part I* (March 1957), 114–18. Also, see Rumsey's textbook of the same title (New York: Academic Press, 1966).

[10]Much of the material in this section is taken from a tutorial paper written by the author in 1962. See "A View of Frequency Independent Antennas," *Microwave Journal*, (1962), pp. 61–68. Copyright 1962 *Microwave Journal*. Reprinted with permission.

[11]For a review of the highlights in the development of this subject, see E. C. Jordan, G. A. Deschamps, J. D. Dyson, and P. E. Mayes, "Developments in Broadband Antennas," *IEEE Spectrum*, 1 (1964), 58–71.

in which C is the angle through which the second antenna must be rotated in order to achieve congruence with the first. Here C depends on K, but neither depends on θ nor ϕ.

Congruence implies that the original antenna would perform exactly the same at the two frequencies, except for a rotation C in the azimuthal coordinate of its radiation pattern as the frequency is changed from ν to ν/K. If it should develop that the range of K is unrestricted, that is, if

$$0 \leq K < \infty \tag{8.35}$$

then the original antenna must have a pattern shape and impedance that are independent of frequency. (The pattern may rotate in ϕ with frequency due to the parameter C, but its shape will be unaltered.)

If (8.35) holds, the nature of the function $F(\theta, \phi)$ can be deduced from (8.34). Differentiation of both sides with respect to C gives

$$\frac{dK}{dC}F(\theta, \phi) = \frac{\partial}{\partial C}F(\theta, \phi + C) = \frac{\partial}{\partial(\phi + C)}F(\theta, \phi + C) \tag{8.36}$$

whereas differentiation of both sides with respect to ϕ gives

$$K\frac{\partial}{\partial\phi}F(\theta, \phi) = \frac{\partial}{\partial\phi}F(\theta, \phi + C) = \frac{\partial}{\partial(\phi + C)}F(\theta, \phi + C) \tag{8.37}$$

Combining these two results, one obtains

$$\frac{dK}{dC}F(\theta, \phi) = K\frac{\partial}{\partial\phi}F(\theta, \phi) \tag{8.38}$$

which can be rewritten, with the aid of (8.32), in the form

$$\frac{1}{K}\frac{dK}{dC} = \frac{1}{r}\frac{\partial r}{\partial\phi} \tag{8.39}$$

Since the left side of (8.39) is independent of θ and ϕ, it follows that

$$r = F(\theta, \phi) = e^{a\phi}f(\theta) \tag{8.40}$$

is a general solution of (8.39), in which $a = (1/K)(dK/dC)$ is a parameter and $f(\theta)$ is a completely arbitrary function.

Equation 8.40 was first derived by V. H. Rumsey[12] and is the central result of the analysis. Any antenna that has surfaces which can be described by functions of the form of (8.40) will have pattern and impedance characteristics that are independent of frequency. Several important classes of such antennas can be identified.

[12]Rumsey, "Frequency Independent Antennas."

PLANAR SPIRALS If one chooses $f(\theta)$ so that

$$\frac{df}{d\theta} = f'(\theta) = A\delta\left(\frac{\pi}{2} - \theta\right) \tag{8.41}$$

with A an arbitrary positive constant and δ the Dirac delta function, then (8.40) becomes

$$r = r_0 e^{a(\phi - \phi_0)} \qquad \text{if } \theta = \frac{\pi}{2} \tag{8.42}$$

$$r = 0 \qquad \text{if } \theta \neq \frac{\pi}{2} \tag{8.43}$$

In (8.42), $r_0 e^{-a\phi_0}$ is a substitution for A. The antenna surface is seen to lie in the XY-plane and (8.42) can be recognized as the equation of an equiangular spiral.

Since the parameter A is arbitrary, it follows that in (8.42) r_0 can be considered as fixed, with ϕ_0 playing the role of a parameter. If ϕ_0 is given the values 0 and π, the antenna of Figure 8.18a results. If ϕ_0 is allowed to take on the values 0, $\pi/2$, π, and $3\pi/2$, four spiral forms occur, as shown in Figure 8.18b, with several symmetrical possibilities for connecting the terminals. If ϕ_0 is allowed to assume *all* values from 0 to ϕ_1, and *all* values from π to $\pi + \phi_1$, with ϕ_1 arbitrary, an antenna of the type shown in Figure 8.18c arises. From these few examples, the variety of possible combinations for the planar spiral case is seen to be endless.

In the ideal theoretical analysis resulting in (8.36), the antenna shapes shown in Figure 8.18 are assumed to be infinite. However, investigation of the current distribution on such antennas reveals that the principal part of the excitation occurs in a resonant region around $r = \lambda/2$. Thus when the planar spiral antennas of Figure 8.18 are truncated at some finite size, one can anticipate that the antenna should perform satisfactorily down to a frequency at which the wavelength is comparable to the antenna size. An upper frequency limit can be expected when the actual antenna terminals no longer behave as a pair of points infinitesimally apart at the origin. Experiments confirm these frequency limits.

An interesting feature of the planar spiral antenna is that Babinet's principle may be applied to it (see Appendix F). With reference to Figure 8.18c, if Z_1 is the input impedance of the antenna for a value $\phi_1 = \alpha$ and Z_2 is the impedance for a

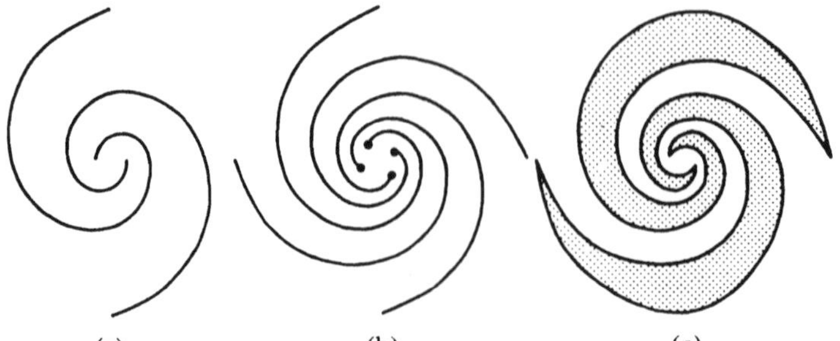

(a) (b) (c)

Fig. 8.18 Some Simple Shapes for Frequency-Independent Planar Antennas value $\phi_1 = \pi - \alpha$, the two antennas form complementary screens and thus $Z_1 Z_2 = \eta^2/4$. For the special case that $\phi_1 = \pi/2$, $Z_1 = Z_2 = 188.5$ ohms. This self-complementary feature was first pointed out by Mushiake.

CONICAL SHAPES If one returns to the generic equation of (8.40), it is apparent that an equally acceptable choice for $f(\theta)$ would result from the requirement that

$$f'(\theta) = A\delta(\beta - \theta) \tag{8.44}$$

in which β is any angle in the range $0 \leq \beta \leq \pi$. The previous discussion of planar spirals can be repeated, except that now the spirals are wrapped on a conical surface.

By symmetry, the *planar* spiral must exhibit the same pattern shape in $0 \leq \theta \leq \pi/2$ as in $\pi/2 \leq \theta \leq \pi$, which severely limits its practical applications. But the conical spiral does not suffer from this limitation, and appropriate selection of the value of β can result in an antenna which produces a single end-fire beam that is circularly polarized.[13]

THE LOG-PERIODIC ELEMENT An interesting and ultimately practical approximation to a frequency independent antenna has been conceived by R. H. DuHamel and D. E. Isbell[14] and is illustrated in Figure 8.19. If successive radii are in the

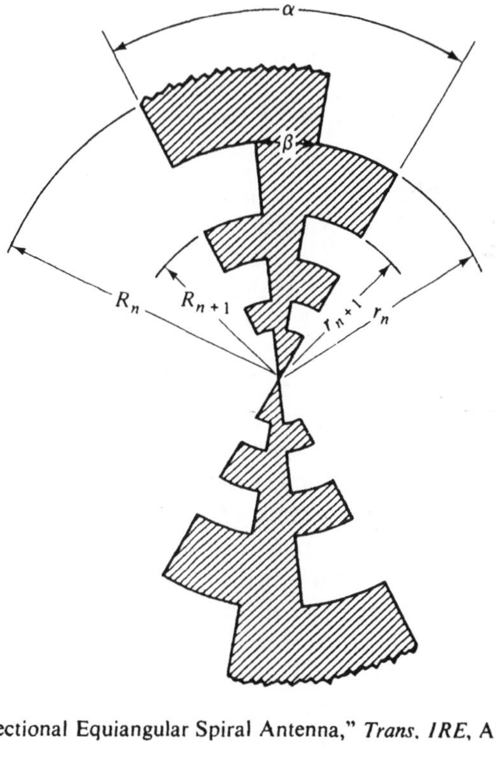

Fig. 8.19 A Logarithmically Periodic Planar Antenna

[13] J. D. Dyson, "The Unidirectional Equiangular Spiral Antenna," *Trans. IRE*, AP-7 (1959), 329–34.

[14] R. H. DuHamel and D. E. Isbell, "Broadband Logarithmically Periodic Antenna Structures," *IRE National Convention Record* (1957), pp. 119–28.

common ratio

$$\frac{r_{n+1}}{r_n} = \frac{R_{n+1}}{R_n} = \tau = e^{2\pi a} \tag{8.45}$$

then if the shape of the original antenna is described by

$$r = f(\theta) \tag{8.46}$$

a new antenna, scaled to give

$$r = Kf(\theta) \tag{8.47}$$

can be made congruent to the original antenna, but only for a restricted discrete set of values for K. These K values are given by

$$K = \tau^m = e^{2\pi m a} \tag{8.48}$$

in which m is an integer. Thus, at any two frequencies in the ratio $e^{2\pi m a}$, the antenna of Figure 8.19 should give the same pattern and impedance. For this reason, the configuration is called a *logarithmically periodic planar antenna.*

If it turns out that the performance does not vary greatly in the frequency range $\nu_1 \leq \nu \leq \nu_2$, with $\nu_2/\nu_1 = \tau = e^{2\pi a}$, then the configuration of Figure 8.19 is broadband. DuHamel and his co-workers found experimentally that some choices of the parameters α, β, and τ for this antenna gave better frequency characteristics than others.

LOG-PERIODIC WIRE ANTENNAS One of the most important practical advances in the subject of frequency-independent antennas was made by DuHamel, who discovered that the fields fell off very sharply with distance from the conductors of antennas of the types shown in Figure 8.18c and 8.19. This suggested that perhaps there was a strong current concentration near the edges of the conductors. If this were so, then removal of most of the material of the antenna of Figure 8.19 should not seriously affect the pattern and impedance characteristics. When this removal is accomplished, the wire antenna of Figure 8.20 results. As anticipated, the performance of this antenna is almost identical to that of its parent.

DuHamel found further that one need not adhere strictly to the shape of Figure 8.20. A wire structure of the form shown in Figure 8.21 is equally suitable. The criterion that must be observed is that the lengths of the transverse elements and their spacings must increase in the same geometric progression.

The antennas of Figures 8.19 through 8.21 suffer from the same deficiency as the planar spiral, in that they create bidirectional patterns, for which the practical applications are limited. But just as Dyson was able to overcome this limitation for spiral antennas by wrapping the spirals on cones, DuHamel found that it is possible to enhance the applicability of log-periodic antennas by folding the two halves so that they lie on the surfaces of a wedge. This is illustrated in Figure 8.22. The result

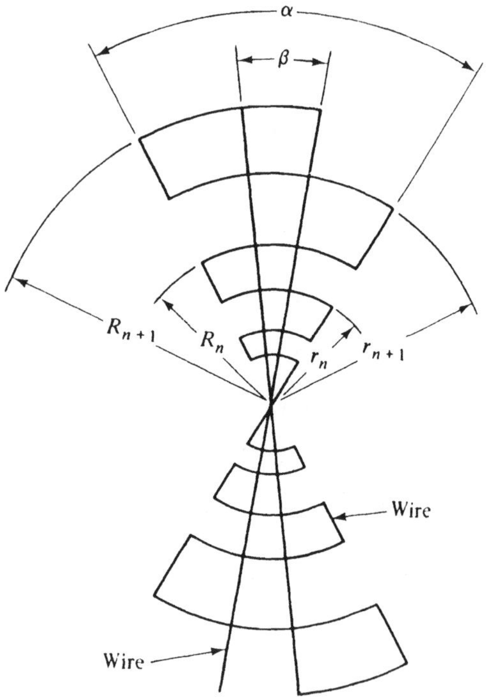

Fig. 8.20 A Wire Evolvement of the Antenna of Fig. 8.19

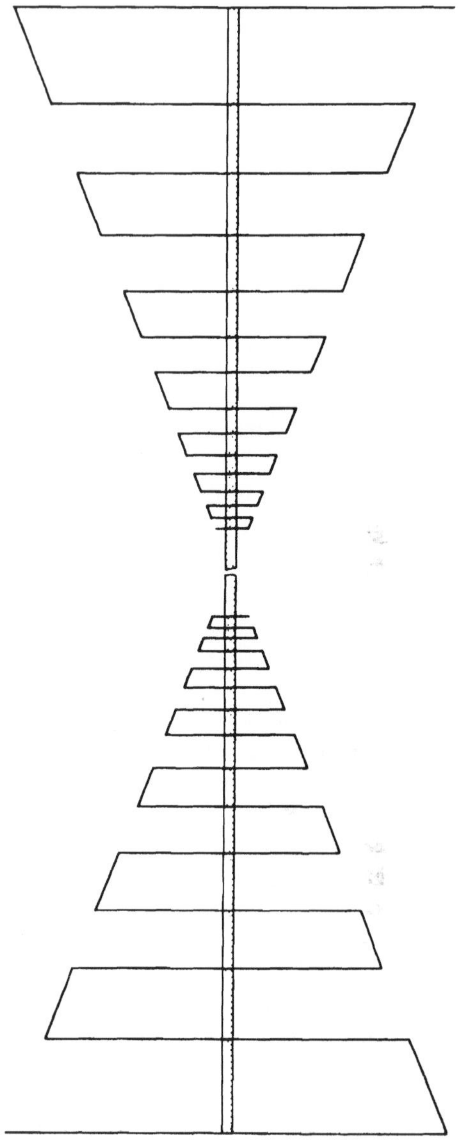

Fig. 8.21 A Wire Variant of the Antenna of Fig. 8.20

is a vee-type antenna which radiates a unidirectional pattern whose main beam points off the tip of the antenna. Pattern and input impedance characteristics are comparable to what Dyson was able to achieve with conical spirals, the principal difference being that linear polarization is obtained with the log-periodic vee whereas circular polarization occurs with the conical spiral.

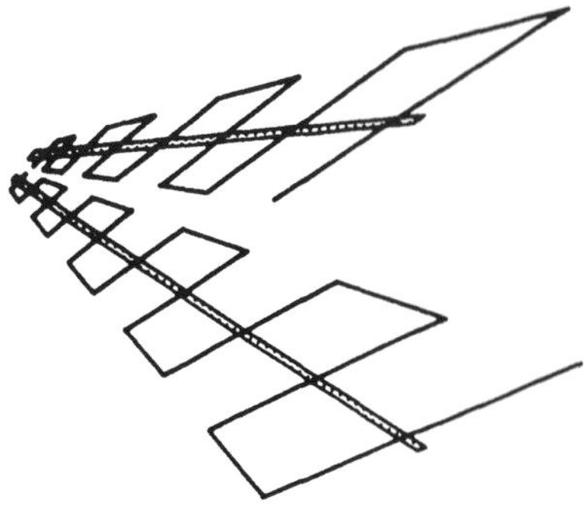

Fig. 8.22 A Log-Periodic Wire Vee Antenna

An example of the performance of a log-periodic vee-wire antenna of the type shown in Figure 8.22 is provided by R. H. DuHamel and F. R. Ore.[15] For a wedge angle of 45°, the *E*-plane and *H*-plane beamwidths were each 66°, the gain was 9.2 dB, and the front-to-back ratio was 12.3 dB. The average input impedance was 110 ohms with a VSWR referred to this value which did not exceed 1.45 over a 10:1 band. Typical patterns for one frequency octave are shown in Figure 8.23.

Two log-periodic vee-antennas can be placed in space quadrature with a common apex, as shown in Figure 8.24. If fed with equal signals that are in time quadrature, they combine to give a single circularly polarized beam when one of the vees is scaled by a quarter of a period. Such an antenna thus becomes competitive with a conical spiral and also offers the possibility of polarization diversity.

The front-to-back ratio of the end-fire pattern of a log-periodic vee-antenna is found to be sensitive to the wedge angle. In the extreme case that the wedge angle approaches zero, the pattern disintegrates badly. This is an interesting result, because in this extreme case the antenna is still a log-periodic structure, but fed by a two-wire line of constant spacing, as suggested by Figure 8.25. The poor performance of this antenna can be traced to the method of feeding. If one assumes that the frequency of operation is such that the nth transverse element is close to its half-wavelength resonance, then the principal excitation of the array involves the $(n - 1)$st, nth, and $(n + 1)$st elements, since the other elements to the right or left of these three are increasingly detuned. Therefore an approximate model for the array is the three-element structure shown in Figure 8.26a. (The skirt wires have been deleted since they contribute little to the performance or to the explanation.)

A study of Figure 8.26a quickly reveals what is wrong. This three-dipole array is similar to a Yagi-Uda array, except that all three elements are driven. But the director, that is, the $(n - 1)$st element, should have a current which lags the current in the nth element. And the reflector, that is, the $(n + 1)$st element, should have a current which leads the current in the nth element. With feeding from the left, the situation is just the opposite from what it should be.

[15] R. H. DuHamel and F. R. Ore, "Logarithmically Periodic Antenna Designs," *IRE National Convention Record* (1958), 139–52.

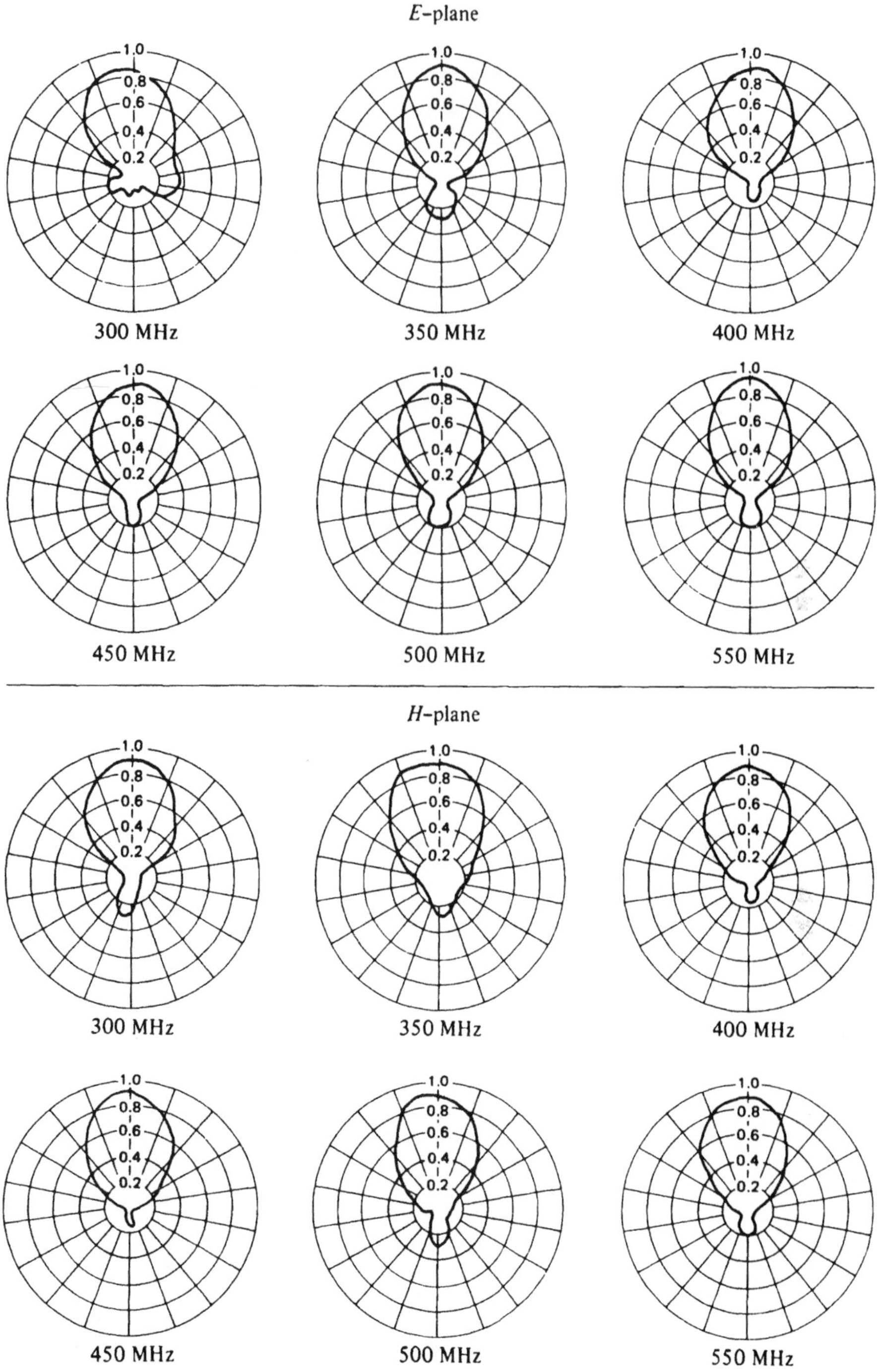

Fig. 8.23 Power Patterns for a Log-Periodic Wire Vee Antenna; Wedge Angle 45°; Polar Plots, Linear Scale (© 1958 IEEE. After Duttamel and Ore, *IRE National Convention Record*, 1958.)

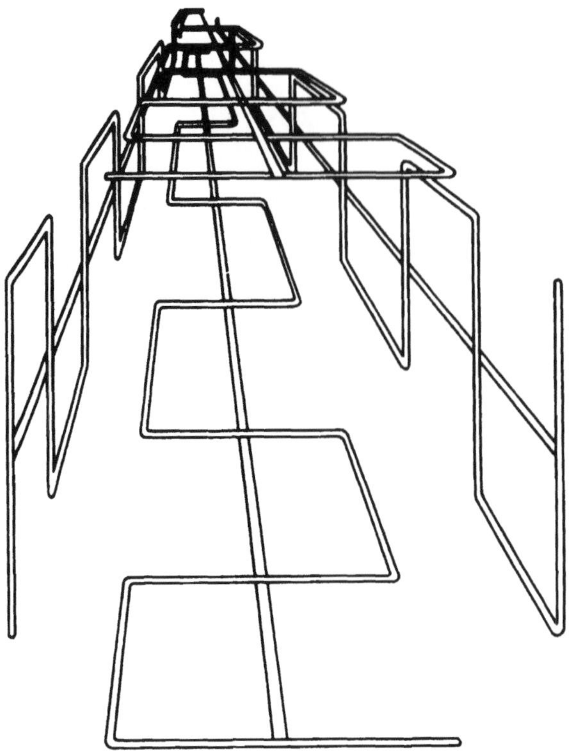

Fig. 8.24 Two Logarithmically Periodic Wire Vee Antennas in Space Quadrature

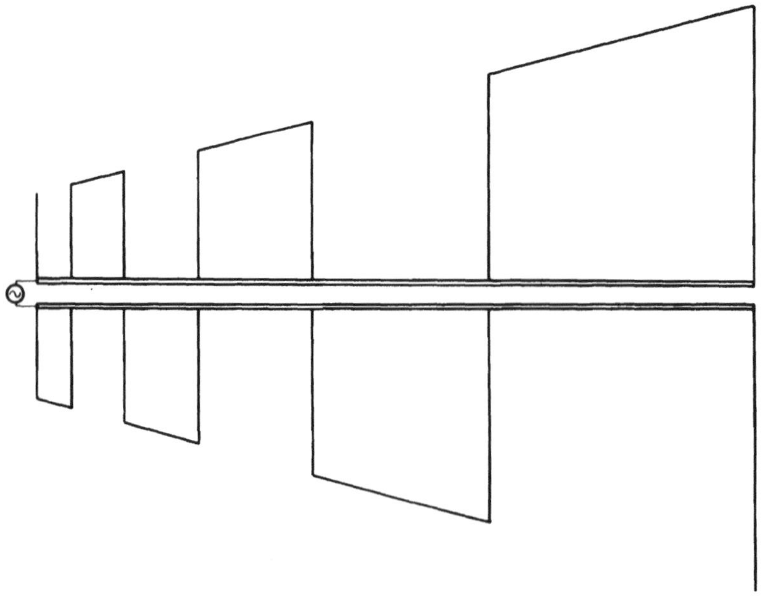

Fig. 8.25 An Improperly Fed Log-Periodic Dipole Array

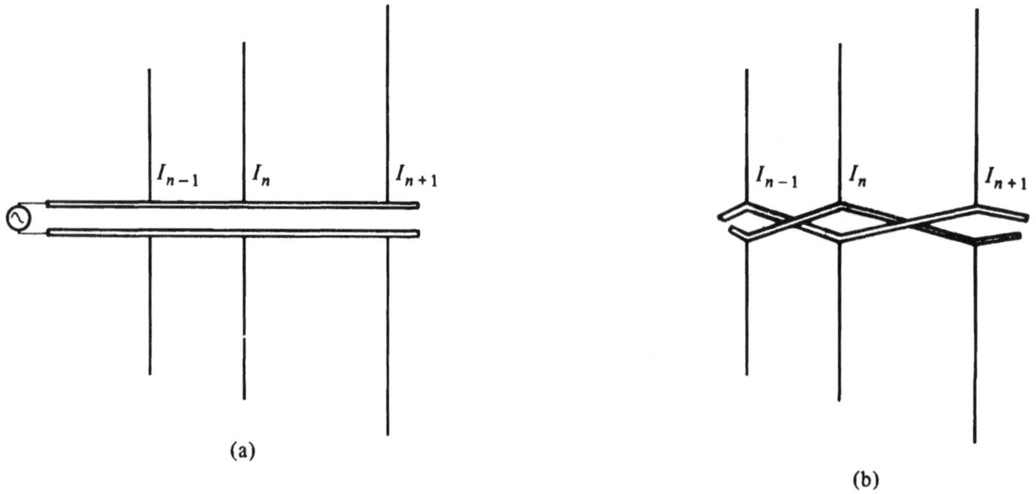

Fig. 8.26 Alternate Methods for Feeding a Three-Dipole Array

The remedy is simple. One need only reverse the feeding at successive junctions of the dipoles with the transmission line, as suggested by Figure 8.26b. Recognition of this fact has led to perhaps the most practical and widely used of all the log-periodic arrays, with typical construction indicated by Figure 8.27. Though offering less directivity than a Yagi-Uda multi-element array, with properly chosen scaling parameters this log-periodic array can be made extremely broadband (multi-octave), both in terms of pattern characteristics and input VSWR.[16]

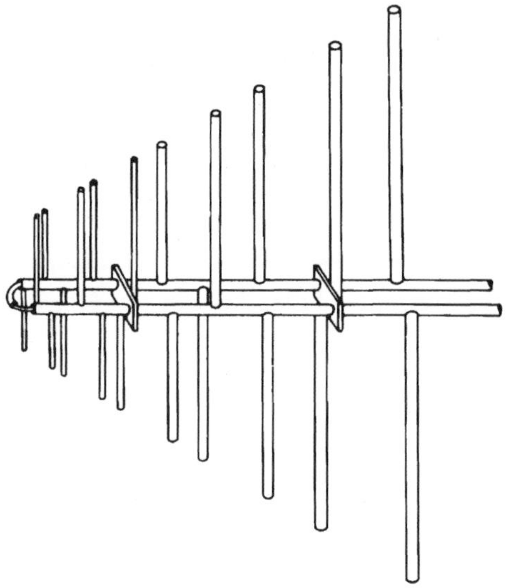

Fig. 8.27 A Properly Fed Log-Periodic Dipole Array

[16] D. E. Isbell, "Log Periodic Dipole Arrays," *IRE Trans. Antennas Propagat.*, AP-8 (1960), 260–67.

8.10 Ground Plane Backed Linear Dipole Arrays

One- and two-dimensional arrays of equispaced dipoles, placed $\lambda/4$ in front of a ground plane in order to confine the radiation essentially to a half-space, find wide application. In very large arrays, all but the elements near the ends or the periphery "see" approximately the same environment, both in physical location of neighboring elements and in the local distribution of dipole excitations. Thus most of the dipoles in these large arrays have roughly the same active impedance, and it is common practice to make this assumption. However, even when the assumption is valid, the active impedance is not the same as the self-impedance of an isolated element. Mutual coupling to nearest neighbors must be taken into account. For small arrays, the "common environment" assumption has to be discarded, and the active impedance may vary widely from element to element. Since the "common" active impedance in large arrays is affected mostly by nearest neighbors, that is, by a small local array, the same technique for determining active impedance in small arrays can be applied with equal success to the large array problem.

Consider first a linear array of similarly oriented dipoles that have a common interelement spacing d, with all the dipoles parallel to, and a common distance h in front of, a ground plane. It will be assumed that the ground plane is composed of a good conductor and extends at least $\lambda/2$ beyond the feed points of the end dipoles. For practical values of h—that is, $\lambda/4$ or less—the image principle can be invoked to good approximation, even for the end dipoles, and thus this antenna is equivalent to a pair of linear arrays a distance $2h$ apart.

If there are N dipoles in the array, one can write

$$V_{pq} = \sum_{r=1}^{2} \sum_{s=1}^{N} I_{r,s} Z_{pq}^{rs} \tag{8.49}$$

in which $r = 1$ identifies the row of dipoles and $r = 2$ identifies the row of images. Since $I_{1,s} = -I_{2,s}$ (the image currents are equal and opposite to the actual dipole currents), Equations 8.49 can be rewritten as

$$V_{pq} = \sum_{s=1}^{N} I_{1,s}(Z_{pq}^{1,s} - Z_{pq}^{2,s}) \tag{8.50}$$

The active (input) impedance of the qth dipole can be obtained from (8.50) by the operation

$$Z_q^a = \frac{V_{1,q}}{I_{1,q}} = \sum_{s=1}^{N} \frac{I_{1,s}}{I_{1,q}}(Z_{1,q}^{1,s} - Z_{1,q}^{2,s}) \tag{8.51}$$

Retention of the notation involving double subscripts and double superscripts is no longer necessary, and one can write

$$Z_m^a = \sum_{n=1}^{N} \frac{I_n}{I_m} Z_{mn} \tag{8.52}$$

with the single index m replacing $1, q$ and the single index n replacing $1, s$. In (8.52) Z_{mn} is a replacement for $Z^{1,s}_{1,q} - Z^{2,s}_{1,q}$. In words, Z_{mn} is the mutual impedance between the mth dipole and the nth dipole minus the mutual impedance between the mth dipole and the image of the nth dipole. Similarly, Z_{mm} is the self-impedance of the mth dipole minus the mutual impedance with its image.

As an illustration of the use of (8.52), consider the problem of the design of the five-element linear dipole array shown in Figure 8.28. Assume that it is desired to excite this array with the equiphase current distribution

$$0.8 \quad 0.9 \quad 1.0 \quad 0.9 \quad 0.8$$

This will produce a sum pattern with a main beam at broadside and symmetrical side lobes at the heights -14.2 dB and -14.8 dB.

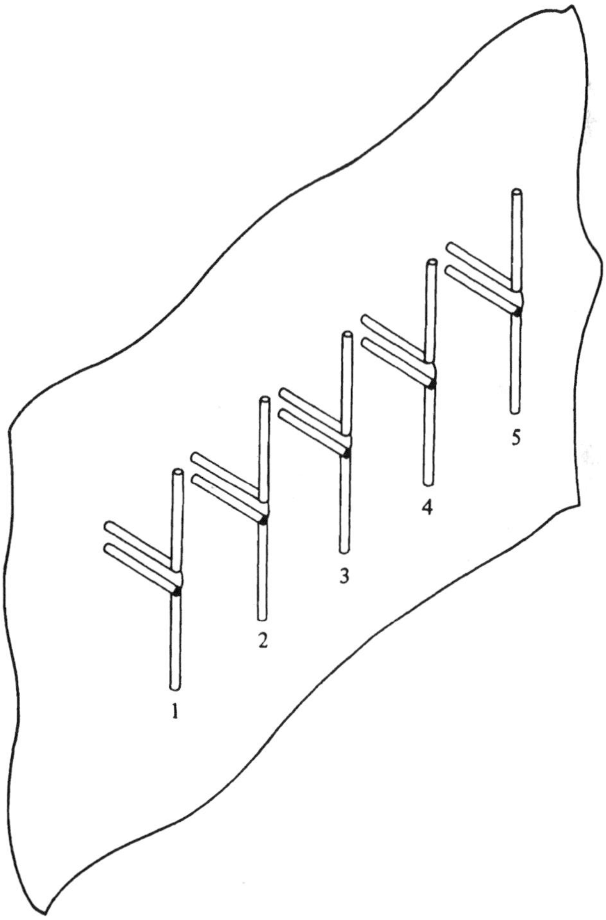

Fig. 8.28 A Five-Element Linear Array of Dipoles Backed by a Ground Plane

Let the interelement spacing d equal $\lambda/2$ and the spacing h off the ground plane be $\lambda/4$. If the starting lengths of all five dipoles are taken to be $\lambda/2$ for the purpose of computing mutual impedance, then use of Equations 7.155 and 7.156 yields the information that

$$Z_{11} = Z_{22} = Z_{33} = Z^{self} - (-12.53 - j29.93) = Z^{self} + 12.53 + j29.93$$

$$Z_{12} = Z_{23} = Z_{34} = Z_{45} = (-12.53 - j29.93) - (-24.64 + j0.78)$$
$$= 12.11 - j30.71$$

$$Z_{13} = Z_{24} = Z_{35} = (4.01 + j17.74) - (13.28 + j9.65) = -9.27 + j8.09$$

$$Z_{14} = Z_{25} = (-1.89 - j12.30) - (-7.21 - j9.39) = 5.32 - j2.91$$

$$Z_{15} = (1.08 + j9.36) - (4.38 + j8.04) = -3.30 + j1.32$$

With the desired current distribution, this gives for the active impedances

$$Z_1^a = Z_5^a = Z_1^{self} + 17.25 + j3.54$$

$$Z_2^a = Z_4^a = Z_2^{self} + 32.21 - j25.98$$

$$Z_3^a = Z_3^{self} + 19.50 - j12.41$$

It is efficient to have these active impedances pure real. If one assumes that the mutual impedance terms will change negligibly as the dipole lengths are adjusted, then the new lengths should be such that

$$X_1^{self} = -3.54 \qquad X_2^{self} = 25.98 \qquad X_3^{self} = 12.41$$

Equation 7.71 can be used to determine l_1, l_2, and l_3. For $a/\lambda = 0.004763$, one finds that

$$2l_1/\lambda = 2l_5/\lambda = 0.466 \qquad 2l_2/\lambda = 2l_4/\lambda = 0.489 \qquad 2l_3/\lambda = 0.478$$

This process could be iterated, using the new lengths to recalculate the mutual impedances, in order to improve on the accuracy. However, one can assume that the values just found are accurate enough and proceed. With the mutual impedances taken to be unchanged as the lengths are trimmed to these new values, it follows that

$$Z_1^a = Z_5^a = R_1^{self} + 17.25 = 76.94$$

$$Z_2^a = Z_4^a = R_2^{self} + 32.21 = 100.67$$

$$Z_3^a = R_3^{self} + 19.50 = 83.82$$

The relative powers radiated by these dipoles are

$$P_1 = P_5 = (0.8)^2(76.94) = 49.24 \qquad P_2 = P_4 = (0.9)^2(100.67) = 81.54$$

$$P_3 = (1)^2(83.82) = 83.82$$

It can be observed that, even though the taper in the current distribution is slight, the effect of mutual coupling causes almost a 2:1 distribution in radiated power.

Imagine that the balun coax characteristic impedances are adjusted so that the active resistances $R_1^a, \ldots, R_5^a$ are transformed to appear as resistances $R_1, \ldots, R_5$ at the ground plane end of each balun. With $R_1, \ldots, R_5$ presented as shunt obstacles $\lambda/2$ apart along a main line coax which runs behind the ground plane (coaxial T-joints are used at each coupling junction), one desires that

$$\sum_{i=1}^{5} G_i = G_0^{ML}$$

in which $G_i = 1/R_i$ and G_0^{ML} is the characteristic conductance of the main line. This will insure an input match to the array. But the same voltage magnitude exists across each of the G_i, and thus $V^2 G_i = P_i$. Therefore $G_i/G_j = P_i/P_j$. For this example

$$\frac{G_1}{G_3} = \frac{G_5}{G_3} = 0.587 \qquad \frac{G_2}{G_3} = \frac{G_4}{G_3} = 0.973$$

as a consequence of which, $\sum_{i=1}^{5} G_i = 4.120\text{G}_3 = G_0^{ML}$, or

$$G_1 = G_5 = 0.143\text{G}_0^{ML} \qquad G_2 = G_4 = 0.236 G_0^{ML} \qquad G_3 = 0.243 G_0^{ML}$$

One needs to select G_0^{ML} at a practical level such that the branch line (balun) characteristic impedances are also at a suitable level. These latter are given by $Z_{0,i}^{BL} = (R_i^a/G_i)^{1/2}$ and thus

$$Z_{0,1}^{BL} = Z_{0,5}^{BL} = \frac{23.20}{\sqrt{G_0^{ML}}} \qquad Z_{0,2}^{BL} = Z_{0,4}^{BL} = \frac{20.65}{\sqrt{G_0^{ML}}} \qquad Z_{0,3}^{BL} = \frac{18.57}{\sqrt{G_0^{ML}}}$$

The choice of a 25-ohm characteristic impedance for the main line ($G_0^{ML} = 0.04$) results in

$$Z_{0,1}^{BL} = Z_{0,5}^{BL} = 116 \qquad Z_{0,2}^{BL} = Z_{0,4}^{BL} = 103 \qquad Z_{0,3}^{BL} = 93$$

which are reasonable values in air-filled coax. The ratio of outer to inner radii for the conductors of these branch coaxial lines can be determined from the formula $Z_0^{BL} = 60 \ln(b/a)$.

One can conclude that, to the extent the dipole lengths have been determined correctly and under the assumption that (7.71) and (7.155) – (7.156) adequately represent the self-impedances and mutual impedances, this transmission line network will provide an input match and insure the desired antenna pattern. In practice, the gap problem may be such that (7.71) is not an acceptable representation of the self-impedance. In such cases, experimental data can be gathered on Z^{self} versus $2l/\lambda$

and an empirical curve fitted to this data can be used in place of (7.71), with the design procedure otherwise unaltered.

8.11 Ground Plane Backed Planar Dipole Arrays

The design of a transmission line harness for a two-dimensional set of dipoles arrayed before a ground plane is basically the same as the procedure detailed for linear arrays in the previous section. For the mnth dipole in the array, the active impedance is given by

$$Z^a_{mn} = \sum_{p=1}^{M} \sum_{q=1}^{N} \frac{I_{pq}}{I_{mn}} Z^{pq}_{mn} \tag{8.53}$$

in which Z^{pq}_{mn} is the mutual impedance between the mnth dipole and the pqth dipole minus the mutual impedance between the mnth dipole and the image of the pqth dipole. The self-impedance of the mnth dipole minus the mutual impedance with its own image is Z^{mn}_{mn}. Equation 8.53 is merely a restatement of (8.52), using double subscript notation because of the shift from linear arrays to planar arrays.

The use of (8.53) in the design of a feeding network will be illustrated for the case of the two-by-three array shown in Figure 8.29. It is assumed that all six dipoles are fed through baluns of the type shown in Figure 8.3c. The dipoles are $\lambda/4$ in front of a large ground plane, are 0.6λ on centers in both directions, and are built of tubular conductors for which $a/\lambda = 0.012$. It is desired that the current distribution be uniform in amplitude and equiphase, which will cause a broadside-broadside sum pattern with a -13.5 dB side lobe level. An input match is desired where the transmission line harness connects to the transmitter (receiver).

By symmetry, all four corner dipoles in the array will be the same as each other, and the two middle dipoles will be the same as each other. Thus attention can be limited to a determination of Z^a_{11} and Z^a_{21}. Once again, one can begin by assuming all dipoles are $\lambda/2$ long for the purpose of computing mutual impedance and use (7.155) and (7.156) to determine that

$$Z^{11}_{11} = Z^{21}_{21} = Z^{self} - (-12.53 - j29.93) = Z^{self} + 12.53 + j29.93$$
$$Z^{12}_{11} = Z^{22}_{21} = (-23.31 - j15.87) - (-20.18 + j10.29) = -3.13 - j26.16$$
$$Z^{21}_{11} = Z^{31}_{21} = (14.67 - j4.01) - (-10.41 - j3.08) = 25.08 - j0.93$$
$$Z^{22}_{11} = Z^{32}_{21} = (-10.55 + j3.43) - (-2.35 + j11.04) = -8.20 - j7.61$$
$$Z^{31}_{11} = (-1.23 + j2.52) - (2.13 + j1.84) = -3.36 + j0.68$$
$$Z^{32}_{11} = (2.92 + j0.76) - (2.93 - j2.00) = -0.01 + j2.76$$

With all desired currents equal, it is found that

$$Z^a_{11} = Z^{self}_{11} + 22.91 - j1.33$$
$$Z^a_{12} = Z^{self}_{21} + 43.16 - j13.31$$

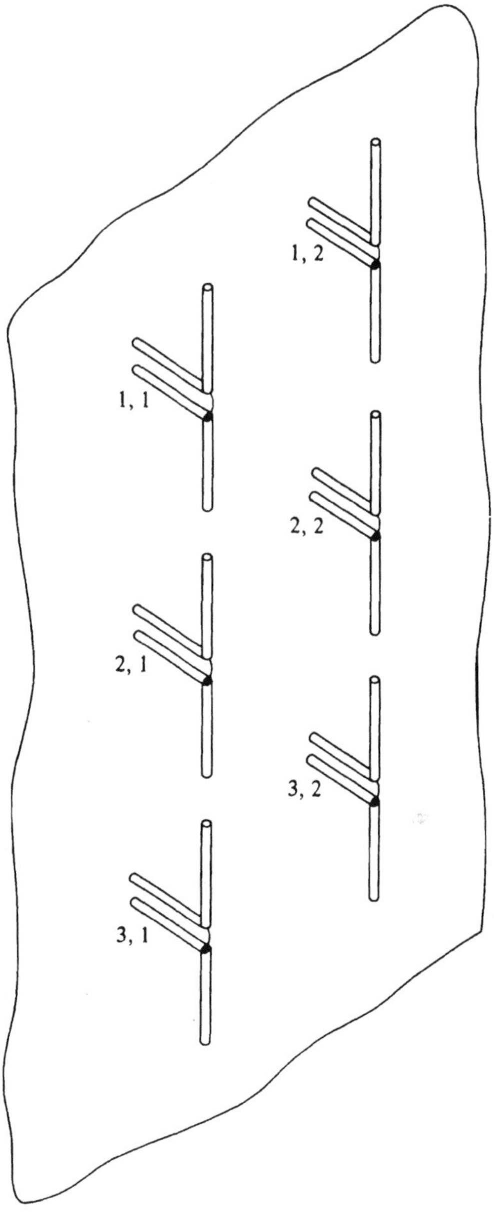

Fig. 8.29 A Two-by-Three Planar Array of Dipoles Backed by a Ground Plane

As before, assume that the dipole lengths will be adjusted so that Z_{11}^a and Z_{21}^a are pure real, and that this adjustment primarily affects Z_{11}^{self} and Z_{21}^{self}. Thus the dipole lengths are sought which cause

$$X_{11}^{self} = 1.33 \quad \text{and} \quad X_{21}^{self} = 13.31$$

to be satisfied. Use of (7.71) yields the values

$$\frac{2l_{11}}{\lambda} = 0.462 \quad \text{and} \quad \frac{2l_{21}}{\lambda} = 0.475$$

and thus

$$R_{11}^a = 57.82 + 22.91 = 80.73 \quad \text{and} \quad R_{21}^a = 62.81 + 43.16 = 105.97$$

If one assumes that the three dipoles (1, 1), (2, 1), and (3, 1) are connected in shunt to a main line coax via the balun quarter-wave sections, the transformed resistances are R_{11}, R_{21}, and R_{31}, these values being governed by the characteristic impedances of the respective branch lines. In this example, the relative radiated powers are

$$P_{11} = (1)^2 R_{11}^a = 80.73 = P_{31} \qquad P_{21} = (1)^2 R_{21}^a = 105.97$$

and these powers are also given by

$$P_{11} = V^2 G_{11} \qquad P_{21} = V^2 G_{21} \qquad P_{31} = V^2 G_{31}$$

in which V is the common voltage magnitude across the three shunt obstacles $G_{11} = 1/R_{11}$, and so on. Thus

$$\frac{G_{11}}{G_{21}} = \frac{G_{31}}{G_{21}} = \frac{P_{11}}{P_{21}} = \frac{80.73}{105.97} = 0.762$$

If G_0^{ML} is the characteristic conductance of the main line feeding the three dipoles, then to match that line one requires that

$$\sum_{i=1}^{3} G_{i1} = G_0^{ML} = G_{21}[1 + 2(0.762)] = 2.524\, G_{21}$$

If, for example, $G_0^{ML} = 0.02$(a 50-ohm coaxial line), then

$$G_{11} = 0.0060 \qquad G_{21} = 0.0079 \qquad G_{31} = 0.0060$$

The needed branch line characteristic impedances are

$$Z_{0,11}^{BL} = Z_{0,31}^{BL} = \left[\frac{(80.73)}{0.0060}\right]^{1/2} = 116 \text{ ohms}$$

$$Z_{0,21}^{BL} = \left[\frac{(105.97)}{0.0079}\right]^{1/2} = 116 \text{ ohms}$$

It is a coincidence peculiar to this example that these two values are the same. They are at a practical level. If this main line coax and its twin (which feeds the other three dipoles) are joined in a T-junction, the combined load is 25 ohms, which can be matched to the transmitter (receiver).

The accuracy of the foregoing procedure can be improved if all the mutual impedances are recalculated using the new set of lengths. As stated before in connection with the linear array application, if the gap problem is such that (7.71) does not accurately represent Z^{self}, then experimental data can be gathered and an empirical formula fitted to the data and used in place of (7.71).

8.12 The Design of a Scanning Array

If a controllable uniform progressive phase can be attached to the current distribution of an array which has been designed to produce a sum pattern, the main beam will *scan*. (See Section 4.3.) This scanning feature unfortunately introduces pattern distortion and input impedance disturbance, both of which usually become more severe as the scan angle is increased. The causes are changes in mutual coupling and in the electrical lengths of those segments of the feeding structure which contain the phaseshifters. Compensation to prevent this performance deterioration can be added to the feeding structure, but only at the cost of increased complexity.

As an illustration of the problems that can be encountered. consider again the one-by-five dipole array of Figure 8.28. Imagine that there is a requirement to scan the sum pattern of this array in the H-plane. One way to accomplish this is to place identical variable phase shifters between successive junctions in the main line feed, as shown in Figure 8.30a. An alternate possibility is to place variable phase shifters in the branch lines, as suggested by Figure 8.30b. (A third method is to cause beam

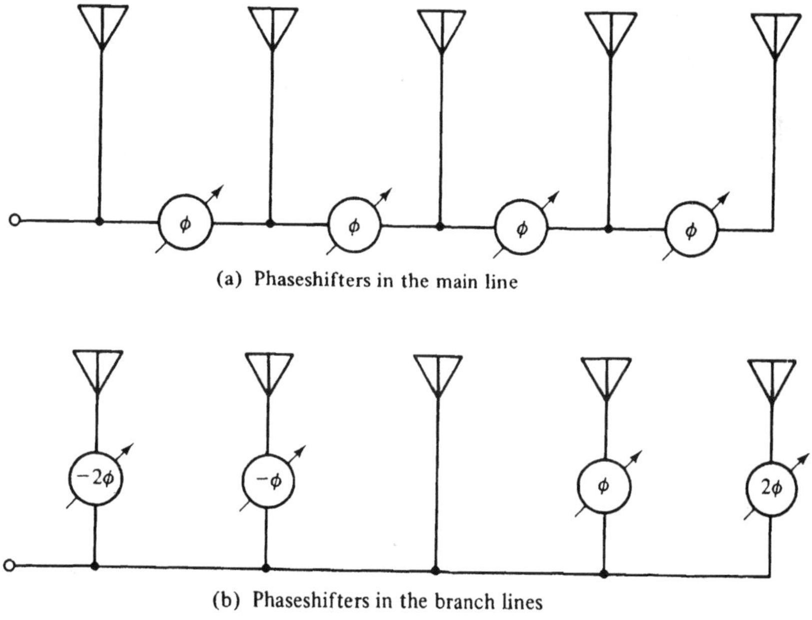

Fig. 8.30 Alternate Feeding Structures for a Scanning Linear Array

scanning by varying the frequency, in effect converting the main line to a traveling wave feed with electrical length that is strongly frequency-dependent. This possibility will be discussed in Chapter 9.) With the phase shifters placed in the branch lines, they will not have identical settings for a given beam position; to attempt to achieve a uniform progressive phase shift in the dipole currents, one must use phaseshifters with settings that are proportional to their positions relative to the array center. This imposes a severe requirement on the phaseshifter design. However, these branch line phaseshifters need not handle the entire transmitter power, unlike the first phaseshifter in the main line placement of Figure 8.30a. Each method of introducing phaseshift has its advantages and disadvantages.

Even if one idealizes the phaseshifters by assuming that they are matched and lossless, their function is affected by the presence of mutual coupling, whether they are inserted in the main line or the branch lines. To see this, suppose that the array shown in Figure 8.28 is to scan $\pm 10°$ about broadside, and that the phaseshifters are perfect and placed in the main line. With all dipole lengths and characteristic impedances optimized for the beam at broadside, the design results of Section 8.10 may be utilized. To recapitulate, it will be assumed that

$$Z_{11} = Z_{55} = (59.69 + j3.54) - (-12.53 - j29.93) = 72.22 + j26.39$$

$$Z_{22} = Z_{44} = (68.46 + j25.98) - (-12.53 - j29.93) = 80.99 + j55.91$$

$$Z_{33} = (64.32 + j12.41) - (-12.53 - j29.93) = 76.85 + j42.34$$

$$Z_{12} = Z_{23} = Z_{34} = Z_{45} = 12.11 - j30.71 \qquad Z_{13} = Z_{24} = Z_{35} = -9.27 + j8.09$$

$$Z_{14} = Z_{25} = 5.32 - j2.91 \qquad Z_{15} = -3.30 + j1.32$$

$$Z_0^{ML} = 25 \qquad Z_{0,1}^{BL} = Z_{0,5}^{BL} = 116 \qquad Z_{0,2}^{BL} = Z_{0,4}^{BL} = 103 \qquad Z_{0,3}^{BL} = 93$$

The voltages and currents at the dipole terminals are connected by the equations

$$V_m = \sum_{n=1}^{5} I_n Z_{mn} \tag{8.54}$$

The voltages V'_m and currents I'_m at the inputs to the balun sections (which are all $\lambda/4$ long) are related to the dipole voltages and currents by[17]

$$V'_m = (-1)^m j I_m Z_{0,m}^{BL} \qquad I'_m = \frac{(-1)^m j V_m}{Z_{0,m}^{BL}} \tag{8.55}$$

The equivalent circuit of the main line with its branching junctions is shown in Figure 8.31. The electrical length between junctions is $180° + \phi$ and thus

[17]Since the main line coupling taps are $\lambda/2$ apart, the signals sent up into the branch lines alternate in phase, requiring a reversal of terminals at successive dipoles. This can be represented mathematically by the factor $(-1)^m$ in Equations 8.55.

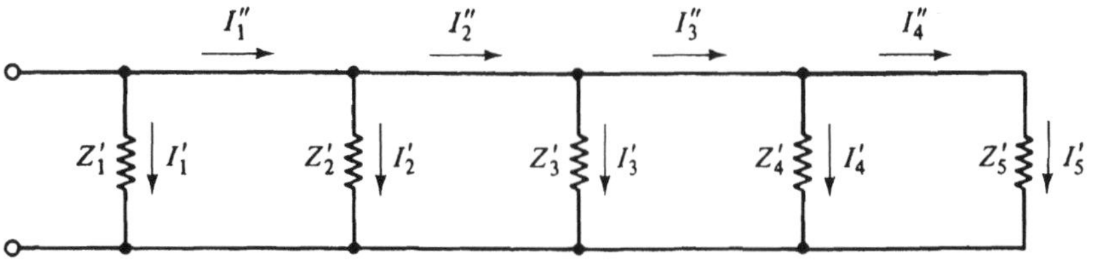

Fig. 8.31 Main Line Equivalent Circuit

$$V'_4 = -V'_5 \cos\phi - jI'_5 Z_0^{ML} \sin\phi \qquad I''_4 = -jV'_5 G_0^{ML} \sin\phi - I'_5 \cos\phi$$

$$V'_3 = -V'_4 \cos\phi - j(I'_4 + I''_4) Z_0^{ML} \sin\phi \qquad I''_3 = -jV'_4 G_0^{ML} \sin\phi - (I'_4 + I''_4)\cos\phi$$

$$V'_2 = -V'_3 \cos\phi - j(I'_3 + I''_3) Z_0^{ML} \sin\phi \qquad I''_2 = -jV'_3 G_0^{ML} \sin\phi - (I'_3 + I''_3)\cos\phi$$

$$V'_1 = -V'_2 \cos\phi - j(I'_2 + I''_2) Z_0^{ML} \sin\phi \qquad I''_1 = -jV'_2 G_0^{ML} \sin\phi - (I'_2 + I''_2)\cos\phi$$

$$I_{IN} = I'_1 + I''_1 \tag{8.56}$$

If (8.55) is used to eliminate V'_m and I'_m from (8.56), the 14 equations in (8.54) and (8.56) can be used to solve for the remaining 14 unknowns. Since the needed uniform progressive phase is $\alpha_z = kd\cos\theta_0$, for $d = \lambda/2$ the phaseshift values $\phi = 0°$, 10°, 20°, and 30° should place the main beam (in the absence of distorting effects) at $\theta_0 =$ 90°, 86.8°, 83.6°, and 80.4°. When these values of ϕ are used in the matrix, the current distributions shown in Table 8.7 result. The input impedance, given by

$$Z_{IN} = \frac{V'_1}{I_{IN}} = \frac{V'_1}{I'_1 + I''_1} = \frac{I_1 (Z_{0,1}^{BL})^2}{V_1 + jI''_1 Z_{0,1}^{BL}} \tag{8.57}$$

is also tabulated. One can observe, even for this modest amount of scanning, that the current distribution quickly departs from what is desired. The input impedance also shows considerable variability.

TABLE 8.7 Dipole currents and input impedance for a scanning five-element linear array

Dipole Number m	Dipole Current, Normalized $\phi = 0°$	$\phi = 10°$	$\phi = 20°$	$\phi = 30°$
1	$0.8\angle 0°$	$0.70\angle 18.3°$	$0.54\angle 65.6°$	$1.17\angle 116.9°$
2	$0.9\angle 0°$	$0.84\angle 7.1°$	$0.67\angle 22.6°$	$0.66\angle 72.7°$
3	$1.0\angle 0°$	$1.00\angle 0°$	$1.00\angle 0°$	$1.00\angle 0°$
4	$0.9\angle 0°$	$0.94\angle -4.0°$	$1.09\angle -9.6°$	$1.46\angle -20.7°$
5	$0.8\angle 0°$	$0.85\angle -5.5°$	$1.03\angle -12.7°$	$1.49\angle -25.6°$
Input Impedance, Ohms	$25\angle 0°$	$18.3\angle -17.3°$	$10.9\angle 10.7°$	$19.8\angle 49.3°$

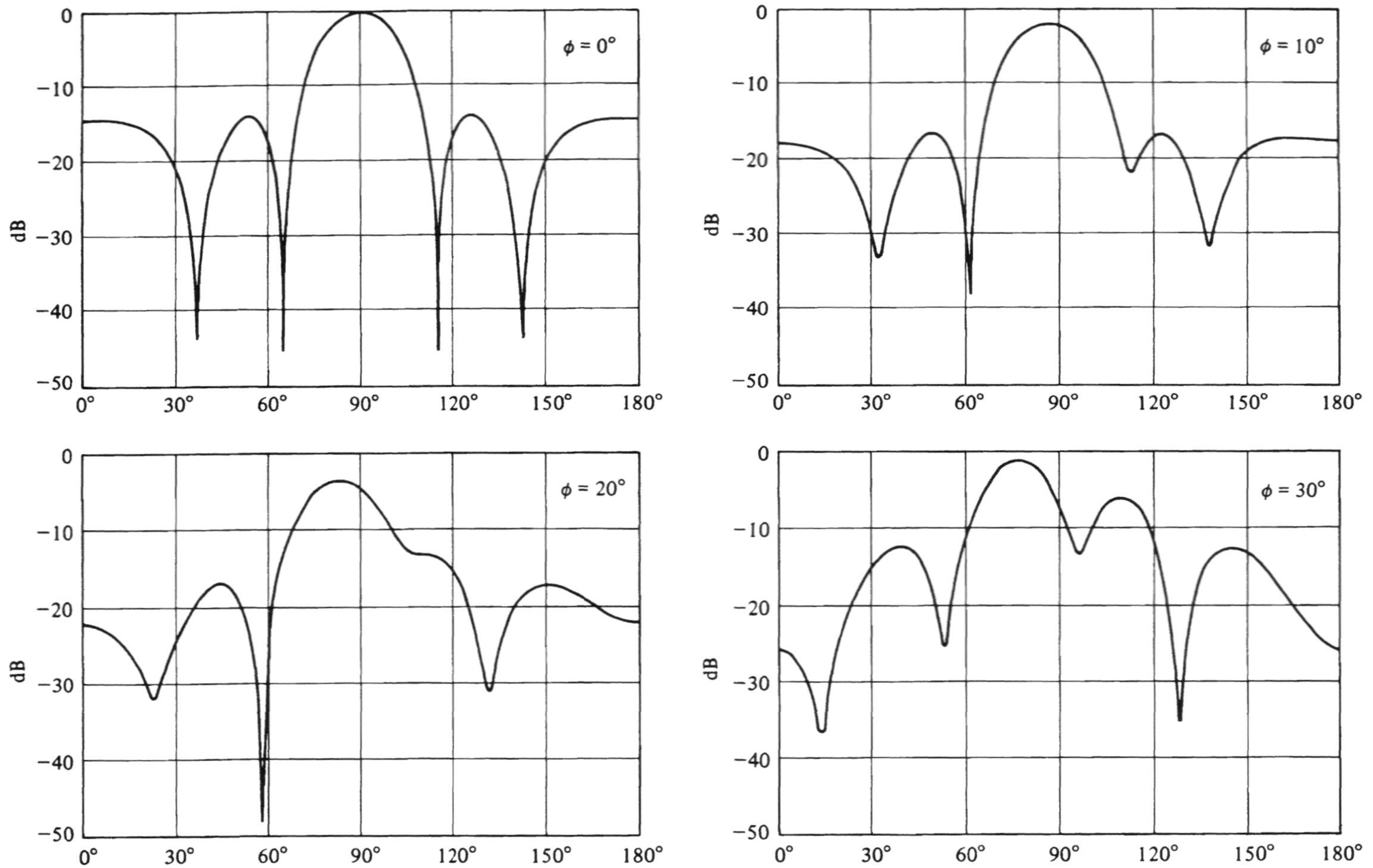

Fig. 8.32 *H*-Plane Sum Patterns of a Five-Element Scanning Linear Array of Dipoles Backed by a Ground Plane; Phase Shifters in the Main Line; $d = 0.5\lambda$, $h = 0.25\lambda$

Array patterns can be calculated from these current distributions and are shown in Figure 8.32. It can be observed that the pattern progressively deteriorates with scan. The height of the main beam is lowered, null-filling occurs, and there is a rise in the side lobe level. The position of the main beam does shift as desired, but not necessarily the proper amount, and attempts to scan the pattern still further are met by unacceptable degradation. Clearly the presence of mutual coupling, plus the nonresonant spacing of coupling junctions along the main line when $\phi \neq 0°$, cause serious problems in the design of this scanning array.

Similar problems of pattern degradation and input mismatch occur with scan for arrays in which the phaseshifters have been placed in the branch lines.[18] For the specific example just discussed (the five-dipole array of Figure 8.28), the interested reader might wish to assume that perfect phase shifters are disposed as in Figure 8.30b and calculate the current distribution, pattern, and input impedance as functions of the uniform progressive phase ϕ.

These scanning problems are less serious for larger arrays, for then the mutual coupling trends toward a common value for all elements and the coupling to the main line per element is lighter, since there are more elements. However, the problems cannot be ignored and their extent can be calculated by the method just outlined.

If one wishes to overcome this degradation of pattern and input impedance with scan, it is possible to determine the transfer characteristics that a set of "phase-shifter/impedance transformer" elements would be required to have when placed in the main line or the branch lines. This is a straightforward but tedious synthesis problem. Physical realizability of such composite elements is a much tougher challenge.

8.13 The Design of Waveguide-Fed Slot Arrays: The Concept of Active Slot Admittance (Impedance)

Waveguide-fed slot arrays differ from two-wire-fed slot arrays in one very important respect. In the latter case, the voltage waves on the two-wire lines which feed the slots can be used to determine *both* the active impedance of each slot and the far-field pattern of the array. (The active impedances are deduced from the positions and relative levels of the maxima and minima of each voltage wave. The pattern is calculable if the voltage wave is known in amplitude and phase at each slot terminal, for then the electric field distribution in each slot is also known in amplitude and phase.)

The situation is more complicated with a waveguide-fed slot array. In that case, the active admittance of each slot can be defined in terms of the propagating waveguide mode incident on the slot, in conjunction with the propagating mode back-scattered by the slot. The positions of the maxima and minima of the sum of these two waves, together with the VSWR, can be used to deduce a normalized active admittance in the usual way. But the sum of these oppositely traveling modes

[18] L. A. Kurtz and R. S. Elliott, "Systematic Errors Caused by the Scanning of Antenna Arrays: Phase Shifters in the Branch Lines," *IRE Trans. Antennas Propagat.* AP-4 (1956), 619–27.

at the waveguide cross section that contains the central point of the slot is not so easily linked to the electric field distribution in the slot as in the case of two-wire feeding. One must determine this linkage in order to design waveguide-fed arrays, and it must be done to include the effects of mutual coupling between slots. The analysis in this and succeeding sections will be concerned with this problem for the special but practical case in which the slots are fed by waveguides of rectangular cross section.

Section 3.5 dealt with the subject of a rectangular waveguide in which a *single* slot was cut to provide a source of radiation to the outside. Three types of slots were illustrated in Figure 3.9, each of which interrupts some of the wall current associated with a TE_{10} mode. This current interruption induces an electric field distribution in the slot which can be viewed as the source of radiation. For a rectangular waveguide with dimensions chosen so that only the TE_{10} mode can propagate, the analysis of Section 3.5 provides a connection between the electric field distribution in the slot and the modal scattering off the slot when an incident TE_{10} mode is the source of excitation. For the longitudinal slot in the broad wall, if a symmetrical standing wave E-field distribution is assumed to exist in the slot, the TE_{10} mode scattering is equivalent to the scattering caused by a shunt element in a two-wire transmission line. Dual analyses undertaken to solve Problems 3.6 and 3.7 at the end of Chapter 3 reveal that the centered inclined slot in the broad wall is equivalent to a series obstacle and that the inclined slot in the narrow wall is equivalent to a shunt obstacle.

Linear arrays of any one of these three slot types can be fabricated by milling a set of equispaced slots in a common wall of a common rectangular waveguide. The lengths and offsets (tilts) of the individual slots must be selected so that the desired electric field intensity, in amplitude and phase, is created in each slot. This will insure the specified pattern. Additionally, it is usually desired that an input match be achieved for the array. How to achieve desired pattern and input impedance is the linear slot array design problem, and it must take into account not only the self-admittance (impedance) of each slot, but also the mutual admittances (impedances), since the slots couple electromagnetically to each other.

Planar slot arrays can be fabricated by placing linear arrays side by side. The design problem is the same in kind, but more complicated because of the two-dimensional nature of the mutual coupling and the relative feeding of waveguides.

Whether the design problem concerns a linear slot array or a planar slot array, it is convenient to define a module of length d, centered around the slot. An example of such a module is shown in Figure 8.33 for the case of a longitudinal slot in the broad wall. A tandem arrangement of such modules will create a linear array, and a parallel arrangement of trains of such modules will result in a planar array.

Inclined slots in a narrow wall are attractive for use as linear slot arrays because of the ease of machining. However, two-dimensional arrays of such slots are not popular. Adjacent waveguides must be spaced so that the wraparound portions of the slots are not shorted out, as they would be if neighboring broad walls were butted up against each other. Also, such arrays are deep ($\sim 3\lambda/4$), which is often undesirable. Therefore planar arrays are usually constructed using slots in the broad wall. Adja-

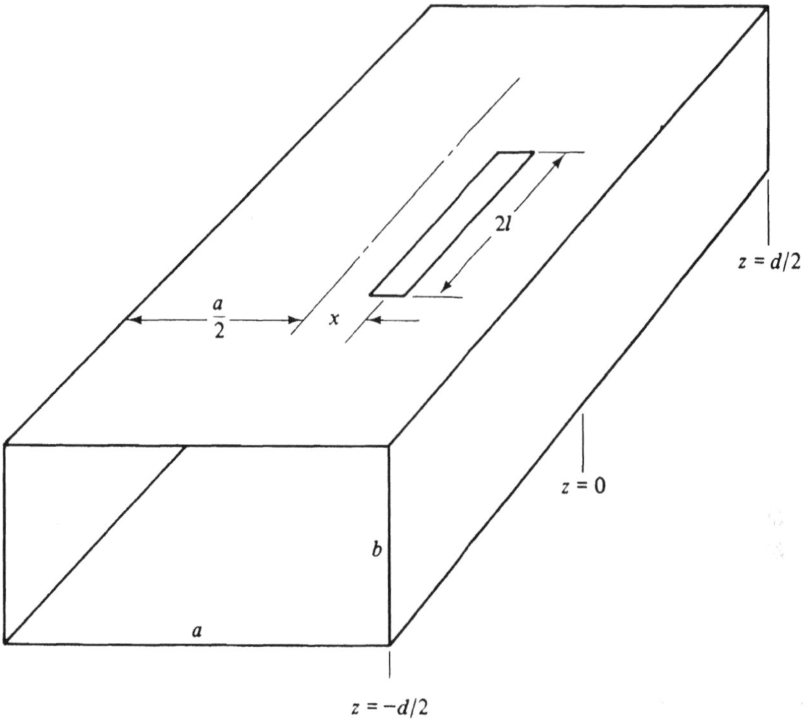

Fig. 8.33 Waveguide-Fed Slot Module

cent waveguides can share a common narrow wall, which is a weight saver, and the box-beam type of structure has mechanical strength advantages. Longitudinal slots have a slight advantage over centered inclined slots in many applications due to the absence of a cross polarized component in the slot field distribution. For this reason, the analysis which follows will focus on the module shown in Figure 8.33. However, the analysis is not limited to arrays of longitudinal broad wall slots; analogous treatments using modules of other slot types are available[19]

If the module length is $d = \lambda_g/2$, with λ_g the guide wavelength for the TE_{10} mode, the slot array is said to be *standing-wave fed.* This is the case that will be considered in what follows. When $d \neq \lambda_g/2$, the slot array is said to be traveling-wave fed. That case will be treated in Chapter 9.

It will be assumed that, in the cross-sectional planes $z = \pm\lambda_g/4$, that is, at the

[19]The analysis presented in this and next three sections is drawn from the article by R. S. Elliott and L. A. Kurtz, "The Design of Small Slot Arrays," *IEEE Trans. Antennas Propagat.*, AP–26(1978), 214–19. (© 1978 IEEE. Reprinted with permission.) Arrays of centered inclined broad wall slots have been analyzed by T. C. Eakins, "Theory of Centered Inclined Slot Arrays with Mutual Coupling" (Master's thesis, University of California, Los Angeles, 1978), and by M. Orefice and R. S. Elliott, Technical Report No. 79–1, Dept. of Electrical Engineering, University of California, Los Angeles, 1979. Also see R. S. Elliott, "An Improved Design Procedure for Small Arrays of Shunt Slots," *IEEE Trans. Antennas and Propagation*, AP–31 (1983), 48–54. Arrays of tilted stripline-fed slots have been treated by P. K. Park (Ph.D. dissertation, University of California, Los Angeles, 1979).

ends of the module, only the propagating TE_{10} mode has a significant value. This implies that all higher-order mode scattering off the slot has decreased to a negligible value when the limits of the module are reached. In this case, the fields at the two ends of the module can be expressed simply by equivalent mode voltages and currents that represent solely the TE_{10} mode.

Let there be N modules in the array, arranged in either a linear or planar lattice. Each module can be viewed as a two-port element, as suggested by Figure 8.34a. For purposes of subsequent notational convenience, the mode voltage and current at one end of the nth module are labeled V'_n and I'_n, and at the other end, V'_{N+n} and I'_{N+n}. The array of N modules is a $2N$-port system, linear and bilateral, and thus the mode voltages and currents can be connected by the equations

$$V'_m = \sum_{n=1}^{2N} I'_n Z_{mn} \qquad m = 1, 2, \ldots, 2N \tag{8.58}$$

If there are no scattering obstacles along the equivalent two-wire line of Figure 8.34a, except possibly in the region $-\epsilon \leq z \leq \epsilon$, with ϵ an infinitesimal, then the standard transmission line formulas give

$$\begin{aligned} V'_n &= jI''_n Z_0 \qquad & V'_{N+n} &= jI''_{N+n} Z_0 \\ I'_n &= jV''_n G_0 \qquad & I'_{N+n} &= jV''_{N+n} G_0 \end{aligned} \tag{8.59}$$

in which $Z_0 = 1/G_0$ is the characteristic impedance of the equivalent transmission line and (V''_n, I''_n), (V''_{N+n}, I''_{N+n}) are defined as shown in Figure 8.34b. In words, V''_n

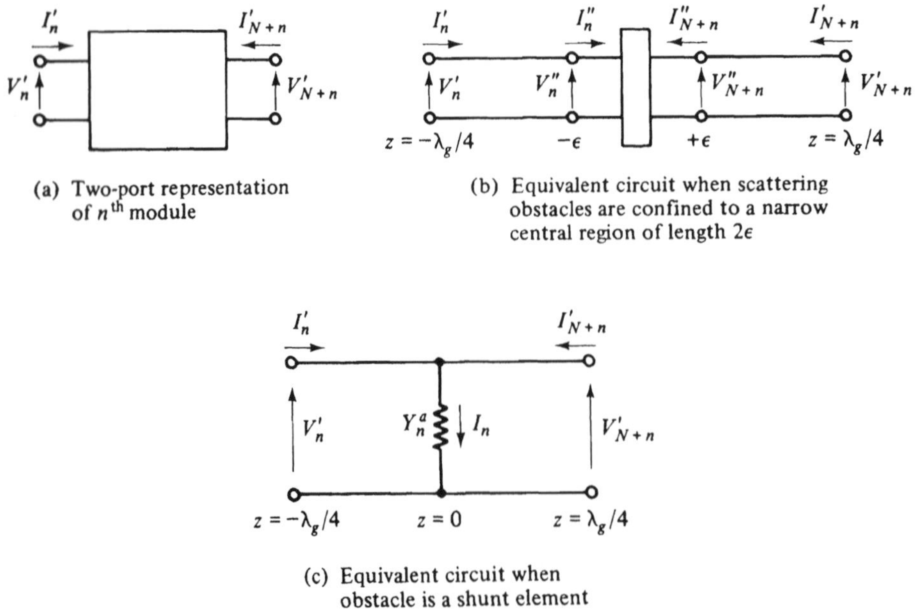

(a) Two-port representation of nth module

(b) Equivalent circuit when scattering obstacles are confined to a narrow central region of length 2ϵ

(c) Equivalent circuit when obstacle is a shunt element

Fig. 8.34 Equivalent Circuits for a Slot Module

and I''_n are the mode voltage and current in the plane $z = -\epsilon$, whereas V''_{N+n} and I''_{N+n} are the mode voltage and current in the plane $z = \epsilon$.

Insertion of (8.59) in (8.58) gives

$$I''_m = \sum_{n=1}^{2N} V''_n Z_{mn}/Z_0^2 \qquad m = 1, 2, \ldots, 2N \tag{8.60}$$

If the electric field distribution in the nth slot is a symmetrical standing wave, regardless of whether that field distribution is caused by internal TE_{10} modes going in either direction past the slot or by external electromagnetic field coupling with other slots or both, the analysis of Section 3.5 shows that the scattering is symmetrical, and thus is equivalent to the scattering off a shunt element in a two-wire line. This assumption of a symmetrical standing wave electric field is usually a good one, particularly when the slots are standing-wave fed and approximately a half-wavelength long and will be made in what follows. Therefore the equivalent circuit of the nth module will be taken as shown in Figure 8.34c, and the active admittance Y_n^a will be defined as representing the shunt element which gives the same scattering in the equivalent circuit as the slot does in the waveguide.

With Figure 8.34c applicable, one can see that $V''_n = V''_{N+n}$ is the mode voltage across the shunt element Y_n^a, and that $I''_n + I''_{N+n}$ is the total current flowing through Y_n^a. In recognition of this, (8.60) can be put in the more useful form

$$I''_m + I''_{N+m} = \sum_{n=1}^{N} \frac{V''_n(Z_{mn} + Z_{N+m,n} + Z_{m,N+n} + Z_{N+m,N+n})}{Z_0^2} \qquad m = 1, 2, \ldots, N \tag{8.61}$$

If one makes the substitutions

$$V_n = V''_n \qquad I_n = I''_n + I''_{N+n}$$

$$Y_{mn} = \frac{Z_{mn} + Z_{N+m,n} + Z_{m,N+n} + Z_{N+m,N+n}}{Z_0^2}$$

then (8.61) becomes

$$I_m = \sum_{n=1}^{N} V_n Y_{mn} \tag{8.62}$$

in which I_m is the total mode current flowing in the equivalent active shunt admittance Y_m^a, V_m is the mode voltage appearing across Y_m^a, and $[Y_{mn}]$ is the matrix that connects the set of mode voltages and the set of mode currents. From (8.62), one can readily deduce that

$$Y_m^a = \frac{I_m}{V_m} = \sum_{n=1}^{N} \left(\frac{V_n}{V_m}\right) Y_{mn}$$

$$= Y_{mm} + \sum_{n=1}^{N}{}' \left(\frac{V_n}{V_m}\right) Y_{mn} \tag{8.63}$$

In (8.63) the prime on the last summation sign serves to indicate that the term $m = n$ is not included. Here, Y_{mm} is generally referred to as the *self-admittance of the* mth *slot* and Y_{mn} is called the *mutual admittance between the* mth *and* nth *slots*. The active admittance is thus seen to be the self-admittance plus the mode-voltage-weighted sum of the mutual admittances. It is sometimes convenient to write (8.63) in the form

$$Y_m^a = Y_{mm} + Y_m^b \tag{8.64}$$

in which

$$Y_m^b = \sum_{n=1}^{N}{}' \left(\frac{V_n}{V_m}\right) Y_{mn} \tag{8.65}$$

is called simply the *mutual coupling term.*

The reader may be puzzled by the part of this analysis in which the mode voltage and current at each end of the module (where clean TE_{10} modes are assumed to exist) were expressed in terms of mode voltages and currents at the center of the module (where clearly a single mode picture is not valid). The reason for this is mathematical convenience. Each module is a two-port element, but this transformation succeeds in treating it as though it can be characterized by the single-mode voltage/current pair (V_n, I_n). This pair should be looked upon as a convenient artifice from which one can deduce the TE_{10} mode presence halfway between slots through use of Equations (8.59).

It can be appreciated from the foregoing development that the self-admittance Y_{nn} and the mutual admittances Y_{mn} are independent of the mode voltages and currents at both ends of the module. In particular, they are independent of $-I'_{N+n}/V'_{N+n}$, that is, of the admittance "seen" looking beyond the module.

8.14 Arrays of Longitudinal Shunt Slots in a Broad Wall of Rectangular Waveguides: The Basic Design Equations

In a development that exactly parallels the one found in Section 3.6, and which extends from Equations 3.48 through Equation 3.52, one can show that the scattering off the shunt element Y_n^a/G_0 shown in Figure 8.34c is symmetrical, and given by

$$B = C = -\frac{1}{2}\frac{Y_n^a}{G_0}V_n \tag{8.66}$$

Equation 8.66 arises because the mode voltage is given by $V(z) = Ae^{-j\beta z} + Be^{j\beta z}$. Since the obstacle Y_n^a/G_0 is taken to be at $z = 0$, it follows that $V_n = V(0) = A + B$.

It was also shown in Section 3.6 that the complex amplitude of the scattered TE_{10} mode is related to the electric field intensity in a longitudinal shunt slot via Equation 3.47. If one takes the origin at the center line of the upper broadwall, as shown in Figure 8.33, and simplifies the notation by letting $\beta = \beta_{10}$, Equation 3.47 can be rewritten in the form

$$B_{10} = C_{10} = \frac{-2V_n^s}{j\omega\mu_0(\beta/k)ab}(\cos\beta l_n - \cos kl_n)\sin\frac{\pi x_n}{a} \tag{8.67}$$

in which $V_n^s = wE_x(x_n, 0, 0)$ is the peak voltage at the center of the slot, with w the slot width and $E_x(x_n, 0, \zeta_n)$ the electric field distribution in the slot aperture.

The back scattered waves B_{10} and B in (8.67) and (8.66) can be related by requiring that the TE_{10} wave and its mode voltage equivalent have the same phase at any cross section z, and that the back scattered power level be the same in both cases. It is a simple matter to show that these conditions will be met if

$$\frac{Y_n^a}{G_0} = \left\{ j\left[\frac{8}{\pi^2 \eta G_0} \frac{(a/b)}{(\beta/k)}\right]^{1/2} (\cos \beta l_n - \cos k l_n) \sin \frac{\pi x_n}{a} \right\} \frac{V_n^s}{V_n} \tag{8.68}$$

Equation 8.68 is the first of two design equations that will be used to determine the lengths and offsets of the slots in an array, such that the desired pattern and input admittance level are achieved.

Under the present assumption that the radiating slots in a common waveguide are resonantly spaced, that is, $\lambda_g/2$ apart, it follows that the mode voltage V_n has a common value (except for an alternation in sign) for all the slots in a common waveguide. If the pattern requirements are such that the slot voltages V_n^s are to have the same phase for all n, then Equation 8.68 indicates that all the active admittances Y_n^a/G_0 should have the same phase. [The alternation in phase of V_n is compensated by an alternation in direction of offset x_n, which causes an alternation in the sign of $\sin(\pi x_n/a)$.]

In such circumstances, the usual choice is to make Y_n^a/G_0 pure real for all n. But a return to (8.63) reveals that, if Y_n^a is to be pure real, in general Y_{nn}, the self-admittance of the nth slot, will *not* be pure real. In other words, when mutual coupling is taken into account in the design of slot arrays, resonant self-conductance data is not sufficient to permit a proper design. Indeed, in many practical applications, the needed value of Y_{nn} will be quite far off resonance. This same effect has already been noticed in the case of some dipole arrays considered earlier in this chapter.

To obtain the design equation that will be companion to (8.68), it is useful to link the waveguide-fed slot array to an equivalent array of dipoles via Babinet's principle. To accomplish this, assume that the waveguide-fed slot array is imbedded in an infinite, perfectly conducting ground plane and radiating into a half-space. Imagine as an interim step the existence of a dual antenna consisting of an identical array of slots, also imbedded in an infinite, perfectly conducting ground plane, and also radiating into a half-space, but center-fed by a network of two-wire lines. If the same electric field distribution is established in corresponding slots in the two arrays, the half-space radiation patterns will be the same. It will be assumed that this is the case. However, the admittance characteristics will *not* be the same. The reason for this is that there is higher-order mode scattering off a waveguide-fed slot, which contributes primarily to the susceptive component of Y_{nn}, and which depends on the slot offset x_n. No such effect exists in the two-wire-fed slot, and to model this higher-order mode scattering one must place a load admittance Y_n^L across the terminals of the corresponding two-wire-fed slot. When this is done, the circuit equations for the two-wire-fed slot array are

$$I'_m = \sum_{n=1}^{N} V^s_n \sin kl_n \, Y'_{mn} \tag{8.69}$$

in which the prime superscripts are used to distinguish this array from the waveguide-fed array, for which (8.62) applies. In (8.69), Y'_{mn} (with $m \neq n$) has its customary meaning, being the mutual admittance between slots m and n when they are two-wire-fed and radiating only into a half-space. However,

$$Y'_{nn} = Y'_n + Y^L_n \tag{8.70}$$

with Y'_n the conventional self admittance of the nth slot when it is two-wire-fed and radiating into a half-space, and with Y^L_n the load admittance representing higher order mode scattering.

Next, consider the complementary array of strip dipoles, center-fed by two-wire lines, and radiating into a full space. The circuit equations for this dipole array are

$$V^d_m = \sum_{n=1}^{N} I^d_n \sin kl_n Z_{mn} \tag{8.71}$$

in which the current distribution in the mth dipole has been assumed to be in the form $I^d_n \sin k(l_n - |\zeta|)$. The superscript d is used to distinguish the fact that the terminal voltage and current (V^d_m, $I^d_m \sin kl_m$) appearing in (8.71) refer to the dipole array.

If the distribution of dipole currents I^d_n matches the distribution of slot voltages V^s_n in the two slot arrays, all three will produce the same radiation pattern in a half-space, except for an interchange of **E** and **H** in the dipole case (See Section 7.16). However, for the dipole array to model the admittance characteristics of the two slot arrays, it is necessary to place a load impedance Z^L_n in series at the terminals of the nth dipole. When this is done,

$$Z_{nn} = Z_n + Z^L_n \tag{8.72}$$

with Z_n the conventional self-impedance of the nth dipole when it is center-fed and radiating into a full space. In (8.71), Z_{mn} (with $m \neq n$) is the conventional mutual impedance, calculable from Equations 7.155 and 7.156.

Booker has shown (see Section 7.16) that the admittances of the unloaded two-wire-fed slots (radiating into a half-space) are related to the impedances of the unloaded complementary strip dipoles (radiating into a full space) by the relations

$$Y'_n = \left(\frac{2}{\eta^2}\right) Z_n \qquad Y'_{mn} = \left(\frac{2}{\eta^2}\right) Z_{mn} \tag{8.73}$$

If, for the loaded complementary arrays, the dipole current distribution is the same as the two-wire-fed slot voltage distribution, then the input admittance to the mth slot is given by

$$Y_m^{a'} = \frac{I'_m}{V_m^s \sin kl_m} = Y'_m + Y_m^L + \sum_{n=1}^{N}{}' \frac{V_n^s \sin kl_n}{V_m^s \sin kl_m} Y'_{mn}$$

$$= \left(\frac{2}{\eta^2}\right) Z_m + Y_m^L + \sum_{n=1}^{N}{}' \frac{I_n^d \sin kl_n}{I_m^d \sin kl_m} \left(\frac{2}{\eta^2}\right) Z_{mn}$$

$$= \left(\frac{2}{\eta^2}\right) \left[Z_m + \left(\frac{\eta^2}{2}\right) Y_m^L + \sum_{n=1}^{N}{}' \frac{I_n^d \sin kl_n}{I_m^d \sin kl_m} Z_{mn} \right] \tag{8.74}$$

The active (or input) impedance of the mth dipole can be deduced from (8.71), as follows:

$$Z_m^a = \frac{V_m^d}{I_m^d \sin kl_m} = Z_m + Z_m^L + \sum_{n=1}^{N}{}' \frac{I_n^d \sin kl_n}{I_m^d \sin kl_m} Z_{mn} \tag{8.75}$$

If the admittance and impedance characteristics of these two loaded complementary arrays are to be similar, comparison of (8.74) and (8.75) leads to the conclusion that

$$Y_m^L = \left(\frac{2}{\eta^2}\right) Z_m^L \tag{8.76}$$

In words, the load admittance Y_m^L, placed across the terminals of the mth two-wire-fed slot in order to model higher-order-mode scattering off the corresponding waveguide-fed slot, and the load impedance Z_m^L, placed in series at the input to the mth dipole in order to model the same effect, are linked to each other by Booker's relation.

If the two-wire-fed slot array and its complementary dipole array are to model the waveguide-fed slot array in admittance characteristics as well as pattern, it is necessary that the complex power flows equate at each element. That is,

$$(\tfrac{1}{2} V_m V_m^* Y_m^{a*}) = (\tfrac{1}{2} V_m^s \sin kl_m I_m'^{*})^* = (\tfrac{1}{2} V_m^d I_m^{d*} \sin kl_m) \tag{8.77}$$

But

$$(V_m^s \sin kl_m I_m'^{*})^* = V_m^{s*} \sin kl_m \, I'_m = V_m^{s*} \sin kl_m \sum_{n=1}^{N} V_n^s \sin kl_n \, Y'_{mn}$$

$$= V_m^s V_m^{s*} \sin^2 kl_m \sum_{n=1}^{N} \frac{I_n^d \sin kl_n}{I_m^d \sin kl_m} \left(\frac{2}{\eta^2}\right) Z_{mn}$$

$$= \left(\frac{2}{\eta^2}\right) Z_m^a V_m^s V_m^{s*} \sin^2 kl_m \tag{8.78}$$

When (8.77) and (8.78) are combined, the result for the nth element can be written in the form

$$V_n V_n^* Y_n^a Y_n^{a*} = \left(\frac{2}{\eta^2}\right) Y_n^a Z_n^a V_n^s V_n^{s*} \sin^2 kl_n \tag{8.79}$$

In (8.79), V_n, V_n^s, and Y_n^a are, respectively, the mode voltage, slot voltage, and active admittance of the nth waveguide-fed slot, and Z_n^a is the active impedance of the corresponding loaded nth dipole.

When (8.68) is multiplied by its complex conjugate, one obtains

$$V_n V_n^* Y_n^a Y_n^{a*} = \frac{8G_0(a/b)}{\pi^2 \eta(\beta/k)} (\cos \beta l_n - \cos k l_n)^2 \sin^2 \left(\frac{\pi x_n}{a}\right) V_n^s V_n^{s*} \tag{8.80}$$

If (8.79) and (8.80) are equated, the result can be arranged to give

$$\frac{Y_n^a}{G_0} = \frac{1}{Z_n^a/73} \left[\frac{4(a/b)}{0.61\pi(\beta/k)} \frac{(\cos \beta l_n - \cos k l_n)^2}{\sin^2 k l_n} \sin^2 \frac{\pi x_n}{a} \right] \tag{8.81}$$

in which the manipulation $0.61\eta/\pi = 73$ ohms has been introduced.

Equation 8.81 is a key result of the analysis and is the second design equation. Together with (8.68), it can be used to determine the lengths and offsets of all longitudinal shunts slots in the broadwall of a set of rectangular waveguides in order to produce a desired pattern and a specified input admittance. This will be demonstrated presently.

The reader familiar with Stevenson's pioneering analysis of the admittance (or impedance) properties of a single resonant slot in a thin-walled rectangular waveguide will recognize the factor inside the brackets in (8.81) as his expression for the normalized conductance of a longitudinal shunt slot that has been tuned to resonance.[20] Thus the interpretation can be put on (8.81) that the normalized active admittance of the nth longitudinal shunt slot in an array is given by Stevenson's expression for the resonant normalized conductance, divided by the active impedance of the corresponding loaded dipole, normalized to 73 ohms.

The single slot case is a simple reduction of (8.81). One obtains

$$\frac{Y}{G_0} = \frac{73}{Z^d + Z^L} \left[\frac{4(a/b)}{0.61\pi(\beta/k)} \frac{(\cos \beta l - \cos k l)^2}{\sin^2 k l} \sin^2 \frac{\pi x}{a} \right] \tag{8.82}$$

in which $Z^d = R^d + jX^d$ is the self-impedance of the complementary dipole, and $Z^L = R^L + jX^L$ is the load impedance in series with it, the presence of which models the effects of internal higher order mode scattering off the slot.

Equation 8.82 is consistent with several experimental observations. If the loaded dipole is shortened below resonance, that is, if $X^d + X^L < 0$, the corresponding waveguide-fed slot has a positive susceptance. (This behavior is opposite to that of a two-wire-fed slot.) At resonance, $X^d = -X^L$, and (8.82) becomes

$$\frac{G_r}{G_0} = \frac{73}{R^d + R^L} \left[\frac{4(a/b)}{0.61\pi\beta/k} \frac{(\cos \beta l_r - \cos k l_r)^2}{\sin^2 k l_r} \sin^2 \frac{\pi x}{a} \right] \tag{8.83}$$

in which G_r/G_0 is the normalized resonant conductance, and $2l_r$ is the resonant length. As the offset of the slot increases, more higher order mode scattering occurs, X^L increases, and it takes a larger value of X^d to tune out X^L. This requires a longer

[20] A. F. Stevenson, "Theory of Slots in Rectangular Waveguides," *J. Appl. Phys.*, 19 (1948), 24–38. (Stevenson assumed $kl = \pi/2$.)

dipole, consistent with the observation that the resonant length of the slot increases with offset.

Equation 8.83 suggests that Stevenson's expression (the factor in brackets) is only approximate and is less accurate as the slot width is increased or the slot offset is increased. This is because R^d is affected by both the length and width of the strip dipole. Only for an infinitesimally wide slot on the centerline (for which $R^L = 0$) would one find that $R^d + R^L = 73$ ohms.

8.15 The Design of Linear Waveguide-Fed Slot Arrays

The two design equations which were developed in the preceding section can be rewritten in the abbreviated forms

$$\frac{Y_n^a}{G_0} = K_1 f_n \sin kl_n \frac{V_n^s}{V_n} \tag{8.84}$$

$$\frac{Y_n^a}{G_0} = \frac{K_2 f_n^2}{Z_n^a} \tag{8.85}$$

in which, by inspection,

$$K_1 = -j\left[\frac{8}{\pi^2 \eta G_0}\frac{(a/b)}{(\beta/k)}\right]^{1/2} \qquad K_2 = \frac{292(a/b)}{0.61\pi(\beta/k)} \tag{8.86}$$

and

$$f_n = \frac{\cos \beta l_n - \cos kl_n}{\sin kl_n} \sin \frac{\pi x_n}{a} \tag{8.87}$$

The active admittance of the nth equivalent loaded dipole is given by

$$Z_n^a = Z_{nn} + Z_n^b \tag{8.88}$$

in which

$$Z_n^b = \sum_{m=1}^{N}{}' \frac{V_m^s \sin kl_m}{V_n^s \sin kl_n} Z_{nm} \tag{8.89}$$

is the mutual coupling, and

$$Z_{nn} = Z_n + Z_n^L \tag{8.90}$$

is the loaded self-impedance of the dipole. Before one can make use of the design equation of (8.85), it is necessary to determine Z_{nn} as a function of the length and offset of the complementary waveguide-fed slot.

If one assumes that the input admittance to the nth slot is the same whether all other slots are (1) present and short-circuited, or (2) absent, then this is equivalent to saying that the input impedance to the nth loaded dipole is the same whether all other loaded dipoles are (1) present and open-circuited, or (2) absent. Experiments show that this is a good assumption. It permits one to infer from (8.85) that

$$Z_{nn} = \frac{K_2 f_n^2}{Y_n/G_0} \tag{8.91}$$

in which $Y_n/G_0\,(x_n, l_n)$ is the isolated self-admittance of the nth slot. Thus if one *measures* Y/G_0 for an isolated slot as a function of its length and offset, (8.91) can be used to deduce the function $Z_{nn}\,(x_n, l_n)$ needed for use in (8.85).[21]

R. J. Stegen has found that the admittance data of an isolated slot can be presented in a universal form that is extremely useful for computational purposes.[22] Using standard X-band brass $RG52/U$ waveguide and a frequency of 9.375 GHz, he measured Y/G_0 with slot offset and length as parameters and assembled the data in a pair of curves, which are reproduced in Figure 8.35. If one lets $y = l/l_r$ represent the abscissa scale, then

$$h(y) = h_1(y) + jh_2(y) \tag{8.92}$$

can symbolize the complex sum of these universal curves. With $g(x)$ taken to mean the normalized resonant conductance as a function of offset, it follows that

$$\frac{Y}{G_0} = g(x)h(y) \tag{8.93}$$

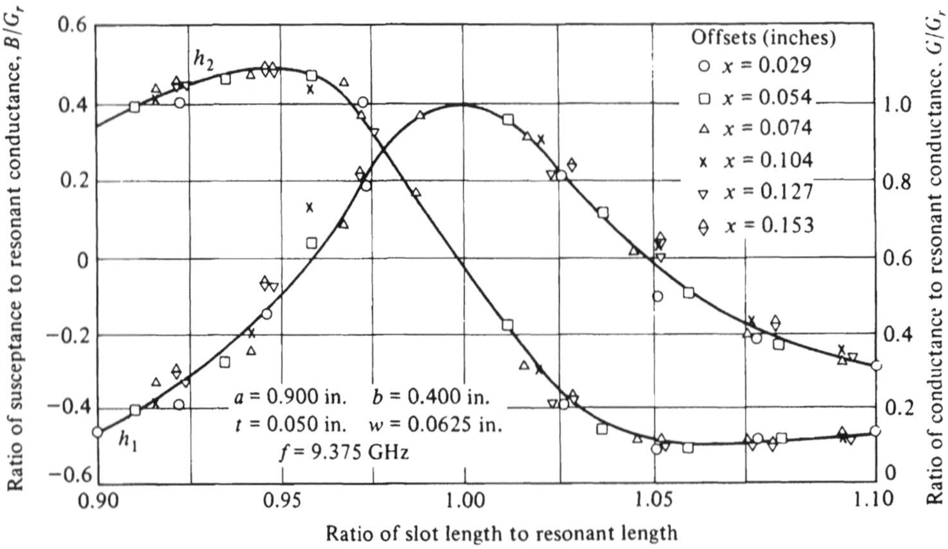

Fig. 8.35 Normalized Self-Admittance Components for a Longitudinal Shunt Slot (After Stegen[22])

[21] Recently it has become feasible, using the method of moments, to make an accurate theoretical determination of Y/G_0 for a rectangular slot, including the effects of slot width and wall thickness. See T. V. Khac, "A Study of Some Slot Discontinuities in Rectangular Waveguides," (Ph.D. dissertation, Monash University, Australia, Nov. 1974).

[22] R. J. Stegen, "Longitudinal Shunt Slot Characteristics," Hughes Technical Memorandum No. 261, Hughes Aircraft Co. (California: Culver City, November 1951). Stegen's curves are reproduced as Figures 9.5, 9.7, 9.9, and 9.10 in *Antenna Engineering Handbook*, ed. H. Jasik (New York: McGraw-Hill Book Co., Inc., 1961).

Stegen's curve of $g(x)$ for this same family of slots is shown in Figure 8.36. To interpret the variable y, one also needs to know the relation between resonant length and offset. This is characterized by the function

$$v(x) = kl_r(x) \tag{8.94}$$

and for Stegen's measurements is shown in Figure 8.37. It is important to note that all four curves, represented by $h_1(y)$, $h_2(y)$, $g(x)$, and $v(x)$ are simple in form and can be easily polyfitted.

With these identifications, one can return to (8.91) and rewrite it in the form

$$Z_{nn}(x_n, y_n) = \frac{K_2 f_n^2(x_n, y_n)}{g(x_n)h(y_n)} \tag{8.95}$$

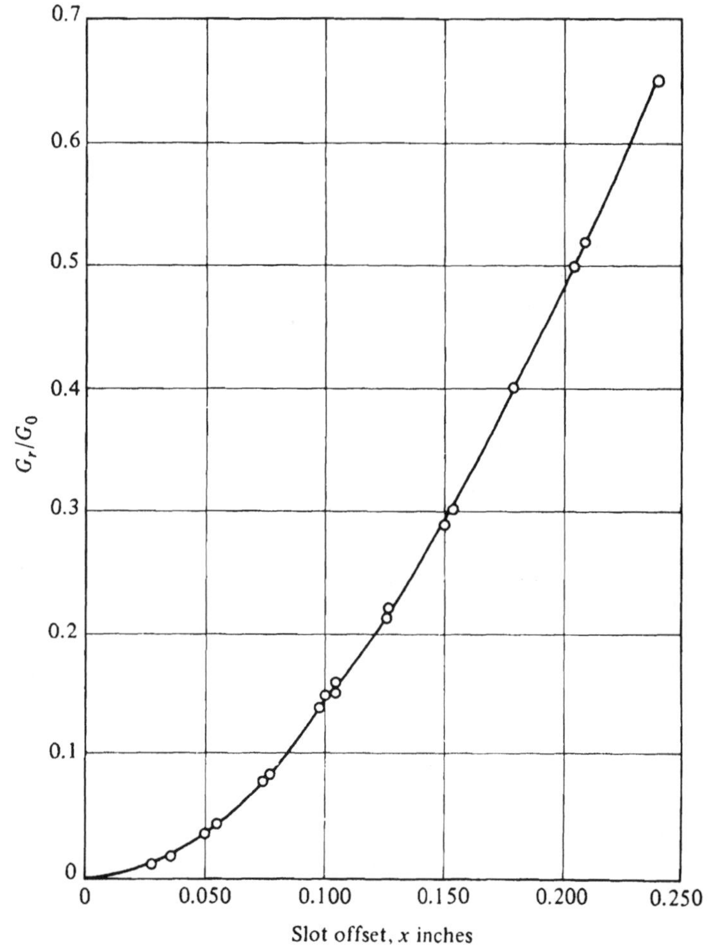

Fig. 8.36 Normalized Resonant Conductance versus Offset for a Longitudinal Shunt Slot; See Fig. 8.35 for Legend (After Stegen[22])

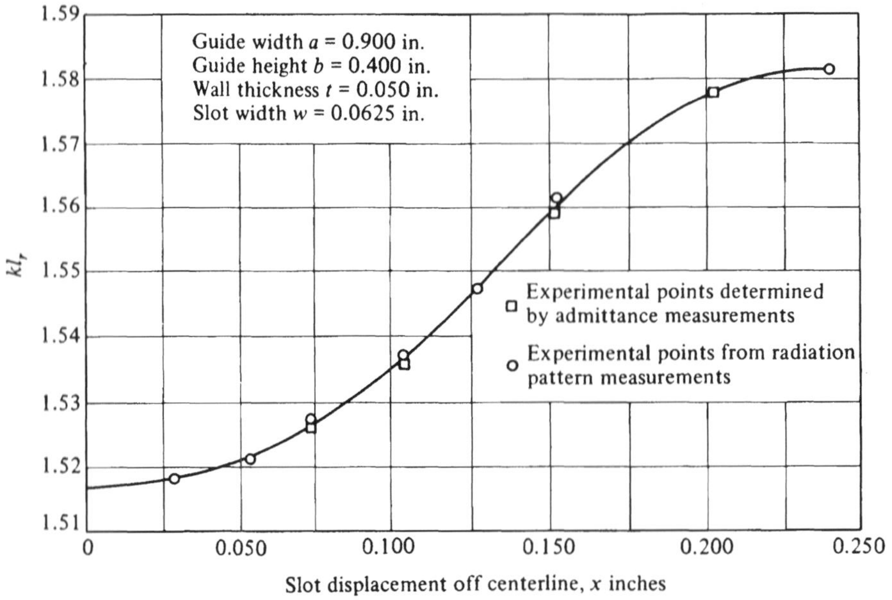

Fig. 8.37 Resonant Length versus Offset for a Longitudinal Shunt Slot; See Fig. 8.35 for Legend (After Stegen[22])

in which

$$f_n(x_n, y_n) = \left\{\frac{\cos[(\beta/k)y_n v(x_n)] - \cos[y_n v(x_n)]}{\sin[y_n v(x_n)]}\right\} \sin\frac{\pi x_n}{a} \tag{8.96}$$

The design of a linear array can now be undertaken. If there are to be N slots, spaced $\lambda_g/2$ apart, and a certain pattern is desired, then the techniques described in Chapter 5 can be used to determine the slot voltage distribution V_n^s. Let it be assumed that this has been done. Next, make an initial estimate of the lengths and offsets of the slots. (This estimate is not critical. One could begin by ignoring mutual coupling, choosing each slot to be self-resonant with the proper distribution of conductances to insure the proper pattern and input match. Or, more simply, one could guess an average slot offset and assign that value and the corresponding resonant length to every slot in the array.) Once a selection has been made of the initial values of slot lengths and offsets, Equations 7.155 and 7.156 can be used to compute all the mutual impedances Z_{mn} between dipoles in the equivalent array. These Z_{mn} values can be placed in (8.89), together with the desired array excitation, to permit computation of a set of starting values for the mutual coupling terms Z_n^b.

In most applications, the desired slot voltage distribution will be equiphase and (8.84) indicates that in such cases the active admittances will have a common phase also. The design procedure is not limited to this situation, but for the purpose of illustration, let us assume this to be the case, and specify further that all the active admittances should be pure real. If the Z_n^b values are not too sensitive to changes in slot length and offset (and experience shows that they are not), one can use the starting values of Z_n^b and impose the condition that

$$X_{nn} = -X_n^b \tag{8.97}$$

Equation 8.95 can be used as the basis for a search for the length/offset combination (x_n, y_n) that will satisfy (8.97). When one conducts this search, the discovery is quickly made that there is a continuum of couplets (x_n, y_n) which satisfy (8.97). There is a corresponding continuum of active admittance values Y_n^a/G_0 that can be calculated from (8.85).

Let (x_n', y_n') be one of the couplets that satisfies (8.97), and let $Y_n^a/G_0\ (x_n', y_n')$ be the corresponding value of active admittance for the nth slot. For any other slot–for example, the mth—there is similarly a continuum of acceptable length/offset combinations and a corresponding continuum of active admittance values. But of all of these, there is only one couplet (x_m', y_m') that can be paired with (x_n', y_n') such that (8.84) gives the proper slot voltage distribution. What one requires is that

$$\frac{Y_n^a/G_0(x_n', y_n')}{Y_m^a/G_0(x_m', y_m')} = \frac{f_n(x_n', y_n')}{f_m(x_m', y_m')}\frac{\sin kl_n}{\sin kl_m}\frac{V_n^s/V_n}{V_m^s/V_m} \tag{8.98}$$

Equation 8.98 serves to identify *sets* of acceptable length/offset combinations such that all members of a set (one for each slot) satisfy (8.98) as well as (8.97).

Of all these sets, the proper one to choose is the one which causes the sum of the normalized active admittances to be unity, since this is the condition for an input match.

It is too costly in computer time, and not even desirable, to identify more than one set of acceptable length/offset combinations. If the set that has been identified gives $\sum Y_n^a/G_0 > 1$, in the next iteration one will know that smaller offsets should be chosen; if $\sum Y_n^a/G_0 < 1$, larger offsets will be needed. But in any event, the process will need to be iterated, because the new lengths and offsets can be used to compute an improved set of Z_n^b values.

For large N, this iterative process can be completely computerized, with the program commanded to stop when the lengths and offsets determined in one iteration differ from those of the previous iteration by amounts less than the machining tolerance that can be specified. The writing of a complete computer program for this iterative process requires care and it is helpful to go through a simple case "by hand" as a background step.

One such case which can be instructive involves the design of a four-slot linear array with the specifications that the slot voltage distribution be equiphase and in the ratio 1: 2: 2: 1 and that an input match be achieved. Let it be assumed that this is to be done for longitudinal shunt slots in standard X-band waveguide, as depicted in Figure 8.38, and at 9.375 GHz, so that Stegen's curves can be used.

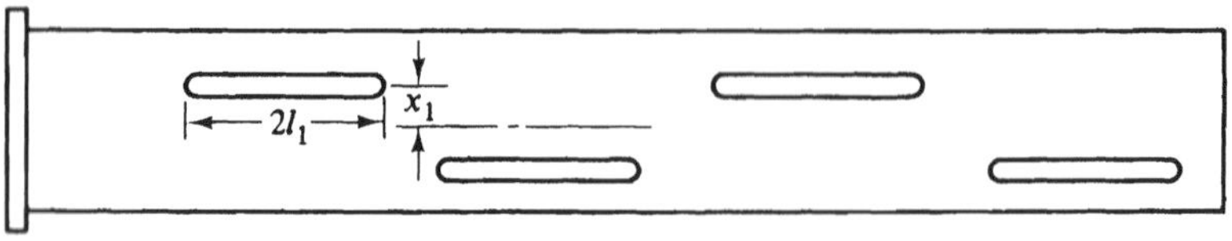

Fig. 8.38 A Four-Element Array of Longitudinal Shunt Slots in the Broad Wall of a Rectangular Waveguide

As a first step, polyfits to the curves shown in Figures 8.35 through 8.37 should be obtained. For the present purposes, it is sufficient to use the simple representations

$$h_1(y) = 1 - 275(y - 1)^2 \tag{8.99}$$

$$h_2(y) = -14(y - 1) \tag{8.100}$$

$$g(x) = 1.177 \sin^2 \frac{\pi x}{a} \tag{8.101}$$

$$v(x) = 1.517 + 1.833x^2 \tag{8.102}$$

The second step involves selection of initial lengths and offsets. This will be done by ignoring mutual coupling (initially). Then all four slots should be self-resonant such that the sum of their normalized conductances is unity and such that the slot voltage distribution is 1: 2: 2: 1. Under these initial assumptions, Equation 8.84 becomes

$$\frac{g(x_n)}{f_n} = \frac{K_1}{V_n} V_n^s \sin v(x_n) = \frac{1.177 \sin (\pi x_n/a) \sin v(x_n)}{\cos [(\beta/k)v(x_n)] - \cos v(x_n)} \tag{8.103}$$

Because of the desired symmetrical slot voltage distribution, slots 1 and 4 will have the same length and opposite offsets, as will slots 2 and 3. The mode voltage V_n will have the same magnitude at each slot, but it will alternate in sign. Thus, when (8.103) is written successively for the first and second slots and a ratio is taken, one obtains

$$\frac{\sin (\pi x_2)/a) \sin v(x_2)}{\cos [(\beta/k)v(x_2)] - \cos v(x_2)} = -2 \frac{\sin (\pi x_1/a) \sin v(x_1)}{\cos [(\beta/k)v(x_1)] - \cos v(x_1)} \tag{8.104}$$

Equation 8.104 will insure the desired slot voltage distribution. In addition, to get an input match,

$$\sum_{i=1}^{4} g(x_i) = 1 = 2[g(x_1) + g(x_2)]$$

or,

$$1.177\left[\sin^2\left(\frac{\pi x_1}{a}\right) + \sin^2\left(\frac{\pi x_2}{a}\right)\right] = \frac{1}{2} \tag{8.105}$$

For $a = 0.900$ inches and $\nu = 9.375$ GHz, one finds that $\beta/k = 0.714$. If one uses these values in (8.104) and (8.105), simultaneous solution gives

$$\begin{aligned} x_1 &= -x_4 = 0.082 \text{ in.} & 2l_1/\lambda &= 2l_4/\lambda = 0.487 & g(x_1) &= g(x_4) = 0.093 \\ x_2 &= -x_3 = -0.180 \text{ in.} & 2l_2/\lambda &= 2l_3/\lambda = 0.502 & g(x_2) &= g(x_3) = 0.407 \end{aligned} \tag{8.106}$$

These are the starting lengths and offsets of the slots in the array. If there were no mutual coupling, they would give the desired slot voltage distribution and an input match.

With these assumed lengths and offsets, one is able to compute Z_{ij} for the equivalent dipole array, using Equations 7.155 and 7.156. It is found that the initial values of mutual impedance are

$$\begin{aligned} Z_{12} &= 0.37 - j8.39 & Z_{13} &= 1.49 + j1.28 \\ Z_{14} &= -0.67 + j0.47 & Z_{23} &= -2.88 - j7.81 \end{aligned} \tag{8.107}$$

From this, one can compute the following initial values of Z_n^b:

$$Z_1^b = [2(0.37 - j8.39) + 2(1.49 + j1.28)]\frac{1.000}{0.999} + (-0.67 + j0.47) = 3.05 - j13.75$$

$$Z_2^b = \left[\frac{1}{2}(0.37 - j8.39) + \frac{1}{2}(1.49 + j1.28)\right]\frac{0.999}{1.000} + (-2.88 - j7.81) = -1.95 - j11.37$$

Since $K_2 = 480$ for this waveguide size and frequency, Equation 8.85 becomes

$$\frac{Y_n^a}{G_0} = \frac{480 f_n^2}{[480 f_n^2 / g(x_n)h(y_n)] + Z_n^b} \tag{8.108}$$

The objective is to select (x_1, y_1) and (x_2, y_2) so that

$$\mathcal{I}_m \frac{480 f_1^2}{g(x_1)h(y_1)} = +13.75 \tag{8.109}$$

$$\mathcal{I}_m \frac{480 f_2^2}{g(x_2)h(y_2)} = +11.37 \tag{8.110}$$

under the pattern restriction that

$$\frac{Y_2^a/G_0}{f_2} = -\frac{V_2^s}{V_1^s}\frac{\sin[y_2 v(x_2)]}{\sin[y_1 v(x_1)]}\frac{Y_1^a/G_0}{f_1} = -2\frac{1.000}{0.999}\frac{Y_1^a/G_0}{f_1} \tag{8.111}$$

and under the input admittance restriction that

$$\frac{Y_1^a}{G_0} + \frac{Y_2^a}{G_0} = \frac{1}{2} \tag{8.112}$$

With the aid of Equations 8.99 through 8.102, a trial-and-error solution of (8.109) through (8.112) is found to be

$$\begin{array}{llll} x_1 = 0.086 \text{ in.} & y_1 = 1.0125 & 2l_1/\lambda = 0.4933 & \dfrac{Y_1^a}{G_0} = 0.0988 \\ x_2 = -0.176 \text{ in.} & y_2 = 1.0099 & 2l_2/\lambda = 0.5058 & \dfrac{Y_2^a}{G_0} = 0.4010 \end{array} \tag{8.113}$$

These results can be iterated. The lengths and offsets in (8.113), when used in (7.155) and (7.156), will produce an improved set of Z_{ij}. However, if this is done, one finds that the new Z_n^b are

$$Z_1^b = 2.95 - j13.73 \qquad Z_2^b = -1.71 - j11.43$$

and these values are so close to the previous set that negligible further change will be found if the iteration process is carried further. Thus (8.113) can be accepted as the design with external mutual coupling taken into account.

A comparison of (8.113) and (8.106) reveals that there is a 5% change in the offset of slots 1 and 4, a $2\frac{1}{2}$% change in the offset of slots 2 and 3, and a 1% lengthening of slots 2 and 4. These changes may seem small enough that one could argue in this application that mutual coupling be neglected. But the effect of these changes on aperture distribution and input admittance are significant, as shall be seen in Section 8.17.

The smallness of these changes can be traced to the fact that these slots, being in a common waveguide, are almost end-fire to each other, so that mutual coupling is lower and falls off faster than when the slots are broadside. One can anticipate a bigger problem with mutual coupling in planar arrays, as will be seen in Section 8.16.

As a final comment on the design of linear slot arrays, the modern trend is to make the b-dimension of the waveguide smaller to save on weight and depth. However, this lengthens the slots and places adjacent ends of successive slots closer together, thus increasing the mutual coupling, and making it even more essential that its effect be included.

8.16 The Design of Planar Waveguide-Fed Slot Arrays

When a family of waveguide-fed linear slot arrays is arranged as shown in Figure 8.39, a planar array results. This introduces a new variable into the design procedure. The individual waveguides containing the radiating slots (hereafter referred to as *branch line waveguides*) may be excited in a variety of ways. Perhaps the most common is to run a main line waveguide transversely across the back of the array and use coupling slots to energize the branch lines. Another method is to use a corporate feed, consisting of a set of T-junctions which serve to split the power in a sequence of steps down to the level of the individual branch lines. But whatever method is used, the mode voltages in the various branch line waveguides can be adjusted in

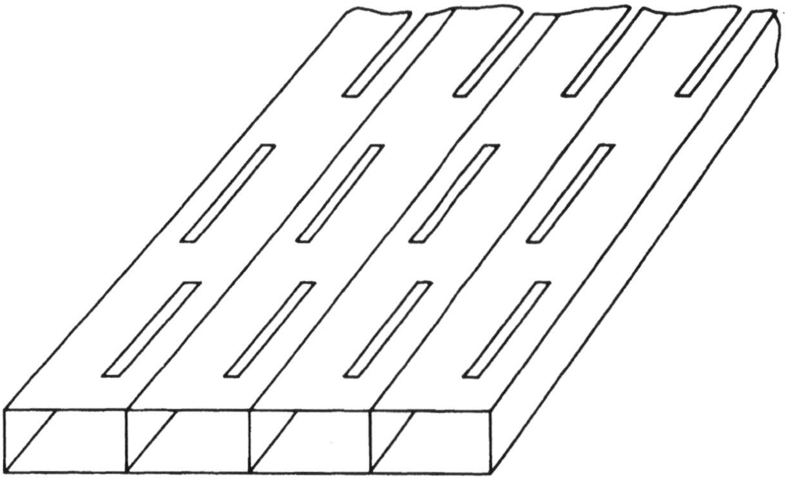

Fig. 8.39 A Planar Array of Longitudinal Shunt Slots

relative level by the coupling mechanism. This is an additional parameter that can be exploited in the design procedure.

It is convenient to go to double subscript notation in the design of planar arrays. Thus the basic design equations become

$$\frac{Y^a_{mn}}{G_0} = K_1 f_{mn} \sin v(x_{mn}) \frac{V^s_{mn}}{V_{mn}} \tag{8.114}$$

$$\frac{Y^a_{mn}}{G_0} = \frac{K_2 f^2_{mn}}{Z^a_{mn}} \tag{8.115}$$

in which K_1 and K_2 are still given by (8.86) and

$$f_{mn} = \left\{ \frac{\cos[(\beta/k) y_{mn} v(x_{mn})] - \cos[y_{mn} v(x_{mn})]}{\sin[y_{mn} v(x_{mn})]} \right\} \sin \frac{\pi x_{mn}}{a} \tag{8.116}$$

in which

$$y_{mn} = l_{mn}/l_r(x_{mn}) \tag{8.117}$$

is the normalized slot length, and $(x_{mn}, 2l_{mn})$ are the offset and length of the mth slot in the nth branch line waveguide.

In most respects, the design of a planar slot array proceeds exactly as for a linear slot array, which was described and illustrated in the previous section. One assumes that every branch line array is resonantly spaced (this restriction will be lifted in Chapter 9). This implies that the mode voltages are given by

$$V_{mn} = (-1)^m V_n \tag{8.118}$$

in which V_n is the reference mode voltage in the nth branch line. For this reason (8.114) can be rewritten in the form

$$\frac{Y^a_{mn}}{G_0} = K_1 \, | f_{mn} | \sin v(x_{mn}) \frac{V^s_{mn}}{V_n} \tag{8.119}$$

since the alternation in direction of offset causes the compensating relation

$$\sin \frac{x_{mn}}{a} = (-1)^m \left| \sin \frac{\pi x_{mn}}{a} \right| \tag{8.120}$$

When (8.115) and (8.119) are used to design a planar array, one must assume an initial set of slot lengths and offsets in order to compute an initial set of Z^b_{mn} values. But one must also assume an initial branch line mode voltage distribution in order to use (8.119). It may be that, as the design proceeds and a series of iterations converges on a final set of slot lengths and offsets, one finds that the set of offsets in a particular branch line is inconveniently small or large (outside the trustworthy range of experimental design data). This can be altered by a change in the V_n distribution. If one wishes to increase the average offset in a branch line without increasing the slot voltage level, this can be accomplished by lowering the coupling to that branch line, which serves to lower that particular branch line mode voltage. Considerable adjusting back and forth is usually needed in the course of the design in order to insure that the final spread of offsets in the branch lines is in an optimum range, and that all the coupling coefficients between the main line feeding structure and the branch lines are also in an optimum range. As a consequence of this adjustment, it is most unlikely that the sum of the normalized active admittances in any branch line is unity; the branch lines do not have to be matched in order to achieve a match in the main line.

A simple illustration of the design of a planar slot array, one which dramatically demonstrates the strong effect of mutual coupling, involves the two-by-four slot antenna shown in Figure 8.40. It was desired to excite this array so that all eight slot

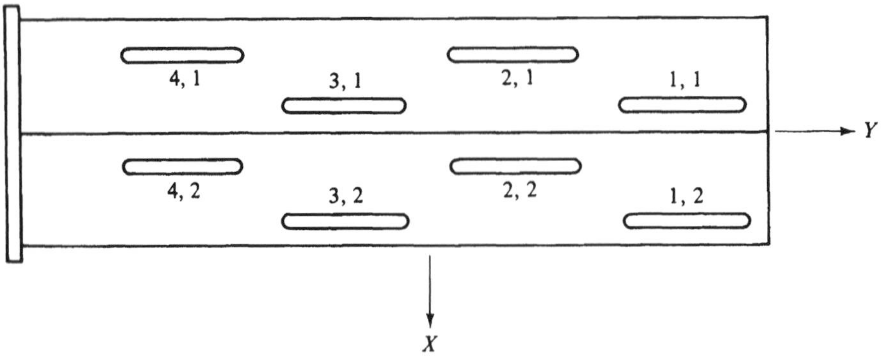

Fig. 8.40 A Two-by-Four Array of Longitudinal Shunt Slots

voltages would be equal and in phase. Additionally, in order to insure that all slot offsets would be in the favorable dynamic range, the condition

$$\sum_{m=1}^{4} \frac{Y_{m,1}^a}{G_0} = \sum_{m=1}^{4} \frac{Y_{m,2}^a}{G_0} = 2 + j0 \tag{8.121}$$

was imposed.

Waveguide dimensions $a = 0.924$ inch, $b = 0.123$ inch, and $t = 0.025$ inch were used, together with a slot width of 0.064 inch. Experimental design data in the form of curves similar to those found in Figure 8.35 through 8.37 was available at $\nu = 8.930$ gigahertz and polynomial expressions were fitted to the data, resulting in expressions akin to Equations 8.99 through 8.102. Equations 8.115 and 8.119 were then used to determine the proper slot lengths and offsets, with $V_1 = V_2$ taken to be the mode voltage distribution. The procedure was identical to the one outlined for the one-by-four array in Section 8.15, except that a computer program was used instead of hand calculations. The results are shown in Table 8.8.

TABLE 8.8 Lengths and offsets for two-by-four slot array

Slot Number mn	Offset x_{mn} (Inches)	Length $2l_{mn}$ (Inches)
1, 1	0.060	0.669
2, 1	−0.122	0.708
3, 1	0.060	0.667
4, 1	−0.099	0.693
1, 2	0.099	0.693
2, 2	−0.060	0.667
3, 2	0.122	0.708
4, 2	−0.060	0.669

A study of this table of slot lengths and offsets reveals several interesting and surprising things. First, there is a 2:1 range in slot offsets. (Were one to ignore mutual coupling or assume it was the same for each slot, all offsets would be the same.) Second, no slot in this array is self-resonant; each slot is detuned appropriately to make the individual *active* admittance resonant. Third, there is a quadrant I to quadrant III and quadrant II to quadrant IV symmetry to the lengths and offsets, but no symmetry about the X-axis nor about the Y-axis. This can be traced to non-symmetrical effects caused by staggering the offsets and always occurs in array designs that yield symmetrical patterns. And fourth, it is clearly evident that the presence of broadside neighbors has substantially increased the effect of mutual coupling.

At first it might seem puzzling, for example, that the required offsets of the 2, 1 slot and the 3, 1 slot are so radically dissimilar when the slot voltages are to be the same. However, a return to Figure 8.40 indicates that the environment of the 2, 1 slot is significantly different from the environment of the 3, 1 slot.

An experimental determination of the active input admittance to each branch line waveguide yielded the results

$$\sum_{m=1}^{4} \frac{Y_{m,1}^{a}}{G_0} = 1.90 + j0 \qquad \sum_{m=1}^{4} \frac{Y_{m,2}^{a}}{G_0} = 1.94 \tag{8.122}$$

which were 5% and 3% from the design values. The experimental H-plane pattern at 8.930 GHz, with the array embedded in an 8-inch by 10-inch ground plane, is shown as the solid curve in Figure 8.41. The theoretical pattern (dotted curve) is also shown for comparison.

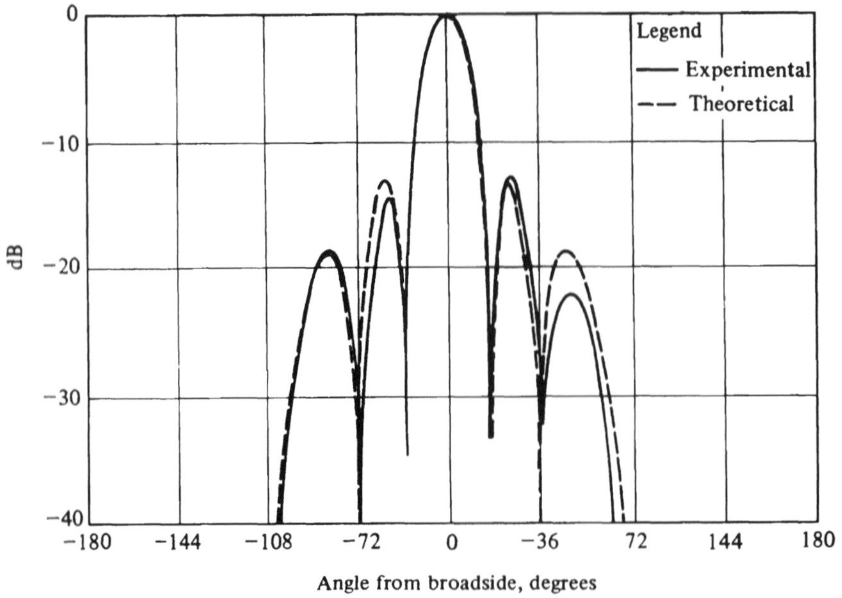

Fig. 8.41 The *H*-Plane Pattern of the Two-by-Four Slot Array Depicted in Figure 8.40; Comparison of Theory and Experiment

8.17 Sum and Difference Patterns for Waveguide-Fed Slot Arrays; Mutual Coupling Included

If Y_n^a/G_0 is eliminated from (8.84) and (8.85), one obtains for a *linear* slot array

$$V_n^s \sin kl_n \, Z_n^a = \left(\frac{K_2}{K_1}\right) V_n f_n \tag{8.123}$$

With the aid of (8.88) and (8.89), this can be rewritten as

$$\sum_{m=1}^{N} V_m^s \sin kl_m \, Z_{nm} = \left(\frac{K_2}{K_1}\right) V_n f_n \tag{8.124}$$

Equations 8.124 can be identified as a set of simultaneous linear equations (with the slot voltages as independent variables) that can be put in the matrix form

$$\begin{bmatrix} Z_{11} & Z_{12} & \cdots & Z_{1N} \\ \cdot & \cdot & & \cdot \\ \cdot & \cdot & & \cdot \\ \cdot & \cdot & & \cdot \\ \cdot & \cdot & & \cdot \\ Z_{N1} & Z_{N2} & \cdots & Z_{NN} \end{bmatrix} \begin{bmatrix} V_1^s \sin kl_1 \\ V_2^s \sin kl_2 \\ \cdot \\ \cdot \\ \cdot \\ V_N^s \sin kl_N \end{bmatrix} = \begin{bmatrix} |f_1| \\ |f_2| \\ \cdot \\ \cdot \\ \cdot \\ |f_N| \end{bmatrix} \tag{8.125}$$

in which it has been recognized that $V_n f_n = V|f_n|$, with V the reference mode voltage in the linear array. The common factor $(K_2/K_1)V$ has been suppressed in going from (8.124) to (8.125).

If the length and offset of every slot is specified, all the impedance terms in the Z-matrix of (8.125) are known, as are all the elements $|f_m|$ in the column matrix. An inversion of (8.125) will give the slot voltage distribution for this set of lengths and offsets.

As an example of the use of (8.125), consider again the four-element linear slot array analyzed in Section 8.15. Because of symmetry considerations, for that case (8.125) takes the form

$$\begin{bmatrix} Z_{11} & Z_{12} & Z_{13} & Z_{14} \\ Z_{12} & Z_{22} & Z_{23} & Z_{13} \\ Z_{13} & Z_{23} & Z_{22} & Z_{12} \\ Z_{14} & Z_{13} & Z_{12} & Z_{11} \end{bmatrix} \begin{bmatrix} V_1^s \sin kl_1 \\ V_2^s \sin kl_2 \\ V_2^s \sin kl_2 \\ V_1^s \sin kl_1 \end{bmatrix} = \begin{bmatrix} |f_1| \\ |f_2| \\ |f_2| \\ |f_1| \end{bmatrix} \tag{8.126}$$

which reduces to

$$\begin{bmatrix} (Z_{11} + Z_{14}) & (Z_{12} + Z_{13}) \\ (Z_{12} + Z_{13}) & (Z_{22} + Z_{23}) \end{bmatrix} \begin{bmatrix} V_1^s \sin kl_1 \\ V_2^s \sin kl_2 \end{bmatrix} = \begin{bmatrix} |f_1| \\ |f_2| \end{bmatrix} \tag{8.127}$$

For the starting lengths and offsets given in (8.106), the mutual impedances are given by (8.107) and

$$\begin{aligned} Z_{11} &= \frac{K_2 f_1^2}{g(x_1)h(y_1)} = \frac{480(0.1184)^2}{0.0938} = 71.74 + j0 \\ Z_{22} &= \frac{K_2 f_2^2}{f(x_2)h(y_2)} = \frac{480(0.2564)^2}{0.4066} = 77.61 + j0 \end{aligned} \tag{8.128}$$

As a consequence of (8.128), the matrix (8.127) becomes

$$\begin{bmatrix} (71.07 + j0.47) & (5.16 - j7.11) \\ (5.16 - j7.11) & (74.73 - j7.81) \end{bmatrix} \begin{bmatrix} V_1^s \sin kl_1 \\ V_2^s \sin kl_2 \end{bmatrix} = \begin{bmatrix} 0.1184 \\ 0.2564 \end{bmatrix} \tag{8.129}$$

Inversion gives

$$V_1^s = 0.001415\underline{|11.6^\circ} \qquad V_2^s = 0.003293\underline{|7.9^\circ} \tag{8.130}$$

so that the slot voltage ratio is

$$\frac{V_2^s}{V_1^s} = 2.33\underline{|-3.7^\circ} \tag{8.131}$$

The reader will recall that the desired ratio is $2.00\underline{|0^\circ}$. One can see that, whereas there is only a 5% or less error in the starting lengths and offsets, this results in a 16.5% magnitude error and a 3.7° phase error in the slot voltage distribution.

For the slot voltages given in (8.130) and the self-impedances and mutual impedances listed in (8.128) and (8.107), one can determine that

$$Z_1^a = 82.00 - j16.85, \quad Z_2^a = 77.14 - j10.71 \tag{8.132}$$

Use of (8.85) reveals that

$$\frac{Y_1^a}{G_0} = \frac{Y_4^a}{G_0} = 0.0788 + j0.0162 \qquad \frac{Y_2^a}{G_0} = \frac{Y_3^a}{G_0} = 0.4014 + j0.0557$$

and thus

$$\sum_{i=1}^{4} \frac{Y_i^a}{G_0} = 0.9604 + j0.1438 = 0.97\underline{|8.5^\circ} \tag{8.133}$$

The input admittance has a susceptive component which is 15% of the conductance. There is a mismatch of 3% in magnitude and 8.5° in phase.

Were one to repeat these calculations for the final offsets and lengths given in (8.113), it would be discovered that the slot voltage distribution is correct in both amplitude and phase. Confirmation of this assertion is left as an exercise.

For a *planar* slot array, elimination of Y_{mn}^a/G_0 from (8.115) and (8.119) gives

$$V_{mn}^s \sin kl_{mn} Z_{mn}^a = \left(\frac{K_2}{K_1}\right) V_n |f_{mn}| \tag{8.134}$$

a result which is identical to (8.123) except for the use of double subscript notation and the added feature that the mode voltage V_n may differ from branch line to branch line. Equation 8.134 can be expanded to give

$$\sum_p \sum_q V_{pq}^s \sin kl_{pq} Z_{mn}^{pq} = V_n |f_{mn}| \tag{8.135}$$

with the constant (K_2/K_1) suppressed.

If the slot lengths and offsets are known, the impedance matrix appearing in (8.135) is known, as are the column matrix elements $V_n|f_{mn}|$. Inversion of (8.135) will give the slot voltage distribution.

A common use of (8.135) is in application to planar arrays, which are divided into four quadrants and fed to produce sum and difference patterns. In such cases it is convenient to use the indexing scheme shown in Figure 8.42. Because of the quadrant I to quadrant III and quadrant II to quadrant IV symmetry in such arrays (already noted for the two-by-four array discussed in Section 8.16), it is unnecessary to apply (8.135) to the entire array. If there are M-by-N slot modules,[23] for the sum pattern $V^s_{pq} = V^s_{M+1-p,N+1-q}$ and (8.135) becomes

$$\sum_{p=1}^{M/2} \sum_{q=1}^{N} V^s_{pq} \sin kl_{pq}(Z^{pq}_{mn} + Z^{M+1-p,N+1-q}_{mn}) = V_n |f_{mn}|$$
$$\left(\Sigma \text{ pattern: } 1 \leq m \leq \frac{M}{2} \qquad 1 \leq n \leq N\right) \tag{8.136}$$

Symmetry conditions indicate that, in (8.36), $V_n = V_{N+1-n}$.

If the slots are assumed to be parallel to the X-axis in Figure 8.42, then the E-plane and H-plane difference patterns correspond to the quadrants being excited as shown in Figure 8.43. For Δ_E (that is, the E-plane difference pattern), $V^s_{pq} =$

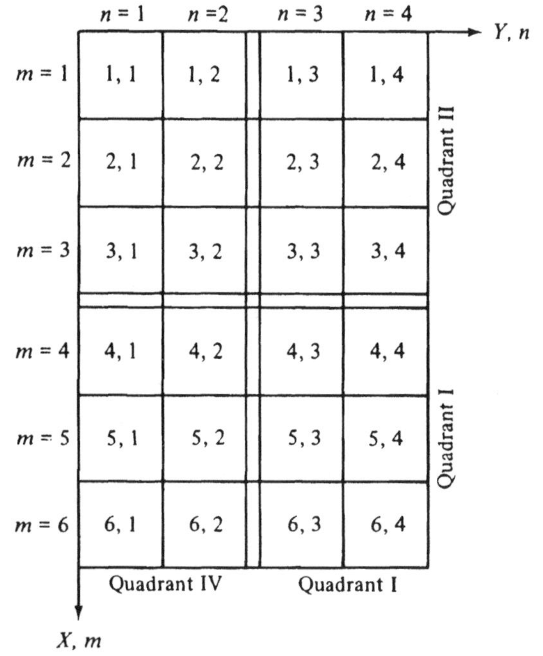

Fig. 8.42 Indexing Notation for Slot Arrays Showing Individual Modules; $M = 6$, $N = 4$

[23]This does not necessarily mean a rectangular array, since some of the modules may be "empty," such as when corner slots are eliminated so that the array will fit in a circular boundary.

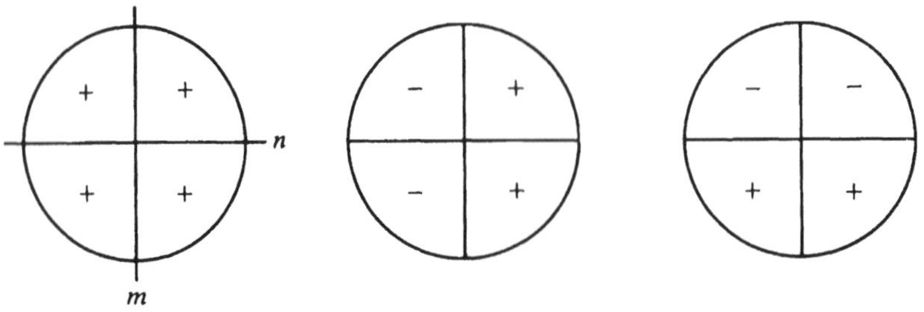

(a) Sum channel (b) E- plane difference channel (c) H-plane difference channel

Fig. 8.43 Quadrant Excitations for Sum and Difference Patterns of Planar Arrays

$-V^s_{M+1-p,N+1-q}$. For this case (8.135) becomes

$$\sum_{p=1}^{M/2}\sum_{q=1}^{N} V^s_{pq}\sin kl_{pq}(Z^{pq}_{mn} - Z^{M+1-p,N+1-q}_{mn}) = V_n|f_{mn}| \qquad (8.137)$$

$$\left(\Delta_E \text{ pattern: } 1 \le m \le \frac{M}{2} \qquad 1 \le n \le N\right)$$

For Δ_H (that is, the H-plane difference pattern), it is also true that $V^s_{pq} = -V^s_{M+1-n,N+1-q}$. For this case (8.135) becomes

$$\sum_{p=1}^{M/2}\sum_{q=1}^{N} V^s_{pq}\sin kl_{pq}(Z^{pq}_{mn} - Z^{M+1-p,N+1-q}_{mn}) = V_n|f_{mn}| \qquad (8.138)$$

$$\left(\Delta_H \text{ pattern: } 1 \le m \le \frac{M}{2} \qquad 1 \le n \le N\right)$$

However, in contrast to the Δ_E case, in (8.138) the relation $V_n = V_{N+1-n}$ applies.

As an illustration of the use of these formulas, if the slot lengths and offsets for the two-by-four array discussed in Section 8.16 are used to compute the Z^{rs}_{ij} and f_{ij} entries needed in (8.136) through (8.138), inversion gives the slot voltage distributions listed in Table 8.9. When these excitations are used to compute the patterns, the

TABLE 8.9 Slot voltage distributions for two-by-four array

Slot No.	Slot Voltage V^s_{mn}		
mn	Σ	Δ_E	Δ_H
1, 1	$1.002 + j0.002$	$0.231 + j0.063$	$1.051 + j0.003$
2, 1	$1.000 - j0.006$	$0.968 + j0.203$	$1.315 - j0.053$
3, 1	$1.002 - j0.005$	$0.031 + j0.139$	$-1.113 + j0.187$
4, 1	$1.004 - j0.003$	$0.791 + j0.128$	$-1.066 + j0.011$
1, 2	$1.004 - j0.003$	$-0.791 - j0.128$	$1.066 - j0.011$
2, 2	$1.002 - j0.005$	$-0.031 - j0.139$	$1.113 - j0.187$
3, 2	$1.000 - j0.006$	$-0.968 - j0.203$	$-1.315 + j0.053$
4, 2	$1.002 + j0.002$	$-0.231 - j0.063$	$-1.051 - j0.003$

result for the sum channel is shown in Figure 8.41. The results for the difference channels are shown in Figure 8.44. Element factors have been included in all these patterns.

The difference patterns exhibited in Figure 8.44 clearly illustrate a basic difficulty in the design of planar arrays for use in sum and difference applications. If the slots are excited by a common feeding structure for all three channels, and if one designs the feeding structure to obtain a good sum pattern, one must accept some poor difference patterns. It is similarly true that, if one were to design the feeding structure in order to produce a good Δ_E pattern, for example, then the resulting Σ and Δ_H patterns are inferior. The only way now known to overcome this deficiency is to use separate feeding structures for the three channels, but this is extremely complicated and costly.

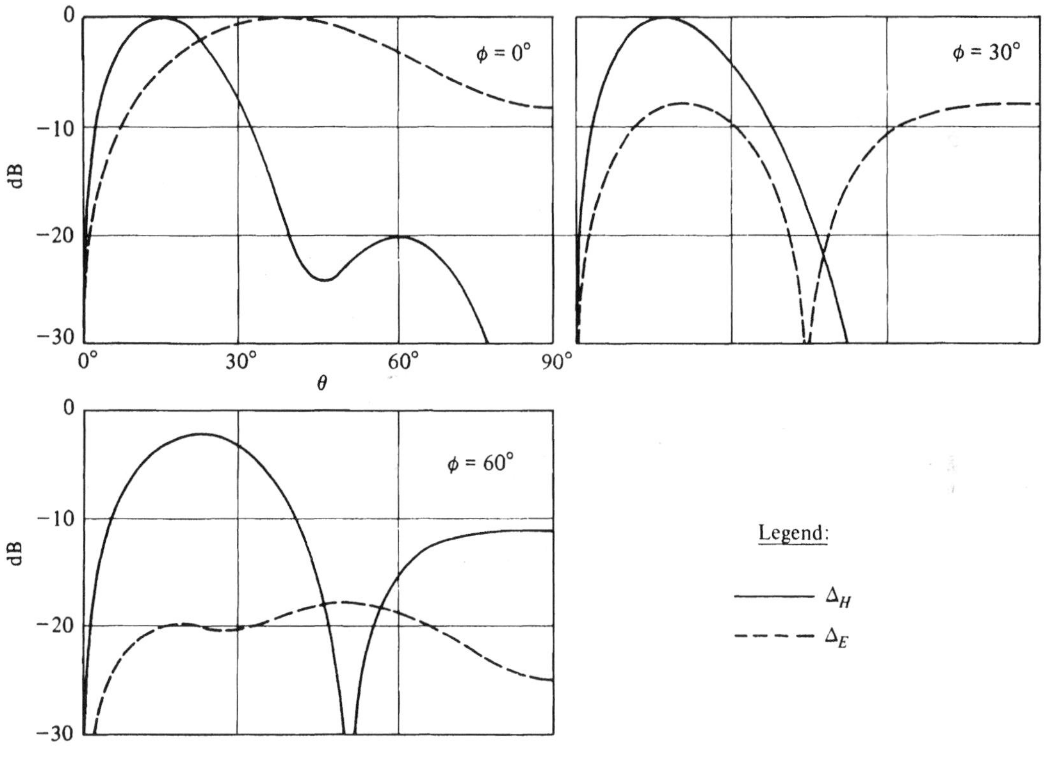

Fig. 8.44 Δ_H and Δ_E Difference Patterns for the Two-by-Four Slot Array of Fig. 8.40

REFERENCES

JORDAN, E. C. and K. G. BALMAIN, *Electromagnetic Waves and Radiating Systems*, 2nd ed. (Englewood Cliffs, New Jersey: Prentice-Hall, 1968), Chapter 15.

KING, R. W. P., *Tables of Antenna Characteristics* (New York: IFI/Plenum, 1971), pp. 7–15.

KRAUS, J. D., *Antennas* (New York: McGraw-Hill Book Co., Inc., 1950), Chapter 7.

MITTRA, R., "Log Periodic Antennas," *Antenna Theory: Part II*, ed. R. E. Collin and F. J. Zucker (New York: McGraw-Hill Book Co., Inc., 1969), Chapter 22.

WOLFF, E. A., *Antenna Analysis* (New York: John Wiley and Sons, Inc., 1966), Sections 5.5, 8.6, and 9.5.

PROBLEMS

8.1 Repeat the monopole design of Section 8.2, assuming that the gap problem can be ignored, and that the Tai-type impedance expression (7.71) is applicable.

8.2 What is the impedance bandwidth (1.5: 1 VSWR criterion) for the monopole designed in Problem 8.1?

8.3 Repeat the dipole-over-a-ground-plane design of Section 8.3, using the improved King-Middleton approximation for self-impedance. Equations 7.110 and 7.111 can be used if a computer is available, or data can be read from Figure 7.18.

8.4 A rectangular slot is cut in a large ground plane of negligible thickness. If the slot is 1 in. wide and is to be center-fed by a two-wire line, what should its length be for resonance at 300 MHz? What is its input resistance at resonance?

8.5 If a cavity is used to box in one side of the slot of Problem 8.4 and if the depth of the cavity is less than $\lambda_g/4$, would you expect to have to lengthen or shorten the slot to reachieve resonance?

8.6 What physical argument would you use to explain the low input VSWR over a wide frequency range for a ground-plane-backed helix?

8.7 Find the optimum spacing of a two-dipole array, with both elements driven, in order to achieve an end-fire array pattern with maximum directivity but only one main lobe. Assume dipole radii $a = 0.0032\lambda$, and determine the lengths of the two dipoles in order to achieve a match with a two-wire line at a reasonable impedance level. Assume series coupling to the line, as in Figure 8.9.

8.8 How would the design of the dipole array in Problem 8.7 be altered if only one element were driven?

8.9 For a three-element Yagi-Uda array, find the optimum lengths of reflector and director if the driven element is 0.475λ long, if the interelement spacing is 0.15λ, and if all three dipoles have a radius $a = 0.0032\lambda$.

8.10 A refinement of the explanation for the need to reverse the feeding at successive elements in a log periodic array of the type shown in Figure 8.27 results from assuming a "locally periodic" behavior along the structure. This permits the identification of transmissive, active, and reflective regions (akin to the director-driven element-reflector model adopted in Section 8.9). Assume this more extensive model and show that the feeding portrayed in Figure 8.26a is incorrect, and that the proper method of feeding is as shown in Figure 8.26b. [Compare with P. E. Mayes, G. A. Deschamps, and W. T. Patton, "Backward-wave Radiation from Periodic Structures and Application to the Design of Frequency-Independent Antennas," *Proc. IRE*, 49 (1961), 962.]

8.11 A linear array of six balun-fed dipoles stands $\lambda/4$ in front of a large ground plane. The dipoles are 0.6λ on centers and in the end-to-end, or tandem orientation. If a -20 dB SLL Dolph-Chebyshev broadside array pattern is desired, together with an input match to 25 ohms, design a coaxial feed for this array. Your design should include a specification of the length of each dipole and the characteristic impedance of each balun section. Assume that all dipoles have a radius $a = 0.0032\lambda$.

8.12 Repeat Problem 8.11 for a six-by-six array.

8.13 Assume that the master input to the six-by-six array of Problem 8.12 is fitted with a perfect magic *T*. Find the *E*-plane difference pattern and the corresponding input impedance.

8.14 Repeat Problem 8.13 for the H-plane difference pattern.

8.15 With the phaseshifters in the branch lines, repeat the analysis of Section 8.12 and show the effect of mutual coupling on pattern and input impedance as the main beam is scanned $\pm 10°$ from broadside.

8.16 Design a resonantly spaced three-element longitudinal shunt slot array in standard *X*-band guide. The frequency of operation is to be 9.375 GHz and the excitation is to be uniform, with an input match.

8.17 Repeat Problem 8.16 for a three-by-three array.

IV continuous aperture antennas

9 traveling wave antennas

9.1 Introduction

A traveling wave antenna is one in which the radiating aperture and feeding structure are intimately contiguous, if not continuously connected. As the name implies, the aperture distribution has features similar to those of a traveling wave; the amplitude of excitation may be tapered, but the phase progression is uniform, or nearly so.

Traveling wave antennas may be one- or two-dimensional. Examples of the former are long wires and their derivatives (Vees and rhombics), polyrods, and leaky waveguides. Examples of the latter include corrugated and dielectric-clad surfaces (both planar and curved) and arrays of leaky waveguides.

This chapter offers an introduction to the analysis and design of some of the practical types of traveling wave antennas. It begins with a discussion of the long wire, followed by an extension to rhombics and Vees. Structures which will support slow waves are then analyzed (notably grounded dielectric slabs and corrugated surfaces) and the launching and termination of these waves is considered, leading to an interpretation of the behavior of slow wave antennas. After this, leaky waveguides are introduced, with particular attention given to long continuous slots in either the narrow or broad wall of a rectangular waveguide (the latter offset from the center line), and to the quasi-continuous case of many closely spaced, nonresonant transverse slots in the broad wall (serrated rectangular waveguide). A design procedure is developed that will yield the aperture geometry necessary to achieve a desired pattern together with an input match.

Trough waveguides are considered next and their beam-scanning capabilities are explored. The chapter closes with a discussion of arrays of longitudinal shunt slots in the broad wall of a rectangular waveguide. The design procedure developed in Chapter 8 for resonantly spaced arrays ($d = \lambda_g/2$) is extended to apply to nonresonant spacing, which results in a traveling wave excitation. Mutual coupling is generally severe in the traveling wave case and this effect is included in the analysis.

Traveling wave antennas typically have good input impedance characteristics (the reflected wave is effectively suppressed by some means) and therefore the emphasis in this chapter will be on pattern characteristics.

9.2 The Long Wire Antenna

One of the simplest of the traveling wave antennas is the long horizontal wire a distance h above the earth, fed at one end against ground and perhaps terminated at the other end in a matched load, as shown in Figure 9.1. If h is not negligible compared to a wavelength, this wire and its image do not behave like a two-wire transmission line, but rather comprise an efficient radiating system. The traveling wave of current proceeding outward along the wire is attenuated due to the radiation, and thus the power absorbed in the matched load may be reduced to an acceptable level by making the wire long enough. Indeed, the leakage may be sufficient to obviate the need for a terminating load. The net current distribution on the wire is then not a standing wave, but is essentially an outward traveling damped wave.

If the earth is highly conductive, the image current lies a distance h below the XY-plane and is also an outward traveling wave, 180° out of phase with the current on the wire. The pattern due to wire plus image can be obtained by multiplying the element pattern of the wire by the array factor $\sin(kh\cos\theta)$.

Assume that the current distribution on the wire can be represented adequately by

$$I(\xi, t) = I_0 e^{j\omega t - \gamma\xi} \tag{9.1}$$

with $\gamma = \alpha + j\beta$ the complex propagation constant. Equations 1.101 and 1.102 can be used to determine the element pattern of the wire, and give

$$\mathcal{A}_{\theta,e}(\theta, \phi) = \cos\theta \cos\phi\, f(\theta, \phi) \tag{9.2}$$

$$\mathcal{A}_{\phi,e}(\theta, \phi) = -\sin\phi\, f(\theta, \phi) \tag{9.3}$$

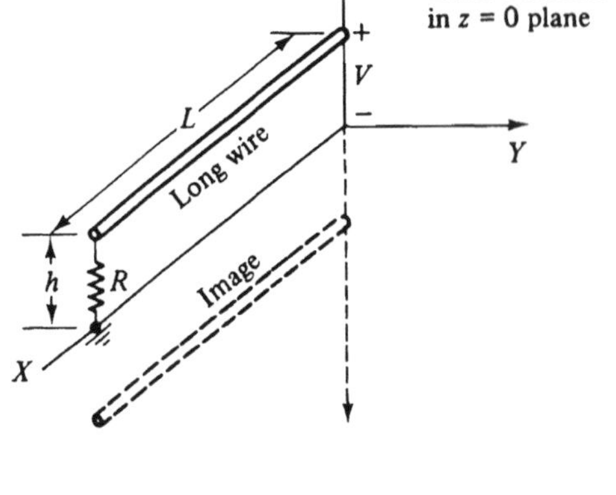

Fig. 9.1 A Long, Terminated Horizontal Wire Antenna; End-Fed (Beverage Antenna)

in which

$$f(\theta, \phi) = I_0 \int_0^L e^{-\gamma\xi} e^{jk\xi \sin\theta \cos\phi} \, d\xi$$
$$= -I_0 \frac{e^{-L(\gamma - jk \sin\theta \cos\phi)} - 1}{(\gamma - jk \sin\theta \cos\phi)} \tag{9.4}$$

This field is a figure of rotation about the wire and is given either by (9.2) for $\phi = 0$ (XZ-plane) or by (9.3) for $\theta = \pi/2$ (XY-plane).

In practical applications (wires composed of good conductors in an air environment), $\beta \cong k$; α is due almost entirely to radiation leakage and due hardly at all to ohmic losses. Further $\alpha \ll k$ (for example, with $L = 10\lambda$ and 10 dB of attenuation in the current level along the wire, α is only 1.8% of k). Thus $f(\theta, \phi)$ is given to good approximation by[1]

$$f(\theta, \phi) = -I_0 \frac{e^{-jkL(1 - \sin\theta \cos\phi)} - 1}{jk(1 - \sin\theta \cos\phi)} \tag{9.5}$$

and its magnitude can be expressed in the form

$$|f(\theta, \phi)| = I_0 L \left| \frac{\sin X}{X} \right| \tag{9.6}$$

with

$$X = \frac{\pi L}{\lambda}(1 - \sin\theta \cos\phi) \tag{9.7}$$

The factor $(\sin X)/X$ can be interpreted as due to a continuous line source which is uniformly excited in amplitude, possessing a uniform progressive phase which places the main beam at end-fire. This result is modified by the multiplicative factors $\cos\theta \cos\phi$ and $-\sin\phi$ in (9.2) and (9.3), as a result of which the actual main beam peak lies off end-fire by an amount which depends on the length of the wire. A typical element pattern is shown in Figure 9.2a for the case $L = 5\lambda$. All lobes shown are conical, since the pattern is a figure of revolution about the wire.

If the earth is a good conductor and $h = \lambda/4$, the array factor of wire plus image is

$$\mathcal{A}_a(\theta) = \sin\left(\frac{\pi}{2} \cos\theta\right) \tag{9.8}$$

and is plotted in Figure 9.2b. This pattern is a figure of revolution about the zenith axis. The field pattern of wire plus image is the product of the two plots shown in Figure 9.2. It is identically zero in the XY-plane and has a shape in the XZ-plane somewhat similar to the upper half of Figure 9.2a, except that the side lobes are raised relative to the main beam because of the weighting of the array factor.

If the earth is not a good conductor, but rather is more appropriately repre-

[1] E. Hallén, "Properties of Long Antennas," *Journ. App. Phys.*, 19 (1948), 1140–47.

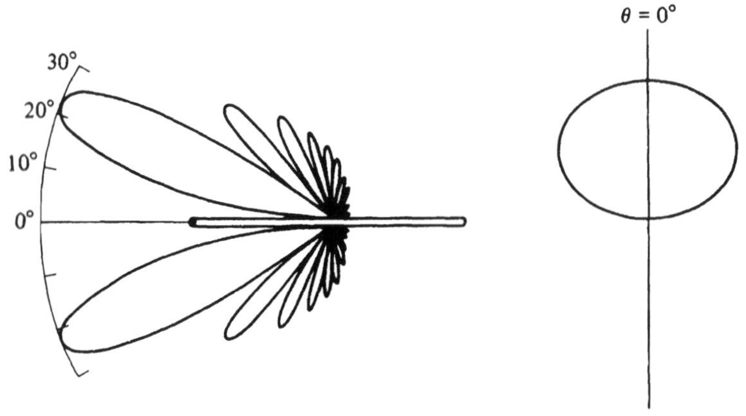

Fig. 9.2 The Element Pattern and Array Pattern of a Traveling Wave Wire Antenna; $L = 5\lambda$, $h = 0.25\lambda$

sented as having a complex dielectric constant $\epsilon' - j\epsilon''$, the height of the wire can be adjusted so that the wave reflected off the earth's surface combines in additive phase with the direct wave from the wire in the direction of maximum radiation.

9.3 Rhombic and Vee Antennas

A rhombic antenna, as its name implies, is composed of four long, straight wires arranged to form a rhombus, as shown in Figure 9.3. It is fed at one corner and terminated in a matched load at the opposite corner. Traveling waves of current, 180°

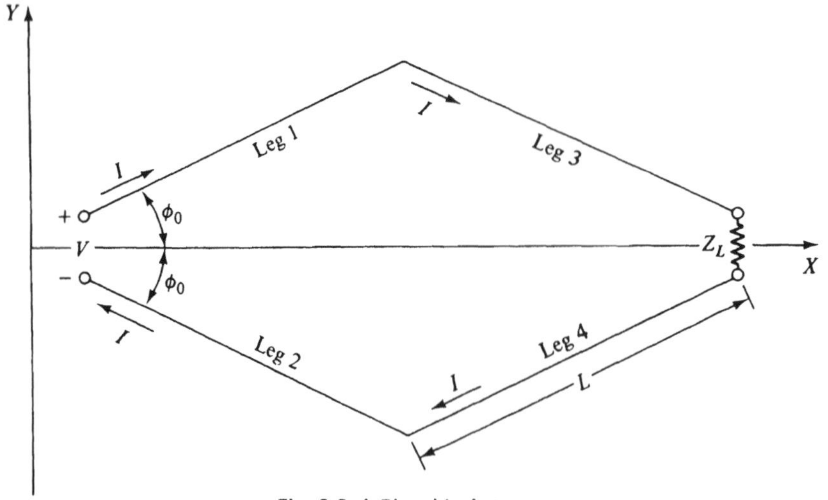

Fig. 9.3 A Rhombic Antenna

out of phase, are launched onto legs 1 and 2. If one neglects corner effects, these traveling waves of current continue outward on legs 3 and 4 and are absorbed in the matched load, thus insuring that no traveling waves are set up in the reverse direction. When the legs of the rhombus are long enough (5λ or more), sufficient radiation has occurred before the traveling waves reach the far corner that an acceptably small portion of the total supplied power is absorbed in the matched load.

If s is a coordinate measured along one of the wires, the traveling wave of current can be expressed in the form $I(s) = I_0 e^{-\gamma s}$ in which $\gamma = \alpha + j\beta$ is the complex propagation constant. In practical applications (wires composed of good conductor in an air environment), $\beta \cong k$; α is due almost entirely to radiation leakage and due hardly at all to ohmic losses. As in the case of the Beverage antenna, discussed in the previous section, $\alpha \ll k$ and therefore the pattern can be determined to a good approximation by assuming $\gamma = jk$.

The rhombic antenna can be viewed as an array of four long wire antennas and the field distribution determined by pattern multiplication. Alternatively, one can return to Equations 1.101 and 1.102 and form the basic expressions for $\mathcal{A}_\theta$ and $\mathcal{A}_\phi$. With the latter course adopted, the position coordinates are given by:

$$\begin{aligned}
&\text{leg 1:}\ \xi = s_1 \cos\phi_0, \quad \eta = s_1 \sin\phi_0 \\
&\text{leg 2:}\ \xi = s_2 \cos\phi_0, \quad \eta = -s_2 \sin\phi_0 \\
&\text{leg 3:}\ \xi = (L + s_3)\cos\phi_0, \quad \eta = (L - s_3)\sin\phi_0 \\
&\text{leg 4:}\ \xi = (L + s_4)\cos\phi_0, \quad \eta = -(L - s_4)\sin\phi_0
\end{aligned}$$

The contribution to $\mathcal{A}_\theta$ from leg 1 is

$$\begin{aligned}
\mathcal{A}_{\theta,1}(\theta,\phi) &= \int_0^L [\cos\theta\cos\phi\cos\phi_0\, I_0 e^{-jks_1} + \cos\theta\sin\phi\sin\phi_0 I_0 e^{-jks_1}] \\
&\quad \cdot e^{jk\sin\theta[s_1\cos\phi_0\cos\phi + s_1\sin\phi_0\sin\phi]}\, ds_1 \\
&= I_0\cos\theta\cos(\phi-\phi_0)\int_0^L e^{-jks_1[1-\sin\theta\cos(\phi-\phi_0)]}\, ds_1 \\
&= I_0\cos\theta\cos(\phi-\phi_0)\frac{1 - e^{-jkL[1-\sin\theta\cos(\phi-\phi_0)]}}{jk[1-\sin\theta\cos(\phi-\phi_0)]}
\end{aligned} \tag{9.9}$$

In like manner, one finds that

$$\mathcal{A}_{\theta,2}(\theta,\phi) = -I_0\cos\theta\cos(\phi+\phi_0)\frac{1 - e^{-jkL[1-\sin\theta\cos(\phi+\phi_0)]}}{jk[1-\sin\theta\cos(\phi+\phi_0)]} \tag{9.10}$$

$$\mathcal{A}_{\theta,3}(\theta,\phi) = -e^{-jkL[1-\sin\theta\cos(\phi-\phi_0)]}\mathcal{A}_{\theta,2}(\theta,\phi) \tag{9.11}$$

$$\mathcal{A}_{\theta,4}(\theta,\phi) = -e^{-jkL[1-\sin\theta\cos(\phi+\phi_0)]}\mathcal{A}_{\theta,1}(\theta,\phi) \tag{9.12}$$

The sum of these four contributions gives

$$|\mathcal{A}_\theta(\theta,\phi)| = \frac{4\pi I_0 L^2}{\lambda}\left|\cos\theta\sin\phi\sin\phi_0\frac{\sin A}{A}\frac{\sin B}{B}\right| \tag{9.13}$$

in which

$$A = \frac{\pi L}{\lambda}[1 - \sin\theta\cos(\phi - \phi_0)] \tag{9.14}$$

$$B = \frac{\pi L}{\lambda}[1 - \sin\theta\cos(\phi + \phi_0)] \tag{9.15}$$

Proceeding in a similar fashion, one finds that

$$|\mathcal{A}_\phi(\theta, \phi)| = \frac{4\pi I_0 L^2}{\lambda}\left|\sin\phi_0(\cos\phi - \sin\theta\cos\phi_0)\frac{\sin A}{A}\frac{\sin B}{B}\right| \tag{9.16}$$

The pattern of principal interest occurs in the XZ-plane. For a rhombic a distance h above a perfectly conducting ground plane, this pattern is given by[2]

$$|\mathcal{A}_\phi(\theta, 0°)| = \frac{4LI_0}{\lambda}\sin\phi_0\sin(kh\cos\theta)\frac{\sin^2\left[\frac{\pi L}{\lambda}(1 - \sin\theta\cos\phi_0)\right]}{\frac{\pi L}{\lambda}(1 - \sin\theta\cos\phi_0)} \tag{9.17}$$

As a function of θ, Equation 9.17 is seen to be the product of the three factors

$$\frac{\sin\pi Y}{\pi Y} \qquad \sin\pi Y \qquad \sin(kh\cos\theta) \tag{9.18}$$

in which $Y = (L/\lambda)(1 - \sin\theta\cos\phi_0)$. The first of these factors has already been plotted in Figure 3.3a. When it is multiplied by $\sin\pi Y$, the result can be displayed as in Figure 9.4a. For the typical case $L/\lambda = 6$, $h/\lambda = 1.5$, and $\phi_0 = 20°$, the factor $\sin(kh\cos\theta)$ appears as Figure 9.4b. Multiplication gives the pattern shown in Figure 9.4c. This pattern is seen to consist of a main beam tilted 10° above the horizon, plus a family of side lobes with an envelope that undulates and decays. The number of side lobes depends on L/λ and the side lobe level is customarily no better than -13 decibels.

The design of a rhombic antenna can be optimized by manipulation of Equation 9.17. First, the height h can be selected by requiring that $\partial\mathcal{A}_\varphi/\partial h = 0$ at θ_0, with θ_0 the desired position of the peak of the main beam. This gives

$$\cos(kh\cos\theta_0) = 0$$

$$kh_m\cos\theta_0 = m(\pi/2) \qquad m = 1, 3, 5, \ldots$$

$$\frac{h_m}{\lambda} = \frac{m}{4\cos\theta_0} \tag{9.19}$$

[2]A. A. de Carvalho Fernandes, "On the Design of Some Rhombic Antenna Arrays," *IRE Trans. Antennas Propagat.*, AP-7 (1959), 39–46. See also E. Bruce, A. C. Beck, and L. R. Lowry, "Horizontal Rhombic Antennas," *Proc. IRE*, 23 (1935), 24–46.

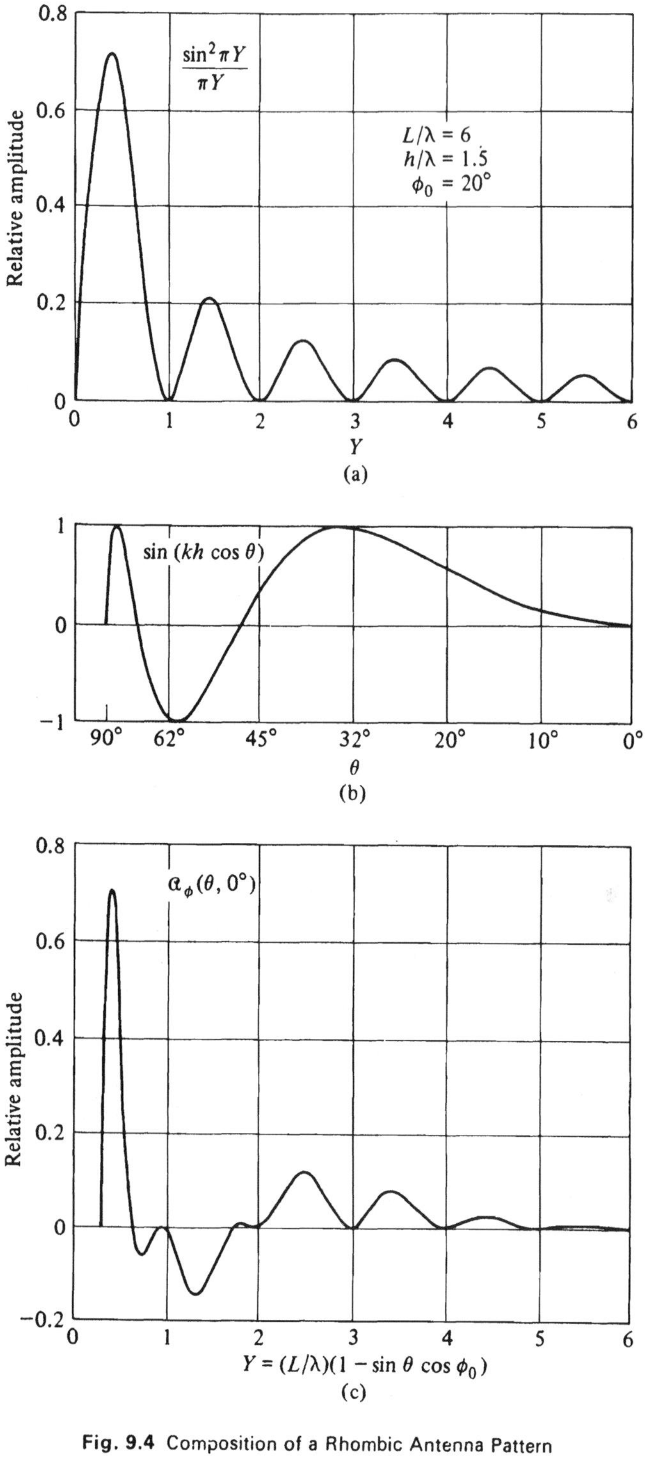

Fig. 9.4 Composition of a Rhombic Antenna Pattern

as a consequence of which the minimum optimum height of the rhombic over a highly conducting earth is

$$\frac{h_1}{\lambda} = \frac{1}{4 \cos \theta_0} \tag{9.20}$$

The position of the peak of the main beam can be found from $\partial \mathcal{A}_\phi / \partial \theta = 0$. This gives

$$kh \sin \theta \cos (kh \cos \theta) \sin Y + kL \cos \theta \cos \phi_0 \sin (kh \cos \theta) \left[\cos Y - \frac{\sin Y}{2Y} \right] = 0$$

which can be rearranged into the form

$$h = \cos \phi_0 \cot \theta \tan (kh \cos \theta) \left\{ \frac{1}{k(1 - \sin \theta \cos \phi_0)} - \frac{L}{\tan [kL/2(1 - \sin \theta \cos \phi_0)]} \right\} \tag{9.21}$$

If h has been chosen so that (9.19) is satisfied, then $\tan(kh \cos \theta_0)$ is infinite, and the only way that the right side of (9.21) can be finite is if

$$\tan \left[\frac{kL}{2} (1 - \sin \theta_0 \cos \phi_0) \right] = kL(1 - \sin \theta_0 \cos \phi_0) \tag{9.22}$$

Equation 9.22 has the solutions

$$\frac{L}{\lambda} (1 - \sin \theta_0 \cos \phi_0) = 0.371,\ 1.465,\ 2.480, \ldots$$

and thus the minimum optimum length of a rhombic arm can be computed from

$$\frac{L}{\lambda} = \frac{0.371}{1 - \sin \theta_0 \cos \phi_0} \tag{9.23}$$

The best choice for corner angle can be determined from

$$\frac{\partial \mathcal{A}_\phi(\theta_0, 0^\circ)}{\partial \phi_0} = 0 \tag{9.24}$$

which leads to the result

$$\cos \phi_0 = \sin \theta_0 \tag{9.25}$$

As an illustration, if the main beam is to be 15° above the horizon, then $\theta^\circ = 75^\circ$ and the corner angle at the feed should be $2\phi_0 = 30^\circ$. The optimum leg length is 5.5λ and the proper height above ground is one wavelength.

The Vee antenna, as one might suspect, is a simplification of the rhombic, with legs 3 and 4 removed. If the Vee is parallel to a ground plane, matched loads can be run to ground from the outer ends of legs 1 and 2, but if the leakage is great enough this is not necessary, and without a ground plane it is not possible. The fields can be deduced from an analysis which parallels what has already been done for the rhombic. For example, $\mathcal{A}_\theta(\theta, \phi)$ is given by the sum of (9.9) and (9.10). The important pattern cut is $|\mathcal{A}_\phi(\theta, 0°)|$; the reader may wish to determine $\mathcal{A}_\phi$ as an exercise.

9.4 Dielectric-Clad Planar Conductors

A flat metallic conductor, onto which a sheet of homogeneous isotropic dielectric has been bonded, as shown in Figure 9.5, is capable of supporting a traveling wave and can thus serve as a transmission line or an antenna, depending on the termination. The electromagnetic behavior of this composite structure can be explained by first assuming that it is infinite in extent in both the X- and Z-directions.[3]

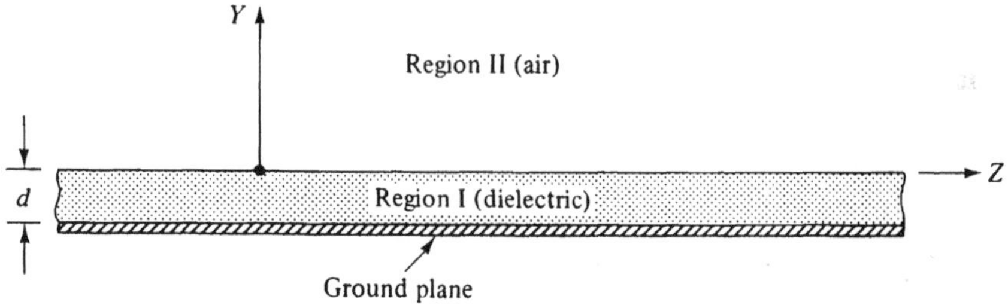

Fig. 9.5 A Dielectric-Clad Ground Plane

Let any field component either in region I (the dielectric) or region II (air) be expressible in the form

$$f(y)e^{j(\omega t - \beta z)} \tag{9.26}$$

This implies that there is a suitable x-independent source at $z = -\infty$ which is causing a wave to travel across the dielectric-clad surface with a propagation constant β. For region I, Maxwell's equations decompose into the following two sets.

$$\begin{array}{ll} \textit{TE} & \textit{TM} \\ E_x = -\dfrac{j\omega\mu_0}{h_1^2}\dfrac{\partial H_z}{\partial y} & H_x = \dfrac{j\omega\epsilon}{h_1^2}\dfrac{\partial E_z}{\partial y} \\ H_y = -\dfrac{j\beta}{h_1^2}\dfrac{\partial H_z}{\partial y} & E_y = -\dfrac{j\beta}{h_1^2}\dfrac{\partial E_z}{\partial y} \\ H_z = f_1(y)e^{j(\omega t - \beta z)} & E_z = g_1(y)e^{j(\omega t - \beta z)} \end{array} \tag{9.27}$$

[3]S. S. Attwood, "Surface Wave Propagation Over a Coated Plane Conductor," *Jour. App. Phys.*, 221 (1951), 504–9.

Here $k^2 = \omega^2\mu_0\epsilon$ and $h_1^2 = k^2 - \beta^2$. Similarly, in region II,

$$\begin{array}{ll} TE & TM \\ E_x = -\dfrac{j\omega\mu_0}{h_2^2}\dfrac{\partial H_z}{\partial y} & H_x = \dfrac{j\omega\epsilon_0}{h_2^2}\dfrac{\partial E_z}{\partial y} \\ H_y = -\dfrac{j\beta}{h_2^2}\dfrac{\partial H_z}{\partial y} & E_y = -\dfrac{j\beta}{h_2^2}\dfrac{\partial E_z}{\partial y} \\ H_z = f_2(y)e^{j(\omega t-\beta z)} & E_z = g_2(y)e^{j(\omega t-\beta z)} \end{array} \tag{9.28}$$

with $k_0^2 = \omega^2\mu_0\epsilon_0$ and $h_2^2 = k_0^2 - \beta^2$.

The longitudinal field components satisfy the wave equation, as a result of which

$$\begin{array}{ll} \left(\dfrac{d^2}{dy^2} + h_1^2\right)f_1(y) = 0 & \left(\dfrac{d^2}{dy^2} + h_1^2\right)g_1(y) = 0 \\ \left(\dfrac{d^2}{dy^2} + h_2^2\right)f_2(y) = 0 & \left(\dfrac{d^2}{dy^2} + h_2^2\right)g_2(y) = 0 \end{array} \tag{9.29}$$

The appropriate solutions, with no sources at $y = +\infty$ and with no wave motion in the $+y$ direction in region II, are

$$\begin{array}{ll} f_1(y) = A_1e^{-jh_1y} + B_1e^{jh_1y} & g_1(y) = C_1e^{-jh_1y} + D_1e^{jh_1y} \\ f_2(y) = A_2e^{-\alpha y} & g_2(y) = C_2e^{-\alpha y} \end{array} \tag{9.30}$$

in which

$$\alpha = jh_2 = \sqrt{\beta^2 - k_0^2} \tag{9.31}$$

is a pure real number.

For TE waves, $H_y \equiv 0$ at $y = -d$, whereas for TM waves, $E_z \equiv 0$ at $y = -d$, and thus

$$B_1 = A_1e^{j2h_1d} \quad \text{and} \quad D_1 = -C_1e^{j2h_1d} \tag{9.32}$$

When this information is used in (9.27) and (9.28), one finds that

$$\begin{array}{ll} TE\text{—Region I} & TE\text{—Region II} \\ E_x = \dfrac{j\omega\mu_0}{h_1}A\sin[h_1(y+d)]e^{j(\omega t-\beta z)} & E_x = \dfrac{-j\omega\mu_0}{\alpha}A_2e^{j(\omega t-\beta z)-\alpha y} \\ H_y = \dfrac{j\beta}{h_1}A\sin[h_1(y+d)]e^{j(\omega t-\beta z)} & H_y = -\dfrac{j\beta}{\alpha}A_2e^{j(\omega t-\beta z)-\alpha y} \\ H_z = A\cos[h_1(y+d)]e^{j(\omega t-\beta z)} & H_z = A_2e^{j(\omega t-\beta z)-\alpha y} \end{array} \tag{9.33}$$

where the substitution $A = 2A_1e^{jh_1d}$ has been made. Similarly,

TM—Region I	*TM*—Region II
$H_x = \frac{j\omega\epsilon}{h_1} C \cos[h_1(y+d)] e^{j(\omega t-\beta z)}$	$H_x = \frac{j\omega\epsilon_0}{\alpha} C_2 e^{j(\omega t-\beta z)-\alpha y}$
$E_y = -\frac{j\beta}{h_1} C \cos[h_1(y+d)] e^{j(\omega t-\beta z)}$	$E_y = -\frac{j\beta}{\alpha} C_2 e^{j(\omega t-\beta z)-\alpha y}$ (9.34)
$E_z = C \sin[h_1(y+d)] e^{j(\omega t-\beta z)}$	$E_z = C_2 e^{j(\omega t-\beta z)-\alpha y}$

with $C = 2jC_1 e^{jh_1 d}$.

The matching of tangential **E** and **H** at the air-dielectric interface gives the following.

TE	*TM*
$A = \frac{A_2}{\cos h_1 d}$	$C = \frac{C_2}{\sin h_1 d}$
$\tan h_1 d = -\frac{h_1}{\alpha}$	$\tan h_1 d = \frac{\epsilon}{\epsilon_0}\frac{\alpha}{h_1}$

(9.35)

When the defining relations for α and h_1 are placed in (9.35), one obtains

$$\tan\left\{\left[\frac{\epsilon}{\epsilon_0} - \left(\frac{\beta}{k_0}\right)^2\right]^{1/2} k_0 d\right\} = -\left[\frac{(\epsilon/\epsilon_0) - (\beta/k_0)^2}{(\beta/k_0)^2 - 1}\right]^{1/2} \quad TE \tag{9.36}$$

$$= \frac{\epsilon}{\epsilon_0}\left[\frac{(\beta/k_0)^2 - 1}{(\epsilon/\epsilon_0) - (\beta/k_0)^2}\right]^{1/2} \quad TM \tag{9.37}$$

Equations 9.36 and 9.37 permit calculation of the normalized propagation constant β/k_0 as a function of the relative permittivity of the dielectric layer and its thickness for *TE* and *TM* waves traveling across the composite surface. A study of these equations reveals important characteristics of this type of wave propagation. First, in the range

$$1 \leq \frac{\beta}{k_0} \leq \sqrt{\frac{\epsilon}{\epsilon_0}} \tag{9.38}$$

the right side of each equation is pure real and has a value which lies in the interval $[-\infty, 0]$ for *TE* waves, and in the interval $[0, \infty]$ for *TM* waves. Therefore positive real values of d can be found, for each value of β/k_0 in the range of (9.38), which will equate the two sides of either (9.36) or (9.37). For *TE* waves, the angle for which the tangent is being taken must lie in the second, fourth, sixth, . . . quadrant, whereas for *TM* waves, it must lie in the first, third, fifth, . . . quadrant.

Second, with the range of β/k_0 prescribed in (9.38), the phase velocity of these waves, which is given by

$$v_{ph} = \frac{\omega}{\beta} = \frac{\omega}{k_0}\frac{1}{(\beta/k_0)} = \frac{c}{(\beta/k_0)} \tag{9.39}$$

is *less* than the speed of light. For this reason, the dielectric-clad ground plane is often referred to as a *slow wave structure*.

Third, since $\alpha = \sqrt{\beta^2 - k_0^2}$, it follows that the larger β/k_0 is in the range of (9.38), the greater is the value of α and the more rapid is the exponential decay of these waves in the direction normal to the surface. The electromagnetic energy being transported is tightly bound to the surface if β/k_0 is even modestly greater than unity. For this reason, these slow waves are sometimes also called *trapped waves.*

The fundamental *TM* mode occurs when d has a value such that the angle

$$\left[\frac{\epsilon}{\epsilon_0} - \left(\frac{\beta}{k_0}\right)^2\right]^{1/2} k_0 d$$

lies in the interval $(0, \pi/2)$. For the fundamental *TE* mode, this angle must be in the interval $(\pi/2, \pi)$, which requires a larger value of d for the same β/k_0. The additional thickness of the dielectric layer lessens the attractiveness of *TE* slow wave propagation.

For a relative dielectric constant $\epsilon/\epsilon_0 = 2.5$, plots of β/k_0 versus d are shown in Figure 9.6 for both the fundamental *TM* and *TE* modes; they illustrate the difference in dielectric thickness requirements for the two types of slow waves.

A discussion of the launching of slow waves on these composite structures, and of the potential use of dielectric-clad planar conductors as transmission lines or antennas, will be deferred to Section 9.7.

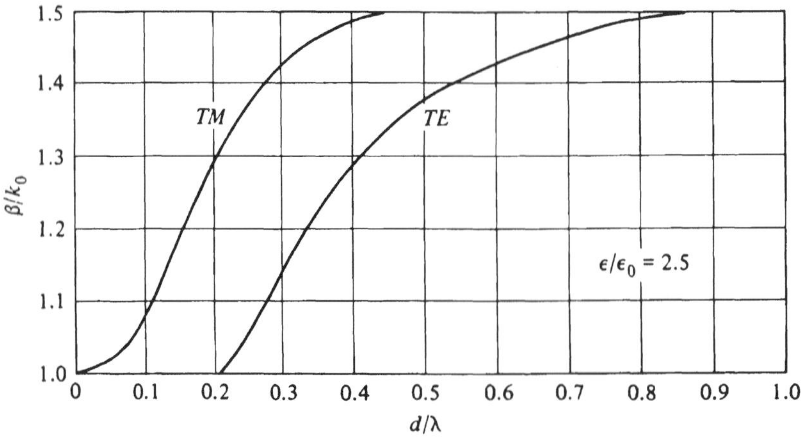

Fig. 9.6 Propagation Constants of Slow Waves on a Dielectric-Clad Ground Plane as Functions of Dielectric Thickness

9.5 Corrugated Planar Conductors

Slow waves of the *TM* type discussed in the previous section can also be supported by corrugated ground planes, with the teeth and gaps that comprise the corrugations running transverse to the direction of propagation.[4] This situation is suggested by

[4]C. C. Cutler, "Electromagnetic Waves Guided by Corrugated Conducting Surfaces," *Bell Telephone Laboratories*, Report MM-44-160-218 (October 1944).

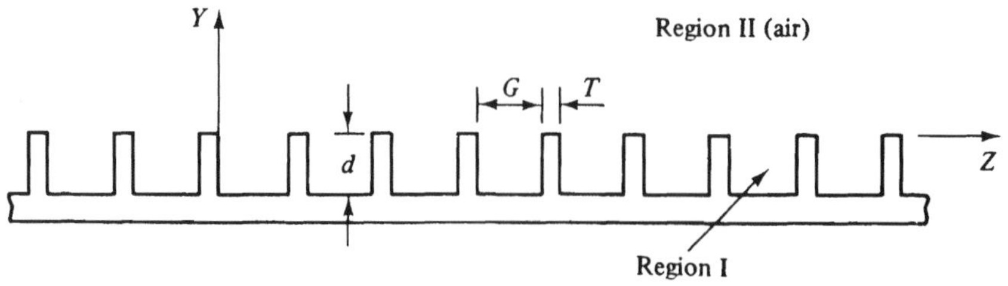

Fig. 9.7 A Planar Conductor with a Corrugated Surface

Figure 9.7, where wave motion is assumed in the Z-direction and the corrugations are parallel to the X-axis.

If $G \gg T$ and if there are many teeth per wavelength, the existence of TM waves in region II (the air-filled half-space above the corrugated structure) will excite standing TEM waves in the gaps which make up region I. In the nth gap, identified by the coordinate z_n, these fields are given by

$$E_z = D_n \sin k(y + d)e^{j\omega t}$$
$$H_x = \frac{j\omega\epsilon}{k} D_n \cos k(y + d)e^{j\omega t} \tag{9.40}$$

where once again $k^2 = \omega^2\mu_0\epsilon$. In (9.40), ϵ is the permittivity of the medium filling the gap (usually air).

The region II fields are given by (9.34) and the matching of tangential **E** and **H** in the plane $y = 0$ gives

$$D_n = \frac{C_2 e^{-j\beta z_n}}{\sin kd}$$
$$\tan kd = \frac{\epsilon}{\epsilon_0}\frac{\alpha}{k} \tag{9.41}$$

When the defining relations for α and k are used in (9.41) the result is that

$$\tan\left[\left(\frac{\epsilon}{\epsilon_0}\right)^{1/2} k_0 d\right] = \left(\frac{\epsilon}{\epsilon_0}\right)^{1/2}\left[\left(\frac{\beta}{k_0}\right)^2 - 1\right]^{1/2} \tag{9.42}$$

Equation 9.42 is similar in form to (9.37). However, the range of β/k_0 for a corrugated ground plane is less restricted than for a dielectric-clad ground plane. One can observe from (9.42) that positive real values of d which cause the angle $(\epsilon/\epsilon_0)^{1/2}k_0 d$ to lie in the interval $[0, \pi/2]$ result in a value of β/k_0 in the range

$$1 \leq \beta/k_0 < \infty \tag{9.43}$$

A plot of β/k_0 versus corrugation depth d is shown in Figure 9.8 for the case $\epsilon = \epsilon_0$. The leveling off of β/k_0 observed in Figure 9.6, due to the finite upper limit

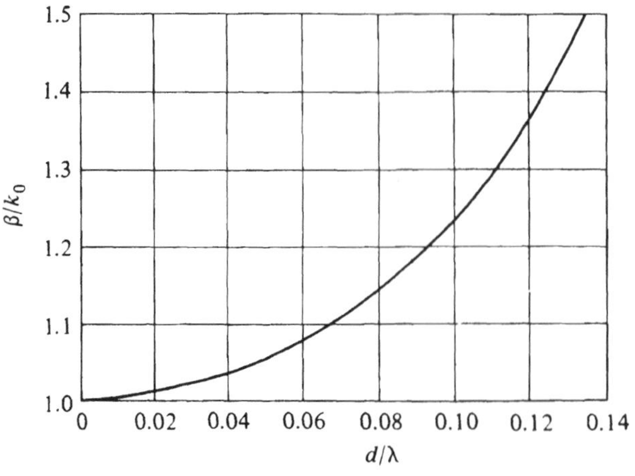

Fig. 9.8 Propagation Constant of *TM* Waves over a Corrugated Surface versus Depth of Corrugations

in (9.38), is absent from the curve of Figure 9.8. However, the values of β/k_0 needed in practical *TM* surface wave antennas can be obtained easily either with a grounded dielectric slab of reasonable thickness and permittivity or with a corrugated surface of reasonable tooth/gap dimensions. This will become evident from the development found in Section 9.7.

9.6 Surface Wave Excitation

For *TM* slow waves supported by a corrugated ground plane, the analysis of the previous section left unanswered the question of how the waves are to be created. If the half-plane $z = 0, y \geq 0$ were singled out in Figure 9.7, on which the current sheets

$$K_y = \frac{j\omega\epsilon_0}{\alpha} C_2 e^{j\omega t - \alpha y} \qquad K_{mx} = -\frac{j\beta}{\alpha\mu_0} C_2 e^{j\omega t - \alpha y} \qquad 0 \leq z < \infty \tag{9.44}$$

could be placed, the fields described by Equations 9.34 would be produced in $z > 0$, as can be seen through recourse to Schelkunoff's equivalence principle and Equations 1.112–1.115. The sources in (9.44) are not physically realizable but they can serve as a useful guide in gaining an introductory appreciation of several types of surface wave launchers.

To see this, consider a rectangular waveguide whose bottom broad wall is corrugated. This structure can be analyzed in a manner similar to the procedure followed in Section 9.5 for the corrugated slab.[5] One finds that the dominant mode is a modified TE_{10} with phase velocity that has been slowed by the presence of the corrugations. If the side walls are permitted to recede to infinity, the solution for waves propagating

[5] R. S. Elliott, "On the Theory of Corrugated Plane Surfaces," *IRE Trans. Antennas Propagat.*, AP-2 (1954), 71–81.

between parallel plates, with one plate corrugated, is obtained.[6] For $G \gg T$ and $G \ll \lambda$, the fields in $0 \leq y \leq b$, that is, in the region above the corrugations and extending to the upper wall (see Figure 9.9), are given by

$$\begin{aligned} H_x &= \frac{j\omega\epsilon_0}{\alpha} C_2' \cosh [\alpha(b - y)] e^{j(\omega t - \beta z)} \\ E_y &= -\frac{j\beta}{\alpha} C_2' \cosh [\alpha(b - y)] e^{j(\omega t - \beta z)} \\ E_z &= C_2' \sinh [\alpha(b - y)] e^{j(\omega t - \beta z)} \end{aligned} \tag{9.45}$$

When these fields are matched at the boundary $y = 0$ to those in the gap regions, the relation

$$\tan kd = \frac{\epsilon}{\epsilon_0} \frac{\alpha}{k} \tanh \alpha b \tag{9.46}$$

can be established.

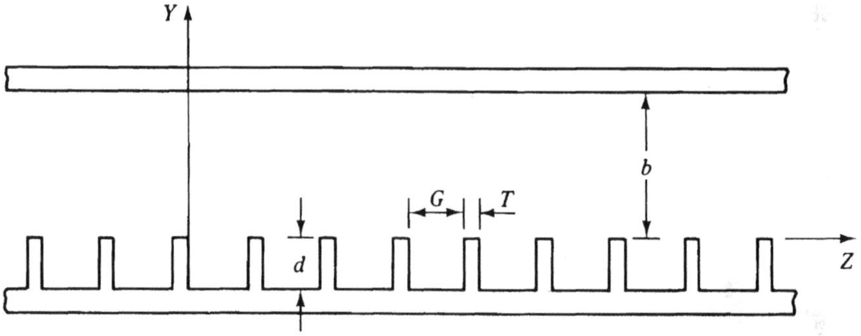

Fig. 9.9 Parallel Plate Transmission Line; Bottom Wall Corrugated

If $e^{-\alpha b} \ll 1$, then $\tanh \alpha b \cong 1$ and (9.46) reduces to (9.35). This means that a given corrugation depth produces the same propagation constant whether the upper plate is present or not. Further, it means that the fields in (9.45) are very close to those described by (9.34). To see this, let $C_2' \sinh \alpha b = C_2$ so that the levels of the two sets of fields are equivalent. Then (9.45c) becomes

$$\begin{aligned} E_z &= C_2 \frac{\sinh [\alpha(b - y)]}{\sinh \alpha b} e^{j(\omega t - \beta z)} \\ &= C_2 \frac{e^{\alpha(b-y)} - e^{-\alpha(b-y)}}{e^{\alpha b} - e^{-\alpha b}} e^{j(\omega t - \beta z)} \\ &\cong C_2[e^{-\alpha y} - e^{-\alpha(2b-y)}] e^{j(\omega t - \beta z)} \\ &\cong C_2 e^{j(\omega t - \beta z) - \alpha y} \qquad 0 \leq y \leq b \end{aligned} \tag{9.47}$$

[6] Ibid.

which agrees with (9.34a) for region II. In like manner, (9.45b) and (9.45c) are approximately equal to their counterparts in (9.34), if $e^{-\alpha b} \ll 1$.

The significance of this result is, for a given corrugation depth d (and thus a given β/k_0 and α), if b is made large enough to cause the condition $e^{-\alpha b} \ll 1$, then it does not matter whether the top wall is present or not. The trapped waves, which are concentrated near the lower (corrugated) wall, would not be seriously affected if the upper wall were suddenly to stop, as suggested by Figure 9.10.

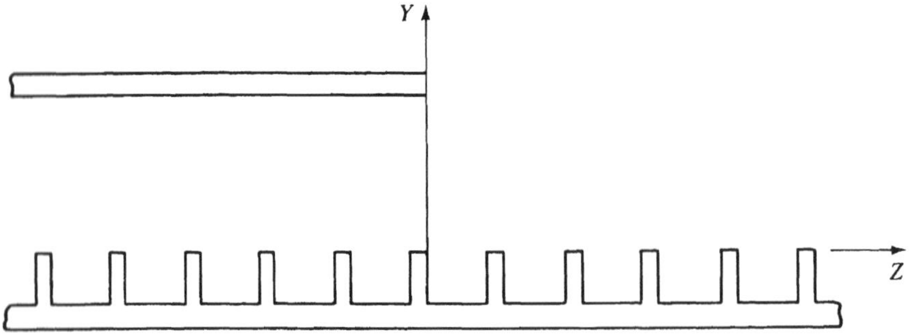

Fig. 9.10 Corrugated Slab Fed by a Corrugated Parallel Plate Transmission Line

This immediately suggests two potentially satisfactory wave launchers. A line source, such as an array of longitudinal shunt slots, could be designed to excite a *TEM* mode in a parallel plate region with the upper wall flared to achieve the proper aperture height b, and with the lower wall corrugated with gradually increasing gap depth until the ultimate depth is achieved. Alternatively a conventional rectangular waveguide could be used to feed a horn with side walls that flare out till they reach the desired separation in the x-dimension, with an upper wall that flares to achieve the proper aperture height b, and with a lower wall corrugated with gradually increasing gap depth until the ultimate depth is achieved. Either of these possibilities is suggested in cross section by Figure 9.11.

An estimate of the effectiveness of *TM* surface wave launchers of these two types can be made by the following argument. Assume that the gradualness of the deepening of the corrugations and the gentleness of the flare of the upper wall (and perhaps

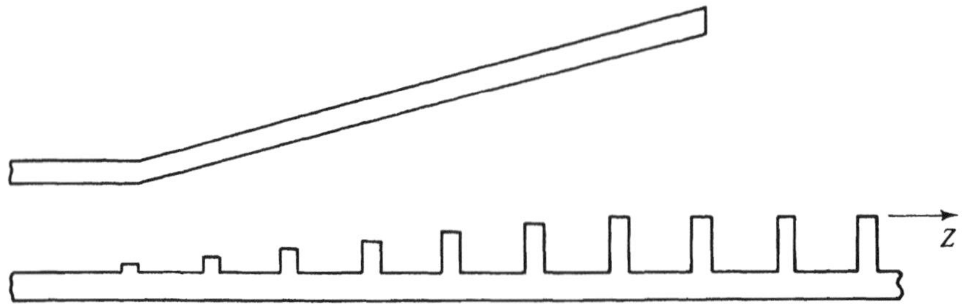

Fig. 9.11 Flared Feed for a Corrugated Surface

the side walls) are sufficient to provide a good match to the line source or rectangular waveguide. Conditions at the mouth of the feed can then be idealized by the parallel plate picture of Figure 9.10. In that picture, if the proper equivalent sources are placed on the half-plane $z = 0, y \geq 0$, the fields in $z \geq 0, y \geq 0$ will be the same as those caused by the actual feed. With $e^{-\alpha b} \ll 1$, the termination of the upper wall sets up a negligible reflection, so the fields in the mouth should be given quite accurately by (9.45). With b/λ large enough to cause $e^{-\alpha b} \ll 1$, the spillover from the feed in $z \leq 0$, $y > b$ should be small enough to permit the assumption that the fields are negligible on $z = 0, y > b$. If this is so, the equivalent sources that should be placed on the half-plane $z = 0, y \geq 0$ are

$$\left.\begin{aligned} K_y &= \frac{j\omega\epsilon_0}{\alpha} C_2' \cosh [\alpha(b - y)] e^{j\omega t} \\ & \qquad\qquad\qquad\qquad 0 \leq y \leq b \\ K_{mx} &= -\frac{j\beta}{\alpha\mu_0} C_2' \cosh [\alpha(b - y)] e^{j\omega t} \\ K_y &\equiv 0 \qquad K_{mx} \equiv 0 \qquad b < y < \infty \end{aligned}\right\} \tag{9.48}$$

The ideal surface wave described by Equations 9.34 requires the sources in (9.44). How close do the sources in (9.48) come to meeting this requirement? An answer can be provided to this question if one compares $\cosh [\alpha(b - y)]/\cosh \alpha b$ to $e^{-\alpha y}$ in $0 \leq y \leq b$ and compares zero to $e^{-\alpha y}$ in $b < y < \infty$.

As an illustration, let $\epsilon/\epsilon_0 = 1$, $\beta/k_0 = 1.10$, and $b/\lambda = 1$. Then $\alpha/k_0 = 0.458$, $e^{-\alpha b} = 0.056$, $\tanh \alpha b = 0.994$, and $d/\lambda = 0.068$. The conditions assumed in the approximations are seen to be met reasonably well even though β/k is only modestly above unity and even though the feed mouth is only one wavelength tall. A comparison of the equivalent sources in (9.44) and (9.48) for this case is provided by the entries in Table 9.1. One can see good agreement over the bottom half of the mouth but the discrepancy grows as $y \rightarrow b$. However, the field values are diminishing as $y \rightarrow b$, which lessens the seriousness of this deviation.

The agreement between (9.44) and (9.48) improves as β/k_0 or b/λ increases. The development in Section 9.7 will show that the value of β/k_0 needed for optimum pat-

TABLE 9.1 Equivalent source distributions for TM surface waves
$\beta/k_0 = 1.10$ $b/\lambda = 1$

y/λ	Sources (9.48)	Sources (9.44)	y/λ	Sources (9.48)	Sources (9.44)
0	1.000	1.000	0.7	0.156	0.133
0.1	0.751	0.750	0.8	0.131	0.100
0.2	0.567	0.562	0.9	0.116	0.075
0.3	0.427	0.422	1.0	0.112	0.056
0.4	0.325	0.316	2.0	0.000	0.003
0.5	0.250	0.237	3.0	0.000	0.000
0.6	0.195	0.178	4.0	0.000	0.000

tern performance is dictated by the length of the corrugated surface and is usually not much greater than unity. This leaves b/λ as the only control parameter to improve the launching efficiency for this type of feed. Practical considerations limit the aperture height and thus the effectiveness of surface wave excitation. However, experimental results to be presented in Section 9.7 will indicate that modest values of b/λ are adequate to cause acceptable surface wave generation.

This entire discussion could be repeated for the launching of either *TM* or *TE* slow waves on a grounded dielectric slab. The conclusion would once again be reached that horn-type feeds can be designed to be satisfactory launchers. As in the corrugated case, the wall flare from line source or waveguide to the feed mouth should be gradual and the dielectric should extend inside the feed, where its thickness is gradually brought to zero to assist in a good match.

What has been presented here is only a brief introduction to surface wave excitation. Other more complex feeding schemes have been devised, and elaborate analyses have been developed for the computation of excitation efficiency. The interested reader should consult the comprehensive review of this subject by F. J. Zucker[7] and the discussions by H. M. Barlow and J. Brown,[8] by R. E. Collin;[9] and by C. H. Walter.[10]

9.7 Surface Wave Antennas

If the feed shown in Figure 9.11 is properly designed, a slow wave will be launched, traveling over the corrugated surface in the $+Z$-direction. Were a mirror-image feed placed some distance to the right in order to absorb this slow wave, the intervening section of corrugated slab could be viewed as a transmission line, albeit an imperfect one. There would be losses: ohmic heating of the conductors, radiation from the first feed in modes other than the surface wave, spreading of the fields in the X-direction, and imperfect reception by the second feed. A serious question could be raised about the necessity for a slab which is wide in the X-dimension. Clearly, this structure does not serve as a good transmission line,[11] but when properly designed it can become a good antenna.

[7]F. J. Zucker, "Surface-Wave Antennas," *Antenna Theory*, *Part II*, ed. R. E. Collin and F. J. Zucker (New York: McGraw-Hill Book Co., Inc., 1969), Chapter 21, pp. 313–20.

[8]H. M. Barlow and J. Brown, *Radio Surface Waves* (Oxford: Clarendon Press, 1962), pp. 92–136.

[9]R. E. Collin, *Field Theory of Guided Waves* (New York: McGraw-Hill Book Co., Inc., 1960), pp. 485–506.

[10]C. H. Walter, *Traveling Wave Antennas* (New York: McGraw-Hill Book Co., Inc., 1965), pp. 282–311.

[11]The cylindrical equivalents of the grounded dielectric slab and corrugated planar conductor, namely the dielectric-sheathed wire and the corrugated rod, *do* make good transmission lines when fed by properly designed conical horns. The dielectric-coated wire has been intensively investigated for such purposes. See G. Goubau, "Surface Waves and Their Application to Transmission Lines," *J. Appl. Phys.* 21 (1950), 1119–28.

As an introduction to the subject of slow wave structures used as antennas, assume that the corrugated surface and feed of Figure 9.11 are infinite in the x-dimension and that all fields are x-independent. Assume further that the regular pattern of corrugations does not persist all the way to $z = +\infty$, but rather terminates a distance L from the feed mouth in a ground plane which *does* extend to infinity. This situation is suggested in Figure 9.12a.

Next imagine a closed surface S consisting of the half-plane $z = 0, y \geq 0$, the half-plane $y = \delta, z \geq 0$, with δ a positive infinitesimal; the cylindrical section of infinite radius which connects these two half-planes at infinity and encloses the three quadrants in which z or y is negative; and the two end-cap surfaces $x = \pm\infty$. The

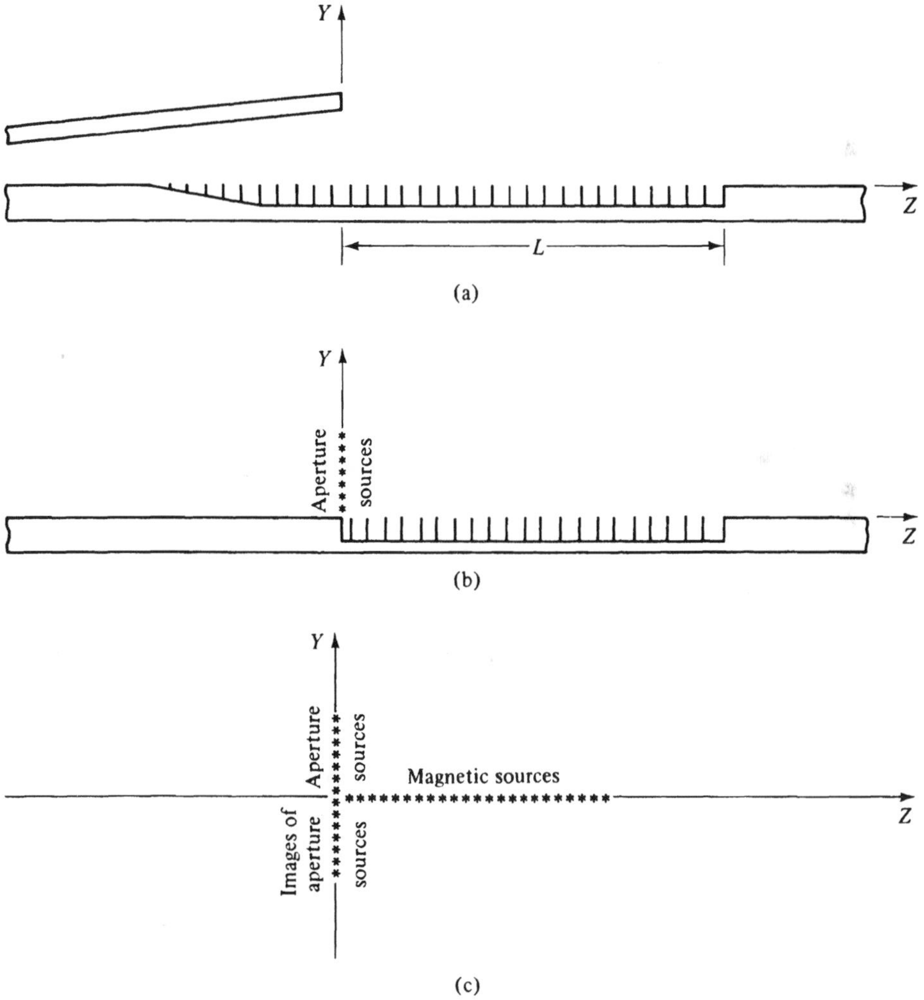

Fig. 9.12 Equivalent Sources for a Corrugated Surface Antenna with an Infinite Ground Plane Extension

volume V inside S contains *all* the sources and Schelkunoff's equivalence principle can be invoked. The proper set of secondary sources on S will produce fields in the quadrant $z \geq 0$, $y \geq 0$ identical to those caused by the actual sources.

It will be assumed that the feed is efficient enough that the true fields in the quadrant $z \leq 0$, $y \geq 0$ are negligible. A useful by-product of this assumption is that a semi-infinite ground plane extending to $z = -\infty$ can replace the actual external feed structure, as pictured in Figure 9.12b. This permits use of the image principle, with the currents in the ground plane accounted for by doubling the equivalent magnetic sources over the corrugated surface, canceling the equivalent electric sources over the corrugated surface and placing images of the aperture sources on the half-plane $z = 0$, $y \leq 0$, as suggested by Figure 9.12c.

The current sheets given by (9.48) will be taken as an adequate approximation to the true secondary sources which should be placed on the half-plane $z = 0$, $y \geq 0$. The images of these current sheets need to be placed on the half-plane $z = 0$, $y \leq 0$. The magnetic sources which are to occupy the position vacated by the corrugated surface can be deduced from (9.34) and are given by

$$K_{mx} = -2\mu_0^{-1}C_2 e^{j(\omega t - \beta z)} \qquad 0 \leq z \leq L \tag{9.49}$$

The far field can be expressed as the sum of two terms, in the form

$$E_\theta(\theta) = E_\theta^s(\theta) + E_\theta^f(\theta) \tag{9.50}$$

with E_θ^s and E_θ^f the contributions from the corrugated surface and feed, respectively, and with θ measured from the Z-axis in the YZ-plane. With the aid of the results of Appendix G, one can show that

$$E_\theta^s(\theta) = 2jkLC_2\mu_0^{-1}e^{-j(\pi L/\lambda)((\beta/k)-\sin\theta)}\frac{\sin \pi L/\lambda\,(\beta/k - \sin\theta)}{\pi L/\lambda(\beta/k - \sin\theta)} \tag{9.51}$$

$$E_\theta^f(\theta) = 2C_2'\mu_0^{-1}\frac{\beta/k + \sin\theta}{(\alpha/k)^2 + \cos^2\theta}\left[\sinh \alpha b + \frac{\cos\theta\sin(kb\cos\theta)}{\alpha/k}\right] \tag{9.52}$$

The part of the field attributable to the corrugated surface is seen to be in the form $\sin \pi Y/\pi Y$, with $Y = (L/\lambda)[(\beta/k) - \sin\theta]$. This general pattern has been plotted in Figure 3.3a, and consists of a main beam and a family of side lobes which decay in height with distance from the main beam. However, a feature of the present application is that $\beta/k > 1$; therefore this pattern does not reach the peak corresponding to $Y = 0$. It is important to be sure that β/k is not so large that all of the main beam of $(\sin \pi Y)/\pi Y$ lies in the invisible range.

W. W. Hansen and J. R. Woodyard have shown[12] that if

$$\left(\frac{\beta}{k} - 1\right)\frac{\pi L}{\lambda} = \frac{\pi}{2} \tag{9.53}$$

[12] W. W. Hansen and J. R. Woodyard, "A New Principle in Directional Antenna Design," *Proc. IRE*, 26 (1938), 333–45.

then patterns of the type in (9.51) will have minimum beamwidth. This condition is usually adopted in the design of corrugated surface antennas and simplifies (9.51) to the form

$$E_\theta^s(\theta) = -2kLC_2\mu_0^{-1}\frac{\sin(\pi L/\lambda)(\beta/k - \sin\theta)}{(\pi L/\lambda)(\beta/k - \sin\theta)} \tag{9.54}$$

which has a main beam peak at end-fire and a family of side lobes, the first of which is at a height −9.5 decibels relative to the main beam.

The feed pattern (9.52) is quite broad and exhibits no nulls in the visible range for practical values of β/k and b/λ. It also makes a maximum contribution at end-fire. The ratio $E_\theta^s(\pi/2)/E_\theta^f(\pi/2)$ is a measure of the feed suppression and depends on the relation between the constants C_2 and C_2', which in turn depends on the aperture height b/λ. If one connects C_2 and C_2' by equating the power emerging from the feed to the power transported by the surface wave, it is a simple matter to compute the feed suppression. This has been done[13] and the results are plotted in Figure 9.13. It

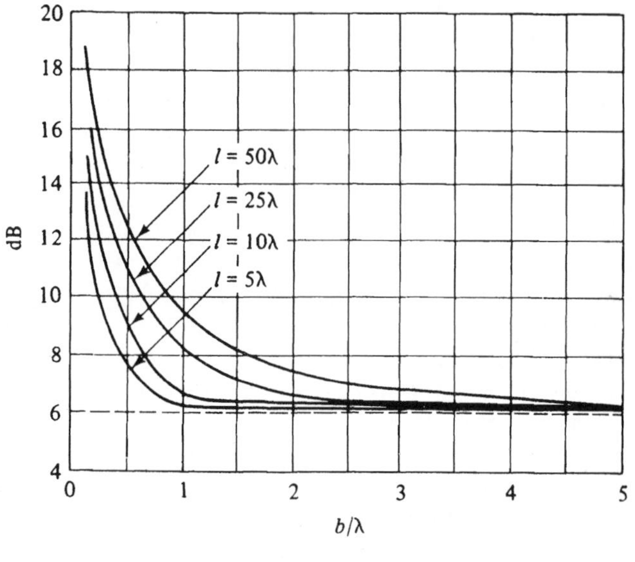

Fig. 9.13 Feed Suppression for a Corrugated Surface Antenna

is evident that, for a given b/λ, there is more feed suppression if L/λ is greater. But it is also evident that, for corrugated surfaces of practical length, the feed suppression rapidly reaches the asymptotic value of 6 decibels as b/λ is increased.

The presence of this asymptote can be understood by returning to (9.52) and realizing that, as b/λ increases, $\sinh \alpha b$ dominates $[\cos\theta \sin(kb\cos\theta)]/(\alpha/k)$ and

[13]Elliott, "Theory of Corrugated Plane Surfaces," p. 76.

$C_2' \sinh \alpha b \longrightarrow C_2$. The limiting value of the feed suppression ratio is therefore

$$\lim_{b/\lambda \to \infty} \frac{E_\theta^s(\pi/2)}{E_\theta^f(\pi/2)} = -\frac{2C_2\mu_0^{-1}kL/(\pi/2)}{2C_2\mu_0^{-1}(\beta/k + 1)/[(\beta/k)^2 - 1]} = -2 \tag{9.55}$$

by virtue of the Hansen-Woodyard relation (9.53).

That it is desirable to suppress feed radiation is apparent from this limiting ratio. If the field attributable to the corrugated surface is only twice the value at end-fire of the field due to the feed and if they are out of phase, then at the peak of the first side lobe of $E_\theta^s(\theta)$, the two contributions are *in* phase. With the feed pattern broad, this can result in a first side lobe as high as the main beam, an unacceptable situation.

But to make b/λ small enough to suppress the feed radiation adequately is to go against the prescription for efficient launching of the surface wave given in Section 9.6, namely, to make b/λ large. Clearly, a tradeoff situation exists. That it is possible to get effective excitation of the surface wave and still have acceptable feed suppression will be demonstrated shortly, when theory and experiment are compared.

First, it is necessary to reconsider several of the assumptions made in this analysis. The source distribution (9.49) tacitly implies the neglect of any reflected surface wave caused at $z = L$ by the juncture of the corrugated surface and the ground plane. This can be justified on the basis of the Hansen-Woodyard relation (9.53), which can be rewritten as

$$\frac{\beta}{k} = 1 + \frac{\lambda}{2L} \tag{9.56}$$

from which one can see that, even for corrugated surfaces that are only 5λ long, β/k should be no greater than 1.1. This in turn implies that the corrugations are shallow and that the transition to ground plane is not severe.

The assumption that the surface is infinite in its x-dimension is a good one if the actual surface has a width greater than or equal to 5λ, for then the finite width has negligible effect on the propagation formula in (9.42) and any of its consequences. The pattern in the YZ-plane can be computed as though the transverse width were infinite. Patterns in planes containing the X-axis will be governed by the x-dependency of the aperture distribution.

Experiments performed by M. J. Ehrlich and L. Newkirk[14] provide a test of the foregoing analysis. They used a corrugated surface which was 2λ wide and 7.33λ long embedded in a ground plane 40λ wide and 70λ long, with the tooth/gap dimensions adjusted to give $\beta/k_0 = 1.07$, consistent with the Hansen-Woodyard relation. Patterns were taken using a receiving horn which was mounted on a rotatable arm 50λ long, thereby simulating the far-field measurements for the case of an infinite ground plane. With $b/\lambda = 0.73$, the experimental pattern which they obtained at 9.840 GHz is shown as the solid line in Figure 9.14. The theoretical pattern computed from (9.54) is shown dotted for comparison.

[14] M. J. Ehrlich and L. Newkirk, "Corrugated Surface Antennas," *IRE Convention Record*, Part 2 (1953), 18–33.

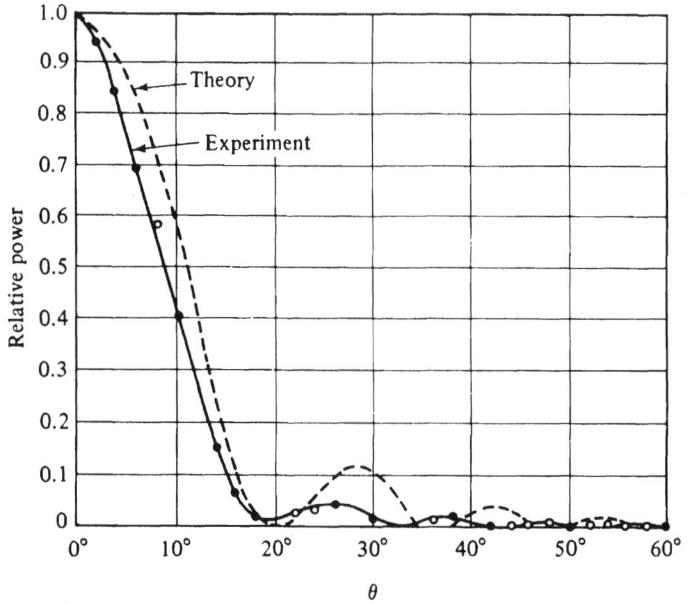

Fig. 9.14 Experimental and Theoretical *E*-Plane Patterns for a Corrugated Surface Antenna; $L = 7.33\lambda$; Large Ground Plane (© 1953 IEEE. Reprinted from M. J. Ehrlich and L. Newkirk, *IRE Convention Record*, 1953.)

The agreement between theory and experiment for this case is good enough to permit the inference that the dominant contribution to the radiation pattern is coming from the corrugated surface and that the feed radiation is adequately suppressed, even though b/λ is only 0.73. However, the null positions and side lobe heights do not quite agree, and the theoretical pattern gives a main beam which is too broad. When the feed radiation is included in the theoretical computation, better correlation is achieved.[15] The principal remaining defect is the assumption that a pure surface wave is traveling across the corrugated surface, and that therefore (9.54) is the proper representation of the equivalent sources. F. J. Zucker offers an example of the measured field distribution immediately above a 4λ surface wave antenna, indicating that the amplitude is far from uniform, particularly near the feed end.[16]

Despite the approximations, the foregoing analysis provides a satisfactory explanation, at an introductory level, of the principal characteristics of corrugated surface antennas. The same approach can be applied equally well to surface wave antennas in which the basic trapping structure is a grounded dielectric slab, supporting either a *TE* or *TM* mode.

Other methods of analysis have been favored by some workers. F. J. Zucker[17] has shown a preference for the "discontinuity radiation" point of view, in contrast to

[15] Elliott, "Theory of Corrugated Plane Surfaces," Figure 5.

[16] Zucker, "Surface-Wave Antennas," Figure 21.5.

[17] Ibid., pp. 302–4 and Problem 21.9.

the foregoing (which he calls the Kirchhoff-Huyghens approach), but argues quite correctly that the two procedures, given the stated approximations, lead to identical results.

Finally, it must be mentioned that the analysis so far has assumed a surface wave antenna of length L followed by an infinitely long ground plane. If the ground plane is finite in length (or nonexistent), the slow wave traversing the antenna will be diffracted at the terminus, resulting in a pattern in which the peak of the main beam is lifted off the surface with some radiation leaking into the lower hemisphere, as typified by the experimental pattern shown in Figure 9.15. An estimate can be made of the angular position of the peak of the main beam as a function of the lengths of the trapping surface and its ground plane.[18]

For some applications of surface wave antennas, it is not desirable to have the main beam tilted up above end-fire. In such situations, a desirable feature of trapped

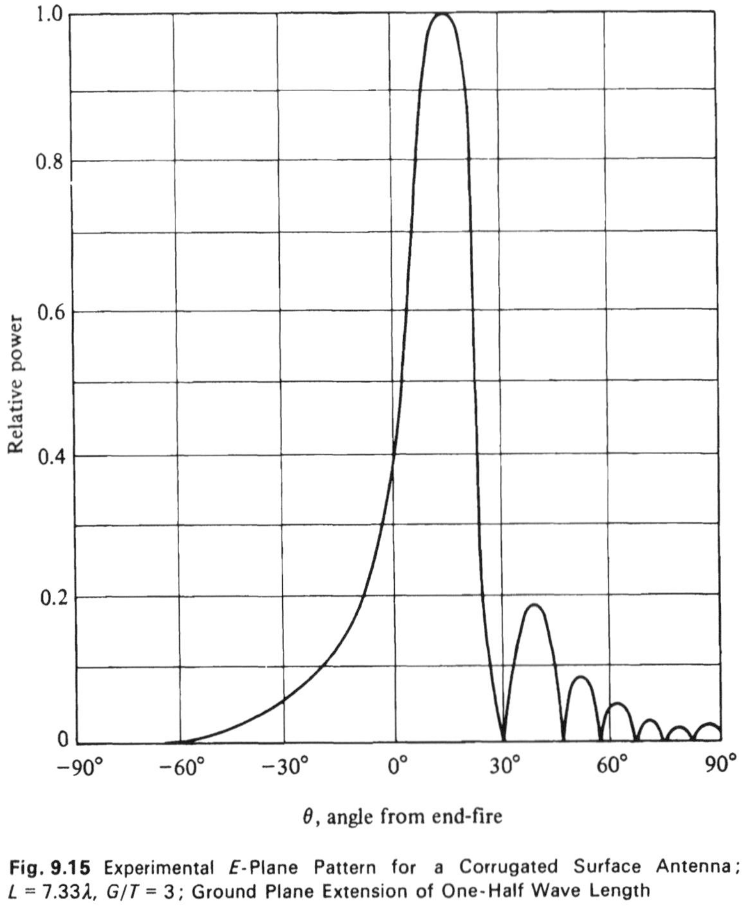

Fig. 9.15 Experimental E-Plane Pattern for a Corrugated Surface Antenna; $L = 7.33\lambda$, $G/T = 3$; Ground Plane Extension of One-Half Wave Length

[18]Elliott, "Theory of Corrugated Plane Surfaces," Figures 6 and 7.

waves can be exploited. Trapped waves will cling to a curved surface, although some leakage occurs (the more so the more radical the curvature). By using a section of a corrugated or dielectric-clad *cylinder* in lieu of a *planar* trapping structure, one can position the main beam at end-fire by proper choice of the length of the arc and the radius of curvature. The leakage of trapped wave causes null-filling in the pattern, which is desirable in some applications.[19] The same effect can be achieved with a spherical cap that is dielectric-clad or corrugated. The resulting ϕ-symmetric pattern, devoid of deep nulls, is useful in beacon antenna applications.[20]

All of the slow wave antennas that were described in this section have the property of a uniform repetitive tooth/gap geometry over the entire aperture, a feature used to justify the assumption of an aperture distribution derivable for dielectric-clad surfaces from (9.33) and (9.34) and for corrugated surfaces from (9.34). The aperture distribution can be modified by modulating the trapping structure. A variety of methods for doing this have been discovered and C. H. Walter can be consulted for a survey and bibliography.[21]

9.8 Fast Wave Antennas

A waveguide mode typically propagates at a phase velocity greater than the speed of light. If the waveguiding structure which supports this mode is properly "opened up," the energy contained in the mode can be leaked to the exterior region, resulting in what is called a *leaky wave antenna.* In practice, one wishes to govern the rate of leakage to achieve a desired apreture distribution. With the aperture many wavelengths long, the leakage rate is everywhere low and the phase velocity of the leaky mode differs but little from the phase velocity of its nonleaky counterpart. As a consequence, there is a quasi-uniform progressive phase distribution to the aperture distribution, corresponding to the passage of a fast wave over the aperture. Thus such structures are also called *fast wave antennas.*

Four examples of fast (leaky) wave antennas are shown in Figure 9.16. The first three give an E-field distribution in the slot that is essentially transverse to the longitudinal Z-axis, which translates into an E_ϕ-polarization in the far field. The last example gives a longitudinal E-field in the aperture, thus causing an E_θ-type far-field pattern. The local value of transverse width of the long continuous slot determines the local rate of leakage for the first two examples. The local value of offset from the center line determines the local leakage rate for the meandering slot, and the local value of transverse slot length determines the local rate of leakage for the serrated waveguide. The first three examples are clearly continuous slots with respect to the longitudinal coordinate, whereas the fourth is only quasi-continuous. However, if the

[19] R. S. Elliott, "Azimuthal Surface Waves on Circular Cylinders," *J. Appl. Phys.*, 26 (1955), 368–76.

[20] R. S. Elliott, "Spherical Surface Wave Antennas," *IEEE Trans. Antennas Propagat.*, AP-4 (1956), 422–28.

[21] Walter, *Traveling Wave Antennas*, pp. 373–84.

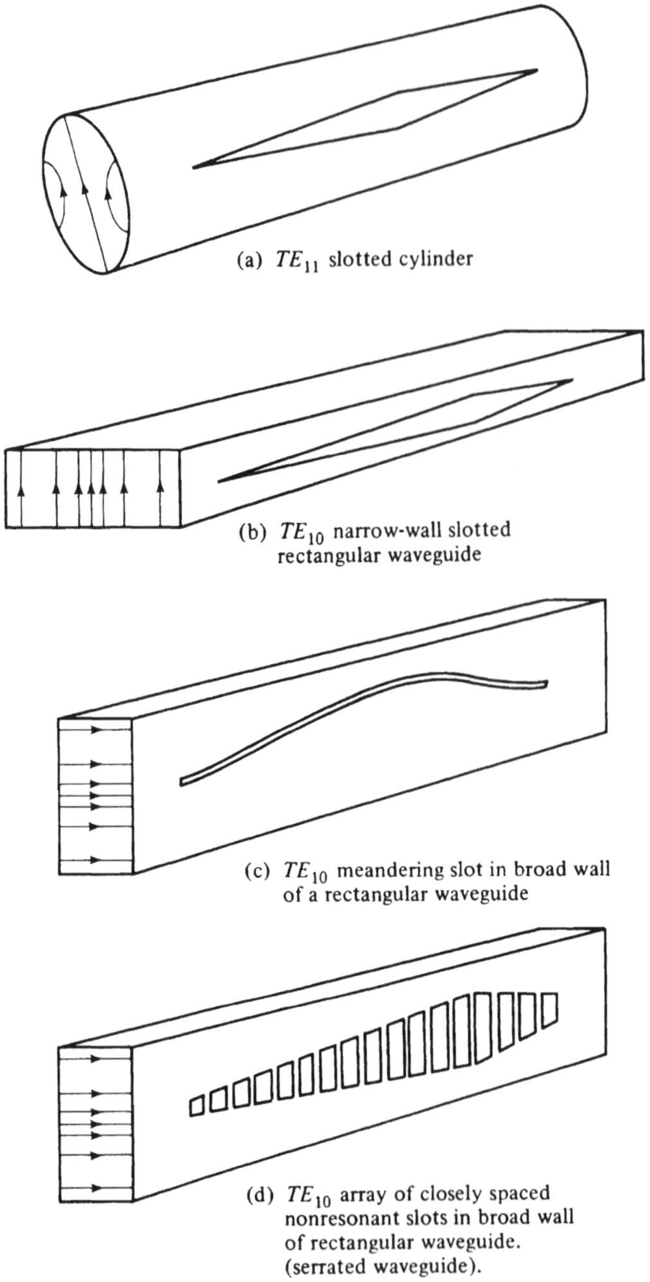

Fig. 9.16 Four Types of Fast (Leaky) Wave Antennas

metal region between adjacent nonresonant slots is narrow and if there are many slots per guide wavelength, this structure may also be treated as though the aperture were continuous.

To see how a desired aperture distribution can be achieved in principle fo structures such as these, let some representative field component in the waveguide b expressed as a decaying wave, such that the power present at the cross section z given by

$$P(z) = B^2 \exp\left[-2\int_0^z \alpha(\zeta)\, d\zeta\right] \quad (9.57)$$

with B the original field amplitude and α the field attenuation per unit length. Becaus it is assumed that the transverse dimension governing leakage rate is controllable, α is a function of longitudinal position ζ.

Differentiation of (9.57) gives

$$\frac{1}{P(z)}\frac{dP(z)}{dz} = -2\alpha(z) \quad (9.58)$$

as the point relation connecting α and P.

If the aperture distribution, collapsed onto the Z-axis, is represented by $A(\zeta)e^{-j\beta\zeta}$ with $A(\zeta)$ adjusted in level so that

$$P(z) = P_{IN} - \int_0^z |A(\zeta)|^2\, d\zeta \quad (9.59)$$

then the input power P_{IN} is given by

$$P_{IN} = \int_0^L |A(\zeta)|^2\, d\zeta + P_{LOAD} \quad (9.60)$$

In (9.60), L is the length of the aperture and P_{LOAD} is the power left inside the wave guide at the end of the aperture, which travels on to be absorbed by an interna matched load which terminates the waveguide.

If one lets

$$P_{LOAD} = f P_{IN} \quad (9.61)$$

so that f is the fraction of the input power absorbed in the load, then (9.60) become

$$P_{IN}(1-f) = \int_0^L |A(\zeta)|^2\, d\zeta \quad (9.62)$$

and as a result (9.59) is converted to

$$P(z) = \frac{1}{1-f}\int_0^L |A(\zeta)|^2\, d\zeta - \int_0^z |A(\zeta)|^2\, d\zeta \quad (9.6$$

Differentiation of (9.63) gives

$$\frac{dP}{dz} = -|A(z)|^2 \quad (9$$

thus (9.58) becomes

$$\alpha(z) = \frac{\frac{1}{2}|A(z)|^2}{\frac{1}{1-f}\int_0^L |A(\zeta)^2\, d\zeta - \int_0^z |A(\zeta)|^2\, d\zeta} \tag{9.65}$$

Equation 9.65 is a useful design result. If the desired pattern is specified, $A(z)$ known and this equation can be used to determine $\alpha(z)$, once a value is chosen for (This will be discussed more fully later.) The task then remains to relate the trans-erse dimensions of the aperture to the newly found $\alpha(z)$.

The connection between the rate of attenuation α and the transverse geometry of a particular type of leaky wave antenna can be determined either theoretically or experimentally. The theoretical approach can be illustrated for the case of an infinitely long slotted circular waveguide for which all cross sections are identical and typified by Figure 9.17. General field expressions can be written for the interior and exterior

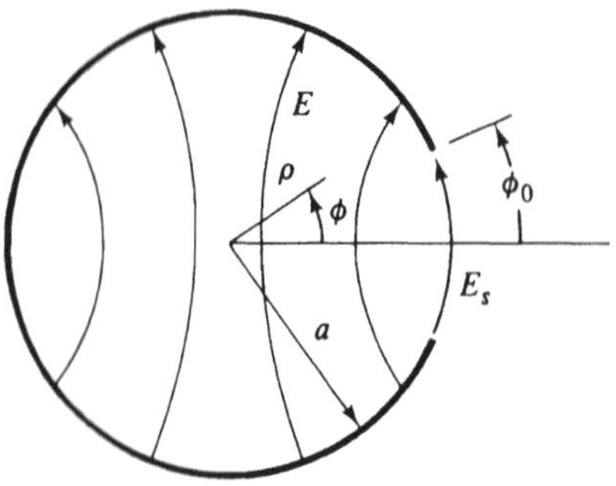

Fig. 9.17 Notation for a Slotted Cylinder Excited by a TE_{11} Mode

regions and matched across the slot boundary. When the internal field is assumed to be basically a TE_{11} mode (that is, to converge to the conventional TE_{11} mode for a circular waveguide when $\phi_0 \longrightarrow 0$), the complex propagation constant $\alpha + j\beta$ can be deduced as a function of ϕ_0 and the normalized cylinder radius.

V. H. Rumsey[22] and R. F. Harrington[23] have used a variational method to btain a numerical solution for this geometry and their results are shown as the solid irves in Figure 9.18. It can be observed that, with a/λ held fixed, α is sensitive to , but β/k is not.

A transverse resonance method has also been employed by Goldstone and ner to obtain a theoretical solution for the propagation constant of a TE_{11} slotted nder[24] and the results are in substantial agreement with those displayed in Figure . The slotted rectangular waveguide shown in Figure 9.16b has been analyzed by

[22]V. H. Rumsey, "Traveling Wave Slot Antennas," *J. App. Phys.*, 24 (1953), 1358–65.

[23]R. F. Harrington, "Propagation Along a Slotted Cylinder," *J. Appl. Phys.*, 24 (1953), 1366– also Walter, *Traveling Wave Antennas*, pp. 163–72.

[24]L. O. Goldstone and A. A. Oliner, "Leaky Wave Antennas II: Circular Waveguides," *IRE ntennas Propagat.*, AP-7 (1959), 280–90. See also Walter, *Traveling Wave Antennas*, pp. 95.

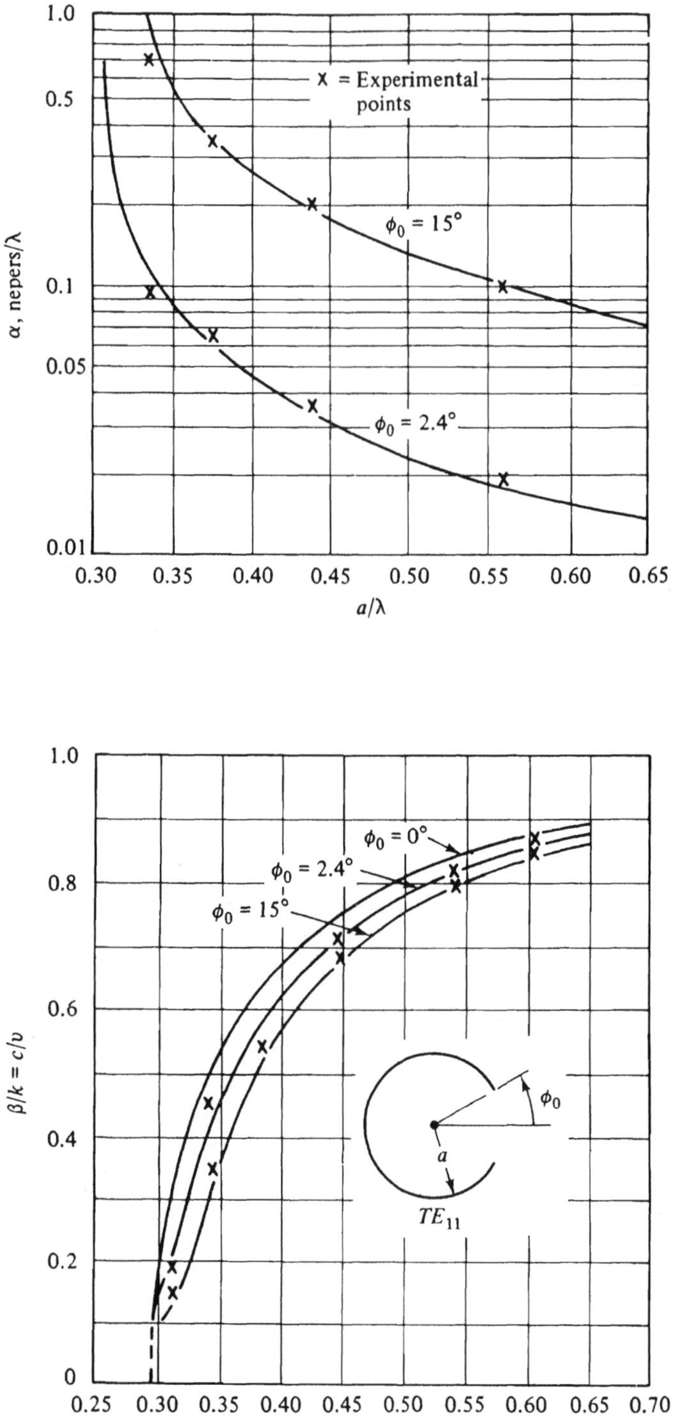

Fig. 9.18 Complex Propagation Constant for a TE_{11} Slotted Cylinder (© 1953 American Institute of Physics. Reprinted from R. F. Harrington, *J. Appl. Phys.*, 1953.)

the same technique, and L. O. Goldstone and A. A. Oliner present theoretical curves for the complex propagation constant versus slot width for standard X-band waveguide.[25]

Experimental determination of $\alpha + j\beta$ versus transverse geometry can be achieved by a variety of methods. Near-field measurements using a small movable probe placed in front of the aperture are practical if α is not too small and the aperture is not too long. An alternative near-field method with the same restrictions employs a scatterer in place of the probe and a precisely constructed magic T.[26]

For α small and the aperture long, a far-field measurement technique can be used to determine the complex propagation constant. To see how this is accomplished, assume a leaky wave antenna for which the transverse cross section is unchanged over the entire aperture length L. With α small, discontinuity effects at the two ends of the aperture can be ignored and with a matched load placed inside the waveguide beyond $z = L$, the collapsed aperture distribution is given by

$$A(\zeta) = Ke^{j(\omega t - \beta\zeta) - \alpha\zeta} \tag{9.66}$$

with K, α, and β *constants*. The far-field pattern corresponding to (9.66) can be deduced easily from (1.128) through (1.131) and, with the element factor suppressed, the normalized radiation intensity is

$$\mathcal{P}(\theta) = \left| \frac{\sin[(\pi L/\lambda)[\cos\theta - \beta/k + (j\alpha/k)]]}{(\pi L/\lambda)[\cos\theta - \beta/k + (j\alpha/k)]} \right|^2 \tag{9.67}$$

With α small, this is a pattern with a main beam pointing at the angle

$$\theta_0 = \arccos\left(\frac{\beta}{k}\right) \tag{9.68}$$

plus a family of side lobes interspersed by fairly deep "nulls." With L large, θ_0 can be determined quite accurately. One can conclude that if any long fast wave antenna of constant cross section and slight leakage is terminated by a matched load, and if the angle off endfire at which the peak of the main beam occurs is measured, (9.68) can be used to deduce the imaginary component of the propagation constant.

This experiment can be improved by replacing the matched load by a short circuit, for then if the generator is matched the aperture distribution becomes

$$A(\zeta) = Ke^{j\omega t - \gamma\zeta} + Ke^{-\alpha L}e^{j\omega t + \gamma\zeta} \tag{9.69}$$

The radiation pattern now takes on the appearance suggested by Figure 9.19, with a rearward main beam as well as a forward main beam. These two beams are sepa-

[25] L. O. Goldstone and A. A. Oliner, "Leaky Wave Antennas I: Rectangular Waveguides," *IRE Trans. Antennas Propagat.*, AP-7 (1959), 307–19. See particularly Figure 7.

[26] Walter, *Traveling Wave Antennas*, pp. 158–59.

Fig. 9.19 Principal Plane Pattern of a Leaky Wave Antenna Terminated in a Short Circuit, Showing Forward and Rearward Main Beams

rated by an angle

$$\theta' = \pi - 2\theta_0 = \pi - 2 \arccos\left(\frac{\beta}{k}\right) \tag{9.70}$$

A measurement of θ' avoids the need to align the leaky wave antenna and permits deduction of β. But additionally, the rearward main beam is N dB below the forward main beam, with N given by

$$N = 10 \log_{10} e^{-2\alpha L} \tag{9.71}$$

and thus α can also be determined from this experiment.

K. C. Kelly[27] has used this method to determine α and β for a family of 12 serrated waveguides. He used standard X-band rectangular waveguide ($a = 0.900$ inch, $b = 0.400$ inch, and $t = 0.050$ inch) with a series of closely spaced nonresonant slots milled into one of the broad walls, as suggested by Figure 9.20. For all members of the family, $G = 2/32$ inch, $T = 3/32$ inch, and $L = 10\frac{5}{16}$ inches. For any one member of the family, l is constant and the members are distinguished by different values of l in the sequence 0.225 inch, 0.250 inch, . . . , 0.500 inch. The values of α and β found

[27]K. C. Kelly and R. S. Elliott, "Serrated Waveguide—Part II: Experiment," *IRE Trans. Antennas Propagat.*, AP-5 (1957), 276–83.

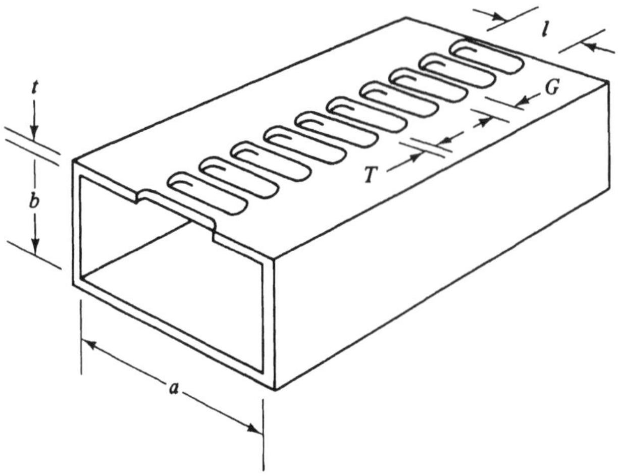

Fig. 9.20 Serrated Waveguide Geometry

by reduction of the pattern data are shown in Figure 9.21, where α varies smoothly with l and has a considerable dynamic range, while β only changes 10% as l increases from zero to 0.500 inch.

R. F. Hyneman[28] has provided a theoretical analysis of a limiting case of the structure shown in Figure 9.20, namely when $G/(G + T) = 1$ and $t \longrightarrow 0$. He assumed square-ended slots and also found that β/k was insensitive to l. His curves for α versus l/a and a/λ are shown in Figure 9.22.

In general, agreement between theory and experiment is excellent for all leaky wave antennas for which the dependence of $\gamma = \alpha + j\beta$ on transverse geometry has been analyzed. The experimental points indicated in Figure 9.18 for the case of the TE_{11} slotted cylinder, and in Figure 9.22 for the limiting case of a serrated waveguide are convincing examples of this.

With the relation between α and transverse dimensions established, one can turn to the design of a leaky wave antenna. As an example, suppose the collapsed aperture distribution

$$A(\zeta) = \left[1 + \sin\left(\frac{\pi\zeta}{L}\right)\right] e^{-j\beta_0\zeta} \tag{9.72}$$

is selected, with β_0 the propagation constant when no leakage occurs and L the length of the aperture. If (9.72) is used in (9.65), one obtains

$$\alpha(z) = \frac{1}{L}\left\{\frac{[1 + \sin(\pi z/L)]^2}{[5.546/(1 - f)] - (4/\pi)[1 + (3\pi z/4L) - \cos(\pi z/L) - \frac{1}{8}\sin(2\pi z/L)]}\right\} \tag{9.73}$$

[28] R. F. Hyneman, "Closely Spaced Transverse Slots in Rectangular Waveguide," *IRE Trans. Antennas Propagat.*, AP-7 (1959), 335–42.

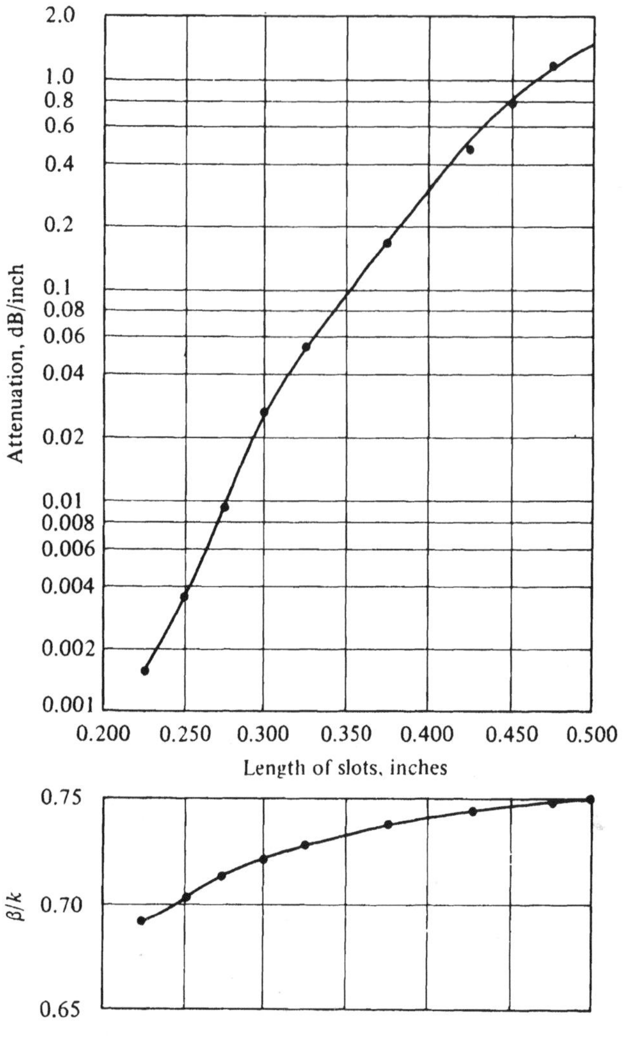

Fig. 9.21 Experimentally Determined Complex Propagation Constant of *X*-Band Serrated Waveguide (© 1957 IEEE. Reprinted from K. C. Kelly and R. S. Elliott, *IRE AP Transactions*, 1957.)

A study of (9.73) reveals that $\alpha(z)$ is inversely proportional to the length of the aperture, but that if L is fixed, a greater dynamic range of $\alpha(z)$ results from the choice of a smaller value of f. This point is illustrated by the curves of Figure 9.23, which are plots of (9.73) for $f = 10\%$ and 20%. Given the finite range of transverse dimensions over which α is a well-behaved function, there is a lower limit on the fraction of the input power which must be delivered to the matched load. Ceteris paribus, the greater the value of L, the smaller f need be.

The proper value of f for a given aperture length and a given type of leaky waveguide can be deduced by trial and error. For example, if the aperture distribution of

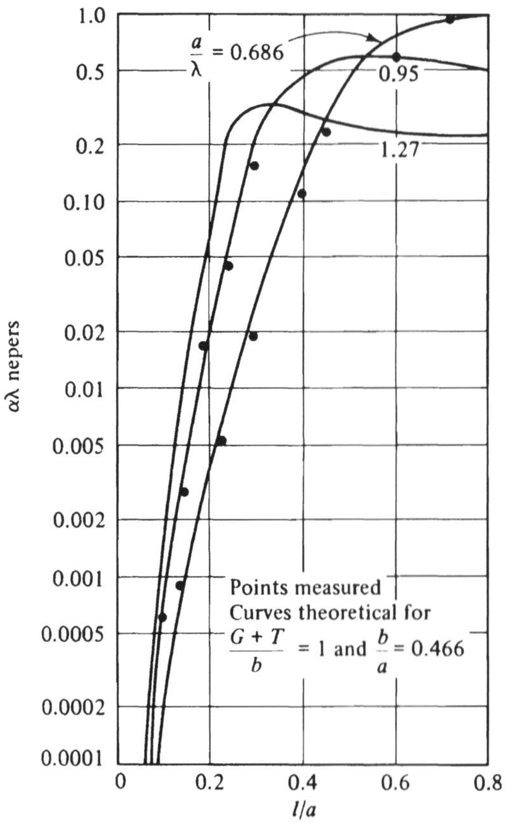

Fig. 9.22 Leakage Rate versus Slot Length for Limiting Case of Serrated Waveguide (© 1959 IEEE. Reprinted from R. F. Hyneman, *IRE AP Transactions*, 1959.)

(9.72) is to be achieved with a serrated waveguide of length $L = 10.140$ inches for which Kelly's data of Figure 9.21 is applicable, choice of $f = 0.1$ results in the design curve shown in Figure 9.24. (Data in the range $0.500 \leq l \leq 0.535$ has been inferred by extrapolation, and the conversion 1 decibel = 8.686 nepers has been used.)

When a serrated waveguide was machined to conform to the design data contained in Figure 9.24 and the principal plane pattern was measured at the design frequency of 9 GHz, the results shown in Figure 9.25 were achieved. Since the theoretical side lobe level for the aperture distribution of (9.71) is −17.7 dB, it can be seen that the design is quite satisfactory. Some null-filling and irregularity in the heights of individual side lobes can be attributed to the fact that β/k is not quite a constant over the aperture.

Leakage rate curves of the type shown in Figure 9.23 can be generated with f as parameter for any desired aperture distribution. This information can then be combined with the knowledge of α versus transverse dimensions for any type of leaky waveguide to produce a design curve similar to Figure 9.24. This procedure has been used successfully with all four leaky waveguide types shown in Figure 9.16.

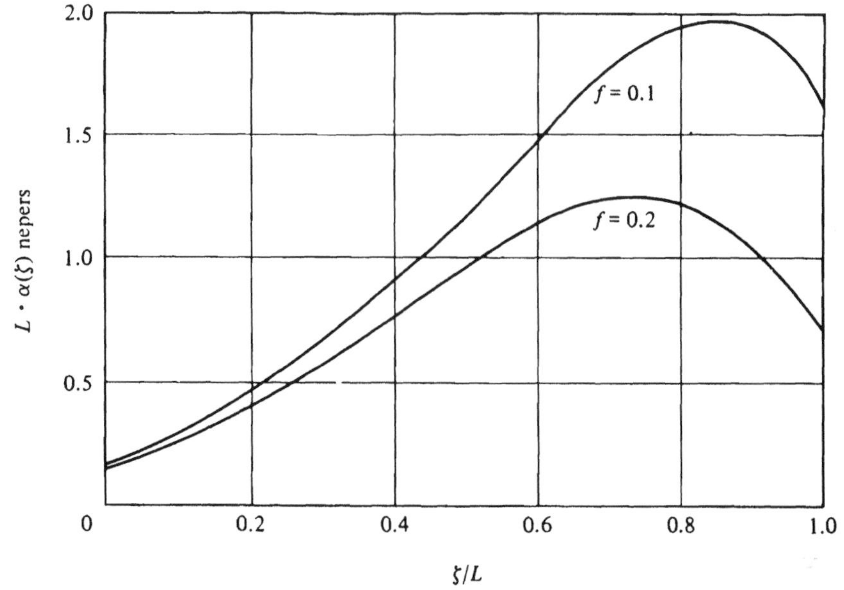

Fig. 9.23 Leakage Rate as a Function of Position in the Aperture for Aperture Distribution of Equation 9.72, Effect of Power Fraction Dissipated in the Load

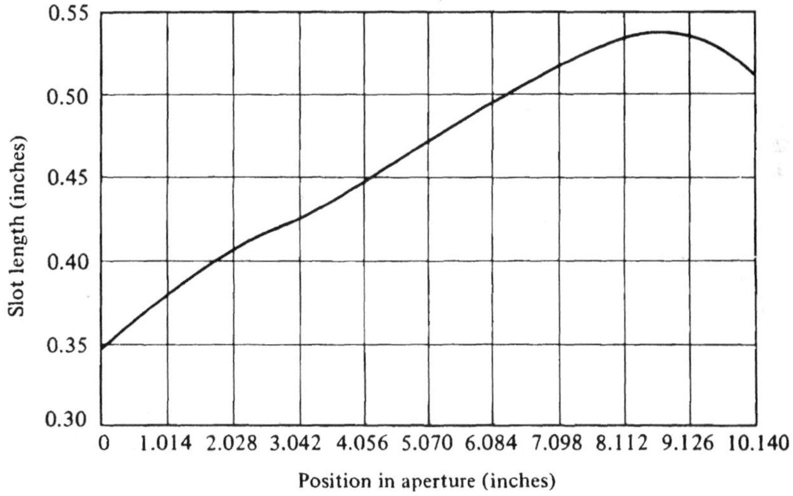

Fig. 9.24 Design Curve for Serrated Waveguide Antenna with Aperture Distribution $= 1 + \sin(\pi\zeta/L)$ (© 1957 IEEE. Reprinted from K. C. Kelly and R. S. Elliott, *IRE AP Transactions*, 1957.)

Leaky waveguides can be placed side by side to form planar arrays. Proper design must account for mutual coupling between waveguides, which is severe for the structures shown in Figures 9.16a, b and c and mild for the structure shown in

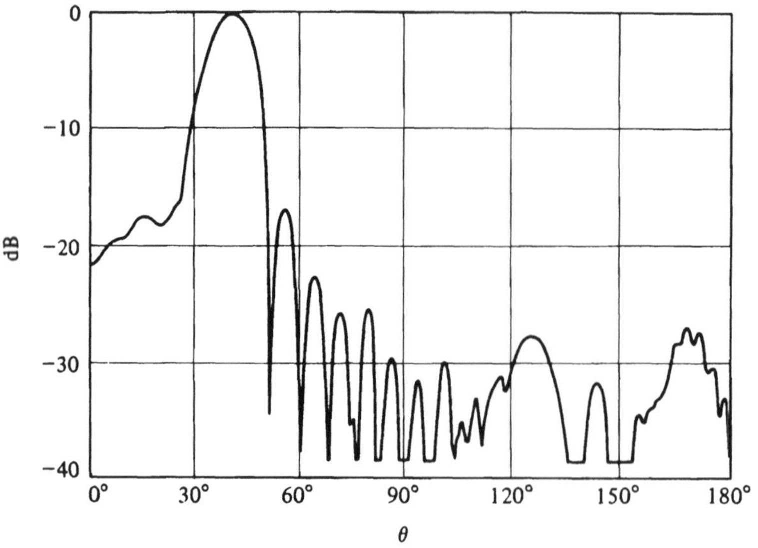

Fig. 9.25 Principal Plane Pattern (Experimental) for a Serrated Waveguide Antenna; $L = 7.75\lambda$, $\nu = 9$ GHz; Aperture Distribution $= 1 + \sin(\pi\zeta/L)$ (© 1957 IEEE. Reprinted from K. C. Kelly and R. S. Elliott, *IRE AP Transactions*, 1957.)

Figure 9.16d. C. H. Walter[29] indicates several methods for including the effects of mutual coupling and K. C. Kelly[30] reports on the design and performance of ten side-by-side serrated waveguides.

9.9 Trough Waveguide Antennas

The trough waveguide is a versatile structure that has modified forms capable of supporting traveling waves with phase velocities above, at, and below the speed of light. In the fast wave case, the leakage is also controllable, making the trough waveguide an attractive candidate for leaky wave antenna applications.

The symmetrical form of a trough waveguide is shown in Figure 9.26. The E-field distribution for the fundamental mode is suggested in the figure. This structure can be viewed as half of a strip transmission line operating in its first higher *TE* mode, a mode for which the plane bisecting the stripline is an electric null plane. Because of the antisymmetry of the E-lines, the symmetrical trough waveguide *cannot* radiate. However, a variety of modifications to the structure, each of which introduces an asymmetry, will cause an unbalance in the E-lines and thus can produce radiation.

One of the common modifications for creating an asymmetry is illustrated by Figure 9.27, in which the trough depth on one side of the center fin has been reduced

[29] Walter, *Traveling Wave Antennas*, pp. 367–71.

[30] Kelly, "Closely Spaced Transverse Slots," p. 282.

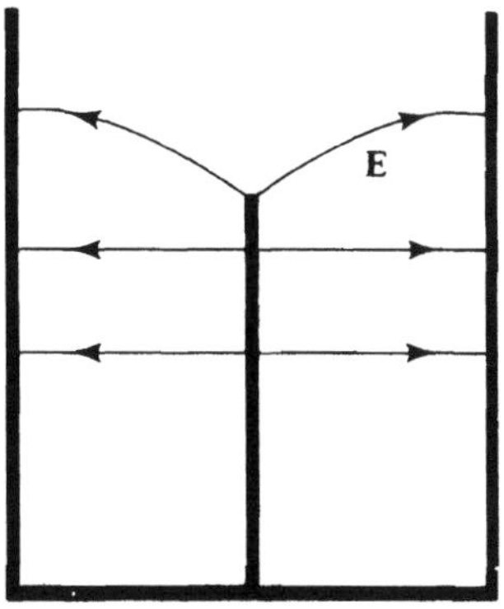

Fig. 9.26 A Symmetrical Trough Waveguide

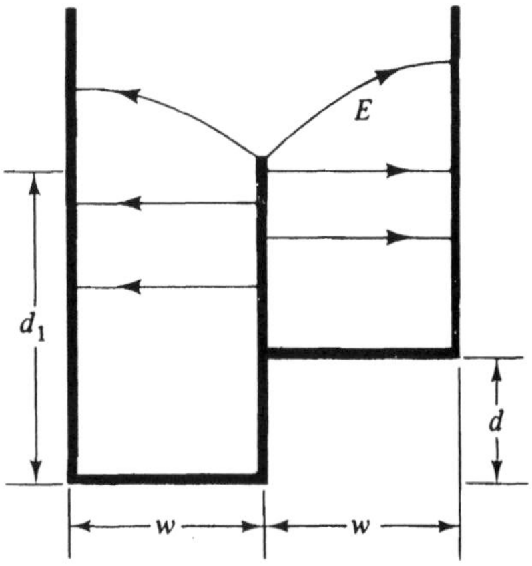

Fig. 9.27 An Asymmetrical Trough Waveguide

by an amount d. W. Rotman and A. A. Oliner[31] have investigated this geometry both theoretically and experimentally and have found the dependence of β/k and α on trough dimensions to be as shown in Figure 9.28. In this modified form, the trough waveguide is a fast wave structure with phase velocity which is relatively independent of d/λ but with a leakage rate which covers a useful dynamic range as d/λ is varied. Agreement between theory and experiment is seen to be quite satisfactory. These curves can be used to design a leaky trough waveguide antenna, employing the technique developed in Section 9.8. It is interesting to observe that if *both* d and d_1 are varied, the leakage rate can be controlled in just the right way to achieve a desired aperture distribution with β held constant over the aperture.

If the center fin of the symmetrical trough waveguide is serrated, as shown in the insert to Figure 9.29, the phase velocity can be modified. A. A. Oliner and W. Rotman[32] have determined the dependence of β/k on the various dimensions of this structure and their curves, which are reproduced in Figure 9.29, show a range that embraces both fast and slow waves. Rotman has been led by this discovery to suggest an arrangement of three serrated center fins, side by side and virtually touching, with only the middle fin movable. As this middle fin is translated longitudinally through one serration period $G + T$, the phase velocity can be swept through a sizeable dynamic range. With an introduced asymmetry, such as that shown in Figure 9.27, leakage can occur and mechanical scanning of the main beam from endfire to broadside is conceptually possible. Alternatively, with a single serrated center fin, frequency scanning of the main beam is possible, as can be seen from the curves of Figure 9.29.

[31] W. Rotman and A. A. Oliner, "Asymmetrical Trough Waveguide Antennas," *IRE Trans. Antennas Propagat.*, AP-7 (1959), 153–62.

[32] A. A. Oliner and W. Rotman, "Periodic Structures in Trough Waveguides," *IRE Trans. Microwave Theory Tech.*, MTT-7 (1959), 134–42.

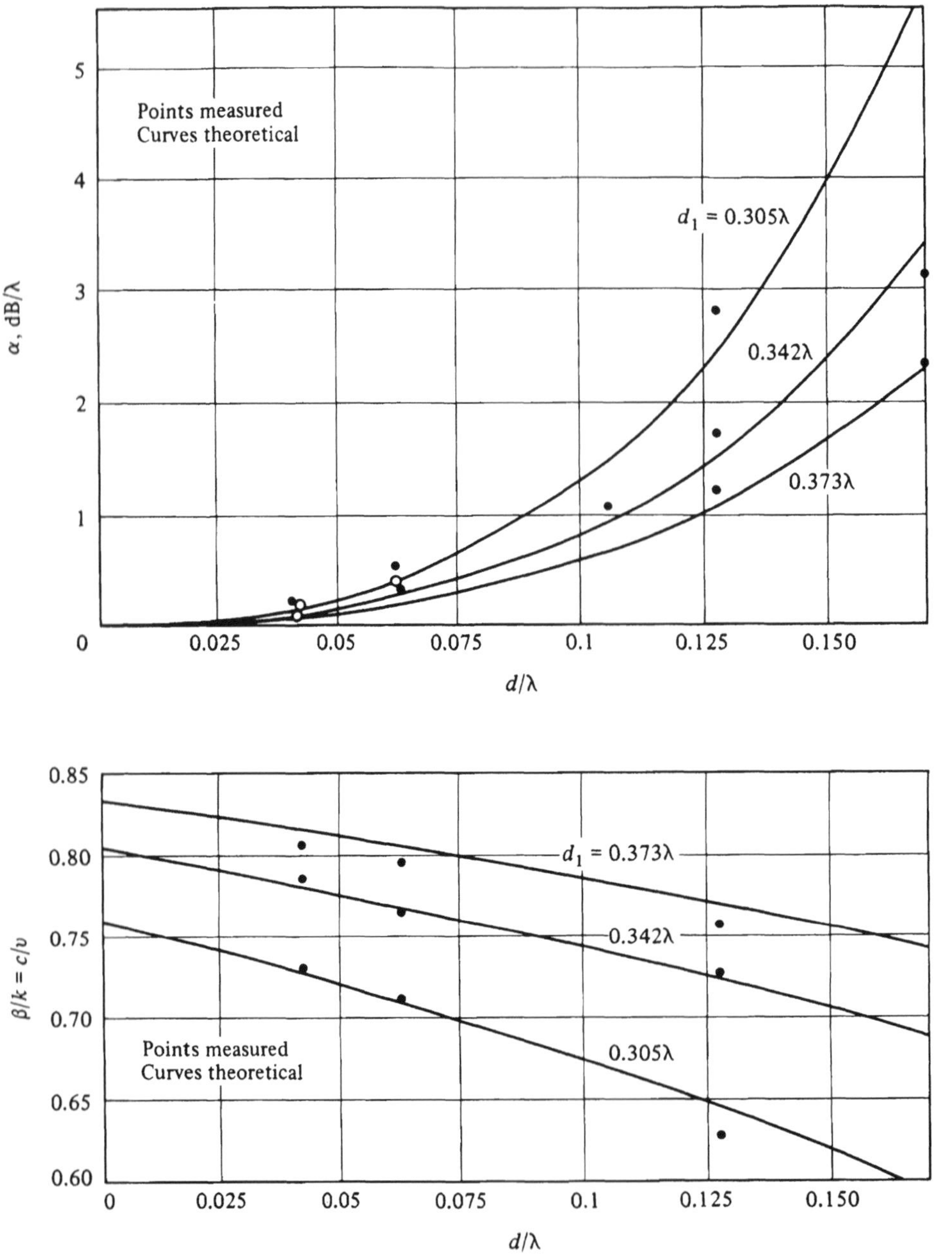

Fig. 9.28 Complex Propagation Constant for the Asymmetrical Trough Waveguide of Figure 9.27. (© 1959 IEEE. Reprinted from W. Rotman and A. A. Oliner, *IRE AP Transactions*, 1959.)

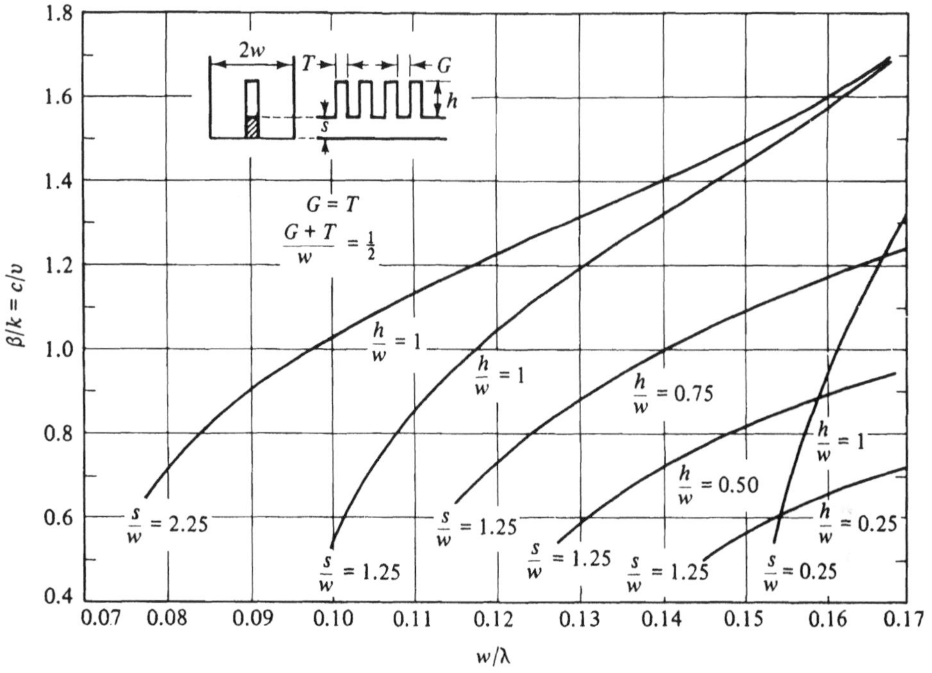

Fig. 9.29 Phase Velocity for a Symmetrical Trough Waveguide with a Serrated Center Fin (© 1959 IEEE. Reprinted from A. A. Oliner and W. Rotman, *IRE AP Transactions*, 1959.)

9.10 Traveling Wave Arrays of Quasi-Resonant Discretely Spaced Slots [Main Beam at $\theta_0 = \arccos(\beta/k)$]

The three types of radiating slots that can be cut in the walls of a rectangular waveguide, and which were described in Section 3.5, can be spaced $\lambda_g/2$ apart and alternately offset (or tilted). This gives rise to a standing wave array for which a design procedure was presented in Section 8.15. However, the slots need not be spaced a half guide-wavelength apart. If the spacing $d \neq \lambda_g/2$, and a matched termination is placed beyond the last slot, a traveling wave array results. The excitation of such arrays quite naturally permits a uniform progressive phase in the aperture distribution. This, together with a properly controlled amplitude taper, will produce a sum pattern with the main beam at some angle other than broadside. There are many practical applications in which this is desirable.

Mutual coupling must be taken into account if precise control of the aperture distribution is to be achieved. For the case of longitudinal shunt slots in the broad wall, this can be accomplished by an extension of the design procedure developed in Section 8.15. It will be recalled that the two basic equations are

$$\frac{Y_n^a}{G_0} = K_1 f_n \frac{V_n^s \sin kl_n}{V_n} \tag{9.74}$$

$$\frac{Y_n^a}{G_0} = \frac{K_2 f_n^2}{[K_2 f_n^2/g(x_n)h(y_n)] + Z_n^b} \tag{9.75}$$

Definitions of the various quantities appearing in (9.74) and (9.75) can be found in Section 8.15.

For a linear array of longitudinal slots in a common waveguide, if $d = \lambda_g/2$, the mode voltage V_n has the same magnitude at all slots, merely alternating in sign as successive slots are reached. But for $d \neq \lambda_g/2$, the situation is more complicated. The equivalent circuit in this more general case is shown in Figure 9.30, with a matched load assumed to exist beyond slot 1.

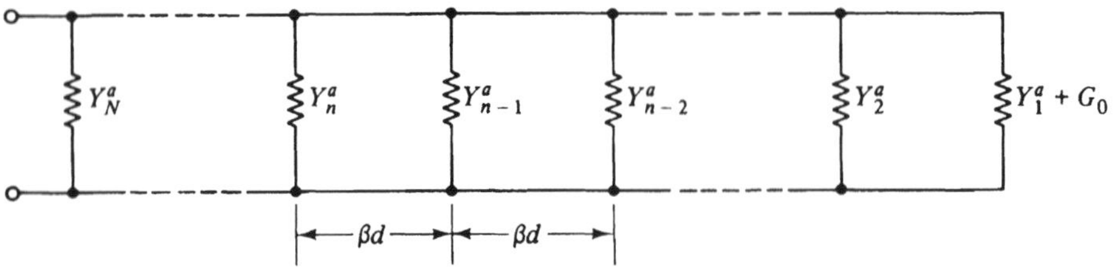

Fig. 9.30 The Equivalent Circuit of a Traveling Wave Array of N Slots

If Y_n is the total admittance seen looking into the nth junction toward the matched load, then

$$\frac{Y_n}{G_0} = \frac{Y_n^a}{G_0} + \frac{(Y_{n-1}/G_0)\cos\beta d + j\sin\beta d}{\cos\beta d + j(Y_{n-1}/G_0)\sin\beta d} \tag{9.76}$$

with β the propagation constant of the TE_{10} mode. The mode voltages at successive junctions are related by

$$\begin{aligned} V_n &= V_{n-1}\cos\beta d + jI_{n-1}Z_0\sin\beta d \\ &= V_{n-1}\left[\cos\beta d + j\left(\frac{Y_{n-1}}{G_0}\right)\sin\beta d\right] \end{aligned} \tag{9.77}$$

When the design equation of (9.74) is written separately for th nth and $(n - 1)$st slots and the ratio is taken, one obtains

$$\frac{Y_{n-1}^a/G_0}{Y_n^a/G_0} = \frac{f_{n-1}}{f_n}\frac{V_{n-1}^s \sin kl_{n-1}}{V_n^s \sin kl_n}\frac{V_n}{V_{n-1}} \tag{9.78}$$

The ratio V_n/V_{n-1} can be eliminated from (9.78) through use of (9.77). Rearrangement gives

$$\frac{Y_n^a/G_0}{f_n \sin kl_n} = \frac{Y_{n-1}^a/G_0}{f_{n-1}\sin kl_{n-1}}\frac{V_n^s/V_{n-1}^s}{\cos\beta d + j(Y_{n-1}/G_0)\sin\beta d} \tag{9.79}$$

Equation 9.79 is a recurrence relation which, in conjunction with (9.75), will permit determination of the length and offset of the nth slot, once the length and offset of the $(n-1)$st slot are known.

A design procedure can now be formulated. One begins by assuming an initial set of lengths and offsets for the slots. This is solely for the purpose of computing initial values of the mutual coupling terms Z_n^b, through use of Equations 8.89, 7.155, and 7.156, with the implicit understanding that the desired slot voltage distribution V_n^s has been specified. It is entirely adequate for this first calculation to take all the slots to be on the center line and a half-wavelength long.

Next, one *guesses* a value for Y_1^a/G. (How to make a judicious guess will be indicated shortly.) Equation 9.75 can then be used, with insertion of the initial values of Z_1^b and Y_1^a/G_0, to determine an improved estimate of (x_1, l_1).

If the slot spacing d has been selected (more about this, too, shortly), the right side of (9.79) is known for the case $n = 2$. Simultaneous solution of (9.75) and (9.79) will yield an estimate of Y_2^a/G_0 and (x_2, l_2). But then the right side of (9.79) becomes known for the case $n = 3$. The process can be repeated to find an estimate of Y_3^a/G_0 and (x_3, l_3), and ultimately an estimate of Y_N^a/G_0 and (x_N, l_N).

This new set of slot lengths and offsets can be tested for adequacy in several ways. First, and most obvious, the maximum and minimum offsets should be in the range in which reliance can be placed on the design data. If one or the other of these extremes is out of range, the next guess for Y_1^a/G_0 will have to be adjusted in the proper direction.

Second, the procedure just described requires the calculation of Y_{n-1}/G_0 at each stage, so the information is available to permit a determination of the normalized input admittance to the array. This is obtained simply by using $n = N$ in (9.76). If Y_N/G_0 is not close enough to a match, this is probably an indication that the spacing d needs to be adjusted. This point will be elaborated later in the discussion.

Third, a computation can be made of the fraction f of the input power that is absorbed in the load. The normalized power radiated by each slot is

$$P_n = \frac{1}{2} \mathcal{R}e\, V_n V_n^* \left(\frac{Y_n^a}{G_0}\right)^* \tag{9.80}$$

Therefore, since the normalized power into the matched load is $\frac{1}{2}V_1 V_1^*$, it follows that

$$f = \frac{\frac{1}{2}V_1 V_1^*}{\frac{1}{2}V_1 V_1^* + \sum P_n} \tag{9.81}$$

An efficient design requires that f be as small as feasible. A raising of the value of Y_1^a/G_0, with the concomitant increase in the offset x_1, will serve to increase all the offsets and all the normalized active admittances, and thus $\sum P_n$, thereby lowering f. This argument leads to the conclusion that the guess for the value of Y_1^a/G_0 should be adjusted so that the maximum offset x_n (usually for the central slot) is at the upper end of the reliable data range, for then f will have its minimum feasible value.

It is highly unlikely that the initial choices of d and (particularly) of Y_1^a/G_0 will lead to a meeting of the design criteria, but even if one were so fortunate as to obtain this outcome, the process should be iterated, because now a more realistic set of slot lengths and offsets is known, permitting an improved set of Z_n^b values to be calculated. A succession of iterations will normally need to be undertaken, with the process terminated when three goals are achieved: (1) the input admittance is acceptably close to a match, (2) the fraction of power dissipated in the matched load has been minimized, and (3) the final set of (x_n, l_n) values is closer to the penultimate set than fabrication tolerances.

Further consideration can now be given to the problem of making an initial guess of the value of Y_1^a/G_0. If there are many slots in the array, the power radiated by slot 1 will be small compared to the total radiated power (particularly so with a tapered aperture distribution). Thus $P_1 \lll \sum P_n$. This means that it is feasible to have P_1 much less than the power dissipated in the load. In quantitative terms,

$$\frac{1}{2} V_1 V_1^* \left(\frac{Y_1^a}{G_0}\right)^* = P_1 \ll P_L = \frac{1}{2} V_1 V_1^* \tag{9.82}$$

This implies that $Y_1^a/G_0 \ll 1$, which is vital, for then the offset of the first slot will be at a level suitably low to prevent the slots in the center of the array from being excessively offset.

If $Y_1^a/G_0 \ll 1$, then

$$\frac{Y_1}{G_0} = \frac{Y_1^a}{G_0} + \frac{G_0}{G_0} \cong 1 \tag{9.83}$$

and a return to Equation 9.77 indicates that $V_2 \cong V_1 e^{j\beta d}$. This argument can be cascaded, because with N large, $Y_2^a/G_0 \ll 1$, and so on, so that (approximately) $V_n = V_{n-1} e^{j\beta d}$. In words, for a traveling wave array of many slots, the mode voltages essentially have a common magnitude and a uniform progressive phase which corresponds to the passage of a wave along the aperture characterized by the function $\exp[j(\omega t - \beta z)]$.

Under these conditions, Equation 9.79 indicates that the same uniform progressive phase can be obtained for the slot voltage distribution if the active admittances are chosen to have approximately the same phase. Thus this type of array is suitable for applications in which a sum pattern is desired with the main beam at an angle θ_0, for this can be achieved by an aperture distribution that gives

$$\mathcal{S}(\theta) = \sum V_n^s e^{jnkd\cos\theta} = \sum |V_n^s| e^{jnd(k\cos\theta - \beta)} \tag{9.84}$$

from which it follows that

$$\frac{\beta}{k} = \cos\theta_0 \tag{9.85}$$

The pattern requirement that the main beam point at an angle θ_0 thus serves to tie down the value of β.[33]

With all of the active admittances essentially at a common phase, and all the mode voltages possessing approximately the same amplitude, one can observe that

$$\frac{P_1}{P_n} = \frac{\frac{1}{2}V_1 V_1^*(Y_1^a/G_0)^*}{\frac{1}{2}V_n V_n^*(Y_n^a/G_0)^*} \simeq \frac{|Y_1^a/G_0|}{|Y_n^a/G_0|} \tag{9.86}$$

But the powers radiated by the individual slots are also roughly proportional to the squares of the slot voltages. Thus

$$\left|\frac{Y_1^a}{G_0}\right| \cong \left|\frac{V_n^s}{V_1^s}\right|^2 \left|\frac{Y_n^a}{G_0}\right| \tag{9.87}$$

Suppose that one identifies the slot with the largest desired slot voltage (call it the nth slot) and ignores mutual coupling for the purpose of making a preliminary calculation. If this slot is given the largest offset consistent with being in the acceptable dynamic range of offsets, and the corresponding resonant length, then $g(x_n)$ is an estimation of its normalized admittance. If $g(x_n)$ is used for $|Y_n^a/G_0|$ in (9.87), one can compute a starting value for $|Y_1^a/G_0|$. The phase that can be attached to this starting admittance is somewhat arbitrary, but probably should not be far from zero degrees.

An initial selection of the element spacing d also needs to be made. An economical design calls for the minimum number of slots consistent with the beamwidth requirement and the desire not to have an additional main beam appearing at reverse end-fire. For N large, the constraint in (4.30) can be used to estimate d. For N modest, placement of the roots on a Schelkunoff unit circle will permit an estimation of the maximum value of d. (See Section 4.4).

The value of d chosen initially may need to be modified slightly as the iterations proceed. The reason for this can be appreciated by considering the factors which influence the input admittance. One can recall from Section 8.14 that the back-scattered wave from the nth slot, referenced at the center of the nth slot, is given by

$$B_n = -\frac{1}{2}V_n \frac{Y_n^a}{G_0} \cong -\frac{1}{2}V_N \frac{Y_n^a}{G_0} e^{-j\beta(N-n)d} \tag{9.88}$$

It follows that

$$B_n' = B_n e^{-j\beta(N-n)d} \cong -\frac{1}{2}V_N \frac{Y_n^a}{G_0} e^{-j2\beta(N-n)d} \tag{9.89}$$

is the wave back-scattered from the nth slot, referenced at the input. One can deduce from (9.89) that there will be a match if

$$\sum_{n=1}^{N} \frac{Y_n^a}{G_0} e^{j2n\beta d} = 0 \tag{9.90}$$

[33]For a distinctly different solution to this problem, in which $\theta_0 \neq \arccos(\beta/k)$, see Section 9.11.

For a given set of active admittance values, there is a unique set of d values that satisfy (9.90). If N is large enough (10 slots or more), one of these values of d will be close to the original selection of d, so that only a minor adjustment need be made. Indeed, for N very large, the left side of (9.90) is small for all values of d in the admissible range.

The theory just presented has been tested by the design of an array of 21 longitudinal shunt slots.[34] Standard X-band waveguide was used ($a = 0.900$ inch, $b = 0.400$ inch, $t = 0.050$ inch) with a central frequency of 9.375 GHz. The slots were 0.063 inch wide and round-ended, and thus Stegen's design data, embodied in Figures 8.35 through 8.37, was applicable. Since $\beta/k = 0.714$ for these dimensions and frequency, it was decided to design for a sum pattern with the main beam at $\theta_0 = \arccos 0.714 = 44.5°$. A Dolph-Chebyshev distribution was specified that would give a -30 dB side lobe level. The interelement spacing was chosen to be $d = 0.545\lambda$, which is comfortably below the constraint (4.30), and thus the uniform progressive phase in the aperture distribution was $\beta d = 140°$. The relative slot voltage magnitudes were taken from the tables of L. B. Brown and G. A. Scharp[35] and are listed in Table 9.2. The theoretical array pattern corresponding to this distribution is shown dotted in Figure 9.31.

TABLE 9.2 Desired slot voltage distribution main beam at 45°; 30-dB Dolph-Chebyshev pattern

n	$\lvert V_n^s \rvert$	n	$\lvert V_n^s \rvert$	n	$\lvert V_n^s \rvert$
1	0.3337	8	0.8829	15	0.7995
2	0.2789	9	0.9465	16	0.7014
3	0.3780	10	0.9864	17	0.5946
4	0.4849	11	1.0000	18	0.4849
5	0.5946	12	0.9864	19	0.3780
6	0.7014	13	0.9465	20	0.2789
7	0.7995	14	0.8829	21	0.3337

An original guess of 0.085 was made for the value of Y_1^a/G_0, and the design procedure was initiated. Three iterations brought the slot offsets to stability within 0.001 inch; the resulting values for lengths and offsets are shown in Table 9.3. The normalized theoretical input admittance was $0.955 + j0.009$, and 12.3% of the power was predicted to be absorbed in the dummy load. These figures could have been improved by adjusting Y_1^a/G_0 to a higher value and by changing d slightly, but since the sole purpose of the design was to validate the theory, it was decided to avoid further computer costs and accept this as an adequate test.

A longitudinal shunt slot array was constructed using the lengths and offsets

[34] R. S. Elliott, "On the Design of Traveling-Wave-Fed Longitudinal Shunt Slot Arrays," *IEEE Trans. Antennas Propagat.*, AP-27 (1979), 717–20.

[35] L. B. Brown and G. A. Scharp, "Chebyshev Antenna Distribution, Beamwidth, and Gain Tables," *Nav. Ord. Report* 4629, (California: Corona, 1958).

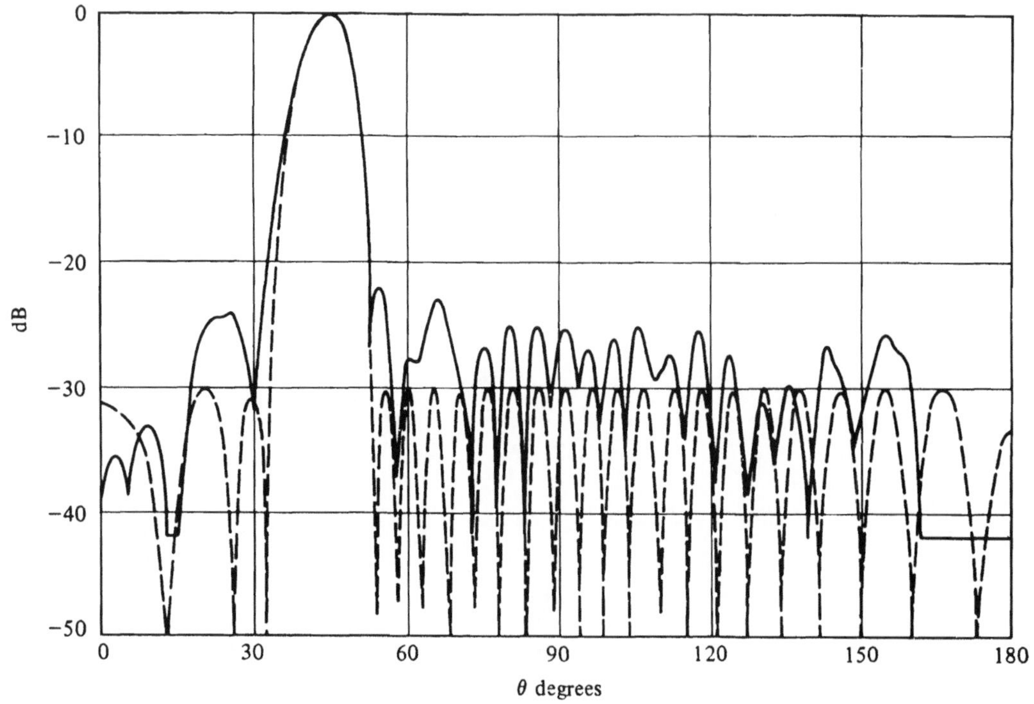

Fig. 9.31 Theoretical and Experimental Patterns for a 21-Element Traveling Wave Array of Longitudinal Shunt Slots; ν = 9.375 GHz (© 1979 IEEE. Reprinted from *IEEE AP Transactions*, 1979.)

listed in Table 9.3. The experimental pattern at 9.375 GHz is shown as the solid curve in Figure 9.31 and the input *VSWR* versus frequency is displayed in Figure 9.32.

The experimental performance can be summarized by noting that the pattern has a well-defined main beam in the proper position and with the proper beamwidth; the side lobe level is poorer than theoretical—the innermost sidelobe on one side is at −22 dB, the three innermost side lobes lie between −22 and −24 dB, and the remainder are all at least 25 dB down from the main beam; however, the outer side-lobes do not fall off as they should considering the element pattern behavior in the

TABLE 9.3 Slot lengths and offsets

n	x_n	$2l_n^{(0)}$	n	x_n	$2l_n^{(0)}$	n	x_n	$2l_n^{(0)}$
1	0.078 in.	0.605 in.	8	0.081 in.	0.612 in.	15	0.055 in.	0.613 in.
2	0.028 in.	0.604 in.	9	0.081 in.	0.613 in.	16	0.030 in.	0.617 in.
3	0.061 in.	0.603 in.	10	0.079 in.	0.613 in.	17	0.027 in.	0.619 in.
4	0.071 in.	0.607 in.	11	0.075 in.	0.612 in.	18	0.026 in.	0.620 in.
5	0.078 in.	0.609 in.	12	0.070 in.	0.612 in.	19	0.024 in.	0.619 in.
6	0.081 in.	0.610 in.	13	0.066 in.	0.613 in.	20	0.022 in.	0.616 in.
7	0.081 in.	0.610 in.	14	0.062 in.	0.613 in.	21	0.025 in.	0.621 in.

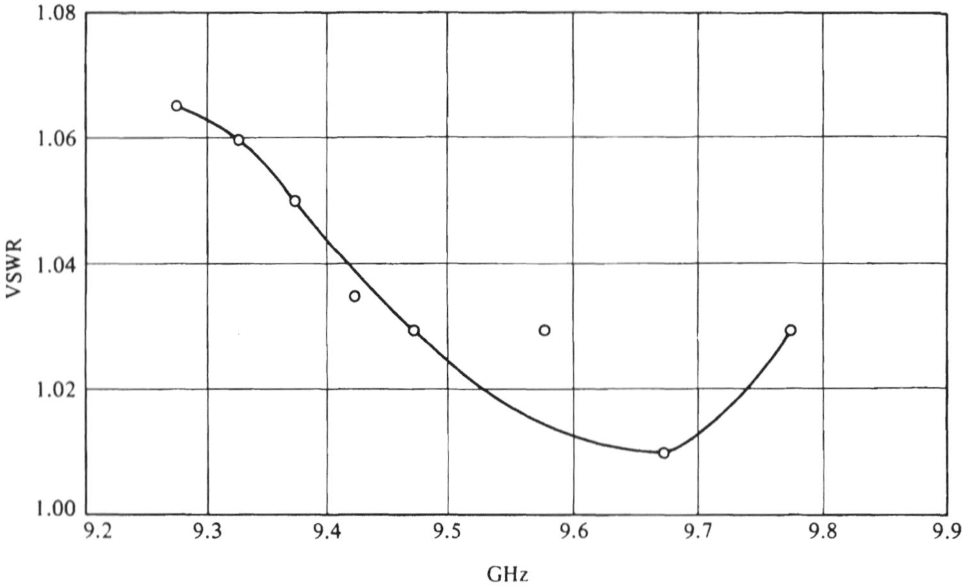

Fig. 9.32 Input VSWR versus Frequency for a 21-Element Traveling Wave Array of Longitudinal Shunt Slots (© 1979 IEEE. Reprinted from *IEEE AP Transactions*, 1979.)

H-plane; the input VSWR is very good over a five percent frequency band, being 1.05 at the design frequency and hitting a low of 1.01 at 9.675 GHz; though not shown, the pattern held up well in this frequency band.

The inability to achieve a -30 dB side lobe level is laid to the presence of *internal* mutual coupling, a factor which was not accounted for in the design. L. A. Kurtz has shown[36] experimentally that the TE_{20} mode scattered off a slot may be strong enough to affect materially the slot voltages of its two adjacent neighbors. In the present design, where the slots are all on the same side of the center line, and where the tip-to-tip spacing is only about one-sixteenth of an inch, it is estimated that this internal TE_{20} mode effect is strong enough to cause as much as a 5% error in some of the slot voltages, an effect which can easily account for the loss of about 5 dB in the side lobe level. The TE_{20} mode problem can also explain why the power into the load was higher than the predicted 12.3 percent. It measured 22.9 percent at 9.375 GHz and hit a low of 17.8% at 9.475 GHz.

9.11 Traveling Wave Arrays of Quasi-Resonant Discretely Spaced Slots (Main Beam Near Broadside)

In the design procedure described in Section 9.10, the argument was made that, for long arrays with small individual slot admittances, a natural phase progression $e^{j\beta d}$ occurs in the ratio of mode voltages at successive slots. As a consequence, if the active

[36]Private communication.

admittances are made cophasal, the slot voltages also progress as $e^{j\beta d}$ and this aperture distribution produces a pattern whose main beam points at the angle $\theta_0 = \arccos(\beta/k)$. The slot offsets in such an array are all on the *same* side of the center line.

There is no fundamental reason why the design procedure should be limited to placing the main beam at $\theta_0 = \arccos(\beta/k)$. However, if it is desired for such arrays to place the main beam at an angle $\theta_1 \neq \theta_0$, one finds that, as θ_1 departs from θ_0, the required active admittances are no longer cophasal. Here, θ_1 need not be far removed from θ_0 before the required phases of the active admittances force the needed self-admittances into the range of unreliable design data. It is therefore more staisfactory to adjust the value of β by changing the a dimension of the waveguide in order to accommodate the requirement of θ_1.

An exception occurs if one is willing to consider the situation in which successive slots are displaced in *alternate* directions from the center line of the broad wall of the waveguide. For example, in the illustration of the traveling wave array of 21 slots presented in Section 9.10, had the offsets listed in Table 9.3 been alternating in sign and had this not affected the mutual coupling, the uniform progression phase in the aperture distribution would have been $\beta d - \pi$ instead of βd. This would have resulted in a main beam pointing somewhere between broadside and *reverse* end-fire, since in that application $\beta d = 140°$. (Specifically, the main beam would have pointed 12° off broadside.) Of course, the mutual coupling *would* have changed, but this argument suggests an alternate design procedure which has practical applications.

With alternating offsets, the main beam will point naturally at an angle θ_0 given by

$$\theta_0 = \arccos \frac{\beta d - \pi}{kd} = \arccos\left(\frac{\beta}{k} - \frac{\lambda}{2d}\right) \tag{9.91}$$

One can see from (9.91) that, if $\beta d = \pi$, the main beam is at broadside and the case of a resonantly spaced array has been recaptured. If $\beta d < \pi$, the main beam points somewhere between broadside and reverse end-fire. If $\beta d > \pi$, the main beam points at an angle which lies between broadside and forward end-fire.

There are practical limitations on the choice of θ_0 and thus of the slot spacing d. If θ_0 is not chosen to be at least one beamwidth removed from broadside, the backscattered waves given in (9.89) will not sum to a negligible value and an input match will not be achieved. If d is *decreased* from the value needed to place the main beam off broadside one beamwidth toward reverse end-fire, it will not take much decrease before the slots overlap, with concomitant serious difficulties in mutual coupling. If d is *increased* from the value needed to place the main beam off broadside one beamwidth toward forward end-fire, it will not take much increase before the slots are sufficiently separated ($d = \lambda$) to introduce unwanted extra main beams. Thus the design of traveling wave arrays for which the placement of the main beam is given by (9.91) is restricted to a practical range near broadside, but excluding one beamwidth either way from broadside.

This restriction does not eliminate traveling wave arrays of this type from consideration for practical applications. Such arrays are more broadband in input impedance and pattern performance than their resonantly spaced counterparts and the low squint angle means high aperture efficiency. The determination of slot lengths and offsets proceeds exactly as outlined in Section 9.10, the only difference being that the aperture distribution is specified to have a uniform phase progression $\beta d - \pi$ rather than βd. This will automatically insure that the offsets alternate in sign. The interested reader might wish to repeat the design of a traveling wave array of 21 slots, described in Section 9.10, with the one change that the main beam is to point one beamwidth off broadside.

Traveling wave arrays of the type that satisfy the beam placement formula (9.91) can also be designed on an incremental conductance basis[37] but the procedure described in Section 9.10, which takes external mutual coupling into account on a discrete basis, will result in a more satisfactory design.

9.12 Frequency Scanned Arrays

Traveling wave arrays of quasi-resonant discretely spaced slots can be constructed in rectangular waveguide using any one of the three types of radiating slots pictured in Figure 3.9. If excitation of the slots is by means of a TE_{10} mode and if the aperture distribution is designed to give a sum pattern, then if the direction of offset (tilt) is *not* alternated, the pointing position of the main beam is given by Equation 9.85, that is,

$$\theta_0 = \arccos\left(\frac{\beta}{k}\right) \tag{9.92}$$

On the other hand, if the direction of offset (tilt) *is* alternated, the pointing position of the main beam is given by Equation 9.91 that is,

$$\theta_0 = \arccos\left(\frac{\beta}{k} - \frac{\lambda}{2d}\right) \tag{9.93}$$

with d the slot-to-slot spacing in the longitudinal direction.

Since for the TE_{10} mode $\beta/k = [1 - (\lambda/2a)^2]^{1/2}$, if the frequency is varied, the main beam position of either type of array will change. Thus frequency can be used as a variable to cause beam scanning.

To gain a feeling for the sensitivity of beam position to changes in frequency, consider again the array of 21 longitudinal shunt slots described in Section 9.10. Standard X-band waveguide was used with $a = 0.900$ inch and a central frequency $\nu_0 = 9.375$ GHz. Equation 9.92 predicted a main beam position of 44.5°, and this was confirmed by experiment. It is a simple matter to determine $\theta_0(\nu)$ for this array from (9.92). The results for a 10% bandwidth are depicted by the lower dotted curve in Figure 9.33. A $\pm 5\%$ variation in frequency has caused a $\pm 6.5\%$ swing in beam position.

[37]V. T. Norwood, "Note on a Method for Calculating Coupling Coefficients of Elements in Antenna Arrays," *IRE Trans. Antennas Propagat.*, AP-3 (1955), 213–14.

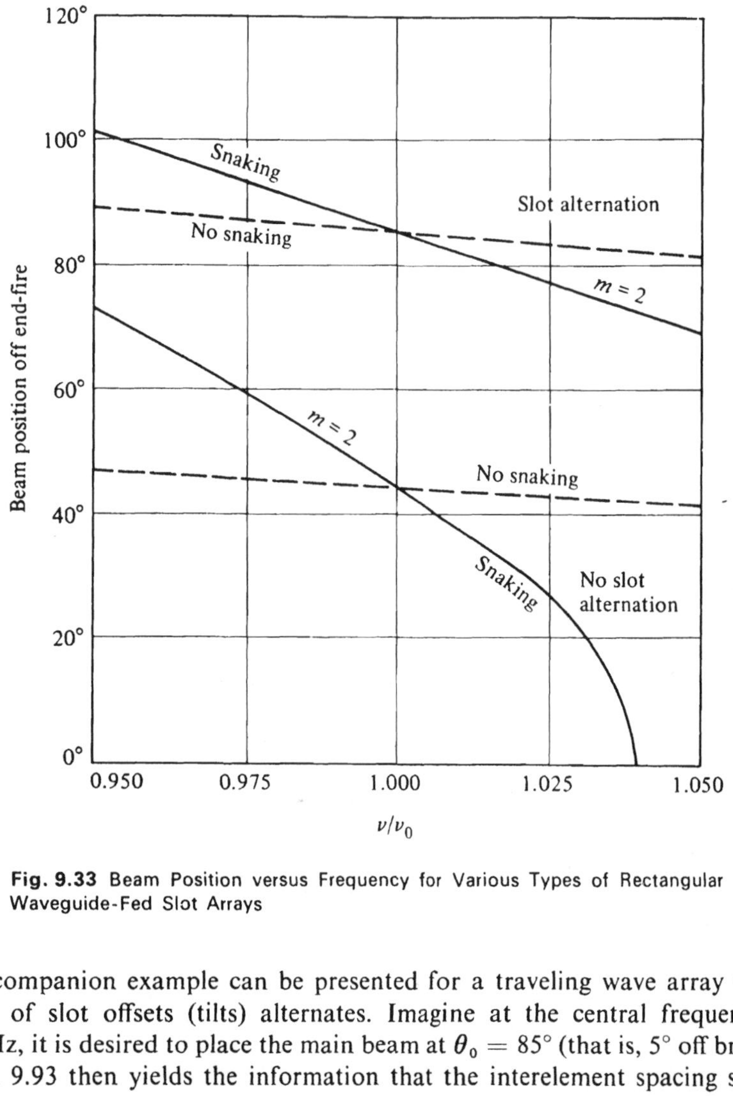

Fig. 9.33 Beam Position versus Frequency for Various Types of Rectangular Waveguide-Fed Slot Arrays

A companion example can be presented for a traveling wave array when the direction of slot offsets (tilts) alternates. Imagine at the central frequency $\nu_0 =$ 9.375 GHz, it is desired to place the main beam at $\theta_0 = 85°$ (that is, 5° off broadside). Equation 9.93 then yields the information that the interelement spacing should be $d = 1.003$ inch. If for this array the frequency is varied, further use of (9.93) produces the upper dotted curve in Figure 9.33. A $\pm 5\%$ variation in frequency has resulted in a $\pm 4.5\%$ movement in beam position.

Clearly, for both array types the amount of beam scanning that has been achieved is modest. In some applications, where the requirement is to maintain beam position constant over a sizeable frequency band, even this small beam swing can be nettling. However, this effect can be augmented and turned to advantage when the need is to scan the main beam over a considerable angular range. The augmentation occurs when the electrical length inside the waveguide from slot to slot is increased by snaking the feed, as shown for an edge slot array in Figure 9.34.

When this serpentine effect is introduced, the array factor becomes

$$\mathcal{A}(\theta) = \sum V_n^s e^{jnkd\cos\theta} = \sum |V_n^s| e^{jn(kd\cos\theta - \beta d')} \tag{9.94}$$

for an array with slots that do not alternate in direction of offset (tilt). Therefore the pointing direction of the main beam is given by

$$\theta_0 = \arccos\frac{\beta d' - 2\pi m}{kd} \tag{9.95}$$

with m an integer. The term $2\pi m$ has been subtracted from $\beta d'$ to reveal the proper uniform progressive phase in the aperture distribution. If the serpentine section is longer, m will be a larger integer. With no snaking of the feed, $m = 0$ and $d' = d$, and Equation 9.95 reduces to (9.92).

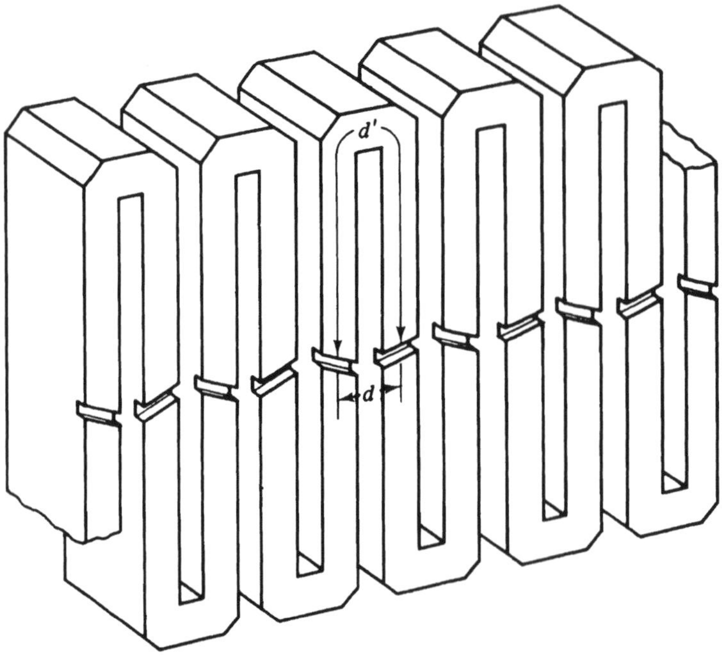

Fig. 9.34 A Serpentine Feed for an Edge Slot Array

As an illustration of the use of (9.95), suppose an edge slot array is to be cut in a snaked X-band waveguide, as shown in Figure 9.34. The transverse dimensions are standard and the main beam is once again to point at 44.5° at the central frequency $\nu_0 = 9.375$ GHz. With d/λ_0 chosen to have a value of 0.545, as in the array discussed in Section 9.10, the interelement spacing is $d = 0.685$ inch. Then for $m = 2$, Equation 9.95 yields the information that $d' = 4.207$ inch. Further use of (9.95) provides $\theta_0(\nu)$, which is plotted as the lower solid curve in Figure 9.34. Now a distinctly larger variation of beam position with frequency can be noted. Indeed, the beam is scanned all the way to end-fire with only a 4% increase in frequency.

If the slots do alternate in direction of offset (tilt), the only modification that needs to be made in the foregoing argument is to subtract an additional π radians from $\beta d'$. The formula for main beam position becomes

$$\theta_0 = \arccos \frac{\beta d' - 2\pi(m + \frac{1}{2})}{kd} \tag{9.96}$$

which is seen to reduce to (9.93) if $m = 0$ and $d' = d$.

For an edge slot array with the tilt angles alternating in direction, an interelement spacing of $d/\lambda = 0.797$ will place the main beam at 85° at $\nu_0 = 9.375$ GHz. With $m = 2$, $d' = 4.073d = 4.086$ inches. With this knowledge, $\theta_0(\nu)$ can be calculated from (9.96). The results are shown as the upper solid curve in Figure 9.34. A $\pm 5\%$ frequency variation results in a beam swing from 69° to 102°. Of course, one would not want to scan the beam through broadside because of the input VSWR difficulties, but for an array long enough to give a 1° beamwidth, a 6% frequency band will scan the beam from 69° to 89°.

These beam swings can be magnified still further by choosing larger values of m (that is, making the serpentine length greater). Obviously, there are penalties associated with doing this. The mechanical complexity increases, as do the bulk, the weight, and the cost of construction. The waveguide losses also increase. One must balance the difficulties of building a source with a greater frequency band of operation to get a wider beam scan with a given serpentine feed, against the difficulties of increasing the serpentine length to get a wider beam scan with a given frequency band of operation of the source.

Frequency scanned arrays have found favor in a variety of practical applications. One use is suggested by Figure 9.35, which shows a serpentined rectangular

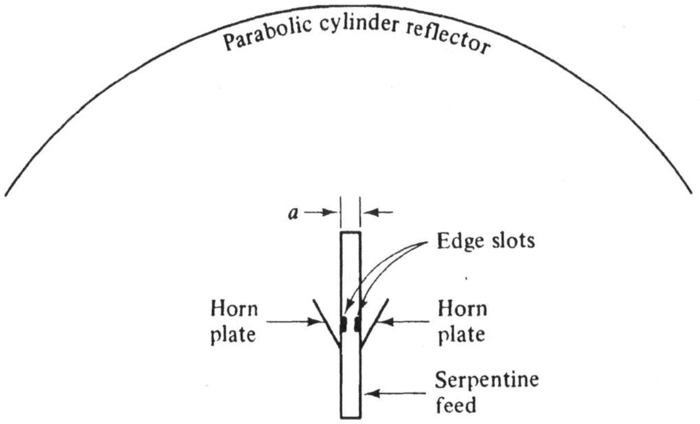

Fig. 9.35 A Frequency Scanned Array Used as the Feed for a Parabolic Cylinder Reflector

waveguide with edge slots cut in *both* narrow walls. Horn-type plates are placed over both arrays so that the beam of each is directed at a parabolic cylinder reflector. The two array patterns combine to illuminate the reflector so that its pattern is a pencil beam which scans in a plane perpendicular to the paper as the frequency is varied. Two-dimensional slot arrays have also been constructed, with frequency scanning in one dimension resulting from use of serpentine feeds.

REFERENCES

BARLOW, H. M. and J. BROWN, *Radio Surface Waves* (Oxford: Clarendon Press, 1962).

COLLIN, R. E., *Field Theory of Guided Waves* (New York: McGraw-Hill Book Co., Inc., 1960).

HARPER, A. E., *Rhombic Antenna Design* (New York: D. Van Nostrand, 1941).

KRAUS, J. D., *Antennas* (New York: McGraw-Hill Book Co., Inc., 1950), pp. 407–413.

LAPORT, E. A., "Long Wire Antennas," *Antenna Engineering Handbook*, ed. H. Jasik (New York: McGraw-Hill Book Co., Inc., 1961), Chapter 4.

WALTER, C. H., *Traveling Wave Antennas* (New York: McGraw-Hill Book., Inc., 1965).

WOLFF, E. A., *Antenna Analysis* (New York: John Wiley and Sons, Inc., 1966), Chapter 8.

PROBLEMS

9.1 Verify the results contained in Equations 9.10 through 9.12.

9.2 Derive the field expressions for a rhombic antenna using pattern multiplication, with the element patterns those of a single long wire.

9.3 Provide the optimum design for a rhombic antenna which is to be parallel to and a distance h above a conducting earth, if the main beam of the pattern is to point 10° above the horizon.

9.4 Determine $\mathcal{A}_\phi(\theta, \phi)$ for a V-antenna and then plot $|\mathcal{A}_\phi(\theta, 0°)|$ for $L/\lambda = 5$, $h/\lambda = 1.5$, and $\phi_0 = 20°$.

9.5 For a parallel plate transmission line, with the bottom wall corrugated, establish from first principles that the fields in the region above the corrugations are given by (9.46) when $G \gg T$ and $G \ll \lambda$.

9.6 Assume that the feed pattern for the corrugated surface antenna of Figure 9.12a is given by Equation 9.52. Show that the ratio of feed radiation at $\theta = 90°$ (end-fire) to feed radiation at $\theta = 0°$ is $20 \log_{10}[(\beta/k)/(\beta/k - 1)]$. With the Hansen-Woodyard relation of (9.53) invoked and (9.54) accepted as valid, show that the end-fire radiation attributable to the corrugated surface is 6 dB above that due to the feed.

9.7 Establish the equations from which the complex propagation constant can be deduced for TE azimuthal waves on dielectric-clad circular cylinders.

9.8 Repeat Problem 9.7 for TM azimuthal waves. Indicate the change in your analysis if the cylinder is corrugated instead of being dielectric-clad.

9.9 Establish the equation from which the complex propagation constant can be determined for *TM* latitudinal waves on a corrugated sphere.

9.10 A continuous line source with a uniform progressive phase and a saw-tooth amplitude distribution of the form $A(\zeta) = 1 - (\zeta/L)$ will produce a pattern which is approximately $\csc^2\theta$ in form. Show this, and then use Equations 9.58 and 9.64 to find $\alpha(z)$ if $f = 10\%$. Then use Kelly's experimental data, contained in Figure 9.21, to design an *X*-band serrated waveguide antenna 7.5λ long for this application. (Compare with Kelly[26] for an experimental validation.)

9.11 A collapsed continuous line-source distribution is desired in the form $A(\zeta) = [2 + \sin(\pi\zeta/L)]e^{-j\beta\zeta}$. Use the theoretical curves of Rotman and Oliner, contained in Figure 9.28, to design an asymmetrical trough waveguide antenna 10λ long to produce this aperture distribution. At first, keep d_1 constant at 0.342λ and find $d(z)$. Then modify your design by also varying d_1 to keep β constant.

9.12 Design a five-element longitudinal shunt slot array in standard *X*-band waveguide with $\nu = 9.375$ GHz in order to have a constant amplitude aperture distribution and a pattern with the main beam at $\theta_0 = 45°$. There should be an input match to the array.

9.13 Repeat Problem 9.12 with the main beam one beamwidth off broadside.

10 reflectors and lenses

10.1 Introduction

A reflector or a lens that is large in wavelengths and fed by a small horn or other elementary source is a relatively simple, inexpensive antenna capable of producing a high-gain pattern with a reasonable side lobe level. Pattern integrity and input impedance can be maintained at a good level over a significant frequency band. Lenses are typically heavier than reflectors, but have inherently less severe aperture blockage problems due to feed placement. Both antenna types are natural candidates for many ground-based, airborne, and ship-based applications, and unfurlable reflectors have become attractive for satellite use. Beam scanning by mechanical rotation of the lens or reflector is common when inertial effects do not prevent practical achievement of the desired rate of scan. In some applications, the inertial effects are eliminated (or alleviated) by keeping the lens or reflector stationary but using multiple feeds (or movable feeds) to produce beam scanning. The result is that both antenna types, and particularly reflectors, are attractive for use in a diverse catalog of situations requiring apertures whose dimensions are large in terms of wavelengths.

The large size of these antennas in terms of wavelengths permits effective use of the principles of optics in their design. Thus this chapter begins with a discussion of the simplifications of solutions to Maxwell's equations that occur when the wavelength becomes small. The geometrical optics field, as the limiting solution is called, has many useful properties which are demonstrated and then employed to deduce several simple reflector shapes that will collimate the rays of the field. These include the parabolic cylinder and the paraboloid. The compound feed that bears Cassegrain's name, consisting of a point source and a hyperboloid, is also analyzed.

Aperture blockage of a reflector is then discussed in general terms, with tradeoffs indicated between feed size and reflector depth and between feed offset and reflector depth. This is followed by the development of a technique for shaping a cylindrical reflector in order to provide an arbitrary secondary pattern, rather than one arising from collimation of all secondary rays. The same procedure is also demonstrated for a doubly curved reflector which is often, in essence, a perturbation of the basic paraboloid, just as the shaped cylindrical reflector is often a perturbation of the basic parabolic cylinder.

Several methods for computing the far field of reflector antennas are introduced. In one approach, the geometric optics field is used to determine the values of equivalent sources in a secondary aperture with and without an allowance for aperture blockage. In another, the geometric optics field is used to deduce an approximation to the current distribution on the reflector surface. Edge effect limitations of both these methods are noted. The treatment of reflector antennas closes with a discussion of Cassegrain-type dual reflector systems that have been modified to improve the aperture efficiency.

The treatment of reflector antennas in this chapter is introductory in nature. The interested reader can gain an overview of the historical development of this antenna type by reading the prefatory article by A. W. Love[1] in a collection of significant papers that he has assembled on the subject. The remainder of the collection and the exhaustive bibliography give a clear indication of the problem areas and advances that have been made in the design of reflector antennas.

The attention given to lens antennas in this chapter is more brief. First to be considered are lenses composed of homogeneous, isotropic dielectric materials. Several conventional shapes (for example, elliptical, hyperbolic) which can transform the diverging rays of a primary source to a collimated secondary field are analyzed. The technique of stepping to save weight is described. Artificial dielectrics are introduced and various design approaches for metal plate lenses are indicated. The chapter concludes with a discussion of the Luneburg lens and its applications.

10.2 Geometrical Optics: The Eikonal Equation[2]

The reflector and lens antennas to be discussed in this chapter, because they are taken to be very large compared to the free-space wavelength λ_0, can be designed with considerable success by invoking the assumptions of geometrical optics. Basically, this involves neglect of the wavelength since, if $\lambda_0 \longrightarrow 0$, solutions to Maxwell's equations can be formulated in geometrical terms. One finds that the energy being transported by the electromagnetic field can then be viewed as traveling along a family of rays. These rays can be traced through dielectric media or while undergoing reflection at metallic boundaries. In this manner, the conversion from the feed radiation impinging on a reflector or lens to the radiated field emerging from the antenna can be understood and controlled.

[1]A. W. Love, ed., *Reflector Antennas* (New York: IEEE Press, 1978), pp. 2–15.

[2]The development in this section follows an excellent exposition by M. Born and E. Wolf, *Principles of Optics* (New York: Pergamon Press, 1959), pp. 109–32. Another useful source has been W. V. T. Rusch and P. D. Potter, *Analysis of Reflector Antennas* (New York: Academic Press, 1970), Chapter 2.

The geometrical optics field is a solution to Maxwell's equations for which each field component has the same family of equiphase surfaces. This implies that, in a nonconducting isotropic medium that is not necessarily homogeneous, **E** and **H** can be expressed in the common form

$$\mathbf{E}(x, y, z, t) = \mathbf{E}_0(x, y, z)e^{j\omega t - jk\psi(x,y,z)} \tag{10.1}$$

$$\mathbf{H}(x, y, z, t) = \mathbf{H}_0(x, y, z)e^{j\omega t - jk\psi(x,y,z)} \tag{10.2}$$

where $k = \omega\sqrt{\mu_0\epsilon_0} = 2\pi/\lambda_0$ is the free-space wave number. The pure real function $\psi(x, y, z)$, when set equal to a constant, defines an equiphase surface. The vector functions $(\mathbf{E}_0, \mathbf{H}_0)$ are generally complex.

In a source-free region, the above fields satisfy Maxwell's equations in the form

$$\begin{aligned}
\nabla \times \mathbf{E} &= e^{-jk\psi}\,\nabla \times \mathbf{E}_0 - jke^{-jk\psi}\,\nabla\psi \times \mathbf{E}_0 = -j\omega\mu\mathbf{H}_0e^{-jk\psi} \\
\nabla \times \mathbf{H} &= e^{-jk\psi}\,\nabla \times \mathbf{H}_0 - jke^{-jk\psi}\,\nabla\psi \times \mathbf{H}_0 = j\omega\epsilon\mathbf{E}_0e^{-jk\psi} \\
\nabla \cdot \epsilon\mathbf{E} &= e^{-jk\psi}\,\nabla \cdot \epsilon\mathbf{E}_0 - jke^{-jk\psi}\,\nabla\psi \cdot \epsilon\mathbf{E}_0 \equiv 0 \\
\nabla \cdot \mu\mathbf{H} &= e^{-jk\psi}\,\nabla \cdot \mu\mathbf{H}_0 - jke^{-jk\psi}\,\nabla\psi \cdot \mu\mathbf{H}_0 \equiv 0
\end{aligned} \tag{10.3}$$

where the time factor $e^{j\omega t}$ has been suppressed. Rearrangement gives

$$\begin{aligned}
\nabla\psi \times \mathbf{E}_0 - \frac{\omega\mu}{k}\mathbf{H}_0 &= \frac{1}{jk}\nabla \times \mathbf{E}_0 \\
\nabla\psi \times \mathbf{H}_0 + \frac{\omega\epsilon}{k}\mathbf{E}_0 &= \frac{1}{jk}\nabla \times \mathbf{H}_0 \\
\nabla\psi \cdot \epsilon\mathbf{E}_0 &= \frac{1}{jk}\nabla \cdot \epsilon\mathbf{E}_0 \\
\nabla\psi \cdot \mu\mathbf{H}_0 &= \frac{1}{jk}\nabla \cdot \mu\mathbf{H}_0
\end{aligned} \tag{10.4}$$

The interest here is in solutions to (10.4) for $k = 2\pi/\lambda_0$ very large. If the spatial derivatives multiplying $1/jk$ on the right sides of these equations are not great, the terms on the right can be neglected, permitting the reduction

$$\begin{aligned}
\nabla\psi \times \mathbf{E}_0 - \frac{\omega\mu}{k}\mathbf{H}_0 &= 0 \\
\nabla\psi \times \mathbf{H}_0 + \frac{\omega\epsilon}{k}\mathbf{E}_0 &= 0 \\
\mathbf{E}_0 \cdot \nabla\psi &= 0 \\
\mathbf{H}_0 \cdot \nabla\psi &= 0
\end{aligned} \tag{10.5}$$

Because the last two of Equations (10.5) can be obtained merely by dotting $\nabla\psi$ into the first two, the essence of this result is equations (10.5a,b). If (10.5a) is solved for $\mathbf{H}_0$ and the result placed in (10.5b), one obtains

$$\nabla\psi \times (\nabla\psi \times \mathbf{E}_0) + n^2\mathbf{E}_0 = 0 \tag{10.6}$$

in which

$$n = \sqrt{\frac{\mu(x, y, z)\epsilon(x, y, z)}{\mu_0\epsilon_0}} \tag{10.7}$$

is the local value of the refractive index. Equation 10.6 can be expanded to the form

$$(\mathbf{E}_0 \cdot \nabla\psi)\nabla\psi - |\nabla\psi|^2\mathbf{E}_0 + n^2\mathbf{E}_0 = 0 \tag{10.8}$$

The first term in (10.8) is zero by virtue of (10.5c). Since $\mathbf{E}_0$ is not identically zero in any nontrivial solution, it follows that

$$|\nabla\psi|^2 = \left(\frac{\partial\psi}{\partial x}\right)^2 + \left(\frac{\partial\psi}{\partial y}\right)^2 + \left(\frac{\partial\psi}{\partial z}\right)^2 = n^2(x, y, z) \tag{10.9}$$

This result (which could have been obtained equally well by eliminating $\mathbf{E}_0$ from 10.5a,b) is basic to geometrical optics and (10.9) is known as the *eikonal equation*; ψ itself is often referred to as the *eikonal function*. The equiphase surfaces or wavefronts with ψ a constant must everywhere adopt a shape that satisfies the point relation in (10.9).

Many other properties of a geometric field may be deduced from (10.5). Principal among these are the following.

1. It is apparent from (10.5a,b) that $\mathbf{E}_0$ and $\mathbf{H}_0$ are perpendicular to each other and to $\nabla\psi$. Thus they both lie in the equiphase surface with ψ a constant and the geometrical optics field is everywhere *TEM*.

2. The time-average electric and magnetic energy densities are given by

$$W_e = \frac{1}{4}\epsilon\mathbf{E} \cdot \mathbf{E}^* = \frac{1}{4}\epsilon\mathbf{E}_0 \cdot \mathbf{E}_0^* = \frac{1}{4c}\mathbf{E}_0 \cdot (\mathbf{H}_0^* \times \nabla\psi) \tag{10.10}$$

$$W_m = \frac{1}{4}\mu^{-1}\mathbf{B} \cdot \mathbf{B}^* = \frac{1}{4}\mu_0\mathbf{H}_0 \cdot \mathbf{H}_0^* = \frac{1}{4c}\mathbf{H}_0^* \cdot (\nabla\psi \times \mathbf{E}_0) \tag{10.11}$$

Therefore $W_e = W_m$ and each is equal to half of the total time-average stored energy density W_{total}.

3. The time-average power flow density is

$$\mathcal{P} = \frac{1}{2}\mathcal{R}e\mathbf{E} \times \mathbf{H}^* = \frac{1}{2}\mathcal{R}e\mathbf{E}_0 \times \mathbf{H}_0^* = \frac{k}{2\omega\mu}\mathcal{R}e\mathbf{E}_0 \times (\nabla\psi \times \mathbf{E}_0^*)$$

$$= \frac{k}{2\omega\mu}(\mathbf{E}_0 \cdot \mathbf{E}_0^*)\nabla\psi = \frac{1}{2n^2}\mathbf{E}_0 \cdot (\mathbf{H}_0^* \times \nabla\psi)\nabla\psi = \frac{2c}{n^2}W_e\nabla\psi$$

If a unit vector $\mathbf{1}_s$ in the direction of $\nabla\psi$ is defined by

$$\mathbf{1}_s = \frac{\nabla\psi}{|\nabla\psi|} = \frac{\nabla\psi}{n} \tag{10.12}$$

this last result can be put in the form

$$\mathcal{P} = \mathbf{1}_s v W_{total} \tag{10.13}$$

in which $v = c/n$ is the local propagation velocity of the wave and $W_{total} = 2W_e = 2W_m$ can be found from either (10.10) or (10.11). In words, Equation 10.13 says that (a) the time-average Poynting vector is locally directed parallel to the normal to the wavefront and (b) its magnitude is equal to the product of the time-average energy density and the local propagation velocity of the wave.

4. The geometrical rays may be defined as a family of curves which are everywhere normal to the geometrical wavefronts $\psi =$ constant. If $\mathbf{r}(s)$ is the position vector of a point on a ray, treated as a function of the distance s along the ray, then $\mathbf{dr}/ds = \mathbf{1}_s$ and the equation of a ray can be written as

$$n\frac{d\mathbf{r}}{ds} = \nabla\psi \tag{10.14}$$

where use has been made of (10.12). Differentiation of (10.14) with respect to s and some manipulation gives

$$\begin{aligned} \frac{d}{ds}n\left(\frac{d\mathbf{r}}{ds}\right) &= \frac{d}{ds}(\nabla\psi) = \left(\frac{d\mathbf{r}}{ds}\cdot\nabla\right)\nabla\psi \\ &= \frac{1}{n}(\nabla\psi\cdot\nabla)\,\nabla\psi = \frac{1}{2n}\nabla|\nabla\psi|^2 \\ &= \frac{1}{2n}\nabla n^2 \end{aligned}$$

as a consequence of which

$$\frac{d}{ds}\left(n\frac{d\mathbf{r}}{ds}\right) = \nabla n \tag{10.15}$$

Equation 10.15 is a particularly useful form in which to cast the differential equation of a ray since it involves only the refractive index.

Some of the characteristics of rays can be revealed by a general consideration of (10.15). Expansion gives

$$\frac{d}{ds}\left(n\frac{d\mathbf{r}}{ds}\right) = \frac{d}{ds}(\mathbf{1}_s n) = n\frac{d}{ds}(\mathbf{1}_s) + \mathbf{1}_s\frac{dn}{ds} = \nabla n \tag{10.16}$$

The vector

$$\mathbf{M} = \frac{d}{ds}(\mathbf{1}_s) \tag{10.17}$$

in (10.16) has a physical interpretation which can be understood with reference to Figure 10.1a. A curved ray *OPQR* (which does not necessarily lie in a plane) is shown,

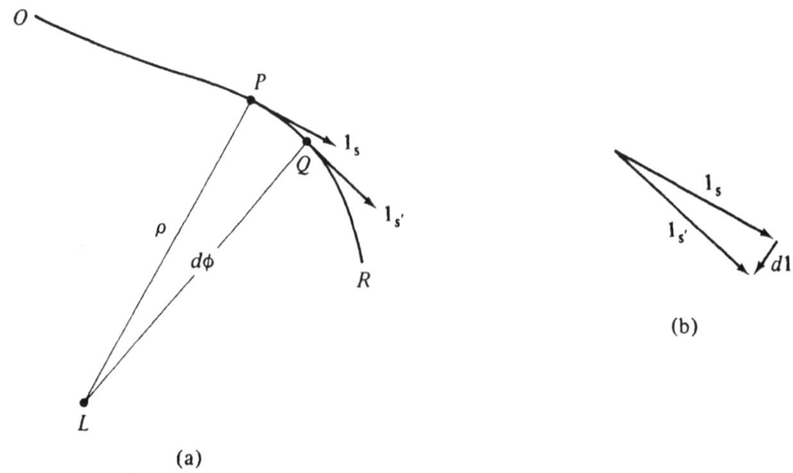

Fig. 10.1 The Geometry of a Curved Ray

with unit tangent vectors indicated at points P and Q. Imagine that planes perpendicular to $\mathbf{1}_s$ and $\mathbf{1}_{s'}$ are erected at P and Q, respectively. These planes intersect in a line LL' (not shown). The intersection of that line with the plane containing $\mathbf{1}_s$ and Q defines the point L. The distance $\rho = \overline{LP}$ is called the *principal radius of curvature* of the curve $OPQR$ at the point P.

From Figure 10.1b it can be seen that

$$d\mathbf{1}_s = \mathbf{1}_{s'} - \mathbf{1}_s = \mathbf{1}_\rho \, d\phi$$

in which $d\phi$ is the angle subtended at L by the curve segment $ds = \overline{PQ}$ and $\mathbf{1}_\rho$ is a unit vector in the direction from P to L. It follows that

$$\mathbf{M} = \frac{d(\mathbf{1}_s)}{ds} = \frac{\mathbf{1}_\rho \, d\phi}{\rho \, d\phi} = \frac{\mathbf{1}_\rho}{\rho} \tag{10.18}$$

From (10.16) and (10.18),

$$n\mathbf{M} = \nabla n - \mathbf{1}_s \frac{dn}{ds} \tag{10.19}$$

If the scalar product is formed of (10.19) and $\mathbf{1}_\rho$, the result is that

$$|\mathbf{M}| = \frac{1}{\rho} = \frac{1}{n}\mathbf{1}_\rho \cdot \nabla n = \mathbf{1}_\rho \cdot \nabla(\ln n) \tag{10.20}$$

This equation provides much useful information about the nature of rays. First, in a region in which n is a constant, $\nabla(\ln n) \equiv 0$ and $\rho \to \infty$. Hence in a homogeneous medium, all rays are *straight lines*.

Second, if $n \neq$ constant, $\nabla(\ln n) \not\equiv 0$, and ρ is finite, that is, the rays are *curved.* Further, since ρ is positive real, *the rays always bend toward the region of higher refractive index.*

5. If a sheath of rays is used as the boundary of a "tube," as shown in Figure 10.2, the result (10.13) can be used to argue that the time-average power flow must be longitudinally through the tube (no power crosses the tube walls), and thus that $\mathcal{P}_1 dS_1 = \mathcal{P}_2 dS_2$, with $\mathcal{P}_1$ the entry power density and $\mathcal{P}_2$ the exit power density, and with dS_1 and dS_2 the transverse cross-sectional areas at the two ends of the tube. More generally, $\mathcal{P} dS$ is seen to remain constant along any tube of rays. This is known as the *intensity law of geometrical optics.*

When the medium is homogeneous, so that the rays are straight lines, the intensity law can be expressed in a different form. With reference to Figure 10.3, consider a tube of rays with transverse cross sections that are rectangular. Projections

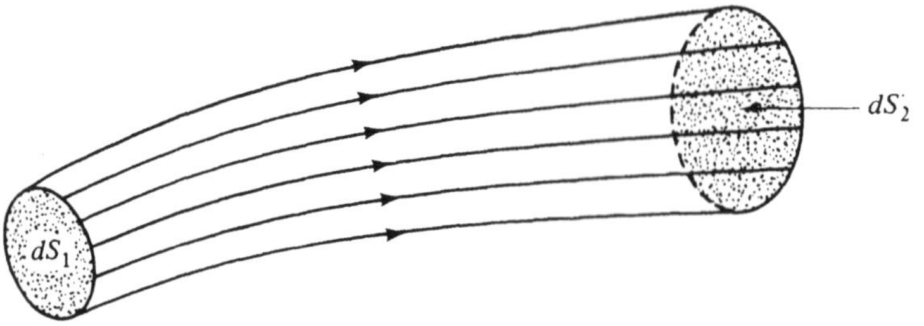

Fig. 10.2 A Tube of Geometrical Rays

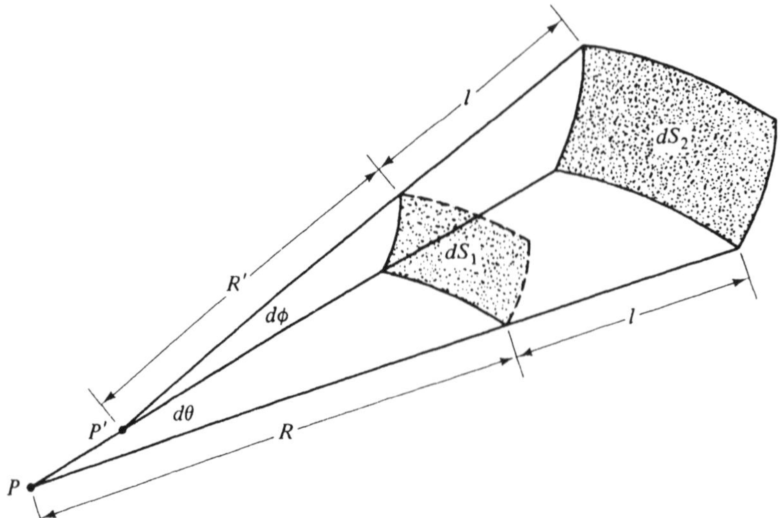

Fig. 10.3 The Intensity Variation of Rectilinear Rays

back toward the source will locate apparent ray centers P and P' for pairs of sides of the tube. It is evident that

$$dS_1 = RR'\,d\theta\,d\phi \quad \text{and} \quad dS_2 = (R + l)(R' + l)\,d\theta\,d\phi$$

and hence

$$\frac{\mathcal{P}_2}{\mathcal{P}_1} = \frac{dS_1}{dS_2} = \frac{RR'}{(R + l)(R' + l)} \tag{10.21}$$

If $l \gg R, R'$, the inverse square law results:

$$\frac{\mathcal{P}_2}{\mathcal{P}_1} = \frac{RR'}{l^2} \tag{10.22}$$

6. The phase delay along a ray from a point P_1 to a point P_2 is given by the integral

$$\Phi = \omega T = \omega \int dt = \omega \int_{P_1}^{P_2} \frac{ds}{v} = \frac{\omega}{c} \int_{P_1}^{P_2} n\,ds = k \int_{P_1}^{P_2} n\,ds \tag{10.23}$$

and for this reason the integral $\int n\,ds$ is often referred to as the optical length of the ray. It follows from (10.12) that

$$\mathbf{1}_s n\,ds = \mathbf{1}_s |\nabla\psi|\,ds = \mathbf{1}_s\,d\psi \tag{10.24}$$

as a consequence of which (10.23) becomes

$$\Phi = k[\psi(P_2) - \psi(P_1)] \tag{10.25}$$

This is an extremely useful result. It states that the phase delay for a geometrical optics field along all rays connecting any two equiphase surfaces is the same.

7. If a uniform plane wave of any given polarization is incident at arbitrary angle i on an infinite perfectly conducting plane reflector, a straightforward application of Maxwell's equations leads to the conclusion that the angle of reflection r equals the angle of incidence.[3] This result is often called Snell's law of reflection. If the reflector is curved, but its radii of curvature are very large compared to the wavelength, it is still an excellent approximation to assume $i = r$. Indeed, if the radii of curvature are locally finite at all points on a reflector, the statement that $i = r$ when $\lambda_0 \longrightarrow 0$ is precisely correct at all points on the reflector except along the boundary. A differential application of Maxwell's equations, in a development that is completely analogous to what is done for the case of a uniform plane wave incident on an infinite plane reflector, serves to establish this. It is as though locally the curved reflector behaves like an infinite plane reflector.

[3]See, for example, S. Ramo, J. R. Whinnery, and T. Van Duzer, *Fields and Waves in Communication Electronics* (New York: John Wiley and Sons, Inc., 1965), pp. 352–55.

Similarly, if a uniform plane wave of arbitrary polarization is incident at arbitrary angle i on an infinite plane interface between two homogeneous, isotropic, lossless dielectric media, the matching of solutions of Maxwell's equations at the boundary leads to Snell's law of refraction,[4] that is, $n_1 \sin i = n_2 \sin r$. In this relation, n_1 and n_2 are the indices of refraction of the medium containing the incident ray and the refracted ray, respectively. If the interface is curved but its radii of curvature are very large compared to a wavelength, it is still an excellent approximation to assume that locally $n_1 \sin i = n_2 \sin r$. These conclusions are also quite valid for low-loss dielectrics.

10.3 Simple Reflectors

Several of the properties of a geometrical optics field, developed in the preceding section, can be used to determine the shapes of some simple but practical and widely used reflector antennas.

1. PARABOLIC CYLINDER Imagine that an equiphase long line source is placed parallel to the Z-axis, as shown in Figure 10.4, and used to illuminate a cylindrical

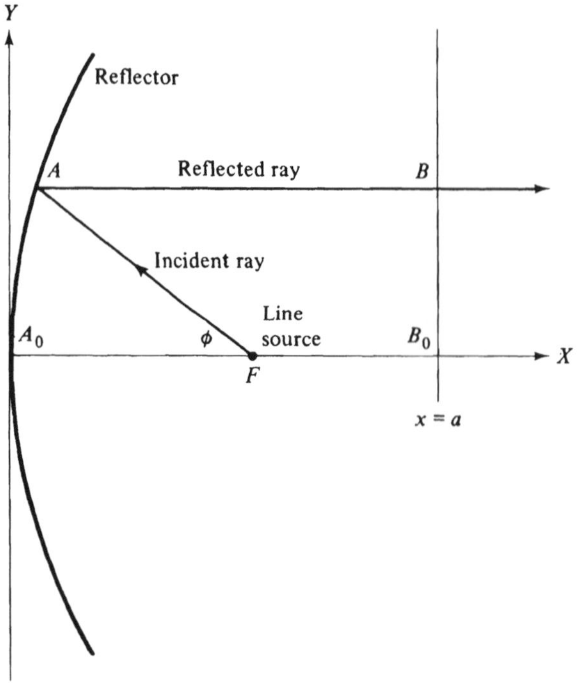

Fig. 10.4 A Parabolic Cylinder Reflector with a Line-Source Feed

[4]Ibid., pp. 358–61.

reflector. With an air medium assumed, the rays from the line source are rectilinear, and one of these rays can be designated as leaving the line source at an angle ϕ. If this and all other rays are to be reflected so that they are collimated into a horizontal bundle, the phase fronts normal to these secondary rays are planes perpendicular to the X-axis, such as the plane $x = a$. Equivalent Huyghens sources, placed in the plane $x = a$, could be used to determine the far field. To the extent that the assumptions of geometrical optics are valid, these Huyghens sources are equiphase; their amplitude distribution depends on the primary radiation pattern of the line source and the effects of reflection. A typical secondary source distribution in the plane $x = a$ can result in a pencil beam pattern with low side lobes.

For a long line source and reflector, the field behavior in a plane parallel to $z = 0$ and not too near either end of the antenna can be deduced by assuming that the structure is infinitely long in the Z-direction. Applying (10.25), one can argue that the phase delay from F to A to B must be the same as the phase delay from F to A_0 to B_0. Since the surrounding medium (air) is homogeneous, this means that the total lengths of the two rays should be equal, that is,

$$\overline{FA} + \overline{AB} = \overline{FA_0} + \overline{A_0B_0} \tag{10.26}$$

If the coordinates of point A are $(x, y, 0)$ and those of point F are $(f, 0, 0)$, Equation 10.26 translates into

$$\sqrt{(x - f)^2 + y^2} + (a - x) = f + a$$

which simplifies to

$$y^2 = 4fx \tag{10.27}$$

Hence the shape that the reflector should have to collimate the rays is that of a parabolic cylinder, with the line source placed at the focus.

The foregoing derivation in effect *assumes* that locally the law of reflection is operative and will alter all ray directions as desired. It is a simple matter to check if this is so. Differentiation of (10.27) yields $2y\,dy = 4f\,dx$ and hence a unit tangent vector to the parabola is given by

$$\mathbf{1}_T = \frac{\mathbf{1}_x\,dx + \mathbf{1}_y\,dy}{\sqrt{(dx)^2 + (dy)^2}} = \frac{\mathbf{1}_x + \mathbf{1}_y(2f/y)}{\sqrt{1 + (2f/y)^2}} \tag{10.28}$$

This vector is shown in Figure 10.5 together with the unit normal vector, obtained by the operation

$$\mathbf{1}_N = \mathbf{1}_T \times \mathbf{1}_z = \frac{-\mathbf{1}_y + \mathbf{1}_x(2f/y)}{\sqrt{1 + (2f/y)^2}} \tag{10.29}$$

The cosine of the angle of incidence is given by

$$\cos i = \mathbf{1}_N \cdot \mathbf{1}_R = \mathbf{1}_N \cdot \frac{\mathbf{1}_x(f - x) - \mathbf{1}_y y}{\sqrt{(f - x)^2 + y^2}} \tag{10.30}$$

and the cosine of the angle of reflection by

$$\cos r = \mathbf{1}_N \cdot \mathbf{1}_x = \frac{2f/y}{\sqrt{1 + (2f/y)^2}} \tag{10.31}$$

Expansion of (10.30) and use of (10.27) quickly reveals that $\cos i = \cos r$ and that the law of reflection is satisfied by the solution $y^2 = 4fx$.

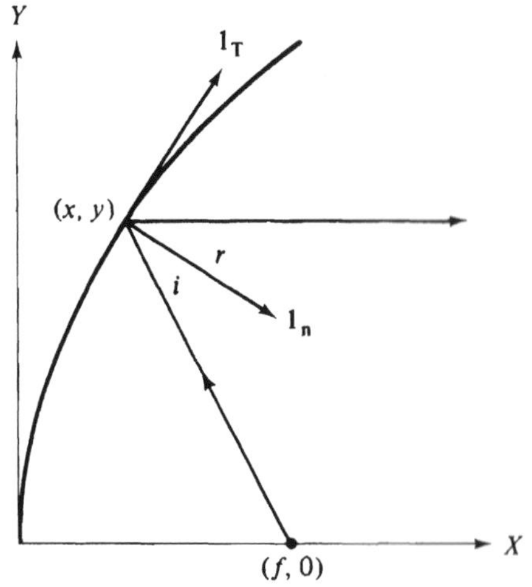

Fig. 10.5 Ray/Surface Relations for a Parabolic Cylinder Reflector

2. PARABOLOID If a *point* source is used to illuminate a reflector, and the requirement is once again to collimate the secondary rays, the analysis is similar to that given above. With reference to Figure 10.6, let the point A have coordinates (x, y, z) and the point F have coordinates $(f, 0, 0)$. Then the requirement that $\overline{FA} + \overline{AB} = \overline{FA_0} + \overline{A_0B_0}$ becomes

$$\sqrt{(x-f)^2 + y^2 + z^2} + (a - x) = f + a$$

This reduces to

$$y^2 + z^2 = 4fx \tag{10.32}$$

which is the equation of a paraboloid. The point source is at the focus.

As with the case of the parabolic cylinder, it is possible to show that primary and secondary rays make equal angles with the normal at any nonedge point on the paraboloid. The proof of this assertion is left as an exercise.

3. HYPERBOLOIDAL REFLECTOR A useful compound feed, first introduced by Cassegrain as part of an optical telescope, is pictured in Figure 10.7. It consists of a point source at G which is assumed to cause a primary geometrical optics field with spherical wave fronts and an associated system of rays which are radial about

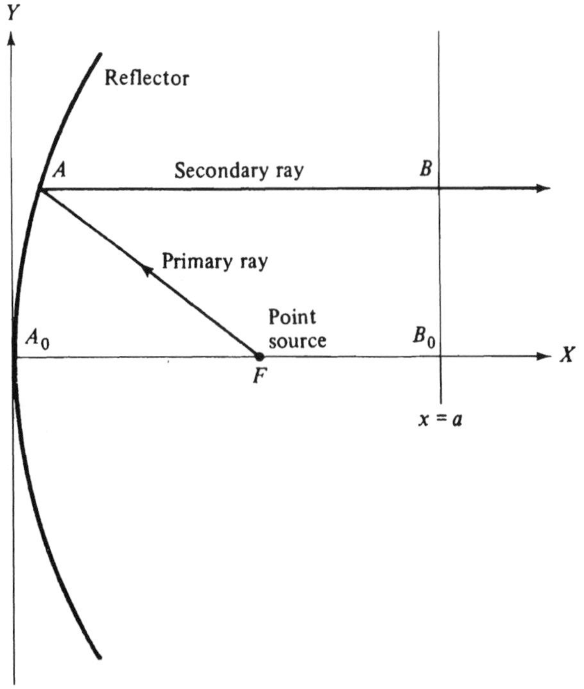

Fig. 10.6 A Paraboloidal Reflector with a Point-Source Feed

G. These rays are intercepted by the reflector and converted to a system of secondary rays that appear to be emanating from the point F. Thus the secondary wavefronts are also spherical. The question to be answered is: What shape must the reflector have to bring this about?

Equating ray lengths, one obtains

$$\overline{GA} + \overline{AB} = \overline{GA_0} + \overline{A_0B_0} = \overline{GO} + \overline{OA_0} + \overline{A_0B_0} \tag{10.33}$$

With the origin taken midway between G and F, $\overline{GO} = \overline{OF}$. And because the reflected waves are to be spherical,

$$\overline{AB} = \overline{FB} - \overline{FA} = \overline{FB_0} - \overline{FA_0}$$

When these substitutions are made in (10.33), rearrangement gives

$$\overline{GA} - \overline{FA} = 2\overline{OA_0} \tag{10.34}$$

This is the equation of a hyperboloid with foci at G and F and eccentricity $e = \overline{OF}/\overline{OA_0}$. This can perhaps be appreciated more readily if the end points of the line segments appearing in (10.34) are assigned the coordinates

$$G(-c, 0, 0) \qquad F(c, 0, 0) \qquad A_0(a, 0, 0) \qquad A(x, y, z)$$

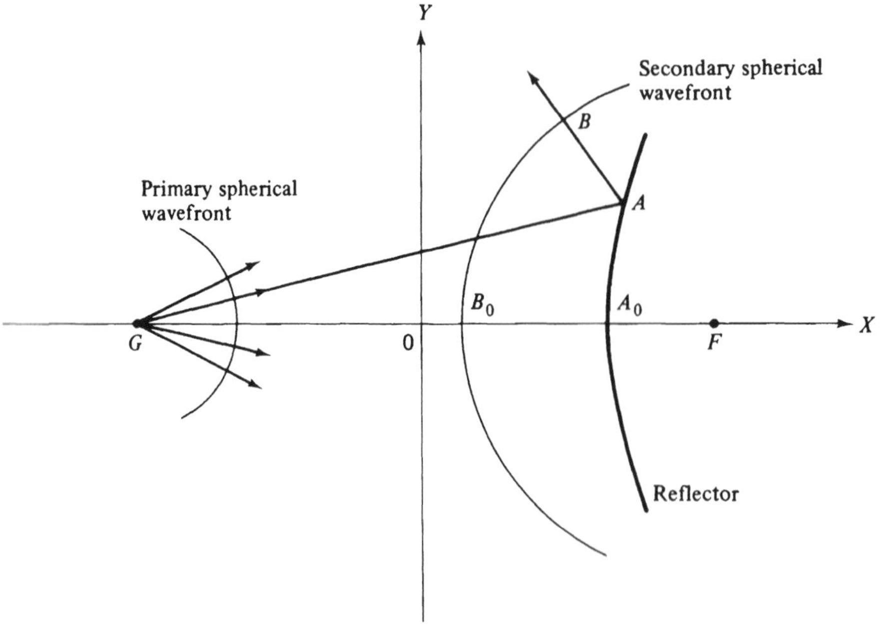

Fig. 10.7 A Cassegrain Feed

which permits (10.34) to be cast in the equivalent form

$$\sqrt{(x+c)^2+y^2+z^2}-\sqrt{(x-c)^2+y^2+z^2}=2a \tag{10.35}$$

Manipulation gives

$$x^2-\frac{y^2+z^2}{e^2-1}=a^2 \tag{10.36}$$

which is a canonical expression for the equation of a hyperboloid.

A Cassegrain feed can be used as shown in Figure 10.8. It has the advantage of permitting the "point source" to be located immediately behind a small opening in the large paraboloidal reflector. This is important in applications that require use of a low-noise transmitter/receiver. A Cassegrain feed also has the advantage of permitting a short focal length paraboloidal reflector to replace one of the same aperture size but much longer focal length, as can be seen from Figure 10.8. The longer focal length equivalent paraboloid shown in Figure 10.8 is much flatter than the actual paraboloid to the left. Since flatter paraboloids have much better defocusing properties (less pattern degradation when a feed is placed away from the focus to produce an extra or scanned beam), by inference a Cassegrain dual reflector system has this desirable feature also.

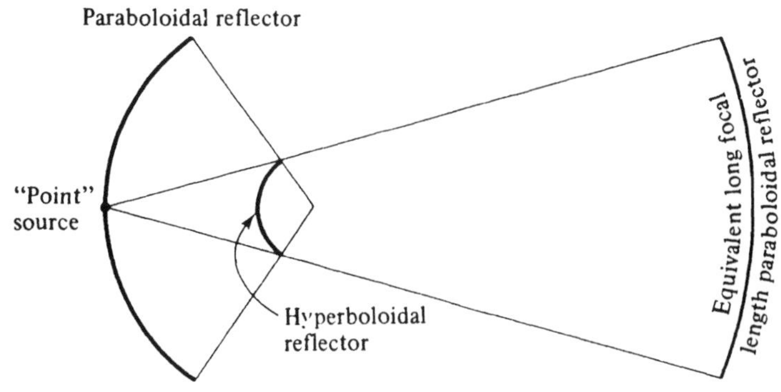

Fig. 10.8 A Cassegrain-Fed Paraboloid and Its Long Focal Length Equivalent

The price paid for these advantages is that the aperture blockage (see Section 10.4) of the hyperboloid reflector is patently greater than that of the primary-focus point source it replaces. For this reason, Cassegrain feeds are usually limited to applications in which the span of the paraboloidal reflector is very large (for instance, greater than 40λ).

10.4 Aperture Blockage

The "line" source of Figure 10.4 and the "point" source of Figure 10.6 in reality have a transverse extent and thus intercept some of the secondary rays, as suggested by Figure 10.9. This causes a perturbation in the secondary aperture distribution (for example in the plane $x = a$) with the consequence that the secondary radiation pattern is affected. An estimate of this effect will be made in Section 10.7. However, there are certain general relations involved in aperture blockage which can be introduced now and which reveal tradeoffs that are available in reflector design.

Consider the parabolic cylinder reflector and line source feed shown in Figure 10.10a. The horn plates must be dimensioned so that the primary aperture height b results in a primary radiation pattern which properly illuminates the reflector. If b is too large, the main lobe of the horn pattern is too narrow and only the central portion of the reflector is excited effectively at a cost in directivity (aperture efficiency). If b is too small, the main lobe of the horn pattern is too broad and some of the primary radiation pattern "spills" past the edges of the reflector, causing unwanted radiation in the back region at a cost in side lobe level and efficiency. Thus for a given focal length f and reflector span D (which combine to determine the subtended angle $2\phi_1$), there is an optimum horn height b. The specification of this optimum is somewhat arbitrary, but a good compromise occurs if the primary radiation at $\phi = \pm\phi_1$ is about $10 dB$ below the value at $\phi = 0°$. Such specification will permit calculation of the proper value for b.

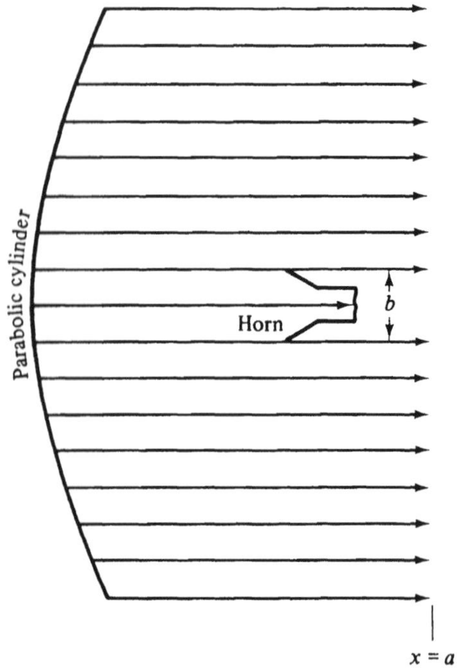

Fig. 10.9 Aperture Blockage Caused by the Finite Extent of the Horn Feed

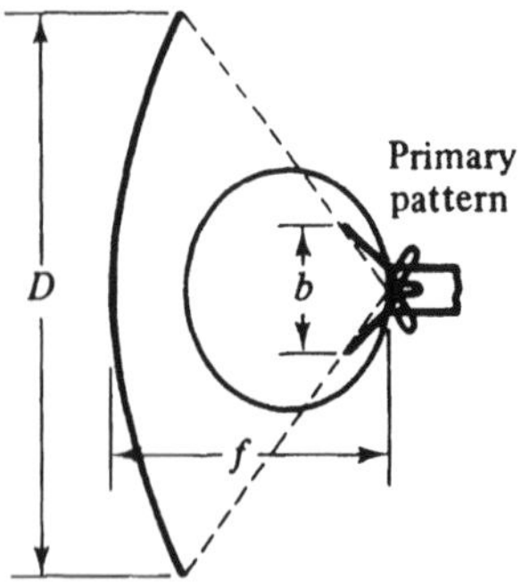

(a) $f/D = 0.50$

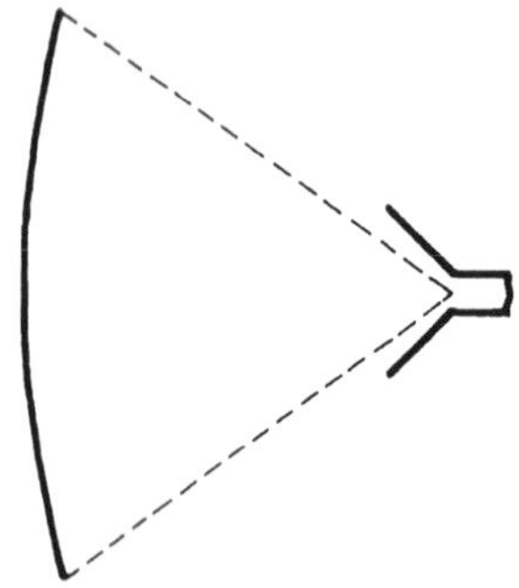

(b) $f/D = 0.75$

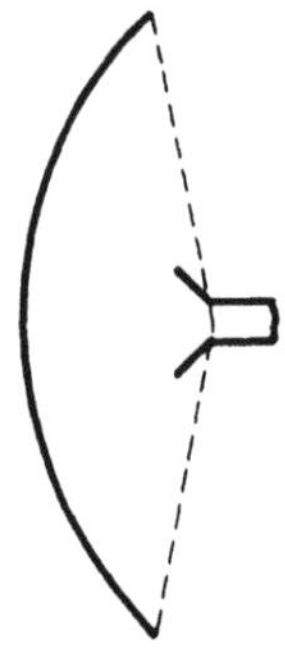

(c) $f/D = 0.33$

Fig. 10.10 The Effect on Reflector Depth and Feed Size of the f/D Ratio

The subtended angle $2\phi_1$ can be deduced by noting that

$$\tan\phi_1 = \frac{y}{f-x} = \frac{y}{f-y^2/4f} = \frac{D/2}{f-D^2/16f} = \frac{0.5}{f/D - D/16f} \tag{10.37}$$

Imagine for the sake of illustration that $f/D = 0.5$ in the reflector/horn assembly shown in Figure 10.9a. Then $\phi_1 = 53°$. Suppose next that a shallower reflector is preferred. This can be accomplished by increasing the f/D ratio. As an illustration, if $f/D = 0.75$, the reflector assumes the flatter shape shown in Figure 10.9b. But now, if the horn feed is to be placed at the focus, half the subtended angle is $\phi_1 = 37°$ and the dimension b must be approximately enlarged by the factor $53°/37° = 1.4$ in order to maintain the same pattern of illumination incident on the reflector. It follows that, for the same secondary aperture height D, the flattening of the reflector has been achieved at the expense of a 40% increase in aperture blockage.

One can go the other way and accept a deeper reflector and the concomitant increases in material and construction cost in order to decrease the aperture blockage. For example, if $f/D = 0.33$, then the reflector takes on the appearance shown in Figure 10.10c, and $\phi_1 = 74°$. Now the b-dimension can be approximately reduced by the factor $53°/74° = 0.7$ and there is a decrease of 38% in the aperture blockage. It can be seen from this simple argument that a tradeoff possibility exists between the amount of interference with the secondary rays caused by feed interception and the depth of the reflector, for a given aperture size D.

These same arguments can be repeated for a paraboloid fed by a pyramidal horn of dimensions a by b. One can conclude that a deeper paraboloid of given span D requires a smaller horn than a shallower paraboloid of the same span D, and thus has less aperture blockage, but at the expense of more weight, more inertia, and poorer defocussing properties. However, it should be noted that aperture blockage is less severe in paraboloidal reflector antennas than it is in parabolic cylinder reflector antennas. The shadowed region, expressed as a fraction of the aperture, is b/D for the parabolic cylinder and $4ab/\pi D^2$ for the paraboloid. For patterns of the same directivity, and thus for apertures of comparable area, the aperture blockage of the paraboloid is less, approximately by the multiplicative factor a/D.

In Cassegrain dual reflector systems, blocking by the subreflector is severe unless D/λ is very large. For this reason, Cassegrains are not often used when $D/\lambda < 40$.

Aperture blockage can be reduced by using an offset feed instead of going to a smaller f/D ratio. This is illustrated in Figure 10.11 and is based on the argument that, as long as the feed is at the focus, there is no fundamental reason why symmetrical portions of the parabola (or paraboloid) need by used. However, the price paid for removing the feed from the path of the secondary rays is once again increased depth of the reflector. This can be seen by comparing Figure 10.11 to Figure 10.10a, which are drawn to the same scale. The two parabolas have the same focal length, but to achieve the same secondary aperture height D, the second reflector is con-

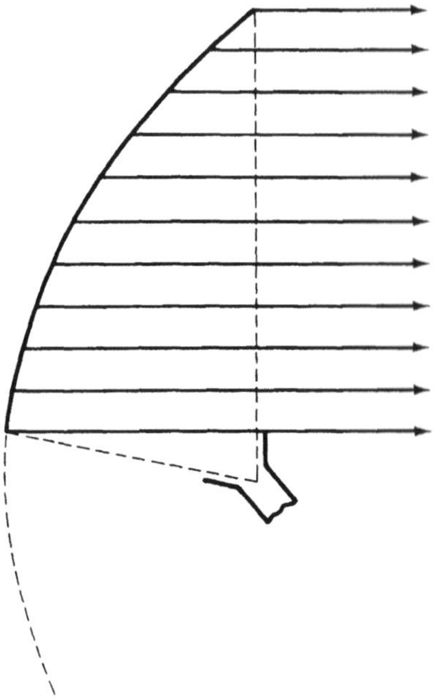

Fig. 10.11 Use of an Offset Feed to Diminish Aperture Blockage

siderably deeper. Here again there is the opportunity to find a suitable compromise between competing desirable design characteristics.

10.5 The Design of a Shaped Cylindrical Reflector

It was shown in Section 10.3 that the diverging primary rays of a long line source, placed at the focus of a parabolic cylinder reflector, were converted to a family of collimated secondary rays. This meant that the equivalent Huyghens sources placed in the secondary aperture $x = a$ were equiphase. If the line source were to radiate a pattern with a single main lobe which centrally illuminated the reflector, the reflector/line-source assembly would produce a sum pattern with a side lobe structure governed by the amplitude taper in the secondary aperture.

Practical applications arise in which it is desirable *not* to collimate the secondary rays, for without collimation a wider variety of secondary patterns is possible. Imagine the situation suggested by Figure 10.12 where it is seen that a primary ray tracing a path at an angle ϕ with respect to the horizontal strikes the cylindrical reflector at a point $P(x, y)$ and, upon reflection, becomes the secondary ray making an angle θ with respect to the horizontal. If $\theta(\phi)$ can be specified, the shape of the reflector can be deduced.

The relation between θ and ϕ depends on knowledge of the feed pattern and the desired secondary pattern. Let it be assumed that the line source and reflector are long enough in the Z-direction that, for the purpose of working with patterns in the central XY-plane, they can be taken as infinitely long. Then $I(\phi)$ in watts per radian-meter can symbolize the primary radiation intensity and $P(\theta)$ in watts per

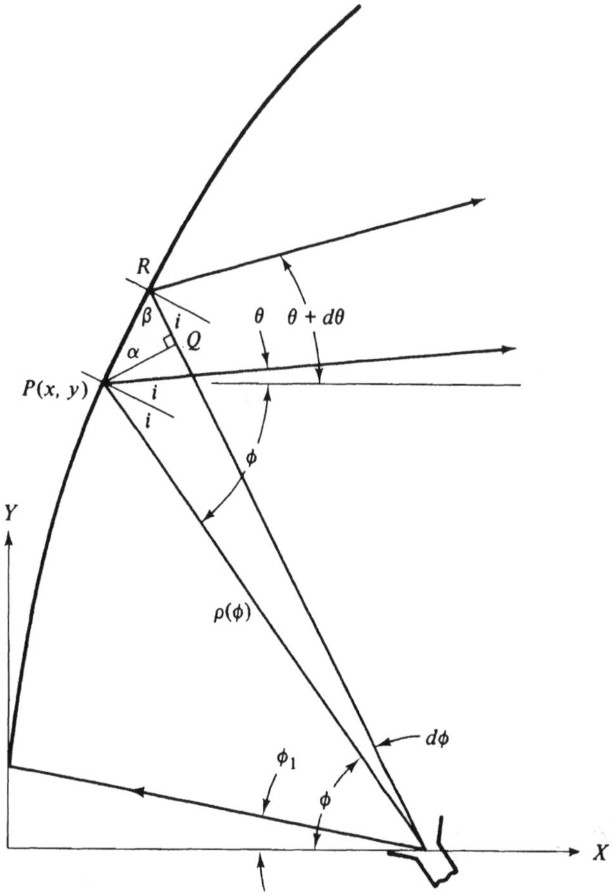

Fig. 10.12 Differential Geometry for a Shaped Cylindrical Reflector

radian-meter can similarly represent the secondary radiation intensity. If use is made of the result obtained in Section 10.2, that radiant energy travels in tubes bounded by sheaths of rays, then it can be concluded that

$$I(\phi)\, d\phi = P(\theta)\, d\theta \tag{10.38}$$

Integration gives

$$\int_{\phi_1}^{\phi} I(\phi')\, d\phi' = \int_{\theta_1}^{\theta} P(\theta')\, d\theta' \tag{10.39}$$

valid over the interval $\phi_1 \leq \phi \leq \phi_2$, $\theta_1 \leq \theta \leq \theta_2$. If $\theta_1 \geq 0$ and if $\phi_1 > 0$, the feed is sufficiently offset to be out of the way of any of the desired secondary rays. If not, there will be aperture blockage. (See Section 10.4 for a discussion of the pros and cons of these alternate feed placements.)

Equation 10.39 can be put in the more useful form

$$\frac{\int_{\phi_1}^{\phi} I(\phi')\,d\phi'}{\int_{\phi_1}^{\phi_2} I(\phi')\,d\phi'} = \frac{\int_{\theta_1}^{\theta} P(\theta')\,d\theta'}{\int_{\theta_1}^{\theta_2} P(\theta')\,d\theta'} \tag{10.40}$$

It is clear from (10.40) that if $I(\phi')$ were replaced by $K_1I(\phi')$, and/or if $P(\theta')$ were replaced by $K_2P(\theta')$, with K_1 and K_2 constants, such insertions would have no effect on the computation of $\theta(\phi)$. This implies that if (10.40) is used instead of (10.39), care need not be taken to normalize the two patterns to a common level.

With $I(\phi)$ and $P(\theta)$ known, Equation 10.40 serves as the vehicle whereby $\theta(\phi)$ can be deduced. Imagine that this has been done. The next task is to find the function $\rho(\phi)$ which defines the shape of the reflector. With reference once again to Figure 10.12, it is clear that

$$\overline{PQ} = \rho d\phi \qquad \overline{RQ} = d\rho$$

and hence

$$\tan\alpha = \frac{d\rho}{\rho\,d\phi} \tag{10.41}$$

It is also evident from a study of this figure that

$$\theta + \phi = 2i \qquad \beta + i = \frac{\pi}{2} = \beta + \alpha$$

from which $\alpha = i = (\theta + \phi)/2$ and thus (10.41) becomes

$$\tan\frac{\theta + \phi}{2} = \frac{d\rho}{\rho\,d\phi} \tag{10.42}$$

Integration gives

$$\ln\frac{\rho(\phi)}{\rho(\phi_1)} = \int_{\phi_1}^{\phi} \tan\frac{\theta' + \phi'}{2}\,d\phi' \tag{10.43}$$

Since $\theta(\phi)$ has already been determined from (10.40), the integrand of (10.43) is known; integration will give the reflector curve $\rho(\phi)$.

As an illustration of the use of this design technique, assume that it is desired to produce a secondary pattern $P(\theta)$ with these features: (a) there is to be a main beam with its peak at $\theta = 3°$, a half-power point at $\theta = 1°$, and a null at $\theta = 0°$; and (b) the other half-power point, which would normally be at $\theta = 5°$, is shifted out to $\theta = 6.5°$ to connect smoothly with a $\csc^2\theta$ curve which extends to $\theta = 20°$, with the pattern dropping to a null at $\theta = 21°$.

A tabulation of the levels of radiation intensity in this secondary pattern is given in Table 10.1 and a plot of the desired pattern is shown in Figure 10.13. Patterns

TABLE 10.1 Desired secondary pattern

θ Degrees	$P(\theta)$ Watts/Radian-Meter	θ Degrees	$P(\theta)$ Watts/Radian-Meter
0	0	11	27
1	75	12	23
2	135	13	20
3	150	14	17
4	143	15	15
5	122	16	13
6	91	17	12
7	68	18	11
8	52	19	10
9	41	20	8
10	33	21	0

of this type are useful in radar systems doing target acquisition or ground-mapping and in airport beacon systems, because the received signal is range-independent (see Appendix H).

Imagine that this pattern is to be produced by a cylindrical reflector whose

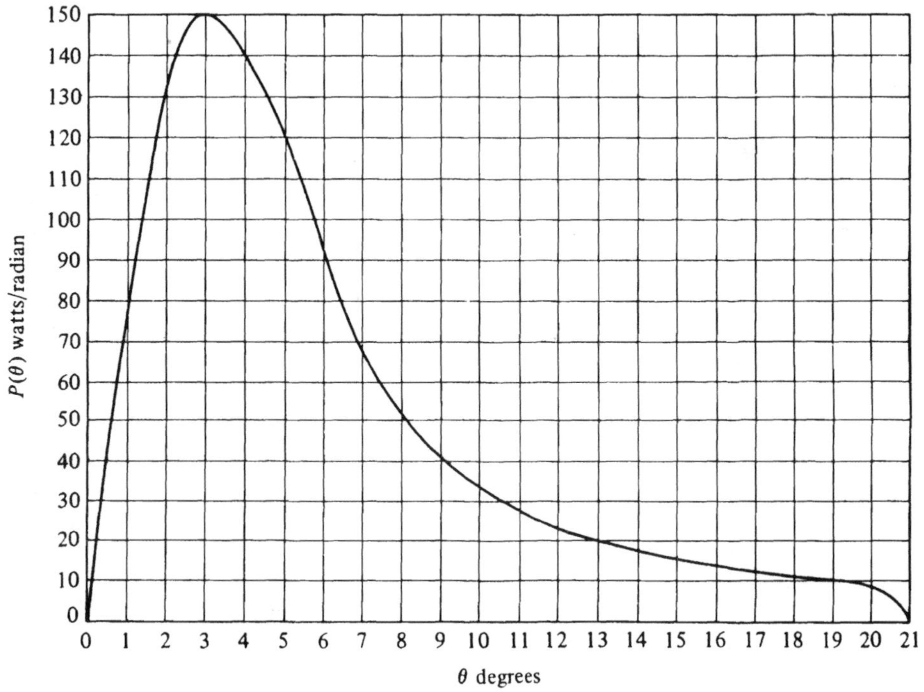

Fig. 10.13 A Desired Secondary Pattern Showing Main Beam and $Csc^2\,\theta$ Roll-Off

extent is such that $\phi_1 = 0^\circ$ and $\phi_2 = 60^\circ$. The primary pattern has been measured and is given to good approximation by $I(\phi) = \cos[3(\phi - 30^\circ)]$. Use of (10.40) gives

$$\sin[3(\phi - 30^\circ)] = 2\frac{\int_{0^\circ}^{\theta} P(\theta')\,d\theta'}{\int_{0^\circ}^{21^\circ} P(\theta')\,d\theta'} - 1 \tag{10.44}$$

Numerical integration of the area under Figure 10.13 permits evaluation of the right side of (10.39) as a function of θ; the results are shown in the second column of Table 10.2. The corresponding values of ϕ can then be deduced using (10.44) and these are entered as the third column of Table 10.2. The computation of $(\theta + \phi)/2$ and its tangent follow readily and these values are listed in the fourth and fifth columns of Table 10.2. A graph of $\tan[(\theta + \phi)/2]$ versus ϕ is shown in Figure 10.14.

Numerical integration of the area under this graph, with use of the proper conversion factor to change degree increments to radian increments gives the data listed in the second column of Table 10.3. The corresponding values of ρ/ρ_0 are given in the third column and the reflector curve is plotted in Figure 10.15. A parabola with the same focal length is shown dotted for comparison. It can be seen that the shaped reflector has less curvature in order to direct the rays at angles $\theta > 0^\circ$.

The reader will observe that the *shape* of the reflector has been determined,

TABLE 10.2 Data for shaped reflector design

θ°	$\sin 3(\phi - 30^\circ)$	ϕ°	$(\theta + \phi)/2$	$\tan[(\theta + \phi)/2]$
0	−1.000	0	0	0
1	−0.930	7.19	4.09	0.072
2	−0.734	14.26	8.13	0.143
3	−0.463	20.81	11.90	0.211
4	−0.187	26.40	15.20	0.272
5	0.062	31.18	18.09	0.327
6	0.260	35.03	20.51	0.374
7	0.409	38.04	22.52	0.415
8	0.521	40.46	24.23	0.450
9	0.607	42.47	25.73	0.482
10	0.676	44.19	27.09	0.512
11	0.732	45.70	28.35	0.540
12	0.779	47.06	29.53	0.566
13	0.819	48.33	30.67	0.593
14	0.854	49.54	31.77	0.619
15	0.884	50.69	32.84	0.646
16	0.910	51.82	33.91	0.672
17	0.933	52.97	34.98	0.700
18	0.954	54.21	36.11	0.729
19	0.974	55.64	37.32	0.762
20	0.991	57.41	38.71	0.801
21	1.000	60.00	40.50	0.854

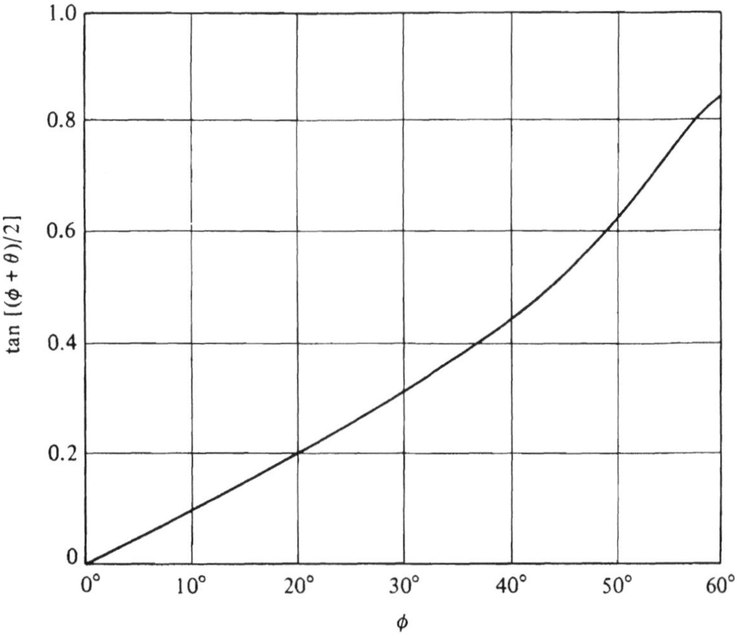

Fig. 10.14 Plot of Integrand for Equation 10.38

TABLE 10.3 Data resulting from integration of Figure 10.13

$\phi°$	$\ln[\rho(\phi)/\rho(0°)]$	$\rho(\phi)/\rho(0°)$
0	0	1.000
10	0.009	1.009
20	0.035	1.036
30	0.080	1.083
40	0.145	1.156
50	0.238	1.268
60	0.368	1.445

but not its *size*, since $\rho(0°)$ has not been specified. A rough estimate of the proper scale factor can be determined by the following argument: Absent the $\csc^2\theta$ pattern shaping, one would be attempting to produce a pattern with a 4° half-power beamwidth. Figure 5.3 indicates that this would require an aperture with a projected length transverse to the $\theta = 3°$ direction of 12.7λ if the amplitude distribution were uniform, perhaps 10% more than this with the taper one would expect from this primary feed; this would give 14λ. If Figure 10.14 is scaled to be consistent with this number, one finds that $\rho(0°) \cong 11.3\lambda$. This estimate is low, since the upper part of the shaped reflector is primarily serving the purpose of pattern-filling in the $\csc^2\theta$ part of the pattern. A value 30% higher than this might not be unreasonable. In

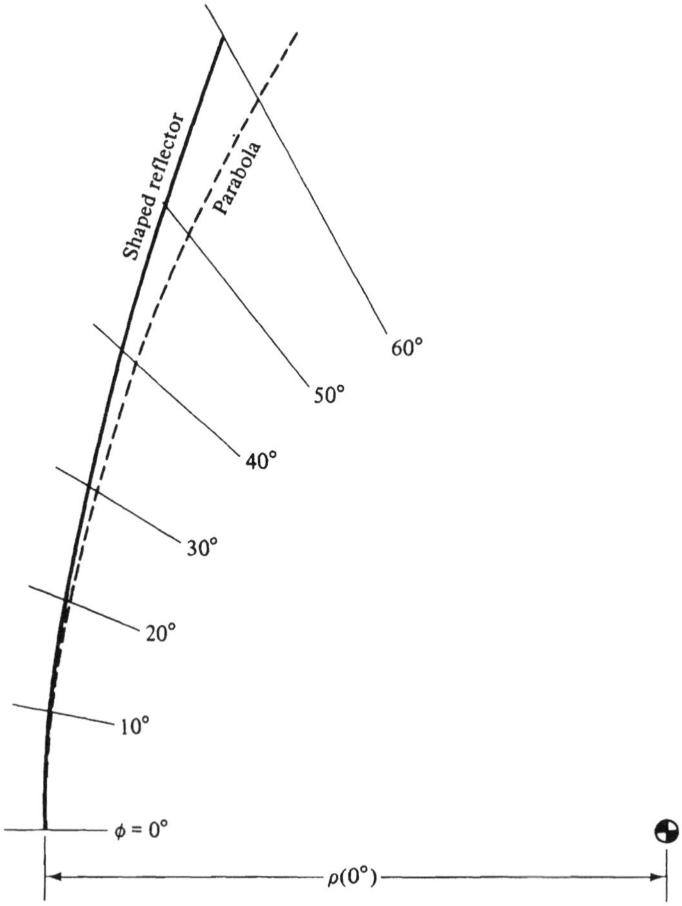

Fig. 10.15 A Cylindrical Reflector Shaped to Produce a $\csc^2 \theta$ Pattern

Section 10.7 a procedure will be presented which permits a much more accurate estimate to be made of $\rho(0°)$, and where it will be found that $\rho(0°)$ should equal 15λ if the desired beamwidth is to be achieved.

10.6 The Design of a Doubly Curved Reflector

It was shown in Section 10.3 that the diverging primary rays of a point source, placed at the focus of a paraboloidal reflector, were converted to a family of collimated secondary rays. It is possible to modify the basic paraboloidal shape so that all the rays are no longer collimated, thus enlarging the class of secondary patterns that can be achieved with a point source and reflector.[5,6] The procedure has many

[5]S. Silver, "Double Curvature Surfaces for Beam Shaping with Point Source Feeds," MIT Radiation Laboratory Report 691 (June 15, 1945).

[6]A. S. Dunbar, "Calculations of Doubly Curved Reflectors for Shaped Beams," *Proc. IRE*, 36 (1948), 1289–96.

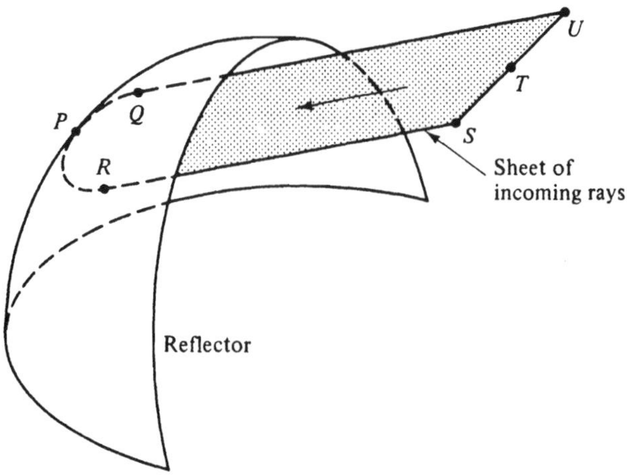

(a)

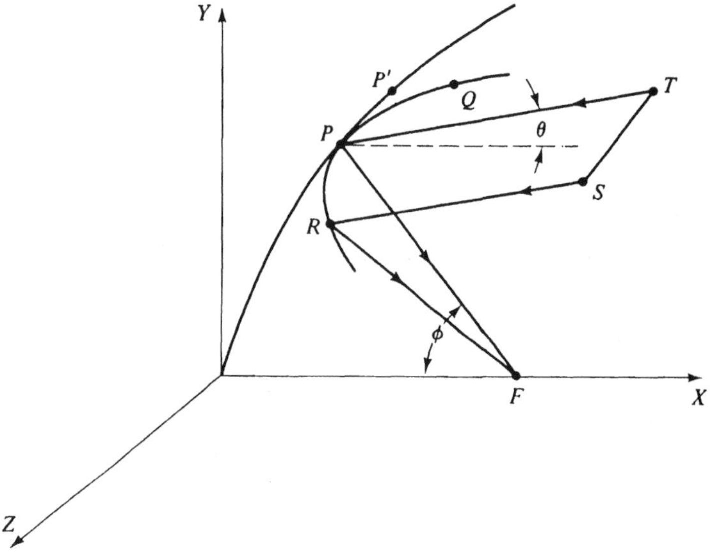

(b)

Fig. 10.16 The Geometry for a Doubly Curved Reflector

similarities to what was done in the analogous problem of shaping a cylindrical reflector (see Section 10.5).

The design of doubly curved reflectors, as these modified paraboloids are called, can best be understood when the antenna is receiving. With reference to Figure 10.16, imagine that a sheet of parallel rays, of which *TP* and *SR* are two members, is incident on the reflector as shown. These rays all lie in a plane which is perpendicular to the XY-plane and which makes an angle θ with the XZ-plane. The reflector and the sheet of rays intersect in a curve RPQ, with P the point which lies in the XY-plane. The reflector surface is to be designed so that all these rays are brought to a focus at the common point F, which lies on the X-axis.

Imagine further that another sheet of parallel rays is incident on the reflector at an angle θ', their intersection being the curve $R'P'Q'$ (not shown), and that all these rays are also to be brought to a focus at F. By extension, if sheets of incoming rays at all angles θ are considered, each having a curve of intersection with the reflector with all rays of all sheets brought to a common focus at F, then the collection of these curves of intersection defines the reflector.

Each curve of intersection has a central point that lies in the XY-plane. The collection of these central points ($P, P', \ldots$) is called the *backbone curve of the reflector*. By anatomical extension, the individual curves of intersection are sometimes referred to as *ribs*.

The equation of a rib can be deduced without difficulty. Let an auxiliary coordinate system $X'Y'Z$ be set up with P as origin and the X'-axis coincident with PT. The line ST, which is part of a wavefront, is characterized by the equation $x' = a'$. Since the phase delay along all ray paths from the wavefront ST to the focal point F is the same, it follows that

$$\overline{TP} + \overline{PF} = \overline{SR} + \overline{RF} \tag{10.45}$$

If $(x', 0, z)$ is the coordinate triplet of the point R in the auxiliary coordinate system $X'Y'Z$, Equation 10.45 becomes

$$a' + \rho = a' - x' + \overline{RF}$$

so that

$$\overline{RF} = \rho + x' \tag{10.46}$$

in which $\overline{PF} = \rho(\phi)$ is the distance from the focus F to the point P on the backbone curve.

Let the projection of the point R in the XY-plane be designated by R^*. The point R^* will lie on the line $\overline{PT}$, as shown in Figure 10.17, and it follows that

$$\begin{aligned} (\overline{RF})^2 &= (\overline{RR^*})^2 + (\overline{R^*F})^2 \\ (RF)^2 &= z^2 + (x')^2 + \rho^2 - 2x'\rho\cos(\theta + \phi) \end{aligned} \tag{10.47}$$

in which it has been recognized that $\angle FPR^* = \theta + \phi$ and the law of cosines has been utilized.

If (10.47) and the square of (10.46) are combined, the result is that

$$\begin{aligned} z^2 &= 2\rho x'[1 + \cos(\theta + \phi)] \\ z^2 &= \left[4\rho \cos^2 \frac{\theta + \phi}{2}\right] x' \end{aligned} \tag{10.48}$$

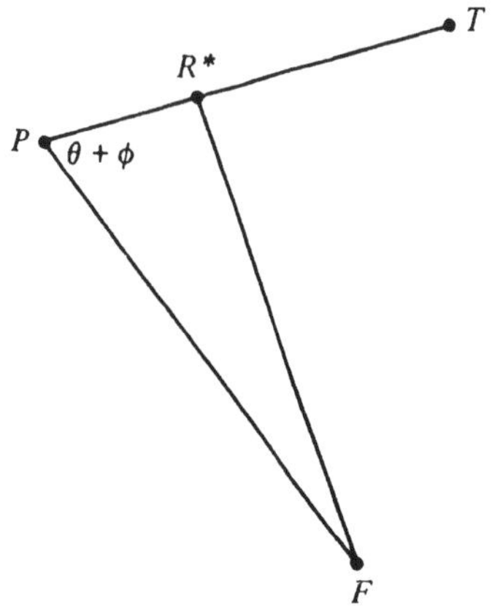

Fig. 10.17 Ancillary Construction for the Design of a Doubly Curved Reflector

Therefore the curve of intersection (rib) is a parabola lying in the $X'Z$ plane and with focal length

$$f = \rho(\phi)\cos^2\frac{\theta(\phi) + \phi}{2} \tag{10.49}$$

Hence if the backbone curve can be found, all the ribs (and thus the reflector surface) can be determined from (10.48).

The shape of the backbone curve is governed by the primary and secondary patterns. To see this, let $I(\phi, \psi)$ be the radiation intensity of the primary pattern in watts per steradian. With the reflector/feed assembly now assumed to be transmitting, $I(\phi, 0°)\,d\phi\,d\psi$ is the power incident on a central element of the reflector. This power is converted to flow outward in a wedge $\rho d\psi$ wide and with a wedge angle $d\theta$. If $P(\theta)$ is the secondary radiation intensity in watts per radian-meter, then

$$I(\phi)\,d\phi\,d\psi = P(\theta)\,d\theta\,\rho\,d\psi \tag{10.50}$$

Integration and normalization gives

$$\frac{\displaystyle\int_{\phi_1}^{\phi}\frac{I(\phi')}{\rho(\phi')}\,d\phi'}{\displaystyle\int_{\phi_1}^{\phi_2}\frac{I(\phi')}{\rho(\phi')}\,d\phi'} = \frac{\displaystyle\int_{\theta_1}^{\theta}P(\theta)\,d\theta}{\displaystyle\int_{\theta_1}^{\theta_2}P(\theta)\,d\theta} \tag{10.51}$$

With $I(\phi)$ known and $P(\theta)$ specified, Equation 10.51 could be used to determine $\theta(\phi)$ if the backbone curve $\rho(\phi)$ were known. But $\rho(\phi)$ is not only not known, it is the goal of the design. Therefore one must proceed iteratively. Usually the desired secondary pattern is a modification of a simple sum pattern, perhaps with $\csc^2\theta$ filling on one side of the main beam. It is then possible to begin by assuming that $\rho(\phi)$ is a parabola with the proper focal length to give the basic unmodified sum pattern. Then (10.51) can be used to get a first approximation to $\theta(\phi)$. This approximation is used in (10.43) to obtain a refined estimation of $\rho(\phi)$, the backbone curve.[7] This new value of $\rho(\phi)$ can be used in (10.51) to determine a refined estimate of $\theta(\phi)$, with the process repeated until an additional iteration causes a negligible change in $\rho(\phi)$. With the backbone curve finalized, the design is complete except for the detailed work of using (10.48) to generate the reflector surface.

The procedure that has just been described is approximate in the sense that, although the family of ribs forms a continuous surface, the slope is not proper at all points. If one attempts to repeat, for the doubly curved reflector, the check undertaken for the parabolic cylinder antenna (see Equations 10.29 through 10.31), it is found that Snell's law of reflection predicts somewhat different paths for the secondary rays than those assumed in deducing the reflector shape. However, if the desired secondary pattern does not differ markedly from a pencil beam, so that the doubly curved reflector is not too different from a paraboloid, this error is not serious.

The reader may have observed that the shaping has only been to decollimate in one dimension. All the rays in a single *sheet* (at angle θ to the XZ-plane) are *parallel*. The *sheets* of rays are no longer parallel, which differs from the case of a paraboloidal reflector. Conceptually, one could envision a reflector surface which would decollimate the rays two-dimensionally, but such designs have limited applicability and the procedure would be quite complicated.

10.7 Radiation Patterns of Reflector Antennas: The Aperture Field Method

The methods for directing secondary rays which have been discussed in previous sections (collimation with a parabolic cylinder or paraboloid Section 10.3; shaped cylindrical reflector, Section 10.5; doubly curved reflector, Section 10.6) can indicate, to the extent that geometrical optics approximations are valid, the *launching* distribution of the secondary field. However, with a finite aperture it is physically impossible to maintain these ray directions throughout all space. Thus another means must be found to calculate the far field.

Conceptually, if the reflector antenna is enclosed in a surface S, and if $\mathbf{E}$ and $\mathbf{H}$ are found everywhere on S, the far field can be computed through use of Equations 1.128 through 1.131. If the antenna radiates primarily in the forward half of space, a convenient selection for the surface S is an infinite plane in front of the reflector[8] closed by an infinite hemisphere, as suggested by Figure 10.18. A geometrical optics approximation to the fields on S consists of assuming that wherever a secondary ray crosses S the field has a value; elsewhere on S it does not. Thus only that part of the infinite plane immediately in front of the reflector is germane; the approximation is obviously better if the infinite plane is as close as possible to the reflector surface.

[7]The reader will appreciate readily that the differential geometry of Figure 10.12 applies equally well to the situation in the XY-plane for the reflector/point-source assembly of Figure 10.15 and thus the backbone curve is also given by (10.43).

[8]An infinite plane is only one of several obvious choices that could be made. Another is to select that part of S which lies in front of the reflector to coincide with a wavefront.

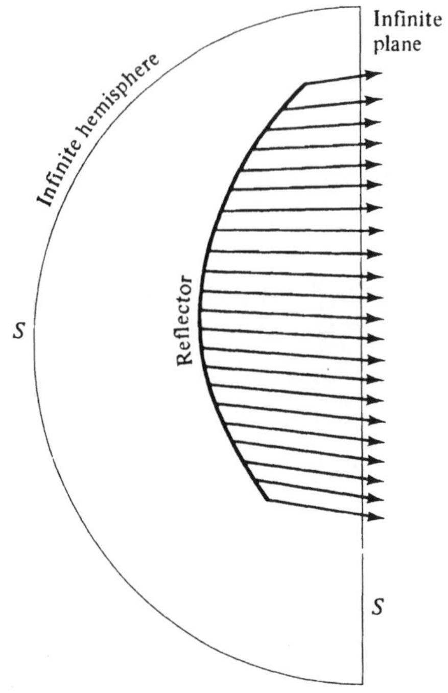

Fig. 10.18 A Reflector Antenna Enclosed by a Secondary Aperture Surface S; Primary Rays and Feed Not Shown

The analyses of Section 10.5 through 10.6 have revealed that, with the primary pattern known and the desired secondary pattern specified, the reflector shape can be determined by geometrical optics means, and thus that the distribution of tubes of radiant energy in the secondary field is calculable in nearby space. Specifically, the energy density distribution crossing the planar surface S can be deduced. This gives the amplitude of the **E** and **H** fields on S. Their polarization is known if the polarization of the feed is specified. Their phase can be determined by calculating the phase delay along each ray from a nearby wavefront to the surface S. With this, one possesses all the knowledge needed to determine the Huyghens sources on S and then to compute an approximation to the far field.

As an example of the use of this method, consider again the parabolic cylinder reflector with a central line source at its focus, as depicted in Figure 10.4. Assume that the reflector extends vertically to $y = \pm D/2$ and thus that the secondary aperture field in the plane $x = a$ will also have a value in the range $-D/2 \leq y \leq D/2$. It is desired to find $P(y)$ in watts per square meter in this range, with $P(y)$ the power density in the secondary field at $x = a$ under geometrical optics assumptions.

It is helpful as a first step to obtain the equation of the parabolic cylinder in polar coordinates. To do this, one can return to Equation 10.43 and set $\theta(\phi) \equiv 0$, then recognize that $\rho(0^\circ) = f$, and obtain

$$\ln \frac{\rho(\phi)}{f} = 2 \int_0^{\phi/2} \tan\left(\frac{\phi'}{2}\right) d\left(\frac{\phi'}{2}\right)$$

from which

$$\rho(\phi) = \frac{f}{\cos^2(\phi/2)} = \frac{2f}{1 + \cos\phi} \tag{10.52}$$

Next, from Figure 10.19, it is evident that, in the geometrical optics approximation, primary and secondary power flows equate such that

$$P(y)\,dy = I(\phi)\,d\phi \tag{10.53}$$

where $I(\phi)$ is the primary pattern in watts per radian-meter. The directed displacement from Q to Q' is given by

$$d\mathbf{s} = \mathbf{1}_T\sqrt{(d\rho)^2 + (\rho d\phi)^2} \tag{10.54}$$

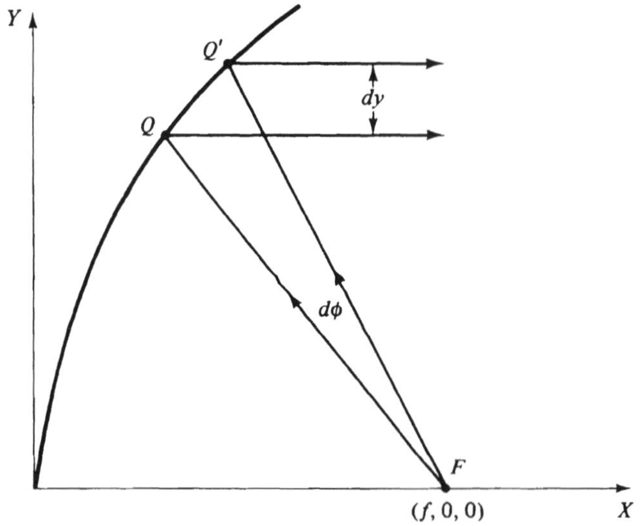

Fig. 10.19 Ray Geometry for a Parabolic Cylinder Reflector

The unit tangent vector $\mathbf{1}_T$ has already been obtained as Equation 10.28. The height of the secondary tube is therefore

$$dy = \mathbf{1}_y \cdot d\mathbf{s} = \sqrt{\rho^2 + \left(\frac{d\rho}{d\phi}\right)^2}\,\frac{2f/y}{\sqrt{1 + (2f/y)^2}}\,d\phi \tag{10.55}$$

When (10.53) and (10.55) are combined and use is made of the fact[9] that

$$\frac{d\rho}{d\phi} = \rho\tan\frac{\phi}{2} \qquad y = \rho\sin\phi$$

[9]The first of these relations is a reduction from (10.42), when $\theta(\phi) = 0$.

the connection between primary and secondary power distributions is found to be given by the simple expression

$$P(y) = \frac{I(\phi)}{\rho(\phi)} = \frac{I(\phi)\cos^2 \phi/2}{f} \tag{10.56}$$

In most practical applications, $I(\phi)$ is a tapered distribution. Equation 10.56 indicates that $P(y)$ is more tapered, due to the factor $\cos^2 \phi/2$. This factor is sometimes called the *space loss*.

As a specific illustration of the use of this result, assume that

$$I(\phi) = \frac{\sin^2 [(\pi b/\lambda) \sin \phi]}{[(\pi b/\lambda) \sin \phi]^2} \tag{10.57}$$

which is the E-plane pattern of a simple, vertically polarized horn of height b (compare with Equation (3.13). If the extent of the reflector is such that $-60° \leq \phi \leq 60°$ and if the illumination at the extremities of the reflector is to be 10 decibels below what it is at the center, then a simple calculation reveals that $b/\lambda = 0.85$. If the reflector is to have a projected height $D/\lambda = 10$, then

$$y_{max} = 5\lambda = \rho(60°)\sin 60°$$

so that $\rho(60°) = 5.774\lambda$. Use of (10.52) reveals that the needed focal length is $f = 4.330\lambda$.

With the dimensions of the reflector and feed horn known, and with $I(\phi)$ given by (10.57), the effective aperture distribution can be found from (10.56). A plot of $f(y) = [P(y)]^{1/2}$ is shown in Figure 10.20. The *field* distribution is $f(y)$; it possesses a smooth taper to an edge value which is 27.5% (−11.2 dB) of the central peak value. An initial specification was that the primary pattern be down 10 dB at $\phi = \pm 60°$; the additional 1.2 dB in the taper of the secondary distribution is due to the presence of the factor $\cos^2 \phi/2$ in Equation 10.56.

The far-field pattern due to this aperture distribution can be found by determining the Huyghens sources from (1.112) through (1.115) and then using Equations G.19 of Appendix G. In this problem the sources are equiphase and, with the element pattern suppressed, the far-field pattern in the XY-plane is given by

$$F_0(\phi) = \int_{-5}^{5} f(p) e^{j2\pi p \sin \phi} \, dp \tag{10.58}$$

where $p = y/\lambda$. For the aperture distribution of Figure 10.20, Equation 10.58 yields the pattern shown in Figure 10.21. Predictably, there is a central main beam with a symmetric side lobe structure that has an envelope which decays with $|\phi|$. The side lobe level is −28 dB, consistent with this amount of aperture taper, and the half-power beamwidth is 6.4°. With reference to Figure 5.3, a *uniform* effective aperture distribution would produce a 5.1° main beam with $D/\lambda = 10$, so the beam broadening factor in this case is 1.26, which is also the loss factor in directivity. For this reason,

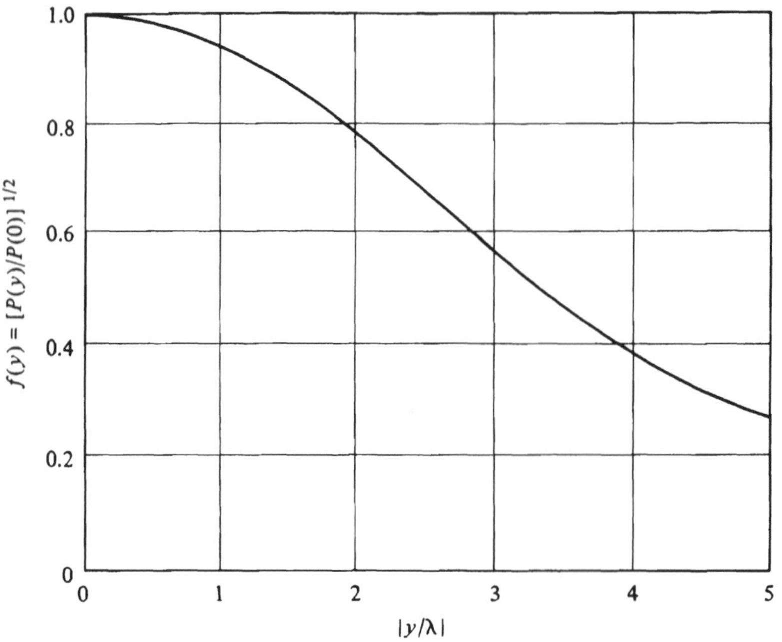

Fig. 10.20 Relative Aperture Distribution for a Parabolic Cylinder Reflector; $D = 10\lambda$, $f = 4.33\lambda$; Primary Pattern Given by Equation 10.57

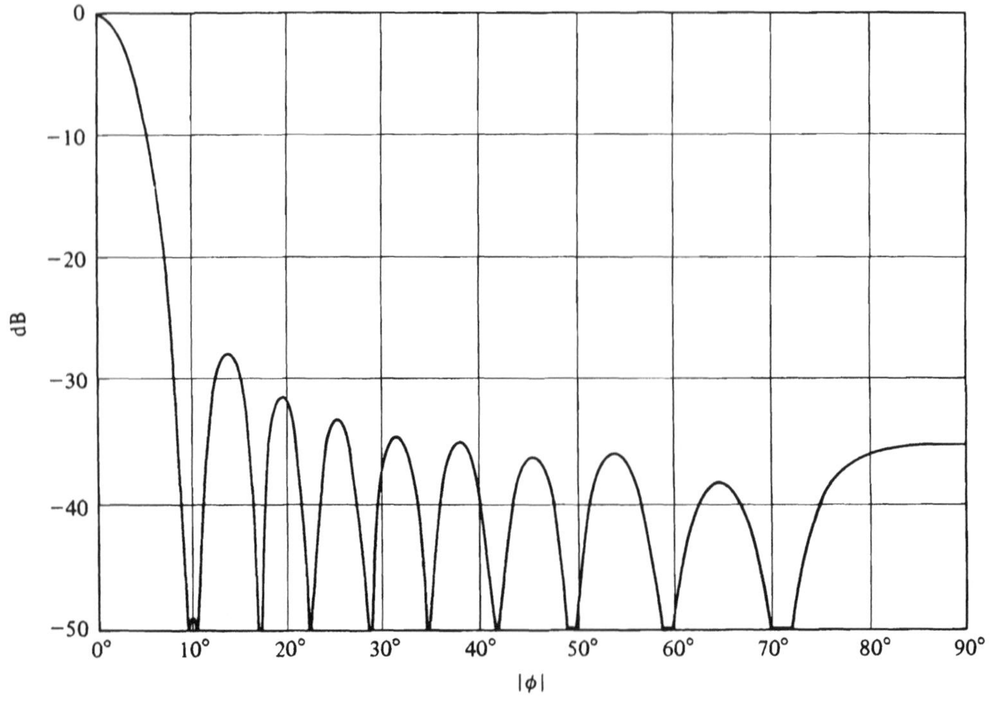

Fig. 10.21 Principal Plane Pattern of a Parabolic Cylinder Reflector Antenna; Aperture Blockage Neglected; $D = 10\lambda$, $f/D = 0.433$, 10 dB Spillover

one can say that the aperture efficiency is $(1.26)^{-1} = 0.79$, or 79%. Figure 5.4 indicates that a Dolph-Chebyshev pattern with a 28 dB side lobe level only occasions a beam broadening factor of 1.13. This illustrates one of the limitations of parabolic cylinder (and paraboloidal) reflectors. Unless one is willing to shape the reflector or use a nonconventional feed, there is an inherent loss in efficiency due to lack of control over the amount and shape of the taper in the secondary aperture distribution. A method which has been devised to overcome this shortcoming, using a modified Cassegrain feed, will be described in Section 10.9.

To this point in the analysis, no attempt has been made to calculate the far field behind the reflector, nor to take into account the presence of the feed. The latter effect (aperture blockage) has been discussed in general terms in Section 10.4. An approximate quantitative estimate of this effect can be obtained by making an assumption that can be understood by returning to Figure 10.9. If when radiating, the primary horn is reasonably well matched to free space, then when receiving, it should also be reasonably well matched. This suggests that the portion of the secondary near field that impinges on the mouth of the horn should be absorbed effectively by the horn, with little further scattering. If this assumption is made, the far-field pattern for the illustrative example under discussion is approximately given by (10.58) *minus* the function

$$F_1(\phi) = \int_{-0.425}^{0.425} f(p)e^{j2\pi p \sin\phi}\, dp \tag{10.59}$$

since the feed height is 0.85λ.

The subtractive pattern of (10.59) is very broad, because b/λ is so small. It is given to good approximation by taking $f(p)$ out from under the integral sign and replacing it by $f(0)$. This produces the result shown in Figure 10.22 which is (as one would expect) a replication of the primary feed pattern.

The functions $F_0(\phi)$ and $F_1(\phi)$ both have their peak values at $\phi = \pi/2$. Therefore the levels of these two patterns are in the same proportion as the areas under the curve of Figure 10.20 in the ranges $0 \le p \le 5$ and $0 \le p \le 0.425$. Numerical integration gives $F_0(0°) = 3.31$ and $F_1(0°) = 0.42$. This means that, with aperture

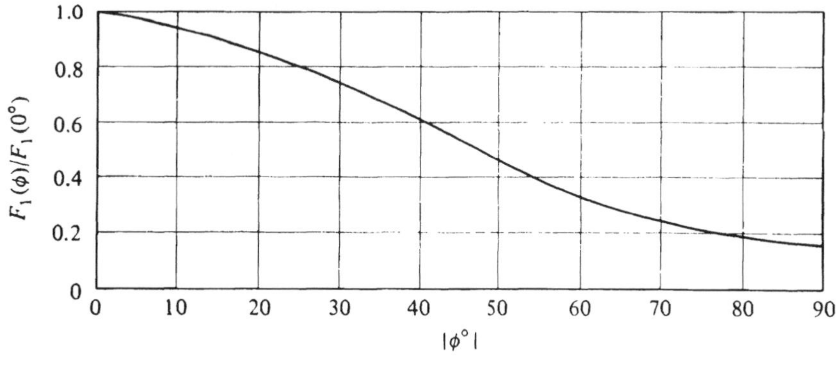

Fig. 10.22 Subtractive Pattern Due to Feed Blockage

blockage included,

$$F_0(0^\circ) - F_1(0^\circ) = 3.31 - 0.42 = 2.89 = \text{height of main beam}$$

Study of Figures 10.21 and 10.22 indicates that $F_0(7^\circ) \cong F_1(7^\circ)$ and thus one would expect a null in the combined pattern at 7°. The first side lobe of Figure 10.21 occurs at 10° and at a height of −48 dB. Thus the first side lobe in the combined pattern should also occur at 10°, with its peak value given by

$$F_0(10^\circ) - F_1(10^\circ) = -\frac{3.31}{251} - (0.42)(0.95) = -0.41$$

From this calculation, one can infer that the height of the first side lobe with aperture blockage included is −17 dB.

There are no more nulls in the combined pattern because $F_0(\phi)$ is never again large enough to cancel $F_1(\phi)$. A dip occurs at 14° because the second side lobe of Figure 10.21 is opposed to the effect of $F_1(14^\circ)$. A similar calculation indicates that this dip achieves a level of −21 dB. The next peak in the combined pattern occurs at the position of the third side lobe peak in Figure 10.21, and is at the level −16 dB. Proceeding in this way, one can sketch in the entire pattern by modifying Figure 10.21 to include the effect of aperture blockage. The total result is shown in Figure 10.23.

It can be seen from this simple exercise that the presence of the feed can have a substantial influence on the secondary pattern. In this case, the main beam has been narrowed somewhat, but the side lobe level has been raised from −28 dB to −16 dB.

The procedure which has just been followed to calculate the far-field pattern of a parabolic cylinder reflector antenna can also be applied to the case of a paraboloidal reflector with a "point" source at its focus. If the closed surface S of Figure 10.18 includes the infinite plane $x = a$ immediately in front of the paraboloid and if spherical coordinates (ρ, ϕ, ψ) are erected with the focus as origin, it is a simple matter to show that the power distribution $P(r, \psi)$ in the aperture $x = a$ is given by

$$P(r, \psi) = \frac{I(\phi, \psi)}{f^2} \cos^4 \frac{\phi}{2} \tag{10.60}$$

In (10.60), $I(\phi, \psi)$ is the primary radiation intensity and (r, ψ) are polar coordinates constructed in the plane $x = a$. This result is seen to be similar to (10.56), obtained earlier for a parabolic cylinder reflector, except that $(\cos^2 \phi/2)/f$ has been replaced by $(\cos^4 \phi/2)/f^2$. Space loss is thus more pronounced for a paraboloidal reflector.

As a specific illustration of the use of Equation 10.60, suppose that a horn that produces a ϕ-symmetric pattern is used to illuminate a paraboloidal reflector of span 10λ. If the horn pattern is given to sufficient accuracy by Equation 10.57, and if the horn size is adjusted so that the edge spillover is −10 dB, the square root of (10.60) gives the equiphase aperture distribution. With linear polarization assumed, the

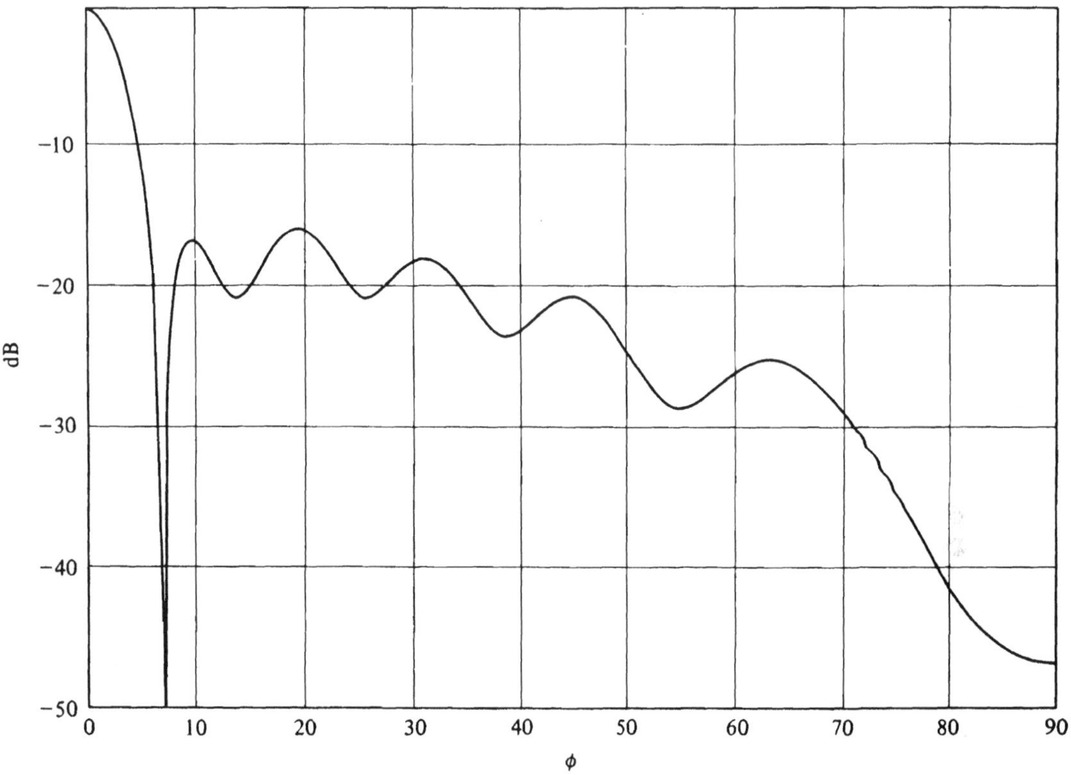

Fig. 10.23 Principal Plane Pattern of a Parabolic Cylinder Reflector Antenna; Aperture Blockage Included; $D = 10\lambda$, $f/D \doteq 0.433$, 10 dB Spillover

equivalent Huyghens sources follow readily and the far-field pattern can be computed with the aid of Equation 6.49. The results are shown in Figures 10.24 and 10.25.

The side lobe level with aperture blockage ignored is seen to be −23 dB. Inclusion of aperture blockage, using the assumption of perfect shadowing, causes the side lobe level to rise, but only to −19 dB. This supports the argument made in Section 10.4 that aperture blockage is less severe with a paraboloid than with a parabolic cylinder. Null filling is also less pronounced, as can be seen by contrasting Figures 10.23 and 10.25.

Of course, in the real-life situation, feed scattering is considerably more complicated than in the idealized examples just given. In the first place, the feed does not simply absorb the blocked rays and leave a well-defined void in the secondary aperture distribution. In the second place, the feed is nowhere near so simple a structure as pictured in Figure 10.9. It has mechanical supports which hold it in place and which also scatter, with the aggregate effect quite complicated. The interested reader is referred to the advanced literature for a deeper treatment of this problem.[10]

[10]For a tutorial introduction to the complexities, see J. Ruze, "Feed Support Blockage Loss in Parabolic Antennas," *Microwave Journal*, 11 (1968), 76–80.

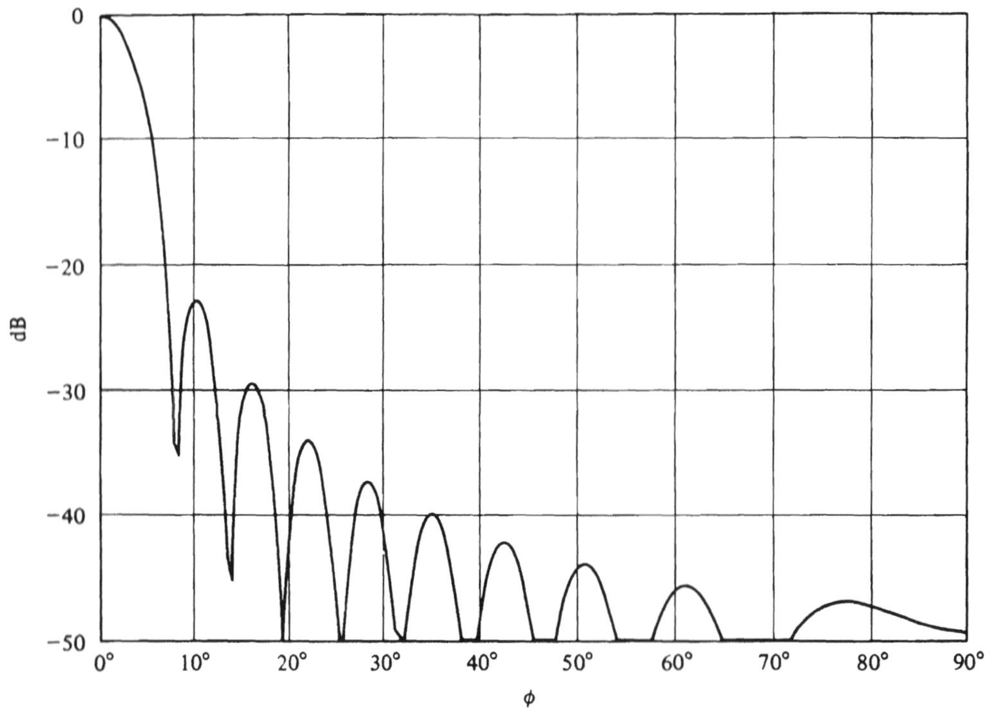

Fig. 10.24 Principal Plane Pattern of a Paraboloidal Reflector Antenna; Aperture Blockage Neglected; $D = 10\lambda$, $f/D = 0.433$, 10 dB Spillover

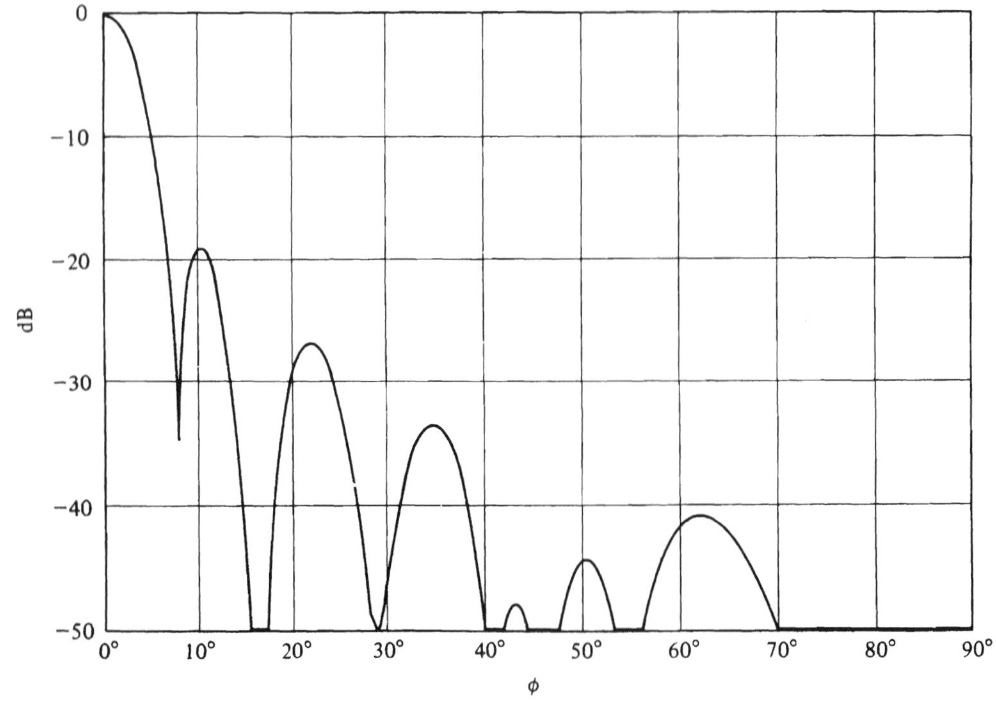

Fig. 10.25 Principal Plane Pattern of a Paraboloidal Reflector Antenna; Aperture Blockage Included; $D = 10\lambda$, $f/D = 0.433$, 10 dB Spillover

A third illustration of the use of the aperture field method concerns the shaped cylindrical reflector discussed in Section 10.5. Here again, the aperture could be chosen to be the plane $x = a$, immediately in front of the reflector. However, since the secondary rays are purposely not collimated, this implies that the aperture field in $x = a$ is not equiphase, meaning that it is necessary to find the total path length from the line source to any specified point in $x = a$. It is easier in this case to choose the surface S depicted in Figure 10.18 to include, as its front portion, a phase front in the geometrical optics secondary field, immediately adjacent to the reflector.

With reference once again to Figure 10.12, consider the primary ray that leaves the line source at angle ϕ, striking the reflector at $P(x, y, 0)$, becoming upon reflection the secondary ray which makes an angle θ with respect to the horizontal. Designate $(\xi, \eta, 0)$, as P' the point in the phase front S pierced by this secondary ray. Then if $\rho' = \overline{PP'}$, S is characterized by

$$\rho + \rho' = K \tag{10.61}$$

with K a constant. If $K = 2\rho_0$, with ρ_0 the distance from the origin to the line source, the equiphase surface S (possibly extended) will go through the position of the line source. This is usually adequate to insure that S is adjacent to the reflector but everywhere in front of it. Then (10.61) becomes

$$\rho' = 2\rho_0 - \rho \tag{10.62}$$

It is evident from Figure 10.12 that

$$\xi = \rho_0 - \rho\cos\phi + \rho'\cos\theta \qquad \eta = \rho\sin\phi + \rho'\sin\theta$$

and these coordinates of a point on S can be normalized to give, with the aid of (10.62),

$$\frac{\xi}{\rho_0} = 1 - \frac{\rho}{\rho_0}\cos\phi + \left(2 - \frac{\rho}{\rho_0}\right)\cos\theta \tag{10.63}$$

$$\frac{\eta}{\rho_0} = \frac{\rho}{\rho_0}\sin\phi + \left(2 - \frac{\rho}{\rho_0}\right)\sin\theta \tag{10.64}$$

Since $\rho(\phi)/\rho_0$ and $\theta(\phi)$ have been found in the course of the design of the shaped reflector (see Section 10.5), it follows that Equations 10.63 and 10.64 can be used to find ξ/ρ_0 and η/ρ_0 as functions of θ. But the desired secondary power distribution $P(\theta)$ is known, and thus the equiphase field distribution on S can be determined. Deduction of the equivalent Huyghens sources and calculation of the far-field pattern then follows rapidly.

For the illustrative problem of Section 10.5, this process results in the far-field pattern shown in Figure 10.26, if $\rho(0^\circ)$ is taken to be 15λ. The desired pattern $P(\theta)$ is shown for comparison by the dashed line. One can observe agreement between the desired and achieved patterns within ± 1 dB out to 17°, beyond which the unrealistic precipitous drop in the desired pattern at 21° causes the two patterns to diverge.

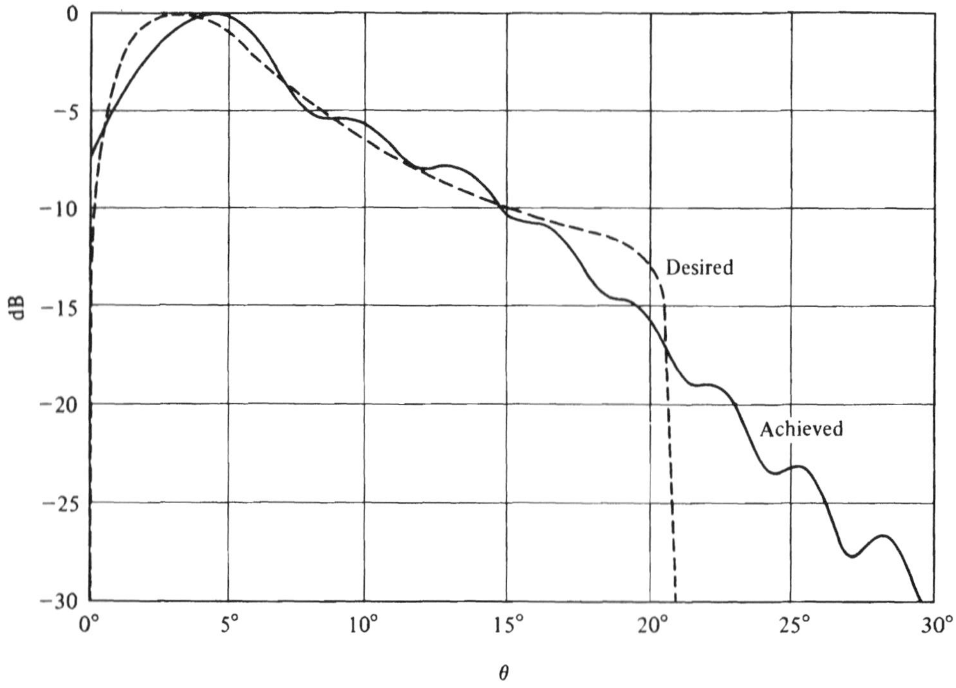

Fig. 10.26 Achieved and Desired Theoretical Far Field Patterns for the Shaped Cylindrical Reflector Antenna of Section 10.5; Aperture Blockage Neglected

The achieved pattern is dependent on the choice of a value for $\rho(0°)$, and it took experimentation to find that 15λ gave the proper 3 dB beamwidth of 5.5°. However, in the desired pattern the 3 dB points occur at 1° and 6.5°, with the peak at 3°. In the achieved pattern the 3 dB points occur at 1.5° and 7°, with the peak halfway between, at 4.25°. This serves to reveal another unrealistic feature of the desired pattern, namely, the placement of the beam peak at a position quite far from midway between the half power points.

Despite these shortcomings (more due to what was desired than to what was achieved) it is clear from Figure 10.26 that the principal goal of shaping the reflector has been achieved. There is a main beam with the proper beamwidth, and null-filling has been accomplished on one side at ± 1 dB of the desired level for a considerable portion of the specified angular range.

10.8 Radiation Patterns of Reflector Antennas: The Current Distribution Method

An alternate method for computing the far-field pattern of a reflector antenna, which in concept is also rigorously correct, involves determination of the current distribution on all parts of the antenna. Were the current distribution known with complete accuracy, Equations 1.101 and 1.102 could be used to obtain a precise expression for the far field. The difficulty in this approach is finding the true current distribution. A useful approximation rests on the assumption that the geometrical optics field representation is valid in the neighborhood of the reflector. That this allows estimation of the reflector currents can be seen from the following.

Let $P(x, y, z)$ be a point on the reflector that is illuminated by the primary radiation, and let $\mathbf{1}_n$ be the normal to the reflector surface at P. With a perfect conductor assumed, the boundary conditions at P are

$$\mathbf{1}_n \times (\mathbf{E}_1 + \mathbf{E}_2) = 0 \qquad \mathbf{1}_n \cdot (\mathbf{H}_1 + \mathbf{H}_2) = 0 \tag{10.65}$$

in which $(\mathbf{E}_1, \mathbf{H}_1)$ is the primary field and $(\mathbf{E}_2, \mathbf{H}_2)$ is the secondary field. It was demonstrated in Section 10.2 that for any geometrical optics field $\mathbf{E}$ and $\mathbf{H}$ are othogonal to each other and that both are transverse to the direction of propagation. In free space, with $\eta = 377$ ohms,

$$\mathbf{E}_1 = -\eta \mathbf{H}_1 \qquad \mathbf{E}_2 = \eta \mathbf{H}_2 \tag{10.66}$$

The difference in signs in (10.66) is due to the reversal of direction of propagation upon reflection. When (10.65) and (10.66) are combined, one obtains

$$H_{1_n} = -H_{2_n} \qquad H_{1_t} = H_{2_t} \tag{10.67}$$

In words, the normal components of the incident magnetic field are equal and opposite, whereas the tangential components are equal and codirected.

From (1.112), the lineal current density at P is given by

$$\mathbf{K} = \mathbf{1}_n \times (\mathbf{H}_1 + \mathbf{H}_2) = \mathbf{1}_n \times 2\mathbf{H}_1 \tag{10.68}$$

If the primary geometrical optics field is known, Equation 10.68 can be used to obtain an approximation to the current distribution on the reflector.

As an illustration of the use of this method, consider again the parabolic cylinder reflector for which a far field was sought in Section 10.7. With the primary power pattern given by (10.57) and vertical polarization assumed, it is clear that

$$\mathbf{H}_1 = \mathbf{1}_z \frac{\sin[(\pi b/\lambda)\sin\phi]}{(\pi b/\lambda)\sin\phi} \frac{e^{-jk\rho(\phi)}}{\sqrt{\rho(\phi)}} \tag{10.69}$$

in which a multiplicative constant has been suppressed. For a parabolic cylinder, $\mathbf{1}_n$ is given by (10.29) and thus

$$\mathbf{K} = \frac{2\sin[(\pi b/\lambda)\sin\phi]}{(\pi b/\lambda)\sin\phi}\left[\frac{\mathbf{1}_x + \mathbf{1}_y(2f/y)}{\sqrt{1 + (2f/y)^2}}\right] \tag{10.70}$$

where $y = \rho \sin\phi$. With the aid of (10.52), this result can be restated in the more useful form

$$\mathbf{K} = \frac{\sin[(\pi b/\lambda)\sin\phi]}{(\pi b/\lambda)\sin\phi}\left[\mathbf{1}_x \frac{\sin\phi}{\cos(\phi/2)} + \mathbf{1}_y 2\cos\frac{\phi}{2}\right] \tag{10.71}$$

To obtain the far-field pattern produced by this current distribution, the cylindrical coordinates equivalent of Equations 1.101 and 1.102 can be used (see Appendix G). In the principal XY-plane, there is only an $\mathcal{A}_\phi$ component and therefore only an E_ϕ-component of the electric field, which is consistent with the assumption of a vertically polarized feed.

For $b/\lambda = 0.85$, $D/\lambda = 10$, and $f/\lambda = 4.33$, which were the dimensions used previously, a plot of $20 \log_{10} |\mathcal{A}_\phi|$ is shown in Figure 10.27. Agreement with Figure 10.21, which was obtained by the aperture field method, is seen to be quite good, particularly in the region of the main beam and the innermost side lobes. It can be argued that, of the two methods (both of which rely on geometrical optics approximations), the current distribution approach should be more precise, since it only relies on the accuracy of the ray assumption for the fields right at the reflector. The aperture field method requires the further assumption that the geometrical optics secondary field is a good approximation to the true field in extension from the reflector out to the surface chosen as the secondary aperture.

However, both methods suffer from the breakdown of geometrical optics at the edges of the reflector. This point has already been noted in Section 10.7 with respect to the aperture field approach. It is even more physically apparent in the

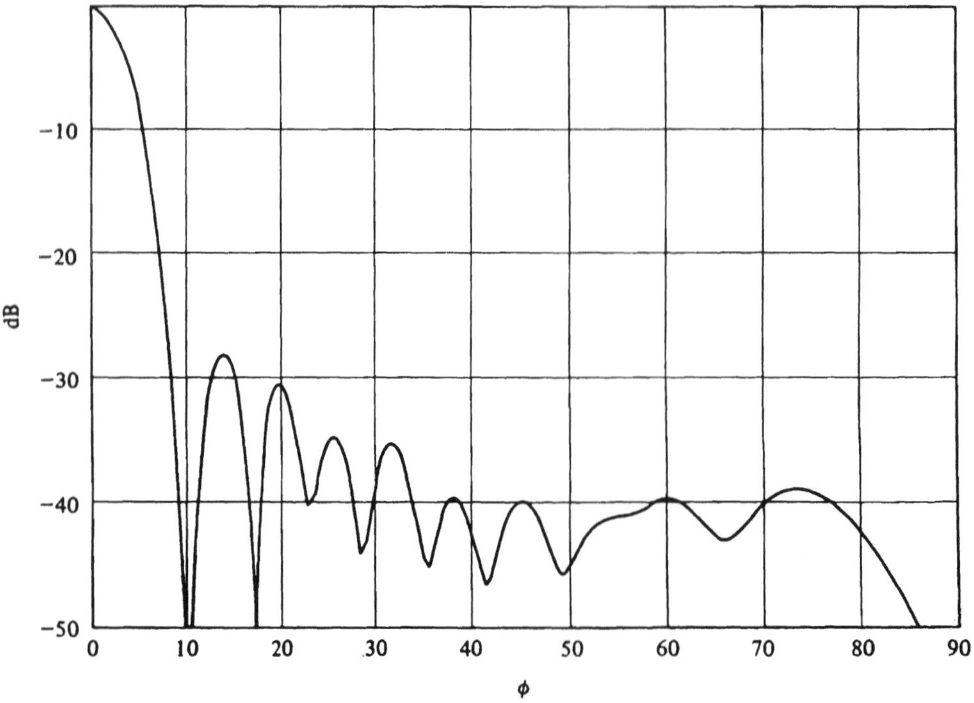

Fig. 10.27 Principal Plane Pattern of a Parabolic Cylinder Reflector Antenna, Obtained by Current Distribution Method. Aperture Blockage Neglected. $D = 10\lambda$, $f/D = 0.433$, 10 dB Spillover

current distribution method as applied to the illustrative example of the parabolic cylinder. A -10 dB spillover at the reflector edges was assumed, indicating that **K**, as computed from (10.71), would not be zero at $\phi = 60$, that is, at the reflector edges. Realistically, one could expect that the current might not quite go to zero at the edge, perhaps wrapping around to the back side before it decayed to a negligible value. However, it certainly should be smaller than the value at the edge given by (10.71). Thus these *edge diffraction effects*, as they are called—which are not taken into account in either of the methods that have been presented—can be expected to have an effect on the far-field secondary pattern.

That the significance of these diffraction effects is greatest when one is computing the field in the region of the outer side lobes can be appreciated by the following argument. For a cylindrical reflector, with the secondary aperture chosen to be in the plane $x = a$, the far field can be expressed as a generalization of (10.58), that is,

$$F(\phi) = \int_{p_1}^{p_2} f(p)e^{jup}\,dp \tag{10.72}$$

with $p = y/\lambda$ and $u = 2\pi \sin \phi$. If this expression is integrated by parts repetitively, one obtains

$$\begin{aligned} F(\phi) &= \sum_{i=0}^{n}(-1)^i \frac{e^{jup}}{(ju)^{i+1}} f^i(p)\Big|_{p_1}^{p_2} \\ &\quad + \frac{(-1)^n}{(ju)^n}\int_{p_1}^{p_2} f^n(p)e^{jup}\,dp \end{aligned} \tag{10.73}$$

Ceteris paribus, this series converges more rapidly when u is greater, that is, as $\phi \rightarrow \pi/2$. Therefore, the zeroth-order term in (10.73), which involves $f(p)$ at the end points, is least important near $\phi = 0°$ and most important near $\phi = 90°$. Said another way, *errors* in $f(p)$ at the edges of the aperture distribution cause the greatest effect in the region of the outer side lobes.

One of the most successful ways found to account for the edge effects involves application of the geometrical theory of diffraction (GTD). The edge is modeled by a semi-infinite ground plane on which a plane wave is incident at a specified angle and off which scattering can be computed exactly. This approach is beyond the scope of the present introductory treatment and the interested reader is referred to the advanced literature.[11]

10.9 Dual Shaped Reflector Systems

The development in Section 10.7 reveals that a parabolic cylinder reflector (or a paraboloidal reflector), illuminated by a feed placed at the focus, has a secondary aperture distribution that is more tapered than the primary radiation. This is evident from Equations 10.56 and 10.60. If one were interested in low side lobes and uncon-

[11]A good start on the significant published work can be made by consulting Part III and the related bibliography in Love, *Reflector Antennas*.

cerned about aperture efficiency and if aperture blockage were not a problem, these tapered distributions would be quite acceptable and would result in impressive patterns such as the one shown in Figure 10.21. However, unless an offset feed is used, aperture blockage *is* a problem, as can be seen from Figure 10.23, and low side lobes are difficult to achieve. Faced with this reality, and desiring higher aperture efficiency, designers of reflector antennas might choose in some applications to de-emphasize low side lobes as a desirable pattern criterion and search for means to control or eliminate the taper.

It was pointed out in Section 6.14 that the maximum directivity that can be achieved from an equiphase planar distribution occurs when the amplitude distribution is uniform. For a paraboloidal reflector fed by a small source at its focus, Equation 10.60 indicates that to achieve this performance, one would need a feed with radiation pattern $I(\phi, \psi)$ proportional to $\sec^4 \phi/2$. This is a physical impracticability with a feed that is typically small in wavelengths.

A different approach to this problem, and one which has proved successful, was first suggested by B. Ye Kinber[12] and then developed by K. A. Green[13] and V. Galindo[14]. It involves use of a Cassegrain-type feed (see Section 10.3) with the shapes of both the hyperboloid and paraboloid modified in order to control the secondary aperture distribution. To understand the approach, refer to Figure 10.28 and assume that a point source is placed at $(f, 0, 0)$ with its radiation intensity $I(\phi)$ rotationally symmetric about the X-axis. The reflectors R_1 and R_2 are also rotationally symmetric. If R_1 is a paraboloid and R_2 is a properly placed hyperboloid with the proper eccentricity, it was shown in Section 10.3 that the rays leaving R_2 and traveling to R_1 all appear to be coming from a virtual phase center (the focus of the paraboloid) and thus that the rays leaving R_1 are collimated. However, these conventional reflector shapes do not result in maximum aperture efficiency. What is needed is to lift the restrictions on reflector shapes so that an arbitrary primary pattern $I(\phi)$ can be transformed to a specified secondary aperture distribution $P(y)$.

To see how this is accomplished, assume that the desired secondary distribution in the plane $x = a$ is to be equiphase. Then

$$r + r' + r'' = \text{constant} = (b - f) + b + a \tag{10.74}$$

From Figure 10.28 it is apparent that

$$r' = \frac{y - r\sin\phi}{\sin\beta} \qquad r'' = a - x$$

[12]B. Ye Kinber, "On Two-Reflector Antennas," *Radio Eng. Electron Phys.*, 7 (1962), 914–21.

[13]K. A. Green, "Modified Cassegrain Antenna for Arbitrary Aperture Illumination," *IEEE Trans. Antennas Propagat.*, AP-11 (1963), 589–90.

[14]V. Galindo, "Design of Dual Reflector Antennas with Arbitrary Phase and Amplitude Distributions," *IEEE Trans. Antennas Propagat.*, AP-12 (1964), 403–8.

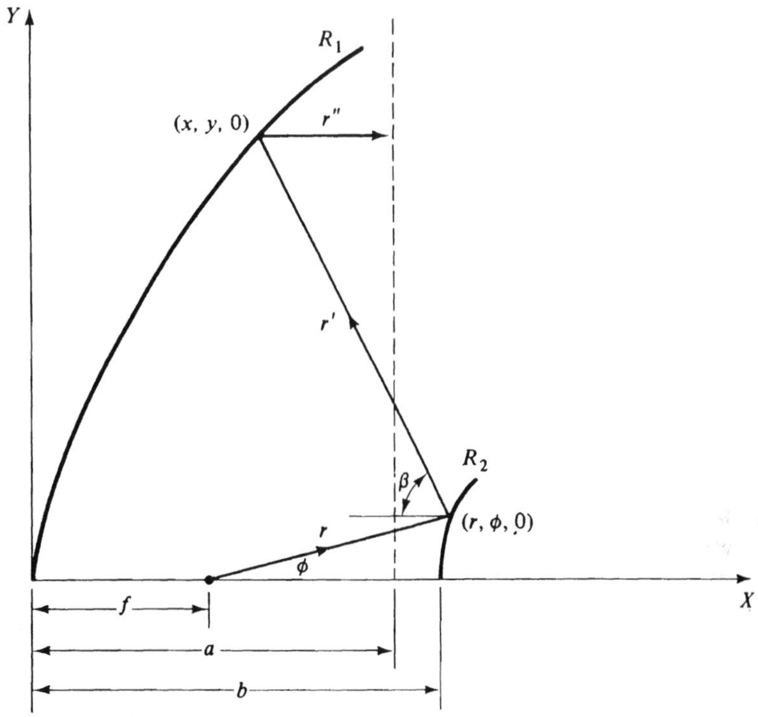

Fig. 10.28 Geometry for a Modified Cassegrain Feed and Reflector

and thus (10.74) becomes

$$r + \frac{y - r \sin \phi}{\sin \beta} - x = 2b - f \tag{10.75}$$

Since the primary pattern and reflectors are rotationally symmetric, the secondary aperture distribution will also have rotational symmetry. As a consequence, the primary power $I(\phi)2\pi \sin \phi \, d\phi$ reappears as the secondary power $P(y)2\pi y \, dy$. When these powers are equated and integrated, normalization gives

$$\frac{\int_0^{\phi} I(\phi') \sin \phi' \, d\phi'}{\int_0^{\phi_{max}} I(\phi') \sin \phi' \, d\phi'} = \frac{\int_0^{y} P(y') y' \, dy'}{\int_0^{y_{max}} P(y') y' \, dy'} \tag{10.76}$$

Maximum directivity (and aperture efficiency) will occur if $P(y)$ is a constant. In that case, (10.76) becomes

$$y^2 = y_{\max}^2 \frac{\int_0^{\phi} I(\phi') \sin \phi' \, d\phi'}{\int_0^{\phi_{max}} I(\phi') \sin \phi' \, d\phi'} \tag{10.77}$$

It was shown in Section 10.5 that the slope of a reflector is measured by the tangent of the angle of incidence of a ray. (See Equation 10.42 as an example of this.) For the reflectors R_1 and R_2, this connection takes the forms

$$\frac{dy}{dx} = \cot \frac{\beta}{2} \tag{10.78}$$

$$\frac{1}{r} \frac{dr}{d\phi} = \tan \frac{\phi + \beta}{2} \tag{10.79}$$

as can be seen from a study of Figure 10.28.

Equations 10.75 and 10.77 through 10.79 comprise a set of four relations from which the dependent variables r, x, y, and β can be determined as functions of ϕ. The function $r(\phi)$ defines the shape of the subreflector R_2 and the parametric set $x(\phi)$, $y(\phi)$ serves to specify the shape of R_1. A computer program of moderate complexity can be written to effect a solution.

The analysis assumes a rotationally symmetric source, which might at first seem to be an invalidating idealization. However, P. D. Potter[15] has described the design of a conical feed horn which satisfies this criterion. W. F. Williams[16] reports the results of an experiment in which such a horn was used to feed a conventional Cassegrain system and a modified Cassegrain of the same size. There was a measured increase in directivity of 1 dB and a change in side lobe level from -23 dB to -17 dB, both figures consistent with the achievement of a secondary aperture distribution uniform in amplitude and phase.

An additional advantage of modified Cassegrain antenna systems is that the spillover at the subreflector can be drastically reduced without penalty to the overall aperture efficiency.

Computer solutions for a modified Cassegrain reveal that R_1 is slightly altered from a paraboloidal shape (see Figure 5 of Williams' paper as an example). However, the change in the shape of the subreflector is more pronounced. A typical result is reported by A. C. Ludwig[17] and is shown in Figure 10.29. It can be observed that the inner region of the modified subreflector is more curved to redirect some of the higher intensity central rays toward the outer region of the high gain reflector R_1.

[15] P. D. Potter, "A New Horn Antenna with Suppressed Sidelobes and Equal Beamwidths," *Microwave Journal*, (1963), pp. 71–8.

[16] W. F. Williams, "High Efficiency Antenna Reflector," *Microwave Journal* (1965), pp. 79–82.

[17] A. C. Ludwig, "Shaped Reflector Cassegrainian Antennas." SPS No. 37–35, Vol. IV (Pasadena, California: Jet Propulsion Laboratories, 1965), 266–8.

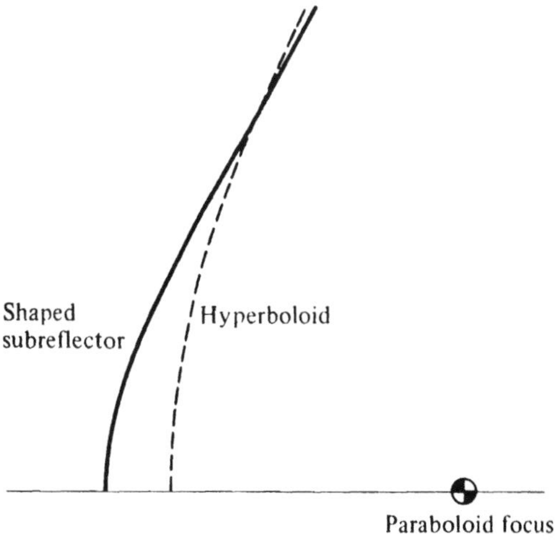

Fig. 10.29 Comparison of Shaped Subreflector and Hyperboloid (After Ludwig[17])

10.10 Single Surface Dielectric Lenses

Attention will now be turned to lens antennas, which share many characteristics with the reflector antennas that have been under discussion in the first part of this chapter. Both antenna types use a low-gain feed (such as a horn) to illuminate a large, high-gain structure (reflector or lens), the latter serving the purpose of redirecting the diverging primary rays so that they become a set of secondary rays that will produce a desirable secondary pattern. The reflector does this by back-scattering the primary rays, the basic process being one of reflection. The lens accomplishes the same result by forward-scattering the primary rays, the basic process being one of diffraction. For this reason, lens antennas have one inherent advantage over reflector antennas—the feed is not in the path of the secondary rays. An offset to this advantage lies in the fact that lens antennas are typically thicker, heavier, and more difficult to construct than reflectors. Despite this, there are applications in which a lens antenna is clearly the superior choice.

A simple introduction to the subject of lens antenna design can be obtained by posing the situation suggested by Figure 10.30. An equiphase line source is placed along the Z-axis and emits a family of diverging rays which, if the line source is long enough, can all be assumed to be parallel to the XY-plane. One of these rays is indicated by the line segment $\overline{OA}$, emerging from the line source at an angle ϕ with respect to the X-axis.

All of space will be assumed to consist of two homogenous, isotropic, lossless media, divided by the single cylindrical interface that intersects the XY-plane in the curve AA_0A'. To the left of this interface, the constitutive parameters are μ_1 and ϵ_1 and to the right of the interface they are μ_2 and ϵ_2. The relative refractive index n will be defined by

$$n = \sqrt{\frac{\mu_2\epsilon_2}{\mu_1\epsilon_1}} \tag{10.80}$$

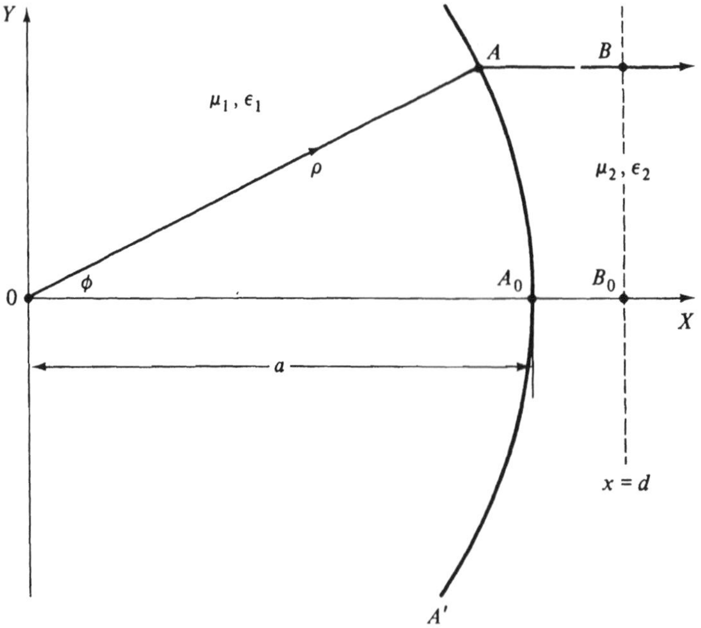

Fig. 10.30 Geometry for a Dielectric Lens

The design problem is to select the shape of the curve AA_0A' so that all the secondary rays are collimated, with the plane $x = d$ representing a secondary wave front.

Equation 10.25 can be applied to this situation, with the argument made that the phase delay along the path OAB should be the same as along the path OA_0B_0. If $\overline{OA} = \rho$ and $\overline{OA_0} = a$, then

$$k_1\rho + k_2(d - \rho\cos\phi) = k_1a + k_2(d - a) \tag{10.81}$$

in which $k_1 = \omega\sqrt{\mu_1\epsilon_1}$ and $k_2 = \omega\sqrt{\mu_2\epsilon_2}$ are the wave numbers in the two media. Equation 10.81 simplifies to

$$\frac{\rho}{a} = \frac{n - 1}{n\cos\phi - 1} \tag{10.82}$$

which is the equation of a conic section of ellipticity n with a focus at the origin.[18] This can be appreciated more fully when (10.82) is reexpressed in cartesian coordinates. If the point A is identified by the triplet $(x, y, 0)$, transformation of (10.82) gives

[18]See, for example, C. E. Love, *Analytic Geometry*, 3rd ed. (New York: Macmillan Co., 1938), pp. 75–99.

$$\frac{\left(x - \frac{n}{n+1}a\right)^2}{\left(\frac{a}{n+1}\right)^2} - \frac{y^2}{\left(\frac{n-1}{n+1}\right)a^2} = 1 \tag{10.83}$$

Two classes of solutions are represented by (10.83).

CASE 1. n < 1 In this case, (10.83) can be written in the form

$$\frac{(x')^2}{(a')^2} + \frac{y^2}{(b')^2} = 1 \tag{10.84}$$

which is the equation of an elliptic cylinder. In (10.84),

$$a' = \frac{a}{1+n} \qquad b' = \sqrt{\frac{1-n}{1+n}}a \qquad b' = a'\sqrt{1-e^2} \tag{10.85}$$

with e the ellipticity. Elimination of a' from the above expressions gives $e = n$, as asserted earlier. The foci are at $x' = \pm c'$, with c' given by

$$c' = \sqrt{(a')^2 - (b')^2} = \frac{n}{n+1}a \tag{10.86}$$

Since $x' = x - (n/n+1)a$, it follows that, in the original coordinate system, the foci are at the positions

$$x = 0, \frac{2n}{n+1}a \tag{10.87}$$

This means that the line source should be placed at the far focus of the elliptic cylinder if the requirement is to collimate the secondary rays.

CASE 2. n > 1 In this case, (10.83) can be written in the form

$$\frac{(x')^2}{(a')^2} - \frac{(y)^2}{(b')^2} = 1 \tag{10.88}$$

which is the equation of a hyperbolic cylinder. In (10.88)

$$a' = \frac{a}{n+1} \qquad b' = \sqrt{\frac{n-1}{n+1}}a \qquad b' = a'\sqrt{e^2-1} \tag{10.89}$$

from which, once again, $e = n$. The foci are at the positions $x' = \pm c'$, with c' given by

$$c' = \sqrt{(a')^2 + (b')^2} = \frac{n}{n+1}a \tag{10.90}$$

which is the same result as in the elliptical case. Hence, the foci are at positions calculable from (10.87), and the line source should once again be placed at the left focus if the secondary rays are to be collimated.

If the line source is replaced by a *point* source at the origin, a simple repetition of the foregoing development leads to the conclusion that, for collimation, the interface should be described by

$$\frac{\left(x - \dfrac{n}{n+1}a\right)^2}{\left(\dfrac{a}{n-1}\right)^2} - \frac{y^2 + z^2}{\left(\dfrac{n-1}{n+1}\right)a^2} = 1 \tag{10.91}$$

Hence, for $n < 1$ the interface is an ellipsoid; for $n > 1$, it is a hyperboloid. In both cases, the point source is at the left focus.

Practical lens antennas can be conceived based on this information. Figure 10.31a shows a dielectric lens cross section in which the inner contour is a circle and the outer contour is an ellipse, with the exterior region free space. The center of the circle and the left focus of the ellipse are coincident at the source point (line) of the primary rays. These rays pass undiffracted through the inner surface, but are diffracted by the outer surface and emerge collimated. A translation parallel to the Z-axis of the contour shown in Figure 10.31a creates an elliptic cylinder lens ($n < 1$), while a rotation of the contour about the X-axis produces an ellipsoidal lens ($n < 1$).

Similarly, Figure 10.31b shows a dielectric lens cross section in which the inner contour is a hyperbola and the outer contour is a straight line. The primary rays are diffracted at the inner surface and are collimated in the lens. They suffer no further

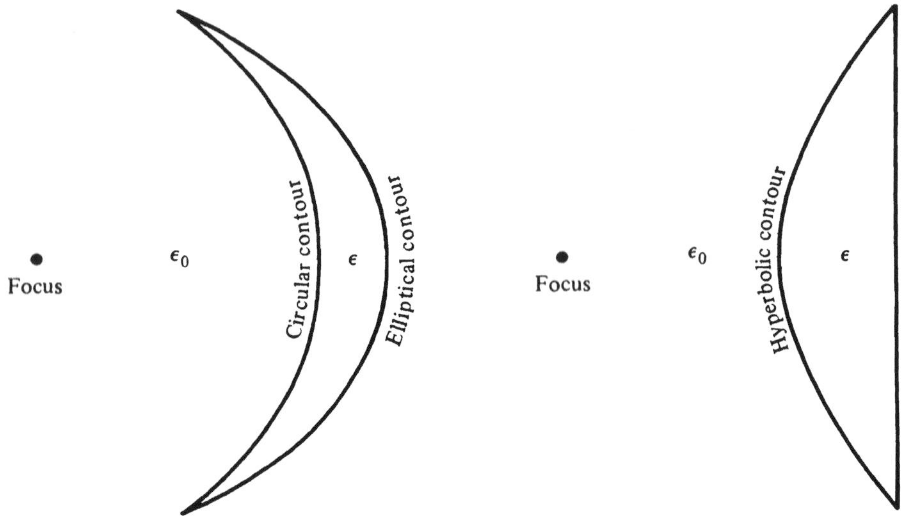

(a) Elliptical lens (b) Hyperbolic lens

Fig. 10.31 Cross Sections of Dielectric Lenses that Can Collimate Secondary Rays diffraction upon emergence from the surface, so they remain collimated. A translation of the contour shown in Figure 10.31b results in a hyperbolic cylinder lens ($n > 1$), while a rotation of the contour generates a hyperboloidal lens ($n > 1$).

There has been an implied assumption in this development that the boundary conditions on the electromagnetic waves at the interface will cause the secondary rays to be horizontal, that is, that the ray trace OAB in Figure 10.30 will satisfy Snell's law of refraction. It is a simple matter to show that, if the equation describing the interface is given either by (10.83) or (10.91), this condition is satisfied. The proof, which is left as an exercise, is similar to what was done in Section 10.3, where it was demonstrated that Snell's law of reflection is obeyed by a parabolic cylinder antenna with a line source placed at its focus.

Lenses of the types shown in Figure 10.31 are called *single-surface* lenses because all of the diffraction occurs at one interface. Two surface lenses can also be devised, but their analysis is somewhat more involved and will not be undertaken here.[19]

10.11 Stepped Lenses

If the angle subtended by the feed from the extremities of the lens is large, as it invariably is in practice, the thickness of the lens (in the x-dimension) varies markedly from center to extremity, as can be seen for the elliptic and hyperbolic lenses shown in Figure 10.31. This causes the lens to be bulky and heavy, a disadvantage which can be overcome by stepping. The basic idea is suggested in Figure 10.32. Zones have been created in either the inner or outer lens surface by stepping the thickness. This must be done so that the rays which pass through different zones are still collimated and in phase in the secondary aperture $x = d$.

If the zones are labeled 0, 1, 2, 3, . . . , counting from the edge of the lens in toward the center, stepping of the different contour segments of the lenses displayed in Figure 10.27 can be accomplished as follows:

(a) ELLIPTICAL LENS—CIRCULAR SEGMENT (FIGURE 10.32a) Let r_0 and r_m be the radii of the zeroth and mth zones. An equiphase secondary aperture distribution will be achieved if

$$k_0 r + k(r_m - r_0) = k_0 r_m + 2\pi m$$

with k_0 and k the wave numbers in free space and in the dielectric. This relation simplifies to

$$r_m - r_0 = \frac{m\lambda_0}{n - 1} \tag{10.92}$$

in which n is the refractive index of the dielectric and λ_0 is the free-space wavelength.

[19]For an introduction to the subject, see E. A. Wolff, *Antenna Analysis* (New York: John Wiley and Sons, Inc., 1966), pp. 468–71.

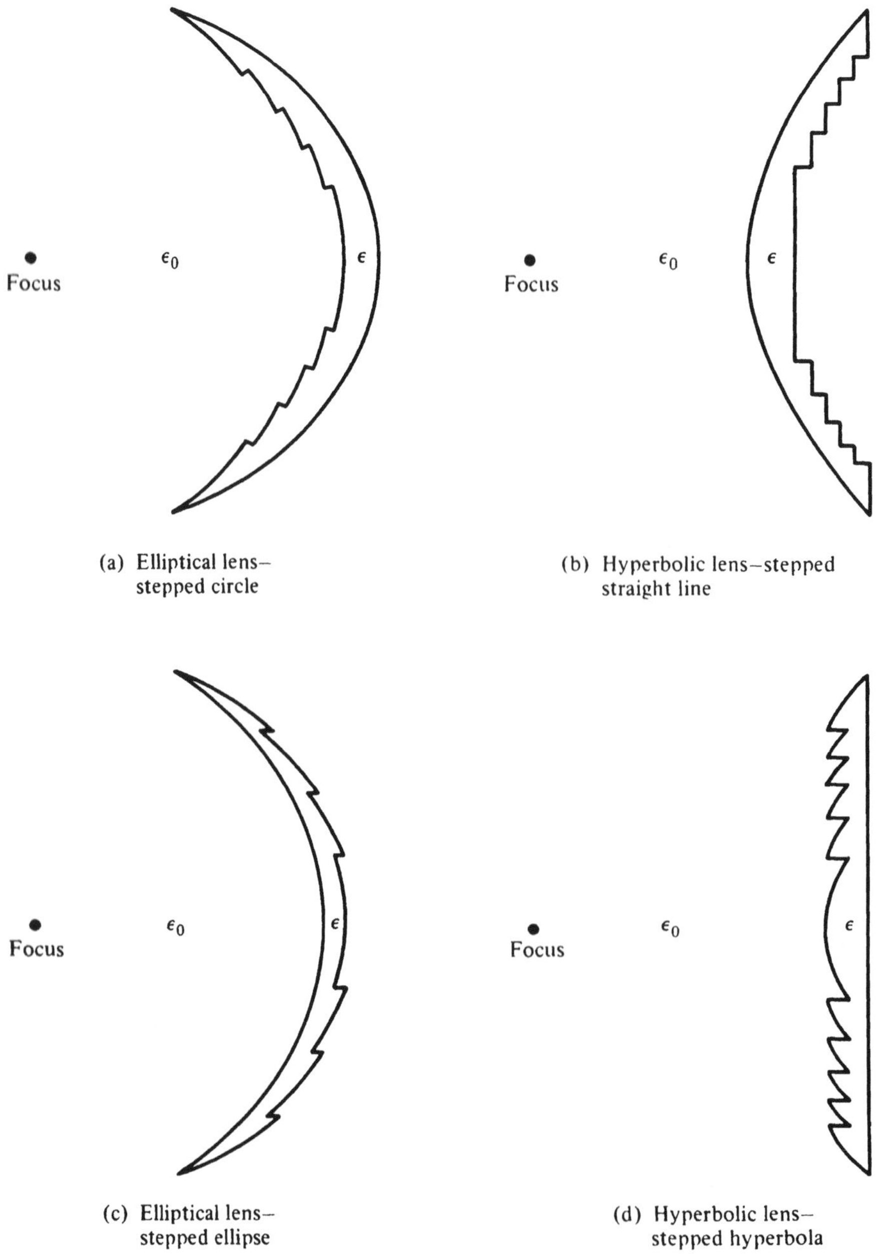

Fig. 10.32 Cross Sections of Stepped Dielectric Lens

Hence the radii steps are equal. The extent of each zone is determined by the minimum and maximum thicknesses required for structural strength.

(b) HYPERBOLIC LENS—STRAIGHT LINE SEGMENT (FIGURE 10.32b) Let x_0 and x_m be the longitudinal dimensions of the zeroth and mth zones. The necessary condition is that

$$k(x_0 - x_m) = k_0(x_0 - x_m) + 2\pi m$$

from which

$$x_0 - x_m = \frac{m\lambda_0}{n - 1} \tag{10.93}$$

Once again the steps are equal and the extents of the different zones are governed by specification of the minimum and maximum thicknesses of the dielectric material.

(c) ELLIPTICAL LENS—ELLIPTICAL SEGMENT (FIGURE 10.32c) For this case, the condition (10.81) needs to be replaced by

$$k_1\rho_m + k_2(d - \rho_m \cos\phi) = k_1 a + k_2(d - a) - 2\pi m$$

which reduces to

$$\rho_m(1 - n\cos\phi) = (1 - n)\left(a - \frac{m\lambda_1}{1 - n}\right) \tag{10.94}$$

with λ_1 the wavelength in medium 1.

Equation 10.94 is in exactly the same form as (10.81), with the intersection of the ellipse and the X-axis occurring at $a - m\lambda_1/(1 - n)$ rather than at a. Hence (10.94) represents a family of confocal ellipses, all with their far foci at the origin and with equal increments in their stepped x-intercepts. The extent of each zone is dictated by the spread of thicknesses specified for the dielectric lens.

(d) HYPERBOLIC LENS—HYPERBOLIC SEGMENT (FIGURE 10.32d) For this case, (10.81) is once again replaced by (10.94). Simplification gives

$$\rho_m(n\cos\phi - 1) = (n - 1)\left(a + \frac{m\lambda_1}{n - 1}\right) \tag{10.95}$$

Comparison with (10.81) reveals that (10.95) is a family of confocal hyperbolas, all with their far foci at the origin, and with equal increments in their x-intercepts. As in the other cases, zonal extents are governed by thickness specifications.

The success of this stepping technique depends on the insignificance of scattering at zone boundaries. Obviously, the greater the extent of each zone in wavelengths, the less serious will be this scattering effect.

The four stepped contours shown in Figure 10.32 can be used to generate cylindrical lenses, for use with line sources, by translation parallel to the Z-axis. Alternatively, they can be used to generate rotationally symmetric lenses, for use with point sources, by rotation about the X-axis.

10.12 Surface Mismatch, Frequency Sensitivity, and Dielectric Loss for Lens Antennas

The analysis of single surface lenses in Section 10.10 dealt with the primary and secondary rays that progress in a forward direction toward the secondary aperture at $x = d$. However, at any interface between dielectric media with different constitutive parameters, reflection can also occur. For the elliptical lens shown in Figure 10.31a, the primary rays are normally incident at the inner lens surface, which is either a portion of a circular cylinder or sphere. Locally, the reflection which occurs is the same as though the interface were planar (geometrical optics approximation). It is shown in many standard texts[20] that the reflection coefficient of a plane wave normally incident on an infinite dielectric/air interface is independent of polarization and given by

$$\Gamma = \frac{1 - n}{1 + n} \tag{10.96}$$

All of the reflected rays from this cylindrical (spherical) lens surface come to a focus at the feed. If the feed is well matched to free space when the lens is absent, the reflection coefficient measured in the feed when the lens is present, due to this one source of reflection, is given by (10.96). For example, if a polystyrene lens for which $\epsilon/\epsilon_0 = 2.56$ is used then $n = 1.16$ and $\Gamma = 0.23$. The input VSWR at the feed terminals would be 1.6, and 5% of the primary power would be returned to the feed under these conditions.

The situation is more complicated than just described because there is also reflection off the elliptical cylinder (ellipsoid) that comprises the second surface of the lens depicted in Figure 10.31a. Because the direction of the normal to this surface varies from point to point on the surface, the aggregate back scattering effect is nonfocused. Reflection from this second surface primarily appears as a contribution to the total field in the half-space behind the feed. Higher-order reflections within the lens can usually be ignored. As in the analogous problem of reflection from a paraboloid, there is some depolarization caused by back-scattering off an ellipsoidal lens surface. This effect is essentially not present in the case of the long line source and elliptical cylinder lens.

An analysis of the surface mismatch for a hyperbolic lens is similar but slightly more complicated. With reference to Figure 10.31b, back-scattering off the hyperbolic cylinder (hyperboloid) is nonfocused and contributes to the total field behind the feed, with some depolarization in the case of the hyperboloid. Reflection off the planar surface is collimated and, if n is not too different from unity, most of it is

[20]See, for example, E. C. Jordan and K. G. Balmain, *Electromagnetic Waves and Radiating Systems*, 2nd ed. (Englewood Cliffs, New Jersey: Prentice-Hall, Inc., 1968), pp. 143–44.

transmitted through the hyperbolic surface and focused at the feed. In such cases the reflection coefficient given by (10.96) is applicable as an approximation in the case of the hyperbolic lens.

When the focused back-scattering caused by dielectric/air mismatch at one of the lens surfaces causes an unacceptably high input VSWR at the feed port, a remedy is to use a matching section in the lens. This consists of a quarter-wavelength thick layer of dielectric with refractive index $n' = n^{1/2}$, bonded to the lens surface which is causing the focused reflected rays. Such mismatch correction is obviously frequency-sensitive.

All of the foregoing remarks are unaltered if surface mismatch is considered for the stepped lenses shown in Figure 10.32. However, stepping introduces another effect which is not present in the basic, unstepped, uncorrected lenses of Figure 10.31. Those prototype lenses are frequency-independent, to the extent that geometrical optics approximations are valid. This is not the case when stepping is introduced. The point is readily appreciated upon return to Equations 10.92 through 10.95, each of which shows that the step increments are $\lambda/(1 - n)$, with λ the wavelength in one or the other of the two media. If the steps are properly dimensioned at the design wavelength λ_d, they will not be correct for a slightly different wavelength $\lambda_1 = \lambda_d + \Delta\lambda$.

When there are M zones in the stepped lens surface, the path length difference for rays that go through the innermost and outermost zones is

$$\Delta L_d = (M - 1)\lambda_d \qquad \Delta L_1 = (M - 1)\lambda_1$$

at the two wavelengths. The *change* in path length difference due to a change in frequency is

$$\delta = \Delta L_1 - \Delta L_d = (M - 1)(\lambda_1 - \lambda_d) = (M - 1)\Delta\lambda$$

and thus a measure of the bandwidth is

$$B = \frac{2\Delta\lambda}{\lambda_d} = \frac{2\delta}{(M - 1)\lambda_d} \tag{10.97}$$

A frequently used criterion for pattern degradation due to phase errors across the aperture is that the wavefront should not have a curvature of more than one-eighth wavelength from center to edge ($\delta = \lambda/8$). In this case,

$$B = \frac{25\%}{M - 1} \tag{10.98}$$

It is clear from this relation that the larger the lens aperture, and thus the more zones needed if stepping is employed, the more narrow band the lens becomes.

Another effect that can be estimated in judging the performance of a dielectric lens is the attenuation in the dielectric. The complex permittivity can be expressed as

$$\epsilon = \epsilon_0(\epsilon' - j\epsilon'') = \epsilon_0\epsilon'\left(1 - j\frac{\epsilon''}{\epsilon'}\right) \tag{10.99}$$

The ratio ϵ''/ϵ', which is small for a good dielectric, is often given in the form

$$\frac{\epsilon''}{\epsilon'} = \tan \delta \tag{10.100}$$

with $\tan \delta$ called the *loss tangent* of the dielectric. Values of ϵ' and $\tan \delta$ can be easily obtained by measurement and are usually provided by the manufacturer over the frequency band of interest.

The complex propagation constant is given by

$$\gamma = \alpha + j\beta = [j\omega\mu_0(j\omega\epsilon_0\epsilon')(1 - j\tan\delta)]^{1/2} = jnk(1 - j\tan\delta)^{1/2} \tag{10.101}$$

with $k = 2\pi/\lambda_0$ the free-space wave number and $n = (\epsilon')^{1/2}$ the refractive index. If the loss tangent is small, the attenuation factor α is given to good approximation by

$$a = \frac{nk}{2}\tan\delta = \left(\frac{\pi}{\lambda}\right) n \tan\delta \quad \text{nepers per wavelength} \tag{10.102}$$

For a zoned lens, the average thickness t is given roughly by

$$\frac{t}{\lambda} = \frac{1}{n-1} \tag{10.103}$$

and hence the attenuation in the dielectric can be estimated by the formula

$$t = 27.3\frac{n}{n-1}\tan\delta \quad dB \tag{10.104}$$

As an example, for a stepped polystyrene lens used at X-band (10 GHz), $\epsilon' = 2.54$ and $\tan\delta = 4.3 \cdot 10^{-4}$, and the loss estimate is 0.02 dB.

10.13 The Far Field of a Dielectric Lens Antenna

The same technique that was used in Section 10.7 to find the far field of a reflector antenna can be employed to determine the far field of a dielectric lens antenna. With reference once again to Figure 10.30, assume a line source and hence a lens boundary formed by translating the curve AA_0A' parallel to the Z-axis. If $I(\phi)$ watts per radian-meter is the primary pattern and $P(y)$ watts per square meter is the secondary power distribution in the aperture plane $x = d$, then $I(\phi)\,d\phi = P(y)\,dy$. But $y = \rho\sin\phi$ and hence

$$dy = \rho\cos\phi\,d\phi + \sin\phi\,d\rho = \left(\rho\cos\phi + \sin\phi\frac{d\rho}{d\phi}\right)d\phi$$

From (10.82),

$$\frac{d\rho}{d\phi} = \frac{an(n-1)\sin\phi}{(n\cos\phi - 1)^2}$$

and thus

$$I(\phi) = \frac{a(n-1)}{n\cos\phi - 1}\left(\cos\phi + \frac{n\sin^2\phi}{n\cos\phi - 1}\right)P(y)$$

from which

$$P(y) = \frac{(n\cos\phi - 1)^2}{a(n-1)(n-\cos\phi)}I(\phi) \tag{10.105}$$

Equation 10.105 applies whether $n < 1$ (elliptic cylinder lens) or $n > 1$ (hyperbolic cylinder lens). The field amplitude of the equiphase distribution in the plane $x = d$ is given by $P^{1/2}(y)$. With the polarization specified, the equivalent Huyghens sources can be determined and the far field computed.

If a point source is used, so that the lens boundary is generated by rotating the curve AA_0A' of Figure 10.30 about the X-axis, a similar analysis can be used to obtain the secondary aperture distribution. With spherical coordinates (ρ, ϕ, ψ) erected at the focus, and $I(\phi, \psi)$ the radiation pattern of the feed, a power balance gives

$$I(\phi, \psi)\sin\phi\, d\phi\, d\psi = P(y, \psi)y\, dy\, d\psi \tag{10.106}$$

where (y, ψ) are polar coordinates in the plane $x = d$ and $P(y, \psi)$ is measured in watts per radian-meter. Since $y = \rho\sin\phi$, Equation 10.106 reduces to

$$I(\phi, \psi)\, d\phi = P(y, \psi)\, \rho dy$$

As before, dy can be related to $d\phi$, the result being that

$$P(y, \psi) = \frac{(n\cos\phi - 1)^3}{a^2(n-1)^2(n-\cos\phi)}I(\phi, \psi) \tag{10.107}$$

Equation 10.107 applies whether $n < 1$ (ellipsoidal lens) or $n > 1$ (hyperboloidal lens). As before, $P^{1/2}(y, \psi)$ gives the field amplitude of the equiphase distribution in the plane $x = d$, so all the information needed to calculate the far-field pattern is embodied in (10.107).

Plots of Equations 10.105 and 10.107, under the assumption of an isotropic primary source, are shown in Figure 10.33 for $n = 0.5$ (elliptical lens) and $n = 2$ (hyperbolic lens). It can be observed that the action of a hyperbolic lens is to impose more taper on the primary distribution, whereas an elliptic lens lessens the taper. Because of the resulting penalty in aperture efficiency, hyperbolic lenses are not attractive for applications in which the lens would subtend an angle at the feed much greater than $2\phi_{max} = 60°$. On the other hand, if an elliptic lens has a subtended angle $2\phi_{max}$ such that $\cos\phi_{max} < n$, the pole in (10.105) and (10.107) can cause design tolerance difficulties unless the taper in the primary distribution dominates this effect.

A specific example of the use of these results is posed in Problems 10.21 and 10.22 at the end of this chapter.

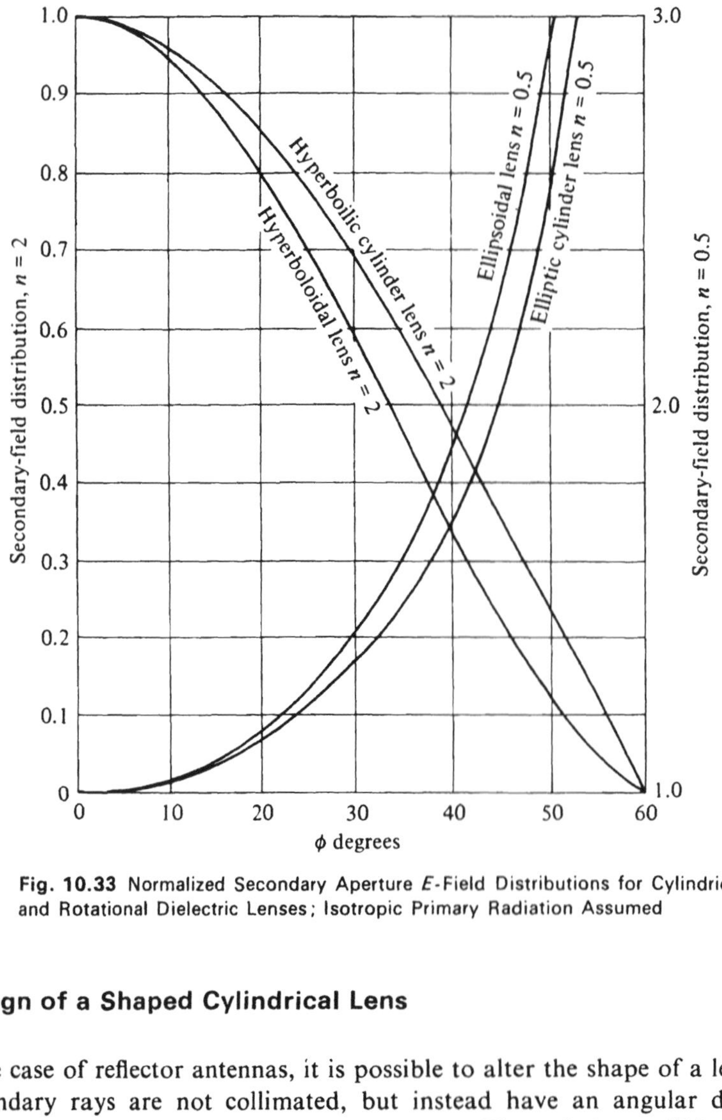

Fig. 10.33 Normalized Secondary Aperture *E*-Field Distributions for Cylindrical and Rotational Dielectric Lenses; Isotropic Primary Radiation Assumed

10.14 The Design of a Shaped Cylindrical Lens

As in the case of reflector antennas, it is possible to alter the shape of a lens so that the secondary rays are not collimated, but instead have an angular distribution corresponding to a more general secondary pattern. The technique will be illustrated for the hyperbolic cylinder lens of Figure 10.31b, with the planar exit surface modified to decollimate the secondary rays in some desired manner.

The situation is suggested by Figure 10.34. A family of horizontal rays travels through the lens and the one shown impinges on the exit surface at the point (x, y), making an angle i with the normal to the surface at that point. The refracted ray departs at an angle $r = i + \theta$ to the normal, with θ the angle this ray makes with the horizontal.

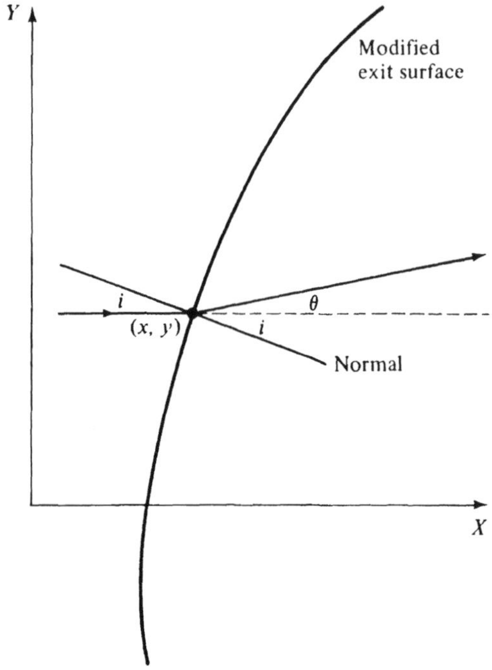

Fig. 10.34 A Hyperbolic Cylinder Lens Modified to Give a Shaped Secondary Pattern

A unit normal vector at (x, y) is given by

$$\mathbf{1}_n = \frac{d\mathbf{s}}{ds} \times \mathbf{1}_z = \frac{\mathbf{1}_x\, dy - \mathbf{1}_y\, dx}{ds} \tag{10.108}$$

with $d\mathbf{s}$ a displacement along the exit contour to the neighboring point $(x + dx, y + dy)$. It follows that

$$\mathbf{1}_z \sin i = \mathbf{1}_n \times \mathbf{1}_x = \mathbf{1}_z \frac{dx}{ds} \tag{10.109}$$

Similarly, if a unit vector $\mathbf{1}_w$ parallel to the exiting ray is defined by

$$\mathbf{1}_w = \mathbf{1}_x \cos\theta + \mathbf{1}_y \sin\theta \tag{10.110}$$

then

$$\mathbf{1}_z \sin r = \mathbf{1}_n \times \mathbf{1}_w = \mathbf{1}_z\left(\sin\theta \frac{dy}{ds} + \cos\theta \frac{dx}{ds}\right) \tag{10.111}$$

With the exit region free space, Snell's law of refraction gives

$$\frac{\sin r}{\sin i} = n = \frac{\sin \theta \, dy + \cos \theta \, dx}{dx}$$

which can be recast in the form

$$dx = \frac{\sin \theta}{n - \cos \theta} dy \tag{10.112}$$

If the connection between θ and y is known, this result can be integrated to give

$$x - x_1 = \int_{y_1}^{y} \frac{\sin \theta(y')}{n - \cos \theta(y')} dy' \tag{10.113}$$

thus establishing the shape of the exit contour as the function $x(y)$.

As before, the function $\theta(y)$ can be deduced from

$$\frac{\int_{y_1}^{y} P(y')\, dy'}{\int_{y_1}^{y_2} P(y')\, dy'} = \frac{\int_{\theta_1}^{\theta} W(\theta')\, d\theta'}{\int_{\theta_1}^{\theta_2} W(\theta')\, d\theta'} \tag{10.114}$$

with $P(y)$ the power distribution in the lens, related to the feed pattern by Equation 10.105; the function $W(\theta)$ watts per radian-meter is the desired far-field secondary pattern.

A design problem illustrating the use of this technique is posed in Problem 10.25 at the end of this chapter.

10.15 Artificial Dielectrics: Discs and Strips

A major practical disadvantage of lens antennas composed of homogeneous, isotropic dielectrics (such as polystyrene) is their weight. Even with stepping, such antennas are so heavy that their use is precluded in all but some ground-based applications. For this reason considerable interest has been shown in the development of inhomogeneous materials with low density and high effective permittivity. Several *artificial dielectrics* (as such materials are called) have been devised with properties that make them well suited for use in lens antennas.

An early candidate was a composite consisting of a lightweight, low permittivity host material (such as polyfoam) in which was imbedded a regular three-dimensional array of conducting spherical particles. This is a medium which can be analyzed with little difficulty[21] and one finds that the equivalent permittivity exceeds ϵ_0 for practical sphere sizes and spacings. However, the equivalent permeability is less than μ_0 and this offset results in a refractive index that rises only to a maximum

[21] See, for example, J. Brown, "Lens Antennas," *Antenna Theory, Part II*, ed. R. E. Collin and F. J. Zucker (New York: McGraw-Hill Book Co., Inc., 1969), Chapter 18, pp. 105–8.

value of 1.27 when the spheres are touching. This value of n is too low for most lens antenna applications. However, the analysis of an array of conducting spheres reveals an important fact. Were the spheres to become spheroids, flattened in the direction of propagation of an electromagnetic wave passing through the medium, the permeability reduction would be lessened. With total flattening, so that the metallic spheres are replaced by metallic discs lying in equispaced planes perpendicular to the direction of propagation, the permeability effect vanishes. Not so the favorable permittivity effect. The increase in permittivity caused by the presence of the metallic discs can be substantial, and refractive indices as high as two or three can be achieved with practical disc sizes and spacings.

The disc dielectric is shown in Figure 10.35a. By symmetry, its refractive index is independent of the polarization of a normally incident electromagnetic plane wave. Because of this it can be used as a lens in conjunction with feeds that are linearly polarized (either horizontally or vertically), circularly polarized, or elliptically polarized.

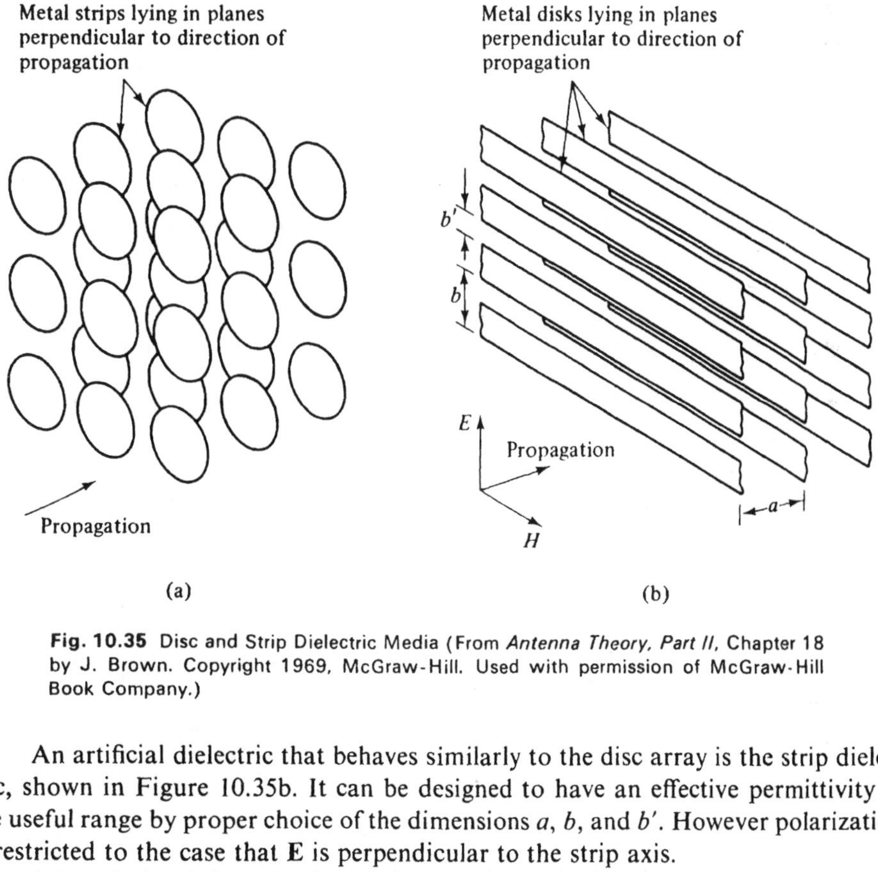

Fig. 10.35 Disc and Strip Dielectric Media (From *Antenna Theory, Part II*, Chapter 18 by J. Brown. Copyright 1969, McGraw-Hill. Used with permission of McGraw-Hill Book Company.)

An artificial dielectric that behaves similarly to the disc array is the strip dielectric, shown in Figure 10.35b. It can be designed to have an effective permittivity in the useful range by proper choice of the dimensions a, b, and b'. However polarization is restricted to the case that **E** is perpendicular to the strip axis.

An analysis of the behavior of a strip dielectric medium relies on several techniques that have been used in earlier chapters. First, if a uniform plane wave is normally incident on the strips, as suggested in Figure 10.35b, the same electric field will be induced in every gap in a common transverse plane. The forward and backward **E**-field scattering from this plane will be *symmetrical* (See Appendix F). This is equivalent to a shunt obstacle in a transmission line (compare with Equation 3.52). The problem is similar to scattering from a capacitive iris in a rectangular waveguide and can be analyzed by a quasi-static method if $b'/\lambda \ll 1$. With losses ignored, the normalized equivalent shunt admittance is found to be purely susceptive and given by[22]

$$\frac{B}{G_0} = \frac{2kb}{\pi} \ln\left[\csc\left(\frac{\pi b'}{2b}\right)\right] \tag{10.115}$$

It follows that the strip dielectric medium of Figure 10.35b is equivalent to a periodically loaded transmission line, with shunt capacitive elements placed a units apart, as shown in Figure 10.36a. That the elements should be capacitive is due to the size restriction on b/λ, which causes the local **E**-field to store energy electrostatically.

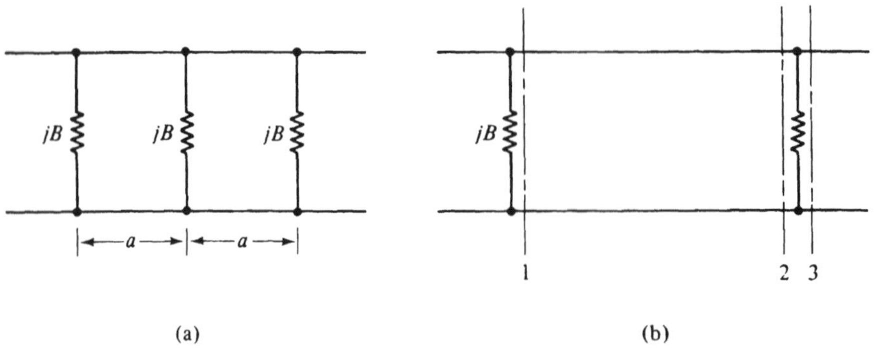

Fig. 10.36 A Periodically Loaded Transmission Line Equivalent to a Strip Dielectric

Because of the lossless periodic loading of the transmission line, the mode voltages at corresponding points in successive sections differ only by a phase constant ϕ. The same is true for the mode currents. Hence, with reference to Figure 10.36b, one can write

$$\begin{aligned} V_2 &= V_1 \cos ka - jI_1 Z_0 \sin ka = V_3 = V_1 e^{-j\phi} \\ I_2 &= -jV_1 G_0 \sin ka + I_1 \cos ka = I_3 + jBV_3 = I_1 e^{-j\phi} + jBV_1 e^{-j\phi} \end{aligned} \tag{10.116}$$

in which (V_1, I_1) are the mode voltage and current at cross section 1, and so on. The cross sections are chosen so that the first is infinitestimally to the right of a shunt element; the second and third straddle the next shunt element, infinitesimally to

[22]See, for example, N. Marcuvitz, *Waveguide Handbook*, Vol. 10 of MIT Rad. Lab. Series (New York: McGraw-Hill Book Co., Inc., 1951), pp. 138–67.

each side of it. Equations 10.116 combine the relations between input and output voltages on an obstacle-free stretch of transmission line and the continuity conditions at a shunt element.

If (10.116a) is solved for I_1 and the result placed in (10.116b), a simple manipulation gives

$$\cos\phi = \cos ka - \frac{B}{2G_0}\sin ka \tag{10.117}$$

With the dimensions a, b, and b' specified, Equation 10.115 can be used to compute B/G_0 and then Equation 10.117 will yield the value for ϕ, the phase delay per section. The equivalent index of refraction can be defined by forming the ratio of this phase delay to the phase delay that would occur in the absence of the strips, namely ka. Hence,

$$n = \frac{\phi}{ka} \tag{10.118}$$

It is useful to consider the form taken by (10.117) when k is small (long wavelength). Then

$$\cos\phi = \cos nka \cong 1 - \tfrac{1}{2}(nka)^2$$

$$\cos ka \cong 1 - \tfrac{1}{2}(ka)^2 \qquad \sin ka \cong ka$$

and substitution in (10.117) gives

$$n^2 \cong 1 + \frac{B/G_0}{k_0 a} \tag{10.119}$$

If B/G_0 is replaced in (10.119) by the right side of (10.115) and $k_0 \to 0$, the relation becomes exact and one obtains

$$n_0^2 = 1 + \frac{2b}{\pi a}\ln\left[\csc\left(\frac{\pi b'}{2b}\right)\right] \tag{10.120}$$

with n_0 the "static" (zero frequency) value of the refractive index. It can be observed that n_0 is calculable from (10.120) once the dimensions of the strip dielectric are specified.

Equations 10.115, 10.118, and 10.120 can be combined and substituted in (10.117) to give the result

$$\cos nka = \cos ka - \tfrac{1}{2}(n_0^2 - 1)ka\sin ka \tag{10.121}$$

which is a particularly useful equation from which to deduce the index of refraction n.

Plots of n versus a/λ for several values of n_0 are displayed in Figure 10.37. Cutoff occurs when $\cos nka$ reaches -1, but it can be seen that values of n significantly greater than unity are achievable with practical strip dimensions.

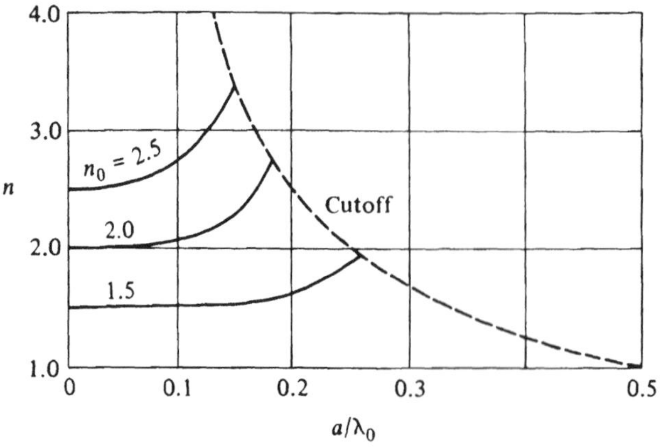

Fig. 10.37 The Refractive Index of Strip Dielectrics

The curves of Figure 10.37 are also applicable to the disc dielectric if the proper value of n_0 is used. The computation of B/G_0 for an array of discs is not simple but static measurements on a sample of the medium, placed in an electrolytic tank, can provide an experimental value.[23]

Since disc and strip dielectric media require normal passage of the electromagnetic wave, and since they present a refractive index greater than unity, they are suitable for use as a lens antenna of the type shown in Figure 10.31b. The strip dielectric is appropriate for the hyperbolic cylinder/line-source case, whereas the disc dielectric can be used either as a hyperbolic cylinder lens or a hyperboloid lens. Stepping is feasible. Both of these artificial dielectrics exhibit loss tangents which are significantly greater than that of a typical homogeneous isotropic dielectric, principally because of the finite conductivity of the obstacles, edge effects if the obstacles are not carefully made, and adhesive losses. However, with controlled construction, the losses are quite acceptable in lens applications and the great savings in weight is a considerable advantage.

10.16 Artificial Dielectrics: Metal Plate (Constrained) Lenses

Another type of artificial dielectric which has been widely used in lens antenna applications is shown in Figure 10.38. It consists of an array of equispaced thin metal plates, each lying in a plane which contains **E** and the direction of propagation. A host material is not needed to hold the plates in place, since they can be connected together by transverse metal rods normal to **E**. Thus a rigid construction is easily obtained, and some of the loss components present in disc and strip dielectrics are eliminated.

[23]S. B. Cohn, "Artificial Dielectrics for Microwaves," *Proc. Symposium on Modern Advances in Microwave Techniques* (Polytechnic Institute of Brooklyn, 1955).

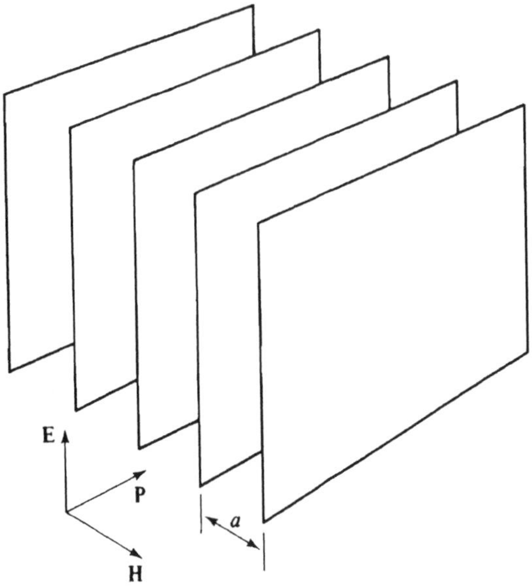

Fig. 10.38 A Metal Plate Artificial Dielectric

Since $\mathbf{E} \equiv 0$ on the walls of the metal plates, for $\lambda/2 < a < \lambda$, the electromagnetic waves which pass between the plates must be in the form of the dominant mode with phase constant β given by the relation

$$\beta = \sqrt{k^2 - \left(\frac{\pi}{a}\right)^2} \tag{10.122}$$

As a consequence, the index of refraction is

$$n = \frac{\beta}{k} = \sqrt{1 - \left(\frac{\lambda}{2a}\right)^2} \tag{10.123}$$

It can be observed from (10.123) that the refractive index of a metal plate dielectric is less than unity, rising from a value of zero when $a = \lambda/2$ to a value of 0.886 when $a = \lambda$. This wide latitude in the choice of a value for n is tempered by the fact that, as n departs further from unity, the surface mismatch between a metal plate dielectric and the adjacent air medium is aggravated. Stated differently, the arriving electromagnetic wave has a locally uniform electric field which must be transformed to a field that is locally a sequence of half-sinusoids. The more of these half-sinusoids there are per unit length in the **H** direction, the greater the mismatch.

Because $n < 1$, the basic lens type to which the metal plate dielectric is easily adapted is the elliptical shape, shown in contour in Figure 10.39. Translation gives an elliptic cylinder lens, suitable for use with a line source; rotation gives an ellipsoidal lens, suitable for use with a point source. Parallel, equispaced slices through

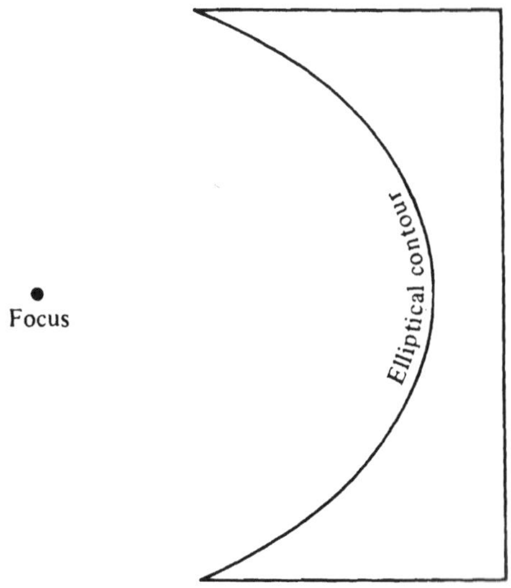

Fig. 10.39 Central Cross Section of a Metal Plate Lens

the solid figures caused by translation or rotation give the templates for the individual metal plates which will comprise the lens.

A second set of metal plates can be added to the structure of Figure 10.39, arranged so that they are perpendicular to the first set. This is suggested in Figure 10.40. Since the second set of plates is transverse to **E**, their presence does not affect the refractive index for vertical polarization. But the second set can be designed to

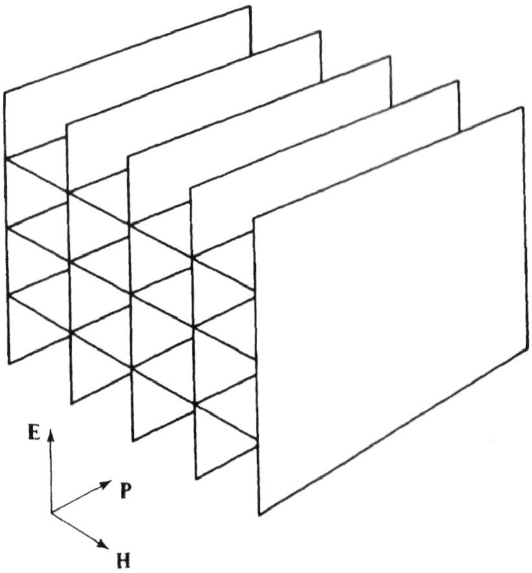

Fig. 10.40 An Eggcrate Artificial Dielectric

achieve the same result with a horizontally polarized wave. The resulting structure is called an *eggcrate dielectric*. With the same plate spacing in both dimensions, collimation of circularly polarized primary radiation, as well as polarization diversity, can be achieved.

Stepping of these metal plate lenses can be accomplished in the manner described in Section 10.11 for homogeneous, isotropic dielectrics. Because the waves are constrained to travel between parallel plates, these structures are often called *constrained lenses*. With an elliptic cylinder or ellipsoidal lens boundary, wave diffraction naturally aids this constraint, but the constraint is present regardless of the shape of the boundary, and other boundary shapes are sometimes employed.

Some depolarization occurs with an ellipsoidal metal plate lens. This effect is absent in the elliptic cylinder case.

10.17 The Luneburg Lens

A class of lens that has proven extremely useful in antenna applications (and also in scattering work) is characterized by a refractive index which is variable, but spherically symmetric. By this one means that if the origin of coordinates is chosen at the center of the lens, then $n = n(r)$. The nature of rays in such a medium can be deduced with the aid of Figure 10.41a. A ray MPQ is shown, and $\mathbf{r}$ is the directed distance from the origin O to P; $\mathbf{1}_s$ is a unit tangent vector at P.

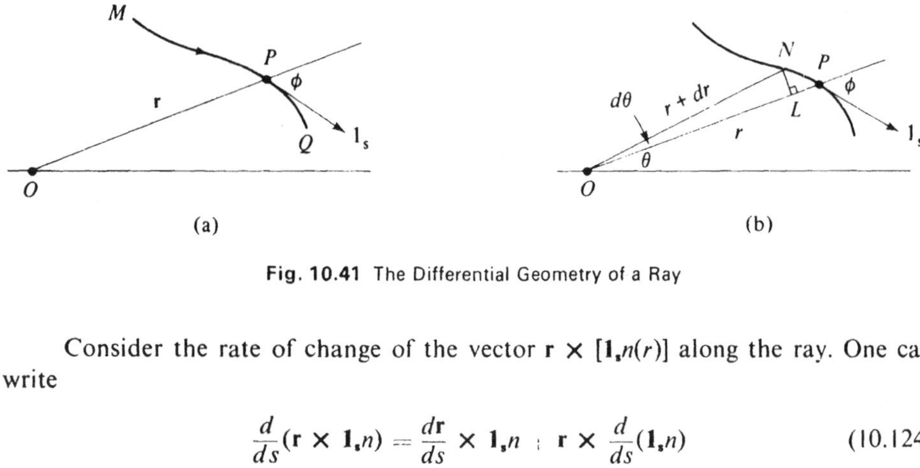

(a) (b)

Fig. 10.41 The Differential Geometry of a Ray

Consider the rate of change of the vector $\mathbf{r} \times [\mathbf{1}_s n(r)]$ along the ray. One can write

$$\frac{d}{ds}(\mathbf{r} \times \mathbf{1}_s n) = \frac{d\mathbf{r}}{ds} \times \mathbf{1}_s n + \mathbf{r} \times \frac{d}{ds}(\mathbf{1}_s n) \tag{10.124}$$

with s a measure of distance on the ray. Because $d\mathbf{r}/ds = \mathbf{1}_s$, the first term on the right in (10.124) vanishes. Also, by virtue of (10.15) the second term becomes $\mathbf{r} \times \nabla \mathbf{n}$. Since $n(r)$ is a spherically symmetric function,

$$\nabla n = \mathbf{1}_r \frac{dn}{dr} \tag{10.125}$$

as a consequence of which the second term on the right side of (10.124) also vanishes. Hence

$$\mathbf{r} \times [1, n(r)] = \text{constant} \tag{10.126}$$

The implication of this result is that each ray is a *plane curve* which lies in a plane containing the origin. Along a ray,

$$rn(r)\,|\sin\phi| = K_1 \tag{10.127}$$

with the angle ϕ defined as in Figure 10.41a and K_1 a constant. Equation 10.127 is often called *Bouquer's formula.*

The angle ϕ can be connected to the differential geometry of a ray with the help of Figure 10.41b. The right triangle LNP is such that

$$\sin\phi = \frac{LN}{NP} = \frac{r\,d\theta}{\sqrt{(dr)^2 + (r\,d\theta)^2}} = \frac{r(\theta)}{\sqrt{r^2(\theta) + \left(\dfrac{dr}{d\theta}\right)^2}} \tag{10.128}$$

The combination of (10.127) and (10.128) gives

$$\frac{dr}{d\theta} = \pm\frac{r}{K_1}\sqrt{r^2n^2(r) - K_1^2} \tag{10.129}$$

Integration yields

$$\theta - \theta_0 = \pm\int_{r_0}^{r} \frac{K_1\,dr}{r\sqrt{r^2n^2(r) - K_1^2}} \tag{10.130}$$

which is the equation of a ray in a medium with a spherically symmetric refractive index.

With this as background, consider the situation suggested by Figure 10.42. A lens of radius a is shown for which $n(r)$ is a monotonically decreasing function of r, with $n(a) = 1$. A ray $P_1Q_1Q_2P_2$ is indicated. It leaves from P_1 at an angle α with respect to the X-axis and travels the straight line path P_1Q_1 in the homogeneous air medium. However, the path Q_1Q_2 is curved because the lens is inhomogeneous. It was demonstrated in Section 10.2, as an interpretation of Equation 10.20, that rays always bend toward the region of higher refractive index. Thus with $n(r)$ increasing toward the lens center, Q_1Q_2 bends as shown. The ray then travels a straight-line path Q_2P_2 in air, intersecting the X-axis at the point P_2. By virtue of (10.126), this ray is a *plane curve.*

Imagine the surface generated by rotating $P_1Q_1Q_2P_2$ about the X-axis. The intersection of this surface with any plane containing the X-axis is also a ray because of the spherical symmetry of the lens. A sheath of rays can thus be envisioned, all of which leave P_1 at the conical angle α and all of which come to a focus at P_2.

Next, imagine another sheath of rays that leave P_1 at a *different* conical angle α'. Will these rays also come to a focus at P_2, or at some other point? Investigation

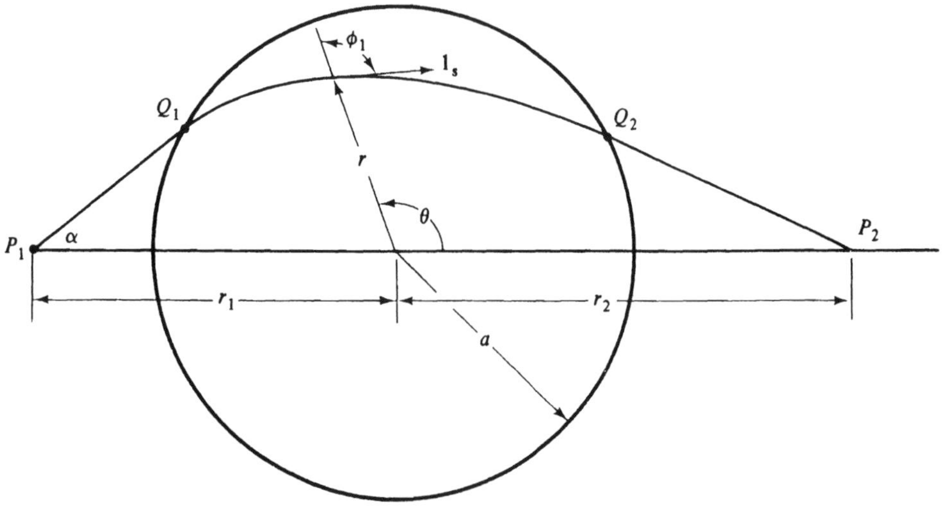

Fig. 10.42 The Cross Section of a Spherically Symmetric Lens

of this question by R. R. Luneburg[24] disclosed that, if $n(r)$ is properly chosen, all rays that leave P_1 and enter the lens can be brought to focus at the common point P_2. This class of lenses appropriately bears his name. In its most common form, the Luneburg lens is designed so that the focal point P_1 is on the surface of the lens and the focal point P_2 is at infinity. This provides the practical advantage that radiation from a point source can be converted to a plane wave emerging from the lens in a direction diametrically opposite to that of the feed.

The form which the refractive index function $n(r)$ must take in order to satisfy the Luneburg condition, namely that all rays which leave P_1 and enter the lens come to a common focus at P_2, can be deduced with the aid of Equations 10.127, 10.129, and 10.130. First, since the point P_1 is outside the lens and thus in a region where $n = 1$, the constant K_1 is given simply by

$$K_1 = r_1 \sin \alpha \tag{10.131}$$

It follows that K_1 has a positive value for all rays that enter the lens, but a value which is dependent on the angle at which the ray departs from P_1. The range of K_1 is over the interval $[0, a]$.

Second, a study of Figure 10.42 reveals that there is a point that has coordinates which can be designated by (r_m, θ_m) where the ray comes closest to the origin, and that this point is inside the lens. For $\theta < \theta_m$, $dr/d\theta < 0$, and for $\theta > \theta_m$, $dr/d\theta > 0$. Hence in Equations 10.129 and 10.130, the minus sign should be used in the range $0 \leq \theta \leq \theta_m$ and the plus sign should be used in the range $\theta_m \leq \theta \leq \pi$. Two special

[24]R. K. Luneburg, *The Mathematical Theory of Optics* (Berkeley: University of California Press, 1964), pp. 164–88. This is a reproduction of Dr. Luneburg's lecture notes at Brown University. The original issue of these notes had mis-spelled his name as Luneberg.

applications of (10.130) then give

$$0 - \theta_m = -\int_{r_m}^{r_2} \frac{K_1\, dr}{r\sqrt{n^2r^2 - K_1^2}} \tag{10.132}$$

$$\pi - \theta_m = \int_{r_m}^{r_1} \frac{K_1\, dr}{r\sqrt{n^2r^2 - K_1^2}} \tag{10.133}$$

The refractive index function $n(r)$ must be chosen so that these two equations hold for all values of the constant K_1 in $[0, a]$. Their difference yields

$$\begin{aligned} \pi = &\int_{r_m}^{a} \frac{K_1\, dr}{r\sqrt{n^2r^2 - K_1^2}} + \int_{a}^{r_1} \frac{K_1\, dr}{r\sqrt{r^2 - K_1^2}} \\ &+ \int_{r_m}^{a} \frac{K_1\, dr}{r\sqrt{n^2r^2 - K_1^2}} + \int_{a}^{r_2} \frac{K_1\, dr}{r\sqrt{r^2 - K_1^2}} \end{aligned} \tag{10.134}$$

Since

$$\int \frac{K_1\, dr}{r\sqrt{r^2 - K_1^2}} = -\arcsin \frac{K_1}{r}$$

it follows that if $f(K_1)$ is defined by

$$f(K_1) = \int_{r_m}^{a} \frac{K_1\, dr}{r\sqrt{n^2r^2 - K_1^2}} \tag{10.135}$$

then (10.134) can be rewritten in the form

$$f(K_1) = \frac{1}{2}\left[\pi + \arcsin \frac{K_1}{r_1} + \arcsin \frac{K_1}{r_2} - 2 \arcsin \frac{K_1}{a}\right] \tag{10.136}$$

The defining relation in (10.135) can be transformed into an integral equation of a known type by the following manipulation[25]: Let

$$r' = \frac{r}{a} \qquad K = \frac{K_1}{a} \qquad \rho(r') = r'n(r')$$

As a result, (10.135) and (10.136) become

$$f(K) = \int_{r'_m}^{1} \frac{K\, dr'}{r'\sqrt{\rho^2(r') - K^2}} \tag{10.137}$$

and

$$f(K) = \frac{1}{2}\left[\pi + \arcsin \frac{K}{r'_1} + \arcsin \frac{K}{r'_2} - 2 \arcsin K\right] \tag{10.138}$$

[25]The development from this point follows closely the original presentation by Luneburg, *Theory of Optics*, pp. 184–87.

Next, introduce the variable τ by the definition

$$\tau = \ln r'$$

which converts (10.137) to the form

$$f(K) = \int_{\tau_m}^{0} \frac{K\,d\tau}{\sqrt{\rho^2(\tau) - K^2}} \tag{10.139}$$

The exact relationship between ρ and τ depends on $n(r)$, but since $n(r)$ is a monotonic function, so too is $\rho(\tau)$. The ranges are $0 \leq \rho \leq 1$ and $-\infty < \tau < 0$. The form of $\rho(\tau)$ is therefore as suggested in Figure 10.43. If the function $T(\rho)$ is defined by

$$T(\rho) = -\tau(\rho) \tag{10.140}$$

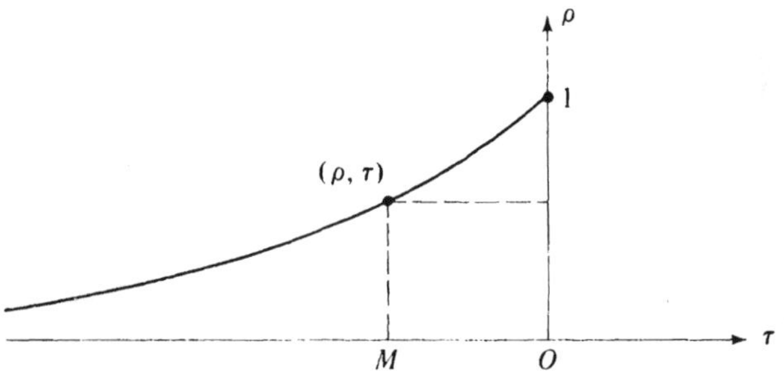

Figure 10.43 The Function $\rho(\tau)$

then $T(\rho)$ is the magnitude of the abscissa $\overline{OM}$ in Figure 10.43. This introduction causes the transformation from (10.139) to

$$f(K) = -K \int_{\rho=K}^{\rho=1} \frac{dT(\rho)}{\sqrt{\rho^2 - K^2}} \tag{10.141}$$

The lower limit on the integral in (10.141) can be explained by returning to (10.127) and noting that

$$\frac{r_1}{a} n(r_1)\,|\sin\phi_1| = \frac{r_1}{a}\sin\alpha = K = \frac{r_m}{a} n(r_m)\,|\sin\phi_m| = \rho_m \tag{10.142}$$

since $\phi_m = \pi/2$.

The integral in (10.141) can be inverted by the following stratagem: Let $g(K)$ be defined by

$$g(K) = -K \int_{K}^{\lambda} \frac{dT(\rho)}{\sqrt{\rho^2 - K^2}} \tag{10.143}$$

in the interval $0 \leq K \leq \lambda$. Multiply (10.143) by $2(K^2 - \rho^2)^{-1/2}$ and integrate with respect to K from ρ to λ. This gives

$$2\int_\rho^\lambda \frac{g(K)}{\sqrt{K^2-\rho^2}}\,dK = -\int_\rho^\lambda \frac{2K}{\sqrt{K^2-\rho^2}}\int_K^\lambda \frac{dT(s)}{\sqrt{s^2-K^2}}\,dK \qquad (10.144)$$

With the order of integration interchanged, (10.144) becomes

$$2\int_\rho^\lambda \frac{g(K)}{\sqrt{K^2-\rho^2}}\,dK = -\int_\rho^\lambda dT(s)\int_\rho^s \frac{2K\,dK}{\sqrt{K^2-\rho^2}\sqrt{s^2-K^2}} \qquad (10.145)$$

Let the integration variable z be defined by

$$K^2 = (s^2-\rho^2)z + \rho^2$$

so that

$$2K\,dK = (s^2-\rho^2)\,dz$$

This converts (10.145) to

$$2\int_\rho^\lambda \frac{g(K)}{\sqrt{K^2-\rho^2}}\,dK = -\int_\rho^\lambda dT(s)\int_0^1 \frac{dz}{\sqrt{z(1-z)}}$$

$$= -\pi\int_\rho^\lambda dT(s)$$

so that

$$T(\rho) - T(\lambda) = \frac{2}{\pi}\int_\rho^\lambda \frac{g(K)}{\sqrt{K^2-\rho^2}}\,dK \qquad (10.146)$$

The result (10.146) can be applied to the function

$$\phi(k) = \frac{\pi}{2} - \arcsin K \qquad (10.147)$$

which can be recognized as part of (10.138). Thus $\phi(K)$ can be written in the integral form

$$\phi(K) = \int_K^1 \frac{K d\rho}{\rho\sqrt{K^2-\rho^2}} = \int_K^1 \frac{K d(\ln\rho)}{\sqrt{K^2-\rho^2}} \qquad (10.148)$$

This equation is of the type of (10.143) with $T(\rho) = -\ln\rho$. Thus from (10.146),

$$T(\rho) - T(1) = -\ln\rho = \frac{2}{\pi}\int_\rho^1 \frac{\phi(K)}{\sqrt{K^2-\rho^2}}\,dK \qquad (10.149)$$

Next, consider the entire function $f(K)$ given alternatively by (10.138) and (10.141). Use of (10.146) yields

$$T(\rho) - T(1) = -\ln \rho + \frac{1}{\pi}\int_\rho^1 \left(\arcsin\frac{K}{r_1'} + \arcsin\frac{K}{r_2'}\right)\frac{dK}{\sqrt{K^2 - \rho^2}} \quad (10.150)$$

where this time $T(\rho) = -\tau(\rho) = -\ln r'$. Since $T(1) = 0$, if the symbolism

$$\omega(\rho, b) = \frac{1}{\pi}\int_\rho^1 \arcsin\left(\frac{v}{b}\right)\frac{dv}{\sqrt{v^2 - \rho^2}} \quad (10.151)$$

is adopted, (10.150) becomes

$$\ln\left(\frac{\rho}{r'}\right) = \ln\left[n\left(\frac{r}{a}\right)\right] = \omega\left(\rho, \frac{r_1}{a}\right)\Big] + \omega\left(\rho, \frac{r_2}{a}\right) \quad (10.152)$$

This equation, together with the defining relation $\rho = nr'$ determines the required refractive index function $n(r')$ in parametric form.

For the special important case that $r_1 = a$ and $r_2 = \infty$, $\omega(\rho, \infty) = 0$ and

$$\omega(\rho, 1) = \frac{1}{\pi}\int_\rho^1 \frac{\arcsin v}{\sqrt{v^2 - \rho^2}}\,dv$$
$$= \frac{1}{2}\ln\left(1 + \sqrt{1 - \rho^2}\right) \quad (10.153)$$

When this result is combined with (10.152), one finds that

$$n\left(\frac{r}{a}\right) = \sqrt{2 - \left(\frac{r}{a}\right)^2} \quad (10.154)$$

This requirement could be satisfied by a medium with permeability everywhere equaling that of free space but with dielectric constant that decreases parabolically from a value of 2 at $r = 0$ to unity at $r = a$. In practice this is achieved by using a family of concentric shells, each with a constant refractive index to approximate (10.154) in steps. At least ten steps need to be used to obtain good performance.

It is difficult to design a useful feed which has a phase center that can be placed directly on the surface of the Luneburg lens. If the first focal point is at $r_1 > a$ but the second focal point is still at $r_2 \rightarrow \infty$, Equation 10.152 gives

$$\ln[n(r')] = \omega(\rho, r') = \frac{1}{\pi}\int_\rho^{r'_1} \frac{\arcsin(K/r_1')}{\sqrt{K^2 - \rho^2}}\,dK - \frac{1}{\pi}\int_1^{r'_1} \frac{\arcsin(K/r_1')}{\sqrt{K^2 - \rho^2}}\,dK$$
$$= \frac{1}{2}\ln\left[1 + \sqrt{1 - (\rho/r_1')^2}\right] - \frac{1}{\pi}\int_1^{r'_1} \frac{\arcsin(K/r_1')}{\sqrt{K^2 - \rho^2}}\,dK \quad (10.155)$$

Numerical solution of (10.155) will yield $n(r')$. E. A. Wolff[26] has provided results of such calculations in the form of a set of curves that are reproduced in Figure 10.44.

[26]Wolff, *Antenna Analysis*, p. 496.

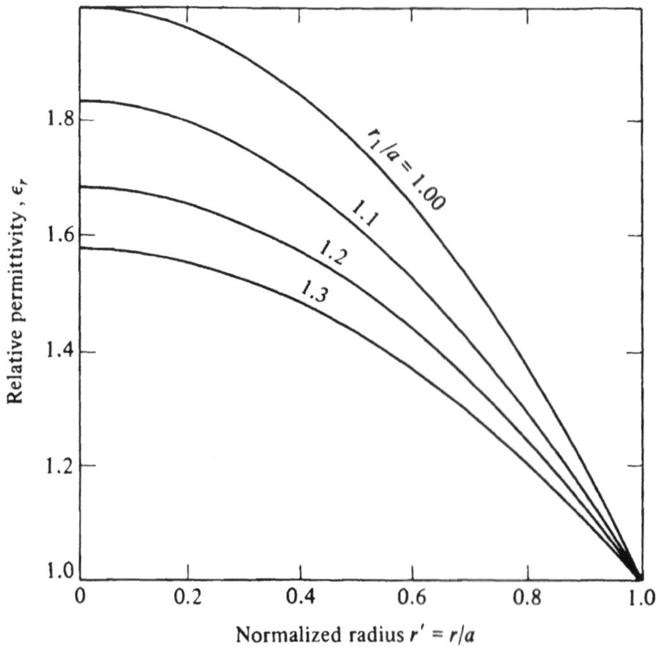

Fig. 10.44 The Required Radial Variation of Dielectric Constant in a Luneburg Lens (From *Antenna Analysis*, by E. A. Wolff. Copyright 1966, John Wiley and Sons, Inc. Used with permission.)

The secondary aperture distribution of a Luneburg lens antenna can be deduced with recourse to Figure 10.45. The ray which leaves the point source at P_1 and enters the lens at Q_1 emerges from the lens at Q_2, parallel to the X-axis. The coordinates of Q_2 are (a, θ). From (10.127) and Figure 10.45,

$$r_1 \sin \alpha = a \sin \phi = a \sin \theta \tag{10.156}$$

Let $I(\alpha, \beta)$ be the radiation intensity of the point source placed at P_1. Then $I(\alpha, \beta) \sin \alpha \, d\alpha \, d\beta$ is the tubular power flow toward the lens. If (y, β) are polar coordinates in the secondary aperture plane $x =$ constant and $P(y, \beta)$ is the power density in that plane, then

$$I(\alpha, \beta) \sin \alpha \, d\alpha \, d\beta = P(y, \beta) y \, dy \, d\beta \tag{10.157}$$

But $y = a \sin \theta$, and use of (10.156) gives

$$y = r_1 \sin \alpha \qquad dy = r_1 \cos \alpha \, d\alpha \tag{10.158}$$

Substitution in (10.157) produces the relation

$$P(y, \beta) = \frac{I(\alpha, \beta)}{r_1^2 \cos \alpha} \tag{10.159}$$

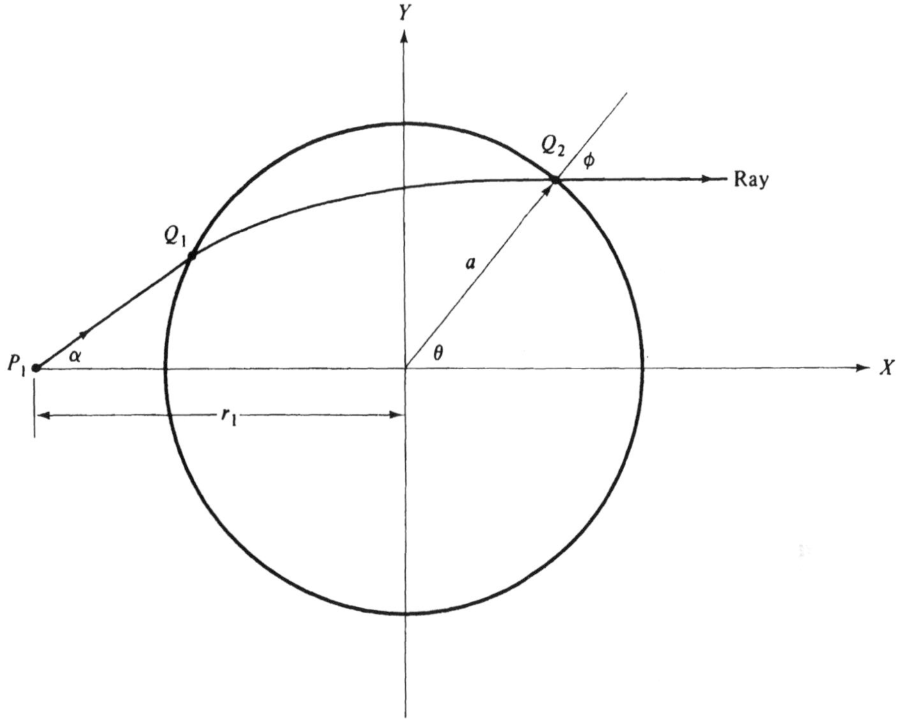

Fig. 10.45 A Luneburg Lens Focused at Infinity

The square root of (10.159) gives the amplitude of the equiphase secondary aperture distribution. From this the equivalent Huyghens sources and the secondary pattern can be deduced.

A Luneburg lens can also be used as an efficient back-scatterer by covering as much as half of its outer surface with a reflector, as suggested by Figure 10.46. This will cause an incoming wave, incident from any direction in a half-space, to be

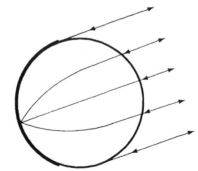

Fig. 10.46 A Luneburg Lens Reflector

reradiated with a secondary pattern with a main beam pointing in the incident direction. Maximum efficiency occurs if the direction of arrival of the incident rays causes them to be brought to a focus at the central point of the reflecting surface.

REFERENCES

BORN, M. and E. WOLF, *Principles of Optics* (New York: Pergamon Press, 1959), Chapter 3.

COLLIN, R. E. and F. J. ZUCKER, ed., *Antenna Theory: Part II* (New York: McGraw-Hill Book Co., Inc., 1969), Chapters 16–18.

DESIZE, L. K. and J. F. RAMSAY, "Reflecting Systems," *Microwave Scanning Antennas*, Vol. I, ed. R. C. Hansen (New York: Academic Press, 1964), Chapter 2.

JASIK, H., ed., *Antenna Engineering Handbook* (New York: McGraw-Hill Book Co., Inc., 1961), Chapters 12, 14, and 25.

LOVE, A. W., ed., *Reflector Antennas*, A Collection of Significant Journal Papers (New York: IEEE Press, 1978).

LUNEBURG, R. K., *Mathematical Theory of Optics* (Berkeley: University of California Press, 1966).

RUSCH, W. V. T. and P. D. POTTER, *Analysis of Reflector Antennas* (New York: Academic Press, 1970), Chapters 2 and 3.

SILVER, S., *Microwave Antenna Theory and Design*, Vol. 12, MIT Rad. Lab. Series (New York: McGraw-Hill Book Co., Inc., 1949), Chapters 4, 5, 11, 12, and 13.

WOLFF, E. A., *Antenna Analysis* (New York: John Wiley and Sons, Inc., 1966), Chapters 7 and 10.

PROBLEMS

10.1 Show that in a homogeneous medium the polarization of the geometric optics field remains constant along a ray.

10.2 Use the results of Section 10.2 to demonstrate Fermat's principle: The ray from a point P_1 to a point P_2 is the curve along which the phase delay is a minimum with respect to infinitesimal variations in path.

10.3 Show that the law of reflection is satisfied for the paraboloidal reflector defined by Equation 10.32 when primary rays emerge from its focal point and all secondary rays are assumed to be collimated parallel to the X-axis.

10.4 Design a shaped cylindrical reflector to produce the secondary pattern shown in Figure 10.13. Let $\phi_1 = 20°$ and $\phi_2 = 80°$ and use the primary pattern $I(\phi) = \cos[3(\phi - 50°)]$.

10.5 Design a shaped cylindrical reflector to produce the secondary pattern shown in Figure 10.13. Let $\phi_1 = 0°$ and $\phi_2 = 60°$ and use the primary pattern $I(\phi) = \cos[3(\phi - 30°)]$. Use a crisscross relationship between primary and secondary rays, that is, $\phi = 0°$ corresponds to $\theta = 21°, \ldots, \phi = 60°$ corresponds to $\theta = 0°$. Do you see an advantage to such a design?

10.6 If $\theta(\phi)$ is a constant so that all secondary rays are collimated, show that Equation 10.43 gives the equation of a parabolic cylinder.

10.7 Design a doubly curved reflector to produce the secondary pattern shown in Figure 10.13. Let $\phi_1 = 0°$ and $\phi_2 = 60°$ and assume $I(\phi) = \cos[3(\phi - 30°)]$. Use an iterative procedure to determine the backbone curve, and terminate when the last iteration gives a $\rho(\phi)$ which differs from the penultimate $\rho(\phi)$ by less than 0.1% everywhere.

10.8 Assume that the feed for the reflector designed in Problem 10.7 is vertically polarized. If the reflector is to be constructed of parallel vertical metallic plates $\lambda/4$ on centers, $\lambda/8$ thick, and $\lambda/2$ deep, find the shape of each plate. Assume that the H-plane pattern is to have a central main beam with a 5° half-power beamwidth and symmetric side lobes.

10.9 Repeat the calculations for the illustrative example of Section 10.7 with the primary radiation down 5, 15, and 20 dB at the reflector edges. Determine thereby the secondary aperture efficiency and aperture blockage as functions of spillover.

10.10 Determine the secondary aperture distribution for the shaped cylindrical reflector of Figure 10.15. Try a sequence of values for $\rho(0°)$ until the half-power beamwidth on the ground side agrees with specification.

10.11 Find the equation of a paraboloid in spherical coordinates centered at the focus.

10.12 Show that, for a paraboloidal reflector, the relation between primary feed pattern $I(\phi, \psi)$ and secondary aperture distribution $P(r, \psi)$ is given by

$$P(r, \psi) = \frac{I(\phi, \psi)}{f^2} \cos^4 \frac{\phi}{2}$$

The geometry of Figure 10.6 applies: ϕ is the angle a primary ray makes with the negative X-axis and (r, ψ) are polar coordinates in the plane $x = a$.

10.13 Assume a paraboloidal reflector with an f/D ratio of 0.5 and with $D = 10\lambda$ is fed by a vertically polarized horn that has a radiation pattern given by Equations 3.8 and 3.9. Find the proper values of a/λ and b/λ to cause a -10 dB spillover at the reflector edge in each principal plane. Then use the result of Problem 10.12 to determine the secondary aperture distribution. From this, find the equivalent Huyghens sources and then compute the far-field pattern in the two principal planes.

10.14 Add in the effect of aperture blockage to the results of Problem 10.13, assuming a perfect void in the secondary aperture distribution due to total absorption by the horn of the secondary field incident on the horn mouth.

10.15 Repeat Problem 10.13 using the reflector current distribution calculated under geometrical optics assumptions.

10.16 For an unmodified Cassegrain antenna system, find the relation between an axially symmetric feed pattern and the secondary aperture distribution.

10.17 Show that Snell's law of refraction is satisfied at the elliptic cylinder surface shown in cross section in Figure 10.31a, with the equation of that surface given by (10.83).

10.18 Show that Snell's law of refraction is satisfied at the hyperbolic cylinder surface shown in cross section in Figure 10.31b, with the equation of that surface given by (10.88).

10.19 Make an accurate scale drawing of the surfaces of an ellipsoidal lens for which $\epsilon/\epsilon_0 =$

2.56 if $a/\lambda = 5$ and $y_{max}/\lambda = 5$. Show two alternate weight-saving designs in which you have stepped one or the other of the lens surfaces.

10.20 Repeat Problem 10.19 for a hyperboloidal lens.

10.21 Consider an elliptic cylinder lens antenna for which $n = 0.5$, with $y_{max}/\lambda = 5$ and $\phi_{max} = 45°$. If the primary radiation intensity is given by

$$I(\phi) = \frac{\sin^2\left(\frac{\pi b}{\lambda}\sin\phi\right)}{\left(\frac{\pi b}{\lambda}\sin\phi\right)}$$

and if b/λ is chosen to have a -10 dB spillover, find the secondary aperture distribution. From this, determine the equivalent Huyghens sources and the secondary pattern.

10.22 Repeat Problem 10.21 for a hyperbolic cylinder lens antenna with $n = 2.0$.

10.23 An ellipsoidal lens antenna for which $n = 0.5$ has a span of 10λ and $\phi_{max} = 28°$. If the primary radiation intensity is rotationally symmetric and given by $I(\phi) = \cos 3\phi$ watts per steradian, find the secondary aperture distribution. Then deduce the equivalent Huyghens sources and the secondary pattern.

10.24 Repeat Problem 10.23 for a hyperboloidal lens antenna with $n = 2.0$.

10.25 Design a shaped cylindrical constrained (metal plate) lens to produce the $\csc^2\theta$ secondary pattern shown in Figure 10.13. Use $n = 0.5$.

10.26 Design a metallic strip artificial dielectric with practical dimensions at S band (3 GHz) to give an equivalent relative permittivity of 4.0.

10.27 Find an expression for the refractive index function of a Luneburg lens when the foci P_1 and P_2 are symmetrically disposed with respect to the lens center. What specific form does this expression take if the foci are on the lens surface?

appendices

A reduction of the vector Green's formula for E

In Chapter 1 the vector Green's theorem is used to establish the relation (1.50), namely,

$$\int_{V'} (\mathbf{E} \cdot \nabla_S \times \nabla_S \times \psi\mathbf{a} - \psi\mathbf{a} \cdot \nabla_S \times \nabla_S \times \mathbf{E})\, dV$$

$$= -\int_{S_1 \ldots S_{N,\Sigma}} (\psi\mathbf{a} \times \nabla_S \times \mathbf{E} - \mathbf{E} \times \nabla_S \times \psi\mathbf{a}) \cdot \mathbf{1}_n\, dS$$

This equation can be transformed in the following manner: Using the ninth vector identity listed on the inside of the back cover, one may write

$$\nabla_S \times \nabla_S \times \psi\mathbf{a} = \nabla_S(\nabla_S \cdot \psi\mathbf{a}) - \nabla_S^2 \psi\mathbf{a} \tag{A.1}$$

However,

$$\nabla_S \cdot \psi\mathbf{a} = \psi\nabla_S \cdot \mathbf{a} + \mathbf{a} \cdot \nabla_S\psi = \mathbf{a} \cdot \nabla_S\psi \tag{A.2}$$

since $\mathbf{a}$ is a *constant* vector. Also

$$\nabla_S^2 \psi\mathbf{a} = \mathbf{a}\nabla_S^2\psi = -k^2\psi\mathbf{a} \tag{A.3}$$

because ψ satisfies the scalar wave equation $\nabla_S^2\psi + k^2\psi = 0$. Thus

$$\nabla_S \times \nabla_S \times \psi\mathbf{a} = \nabla_S(\mathbf{a} \cdot \nabla_S\psi) + k^2\psi\mathbf{a} \tag{A.4}$$

Employing both (A.4) and (1.46), one obtains

$$\mathbf{E} \cdot \nabla_S \times \nabla_S \times \psi\mathbf{a} - \psi\mathbf{a} \cdot \nabla_S \times \nabla_S \times \mathbf{E} = \mathbf{E} \cdot \nabla_S(\mathbf{a} \cdot \nabla_S\psi) + \mathbf{a} \cdot \left(j\omega\psi \frac{\mathbf{J}}{\mu_0^{-1}}\right)$$

Use of the third vector identity (inside of back cover) gives

$$\mathbf{E} \cdot \nabla_S(\mathbf{a} \cdot \nabla_S\psi) = \nabla_S \cdot [\mathbf{E}(\mathbf{a} \cdot \nabla_S\psi)] - (\mathbf{a} \cdot \nabla_S\psi)\nabla_S \cdot \mathbf{E}$$
$$= \nabla_S \cdot [\mathbf{E}(\mathbf{a} \cdot \nabla_S\psi)] - \frac{\rho}{\epsilon_0}\mathbf{a} \cdot \nabla_S\psi$$

so that the left side of (1.50) becomes

$$\int_V \left\{\nabla_S \cdot [\mathbf{E}(\mathbf{a} \cdot \nabla_S\psi)] - \frac{\rho}{\epsilon_0}(\mathbf{a} \cdot \nabla_S\psi) + \mathbf{a} \cdot \left(j\omega\psi \frac{\mathbf{J}}{\mu_0^{-1}}\right)\right\} dV$$
$$= \mathbf{a} \cdot \int_{V'} \left(j\omega\psi \frac{\mathbf{J}}{\mu_0^{-1}} - \frac{\rho}{\epsilon_0}\nabla_S\psi\right) dV - \mathbf{a} \cdot \int_{S_1 \ldots S_N, \Sigma} (\mathbf{1}_n \cdot \mathbf{E})\nabla_S\psi \, dS \tag{A.5}$$

in which the divergence theorem has been employed.

The constant vector **a** may also be taken out in front of the integral sign on the right side of (1.50). Since, with the aid of the fifth vector identity (inside of back cover) and the triple scalar product, one can write

$$[\mathbf{E} \times (\nabla_S \times \psi\mathbf{a})] \cdot \mathbf{1}_n = [\mathbf{E} \times (\nabla_S\psi \times \mathbf{a})] \cdot \mathbf{1}_n = [(\mathbf{1}_n \times \mathbf{E}) \times \nabla_S\psi] \cdot \mathbf{a}$$
$$[\psi\mathbf{a} \times \nabla_S \times \mathbf{E}] \cdot \mathbf{1}_n = -j\omega\psi(\mathbf{a} \times \mathbf{B}) \cdot \mathbf{1}_n = j\omega\psi\mathbf{a} \cdot (\mathbf{1}_n \times \mathbf{B})$$

it follows that

$$-\int_{S_1 \ldots S_N, \Sigma} (\psi\mathbf{a} \times \nabla_S \times \mathbf{E} - \mathbf{E} \times \nabla_S \times \psi\mathbf{a}) \cdot \mathbf{1}_n dS$$
$$= -\mathbf{a} \cdot \int_{S_1 \ldots S_N, \Sigma} [j\omega\psi(\mathbf{1}_n \times \mathbf{B}) - (\mathbf{1}_n \times \mathbf{E}) \times \nabla_S\psi] \, dS \tag{A.6}$$

But (A.5) and (A.6) are modified forms of the left and right sides of (1.50), so they are equal to each other. And since this is true for any arbitrary constant vector **a**, it follows that the integrals themselves must be equal. Thus

$$\int_{V'} \left(j\omega\psi \frac{\mathbf{J}}{\mu_0^{-1}} - \frac{\rho}{\epsilon_0}\nabla_S\psi\right) dV$$
$$- \int_{S_1 \ldots S_N} [(\mathbf{1}_n \cdot \mathbf{E})\nabla_S\psi + (\mathbf{1}_n \times \mathbf{E}) \times \nabla_S\psi - j\omega\psi(\mathbf{1}_n \times \mathbf{B})] \, dS$$
$$= \int_\Sigma [(\mathbf{1}_n \cdot \mathbf{E})\nabla_S\psi + (\mathbf{1}_n \times \mathbf{E}) \times \nabla_S\psi - j\omega\psi(\mathbf{1}_n \times \mathbf{B})] \, dS \tag{A.7}$$

where, for convenience, the surface integral over the sphere Σ has been split off.

Consider this integral. On the surface of the sphere Σ one has

$$(\nabla_S\psi)_\delta = \mathbf{1}_n\left[\frac{d}{dR}\left(\frac{e^{-jkR}}{R}\right)\right]_\delta = -\mathbf{1}_n\left(jk + \frac{1}{\delta}\right)\frac{e^{-jk\delta}}{\delta} \tag{A.8}$$

If $d\Omega$ is the element of solid angle subtended at P by a surface element dS on Σ, then the right side of (A.7) can be written

$$I = \int_0^{4\pi} \left[-(\mathbf{1}_n \cdot \mathbf{E})\mathbf{1}_n \left(jk + \frac{1}{\delta}\right) \frac{e^{-jk\delta}}{\delta} - (\mathbf{1}_n \times \mathbf{E}) \times \mathbf{1}_n \left(jk + \frac{1}{\delta}\right) \frac{e^{-jk\delta}}{\delta} - j\omega(\mathbf{1}_n \times \mathbf{B}) \frac{e^{-jk\delta}}{\delta} \right] \delta^2 \, d\Omega$$

$$\begin{aligned} &= -\delta e^{-jk\delta} \int_0^{4\pi} [jk(\mathbf{1}_n \cdot \mathbf{E})\mathbf{1}_n + jk(\mathbf{1}_n \times \mathbf{E}) \times \mathbf{1}_n + j\omega(\mathbf{1}_n \times \mathbf{B})] \, d\Omega \\ &\quad - e^{-jk\delta} \int_0^{4\pi} [(\mathbf{1}_n \cdot \mathbf{E})\mathbf{1}_n + (\mathbf{1}_n \times \mathbf{E}) \times \mathbf{1}_n] \, d\Omega \end{aligned} \tag{A.9}$$

Since both integrals in (A.9) are well behaved at P, it follows that

$$\lim_{\delta \to 0} I = -\lim_{\delta \to 0} e^{-jk\delta} \int_0^{4\pi} \mathbf{E} \, d\Omega = -\mathbf{E}(P) \int_0^{4\pi} d\Omega = -4\pi \mathbf{E}(P) \tag{A.10}$$

Next consider the limit, as $\delta \longrightarrow 0$, of the *left* side of (A.7). Obviously the surface integrals are well behaved because P is restricted to be a point within V and thus is not on any of the bounding surfaces S_i. As $V' \longrightarrow V$, the volume integral is also well behaved. To see this, spherical coordinates may be introduced centered at P. Then $dV = R^2 \sin\theta \, dR \, d\theta \, d\phi$. Since ψ and $\nabla_S \psi$ contain terms involving R^{-1} and R^{-2} only, the contribution of the volume element at $R = 0$ to the volume integral in (A.7) is finite. Therefore the limiting value of (A.7) is

$$\begin{aligned} \mathbf{E}(x, y, z) &= \frac{1}{4\pi} \int_V \left(\frac{\rho}{\epsilon_0} \nabla_S \psi - j\omega\psi \frac{\mathbf{J}}{\mu_0^{-1}} \right) dV \\ &\quad + \frac{1}{4\pi} \int_{S_1 \ldots S_N} [(\mathbf{1}_n \cdot \mathbf{E}) \nabla_S \psi + (\mathbf{1}_n \times \mathbf{E}) \times \nabla_S \psi - j\omega\psi(\mathbf{1}_n \times \mathbf{B})] \, dS \end{aligned} \tag{A.11}$$

in which (x, y, z) are the coordinates of the point P, and it is to be remembered that a time factor $e^{j\omega t}$ has been suppressed.

B the wave equations for A and Φ

In Chapter 1 the potential functions **A** and Φ were introduced by the defining relations (1.80) and (1.81):

$$\mathbf{A}(x, y, z, t) = \int_V \frac{\mathbf{J}(\xi, \eta, \zeta)e^{j(\omega t - kR)}}{4\pi\mu_0^{-1}R}\,dV \tag{B.1}$$

$$\Phi(x, y, z, t) = \int_V \frac{\rho(\xi, \eta, \zeta)e^{j(\omega t - kR)}}{4\pi\epsilon_0 R}\,dV \tag{B.2}$$

Upon taking the divergence of (B.1) one obtains

$$\nabla_F \cdot \mathbf{A} = \int_V \frac{\mathbf{J}e^{j\omega t}}{4\pi\mu_0^{-1}} \cdot \nabla_F\left(\frac{e^{-jkR}}{R}\right) dV \tag{B.3}$$

since **J** is not a function of (x, y, z). But

$$\mathbf{J} \cdot \nabla_F\left(\frac{e^{-jkR}}{R}\right) = -\mathbf{J} \cdot \nabla_S\left(\frac{e^{-jkR}}{R}\right) = \frac{e^{-jkR}}{R}\nabla_S \cdot \mathbf{J} - \nabla_S \cdot \left(\frac{\mathbf{J}e^{-jkR}}{R}\right)$$

in which use has been made of the third vector identity listed on the inside of the back cover. If $\nabla_S \cdot \mathbf{J}$ is replaced by $-j\omega\rho$ in accordance with (1.45), (B.3) may be written

$$\nabla_F \cdot \mathbf{A} = -j\omega \int_V \frac{\rho(\xi, \eta, \zeta)e^{j(\omega t - kR)}}{4\pi\mu_0^{-1}R}\,dV - \int_S \frac{\mathbf{J}(\xi, \eta, \zeta)e^{j(\omega t - kR)}}{4\pi\mu_0^{-1}R} \cdot d\mathbf{S} \tag{B.4}$$

after the divergence theorem has been employed. Since S may be made large enough to encompass all the sources without containing any of them in its surface, the second integral in (B.4) vanishes and one is left with the conclusion that

$$\nabla \cdot \mathbf{A} = -\frac{j\omega}{c^2}\Phi = -\frac{1}{c^2}\dot{\Phi} \tag{B.5}$$

Through application of the Fourier integral, if **J** and ρ are general functions of time, one sees that

$$\mathbf{A}(x, y, z, t) = \int_V \frac{\mathbf{J}(\xi, \eta, \zeta, t - R/c)}{4\pi\mu_0^{-1}R}\, dV \tag{B.6}$$

$$\Phi(x, y, z, t) = \int_V \frac{\rho(\xi, \eta, \zeta, t - R/c)}{4\pi\epsilon_0 R}\, dV \tag{B.7}$$

these integrals being natural extensions of (B.1) and (B.2). Because linear superposition has been employed, it follows that **A** and **Φ** as given by (B.6) and (B.7) also satisfy (B.5). Further, the fields **E** and **B** arising from the sources $\mathbf{J}(\xi, \eta, \zeta, t)$ and $\rho(\xi, \eta, \zeta, t)$ satisfy

$$\mathbf{E} = -\nabla\Phi - \dot{\mathbf{A}} \tag{B.8}$$

$$\mathbf{B} = \nabla \times \mathbf{A} \tag{B.9}$$

These equations are restatements of (1.82) and (1.83) but for the more general potential functions (B.6) and (B.7).

If one takes the divergence of (B.8) and the curl of (B.9), the result is

$$\nabla \cdot \mathbf{E} = -\nabla^2\Phi - \nabla \cdot \dot{\mathbf{A}} = \frac{\rho}{\epsilon_0}$$

$$\nabla \times \mathbf{B} = \nabla \times \nabla \times \mathbf{A} = \nabla(\nabla \cdot \mathbf{A}) - \nabla^2\mathbf{A} = \frac{\mathbf{J}}{\mu_0^{-1}} + \frac{1}{c^2}\mathbf{E}$$

which, with the aid of (B.5) and (B.8) become

$$\nabla^2\Phi - \frac{1}{c^2}\ddot{\Phi} = -\frac{\rho}{\epsilon_0} \tag{B.10}$$

$$\nabla^2\mathbf{A} - \frac{1}{c^2}\ddot{\mathbf{A}} = -\frac{\mathbf{J}}{\mu_0^{-1}} \tag{B.11}$$

Thus both **A** and **Φ** satisfy the same type of differential equation, the solutions being given formally by (B.6) and (B.7).

C derivation of the Chebyshev polynomials

Chebyshev's differential equation is

$$(1 - u^2)\frac{d^2T_m}{du^2} - u\frac{dT_m}{du} + m^2T_m = 0 \tag{C.1}$$

To find a solution, assume $T_m(u)$ can be expressed as a power series, namely,

$$T_m(u) = \sum_{n=0}^{\infty} a_n u^n \tag{C.2}$$

and then

$$T'_m(u) = \sum_{n=1}^{\infty} n a_n u^{n-1} \quad \text{and} \quad T''_m(u) = \sum_{n=2}^{\infty} n(n-1)a_n u^{n-2}$$

Substitution in (C.1) gives

$$\sum_{n=2}^{\infty} n(n-1)a_n u^{n-2} - \sum_{n=2}^{\infty} n(n-1)a_n u^n - \sum_{n=1}^{\infty} n a_n u^n + m^2 \sum_{n=0}^{\infty} a_n u^n = 0 \tag{C.3}$$

When the coefficients of u raised to the various powers are separately equated to zero, one obtains

$$2a_2 + m^2 a_0 = 0 \qquad n = 0 \tag{C.4}$$

$$6a_3 + (m^2 - 1)a_1 = 0 \qquad n = 1 \tag{C.5}$$

$$(n+2)(n+1)a_{n+2} + (m^2 - n^2)a_n = 0 \qquad n \geq 2 \tag{C.6}$$

From (C.6), the recursion relation

$$a_{n+2} = -a_n \frac{(m+n)(m-n)}{(n+2)(n+1)} \tag{C.7}$$

arises, and is seen to truncate for $n = m$ if m is a positive integer.

For m an even integer, (C.4) and (C.7) in conjunction indicate that all the even coefficients can be expressed in terms of a_0, with the highest nonzero coefficient being a_m. If in such circumstances a_1 is arbitrarily set equal to zero all the odd coefficients are zero by virtue of (C.7). The result is a solution to (C.1) that is a polynomial of degree m containing one arbitrary constant a_0 and only even powers of u.

Similarly, if m is an odd integer, (C.5) and (C.7) taken together reveal that all the odd coefficients can be expressed in terms of a_1, with the highest nonzero coefficient being a_m. If in such circumstances a_0 is arbitrarily set equal to zero all the even coefficients are zero by virtue of (C.7). The result is a solution to (C.1) that is a polynomial of degree m containing one arbitrary constant a_1 and only odd powers of u.

For m an even positive integer, (C.4) and (C.7) give

$$a_2 = -a_0 \frac{m \cdot m}{2 \cdot 1}$$

$$a_4 = a_0 \frac{(m+2)m \cdot m(m-2)}{4 \cdot 3 \cdot 2 \cdot 1}$$

$$\vdots$$

$$a_n = (-1)^{n/2} a_0 \frac{m(m+2)\cdots(m+n-2)m(m-2)\cdots(m-[n-2])}{n!} \tag{C.8}$$

Manipulation of (C.8) yields

$$a_n = (-1)^{n/2} a_0 \frac{2^n\left(\frac{m}{2}\right)\left(\frac{m}{2}+1\right)\cdots\left(\frac{m}{2}+\frac{n}{2}-1\right)\left(\frac{m}{2}\right)\left(\frac{m}{2}-1\right)\cdots\left(\frac{m}{2}-\frac{n}{2}+1\right)}{n!}$$

If $m = 2N$ and $n = 2n'$, then

$$a_{2n'} = (-1)^{n'} a_0 \frac{2^{2n'}}{(2n')!}[N(N+1)\cdots(N+n'-1)N(N-1)\cdots(N-n'+1)]$$

$$a_{2n'} = (-1)^{n'} a_0 \frac{2^{2n'}}{(2n')!} \cdot \frac{(N+n'-1)!}{(N-1)!} \cdot \frac{N!}{(N-n')!} \tag{C.9}$$

From (C.2),

$$\begin{aligned} T_{2N}(u) &= \sum_{n'=0}^{N} a_{2n'} u^{2n'} \\ &= a_0 \sum_{n'=0}^{N} (-1)^{n'} 2^{2n'} \frac{N(N+n'-1)!}{(2n')!(N-n')!} u^{2n'} \\ &= a_0 \sum_{n'=0}^{N} (-1)^{n'} \frac{N}{N+n'} \cdot \frac{(N+n')!}{(2n')!(N-n')!} (2u)^{2n'} \\ &= a_0 \sum_{n'=0}^{N} (-1)^{n'} \frac{N}{N+n'} \binom{N+n'}{2n'} (2u)^{2n'} \end{aligned} \tag{C.10}$$

If one chooses a_0 so that $T_{2N}(0) = (-1)^N$ then from (C.10)

$$T_{2N}(0) = (-1)^N = a_0(2)^{-1}\frac{2N}{N}\cdot\frac{N!}{0!N!} = a_0$$

When this value of a_0 is inserted in (C.10), the result is

$$T_{2N}(u) = \sum_{n=0}^{N}(-1)^{N-n}\frac{N}{N+n}\binom{N+n}{2n}(2u)^{2n} \tag{C.11}$$

Equation C.11 is the general expression for an even-degree Chebyshev polynomial.

Similarly, if m is an odd positive integer,

$$a_3 = -a_1\frac{(m+1)(m-1)}{3\cdot 2}$$

$$a_5 = a_1\frac{(m+3)(m+1)(m-1)(m-3)}{5\cdot 4\cdot 3\cdot 2}$$

$$\vdots$$

$$a_n = (-1)^{-(n-1)/2}a_1\frac{(m+n-2)\cdots(m+3)(m+1)(m-1)(m-3)\cdots(m-[n-2])}{n!} \tag{C.12}$$

If the substitutions $m = 2N - 1$ and $n = 2n' - 1$ are made, manipulation of (C.12) gives

$$a_{2n'-1} = (-1)^{-(n'-1)}a_1\frac{2^{2n'-2}}{N+n'-1}\binom{N+n'-1}{2n'-1} \tag{C.13}$$

From (C.2),

$$T_{2N-1}(u) = a_1\sum_{n=1}^{N}(-1)^{-(n-1)}\frac{2^{2n-2}}{N+n-1}\binom{N+n-1}{2n-1}u^{2n-1} \tag{C.14}$$

It is customary to choose a_1 so that the slope of $T_{2N-1}(u)$ is $(2N-1)(-1)^{N-1}$ at the origin. When this done,

$$T_{2N-1}(u) = \sum_{n=1}^{N}(-1)^{N-n}\frac{2^{2n-2}(2N-1)}{N+n-1}\binom{N+n-1}{2n-1}u^{2n-1} \tag{C.15}$$

Equation C.15 is the general expression for an odd-degree Chebyshev polynomial.

D a general expansion of $\cos^m v$

The development will be restricted to the case in which m is an integer. Since

$$\cos^m v = \left(\frac{e^{iv} + e^{-iv}}{2}\right)^m = 2^{-m}(e^{iv} + e^{-iv})^m \tag{D.1}$$

The binomial expansion gives

$$\cos^m v = 2^{-m} \sum_{n=0}^{m} \binom{m}{n} e^{i(m-2n)v} \tag{D.2}$$

If m is *odd*, there is an even number of terms in this sum, occurring in pairs, such that

$$\cos^m v = 2^{-(m-1)} \sum_{n=0}^{(m-1)/2} \binom{m}{n} \cos (m - 2n)v \tag{D.3}$$

With the substitution $m = 2n' - 1$, (D.3) becomes

$$\cos^{2n'-1} v = \frac{1}{2^{2n'-2}} \sum_{n=0}^{n'-1} \binom{2n' - 1}{n} \cos (2n' - 2n - 1)v \tag{D.4}$$

Finally, with the additional substitution $l = n' - n$, one obtains

$$\cos^{2n'-1} v = \frac{1}{2^{2n'-2}} \sum_{l=1}^{n'} \binom{2n' - 1}{n' - l} \cos (2l - 1)v \tag{D.5}$$

Equation D.5 is the general expansion of $\cos^m v$ for m an odd positive integer.

If one returns to (D.2) and asserts that $m = 2n'$ is an even integer, then

$$\cos^{2n'} v = \frac{1}{2^{2n'}} \sum_{n=0}^{2n'} \binom{2n'}{n} e^{i2(n'-n)v} \tag{D.6}$$

Now there is an odd number of terms in the sum, composed of pairs plus a single term, such that

$$\cos^{2n'} v = \frac{1}{2^{2n'}}\binom{2n'}{n} + \frac{1}{2^{2n'-1}}\sum_{n=0}^{n'-1}\binom{2n'}{n}\cos 2(n'-n)v \tag{D.7}$$

If the substitution $l = n' - n$ is once again used, then

$$\cos^{2n'} v = \frac{1}{2^{2n'}}\sum_{l=0}^{n'}\epsilon_l\binom{2n'}{n'-l}\cos 2lv \tag{D.8}$$

in which $\epsilon_0 = 1$, $\epsilon_l = 2$ for $l \geq 1$. Equation D.8 is the general expansion of $\cos^m v$ for m an even positive integer.

E approximation to the magnetic vector potential function for slender dipoles[1]

In Chapter 7, the potential function

$$\mathcal{A}(z) = \int_{-l}^{l}\int_{0}^{2\pi} \frac{e^{-jkR}}{4\pi R} K_z(z')a \, d\phi \, dz' \tag{E.1}$$

is encountered in Equation 7.18. Here $K_z(z')$ is the lineal current density flowing on the outer cylindrical surface of the dipole (which is assumed to be composed of a perfect conductor and to have a length $2l$). Because of the circular cross section and the ϕ-symmetric method of feeding the dipole, K_z is not a function of ϕ. The quantity R that occurs in (E.1) is the distance between the source point (x', y', z') and the field point (x, y, z), both of which lie on the outer cylindrical surface of the dipole. If the source point and field point are expressed in cylindrical coordinates by (a, ϕ, z') and (a, β, z), respectively, with a the dipole radius, then

$$R = [2a^2 - 2a^2 \cos(\phi - \beta) + (z - z')^2]^{1/2} \tag{E.2}$$

By symmetry, $\mathcal{A}(z)$ is independent of β, so no loss in generality accrues from setting $\beta = 0$ in (E.2).

It is desired to determine how closely (E.1) is approximated by

$$\mathcal{A}_1(z) = \int_{-l}^{l} \frac{e^{-jkr}}{4\pi r} I(z') \, dz' \tag{E.3}$$

in which $I(z') = 2\pi a \, K_z(z')$ is the total axial current and

$$r = [a^2 + (z - z')^2]^{1/2} \tag{E.4}$$

[1]The proof presented here is patterned after one which can be found in J. Galejs, *Antennas in Inhomogeneous Media* (Oxford: Pergamon Press, 1969), §2.5.

Equation E.3 represents the magnetic vector potential function evaluated at the field point $(a, 0, z)$ when the current is concentrated on the Z-axis.

With $ka \ll 1$ and $a \ll l$, Functions E.1 and E.3 are approximated very well by

$$\begin{aligned}\mathcal{A}(z) = &\int_{-l}^{z-b}\int_{0}^{2\pi}\frac{e^{-jk|z-z'|}}{4\pi|z-z'|}K_z(z')a\,d\phi\,dz'\\ &+\int_{z-b}^{z+b}\int_{0}^{2\pi}\frac{e^{-jkR}}{4\pi R}K_z(z')a\,d\phi\,dz'\\ &+\int_{z+b}^{l}\int_{0}^{2\pi}\frac{e^{-jk|z-z'|}}{4\pi|z-z'|}K_z(z')a\,d\phi\,dz'\end{aligned} \tag{E.5}$$

and

$$\begin{aligned}\mathcal{A}_1(z) = &\int_{-l}^{z-b}\frac{e^{-jk|z-z'|}}{4\pi|z-z'|}I(z')\,dz'\\ &+\int_{z-b}^{z+b}\frac{e^{-jkr}}{4\pi r}I(z')\,dz'\\ &+\int_{z+b}^{l}\frac{e^{-jk|z-z'|}}{4\pi|z-z'|}I(z')\,dz'\end{aligned} \tag{E.6}$$

in which $b \gg a$ (for example, $b = 10a$).

The first and third integrals respectively of these two expansions are equal and thus the two potential functions differ primarily because of the second integrals, that is,

$$\begin{aligned}\mathcal{A}(z) - \mathcal{A}_1(z) \cong &\int_{z-b}^{z+b}\int_{0}^{2\pi}\frac{e^{-jkR}}{4\pi R}K_z(z')a\,d\phi\,dz'\\ &-\int_{z-b}^{z+b}\frac{e^{-jkr}}{4\pi r}I(z')\,dz'\end{aligned} \tag{E.7}$$

In the range $z - b \le z' \le z + b$, kR and kr are small and e^{-jkR} and e^{-jkr} can be replaced by $1 - jkR$ and $1 - jkr$. The integrals associated with the factors $-jkR$ and $-jkr$ in (E.7) cancel, leaving

$$\begin{aligned}\mathcal{A}(z) - \mathcal{A}_1(z) \cong &\int_{z-b}^{z+b}\int_{0}^{2\pi}\frac{K_z(z')}{4\pi R}a\,d\phi\,dz'\\ &-\int_{z-b}^{z+b}\frac{I(z')}{4\pi r}dz'\end{aligned} \tag{E.8}$$

It is reasonable to assume that $K_z(z')$ is slowly varying in the interval $2b$ and thus

$$\mathcal{A}(z) - \mathcal{A}_1(z) \cong \frac{I(z)}{4\pi}\left[\int_{0}^{2\pi}\frac{d\phi}{2\pi}\int_{z-b}^{z+b}\frac{dz'}{R} - \int_{z-b}^{z+b}\frac{dz'}{r}\right] \tag{E.9}$$

Execution of the z'-integrations gives

$$\mathcal{A}(z) - \mathcal{A}_1(z) \cong \frac{I(z)}{4\pi}\left\{\int_0^{2\pi} \frac{d\phi}{2\pi} 2\left[\ln\left(y + \sqrt{y^2 + 2a^2(1 - \cos\phi)}\right)\right]_0^b - 2\left[\ln\left(y + \sqrt{y^2 + a^2}\right)\right]_0^b\right\} \tag{E.10}$$

When the ϕ integration is performed on the term $y = 0$ in the first part of (E.10), one finds that the result exactly cancels the $y = 0$ term in the second part of (E.10). Thus

$$\mathcal{A}(z) - \mathcal{A}_1(z) \cong \frac{I(z)}{2\pi}\left\{\frac{1}{2\pi}\int_0^{2\pi} \ln\left(b + \sqrt{b^2 + 2a^2(1 - \cos\phi)}\right) d\phi - \ln\left(b + \sqrt{b^2 + a^2}\right)\right\} \tag{E.11}$$

Equation E.11 can be put in the more compact form

$$\mathcal{A}(z) - \mathcal{A}_1(z) \cong \frac{I(z)}{4\pi^2}\int_0^{2\pi} \ln\frac{b + \sqrt{b^2 + 2a^2(1 - \cos\phi)}}{b + \sqrt{b^2 + a^2}}\, d\phi \tag{E.12}$$

from which it can be recognized that the integrand consists of the logarithm of a number that is never far from unity. Indeed, a power series expansion gives

$$\ln\frac{b + \sqrt{b^2 + 2a^2(1 - \cos\phi)}}{b + \sqrt{b^2 + a^2}} = \ln\left[1 + \frac{a^2}{4b^2}(1 - 2\cos\phi) + \cdots\right] \cong \frac{a^2}{4b^2}(1 - 2\cos\phi) \tag{E.13}$$

When (E.13) is used in (E.12), the integration is simple and gives

$$\mathcal{A}(z) - \mathcal{A}_1(z) \cong \frac{a^2 I(z)}{8\pi b^2} \tag{E.14}$$

If one returns to (E.6), it is evident that $\mathcal{A}_1(z)$ receives its principal contribution from the second integral and that

$$\mathcal{A}_1(z) \cong \frac{I(z)}{4\pi}\int_{z-b}^{z+b} \frac{dz'}{r} \cong \frac{I(z)}{2\pi}\ln\frac{2b}{a} \tag{E.15}$$

A good measure of the adequacy of the approximation is the ratio

$$\frac{\mathcal{A}(z) - \mathcal{A}_1(z)}{\mathcal{A}_1(z)} \cong \frac{(a/b)^2}{4\ln(2b/a)} \tag{E.16}$$

If $b = 10a$, this ratio has the value 0.0008. Even for $b = 4a$ it is only 0.0075, less than a 1% error. Thus it is acceptable to use (7.25) as an approximation for (7.24) if the computations include an integration that extends over a length of the dipole of at least plus and minus several wire diameters. This is possible for all values of z except those close to $z = \pm l$, but in those small regions $I(z)$ is negligible and the value of $\mathcal{A}(z)$ is small, so the error is not serious.

F diffraction by plane conducting screens: Babinet's principle

Wire grids or arrays of slots in a ground plane excited by primary radiators such as horns, or even by plane waves caused by remote sources, can be designed to have useful antenna characteristics. A powerful integral equation technique which permits deduction of the scattering off such planar obstacles when excited by very general primary sources has been formulated by E. T. Copson.[1] An important result arising from Copson's formulation is a rigorous statement of Babinet's principle for complementary conducting screens.

Consider an infinite, perfectly conducting ground plane of negligible thickness in which an arbitrary collection of arbitrarily shaped holes has been cut, as suggested by the first screen in Figure F.1. The holes have surface areas $S_1, S_2, \ldots, S_N$, to which reference will be made by the abbreviation Σ_1. The metallic portion of the screen will be designated by Σ_2. Imagine that sources in $z < 0$ cause an electromagnetic field distribution throughout space (in the absence of the screen) designated by $(\mathbf{E}^i, \mathbf{H}^i)$. The sources induced in the screen cause a scattered field $(\mathbf{E}^s, \mathbf{H}^s)$ and the total field at any point in space is given by

$$\mathbf{E} = \mathbf{E}^i + \mathbf{E}^s, \qquad \mathbf{H} = \mathbf{H}^i + \mathbf{H}^s \tag{F.1}$$

The total field in $z > 0$ (see Section 1.12) can be determined from the equivalent sources

$$\begin{aligned} \mathbf{K} &= \mathbf{1}_z \times \mathbf{H}(\xi, \eta, 0+) \qquad & \mathbf{K}_m &= -\mu_0^{-1}\mathbf{1}_z \times \mathbf{E}(\xi, \eta, 0+) \\ \rho_s &= \epsilon_0 \mathbf{E}_z(\xi, \eta, 0+) \qquad & \rho_{sm} &= \mathbf{H}_z(\xi, \eta, 0+) \end{aligned} \tag{F.2}$$

with $(\xi, \eta, 0+)$ any point in a plane in $z > 0$, parallel to and infinitesimally close to the screen. The potential functions for these equivalent sources are (see Section 1.12)

[1] E. T. Copson, "An Integral Equation Method for Solving Plane Diffraction Problems," *Proc. Roy. Soc. London*, 186A (1946), 100–18.

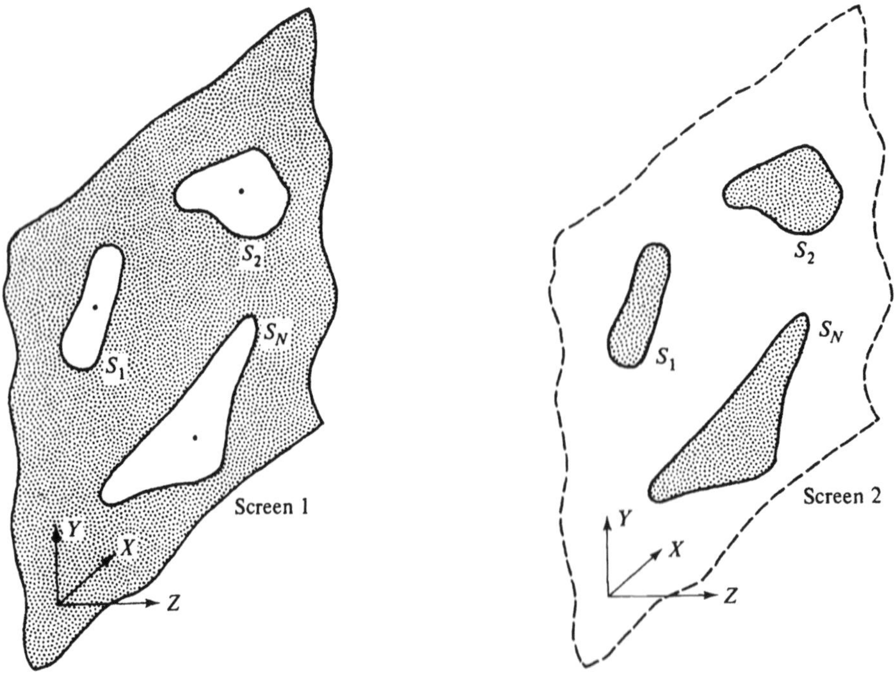

Fig. F.1 Complementary Thin Conducting Screens

$$
\begin{aligned}
\mathbf{A}(x, y, z) &= \frac{1}{4\pi\mu_0^{-1}} \iint_{\Sigma_1+\Sigma_2} [\mathbf{1}_y\mathbf{H}_x(\xi, \eta, 0+) - \mathbf{1}_x\mathbf{H}_y(\xi, \eta, 0+)]\psi \, d\xi \, d\eta \\
\mathbf{F}(x, y, z) &= \frac{1}{4\pi} \iint_{\Sigma_1} [\mathbf{1}_x\mathbf{E}_y(\xi, \eta, 0+) - \mathbf{1}_y\mathbf{E}_x(\xi, \eta, 0+)]\psi \, d\xi \, d\eta \\
\Phi(x, y, z) &= \frac{1}{4\pi} \iint_{\Sigma_1+\Sigma_2} \mathbf{E}_z(\xi, \eta, 0+)\psi \, d\xi \, d\eta \\
\Phi_m(x, y, z) &= \frac{1}{4\pi\epsilon_0} \iint_{\Sigma_1} H_z(\xi, \eta, 0+)\psi \, d\xi \, d\eta
\end{aligned}
\tag{F.3}
$$

in which $\psi = e^{-jkR}/R$ with $R = [(x - \xi)^2 + (y - \eta)^2 + z^2]^{1/2}$.

Equations 1.106 and 1.107 can be used to establish that, for any point (x, y, z) in $z > 0$,

$$
\begin{aligned}
E_x &= \frac{\partial F_y}{\partial z} - j\omega A_x - \frac{\partial \Phi}{\partial x} & H_x &= -\mu_0^{-1}\frac{\partial A_y}{\partial z} - j\omega\epsilon_0 F_x - \epsilon_0 \frac{\partial \Phi_m}{\partial x} \\
E_y &= -\frac{\partial F_x}{\partial z} - j\omega A_y - \frac{\partial \Phi}{\partial y} & H_y &= \mu_0^{-1}\frac{\partial A_x}{\partial z} - j\omega\epsilon_0 F_y - \epsilon_0 \frac{\partial \Phi_m}{\partial y} \\
E_z &= -\frac{\partial \Phi}{\partial z} - \frac{\partial F_y}{\partial x} + \frac{\partial F_x}{\partial y} & H_z &= -\epsilon_0 \frac{\partial \Phi_m}{\partial z} + \mu_0^{-1}\frac{\partial Ay}{\partial x} - \mu_0^{-1}\frac{\partial A_x}{\partial y}
\end{aligned}
\tag{F.4}
$$

Since all of the potential functions are even in z, their first derivatives with respect to z must be odd in z. When the field components in (F.4) are evaluated at $(x, y, -z)$, the result must be zero, because these equivalent sources were chosen so as to give null fields in $z < 0$. But this means that, for any point (x, y, z) in $z > 0$, the first term on the right side of any equation in (F.4) is equal to the sum of the following two terms. For this reason (F.4) can be rewritten in either of two distinct forms.

$$E_x = 2\frac{\partial F_y}{\partial z} \qquad E_y = -2\frac{\partial F_x}{\partial z} \qquad E_z = -2\frac{\partial F_y}{\partial x} + 2\frac{\partial F_x}{\partial y}$$

$$H_x = -2j\omega\epsilon_0 F_x - 2\epsilon_0\frac{\partial \Phi_m}{\partial x} \quad H_y = -2j\omega\epsilon_0 F_y - 2\epsilon_0\frac{\partial \Phi_m}{\partial y} \quad H_z = -2\epsilon_0\frac{\partial \Phi_m}{\partial z} \qquad \text{(F.5)}$$

or

$$E_x = -2j\omega A_x - 2\frac{\partial \Phi}{\partial x} \quad E_y = -2j\omega A_y - 2\frac{\partial \Phi}{\partial y} \quad E_z = -\frac{\partial \Phi}{\partial z} \qquad \text{(F.6)}$$

$$H_x = -2\mu_0^{-1}\frac{\partial A_y}{\partial z} \qquad H_y = 2\mu_0^{-1}\frac{\partial A_x}{\partial z} \qquad H_z = 2\mu_0^{-1}\frac{\partial A_y}{\partial x} - 2\mu_0^{-1}\frac{\partial A_x}{\partial y}$$

It can be observed that (F.5) gives the field components entirely in terms of the equivalent magnetic sources, whereas (F.6) involves only the electric sources.

FORMULATION FOR Σ_1/Σ_2 SMALL If the collection of holes Σ_1 is a small fraction of the entire screen, it is clear from a study of (F.3) that it is advantageous to select the set in (F.5) because then the pertinent potential functions require only integration over Σ_1. It will be assumed in what follows that screen 1 fits this description and that (F.5) will be used to represent the fields in $z > 0$.

With the primary sources in $z < 0$, the determination of the total fields in $z < 0$ cannot be achieved quite so directly. However, one can observe that

$$\begin{aligned} E_x(\xi, \eta, 0-) &= E_x^i(\xi, \eta, 0) + E_x^s(\xi, \eta, 0-) \\ E_y(\xi, \eta, 0-) &= E_y^i(\xi, \eta, 0) + E_y^s(\xi, \eta, 0-) \\ H_z(\xi, \eta, 0-) &= H_z^i(\xi, \eta, 0) + H_z^s(\xi, \eta, 0-) \end{aligned} \qquad \text{(F.7)}$$

The scattered field components that occur in (F.7) are even in z, that is,

$$E_x^s(\xi, \eta, 0-) = E_x^s(\xi, \eta, 0+), \qquad E_y^s(\xi, \eta, 0-) = E_y^s(\xi, \eta, 0+),$$
$$H_z^s(\xi, \eta, 0-) = H_z^s(\xi, \eta, 0+)$$

as a consequence of which the total field components appearing in (F.7) are also even in z. Therefore the scattered fields anywhere in $z < 0$ can be deduced through use of the equivalent magnetic sources

$$\mathbf{K}_m = \mu_0^{-1}\mathbf{1}_z \times \mathbf{E}^s(\xi, \eta, 0-) \qquad \sigma_m = -H_z^s(\xi, \eta, 0-) \qquad \text{(F.8)}$$

which give rise to the potential functions

$$
\begin{aligned}
F'_x(x, y, z) &= -\frac{1}{4\pi} \iint_{\Sigma_1+\Sigma_2} E^s_y(\xi, \eta, 0-)\psi \, d\xi \, d\eta \\
&= -\frac{1}{4\pi} \iint_{\Sigma_1} E_y(\xi, \eta, 0+)\psi \, d\xi \, d\eta + \frac{1}{4\pi} \iint_{\Sigma_1+\Sigma_2} E^i_y(\xi, \eta, 0)\psi \, d\xi \, d\eta \\
&= -F_x(x, y, z) - F^r_x(x, y, z)
\end{aligned} \tag{F.9}
$$

$$
\begin{aligned}
F'_y(x, y, z) &= \frac{1}{4\pi} \iint_{\Sigma_1+\Sigma_2} E^s_x(\xi, \eta, 0-)\psi \, d\xi \, d\eta \\
&= \frac{1}{4\pi} \int_{\Sigma_1} E_x(\xi, \eta, 0+)\psi \, d\xi \, d\eta - \frac{1}{4\pi} \iint_{\Sigma_1+\Sigma_2} E^i_x(\xi, \eta, 0)\psi \, d\xi \, d\eta \\
&= -F_y(x, y, z) - F^r_y(x, y, z)
\end{aligned} \tag{F.10}
$$

$$
\begin{aligned}
\Phi'_m(x, y, z) &= -\frac{1}{4\pi} \iint_{\Sigma_1+\Sigma_2} H^s_z(\xi, \eta, 0-)\psi \, d\xi \, d\eta \\
&= -\frac{1}{4\pi} \int_{\Sigma_1} H_z(\xi, \eta, 0+)\psi \, d\xi \, d\eta + \frac{1}{4\pi} \iint_{\Sigma_1+\Sigma_2} H^i_z(\xi, \eta, 0)\psi \, d\xi \, d\eta \\
&= -\Phi_m(x, y, z) - \Phi^r_m(x, y, z)
\end{aligned} \tag{F.11}
$$

The potential functions F^r_x, F^r_y, and Φ^r_m in (F.9) through (F.11) are those that would apply for the back-scattered fields in $z < 0$ if the screen contained no holes, for then

$$
E^r_x(\xi, \eta, 0) = -E^i_x(\xi, \eta, 0), \qquad E^r_y(\xi, \eta, 0) = -E^i_y(\xi, \eta, 0),
$$
$$
H^r_z(\xi, \eta, 0) = -H^i_z(\xi, \eta, 0)
$$

The total field in $z < 0$ can be found by operating on the potential functions F'_x, F'_y, and Φ'_m as prescribed by Equations F.5 in order to get the scattered fields and then adding the components of the incident field. This gives

$$
\begin{aligned}
E_x &= E^0_x - 2\frac{\partial F_y}{\partial z} & E_y &= E^0_y + 2\frac{\partial F_x}{\partial z} & E_z &= E^0_z + 2\frac{\partial F_y}{\partial x} - 2\frac{\partial F_x}{\partial y} \\
H_x &= H^0_x + 2j\omega\epsilon_0 F_x + 2\epsilon_0 \frac{\partial \Phi_m}{\partial x} & H_y &= H^0_y + 2j\omega\epsilon_0 F_y + 2\epsilon_0 \frac{\partial \Phi_m}{\partial y} & & \\
& & & & H_z &= H^0_z + 2\epsilon_0 \frac{\partial \Phi_m}{\partial z}
\end{aligned} \tag{F.12}
$$

In (F.12), $\mathbf{E}^0 = \mathbf{E}^i + \mathbf{E}^r$, $\mathbf{H}^0 = \mathbf{H}^i + \mathbf{H}^r$ is the total field in $z < 0$ for the case $\Sigma_1 = 0$.

To summarize the results to this point, one can say that the total field in $z > 0$ can be obtained in terms of the potential functions F_x, F_y, and Φ_m via (F.5), and that the total field in $z < 0$ can be obtained in terms of the potential functions F_x, F_y, and Φ_m via (F.12) if one adds the total field that would exist in $z < 0$ if the screen were closed. There still remains the task of finding the aperture distribution $E_x(\xi, \eta, 0+)$, $E_y(\xi, \eta, 0+)$, and $H_z(\xi, \eta, 0+)$ so that the potential functions are known.

Since E_z, H_x, and H_y must be continuous across the holes Σ_1, it follows from (F.5) and (F.12) that if (x, y, z) is a point in the screen occupied by a hole, then

$$\begin{aligned} -2\frac{\partial F_y}{\partial x} + 2\frac{\partial F_x}{\partial y} &= E_z^0 + 2\frac{\partial F_y}{\partial x} - 2\frac{\partial F_x}{\partial y} \\ -2j\omega\epsilon_0 F_x - 2\epsilon_0\frac{\partial \Phi_m}{\partial x} &= H_x^0 + 2j\omega\epsilon_0 F_x + 2\epsilon_0\frac{\partial \Phi_m}{\partial x} \\ -2j\omega\epsilon_0 F_y - 2\epsilon_0\frac{\partial \Phi_m}{\partial y} &= H_y^0 + 2j\omega\epsilon_0 F_y + 2\epsilon_0\frac{\partial \Phi_m}{\partial y} \end{aligned} \tag{F.13}$$

Because

$$E_z^r(x, y, 0) = E_z^i(x, y, 0), \qquad H_x^r(x, y, 0) = H_x^i(x, y, 0), \qquad H_y^r(x, y, 0) = H_y^i(x, y, 0)$$

Equations F.13 reduce to

$$\begin{aligned} -2\frac{\partial F_y}{\partial x} + 2\frac{\partial F_x}{\partial y} &= E_z^i \\ -2j\omega\epsilon_0 F_x - 2\epsilon_0\frac{\partial \Phi_m}{\partial x} &= H_x^i \\ -2j\omega\epsilon_0 F_y - 2\epsilon_0\frac{\partial \Phi_m}{\partial y} &= H_y^i \end{aligned} \tag{F.14}$$

and thus the integral equations linking the unknown aperture field to the incident field are

$$\begin{aligned} \iint_{\Sigma_1}\left[E_x(\xi, \eta)\frac{\partial \psi_0}{\partial x} + E_y(\xi, \eta)\frac{\partial \psi_0}{\partial y}\right] d\xi\, d\eta &= 2\pi E_z^i(x, y) \\ \iint_{\Sigma_1}\left[j\omega\epsilon_0 E_y(\xi, \eta)\psi_0 + H_z(\xi, \eta)\frac{\partial \psi_0}{\partial x}\right] d\xi\, d\eta &= -2\pi H_x^i(x, y) \\ \iint_{\Sigma_1}\left[j\omega\epsilon_0 E_x(\xi, \eta)\psi_0 - H_z(\xi, \eta)\frac{\partial \psi_0}{\partial y}\right] d\xi\, d\eta &= 2\pi H_y^i(x, y) \end{aligned} \tag{F.15}$$

in which

$$\psi_0 = \frac{e^{-jkr}}{r} \qquad r = [(x - \xi)^2 + (y - \eta)^2] \tag{F.16}$$

with both the source point $(\xi, \eta, 0)$ and the field point (x, y, z) lying in Σ_1.

FORMULATION FOR Σ_1/Σ_2 LARGE If the collection of holes Σ_1 is a large fraction of the entire screen, Equations F.15 are difficult and costly to solve for the aperture distribution. It then proves useful to return to the alternate expressions for the fields, embodied in (F.6), and proceed as follows.

Let the sources of the incident field be in $z < 0$. The total field in $z > 0$ is $(\mathbf{E}^i + \mathbf{E}^s, \mathbf{H}^i + \mathbf{H}^s)$, and the scattered field in $z > 0$ can be found from (F.6) if

$$A_x(x, y, z) = -\frac{1}{4\pi\mu_0^{-1}} \iint_{\Sigma_2} H_y^s(\xi, \eta, 0+)\psi \, d\xi \, d\eta$$

$$A_y(x, y, z) = \frac{1}{4\pi\mu_0^{-1}} \iint_{\Sigma_2} H_x^s(\xi, \eta, 0+)\psi \, d\xi \, d\eta \tag{F.17}$$

$$\Phi(x, y, z) = \frac{1}{4\pi} \iint_{\Sigma_2} E_z^s(\xi, \eta, 0+)\psi \, d\xi \, d\eta$$

It is important to note that the integration in (F.17) extends only over Σ_2, that is, the material portion of the screen. This is a consequence of the fact that, because the screen has negligible thickness, H_x^s, H_y^s, and E_z^s are identically zero in Σ_1.

The total fields in $z > 0$ are therefore given by

$$E_x = E_x^i - 2j\omega A_x - 2\frac{\partial \Phi}{\partial x} \qquad E_y = E_y^i - 2j\omega A_y - 2\frac{\partial \Phi}{\partial y} \qquad E_z = E_z^i - \frac{\partial \Phi}{\partial z}$$

$$H_x = H_x^i - 2\mu_0^{-1}\frac{\partial A_y}{\partial z} \qquad H_y = H_y^i + 2\mu_0^{-1}\frac{\partial A_x}{\partial z} \qquad H_z = H_z^i + 2\mu_0^{-1}\frac{\partial A_y}{\partial x} - 2\mu_0^{-1}\frac{\partial A_x}{\partial y} \tag{F.18}$$

Upon reflection, one can conclude that (F.18) also applies in $z < 0$. The reason for this can be seen by examining one of the potential functions. The scattered field in $z < 0$ can be found in terms of the equivalent sources

$$\mathbf{K} = -\mathbf{1}_z \times \mathbf{H}^s(\xi, \eta, 0-) \qquad \rho_s = -\epsilon_0 E_z^s(\xi, \eta, 0-)$$

which gives rise to potential functions such as

$$A_x'(x, y, z) = \frac{1}{4\pi\mu_0^{-1}} \iint_{\Sigma_2} H_y^s(\xi, \eta, 0-)\psi \, d\xi \, d\eta \tag{F.19}$$

But $H_y^s(\xi, \eta, 0-) = -H_y^s(\xi, \eta, 0+)$ anywhere on Σ_2, and therefore $A_x' = A_x$. Similarly, $A_y' = A_y$ and $\Phi' = \Phi$, and thus (F.18) is valid everywhere, both in $z > 0$ and $z < 0$.

The problem still remains that the aperture distribution of the scattered field is unknown. But E_x, E_y, and H_z must vanish everywhere on Σ_2, and thus (F.18) yields the integral equations

$$\iint_{\Sigma_2}\left[j\omega\mu_0 H_y^s(\xi,\eta)\psi_0 - E_z^s(\xi,\eta)\frac{\partial\psi_0}{\partial x}\right]d\xi\,d\eta = -2\pi E_x^i(x,y)$$

$$\iint_{\Sigma_2}\left[j\omega\mu_0 H_x^s(\xi,\eta)\psi_0 + E_z^s(\xi,\eta)\frac{\partial\psi_0}{\partial y}\right]d\xi\,d\eta = 2\pi E_y^i(x,y) \qquad \text{(F.20)}$$

$$\iint_{\Sigma_2}\left[H_x^s(\xi,\eta)\frac{\partial\psi_0}{\partial x} + H_y^s(\xi,\eta)\frac{\partial\psi_0}{\partial y}\right]d\xi\,d\eta = -2\pi H_z^i(x,y)$$

Just as (F.15) is the preferred form to use when finding the aperture distribution if Σ_1/Σ_2 is small, Equations F.20 are preferable when Σ_2/Σ_1 is small.

BABINET'S PRINCIPLE An interesting coupling of (F.15) and (F.20) can be made when two screens are complementary. This situation is suggested by Figure F.1, which shows a second screen with holes and material that are an exact interchange of what one finds for the first screen. Thus Σ_1 for the first screen is matched by Σ_2 for the second screen. In such cases, if (F.15) is used for screen 1 and (F.20) for screen 2, the integrations are over exactly the same regions. Suppose further that, throughout all space,

$$\mathbf{E}_1^i = \kappa\mathbf{H}_2^i, \qquad \mathbf{H}_1^i = -\frac{\kappa}{\eta^2}\mathbf{E}_2^i \qquad \text{(F.21)}$$

in which (E_1^i, H_1^i) is the primary field which excites screen 1, (E_2^i, H_2^i) is the primary field which excites screen 2, and $\kappa = 1$ ohm; $\eta = \sqrt{\mu_0/\epsilon_0}$ is the impedance of free space.[2]

When (F.21) is used in (F.15) and the result is compared to (F.20), one finds that

$$E_{x,1}(\xi,\eta) = -\kappa H_{x,2}^s(\xi,\eta) \qquad E_{y,1}(\xi,\eta) = -\kappa H_{y,2}^s(\xi,\eta) \qquad H_{z,1}(\xi,\eta) = \frac{\kappa}{\eta^2}E_{z,2}^s(\xi,\eta) \qquad \text{(F.22)}$$

which means

$$F_x(x,y,z) = \kappa\mu_0^{-1}A_x(x,y,z) \qquad F_y(x,y,z) = \kappa\mu_0^{-1}A_y(x,y,z)$$
$$\Phi_m(x,y,z) = \kappa\mu_0^{-1}\Phi(x,y,z) \qquad \text{(F.23)}$$

If the relations given in (F.23) are used in (F.5) and the results are compared to (F.18), it is discovered that

$$\mathbf{E}_1 + \kappa\mathbf{H}_2 = \mathbf{E}_1^i \quad \text{and} \quad \mathbf{H}_1 - \frac{\kappa}{\eta^2}\mathbf{E}_2 = \mathbf{H}_1^i \qquad z > 0 \qquad \text{(F.24)}$$

[2] Conceptually, the relationship between the two primary fields given by (F.21) can be achieved as follows: Let $(\mathbf{J}_1^i, \rho_1^i)$ be the sources for $(\mathbf{E}_1^i, \mathbf{H}_1^i)$. Replace these sources by a fictitious set $(\mathbf{J}_m, \rho_m)$ which give the same field $(\mathbf{E}_1^i, \mathbf{H}_1^i)$ everywhere. Finally, replace the magnetic sources by electric sources $(\mathbf{J}_2^i, \rho_2^i)$ such that $\mathbf{J}_2^i = -\kappa\epsilon_0\mathbf{J}_m$ and $\rho_2^i = -\kappa\epsilon_0\rho_m$. The sources $(\mathbf{J}_2^i, \rho_2^i)$ will cause a field $(\mathbf{E}_2^i, \mathbf{H}_2^i)$ which satisfies (F.21).

Similarly, when (F.23) is substituted in (F.12), comparison with (F.18) establishes that

$$\mathbf{E}_1 - \kappa \mathbf{H}_2 = \mathbf{E}_1^{\text{r}} \quad \text{and} \quad \mathbf{H}_1 + \frac{\kappa}{\eta^2}\mathbf{E}_2 = \mathbf{H}_1^{\text{r}} \qquad z < 0 \tag{F.25}$$

Equations F.24 and F.25 are a rigorous statement of Babinet's principle for thin, perfectly conducting plane complementary screens. The scattered fields are complementary in the sense that if the incident fields on the two screens are related by (F.21), then the total fields on the two sides of the screens are connected in the manner given by (F.24) and (F.25). This result is particularly useful when an incident plane wave is assumed, for then (F.21) indicates that a rotation of 90° in the polarization is all that is needed to obtain the incident field for the second screen. The complementary fields are then such that their vector sum is a uniform plane wave; the (θ, ϕ)-dependencies of the total fields are compensatory.

G the far-field in cylindrical coordinates

The situation occasionally arises in antenna analysis when it is convenient to assume that an aperture is infinitely long in one dimension and that the sources and fields are independent of that coordinate. This reduces the analysis to a two-dimensional problem, and as a result cylindrical coordinates become the natural choice as framework for the mathematical development. It is therefore desirable to establish expressions for the far field in cylindrical coordinates due to aperture distributions of this type.

One can begin with the general forms of (1.110) for the retarded vector potential functions. With the aperture assumed to be a cylindrical surface S, infinitely long in the Z-direction (see Figure G.1), and with the aperture distribution z-independent, these expressions become

$$\begin{aligned} \mathbf{A}(x, y, 0, t) &= \int_S \frac{\mathbf{K}(\xi, \eta)e^{j(\omega t - kR)}}{4\pi\mu_0^{-1}R}\,dS \\ \mathbf{F}(x, y, 0, t) &= \int_S \frac{\mathbf{K_m}(\xi, \eta)e^{j(\omega t - kR)}}{4\pi\mu_0^{-1}R}\,dS \end{aligned} \tag{G.1}$$

with (x, y, z) the field point and (ξ, η, ζ) the source point. Because all the sources are z-independent, no loss in generality results from taking the field point to be in the XY-plane, and this has been done in (G.1). Thus

$$R = [(x - \xi)^2 + (y - \eta)^2 + \zeta^2]^{1/2} \tag{G.2}$$

The two integrals in (G.1) are mathematically similar and the remainder of the treatment applies equally well to either. Proceeding with $\mathbf{A}$, one can write

$$\mathbf{A}(x, y, t) = \oint_C \frac{\mathbf{K}(\xi, \eta)e^{j\omega t}}{4\pi\mu_0^{-1}}\left[\int_{-\infty}^{\infty} \frac{e^{-jkR}}{R}\,d\zeta\right]dl \tag{G.3}$$

with dl an increment in length along the transverse contour C.

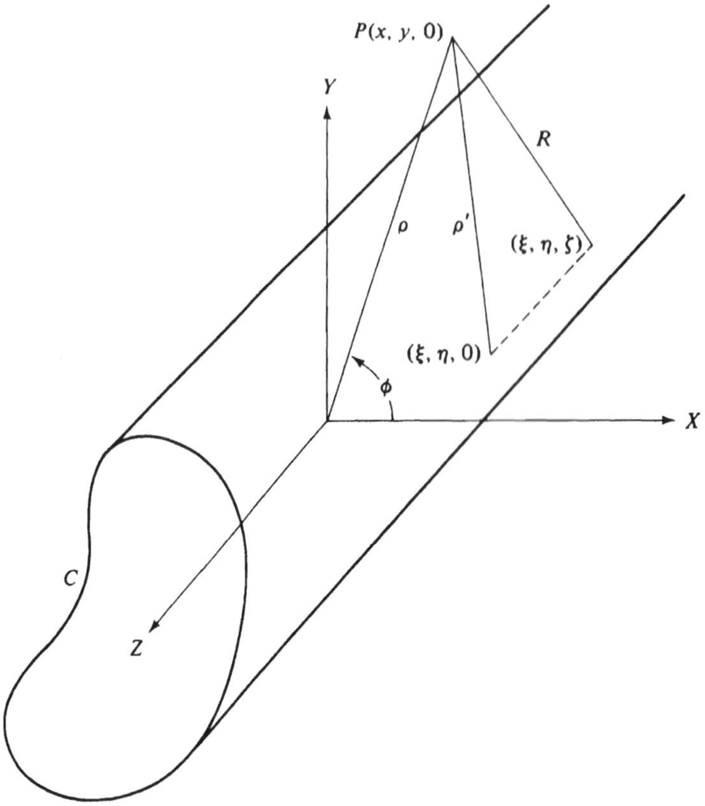

Fig. G.1 An Infinitely Long Cylindrical Aperture

The integral

$$I = \int_{-\infty}^{\infty} \frac{e^{-jkR}}{R}\, d\zeta \tag{G.4}$$

with R given by (G.2), can be evaluated in the following manner.

Hankel functions of the first and second kind can be defined in terms of contour integrals in the complex plane which take a variety of forms. Of pertinence to the present development is the relation[1]

$$H_0^{(2)}(u) = -\frac{1}{\pi j}\int_{-\infty}^{\infty} e^{-ju\cosh v}\, dv \tag{G.5}$$

in which $H_0^{(2)}(u)$ is the Hankel function of zero order and the second kind. In (G.5), integration is along the axis of reals.

[1]See, for example, G. N. Watson, *A Treatise on the Theory of Bessel Functions*, 2nd Ed. (London: Cambridge University Press, 1952), p. 180.

If the substitutions

$$\rho' = [(x-\xi)^2 + (y-\eta)^2]^{1/2} \qquad u = k\rho' \qquad R = \rho' \cosh v \tag{G.6}$$

are introduced, (G.5) becomes

$$H_0^{(2)}(k\rho') = -\frac{1}{\pi j}\int_{-\infty}^{\infty} \frac{e^{-jkR}}{R}\, d\zeta \tag{G.7}$$

Comparison with (G.4) reveals that

$$I = -\pi j H_0^{(2)}(k\rho') \tag{G.8}$$

Since the interest here is in the far field, $k\rho' \gg 1$ and the asymptotic form can be used for the Hankel function,[2] that is,

$$H_0^{(2)}(k\rho') \cong \sqrt{\frac{2}{\pi k\rho'}}\, e^{-j(k\rho'-\pi/4)} \tag{G.9}$$

Further, from (G.6),

$$\begin{aligned}\rho' &= [x^2 + y^2 - 2x\xi - 2y\eta + \xi^2 + \eta^2]^{1/2} \\ &= \rho - \xi\cos\phi - \eta\sin\phi + \frac{\xi^2+\eta^2}{2\rho} + \cdots\end{aligned} \tag{G.10}$$

where polar coordinates have been introduced via the transformation

$$x = \rho\cos\phi \qquad y = \rho\sin\phi \tag{G.11}$$

Thus

$$H_0^{(2)}(k\rho') \cong \sqrt{\frac{2}{\pi k\rho}}\, e^{-j(k\rho-\pi/4)} e^{jk(\xi\cos\phi+\eta\sin\phi)} \tag{G.12}$$

When this far-field approximation is placed in (G.3), one obtains

$$\mathbf{A}(\rho, \phi, t) = \frac{e^{j(\omega t - k\rho)}}{2\mu_0^{-1}\sqrt{2\pi jk\rho}} \oint_C \mathbf{K}(\xi, \eta) e^{jk(\xi\cos\phi+\eta\sin\phi)}\, dl \tag{G.13}$$

Therefore $\mathbf{A}$ can be viewed as the product of the outgoing cylindrical wavefactor

$$\frac{e^{j(\omega t - k\rho)}}{2\mu_0^{-1}\sqrt{2\pi jk\rho}} \tag{G.14}$$

and the weighting function

$$\mathcal{A}(\phi) = \oint_C \mathbf{K}(\xi, \eta) e^{jk(\xi\cos\phi+\eta\sin\phi)}\, dl \tag{G.15}$$

[2]Ibid., pp. 196–98.

Similarly, **F** in the far field is the product of the cylindrical wavefactor (G.14) and the weighting function

$$\mathfrak{F}(\phi) = \oint_C \mathbf{K}_m(\xi, \eta) e^{jk(\xi \cos\phi + \eta \sin\phi)} \, dl \tag{G.16}$$

If the operations indicated by (1.119) are performed and only terms in $\rho^{-1/2}$ are retained, one finds that

$$\mathbf{E} = -j\omega \mathbf{A}_T + jk(\mathbf{l}_\rho \times \mathbf{F}_T) \tag{G.17}$$

which is the same result as (1.123), obtained earlier for the spherical wave case. With suppression of the cylindrical wavefactor,

$$\begin{aligned} E_\phi &= -j\omega \mathcal{A}_\phi - jk\mathfrak{F}_z \\ E_z &= -j\omega \mathcal{A}_z + jk\mathfrak{F}_\phi \end{aligned} \tag{G.18}$$

wherein

$$\begin{aligned} \mathcal{A}_\phi(\phi) &= \oint_C [-\sin\phi K_x(\xi, \eta) + \cos\phi K_y(\xi, \eta)] e^{jk\mathcal{L}} \, dl \\ \mathcal{A}_z(\phi) &= \oint_C K_z(\xi, \eta) e^{jk\mathcal{L}} \, dl \\ \mathfrak{F}_\phi(\phi) &= \oint_C [-\sin\phi K_{xm}(\xi, \eta) + \cos\phi K_{ym}(\xi, \eta)] e^{jk\mathcal{L}} \, dl \\ \mathfrak{F}_z(\phi) &= \oint_C K_{zm}(\xi, \eta) e^{jk\mathcal{L}} \, dl \end{aligned} \tag{G.19}$$

with $\mathcal{L} = \xi \cos\phi + \eta \sin\phi$. Equations G.18 and G.19 can be used to determine the far field for cylindrical apertures that are infinite in extent in the Z-direction and contain sources that are z-independent.

H the utility of a $\csc^2 \theta$ pattern

Consider an antenna A whose radiated pattern is $\mathcal{P}(\theta, \phi)$ watts per steradian. Let this pattern be illuminating a target whose transverse radar cross section is σ. If the target is flying at a constant height h, as shown in Figure H.1, when the range is r the elevation angle θ is given by

$$\csc \theta = \frac{r}{h} \tag{H.1}$$

Assume that the target is flying in a straight line path in the half-plane $\phi = 0°$. Then the intercepted power at range r is

$$\mathcal{P}(\theta, 0°) \frac{\sigma}{r^2} \text{ watts} \tag{H.2}$$

It will be assumed that this power is reradiated isotropically.

If the target is at a different range r', corresponding to a different elevation angle θ', and if the radar cross section σ is unchanged, the incident power is

$$\mathcal{P}(\theta', 0°) \frac{\sigma}{(r')^2} \text{ watts} \tag{H.3}$$

It will be assumed that this power is also reradiated isotropically. It is equivalent to an isotropic source at $(r, \theta', 0°)$ which radiates

$$\mathcal{P}(\theta', 0°) \frac{\sigma}{(r')^2} \frac{r^2}{(r')^2} \text{ watts} \tag{H.4}$$

The signals detected by antenna A when in its receive mode are in the power ratio

$$\frac{P(\theta')}{P(\theta)} = \frac{\mathcal{P}(\theta', 0°)(r^2\sigma/(r')^4)}{\mathcal{P}(\theta, 0°)(\sigma/r^2)} \frac{\mathcal{P}(\theta', 0°)}{\mathcal{P}(\theta, 0°)} \tag{H.5}$$

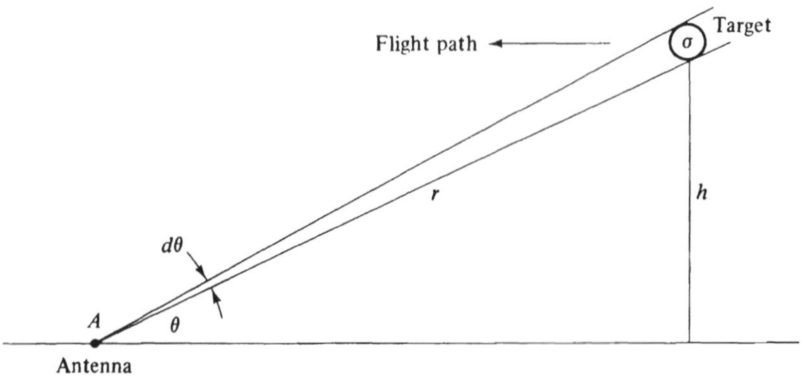

Fig. H.1 Radar Acquisition of a Target with a Horizontal Flight Path

If it is desired to have $P(\theta') = P(\theta)$, then

$$\frac{\mathcal{P}^2(\theta', 0^\circ)}{(r')^4} = \frac{\mathcal{P}^2(\theta, 0^\circ)}{r^4} \tag{H.6}$$

or

$$\frac{\mathcal{P}(\theta, 0^\circ)}{r^2} = \text{constant} \tag{H.7}$$

When (H.1) and (H.7) are combined, the result is that

$$\mathcal{P}(\theta, 0^\circ) \sim \csc^2 \theta \tag{H.8}$$

Thus if the principal plane pattern is designed to be proportional to the square of the cosecant of the elevation angle, the received signal from a target flying at constant height has a strength independent of range. Such patterns have obvious application in target-seeking radar systems. For the same basic reason they are also useful in ground-mapping radars and airport beacons. One should note that this result depends on the premise that σ is range-independent. Modifications of the $\csc^2 \theta$ formula are obtained when other assumptions are made about the radar cross section.

index

AUTHOR INDEX

SUBJECT INDEX

About the Author

Robert S. Elliott (S'46—A'52—SM'54—F'61—LF'87) was born in Brooklyn, NY, on March 9, 1921. He received the A.B. degree in English literature in 1942 and the B.S. degree in electrical engineering in 1943, both from Columbia University, New York, NY: There followed three years in the service of the Navy, first at the Naval Ordnance Laboratory and then at the Applied Physics Laboratory, both in the Washington, DC, area, where he worked successively on acoustic mines, the proximity fuze, and radar.

With the war over, he moved to Champaign-Urbana, IL, to become a graduate student in electrical engineering. Because of the enormous number of returning G.I. 's, all ten members of that first postwar graduate class were pressed into teaching, which proved to be one of the most favorable turning points in his career. Though it took six years to earn the M.S. and Ph.D. degrees, the full-time teaching experience solidified his basic understanding and established his priorities. He still feels that his Ph.D. dissertation, "On the Maximum Frequency of Coherent Oscillations," is one of his most important research contributions.

During the Korean war, the newly married, newly graduated Dr. Elliott became Lt. Elliott on active duty, once again at the Naval Ordnance Laboratory. The election of President Eisenhower and the settling of that conflict meant that the reservists were discharged in early 1953, and academic positions were in short supply. Dr. Elliott accepted an offer to work in the Antenna Research Section of the Microwave Laboratory of Hughes in Los Angeles. Those were golden years, with the Laboratory under the creative guidance of Dr. Lester Van Atta, and with a brilliant group of co-workers that included Tom Taylor and Sam Sensiper. It was during this period that Dr. Elliott made his contributions to the literature on surface waves, worked with Ken Kelly on leaky wave antennas, and began his long collaboration with Lou Kurtz on the analysis and design of slot arrays.

In private life, Barbara and Bob Elliott were building a home and raising their four children. Paula is a nurse, married, and has two children. Laura is a legal secretary. Jeff and Greg both earned their doctorates in solid state physics.

While at Hughes, Dr. Elliott began to offer graduate courses on antennas at night at UCLA, and seriously entertained the idea of joining their regular faculty. But Bob Krausz intervened with the appealing idea that, together with Al Clavin and Lou Kurtz, they should start their own small company. Thus Rantec was born, struggled, succeeded, and ultimately became a Division of Emerson Electric. Long before that happened, Dr. Elliott returned to his first love, teaching, and joined UCLA, where he has now spent 30 productive years.

In 1957 the School of Engineering at UCLA was small; it had only a handful of Electrical Engineering professors. Thus the opportunity existed to develop a modem curriculum and build a respected faculty. Dr. Elliott participated in both of these efforts and served as the first Chair of Electrical Engineering when the School was departmentalized.

His interests are broad, a fact easily gleaned from a perusal of his publications. In 1966 he authored a highly original book, entitled *Electromagnetics*, which uses special relativity to derive a complete electromagnetic theory, with Coulomb's law as the sole postulate. Each chapter is enriched by a historical section describing discoveries pertinent to the subject under discussion. He has championed a redefinition of all electrical quantities and a simplification of the teaching of electromagnetic theory. In the mid- 1970's, he took a hiatus from engineering and returned to school to obtain an M.A. degree in economics.

Dr. Elliott has undertaken research, together with his graduate students and colleagues (notably Professors Alexopoulos and Orchard) and industry associates (such as George Stern and Lou Kurtz) on a variety of topics, but principally antenna pattern synthesis and the design of microwave arrays in the presence of mutual coupling. His 80 journal papers on these subjects culminated in the textbook *Antenna Theory and Design*. Dr. Elliott's honors include a half-dozen Best Teacher Awards, two Best Paper Awards from the IEEE, election to Tau Beta P1, Sigma Xi, the New York Academy of Sciences, and the National Academy of Engineering. He became a Fellow of the IEEE in 1961.

Dr. Elliott has served the IEEE and APS in many capacities. He spearheaded the successful efforts in both 1971 and 1981 to bring the AP/URSI International Symposium to Los Angeles and headed the Coordinating Committee in 1981. He was an early Chapter President for APS in Los Angeles and served on the Antennas and Propagation Society Administrative Committee during 1980-1983. He has given talks worldwide as a Distinguished Lecturer for APS and was a Plenary Session speaker at both the MTT Symposium in May 1988 in New York and at the AP- URSI Symposium that followed in June in Syracuse, speaking on the subject "The History of Electromagnetics as Hertz Would Have Known It" as part of the Hertz Centennial.

The Antenna and Propagation Society Distinguished Achievement Award was established in 1985 to recognize outstanding technical achievement and meritorious service to the Antennas and Propagation Society. In 1988 the Award was presented to Robert S. Elliott with the citation:

> "For his original contributions to antenna pattern synthesis, the design of slot arrays, the relativistic derivation of Maxwell's equations, and for his skill as a teacher, not only of electromagnetics, but also the history of science, and for his participation in the activities of the IEEE."

In 1992 Professor Elliott's career was honored by UCLA when he was chosen to be the first person to hold the newly created Hughes Distinguished Chair in Electromagnetics.

In 2000 Dr. Elliott was the recipient of an IEEE Third Millennium Medal.

SUMMARY OF IMPORTANT VECTOR RELATIONS

Cartesian Coordinates

$$\nabla\Phi = \mathbf{1}_x\frac{\partial\Phi}{\partial x} + \mathbf{1}_y\frac{\partial\Phi}{\partial y} + \mathbf{1}_z\frac{\partial\Phi}{\partial z}$$

$$\nabla\cdot\mathbf{A} = \frac{\partial A_x}{\partial x} + \frac{\partial A_y}{\partial y} + \frac{\partial A_z}{\partial z}$$

$$\nabla^2 = \frac{\partial^2}{\partial x^2} + \frac{\partial^2}{\partial y^2} + \frac{\partial^2}{\partial z^2}$$

$$\nabla\times\mathbf{A} = \begin{vmatrix} \mathbf{1}_x & \mathbf{1}_y & \mathbf{1}_z \\ \dfrac{\partial}{\partial x} & \dfrac{\partial}{\partial y} & \dfrac{\partial}{\partial z} \\ A_x & A_y & A_z \end{vmatrix}$$

Cylindrical Coordinates

$$\nabla\Phi = \mathbf{1}_r\frac{\partial\Phi}{\partial r} + \frac{\mathbf{1}_\phi}{r}\frac{\partial\Phi}{\partial\phi} + \mathbf{1}_z\frac{\partial\Phi}{\partial z}$$

$$\nabla\cdot\mathbf{A} = \frac{1}{r}\frac{\partial}{\partial r}(rA_r) + \frac{1}{r}\frac{\partial A_\phi}{\partial\phi} + \frac{\partial A_z}{\partial z}$$

$$\nabla^2 = \frac{1}{r}\frac{\partial}{\partial r}\left(r\frac{\partial}{\partial r}\right) + \frac{1}{r^2}\frac{\partial^2}{\partial\phi^2} + \frac{\partial^2}{\partial z^2}$$

$$\nabla\times\mathbf{A} = \begin{vmatrix} \dfrac{\mathbf{1}_r}{r} & \mathbf{1}_\phi & \dfrac{\mathbf{1}_z}{r} \\ \dfrac{\partial}{\partial r} & \dfrac{\partial}{\partial\phi} & \dfrac{\partial}{\partial z} \\ A_r & rA_\phi & A_z \end{vmatrix}$$

Spherical Coordinates

$$\nabla\Phi = \mathbf{1}_r\frac{\partial\Phi}{\partial r} + \frac{\mathbf{1}_\theta}{r}\frac{\partial\Phi}{\partial\theta} + \frac{\mathbf{1}_\phi}{r\sin\theta}\frac{\partial\Phi}{\partial\phi}$$

$$\nabla\cdot\mathbf{A} = \frac{1}{r^2}\frac{\partial}{\partial r}(r^2A_r) + \frac{1}{r\sin\theta}\frac{\partial}{\partial\theta}(\sin\theta\, A_\theta) + \frac{1}{r\sin\theta}\frac{\partial A_\phi}{\partial\phi}$$

$$\nabla^2 = \frac{1}{r^2}\frac{\partial}{\partial r}\left(r^2\frac{\partial}{\partial r}\right) + \frac{1}{r^2\sin\theta}\frac{\partial}{\partial\theta}\left(\sin\theta\frac{\partial}{\partial\theta}\right) + \frac{1}{r^2\sin^2\theta}\frac{\partial^2}{\partial\phi^2}$$

$$\nabla\times\mathbf{A} = \begin{vmatrix} \dfrac{\mathbf{1}_r}{r^2\sin\theta} & \dfrac{\mathbf{1}_\theta}{r\sin\theta} & \dfrac{\mathbf{1}_\phi}{r} \\ \dfrac{\partial}{\partial r} & \dfrac{\partial}{\partial\theta} & \dfrac{\partial}{\partial\phi} \\ A_r & rA_\theta & r\sin\theta A_\phi \end{vmatrix}$$

VECTOR IDENTITIES

$$\nabla(\Phi\Psi) = \Psi\nabla\Phi + \Phi\nabla\Psi$$
$$\nabla(\mathbf{A} \cdot \mathbf{B}) = (\mathbf{A} \cdot \nabla)\mathbf{B} + (\mathbf{B} \cdot \nabla)\mathbf{A} + \mathbf{A} \times (\nabla \times \mathbf{B}) + \mathbf{B} \times (\nabla \times \mathbf{A})$$
$$\nabla \cdot (\Phi\mathbf{A}) = \Phi\nabla \cdot \mathbf{A} + \nabla\Phi \cdot \mathbf{A}$$
$$\nabla \cdot (\mathbf{A} \times \mathbf{B}) = \mathbf{B} \cdot \nabla \times \mathbf{A} - \mathbf{A} \cdot \nabla \times \mathbf{B}$$
$$\nabla \times \Phi\mathbf{A} = \Phi\nabla \times \mathbf{A} + \nabla\Phi \times \mathbf{A}$$
$$\nabla \times (\mathbf{A} \times \mathbf{B}) = \mathbf{A}\nabla \cdot \mathbf{B} - \mathbf{B}\nabla \cdot \mathbf{A} + (\mathbf{B} \cdot \nabla)\mathbf{A} - (\mathbf{A} \cdot \nabla)\mathbf{B}$$
$$\nabla \cdot (\nabla \times \mathbf{A}) \equiv 0$$
$$\nabla \times (\nabla\Phi) \equiv 0$$
$$\nabla \times \nabla \times \mathbf{A} = \nabla(\nabla \cdot \mathbf{A}) - \nabla^2\mathbf{A}$$

$$\int_V \nabla \cdot \mathbf{A}\,dV = \oint_S \mathbf{A} \cdot d\mathbf{S}$$
$$\int_S \nabla \times \mathbf{A} \cdot d\mathbf{S} = \oint_C \mathbf{A} \cdot d\mathbf{l}$$